Metal Sprays and Spray Deposition

Hani Henein • Volker Uhlenwinkel
Udo Fritsching
Editors

Metal Sprays and Spray Deposition

Springer

Editors
Hani Henein
Advanced Materials and Processing Lab
University of Alberta
Edmonton, AB, Canada

Volker Uhlenwinkel
Foundation Institute of Materials Science
University of Bremen
Bremen, Germany

Udo Fritsching
Foundation Institute of Materials Science
University of Bremen
Bremen, Germany

ISBN 978-3-319-52687-4 ISBN 978-3-319-52689-8 (eBook)
DOI 10.1007/978-3-319-52689-8

Library of Congress Control Number: 2017944585

Printed on acid-free paper

This Springer imprint is published by Springer Nature
The registered company is Springer International Publishing AG
The registered company address is: Gewerbestrasse 11, 6330 Cham, Switzerland

Preface

This compendium of knowledge in metal spray and spray deposition processes summarizes the technical and scientific state-of-the-art metal treatment and manufacturing via droplet and spray processes to form near-net-shaped components. It is hoped that established production areas and fields such as spray forming (SF), and also emerging fields like additive manufacturing (AM), may be inspired in an economical and ecological sense in developments of new technical approaches.

We are grateful for the contributions of all authors of the chapters in this book. The sharing of their knowledge and experiences in this field is acknowledged. Without them, this book would not have been possible. This book would not have seen the light of day without the assistance of some colleagues and students. In particular, the assistance of Paul Gronau, Evan Chow, and Sining Li was invaluable.

We hope you will enjoy reading and find this book of value for years to come.

Edmonton, AB, Canada — Hani Henein
Bremen, Germany — Volker Uhlenwinkel
Bremen, Germany — Udo Fritsching

Contents

Chapter 1
Introduction

Diran Apelian, Hani Henein, and Udo Fritsching

Near net shape processing or net shape processing has been and continues to be a pursuit of the Materials Science and Engineering community. Net shape processing is a type of manufacturing that produces a product that does not require any further treatment. Near net shape processing is similar except that minor treatment of the product is considered necessary. There are many motivations for developing such routes. Processing metallic and metallic based composite products are capital intensive operations; thus any process that generates a product closer to its final form using less processing steps will require less capital equipment and result in reduced capital investments. Concomitant with the reduction in process steps is the requirement that superior product performance and properties be achieved while reducing the waste generated in processing the part. It is desired to process complex shaped parts with significant throughput and the ability to apply automation in processing. This increases the reliability of products while achieving high volume production. An additional advantage of these processing routes is that they are considered to be green processes.

In the last decade we have seen much interest in green processing or in processes that we term as being sustainable. It may be useful to lay out some basic principles for green processing as it relates to spray forming or droplet consolidation processes. In the most simplistic sense, processes that reduce waste are sustainable processes. Metaphorically speaking, the most sustainable organism we have to

D. Apelian (✉)
Metal Processing Institute, Worcester Polytechnic Institute, Worcester, MA 01609, USA
e-mail: dapelian@wpi.edu

H. Henein
Advanced Materials and Processing Lab, University of Alberta, Edmonton, AB, Canada
e-mail: hhenein@ualberta.ca

U. Fritsching
Foundation Institute of Materials Science, University of Bremen, Bremen, Germany
e-mail: ufri@iwt.uni-bremen.de

H. Henein et al. (eds.), *Metal Sprays and Spray Deposition*,
DOI 10.1007/978-3-319-52689-8_1

learn from is nature. Nature has been around for a long time and much can be learned from nature. Nature is cyclic and there is no waste; furthermore, nature uses a few elements (unlike the scenario we are witnessing in the twenty-first century where most of the elements of the periodic table are being utilized). Waste can be further classified as *production waste* or *post-consumer waste*. The former has much to do with green processing, whereas the latter has to do with manufacturing products that can be disassembled (and repaired and reused) as well as creating value out of scrap. A good example of production waste is red mud during Al production; for every kg of Al produced, 3 kg of red mud is also produced. An example of post-consumer waste is the fact that only ~50% of beverage cans in North America are recycled. On another front, and one that has huge promise is the work that is being carried out by Melt Cognition LLC in developing a mini mill for Al production where the starting material is 100% mixed scrap Al, which is intelligently sorted into the various Al alloys (based on chemistry—XRF and LIBS technologies), and subsequently melted and cast [1]. The process is called AIM (Al Integrated Mini-mill) and creates value out of scrap. In other words, one upcycles rather than downcycles in an effort to create value and attain sustainable processing or green processing. The processing of AIM is in principle similar to that of spray forming. Eliminating processing steps such as homogenization and hot working while reducing machining scrap. For these attributes, spray forming is a green process.

To add to the economic and green processing benefits derived from near or near net shape processing, there are microstructural benefits one can obtain by near net shape processing. From a scalar perspective, processes that reduce the diffusion distance between heterogeneities in the final structure are most desirable, as they enable one to attain better properties and performance in the final component. Microstructural refinement of the end product has been and continues to be a goal in metal processing. For centuries mankind has been making components via casting where the solidification rates are small, and the diffusion distances between heterogeneities are large. Castings are heat treated (call it post-solidification processing) specifically to attain microstructural homogeneity and refinement. This is a good example of how spray forming through a droplet consolidation mechanism circumvents all of the post processing as each droplet has a starting refined microstructure. Spray forming can be thought of assembling these individual droplets into a whole. There are numerous near net shape and net shape processing routes developed and continue to be proposed and tested. These include processes such as strip casting, high pressure die casting, powder metallurgy routes, plasma deposition, cold spray, and melt infiltration.

To establish some context, one can describe metal processing through the phase changes that accompany the process. Figure 1.1 illustrates conceptually a classification based on the phases that are being processed. For example, in sand casting, the cavity is filled with a liquid which undergoes solidification. Depending on the size of the casting, local solidification times vary, but in general the solidification rates are quite slow in the order of several degrees per minute. Whereas in deformation processing, there is no phase change, and all of the processing takes

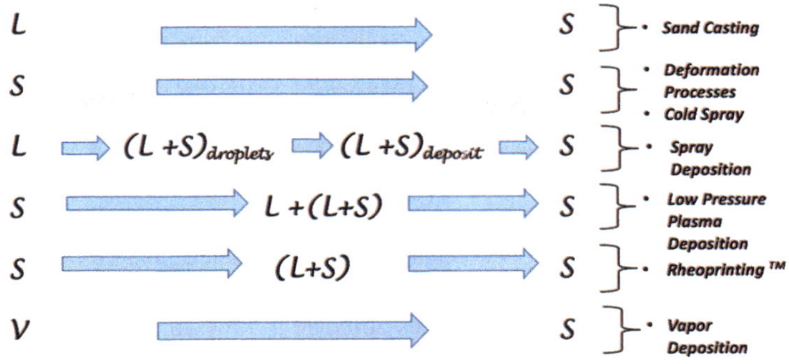

Fig. 1.1 Classification of metal processing based on phase changes in the process

place in the solid state. Cold spray processing and forging are quite dissimilar processes, but what they have in common is that the processing occurs in the solid state. In forging it is the kinetic energy imparted on the workpiece, whereas in cold spray it is the velocity of the powder particles that aid the impact of the powders onto the substrate to form an integral bond. In low pressure plasma deposition, powders are injected into the nozzle and upon exiting one will have a mixture of liquid droplets as well as semi-solid (L + S) droplets impacting the substrate. More recently, with the advent of Additive Manufacturing (AM), we have seen much interest in powder production as most of the AM processes use powder as the starting material. The Rheoprinting™ technology, developed at the Metal Processing Institute, circumvents the use of powders in AM, as the starting material is an ingot and what emerges from the nozzle is a thixotropic metal in the mushy zone [2]. It is AM via control and manipulation of the viscosity of the thixotropic alloy through the nozzle of the printer. Lastly, processing can be done in the vapour phase such as in CVD, PVD, and other related processes.

An important commercial near net shape process is the spray deposition/forming process. It has been nearly half a century since the first publication by Singer [3, 4] described a new method by which atomised droplets are deposited onto a substrate before they are fully solidified. One practice of this process is described here for illustration. The principle behind this process is that molten metal is poured through a nozzle of controlled diameter into a chamber containing inert gas with gas jets directed at the stream of molten metal. There are numerous approaches to the atomization process for molten metals. The liquid melt stream is broken up into droplets. These droplets flow with the atomizing gas exchanging and losing heat while partly or completely solidifying. In spray forming, most droplets trajectories while semi-solid are interrupted by falling onto a substrate. The remaining liquid in the droplets together with some larger liquid droplets aid in filling the pores between deposited droplets. Most droplets in the deposit likely remain separated even by a tiny oxide layer. This mechanism has been described by analogy to a series of balloons filled with ice and water landing onto the substrate by researchers at the University of Alberta. This model would explain why precipitates in a spray formed part are fine in size and homogeneously distributed throughout the deposit despite the very low solidification rate of the deposited ingot and the coarse grain size. Thus, a part in spray forming is built layer by layer as more droplets land on the substrate and subsequently the deposit. Process description and modelling, material evolution models and theories, as well as the current state of the art with various alloy systems are clearly discussed in this book.

There have been great efforts in academia, government and industry to develop the spray forming process and generate unique cost effective products with it. In 1985 the very first Osprey unit was installed in North America at Drexel University (Apelian, Lawley, Doherty); many doctoral theses were published and much of the fundamentals of spray deposition were established [5–16]. In the nearly 50 years of practice of this process, there have been numerous efforts to present article reviews on the status of research and development on the process [17–23]. In addition, in the mid 1990s Lavernia and Wu [21] published a book describing the then state of the art in spray forming. The fundamentals of Spray Forming have also been collected in the Chemical Engineering basic Ullmanns Encyclopedia [22]. As the potential of numerical modelling and simulation has further and further increased in these days, a summarize on Spray Simulation: Modeling and Numerical Simulation of Spray forming Metals has been published in 2004 [23]. Research papers and publications on Spray Forming continue to grow as new knowledge and approaches to practice the process continue to be invented, developed and practiced [24, 25]. Figure 1.2 provides a view of the number of publications that are published as a function of the year of publication. The search was carried out on Web of Science covering the years since 1970, Singer's first publication on the process. Figure 1.3 shows the citations on Spray Forming as a function of year also since 1970. It is clear from both of these plots that activity in this field remains strong. In fact the area of Spray Forming has an H index of 37 indicating that it remains an important area of activity in the field of materials science and engineering. A review of the papers that

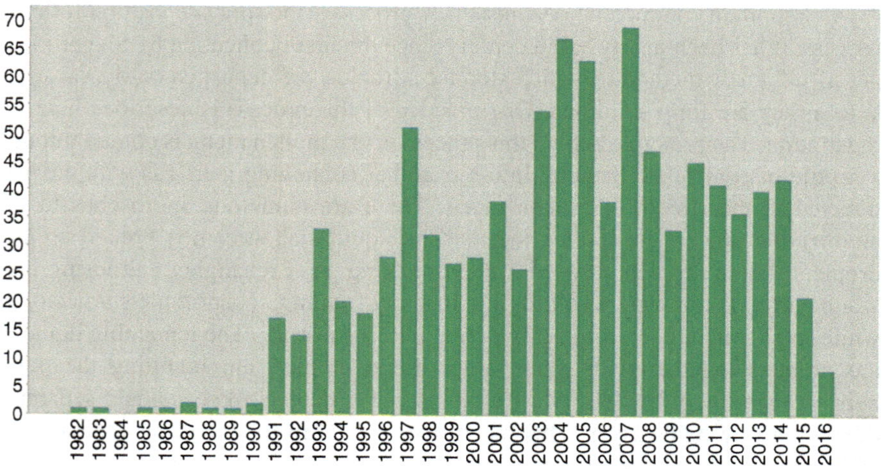

Fig. 1.2 Number of publications on Spray Forming as a function of publication year. Source: Web of Science, June 29, 2016

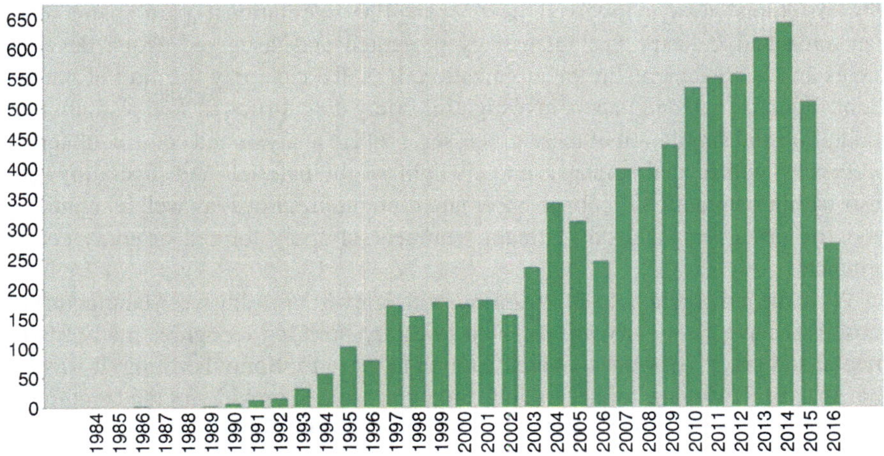

Fig. 1.3 Citations on Spray Forming as a function of year. Source: Web of Science, June 29, 2016

Table 1.1 Frequency of publications on Spray Forming by country

Peoples Republic of China	27%
United States of America	18%
Germany	13%
England	8%
India	7%
Brazil	7%
South Korea	4%
Taiwan	3%
France	2%
Spain	2%
Japan	1%
Wales	1%
Canada	1%
Denmark	1%
Others	6%

Source: Web of Science, June 29, 2016

continue to be published in this field clearly shows that interest in Spray Forming is indeed a worldwide activity. Table 1.1 lists the countries reported to have published on Spray Forming as well as the frequency of such publications.

There have been intense research activities over the past two decades by Bauckhage and the research group of the Collaborative Research Center on Spray Forming at the University of Bremen in Germany. A series of conferences as “International Conference on Spray Deposition and Melt Atomization—SDMA” took place five times between 2000 and 2013 at the University of Bremen [26], sometimes in cooperation with the “International Conference on Spray Forming”.

The meetings always involved more than 100 international participants from academia and industry that intensively presented and discussed actual developments and achievements in the atomization of melts and spray forming of metals. Contributions included papers investigating spray diagnostics as well as numerical modelling and simulation of these processes, process analysis and control, materials properties with special emphasis on new lightweight materials and superalloys and also within conventional copper, steel and aluminium alloys, as well as contributions on processing and downstream treatment of spray formed or spray coated products.

With the emergence of 3D printing with metals or Additive Manufacturing, those that have been active in the field of Spray Forming recognize many fundamental and practical issues that are common to those for Spray Forming. It was felt that an overview on the topic would be timely in order to provide for the community one reference point on the latest developments in Spray Forming. Hence this book is organized into areas of fundamentals in the early chapters. These start with a description of the fundamentals of single fluid atomization. Several techniques are described though not all of them may be easily conducive to Spray Forming. In these techniques, the use of mechanical or electrical energy is used to break up a melt stream. This provides more controlled melt stream break-up conditions, reduced gas consumption, narrower droplet size distribution, and lower overspray powders, while yielding rapidly solidified structures in the spray formed parts. In Chaps. 3, 4, 5 and 7 two fluid atomization is described in terms of its fundamentals along with the fundamentals of Spray Forming with respect to impingement of droplets onto a substrate or deposit as well as transport phenomena governing the process. Diagnostics measurements taken in-situ during atomization is presented in Chap. 6. Chapter 9 described the techniques developed and used to characterize spray formed products. Chapters 10–16 address the state of the art for different alloy systems including aluminium, titanium, and copper and their alloys, steels and superalloys are also addressed. Finally some of the applications of Spray Forming to novel materials such a bulk metallic glasses are outlined.

It is hoped that this compendium of knowledge will spur further activity in this area as well as inspire practical and high throughput approaches to new developments in Additive Manufacturing.

References

1. Melt Cognition LLC. ARPA-E Award No. DE- AR0000417. https://arpa-e.energy.gov/?q=slick-sheet-project/integrated-minimill-produce-aluminum-scrap. 28 Sep 2016.
2. Rheoprinting™. Metal Processing Institute Reports (15-02, 16-01), MPI, WPI, Worcester, MA 01609 USA.
3. Singer, A. R. E. (1982). The challenge of spray forming. *Powder Metallurgy, 25*(4), 195–200.
4. Singer, A. R. E. (1985). Recent developments in the Spray forming of metals. *International Journal of Powder Metallurgy, 21*(3), 219.

5. Mathur, P., & Apelian, D. (1992). Spray casting: A review of technological and scientific aspects. In I. Jenkins & J. V. Wood (Eds.), *Powder metallurgy—An overview* (pp. 22–44). London: Inst. of Metals.
6. Mathur, P., Annavarapu, S., Lawley, A., & Apelian, D. (1991). Spray casting: An integral model or processs understanding and control. *Materials Science and Engineering: A, 142*, 261–276.
7. Mathur, P., Apelian, D., & Lawley, A. (1991). Fundamentals of spray deposition via Osprey processing. *Powder Metallurgy, 34*(2), 109–112.
8. Annavarapu, S., Apelian, D., & Lawley, A. (1990). Spray casting of steel strip: Process analysis. *Metallurgical Transactions A, 21*(12), 3237–3256.
9. Mathur, P., Apelian, D., & Lawley, A. (1989). Analysis of the spray deposition process. *Acta Metallurgica, 37*(2), 429–443.
10. Mathur, P., Annavarapu, S., Apelian, D., & Lawley, A. (1989). Process control, modeling and applications of spray casting. *Journal of the Minerals, Metals and Materials, 41*(10), 23–28.
11. Annavarapu, S., Apelian, D., & Lawley, A. (1988). Processing effects in the spray casting of steel strip. *Metallurgica Transsactions A, 19*, 3077–3086.
12. Apelian, D., Wei, D., & Smith, R. W. (1988). Particle melting and droplet consolidation during low pressure plasma deposition. *Powder Metallurgy International, 20*(2), 7–10.
13. Apelian, D., Lawley, A., Mathur, P. C., & Luo, X. (1988). Fundamentals of droplet consolidation during spray deposition. In P. U. Gummeson & D. A. Gustafson (Eds.), *Modern developments in powder metallurgy* (Vol. 19, p. 397). Princeton, NJ: Metal Powder Industries Federation.
14. Apelian, D., Gillen, G., & Leatham, A. (1987). Near net shape manufacturing via the Osprey process. In F. H. Froes & S. J. Savage (Eds.), *Processing of structural metals by rapid solidification* (pp. 107–120). Metals Park, OH: ASM.
15. Apelian, D., & Gillen, G. (1986). Spray deposition via the Osprey process. *Journal of Metals, 38*(12), 44.
16. Apelian, D., & Kear, B. H. (1985). Plasma deposition processing. In *Plasma processing of materials* (pp. 79–104). Publication NMAB-415. Washington, DC: National Academy Press.
17. Ojha, S. N. (1992). Spray forming—Science and technology. *Bulletin of Materials Science, 15* (6), 527–542.
18. Leatham, A. G., & Lawley, A. G. (1993). The Osprey process—Principles and applications. *International Journal of Powder Metallurgy, 29*(4), 321.
19. Widmark, H. (1993). 30 years of stainless steel development. *Scandanavian Journal of Metallurgy, 22*(3), 156–164.
20. Grant, P. S. (1995). Spray forming. *Progress in Materials Science, 39*(4–5), 497–545.
21. Lavernia, E. J., & Wu, Y. (1996). *Spray atomization and deposition*. Chichester: Wiley.
22. Fritsching, U., & Bauckhage, K. (1999). Spray forming of metals. In *Ullmann's encyclopedia of industrial chemistry* (6th ed.). Wiley: Weinheim.
23. Fritsching, U. (2004). *Spray simulation: Modeling and numerical simulation of sprayforming metals*. Cambridge: Cambridge University Press.
24. Leatham, A. G. (1996). Spray forming technology. *Advanced Materials and Processes, 150*(2), 31–34.
25. Leatham, A. G., & Lawley, A. 1999. Spray forming commercial products: Principles and practice. In *Advanced powder technology*. Materials science forum (Vol. 299–300, pp. 407–415).
26. K. Bauckhage, U. Fritsching, V. Uhlenwinkel, J. Ziesenis, A. Leatham (Eds.). (2000, 2003, 2009, 2010 and 2013). Proceedings of international conference on spray deposition and melt atomization SDMA (Vol. 1–5), Bremen, Germany.

Chapter 2
Single Fluid Atomization Fundamentals

Abdoul-Aziz Bogno, Hani Henein, Volker Uhlenwinkel, and Eric Gärtner

2.1 Introduction

Atomization is simply defined as the breakup of a liquid stream into droplets. It can be achieved in many ways including spraying through a nozzle, pouring on to a rotating disc, etc. Atomization practice and research usually involve materials processing in their liquid state either at or near room temperature (oil-based liquids, paint spraying, aerosol sprays, etc.) or at high temperature (metal melts). Most of the literature describing atomization mechanisms pertains to two fluid atomization in which a second fluid is applied to break up a melt stream into droplets. Two fluid atomization techniques for molten metals are described in Chap. 3.

In view of the requirement of high liquid/solidification cooling rate, high undercooling but also controllable droplets size, shape and solidification microstructures, single fluid atomization (SFA) has established itself as the atomization technique of choice. It is a containerless solidification technique [1] which consists in the transformation of a bulk liquid into a spray of droplets that generally fall and solidify rapidly by losing heat to a surrounding gas of choice (N_2, Ar or He are commonly used). The bulk liquid is produced by heating a material above its melting point and the droplets, generally of narrow size distribution, are either collected after complete solidification as powders or are deposited in a semi-solid

A.-A. Bogno (✉) • H. Henein
Advanced Materials and Processing Lab, University of Alberta, Edmonton, AB, Canada
e-mail: bogno@ualberta.ca; hhenein@ualberta.ca

V. Uhlenwinkel
Foundation Institute of Materials Science, University of Bremen, Bremen, Germany
e-mail: uhl@iwt.uni-bremen.de

E. Gärtner
University Bremen, Bremen, Germany
e-mail: e.gartner@iwt.uni-bremen.de

H. Henein et al. (eds.), *Metal Sprays and Spray Deposition*,
DOI 10.1007/978-3-319-52689-8_2

state onto a substrate to form a strip or as a spray coating. Controllable droplet shape and size and narrow size spectra are important for many technical applications including spray coating and printing. It has been reported that spherical solar cells produced by dropping method for photovoltaic power generation give better performance due to an improved overall microstructures as compared to the bulk Si solar cells made from Si ingots [2, 3].

In recent years, SFA has become the subject of attention of many researchers through the development of several techniques such as the drop-on-demand [4–6], the pulsated orifice ejection method (POEM) [2], the jet-splitting method [7], the flat-fan and pressure swirl methods [8], the centrifugal atomization [9], the ultrasonic atomization [10–12] and the Impulse Atomization (IA) technique [13].

In this chapter, SFA fundamentals will be discussed based on metals atomization. The melt stream break-up mechanism will be emphasized and the controlling parameters of the mass/heat flux and size distribution will be analyzed based on IA, a typical SFA technique developed at the Advanced Materials and Processes Laboratory (AMPL) of the University of Alberta. Finally, microstructures characterization of droplets obtained by IA will be discussed and a new quantitative method to estimate the primary and secondary nucleation undercoolings during rapid solidification of droplets will be presented.

2.2 Droplet Formation

2.2.1 Mechanism of Stream Breakup

Atomization is defined as the break-up of a liquid stream into droplets. Understanding this breakup mechanism is therefore very important in order to optimize the design and improve the performance of SFA systems. As described by Henein [13] and Yuan [14] the breakup mechanism is controlled by two forces: the potential force induced by the head of liquid above the orifice and an external force (disturbance) applied to the melt. Indeed, a liquid sitting over an orifice will flow through it to form a stream when the gravity force is significantly greater than the surface tension and drag force. While the melt head induces an inertial force to drain the melt through the orifices, effective melt flow requires (1) viscous dissipations through the orifices and (2) overcoming of the force induced by the surface tension of the melt which acts opposite to the direction of flow, as the stream exits the orifice [15]. Thus, the melt requires an external force not only to push it through the orifices but also to act as a disturbance that triggers the stream break-up especially for a small orifice size.

Figure 2.1 shows a schematic of a liquid ligament emanating from an orifice. When the liquid ligament emerges from the orifice as a continuous body of cylindrical form (as shown by the schematic) there occurs a competition on the surface of the ligament between the cohesive and disruptive forces. This competition gives rise to

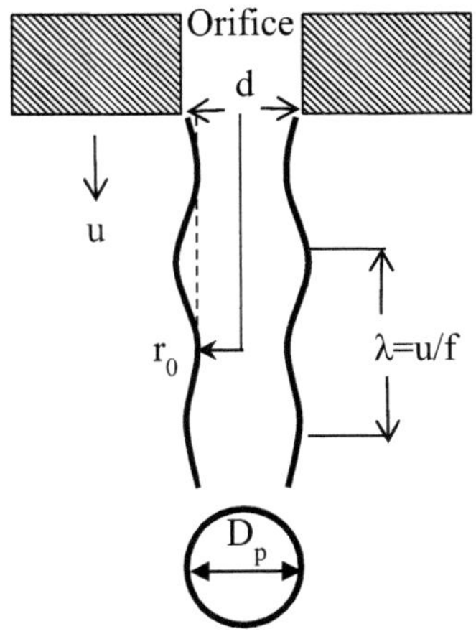

Fig. 2.1 Liquid ligament with periodic perturbations falling with a velocity u from a nozzle orifice of radius r_0 upon an applied impulse of frequency f

oscillations and perturbations of a wavelength λ, which under induced effects of an external disturbance (e.g. impulses) are amplified and the liquid body breaks up into spherical droplets.

According to Rayleigh instability [16], the minimum theoretical wavelength of a perturbation required to break up a liquid ligament of length *L* is:

$$\lambda = 2\pi r_0 \tag{2.1}$$

where r_0 is the radius of the ligament, which is assumed to be equal to the orifice radius. Therefore, in terms of frequency and velocity, the maximum frequency f_{max} that must be applied for the applied perturbation to induce break-up of the liquid ligament is:

$$f_{max} = \frac{u}{2\pi r_0} \tag{2.2}$$

where u is the velocity of the liquid ligament and f the frequency of the applied perturbation. Thus, if for a given ligament of length L and circumference C, the condition $L/C < 1$ is fulfilled then the ligament is expected to form only one droplet, otherwise if $L/C > 1$, the ligament will break up into several droplets depending on its L and C which are determined by the force applied to push the liquid through the nozzle orifice and the nozzle orifice size [14]. The force used to generate the melt stream varies by process type. For example, in the jet-splitting method, the flat-fan and pressure swirl method and the ultrasonic atomization, the pressure is applied by a gas overpressure being applied to the surface of a melt in a crucible. For the POEM and the drop-on-demand methods, this force is being applied using a piezo-electric crystal. For centrifugal atomization the orifice at the bottom of a crucible holding the melt is sufficiently large as to allow the melt to flow out freely under gravity. Finally, for IA the force that is used can be a combination of both gas overpressure and mechanical pressure.

The resulting droplet diameter D_p can be calculated using Eq. (2.3) where it is assumed that droplets generated from the applied perturbation have the same volume as that within one wavelength of the liquid stream [17].

$$\frac{1}{6}\pi D_p^3 = \pi r_0^2 \lambda \tag{2.3}$$

Expressing Eq. (2.3) in terms of f and u:

$$\frac{1}{6}\pi D_p^3 = \pi r_0^2 \left(\frac{u}{f}\right) \tag{2.4}$$

And, rearranging in terms of D_p:

$$D_p = \left(\frac{6r_0^2 u}{f}\right)^{1/3} \tag{2.5}$$

While the frequency f is operator dependent, u is determined by dividing the liquid flowrate Q by the cross sectional area of the orifice πr_0^2 [17] yielding Eq. (2.6).

$$u = \frac{Q}{\pi r_0^2} \tag{2.6}$$

2.2.2 Boundary Between Stream and Dripping Formation

The quantification of the transition from free stream flow to dripping is important in modeling melt atomization. Based on a model initially used to determine physical properties of liquids, the liquid flow from the bottom of a crucible is given by Eq. (2.7) [17]. The flowrate Q is related to the surface tension (σ), the potential force induced by the liquid head (h), the gravity acceleration constant (g), the liquid density (ρ) as well as the discharge coefficient, C_D across the orifice of cross-sectional radius r_o [15].

$$Q = C_D \pi r_0^2 \sqrt{2g\left(h - \frac{\sigma}{\rho g r_0}\right)} \tag{2.7}$$

Combining Eqs. (2.6) and (2.7), the liquid velocity can be expressed in terms of potential and surface forces [Eq. (2.8)].

$$u^2 = C_D^2 2g\left(h - \frac{\sigma}{\rho g r_0}\right) \tag{2.8}$$

Equation (2.8) may be written in dimensionless form by introducing the Froude number, Fr, and Bond number, B_o, as follows.

$$\frac{F_r}{C_D^2} + \frac{1}{B_0} = 1 \tag{2.9}$$

Where

$$F_r = \frac{u^2}{2gh} \tag{2.10}$$

And

$$B_0 = \frac{\rho g r_0 h}{\sigma} \tag{2.11}$$

When h decreases i.e. the level of the melt becomes low (degree of vorticity negligible) the velocity of the stream decreases and the stream will approach laminar flow conditions characterized by a low Reynolds number:

$$Re = \frac{2\rho u r_0}{\mu} \tag{2.12}$$

At low Re it has been shown that there is a linear relationship between C_D and Re as seen in Eq. (2.13), so that a decrease of u (therefore Re) consequent to a decrease in h will lead to a decrease of C_D [15].

$$C_D = a + bRe \tag{2.13}$$

where a and b are respectively the intercept and the slope of the C_D vs Re regression line for low Re values. The melt will stop flowing as a stream when its velocity tends to zero.

When $u \to 0$, $B_0 \to 1$ so that the lower critical head height h_{min} for stream formation is expressed by Eq. (2.14) as follows:

$$h_{min} \geq \frac{\sigma}{\rho g r_0} \tag{2.14}$$

When $h < h_{min}$ flow from the orifice would continue by dripping, forming droplets, until h further decreases and the fluid surface tension keeps it at the orifice. Atomization under these conditions typically produces large droplets at low production rates. In fact, under dripping conditions, the formulations for droplet formation following the stream breakup mechanism described above do not apply. The lower critical velocity for stream formation can be expressed in terms of the Weber number, *We* as follows [18]:

$$We = \frac{2\rho r_0 u^2}{\sigma} > 4 \tag{2.15}$$

2.2.3 Stream Breakup Regimes

Figure 2.2 shows a schematic description of droplet formation by different break-up mechanisms depending on the velocity of the liquid stream.

When u is large enough consequent to an increase of Q and therefore h, the kinetic energy overcomes the surface tension and a continuous liquid stream forms. Droplet formation in this case occurs by Rayleigh instability [16] in the so called "Rayleigh regime" as described earlier. At larger liquid velocity consequent to a larger head height, the relative velocity between the stream and the atomization atmosphere becomes remarkable inducing aerodynamic effects that accelerate the break-up process and shortening of the ligaments lengths is observed. This regime is referred to as "the 1st Wind Break-up" [16]. At a sufficiently high velocity, the static pressure induced on the surface may result in the "whiplash mode" [16, 19]. This mechanism is referred to as "2nd Wind Break-up" [16]. If the velocity of the stream is higher still, "atomization" is observed [16, 19]. Indeed, while viscosity has a damping effect on the growth of disturbances on the liquid surface and surface tension tends to pull the liquid together, aerodynamic forces tend to promote the growth of disturbances. The overall contributions of these forces can be characterized by the Ohnesorge (Oh) non-dimensional number, ratio of viscous forces over inertia and surface tension forces [Eq. (2.16)].

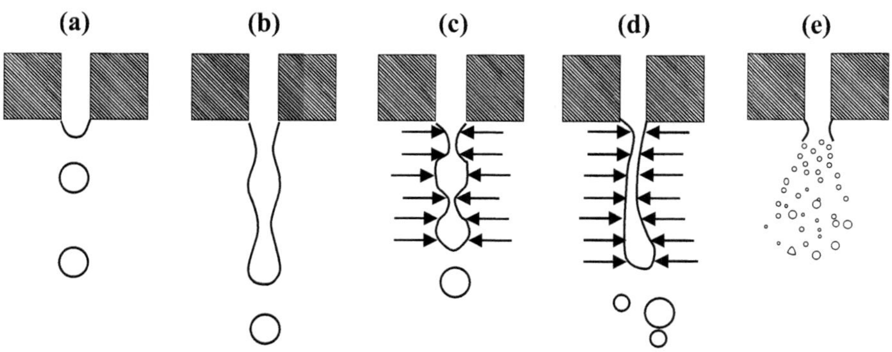

Fig. 2.2 Schematic description of different droplet formation mechanisms (**a**) drip off, (**b**) Rayleigh regime, (**c**) 1st Wind Breakup, (**d**) 2nd Wind Breakup and and (**e**) Atomization

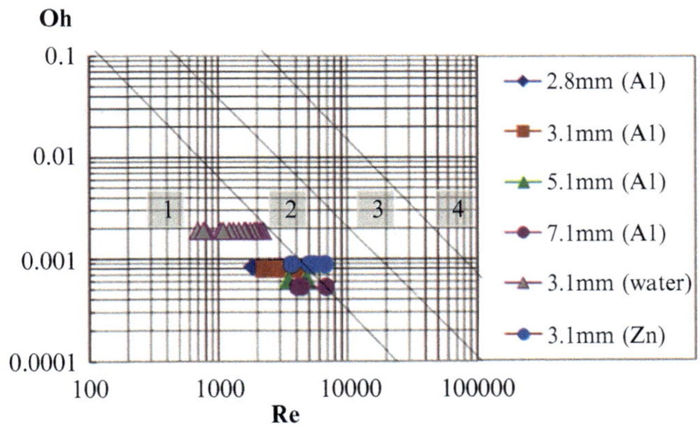

Fig. 2.3 Breakup regimes including examples of stream of Al, water and zinc at various sizes (**1**) Rayleigh regime, (**2**) first wind-induced, (**3**) second wind induced and (**4**) atomization breakup (from [21] with permission)

$$Oh = \frac{\mu}{\sqrt{2\rho\sigma r_0}} \tag{2.16}$$

Or,

$$Oh = \frac{\sqrt{We}}{Re} \tag{2.17}$$

Reitz [20] developed the original work of Ohnesorge [21] to propose a chart of break-up regimes by plotting *Oh* vs *Re* as shown in Fig. 2.3. This was obtained using oils and water. A large *Oh* indicates that viscous effects are more dominant in the liquid. Example data for aluminum, water and zinc processed with various orifice sizes are also shown in the figure.

The data show that water as well as metals melts (Al, Zn) processed at the same orifice sizes remain within the Rayleigh and first wind induced breakup regimes. With a higher flowrate Q (higher *Re*) as one moves from left to right on Fig. 2.3. and the droplet size becomes smaller.

2.2.4 *Spheroidization*

Spheroidization which is the last stage of droplet formation depends on τ the ratio of times taken to dissipate internal mechanical energy or simply the spherodization time (t_{SP}) to the time taken to dissipate thermal energies t_{th} [Eq. (2.18)].

$$\tau = \frac{t_{SP}}{t_{th}} \tag{2.18}$$

If $\tau < 1$, under the influence of surface tension, the droplets have the time to spheroidize before solidification is completed. And if $\tau > 1$, i.e. solidification completes before spheroidization, the droplets will have the shape of ligaments [22].

Indeed, t_{SP} is the time for any oscillations in a spherical shape to be damped by internal viscous stresses characterized by the viscosity coefficient μ. It can be shown that this time should be proportional to $\left(\rho D_p^2/\mu\right)$ [23] reported that a 200 μm kerosene droplets moving at 10 ms^{-1} in an air-atomized spray retained sphericity after approximately 5 ms, which leads to the following expression of t_{SP}:

$$t_{SP} = \frac{0.1\rho D_p^2}{\mu} \tag{2.19}$$

The selection of gas atmosphere can affect the resultant shape of a droplet. When using inert gas, Eqs. (2.18) and (2.19) would result in droplets spheroidizing. However, in the presence of oxygen in the gas atmosphere, an additional force is introduced. The presence of oxygen will result in the oxidation of the surface of the melt ligament. The time of oxidation of molten metals is generally faster than t_{SP}. If this oxide is strong and adherent to the melt, then the droplets will retain and solidify with the shape of the ligament. The required oxygen partial pressure in the gas atmosphere to accomplish this non-spheroidization will vary with melt composition and temperature of atomization.

2.3 Theoretical Energy Requirement

One of the main advantages of SFA is the low energy requirement as compared to two fluid atomization. For instance, there is no need to compress the gas used for atomizing a liquid as it is the case in a gas atomizer. However, like in all melt atomization processes, there is a compulsory energy requirement for superheating the material above its melting point. And, as there is generally no heat recovery during a SFA process, this heat energy goes to waste as it is removed from the atomization chamber to produce powder.

Besides heating the material to melt it, additional energy input is required to break up the melt into droplets. That energy would be the one needed to create the surface area of atomized droplets. To create a surface area S of droplets atomized from a melt of mass M per unit time, the energy E_σ required is the product of S and the surface tension σ [23].

$$E_\sigma = \sigma S = \sigma \sum_{i=1}^{i=N} \pi D_{pi}^2 \tag{2.20}$$

Where N is the total number of droplets produced from the mass M per unit time (1 s).

Thus, the theoretical energy required to atomize unit mass (E_σ/M) is given by:

$$E_\sigma/M = 6\sigma\rho^{-1}D_P^{-1} \tag{2.21}$$

Where M is the product of density by volume of the liquid droplets:

$$M = \rho\pi \sum_{i=1}^{i=N} D_{Pi}^3/6 \tag{2.22}$$

From Eq. (2.21) it is clear that less energy is required for denser liquid and/or larger orifice size (or droplets size). The theoretical power requirement for atomization of different droplets size of a variety of materials is found to vary from 100 to 1000 times lower than the power required for melting [23]. However, the atomization efficiency (theoretical minimum input power/actual input power) which is generally low is found to be less than 1% so that the true power requirement for melting and atomization are in the same range for all materials [23].

Atomization efficiency is affected by different physical properties of liquid material and cooling gas. Indeed, during atomization process, liquid kinetic energy is converted into droplets surface energy and droplets movement inside the cooling gas. For a fixed liquid flow rate, an increase in the gas density can improve atomization, however an increase of the drag force due to the increase of density should be avoided so that the relative velocity between the liquid ligament and the cooling gas is not reduced.

The atomization of a liquid with higher surface tension (viscosity and density being constant) requires more energy but yield a better efficiency. It is possible that the high surface tension of a liquid acts againts coalescence of droplets after break up.

Viscosity is another liquid physical property that resists break up into droplets and consequently requires more energy. The effect of viscosity is generally minimized by superheating the liquid.

2.4 Single Fluid Atomization Techniques

2.4.1 Overview of Existing Techniques

Single fluid atomization techniques have been developed based on the limitations of two fluid atomization techniques such as gas atomization (GA) and water atomization (WA). Formation and break-up of sheet-ligament during these atomization techniques lead somewhat to irregular droplets formation [24] except for when GA is carried out in inert gas atmospheres. If consistent ligaments size and spherical droplets are required, single fluid atomization techniques are more attractive. In addition performance to requirements characterized by energy consumption, spray quality, mass flux across the spray and size distribution have become a real concern in the production of powder for many applications including additive manufacturing. Two parameters are usually considered in describing size distribution (usually log-normal), the mass median droplet size D_{50} and log-normal or geometric standard deviation σ_{LN}. Almost all applications of powders require a specific droplet size or distribution to be supplied. Although two fluid atomization processes have the advantage of yielding very fine droplets with higher throughput (but usually at low yield), their energy consumption and size distribution present considerable limitations. Thus, based on the design and performance of two fluid atomizers, technology transfer has given rise to a variety of single fluid atomization processes including Centrifugal atomization (CA), Drop on Demand Techniques (POEM, PDOD, StarJet), Continuous uniform droplet generation (UDG) and IA.

2.4.2 Centrifugal Atomization

Centrifugal atomization (CA) has been used for several decades in the metal powder industry. Centriguation of molten metal streams is an efficient method for producing high quality powders. It offers many advantages in terms of a relatively small particle size range (50–250 μm), narrow particle size distribution, spherical shape, processing strategy and flexibility, operating costs, high production rate, and high yield [25]. One of the greatest benefits of centrifugal atomization is the narrow particle size distribution. A narrow distribution is increasing the yield of the final product in the desired particle size range. Hence it may be favored over other common techniques such as gas atomization. These advantages allow for the application of centrifugal atomization in the production of common materials such as: Sn, Pb, Al, Mg, Zn, Ti, Co, and corresponding alloys [26]. Despite these advantages and the flexibility of the process, the application on the industrial scale remains relatively limited. The overall production quantities are estimated to be 100,000 t/year [25]. Hence, it is reported of playing a minor role, in comparison to other well-known technologies, such as gas and water atomization. This is attributed to the lack of scientific knowledge in this field [27]. From a scientific point of view, more fundamental research regarding process design and the atomization mechanism need to be conducted in order to improve the process and increase profitability and practicality. It is a question as to whether industry will improve its applications of centrifugal atomization and thus increase the use of the process for powder production. However, it is not yet able to produce particle sizes below 50 μm. Major problems arise concerning the spray chamber dimensions and cleaning of the device. The startup process in centrifugal atomization is

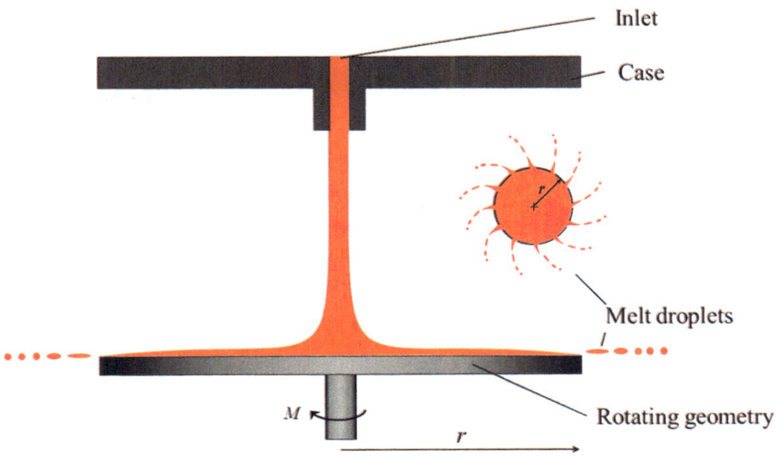

Fig. 2.4 Schematic drawing of a centrifugal atomizer

critical. Pre-solidification on the spinning disc needs to be prevented, and special care must be taken in order to guarantee the stability of the actual atomization process.

2.4.2.1 Melt Flow Disintegration

Centrifugal atomization (also known as spinning disc, rotary, or spinning cup atomization) has received increasingly more attention over the last few decades [25]. In the most standard procedure, a molten metal stream is fed centrally at the top of a rapidly spinning geometry (disc, plate, or cup) forming a liquid film, which is sheared off from the rotating body tangentially (see Fig. 2.4). The underlying mechanism of atomization occurs at the threshold, where the centrifugal force exceeds the viscous force and the surface tension of the melt film at the edge of the disc.

Depending on the melt flow rate on the spinning geometry and the rotational speed, several distinct droplet formation modes can be observed [28]. Figure 2.5 shows the predominant modes when observing a rotating flat disk process: (a) Direct Droplet Formation, (b) Direct Droplet and Ligament Formation, (c) Ligament Formation and (d) Sheet/Film Formation. These regimes may occur on top of the geometry, around, or beyond the edge of the geometry in the horizontal axis. The transition between the different states is due to an increasing liquid flow rate [29]. At relatively small liquid feed rates and rotating speed, the Direct Droplet Formation mode occurs. In this mode, the outer rim of the liquid film begins to alter into a non-uniform shape under the applied centrifugal force. Once the liquid film is unable to maintain its natural shape, the surface tension is overcome by the centrifugal force and a distinct part of the liquid volume detaches from the original body tracing a fine ligament behind it. If the droplet separates

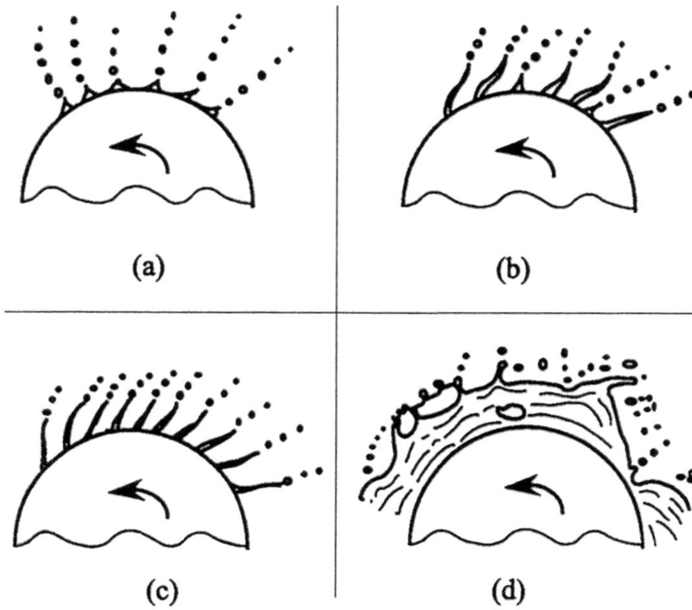

Fig. 2.5 Regimes rotary atomization liquids (**a**) Droplet Formation, (**b**) Droplet and Ligament Formation, (**c**) Ligament Formation and (**d**) Film or Sheet Formation (from [28] with permission)

from the ligament, the ligament will disperse in a series of fine droplets of near uniform size.

Further increasing the flow rate progresses the disintegration regime into the Ligament Formation mode. The volume separation step moves towards the periphery of the rotation disc or cup generating larger ligaments compared to the Direct Droplet mode. Similar to the Direct Droplet mode, ligaments disperse in small volume units-droplets. An even greater melt flow rate will create a continuous film or sheet beyond the geometry rim, forcing the flow disintegration zone off the geometry. Consequently, this regime is addressed as the Film or Sheet Formation mode. The sheet break-up mechanism, regarding the equilibrium state of contraction energy and surface tension at the sheet edge, was thoroughly investigated by Fraser et al. [30].

Figure 2.6a, b shows two in-process pictures of centrifugal melt atomization, using a cup as the rotational geometry to disintegrate the liquid stream into small droplets. Figure 2.6a shows the overall process concept: Metal flow is fed at the top of a rotating geometry. In this case, water is used to increase the cooling rate of the individual metal particles.

Having a closer look at the rotational unit (Fig. 2.6b), one can see the different volume separation modes described earlier. The melt flow is injected from the top of the process chamber onto the cup. The cup is completely filled and ligaments extend from the rim of the cup, disintegrating into small particles.

Fig. 2.6 (**a**) 30 t/h centrifugal atomization of steel with water quenching and (**b**) top view on large cup with metal flow (from [25] with permission)

2.4.2.2 Droplet Generation and Parameters

Droplet separation or atomization will only take place if the centrifugal force exceeds the restoring surface tension of the liquid metal. With the help of a simple force balance, the mean particle size of the produced metal particles is available [29]. The equilibrium condition when interpreting the droplet generation is the following:

$$F_C = F_S \tag{2.23}$$

The centrifugal force F_C and the surface force F_S are in balance. The centrifugal force of a particle on a rotating path is defined by its mass m, the radius of the geometry r_0, and the angular velocity ω.

$$F_C = m r_0 \omega^2 \tag{2.24}$$

The mass of a single liquid droplet can be calculated according to Eq. (2.25) by its density ρ and diameter d.

$$m = \frac{\rho \pi d^3}{6} \tag{2.25}$$

Using the surface force F_S, which is calculated from the surface tension σ and the particle diameter d,

$$F_S = \sigma \pi d \tag{2.26}$$

and integrating both forces in the aforementioned balance, one yields the following expression defining the mean particle diameter produced by centrifugal atomization.

$$d = \sqrt{\frac{6\sigma}{\rho r_0 \omega^2}} \tag{2.27}$$

This simple expression is only applicable for Direct Drop Formation and fails for high metal feed rate where Ligament and Sheet Formation become predominant disintegration modes. Its failure arises from not taking into account the complexity of the bulging film, or aerodynamic forces, or slippage on the atomizer itself [25]. The mean particle size currently attainable by centrifugal atomization varies between 50–250 μm depending on process parameters, especially rotation speed but also material parameters (melt density and surface tension) [31].

2.4.2.3 Rotating Speed and Material Properties

Plookphol et al. [27] used the above expression [Eq. (2.27)] for a series of experiments proving the dependence of the mean particle size on the atomizer size and shape, oxygen level, and feed rate of an SnAgCu alloy. A significant discrepancy has been found when comparing the actual particle size from the process and theoretically calculated. The greatest deviations from the equation were examined in the low rotation regime below <15,000 rpm. However, the force balance in centrifugal atomization is able to predict the process outcome from the particle point of view.

Figure 2.7 shows theoretically possible particle sizes depending on the rotation speed for various pure metals at their liquidus temperatures. Generally, it can be seen that with increasing rotation speed a smaller particle size results due to greater centrifugal forces applied to the melt. Aside from the mechanics, there is a clear influence of the material parameters: density and surface tension. The inner diagram compares the square rooted ratio of surface tension to density of all chosen pure metals. Generally, smaller particles result from lower restoring mechanisms of density and surface tension.

2.4.2.4 Atomizer Size and Shape

A number of authors [27, 32, 33] experimentally confirmed the influence of size and shape of the rotating unit. Generally, it was shown that changing the geometry from disc to cup will create significantly smaller particles. The cup geometry favors the disintegration mechanism of the melt flow compared to a simple disc. A variation of the cup wall angle did prove to provide further enhancement in terms of final

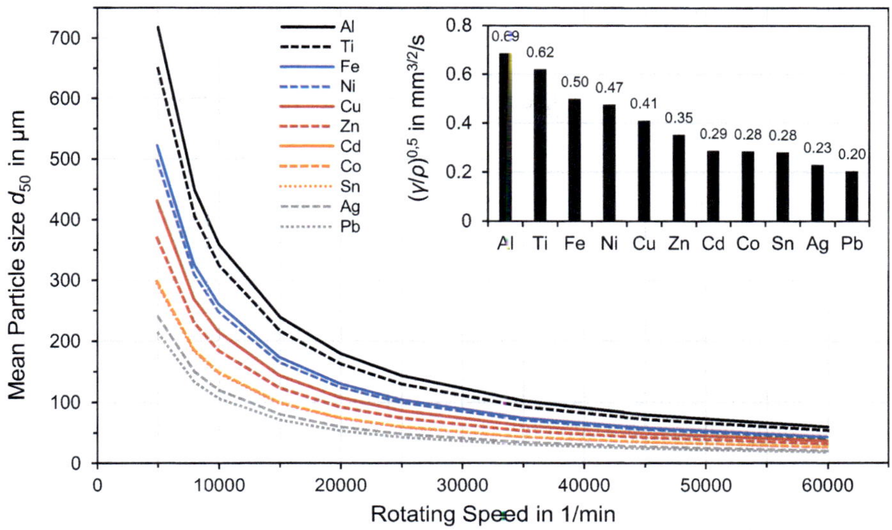

Fig. 2.7 (**a**) Theoretical particle size dependent on the rotation speed of various pure metals r = 20 mm and (**b**) square rooted surface tension to density ratio

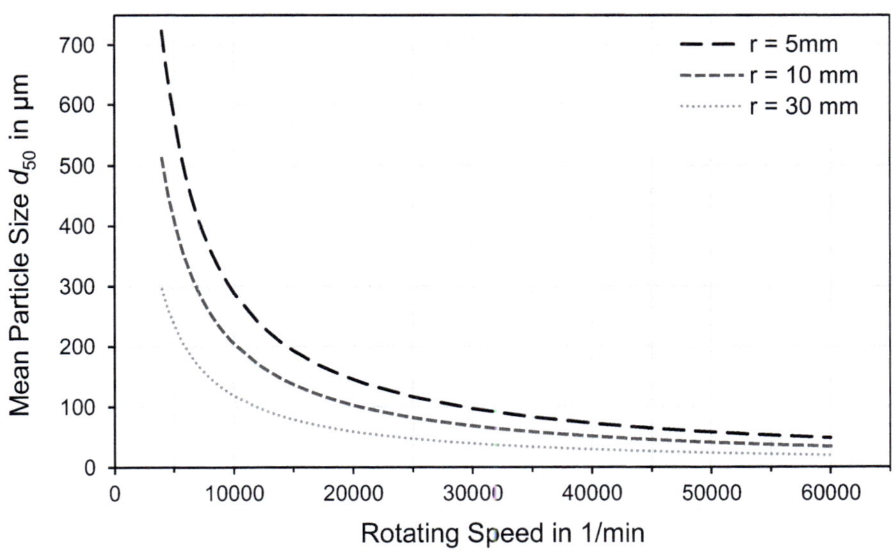

Fig. 2.8 Theoretical particle size dependent on the rotation speed and variation of disc diameter from 5 to 30 mm of SnAgCu alloy calculated after Plookphol [27]

particle size. Figure 2.8 is derived from the theoretical expression of Eq. (2.27) and does not account for technical boundaries. One can see that an increasing disc diameter improves the rotational force at the geometry rim generating smaller droplet units at constant melt flow rate.

2.4.2.5 Feed Rate, Wettability and Oxygen Content

Melt feed rate has been reported as an important process parameter having a major impact not only on mean particle size, but also particle size distribution [27]. An increasing of melt flow will inevitably affect the apparent disintegration mode. Melt layer thickness may increase at the rim of the geometry and coarser particles result. On the other hand, an unwanted premature solidification may occur when choosing a very low melt flow rate. Hence, careful evaluation and adjustment of the melt flow rate will support the process outcome in terms of particles size and throughput.

Another very important system property affecting the continuity and yield of the atomization process is the wettability of the atomizer material [32]. In terms of contact angle of the melt, and also surface roughness of the geometry, it becomes evident when improving wettability, slippage of the melt can be reduced, resulting in steady atomization [27]. Often the atomizer is pre-coated with the melt and mechanically roughened before running the process.

One parameter recently receiving more attention is the amount of oxygen in the processing chamber [33]. Theoretically, oxygen will support the surface oxidation form metal oxide layers. As a result the restoring surface tension of the melt flow increases hence, hindering the capability of secondary disintegration of droplets. Yet, this phenomenon remains unclear. Process safety (i.e. explosion in aluminum atomization) and also consumer specifications demand that the oxygen level be as low as possible [27].

2.4.2.6 Process Design

The design of centrifugal atomizers is mainly driven by the centrifugal forces necessary for the materials to be atomized. The most crucial dimension of an atomizing chamber is the diameter required for sufficient centrifugal radial path. Figure 2.9 shows a common type of industrial atomizer. The melting unit is attached on top of the spray chamber consisting of a crucible, induction coil, furnace, stopper, and motor for the crucible. The melt is introduced through the nozzle onto the rotational geometry. The rotational geometry is powered by a motor, which rotates the disk.

Process monitoring via a camera or a window allows for the process to be controlled and regulated manually, which is especially important at the beginning and the end of a batch. As discussed earlier, surrounding gas may have an influence on the stability of the process and the final product. Commonly, an air supply unit is mounted to the spray chamber. The granular product can be extracted from the unit either continuously or discontinues at the bottom of the metal tank.

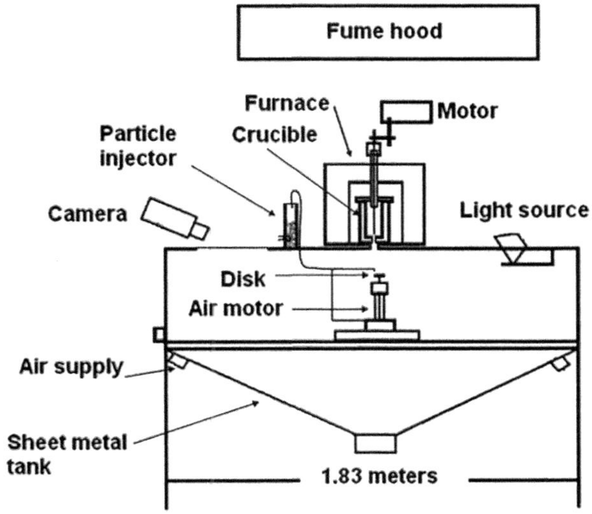

Fig. 2.9 Exemple of a process scheme of a centrifugal atomizer (from [34] with permission)

2.4.3 Drop on Demand Techniques (POEM, PDOD, StarJet)

Drop on demand techniques for molten metal applications have been developed for many purposes. Most of these techniques can generate both single and multiple droplets. In fundamental research the drop on demand technique is often used to study the correlation between the cooling rate and the microstructure of a solidified droplet. The cooling rate of a single droplet in static gas can be calculated and measured accurately. The cooling rate varies greatly by changing the droplet diameter, the ambient gas, and the gas pressure. Drop on demand techniques are also used for droplet deposition and 3D printing [35] Typically, the droplet size is in the range of 100–500 μm. Particularly for metals with high melting points it is challenging to generate particles below 100 μm. Often high melting point alloys have higher surface tensions which hinders the droplet formation. An increase in melt viscosity also affects the droplet formation process because of increasing viscous forces. Below, some drop-on-demand processes are introduced.

2.4.3.1 Pulsated Orifice Ejection Method (POEM)

The Pulsated Orifice Ejection Method (short POEM) was developed at the end of the last century to generate single droplets on demand. The experimental set-up is based on the Rayleigh break-up system (Fig. 2.1). A pressure pulse generated by a piezoelectric actuator is applied to produce a droplet by the displacement of the cylindrical rod (Fig. 2.10). Here, only a few micrometer of displacement is necessary to generate a droplet.

Monodispersed copper particles were generated by Takagi et al. [36]. The gas pressure (0–50 kPa) and the cylindrical rod displacement (approx. 7 μm) have been

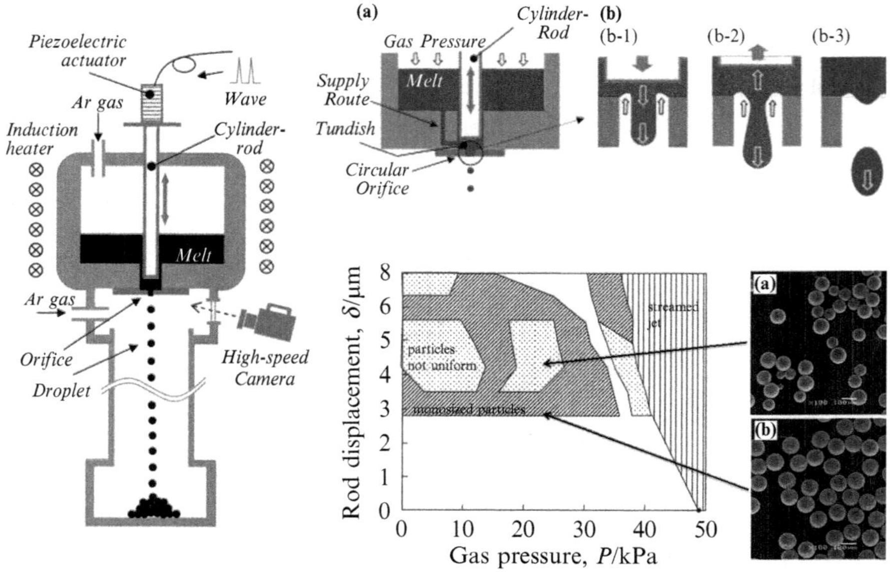

Fig. 2.10 Principle of POEM (*left*), Formation of monosized particles on the pulsated orifice ejection method (*upper right*), Particle formation map resulting from variation of rod displacement and gas pressure and micrographs of the monosized particles obtained under the corresponding conditions on the map (*lower right*) (from [36] with permission)

varied in order to achieve droplet sizes between 75 and 120 μm with a standard deviation of 2 μm. The nozzle diameter was approximately 100 μm. Parameters were adjusted to a frequency of 10 Hz for shaft, a superheat of only 16 K, a shaft velocity of 4.7 cm/s (down) and 0.35 cm/s (up), respectively. As the wettability of the orifice and the melt is important for the process, four different materials with different contact angles were tested: 3 mol% Y_2O_3-ZrO_2 (contact angle 115°), SiC (contact angle135°), Al_2O_3 (contact angle 142°), and graphite (contact angle 163°).

Mui et al. [37] generated Fe-Co-based droplets in an argon atmosphere with a resulting diameter range between 250 and 400 μm. This glass forming alloy has critical cooling rates (less cooling rate causes crystallization) between 700 and 900 K/s. With POEM the final particles size can be adjusted by the gas pressure on the melt surface (2–10 kPa) and the displacement of the cylindrical rod (6–10 μm).

2.4.3.2 Pneumatic Drop-on-Demand Technique (PDOD)

The pneumatic Drop-on Demand generator was first mentioned in a patent [5]. It can operate under two different modes: single droplet and multiple droplets. The principle experimental set-up is given in Fig. 2.11. Compressed gas passes a solenoid valve when it is open for a few milliseconds and the gas flow is spitted in two directions at a T-junction. Thus, the gas partly flows to a crucible and

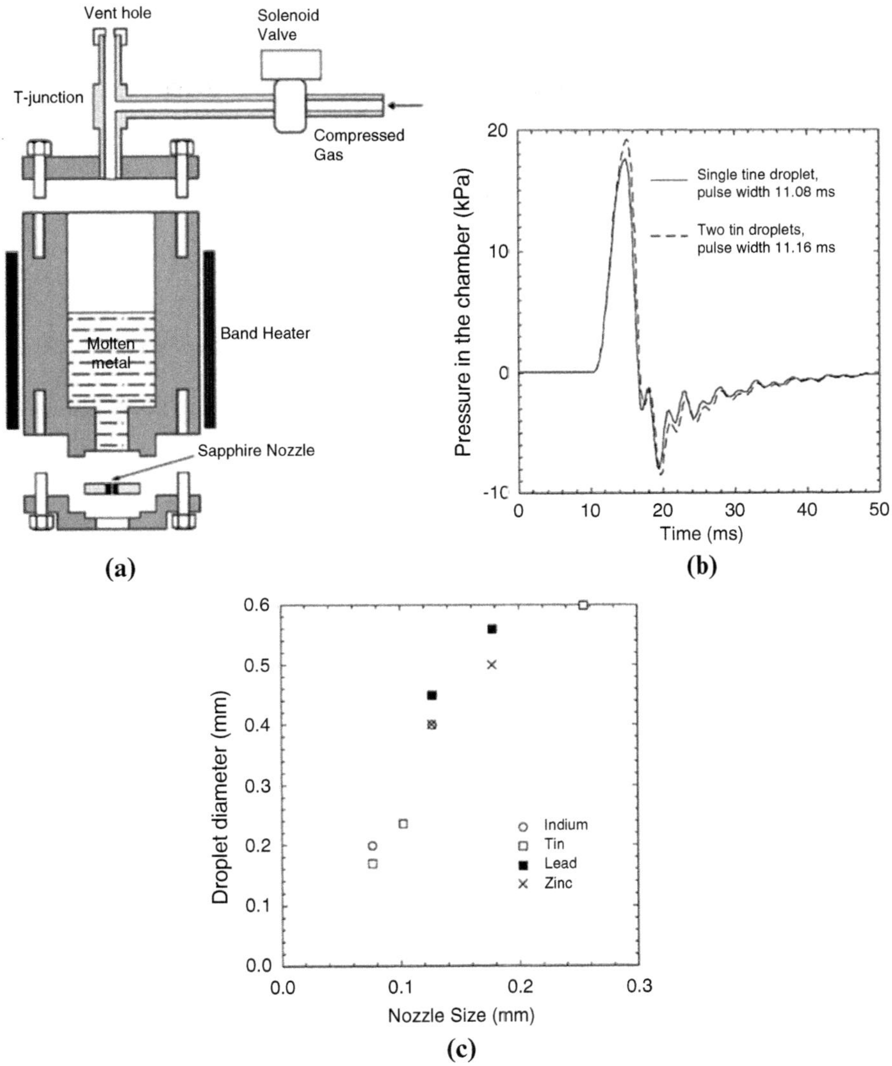

Fig. 2.11 (**a**) Principle of the PDOD, (**b**) typical pressure wave on the surface of the molten metal and (**c**) correlation between nozzle size and droplet size (from [35] with permission)

increases the pressure on the molten metal surface. The other part of the gas leave the system directly through a vent hole. Consequently, this configuration leads to a pressure peak on the melt surface as shown in Fig. 2.11 as well. This peak must be high enough to push one single droplet through the nozzle at the bottom of the crucible. The pressure can be increased by different parameter like the pulse width of the solenoid valve or the pressure of the compressed gas. This technique needs no moving parts in contact with the melt. Therefore, this system is cheaper, robust and convenient to use in high temperature applications. In the beginning this technique was only available for low melting elements like tin, lead, zinc and indium [35] but later a high temperature PDOD was developed to study the rapid solidification of iron based metallic glasses [38] and other alloy systems [39]. Typically, in the single droplet mode all diameters are similar and a little larger than the nozzle diameter. But smaller droplet diameters (smaller than the nozzle diameter) can be generated as well as described by Amirzadeh et al. [6].

2.4.3.3 StarJet Technology

The StarJet droplet generator [40, 41] can operate in three different modes which are the continuous droplet generation, drop on demand, and spray. The principle is similar to the pneumatic droplet generator. The gas flow through the nozzle is modulated by a pressure peak forcing small amounts of melt into a star-shaped channel. Principally, the generator consists of a bronze housing and a star-shaped silicon chip as shown in Fig. 2.12. A continuous inert gas flow is circulated around the melt which flows through the nozzle to prevent oxidation. A small amount of melt is pressed into the star-shaped nozzle and flows through a number of 12–16 channels with a typical width of 20–40 μm. The star-shaped nozzle is exhibited in Fig. 2.12. The melt only touches the surface of the v-shaped silicon tips. This kind of shape minimizes the wall contact of the liquid metal and reduces contact friction.

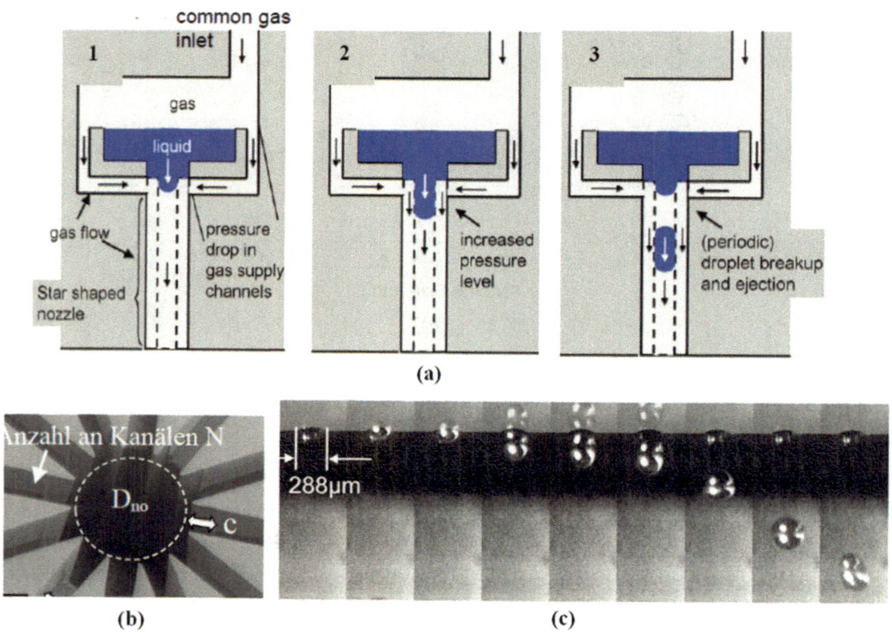

Fig. 2.12 Principle of a StarJet droplet generator (*above*), Detail of the StarJet silicon chip (*left below*), SnPb droplets at the nozzle exit (*right below*), adapted from [40]

At the end of the channel the droplets leave the nozzle as shown in Fig. 2.12. The special production technique of the silicon chip is described in the literature [40].

Different designs of the StarJet nozzles were tested and the nozzle diameter of the silicon chip was varied between 50 and 306 μm resulting in droplet diameters between 48 and 360 μm, respectively. The droplets were generated with a frequency of 0.5–50 Hz in the drop-on-demand mode and 5–4000 Hz in the continuous mode. The temperature of the melt is limited to a maximum temperature of 500 °C. Therefore, only low melting point alloys have been processed (Sn95-Ag4-Cu1, T_m = 210 °C and Zn96-Al4, T_m = 420 °C). It is possible to vary the droplet diameter not only by the nozzle diameter but also by the activation frequency. Because of the relatively long channel (approx. 500 μm), the droplets show very precise trajectories (less than 0.3° off the symmetry axis of the nozzle) which applies to the printing of 3D metal structures.

2.4.4 *Continuous Uniform Droplet Generation (UDG) Based on Rayleigh Instability*

The break-up phenomenon of liquid flow in air at room temperature was first reported by Lord Rayleigh [16]. His observations apply as well to molten metal streams [42]. Generally, a molten metal stream breaks up into unit volumes of varying droplet sizes. A chain of uniform droplets can be generated when a stream is disturbed by a characteristic frequency (Fig. 2.13). The frequency and the droplet size are predictable as the droplet rate is the reciprocal of the frequency. Typically, the disturbance is introduced by a vibrating shaft located in the melt in the vicinity

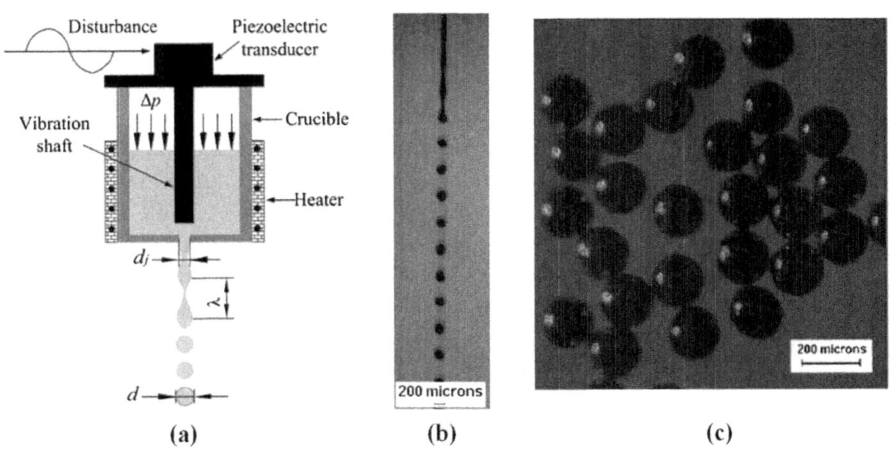

Fig. 2.13 Principle of uniform droplet spray process (*left*), Sn63-Pb37 droplet stream generated with a 100 μm orifice diameter, a frequency of 2600 Hz, a gas pressure of 40 kPa, and oxygen concentration less than 10 ppm (*middle*), Sn63-Pb37 particles with average diameter 192 μm (*right*) (from [43] with permission)

of the nozzle hole. The vibrating shaft consists of a refractory material with good thermal shock resistance and wear properties. This process has been used to investigate droplet deposition and deposit generation (s. Chap. 8). Common particle sizes vary between 75 and 500 μm. Achieving smaller particle sizes is difficult due to the high required initial pressure for smaller nozzle diameters below 50 μm. Because of impurities in the melt, the small nozzle may be susceptible to clogging, which affects the stability of the process.

The Rayleigh break-up has been used for basic scientific applications. Low melting point alloys have been used widely in the beginning and later it was applied to higher melt temperature alloys (Sn-Pb [43], Zn-Sn [42], Al-Alloys, Si) as well.

2.4.5 Impulse Atomization Process (IA)

IA is another droplet generation technique based on Rayleigh break-up which has proven to provide a unique experimental setup for the study of different atomization regimes in a controlled atmosphere [13, 17, 44]. It was developed and patented at the Advance Materials and Processes Laboratory at the University of Alberta in Canada [14, 45]. IA consists in generating one or several melt streams by mechanically pushing a bulk liquid produced by heating a material above its melting point, through orifices of known size and geometry. At the exit of the orifices, the generated discrete lengths of streams (and provided with the required instability by means of impulses) break-up into droplets that spheroidize (depending on the type of gas used) quickly and lose heat to a surrounding stagnant gas of choice while falling through a 4 m long drop tube as shown in Fig. 2.14a. This figure shows an impulse applicator/plunger (2) through which generated impulse is applied to the melt, a crucible (4) containing the melt with a nozzle plate through which the melt is pushed to form liquid ligaments that breakup into droplets Fig. 2.14b.

The droplets are accelerated by gravity during their fall so that no collision occurs despite a narrow spray angle of ~5°. Spherical powders and granules with smooth high quality surfaces have been produced with a mass mean droplets size (D_{50}) ranging from 100 μm to 1000 μm and controlled log-normal standard deviations (STD = D_{84}/D_{50}) of $1.1 \leq \text{STD} \leq 1.6$ [13, 46, 47]. Table 2.1 shows a summary of the type of materials and range of droplets sizes obtained by IA to date.

Figure 2.15 shows an example of droplets morphologies obtained by IA of *Sn-0.7wt% Cu*. Under the same atomization conditions, D_{50} can vary with alloy composition and orifice size. Figure 2.16 shows the variation of D_{50} with orifice diameter for different metals/alloys IA through 2.5 cm nozzle plates with various numbers of orifices in a nozzle plate from 1 to 97.

A critical requirement for processing via sintering or for some processes in Additive Manufacturing is knowledge of the solidified microstructure of the powder. IA has offered the unique capability to provide an environment where powder can be produced in a reproducible, clean, controlled environment that is quantifiable. This enables our ability to model the inter-dependent transport phenomena

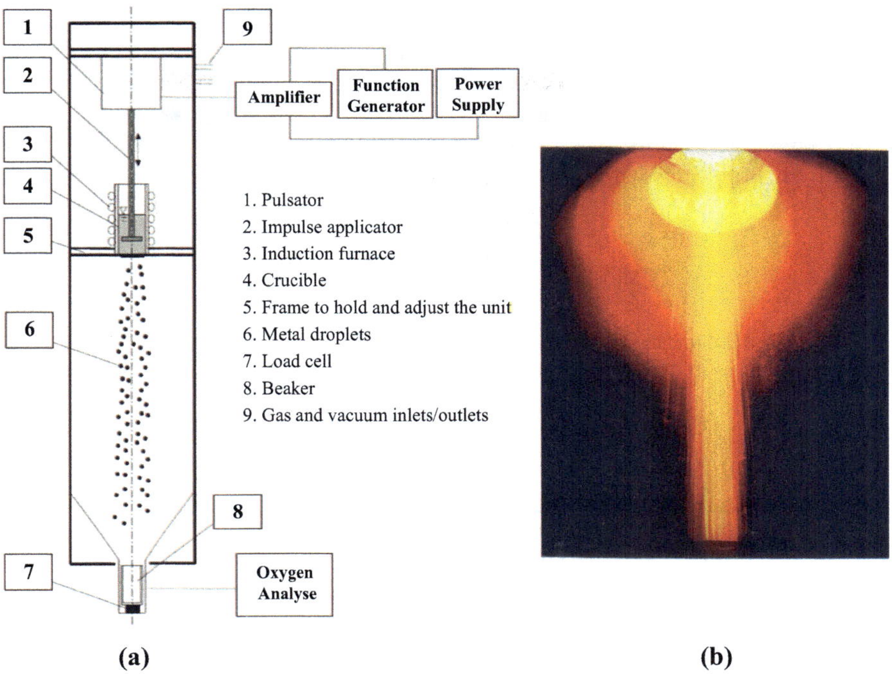

Fig. 2.14 Schematic of the Impulse Atomization (IA) setup

Table 2.1 List of materials and range of mass mean droplets size generated by IA

Metal or alloy	D50 (μm)	Metal or alloy	D50 (μm)
Aluminum		*Tin*	
Al	250–700	Sn-0.7 Cu	250–300
Al-5%Cu	364–517	Sn-0.7%Cu 3% Ag	268–303
Al-10%Cu	364–517	*Magnesium*	
Al-17%Cu	381–515	Mg	350–850
Al-24%Cu	377–521	Mg-9% Al-1%Zr	350–850
Al 6061	250–850	*Steel*	
Al-10%Sr	377–521	Fe-W	425–850
Al24%Sr	377–521	1040	425–850
Al 357	549–692	4140	354–800
Al-4.5%Cu-0.4%Sc	541–712	4140	364–881
Al-1.5%Mg-0.2%Sc	400–800	H-13	538–850
Al-3%Mg-0.2Sc	400–800	D2	385–810
Copper		*Neodynium*	
Cu	200–700	NdFeB	100–700
Bronze (90/10)	400–800	*BMG*	
Lead		FeCo-BMG	310–569
Pb-10% Sn	250–800	CoCr-BMG	585–921
Pb-12% Sn	130–500		
Zinc			
Zn-500ppm Pb	250–700		

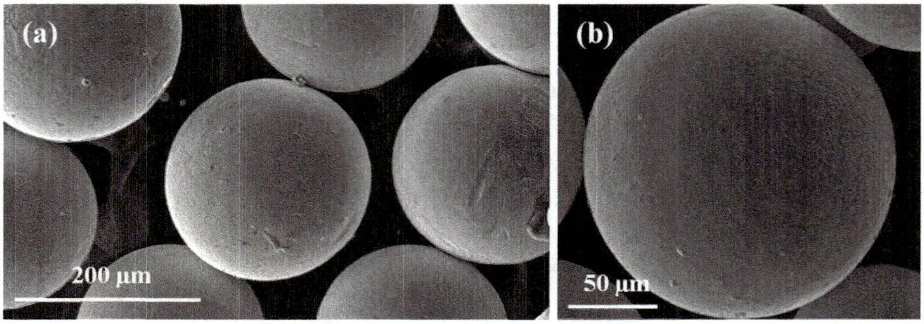

Fig. 2.15 SE-SEM images of the as-atomized Sn-0.7wt% Cu droplets of (**a**) Low magnification image, size range (in μm) 212–250 and (**b**) A higher magnification image, size range (in μm) 125–150

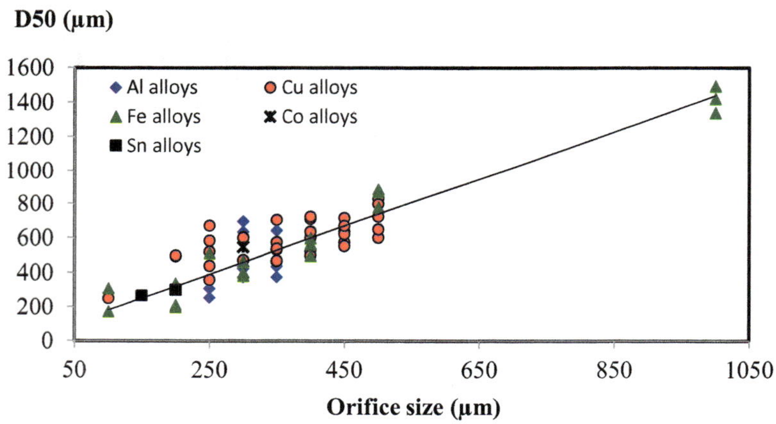

Fig. 2.16 Variation of the mass mean (D_{50}) with orifice diameter for different IA materials droplet cooling

taking place: heat transfer, fluid flow, and solidification. In the following sections, the methodology developed for this quantification of microstructure evolution of solidified droplets is presented.

2.4.5.1 Heat Transport

The solidification microstructure of an atomized droplet is highly dependent on the heat flow conditions. But, a direct measurement of the temperature history of droplets during IA is difficult to achieve for the simple reason that they cool very fast while falling in a sealed chamber filled with inert atmosphere. However, numerical models (based on the quantification of heat exchange between the droplet and gas in an environment with high temperature gradient) of droplet cooling

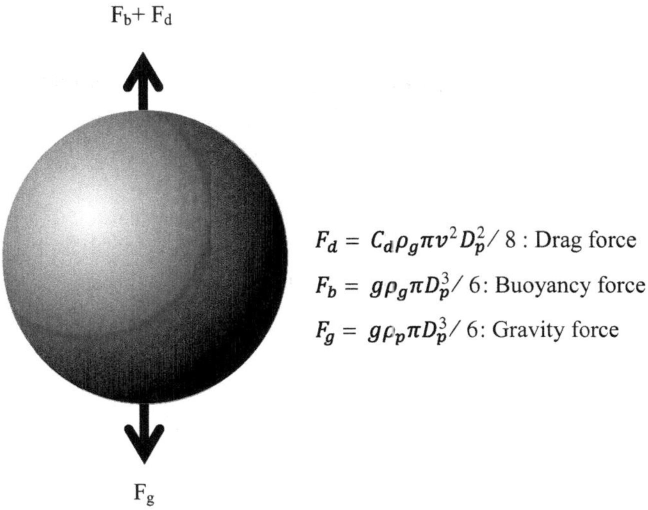

Fig. 2.17 Schematic of the forces acting on IA droplet during its fall

during IA have been developed from Wiskel's heat transfer model formulation [48] and validated [49–52].

Since the cooling of a droplet happens during its flight after atomization, the temperature history is a function of time and displacement down the droplet trajectory. After exiting from an orifice, the trajectory of a droplet is determined by its initial velocity, u, and the different forces acting upon it. These are the gravity force F_g, the buoyancy force F_b and the drag force F_d (Fig. 2.17) [48].

The net force (instantaneous acceleration) on the droplet can be written as:

$$\frac{dv}{dt} = \frac{\rho_p - \rho_g}{\rho_p} g - 0.75 \frac{\rho_g C_d}{\rho_p D_p} v^2 \tag{2.28}$$

Where v represents the relative velocity between the droplet and the atomization fluid which is stagnant ($v_{gas} = 0$), ρ_p and ρ_g are respectively the densities of the droplet and atomization atmosphere. D_p is the droplet diameter and C_d is the drag coefficient [23].

$$C_d = \frac{18.5}{Re^{0.6}} \tag{2.29}$$

Thus, as a result of the sum of these forces, the relative velocity between the droplet and the gas changes during the fly until the droplet reaches its terminal velocity. Consequently the heat transfer from the droplet to the gas occurs under transient conditions so that the heat transfer coefficient between the droplet and the surrounding gas is given by the following expression of heat flow q:

$$q = h_{eff} S \left(T_S - T_g\right) \tag{2.30}$$

where h_{eff} is the effective heat transfer coefficient consisting of the combination of convection, conduction & radiation heat transfer mechanisms, S is the contact area between the atomization atmosphere and the droplet, T_S and T_g are respectively the temperatures of the droplet and atomization atmosphere.

Radiation heat transfer is neglected so that h_{eff} is evaluated as an additive contributions of convection and conduction heat transfer coefficients. Therefore, a relationship with the Nusselt number Nu must be used.

$$h_{eff} = \frac{Nu k_g}{D_p} \tag{2.31}$$

where k_g *is* the gas conductivity at the interface with the droplet of temperature T_S. Indeed, Nu indicates the ease of convective heat transfer from an object and its value depends upon the shape of the object, the relative gas-droplet velocity and the properties of the gas.

Based on the correlation proposed by Whitaker [53], a modified expression of Nu is given by Wiskel et al. [48] as:

$$Nu = 2\frac{B}{k_s(m+1)}\frac{\left(T_S^{m+1} - T_g^{m+1}\right)}{\left(T_S - T_g\right)} + \left(0.4Re^{1/2} + 0.06Re^{2/3}\right)Pr^{0.4}\left(\frac{\mu_g}{\mu_S}\right)^{1/4} \tag{2.32}$$

Where B, m, k_s, μ_g and μ_s are respectively the pre-power coefficient, the power coefficient, the gas conductivity at the droplet surface temperature, the viscosity evaluated at the free stream fluid temperature and the viscosity evaluated at the droplet surface temperature. While Re is already given by Eq. (2.12) the Prandtl number Pr of the gas is expressed as:

$$\mathrm{Pr} = \frac{C_{pg}\mu_g}{k_g} \tag{2.33}$$

The expression given by Eq. (2.33) suggests that the properties of the gas must be evaluated at the droplet temperature as the droplet is falling through the stagnant gas. Note that validation experiments were carried out by Wiskel et al. [48] using the Whitaker [53] and the Ranz-Marshall [54] correlations. The experimental conditions under which the Ranz-Marshal correlation was deveoped is not suitable to be used for the large temperature gradients between melt droplets and gas tempearture. The experimental results performed by Wiskel et al. [48] clearly showed that only the Whitaker correlation should be used as modified by the temperature dependent gas properties.

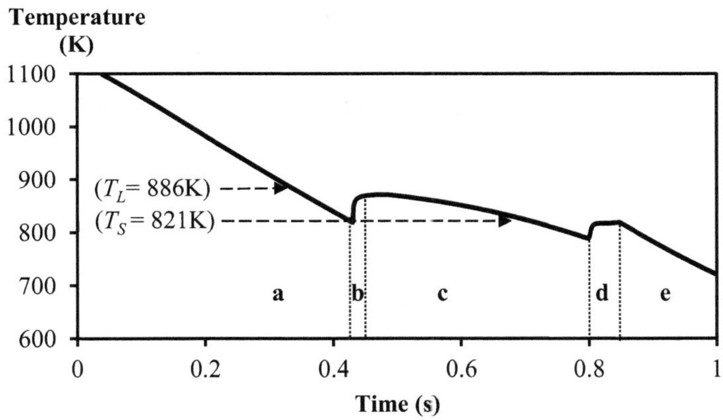

Fig. 2.18 Temperature profile of a 196 mm impulse atomized Al-17wt% Cu droplet undercooled by 70 K, (**a**) liquid state before primary nucleation temperature is reached, (**b**) recalescence of the undercooled melt, (**c**) post recalescence/segregated solidification, (**d**) eutectic solidification and finally and (**e**) cooling in the solid state

2.4.5.2 Cooling Rate

During the fall in the atomization chamber, droplet solidification is determined by continuous cooling in five successive stages as shown in Fig. 2.18. These include: (a) liquid state before primary nucleation temperature is reached, (b) recalescence of the undercooled melt, (c) post recalescence/segregated solidification, (d) eutectic solidification and finally and (e) cooling in the solid state.

Thus, two definitions of cooling rate can be distinguished [55]: (1) Slope of the Time-Temperature curve or *liquid cooling* rate $\dot{T}_L$. And (2) the ratio of solidification temperature interval ΔT_{SL} (primary nucleation temperature till eutectic nucleation temperature) to the local solidification time t_{SL} or the *solidification cooling* rate $\dot{T}_{SL}$:

$$\dot{T}_{SL} = \frac{\Delta T_{SL}}{t_{SL}} \tag{2.34}$$

Solidification cooling rate is obviously controlled by the heat exchange between the stagnant gas and the atomized droplets and therefore depends upon the nature of the gas in the atomization chamber (helium, argon or nitrogen) and the droplets initial falling velocity u and size D_p. For IA droplets of size range 100–1000 μm, u is found to be 0.5 ms^{-1} [14].

The most widely used cooling rates in microstructure predictions are the liquid cooling rate as defined in (1) and the solidification cooling rate as defined in (2). While both cooling rates are atomization gas dependent, the former is a function of droplet size and the latter is closely related to the size of the solidification microstructure. Figure 2.19a shows cell spacing variation with $\dot{T}_{SL}$ and Fig. 2.19b shows liquid cooling rate variation with average droplet size.

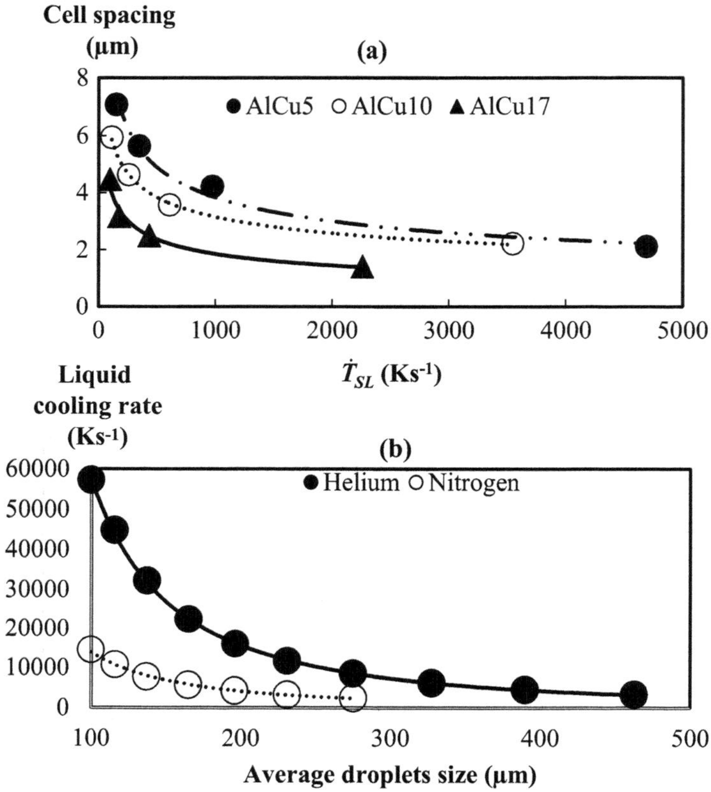

Fig. 2.19 (**a**) Cell spacing variation with solidification cooling rate and (**b**) liquid cooling rate as a function of average droplet size

2.4.5.3 Microstructures

Microstructures characterization of IA droplets is achieved by optical or Scanning Electron Microscopy (SEM), Electron Backscatter Diffraction (EBSD) and X-rays Diffraction (XRD). But also by the more advanced techniques such as Neutron diffraction and X-ray micro-tomography. Indeed, optical microscopy, SEM or EBDS show images of a random cross-section of the droplet obtained by grinding, polishing and sometimes etching the surface of a droplet mounted in an epoxy resine [56, 57]. As shown by SEM (in back scattered electron mode) image in Fig. 2.20, droplets of the same size range at random cross sections can show a non-uniformly distributed microstructures characterized by different fractions of eutectic and different length scales. It is therefore clear that a 3D imaging is required to fully characterize a droplet microstructural features such as nucleation points, dendrites morphology as well as porosity and phase fractions.

Figure 2.21 shows a 3D vizualization of a droplet microstructure consisting of dendrites network orginating from a single nucleation point (Fig. 2.21a) and growing in different planes (Fig. 2.21b). The images were obtained by synchrotron

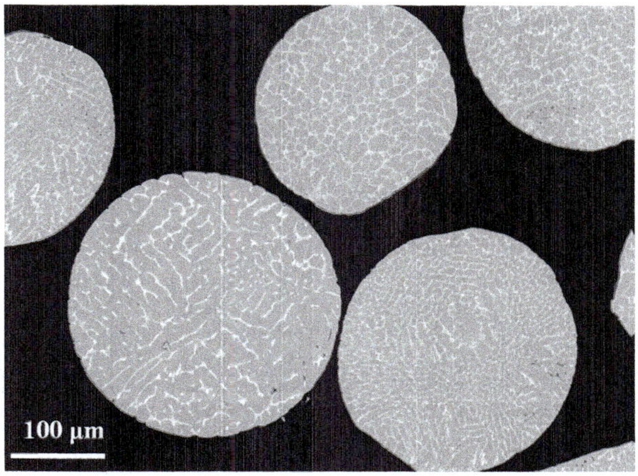

Fig. 2.20 Non-uniformly distributed microstructures of Impulse Atomized Al-4.5wt% Cu droplets. Size range: 212–250 μm, Cooling gas: Argon

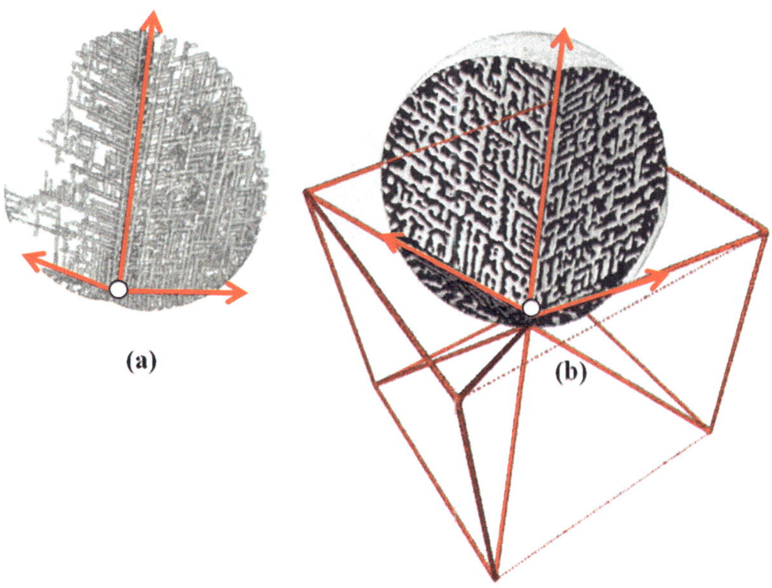

Fig. 2.21 Synchrotron X-ray micro-tomography result showing (**a**) a droplet cross-section and (**b**) the corresponding 3D-reconstruction with characteristic planes and the position of the nucleation centre (*white dot*) in a 290 μm diameter Al-4.5wt% Cu droplet atomized in helium

X-rays micro-tomography at the European Synchrotron Radiation Facility (ESRF) [44].

3D reconstructions by X-ray micro-tomography allows to visualize cross sections of the droplets in every spatial directions. Visualization of microstructures in any direction makes it possible to observe dendrites morphology within a droplet so that the process variables can be inferred.

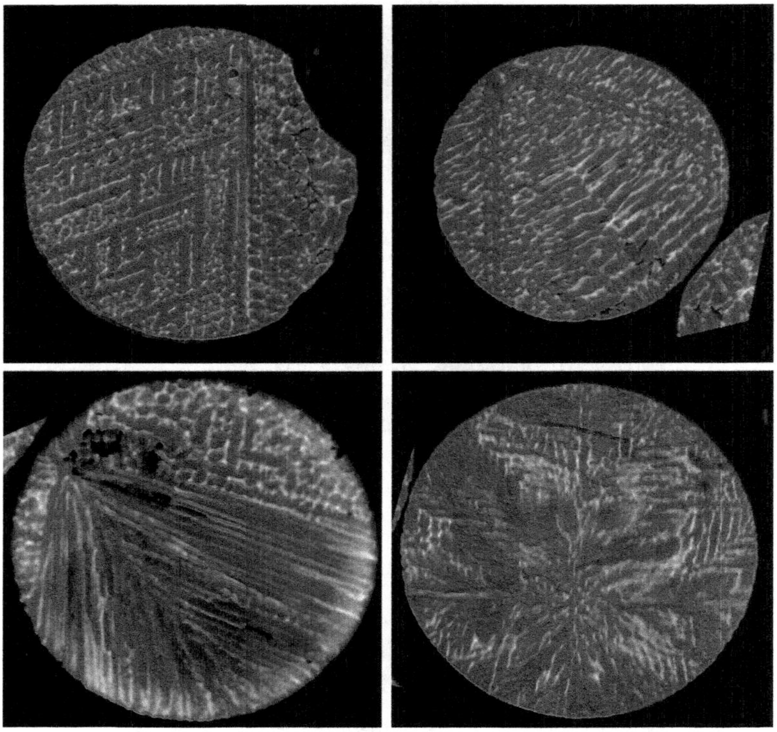

Fig. 2.22 IA Al-4.5wt% Cu Droplets cross-section obtained by synchrotron X-ray micro-tomography showing a wide range of dendrite morphologies within droplets of similar sizes and produced under the same atomization conditions

An analysis of IA Al-4.5wt% Cu droplets by micro-tomography (Fig. 2.22) evidenced a wide range of dendrite morphologies within droplets of similar sizes and produced under the same atomization conditions [58]. The same analysis showed that in most droplets, dendrites growth occurs along the <111> crystallographic direction instead of the <100> growth direction observed in conventional castings.

X-ray micro-tomography allowed not only to quantify but also to analyze porosity formation and curvature in IA droplets. Porosity, mostly cylindrical is found to be due to solidification shrinkage (Fig. 2.23).

It is also possible to carry out microstructural phase identification and quantification of the droplets by the means of X-rays or Neutron Scattering. This is achievable by Rietveld refinement analysis of powder diffraction patterns. The analysis enables to quantify microsegregation during rapid solidification of the droplets. Figure 2.24 shows an example of a diffraction pattern analysis of Al-4.5wt% Cu IA-droplets average size 275 μm analyzed by the Rietveld refinement method.

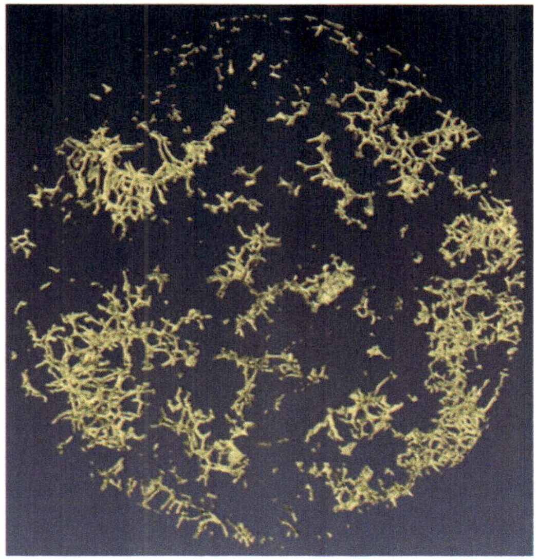

Fig. 2.23 Synchrotron X-ray micro-tomography result showing (**a**) a droplet cross-section and (**b**) the corresponding 3D-reconstruction with characteristic planes and the position of the nucleation point

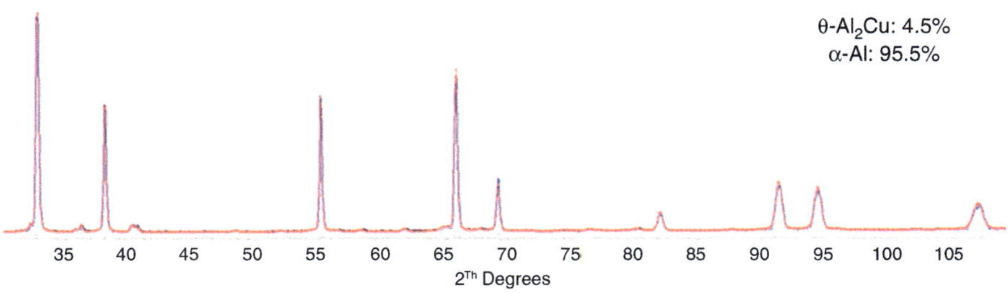

Fig. 2.24 Rietveld refinement fit of neutron diffraction spectra of IA Al-4.5 wt% Cu droplets (size range from 250 μm to 300 μm). Experiment conducted at ILL-Grenoble France

This analysis shows that rapid solidification by IA yields microstructures with reduced microsegregation as the fraction of θ-Al_2Cu (4.5%) is much lower than the Gulliver-Scheil prediction (~5.4%).

2.4.5.4 Quantification of Dendritic and Eutectic Nucleation Undercoolings

Due to the difficulty to monitor the temperatures of droplets moving relatively fast, in-situ measurements of nucleation temperatures cannot be achieved during IA. Consequently a post-mortem method to determine the nucleation temperatures of primary and higher order phases must be developed. Recently, such a method has been developed for binary Al-Cu and D2 steel whose secondary nucleation is a eutectic so that the nucleation tempeature of the primary phase $T_{P,}$ and of the eutectic T_E were calculated [59–61]. This method consists in using experimental results obtained by Rietveld refinement analysis of Neutron scattering data to determine the eutectic fraction (F_E) and eutectic undercooling (ΔT_E). Then, ΔT_P is determined by using ΔT_E in combination with semi-empirical coarsening models of secondary dendrite arms.

2.4.5.5 Estimation of Eutectic Fraction and Undercooling

Phase fraction results obtained by Rietveld refinement analysis of neutron scattering data is used to estimate the eutectic fraction (F_E). To achieve this, Gulliver-Scheil (GS) model of microsegregation is used to determine F_E at different undercoolings following the extended solidus and liquidus lines of the phase diagram so that the temperature at which the calculated F_E equals the the fraction obtained by Rietveld refinement corresponds to the eutectic undercooling ΔT_E of the considered droplet. Figure 2.25 shows a metastable extension of solidus and liquidus lines of the aluminum rich hypoeutectic region of Al-Cu phase diagram obtained by suspension of θ-Al_2Cu by Thermo-Calc© [62].

Considering that rapid solidification results from high cooling rate and undercooling, it does not promote time dependent processes such as diffusion. Therefore, a model that assumes no diffusion in the solid and complete mixing in the liquid such as the GS model of microsegregation is considered to be more suitable for the determination of eutectic fractions in rapidly solidified droplets. F_E *and* ΔT_E obtained for Al-Cu and D2 steel droplets are shown in Fig. 2.26.

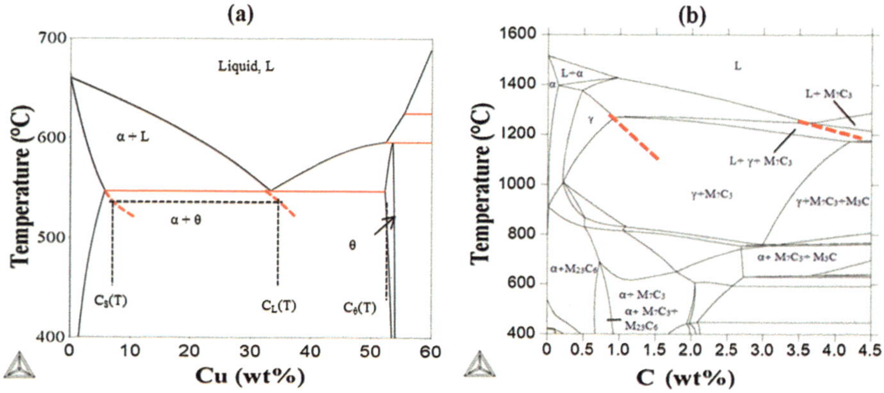

Fig. 2.25 Thermo-Calc calculations of: (**a**) Al-rich corner of Al-Cu phase diagram and (**b**) D2 tool steel pseudo-binary phase diagram. Metastable extension of solidus and liquidus lines are indicated by *dashed lines*

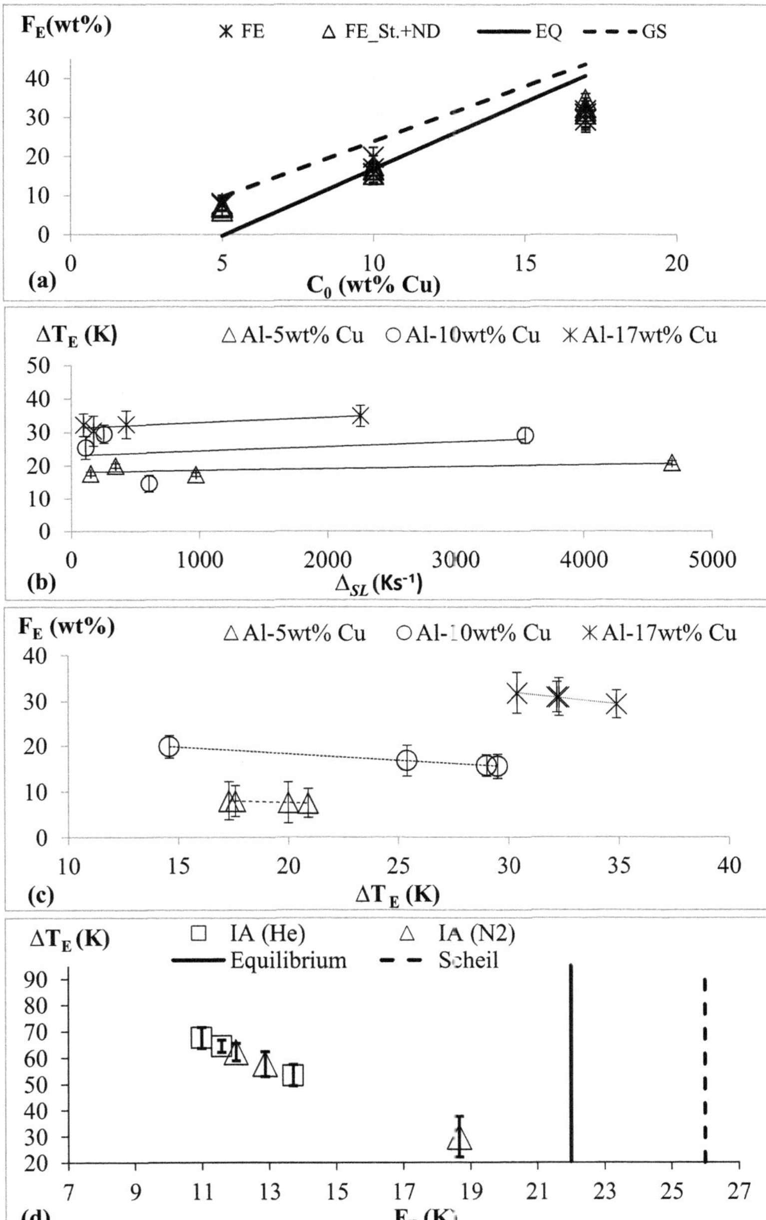

Fig. 2.26 (**a**) Weight percent eutectic variation with Al-Cu alloy nominal composition, (**b**) Eutecticc nucleation undercooling variation with cooling rate for different Al-Cu compositions, (**c**) Weight percent eutectic variation with eutectic nucleation undercooling for different Al-Cu compositions and (**d**) Weight percent eutectic variation with eutectic nucleation undercooling for D2 steel

2.4.5.6 Estimation of Primary Undercooling

Primary dendritic nucleation undercooling ΔT_P is defined as the difference between the equilibrium liquidus temperature T_L and the nucleation temperature T_P of the primary dendritic α-phase.

$$\Delta T_P = T_L - T_P \tag{2.35}$$

It is assumed that (1) α-phase forms as a result of nucleation, growth and coarsening up to the nucleation of the eutectic and that following the nucleation of θ-Al_2Cu phase no primary solid formation takes place, (2) there is no primary α-phase re-melting due its recalescence and (3) most of the dendrites that have formed did so during coarsening so that semi-empirical relationships such as the one reported by Kurz and Fisher [63] can be used to describe the secondary dendrite arms spacing λ_2 as a function of solidification time:

$$\lambda_2 = 5.5(\bar{M}t_{SL})^{1/3} \tag{2.36}$$

where t_{SL} is the local solidification time and $\bar{M}$ (varying from the nucleation temperature of primary α-phase T_P to the eutectic nucleation temperature T_E) is the average value of the coarsening parameter M.

$$\bar{M} = \frac{1}{T_P - T_E'} \int_{T_E'}^{T_P} M(T)dT \tag{2.37}$$

The mathematical formula of the coarsening mechanism proposed by Kattamis [64] has been modified by Roosz and co-workers [65] to account for the temperature dependent physical parameters so that M is defined as:

$$M = \frac{\gamma D_L(T)T}{m(1-k)\Delta H C_L} \tag{2.38}$$

where γ is the energy of solid/liquid interface, ΔH is the latent heat of solidification, m is the liquidus slope, k is the solute partition coefficient, D_L is the solute diffusion coefficient in liquid phase and C_L is the solute composition in liquid phase.

By plotting the measured λ_2 as a function of $\dot{T}$ (estimated by a Thermal model [49]), λ_2 is shown to be related to $\dot{T}$ by a power law.

$$\lambda_2 = A\left(\dot{T}\right)^{-n} \tag{2.39}$$

A and n are alloy-dependent parameters obtained from the best fitting curves ($R^2 > 0.90$) of λ_2 *Vs* $\dot{T}$ [49]. The parameter A, described as a "composition-sensitive" coefficient by Eskin et al. [55] is found to be decreasing with increasing

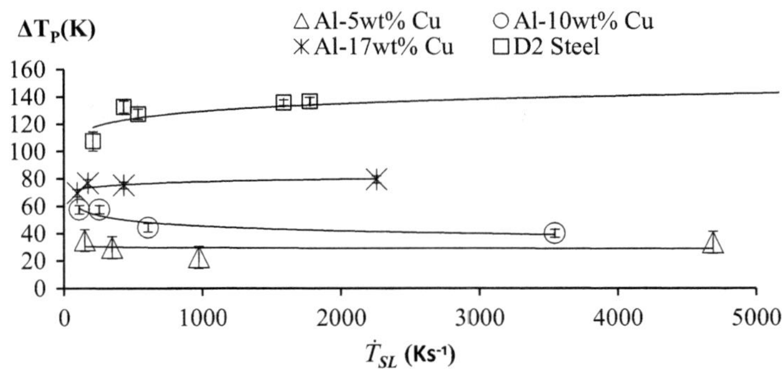

Fig. 2.27 Primary phase nucleation undercooling variation with cooling rate for different Al-Cu compositions and D2 Steel

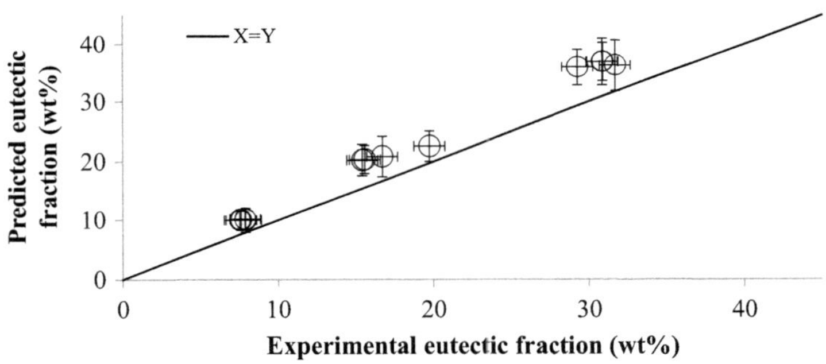

Fig. 2.28 Comparison of experimental results with models output results of weight percent eutectic for the Al-Cu droplets

nominal alloy composition C_0 (A is equal to 43. 96, 22.66 and 20.92 respectively as C_0 varies from 5wt%, 10wt% and 17wt%) while coefficient n is less sensitive to the alloy composition ($n \simeq 1/3$ for the three investigated alloys). Thus, Eqs. (2.36) and (2.39) can be reduced to Eq. (2.40) after substitution of $\frac{\Delta T_{SL}}{t_{SL}}$ for $\dot{T}$.

$$\bar{M}\Delta T_{SL} = \left(\frac{A}{5.5}\right)^3 \tag{2.40}$$

Thus values of $\bar{M}$ and ΔT_{SL} are calculated at different temperatures T varying from T_L to T_E until Eq. (2.40) is solved. The corresponding value of T at which Eq. (2.40) is solved is therefore the temperature of interest T_P and ΔT_P is deduced by subtraction of T_P from T_L. Variation of ΔT_P with cooling rate for Al-Cu and D2 Steel droplets are shown in Fig. 2.27.

ΔT_P and ΔT_E for Al-Cu alloys are then used as input variables to run a microsegregation model for binary alloys [50]. The output results were found to be in agreement with the theory of a spherical droplet solidifying in a quiescent gas. As shown in Fig. 2.28, the fractions of eutectic computed by the microsegregation model compare very favourably with the experimental results.

2.5 Summary

The fundamentals of Single Fluid Atomization (SFA) have been discussed in details with illustrations of common SFA techniques. The advantages of these techniques have been demonstrated through the break up and cooling mechanism of droplets. Rapid solidification microstructures formation and characterization have been discussed through atomized droplets of a wide range of materials by Impulse Atomization (IA). A novel method of quantification of nucleation undercooling has been described. It is clear that SFA techniques provide clean environments where powder production is reproducible, controlled and quantifiable characterization is possible. Thus, direct information on the solidification path and thermal history controlling microstructures development during rapid solidification is accessible through SFA. Analysing IA droplets can allow to map out a wide range of solidification microstructures so that their processing history can be inferred from the knowledge of nucleation undercooling and cooling rate. Knowledge of the solidified microstructure of the powder is critical requirement for processing via sintering or for some processes in Additive Manufacturing.

2.6 List of Symbols

2.6.1 Latin

Symbole	Description
A	Composition sensitive coefficient by Eskin et al.
a	Intercept of the C_D *Vs* Re regression line
b	Slope of the C_D *Vs* Re regression line
B_o	Bond number
c	Circumference of a liquid ligament
d	Diameter of a liquid ligament
E	Energy
f	Frequency
F	Force
Fr	Froude number
g	Gravity acceleration constant
h	Liquid head
k	Conductivity
L	Length
M	Mass
$\bar{M}$	Average value of the coarsening parameter M
n	Alloy-dependent parameter
Nu	Nusselt number
Oh	Ohnesorge number

(continued)

Symbole	Description
Pr	Prandtl number
Q	Liquid flow rate
r	Radius
Re	Reynolds number
S	Surface area
T	Temperature
t	Time
T_E	Eutectic nucleation temperature
u	Velocity
We	Weber number
B	Pre-power coeffient

2.6.2 Greek

Symbole	Description
γ	Energy of solid/liquid interface
Δ	Difference
λ_2	Secondary dendrite arms spacing
ρ	Liquid density
σ	Surface tension
λ	Wave length
μ	Viscosity coefficient
τ	Ratio of time taken to spherodize (t_{sp}) and to dissipate thermal energy (t_{th})
ω	Angular velocity

2.6.3 Indicies

Symbole	Description
C	Centrifugal
eff	Effective
i	Index number
l	Liquid
max	Maximum
p	Droplet
s	Surface
SL	Solid-Liquid
Sp	Spheroidization

References

1. Herlach, D. M. (2012). Containerless undercooling of drops and droplets. In *Solidification of containerless undercooled melts* (pp. 1–30). Weinheim: Wiley.
2. Masuda, S., Takagi, K., Dong, W., Yamanaka, K., & Kawasaki, A. (2008). Solidification behavior of falling germanium droplets produced by pulsated orifice ejection method. *Journal of Crystal Growth, 310*(11), 2915–2922.
3. Ishikawa, A. (1999). Spherical shaped semiconductor integrated circuit, 5,955,776.
4. Chandra, S., & Jivraj, R. (2002). Apparatus and method for generating droplets. *US patent No. 6,446,878 B1*.
5. Kempf, B., Ptaschek, G., Ringelstein, H.-M., Fuchs, R., & DiVicenzo, C. (2007). Method and device for producing spherical metal particles. *US patent No. 7297178 B2*.
6. Amirzadeh Goghari, A., & Chandra, S. (2006). Producing droplets smaller than the nozzle diameter by using a pneumatic droplet generator. In *19th Annual Conference on Liquid Atomization and Spray Systems* (pp. 23–26).
7. Kuzuoka, Y., Isomae, S., & Yamaguchi, Y. (2007). Crystal morphology of spherical silicon particles produced by jet-splitting method. *Journal of Crystal Growth, 304*(2), 487–491.
8. Stelter, M., Brenn, G., & Durst, F. (2002). The influence of viscoelastic fluid properties on spray formation from flat-fan and pressure-swirl atomizers. *Atomization and Sprays, 12*(1–3), 299–327.
9. Lavernia, E. J., & Srivatsan, T. S. (2010). The rapid solidification processing of materials: Science, principles, technology, advances, and applications. *Journal of Materials Science, 45*(2), 287–325.
10. Giulio Caccioppoli, H. H., Clausen, B., & Bonjour, C. (2004). Ultrasonic atomization of metallic melts: Modelling and case studies. In *Powder manufacturing and processing* (p. 59), Vienna, Austria.
11. Lierke, E. G., & Griesshammer, G. (1967). The formation of metal powders by ultrasonic atomization of molten metals. *Ultrasonics, 5*(4), 224–228.
12. Wisutmethangoon, S., Plookphol, T., & Sungkhaphaitoon, P. (2011). Production of SAC305 powder by ultrasonic atomization. *Powder Technology, 209*(1), 105–111.
13. Henein, H. (2002, March). Single fluid atomization through the application of impulses to a melt. *Materials Science and Engineering: A, 326*(1), 92–100.
14. Yuan, D. (1997). *The novel impulse atomization process*. Edmonton: University of Alberta.
15. Roach, S. (2001). *Determination of the physical propoerties of melts*. Edmonton: University of Alberta.
16. Strutt, L. R. J. W. (1878). On the instability of Jets. *London Mathematical Society, 10*, 4–13.
17. Roach, S. J., & Henein, H. (2003). A dynamic approach to determining the surface tension of a fluid. *Canadian Metallurgical Quarterly, 42*(2), 175–186.
18. van Hoeve, W., Gekle, S., Snoeijer, J. H., Versluis, M., Brenner, M. P., & Lohse, D. (2010). Breakup of diminutive Rayleigh jets. *Physics of Fluids, 22*(12), 122003.
19. Aldinger, N., Linck, F., & Claussen, E. (1977). Melt-drop technique for the production of high-purity metal powder. *Modern Developments in Powder Metallurgy, 9*(19), 141–151.
20. Reitz, R. D. (1987). Modeling atomization processes in high-pressure vaporizing sprays. *Atomization and Sprays, 3*, 309–337.
21. Ohnesorge, W. V. (1936). Formation of drops by nozzles and the breakup of liquid jets. *Journal of Applied Mathematics and Mechanics, 16*, 355–358.
22. Jiang, G., Henein, H., & Siege, M. W. (1988). Intelligent sensors for atomization processing of molten metals and alloys. *Analysis, June*, 3.
23. Yule, A. J., & Dunkley, J. J. (1994). *Atomization of melts for powder production and spray deposition*. Oxford series on advanced manufacturing. Oxford: Clarendon Press.
24. Dunkley, J. J. (1978). Production of metal powders by water atomization. *Powder Metallurgy International, 10*(1), 38–41.

25. Aderhold, D., & Dunkley, J. J. (2007). Centrifugal atomization of metal powders. In *Powder metallurgy and particulate materials* (pp. 26–31).
26. Sungkhaphaitoon, S. W. P., & Plookphol, T. (2012). Design and development of centrifugal atomizer for producing zinc metal powder. *International Journal of Applied Physics and Mathematics, 2*(2), 77–82.
27. Plookphol, T., Wisutmethangoon, S., & Gonsrang, S. (2011). Influence of process parameters on SAC305 lead-free solder powder produced by centrifugal atomization. *Powder Technology, 214*(3), 506–512.
28. Liu, H. (2000). *Science and engineering of droplets—Fundamentals and applications.* Westwood, USA: Noyes Publications.
29. Liu, Q. G. J., & Yu, Q. (2012). Experimental investigation of liquid disintegration by rotary cups. *Chemical Engineering Science, 73*, 44–50.
30. Fraser, R. P., & Eisenklam, P. (1953). Research into the performance of atomizers for liquids. *Imperial College Chemical Engineering Society Journal, 7*, 52–53.
31. Grant, P. S. (1995). Spray forming. *Progress in Materials Science, 39*(95), 497–545.
32. Dunkley, J. J., Xie, J. W., & Zhao, Y. Y. (2004). Effects of processing conditions on powder particle size and morphology in centrifugal atomization of tin. *Powder Metallurgy, 47*(2), 168–172.
33. Sheikhalieva, Z. I., & Dunkley, J. J. (2007). Centrifugal atomization of melt. Influence of chamber oxygen on the size and shape of aluminium powder particles. *Euro PM.*
34. Eslamian, M., Rak, J., & Ashgriz, N. (2008). Preparation of aluminum/silicon carbide metal matrix composites using centrifugal atomization. *Powder Technology, 184*(1), 11–20.
35. Cheng, S. X., Li, T., & Chandra, S. (2005). Producing molten metal droplets with a pneumatic droplet-ondemand generator. *Journal of Materials Processing Technology, 159*(3), 295–302.
36. Masuda, K. T. S., Suzuki, H., & Kawasaki, A. (2006). Preparation of monosized copper micor particles by pulsated orifice ejection method. *Materials Transactions, 45*(5), 1380–1385.
37. Miura, A. K. A., Dong, W., Fukue, M., Yodoshi, N., & Takagi, K. (2011). Preparation of Fe-based monodisperse spherical with fully glassy phase. *Journal of Alloys and Compounds, 509*, 5581–5586.
38. Ellendt, L. M. N., Ciftci, N., Goodreau, C., & Uhlenwinkel, V. (2016). Solidification of single droplets under combined cooling conditions. In *Fourth international CONFERENCE on advances in solidification processes (ICASP-4), Materials science and engineering* (Vol. 117, p. 012057).
39. Uhlenwinkel, H., Ellendt, V., Lehmann, N., Naujoks, R., & Bentz, M. (2014). High temperature drop-on-demand system. In *World Congress on powder metallurgy and particulate materials (PM2014).*
40. Metz, T., Birkle, G., Zengerle, R., & Koltay, P. (2009). Starjet: Pneumatic dispensing of nano-To picoliter droplets of liquid metel. In *Proceedings of the IEEE International Conference on Micro Electro Mechanical Systems (MEMS)*, pp. 43–46.
41. Lass, N., Tropmann, A., Metz, T., Zengerle, R., & Koltay, P. (2011). 3D Rapid prototyping by direct printing of liquid metal using the StarJet technology. *Mikrosystemtechnik Kongress.*
42. Cherng, J.-P., & Chun, J.-H. (2002). In flight solidification behaviour of aluminium alloy droplets produced by the uniform droplet spray process. In *SFB372-Kolloquium Band 6.*
43. Xiao-Shan Jiang, J.-M. Z., Qi, L.-H., Luo, J., & Huang, H. (2010). Research on accurate droplet generation for micro-droplet deposition manufacture. *The International Journal of Advanced Manufacturing Technology, 49*(5), 535–541.
44. Prasad, A., & Henein, H. (2008). Droplet cooling in atomization sprays. *Journal of Materials Science, 43*(17), 5930–5941.
45. Huan, D., Henein, H., & FallaVollita, J. A. (1997). Method for producing droplets. *US patent No. 5,609,919.*
46. Mejia, J., Henein, H., Morin, L. C., & Reider, M. (1996). Impulse atomization: A novel technique to economically produce powders with a desired size distribution. In *Advances in powder metallurgy and particulate materials* (pp. 1–185). Princeton, NJ: APMI.

47. Olsen, K., Sterzik, G., & Henein, H. (1995). Upgrading scrap automotive aluminum alloys with the impulse atomization and quench technique. In *International symposium recycling of metals and engineered materials.3 by the Minerals, Metals and Materials Society (TMS), Warrendale* (pp. 67–83).
48. Wiskel, J. B., Henein, H., & Maire, E. (2002). Solidification study of aluminum alloys using impulse atomization: Part I—Heat transfer analysis of an atomized droplet. *Canadian Metallurgy Quarterly, 41*(1), 97–110.
49. Prasad, A., Mosbah, S., Henein, H., & Gandin, C.-A. (2009). A solidification model for atomization. *ISIJ International, 49*(7), 992–999.
50. Gandin, C.-A., Mosbah, S., Volkmann, T., & Herlach, D. M. (Aug. 2008). Experimental and numerical modeling of equiaxed solidification in metallic alloys. *Acta Materialia, 56*(13), 3023–3035.
51. Tourret, D., Reinhart, G., Gandin, C.-A., Iles, G. N., Dahlborg, U., Calvo-Dahlborg, M., et al. (Oct. 2011). Gas atomization of Al–Ni powders: Solidification modeling and neutron diffraction analysis. *Acta Materialia, 59*(17), 6658–6669.
52. Tourret, D., & Gandin, C.-A. (Apr. 2009). A generalized segregation model for concurrent dendritic, peritectic and eutectic solidification. *Acta Materialia, 57*(7), 2066–2079.
53. Whitaker, S. (1972). Forced convection heat transfer correlations for flow in pipes, past flat plates, single e cylinders, single spheres, and for flow in packed beds and tube bundles. *AIChE Journal, 18*(2), 361–371.
54. Ranz, W. E., & Marshall Jr., W. R. (1952). Evaporation from drops. Parts I & II. *Chemical Engineering Progress, 48*(141–146), 173–180.
55. Eskin, D., Du, Q., Ruvalcaba, D., & Katgerman, L. (2005). Experimental study of structure formation in binary Al-Cu alloys at different cooling rates. *Materials Science and Engineering: A, 405*(1–2), 1–10.
56. Prasad, A. (2006). *Microsegregation studies of rapidly solidified binary Al-Cu alloys*. Edmonton: University of Alberta.
57. Ilbagi, A. (2012). *Non-equilibrium containerless solidification of Al-Ni alloys*. Edmonton: University of Alberta.
58. Bedel, M., Reinhart, G., Bogno, A.-A., Gandin, C.-A., Jacomet, S., Boller, E., et al. (2015). Characterization of dendrite morphologies in rapidly solidified Al–4.5wt.%Cu droplets. *Acta Materialia, 89*, 234–246.
59. Henein, H., Gandin, C.-A., Bogno, A.-A., & Delshad Khatibi, P. (2016). Quantification of dendritic and eutectic nucleation undercoolings in rapidly solidified hypo-eutectic Al-Cu droplets. *Metallurgical and Materials Transactions A, 47*, 4606.
60. Bogno, A., Delshad Khatibi, P., Henein, H., & Gandin, C.-A. (2013). Quantification of primary and eutectic undercoolings of impulse atomized Al-Cu droplets. In *Materials science & technology 2013*.
61. Khatibi, P. (2014). *Microstructural investigation of D2 tool steel during rapid solidification*. Edmonton: University of Alberta.
62. Thermo-Calc (2008) Database, TTAL7.
63. Kurz, W., & Fisher, D. J. (1998). *Fundamentals of solidification* (4th ed.). Boca Raton: CRC Press.
64. Coughlin, J. C., Flemings, M. C., & Kattamis, T. Z. (1967). Influence of coarsening on dendrite arm spacing of aluminum-copper alloys. *Transactions of the Metallurgical Society of AIME, 239*, 1504.
65. Roosz, H. E., Halder, A., & Exner, E. (1986). Numerical calculation of microsegregation in coarsened dendritic microstructures. *Materials Science and Technology, 2*, 1149–1155.

Chapter 3
Two Fluid Atomization Fundamentals

Iver E. Anderson and Lydia Achelis

3.1 Introduction

Gas atomized molten metal sprays are the precursor to many types of special metal powder-based materials, where understanding of the process can permit efficient access to desirable types of solidification microstructures to allow tailoring of materials consolidated from such powders or can promote production of a desired powder size class to minimize or eliminate the need for costly sub-sieve size classification [1]. Alternatively, gas atomized molten metal sprays are deposited directly, before complete solidification, on mandrels or substrates to generate pre-form shapes for deformation processing or near-final shapes that are machined to finished form. In the spray deposition process, the pre-forms or near-final shapes also can benefit from close control of droplet size, in terms of deposition efficiency and solidification microstructure control, especially to minimize trapped porosity [2]. To help accomplish the overall goals for our book, this chapter will present insights into the physics of melt breakup and spray formation processes during "two-fluid" atomization, both for well-practiced and emerging types of melt disintegration methods. Due to the complexity and chaos of experiments on molten metal break-up, many researchers have found it most expedient to bypass detailed calculations or extensive modeling of the full atomization process that utilize momentum, surface energies, and energy transfer parameters. Instead, much of

I.E. Anderson (✉)
Ames Laboratory (USDOE), Division of Materials Sciences and Engineering and Materials Science and Engineering Department, Iowa State University, Ames, Iowa 50011, USA
e-mail: andersoni@ameslab.gov

L. Achelis
Department of Production Engineering, University of Bremen and Foundation Institute of Materials Science (IWT), 28359 Bremen, Germany
e-mail: achelis@iwt.uni-bremen.de

H. Henein et al. (eds.), *Metal Sprays and Spray Deposition*,
DOI 10.1007/978-3-319-52689-8_3

the metallic melt atomization research community relies on analysis of high-speed visualization of the melt spray (primarily focused on melt filming or filamentation and initial droplet formation), on calculation of gas-only flows that are verified by gas flow visualization, and on characterization and correlation of the results of each atomization experiment, in terms of the metal powder produced, especially size distribution analysis. Indeed, these types of results will be presented in this chapter. However, it is likely that the growth of capabilities for modeling and simulation of similar atomization processes for other "simpler" fluids, e.g., fuel sprays and aqueous solutions, by multiphase flow analysis will have greater application in the molten metal atomization research community as the demand grows for more efficient and precise powder production and improved spray deposition.

3.2 Gas Atomization Configurations

3.2.1 Free-Fall Gas Atomization (FFA)

3.2.1.1 Introduction

Free-fall gas atomizers were some of the first two fluid atomizer designs to be used for molten metal atomization. In a simple open (unconfined stream) design a melt stream falls from a tundish exit via gravity into the convergence of focused atomization gas jets where it is disintegrated. The distance between the melt stream and exit point of each gas jet (usually in a circular array) is fairly large (about 5 cm) to protect the jets from damage due to melt splashing or impingement from stream "wandering." Also, the melt stream falls a certain distance (perhaps 5–10 cm) before the gas flow impacts it. This makes the free-fall atomizers much less problematic in terms of freezing and clogging when compared to close-coupled gas atomizers, but requires that the gas jets travel a relatively long way before hitting the melt. This reduces their velocity and, consequently, the resulting yields of fine powder for a given gas to metal mass flow ratio (GMR) and, as a result, the convective cooling of the melt due to the cold gas (from expansion chilling) occurs at a later position and with less effectiveness. The free-fall atomizer concept has several applications in liquid droplet/spray processing. Free-fall atomizers are applied in spray drying, powder production, and spray forming processes [3, 4]. In the spray forming process the free-fall atomizer provides an additional advantage, namely the simplified possibility of controlled mechanical or pneumatic scanning and, therefore, oscillating the gas atomizer with respect to one axis. This scanning technique enables a potential influence on control and regulation of the mass flux distribution of droplets in the spray, which is necessary for processes that promote flat product spray forming and twin atomization. This important capability can only be controlled in other atomizer nozzle systems (within a running process) by modulating the atomization gas pressure or flow rate (e.g., in a split manifold).

3.2.1.2 Dual Nozzle Improvement

An improved free-fall atomizer was designed with two gas nozzles arranged above each other as shown in Fig. 3.1. The primary gas nozzle is located on the top of the arrangement such that the primary gas is co-flowing along the axis of the liquid jet. This is intended to provide a gas flow field without any internal recirculating vortex within the primary nozzle, promoting a guided liquid jet into the atomization area and suppressing the weak recirculation gas flow within the secondary nozzle.

The secondary nozzle is the main atomization unit [5]. The concentric gas jets from the secondary nozzle impinge onto the central liquid jet within the atomization area (see Fig. 3.1), causing its disintegration via instabilities from the shearing action of the secondary gas flow and its relative (mismatched) gas velocity [6]. The disintegration process of the liquid jet in the atomization zone is located underneath the secondary gas nozzle [7, 8]. Because of the separation between the atomization area and the atomizer body and the tolerance for wide variations of melt and gas parameters, the free-fall atomizer is mainly used for atomization of melts or viscous high temperature liquids without narrow size distribution requirements for the resulting droplets or powders.

Due to the inclination of the secondary gas flow, a recirculation gas flow effect (see Fig. 3.1) may occur underneath the secondary gas nozzle, depending on the atomizer design and operating conditions. If the intensity of the recirculation flow is reasonably high, liquid ligaments or droplets can be transported back (upstream/outward) to the atomizer body as seen in Fig. 3.2 (left image). This melt flow effect can produce sticking/welding of melt fragments that may clog the gas orifices (or even the melt orifice) and can negatively influence or even stop the atomization

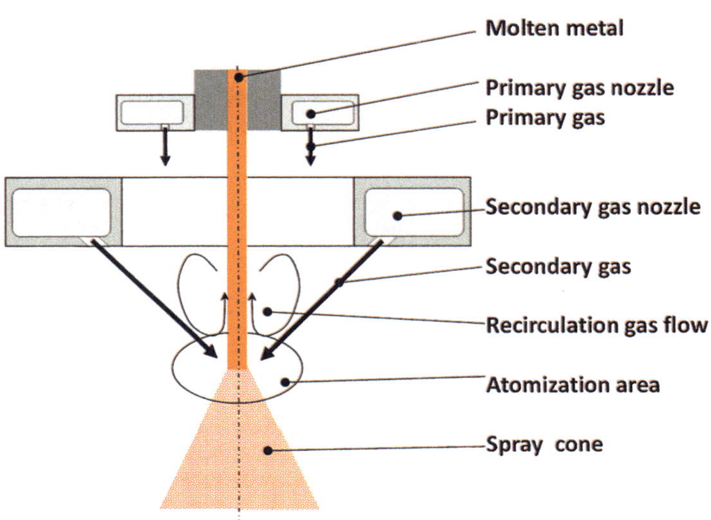

Fig. 3.1 Main components with gas and melt flow regions of an improved version of the conventional free-fall gas atomizer

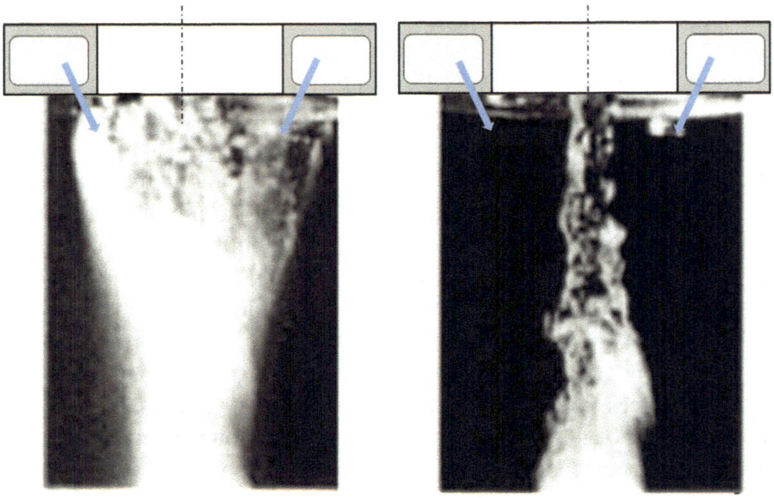

Fig. 3.2 Water disintegration without primary gas flow (*left*), and with primary gas flow (*right*)—adapted from [9]

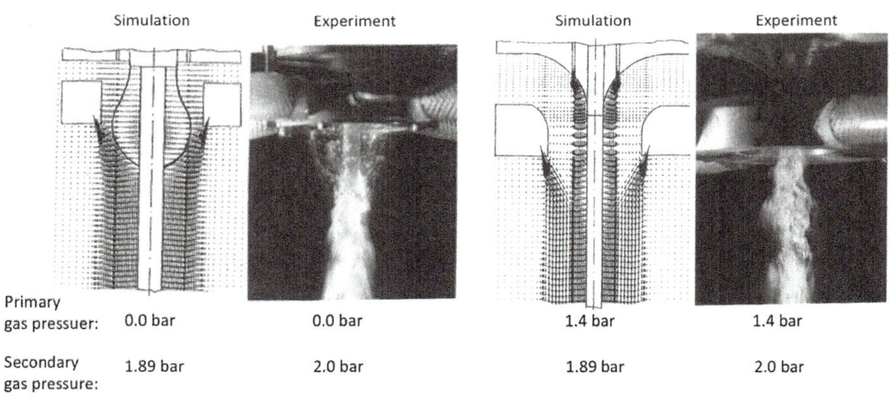

Fig. 3.3 Comparison of simulated and experimental results; without primary gas (*left image*), and with primary gas (*right image*)—adapted from [3, 5]

process. In common improved free-fall atomizer designs, the recirculation gas flow can be suppressed (Fig. 3.2, right image) by the use of a primary (upper) gas nozzle (see Fig. 3.1).

Comparisons of simulated gas velocity vector plots in the vicinity of the free-fall atomizer and images of water disintegration in a similar nozzle are shown in Fig. 3.3. The left comparison shows a calculated flow field and experiment for atomization without the use of primary gas. The gas from each gas jet exits the secondary nozzle by an inclination angle from the central axis of 10° but contacts the liquid stream more intensely (at a larger angle) than indicated by that angle due to free expansion (particularly internally). Thus, the point where the gas hits the melt steam is slightly closer to the nozzle body than the geometric point of jet impingement (theoretical atomization point). External gas is accelerated into the inner flow area due to gas entrainment from the edge of the circular jet. A high proportion of the entrainment gas flows through the gap between primary gas nozzle body and secondary gas nozzle, where a maximum gas velocity of approximately 10 m/s is achieved, according to the calculation. In the inner region of the gas flow field a large recirculation area can be observed indicated by a solid line marking the streamline boundary. Inside the vortex the maximum velocity of the upward-directed (counter flowing along the melt stream exterior) gas is 45 m/s, resulting in upwards and outwards accelerated particles, directed towards the surrounding nozzle body. These recirculating particles may hit the nozzle close to the melt exit, as visualized in the left water spray picture (labeled Experiment in the left image of Fig. 3.3) as a thin veil of spray that is external to the main stream on the central axis. This result is validated by visual observation of water disintegration by an almost identical process. Several droplets can be seen above the theoretical atomization point, accelerated upwards against the main gas flow towards the nozzle body and tundish exit.

In the second comparison on the right side of Fig. 3.3, an improved free-fall atomization unit that utilizes the stabilizing support of primary gas with several other modifications is shown. The modified gas nozzle arrangement uses primary gas nozzles located as close as possible to the melt exit to suppress primary gas flow recirculation and a smooth nozzle contour to prevent detachment of flow, creating a "sheath" gas flow to stabilize the melt stream. In addition the vertical gap between primary and secondary nozzles is increased to maximize volumetric flow of entrainment gas from the spray chamber environment. Process parameters for calculating the primary and secondary gas flows were set at 1.40 and 1.89 bars, respectively. A gas flow field without a recirculating vortex in the vicinity of the nozzle is observed in the simulation results. However, in the experiment there is an apparent extension of the undisrupted stream, which indicated that acceleration of the fluid has occurred. This is because of the close proximity of the primary gas to the liquid jet, where the gas velocity values in the boundary layer between the fluid/gas interface is increased. Unfortunately, the stabilized liquid stream extension does not lead to improved droplet spray formation, which seems to require an additional downstream atomization mechanism.

3.2.1.3 Discussion

During operation of the improved free-fall atomizer, the gas mass flow rates of both gas nozzles (in Fig. 3.1) are controlled by the primary and secondary gas pressures [10]. Problems may occur during atomizer operation if the pressure ratio between the primary and the secondary nozzle is improperly adjusted and an intense recirculation flow is generated. The maximum applicable gas pressure of the primary gas nozzle is limited by initial disturbances that may be generated on the liquid stream before reaching the atomization area. In the same way, the secondary nozzle gas pressure is also limited because the secondary gas mass flow rate determines the necessary primary gas mass flow rate for prevention of recirculation. This coupling of the two nozzle systems limits the applicability (especially for high pressure atomization) of this type of improved free-fall atomizer [3, 11].

New and modified free-fall atomizers have been developed to improve the gas flow, limit gas recirculation, and improve produced powder quality in both size and shape. To improve the operating conditions of the free-fall atomizer it is necessary to increase the gas pressures while still suppressing gas recirculation. One method to achieve this is through using a cylindrical ring device to influence the local atomization gas direction via the Coanda effect to improve the atomization performance [12]. Experimental and numerical studies conducted by other workers [13] have shown that it is possible to achieve a recirculation free flow and very stable atomizer operation without applying any primary gas by using a Coanda-flow device. In addition, the droplet diameter was halved when compared to the conventional free-fall atomizer results. Further experimental studies have focused on narrowing the particle size distribution and improving particle quality [14]. This optimized free fall atomizer arrangement has produced high quality powder with a narrow particle size distribution of 1.6 ($d_{84.3}/d_{50.3}$), and high sphericity.

3.2.2 *Close-Coupled Gas Atomization (CCGA)*

3.2.2.1 Introduction

For the enhanced control of gas atomization in high temperature, high surface tension fluids (molten metals) it is preferred to use more complex two fluid nozzles, termed close-coupled gas atomization (CCGA) nozzles. As illustrated in Fig. 3.4, the atomization gas exits from a gas jets (or a gas slit) that is connected by a

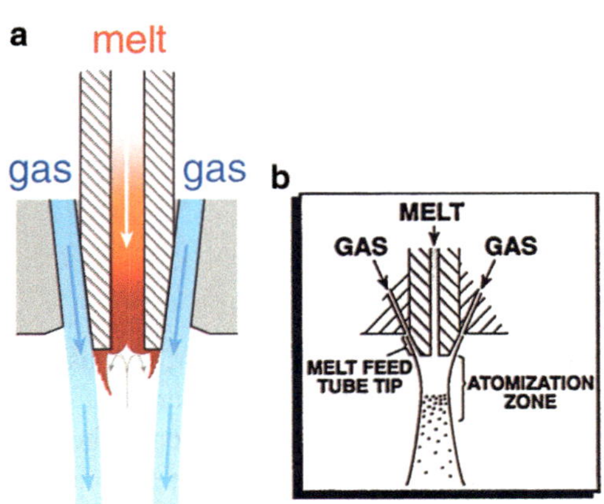

Fig. 3.4 Schematic of interacting melt and atomization gas flows in a close-coupled gas atomization nozzle, showing: (**a**) the gas recirculation effect that promotes melt stream splitting and filming and, (**b**) the nomenclature that is used to describe the CCGA nozzle features

distribution manifold and the gas collides almost immediately with melt from a confined stream that has been split near the tube exit and is distributed as a film across the base of the melt feed tube tip by the action of a strong local gas recirculation effect. The local gas recirculation flow feature will be discussed later in this chapter with regards to the type of gas flow field that is selected, either open wake or closed wake.

On the other hand, CCGA nozzles do have a tendency for melt "freeze-off" events if the surfaces chilled by the rapidly expanding atomization gas are not thermally isolated from the melt, especially when feeding melt into the atomization zone. This is typically overcome by the use of an elevated superheat for the melt (perhaps 100–300 °C) in many situations. Probably, the greatest difficulty for operation of CCGA nozzles is with atomization gas flows that exhibit a high gas (jet or annular slit) apex angle (e.g., 45°) relative to the central axis (see Fig. 3.5a) that can achieve wake closure. These types of nozzles (see Fig. 3.5b) often exhibit a

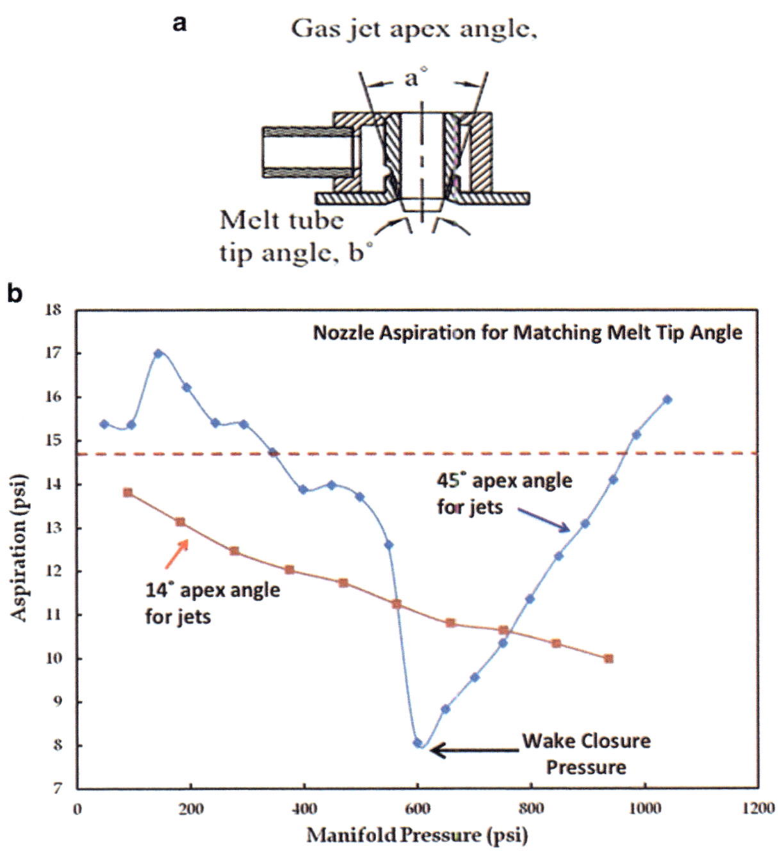

Fig. 3.5 Illustration of the angular relationship of the atomization gas jets and the melt pour tube tip chamfer in (**a**) and the dependence of aspiration pressure (measured with gas-only) for two types of CCGA nozzles that operate only in open wake (14° apex angle) or in (**b**) that can achieve an intense closed wake flow pattern at elevated pressure (45° apex angle)

high sensitivity for stable melt feeding conditions (where a negative aspiration pressure promotes suction and stable melt feeding) on the atomization gas supply manifold pressure (that directly controls the gas flow velocity). This sensitivity is particularly apparent when a nozzle is operated in maximum energy transfer conditions (needed to achieve supersonic velocity gas that contacts the melt), when the gas jet apex angle is equal to the angle of the chamfered surface of the melt feed tube tip (see Fig. 3.5a). In fact, for some CCGA nozzle geometries with large apex angles (see Fig. 3.5b) there can be low and high gas supply manifold pressure (gas velocity) regions when a positive pressure at the melt feed tube tip can promote positive (back) pressure on the melt stream and stop the melt flow, even erupting gas bubbles from the melt surface. On the other hand, for CCGA nozzle geometries with small (e.g., 14°) apex angles (see Fig. 3.5b) there appear to be no manifold pressure (gas velocity) regions without a negative pressure at the melt feed tube tip, so only stable atomization will occur. Of course, since no steep aspiration pressure drop appears (that indicates a closed wake gas flow pattern) for the 14° nozzle, one should not expect to utilize a second disintegration mechanism for highly efficient fine powder making. When first offered the chance to integrate a CCGA nozzle into a typical commercial powder making operation, atomization engineers are most comfortable with small apex angle nozzles, since it is most tolerant of manifold pressure adjustments.

3.2.2.2 Discussion

It should be apparent that CCGA nozzles are capable of more energy intensive disintegration in the near-field "atomization zone" because of the closer proximity of the high energy atomization gas to the initial melt exit, compared to free-fall atomization nozzles, and, thus, should produce higher yields of fine powder. Commercial powder producers have preferred CCGA nozzles for fine powder production for many years. As will be discussed in this chapter, the close coupling arrangement of the melt to the gas also should permit responsive "tuning" of the powder size distribution by the energy level of the atomization gas, if process uniformity is maintained. This size control capability of CCGA, if well developed, could also have applications in droplet spray deposition and for powder making of specific powder size ranges, which is of current interest for additive manufacturing and for thermal spray deposition.

In practice, high speed visualization and size distribution analysis of the powders resulting from molten metal atomization experiments using these CCGA nozzles revealed that the typical supersonic gas flows act to effectively pre-film and disrupt a "thick" melt stream or sheet (see Fig. 3.6). Unsurprisingly, comparisons of the data show that the velocity mismatch or shear between the liquid and gas phase promotes (in effect) the transformation of kinetic energy to surface energy that is required for melt breakup to occur [15]. As will be discussed later in this chapter, there has been some success (to be discussed below) in controlling powder size distribution results that makes use of this rough description of melt breakup physics.

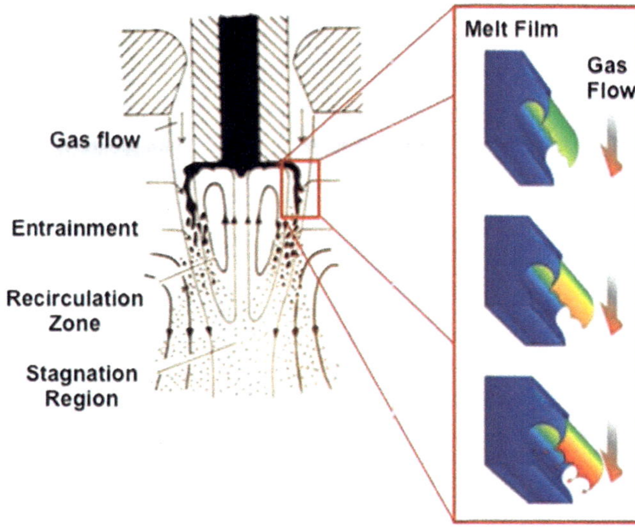

Fig. 3.6 Schematic of the flow patterns of the melt and gas that lead to molten metal droplet formation by the action of a CCGA nozzle

Specifically, this control has been exercised over average particle size (d_{50}) and powder size standard deviation (d_{84}/d_{50}) by variation of atomization gas pressure in the regime of velocity below and slightly above Mach 1.0, where the velocity is most sensitive to supply pressure.

3.2.2.3 Control of Primary Atomization in CCGA

Introduction

A grand challenge for the technology of two-fluid gas atomization of melts is to demonstrate a significant range of droplet size control based on shear velocity, which can be related to a specific melt break-up model [4]. In an attempt to gain control of the process and to limit broadening of atomized particle size distributions, close-coupled two-fluid gas atomization nozzles have been developed in laboratory systems [1] and are in commercial practice [16]. However, truly precise control of close-coupled gas atomization processes for molten metals is complicated by the need to maintain a high temperature melt flow immediately adjacent to a cold, rapidly expanding gas flow without melt stream freezing. This typically militates the use of high melt superheats that can promote elemental losses/contamination from surface reactions or excessive melt flows that can overwhelm the primary atomization effect of the available atomization gas and can broaden the size distribution [4]. One example of excess melt flow can be seen in a high-speed video still image in Fig. 3.7 that shows the persistence of melt sheet fragments far

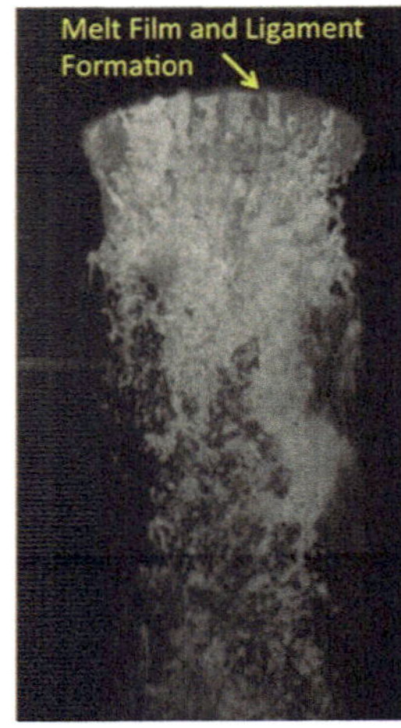

Fig. 3.7 Still image from high-speed (4000 fps) video that shows atomization spray from Fe–Cr alloy, poured at 1750 °C

downstream of the primary atomization zone where only secondary breakup by a much slower gas flow will be active. As a consequence, even laboratory CCGA systems may produce powders with a broad size distribution, demonstrating that unambiguous primary atomization has not yet been achieved in these interacting fluid flow fields [17].

The dominance of uncontrolled secondary break-up of melt fragments among the droplets from primary break-up in an atomized spray probably is most responsible for the common observation of broad size distributions [17], even in CCGA systems. Thus, the overwhelming influence of secondary processes has frustrated most attempts to gain true shear velocity control of the droplet diameter. A major problem lies within the complex two-phase flow field of a typical close-coupled atomizer where dense packets of melt are formed and thinned to yield droplets by the shearing action of high energy gas flows. In the primary atomization zone, the liquid packets can be exposed to strong temperature gradients, internal and external gas recirculation zones, and high velocity shear and internal shock (Mach disk) disruption, if closed wake gas flow conditions (Fig. 3.1a) are utilize and maintained [18]. It should be noted that the internal shear surface that extends downstream from the edge of the melt tip (see Fig. 3.8) for both open and closed wake gas flow patterns provides an envelope for the primary atomization zone. Also, this same melt tip edge is the location of the most intense pure shear disintegration forces and is the location where the melt first encounters the gas. If liquid packets move downstream beyond the near-field primary break-up zone, they can become subject to secondary break-up processes from gas flows of greatly reduced energy that add coarser particles to the yield and broaden the size distribution [4, 17]. Therefore, it is only in the primary break-up zone that high efficiency atomization can be closely controlled because the gas velocity is at a maximum, the melt flow is essentially static, and the pre-filmed melt thickness can be influenced by the melt tube orifice design [1].

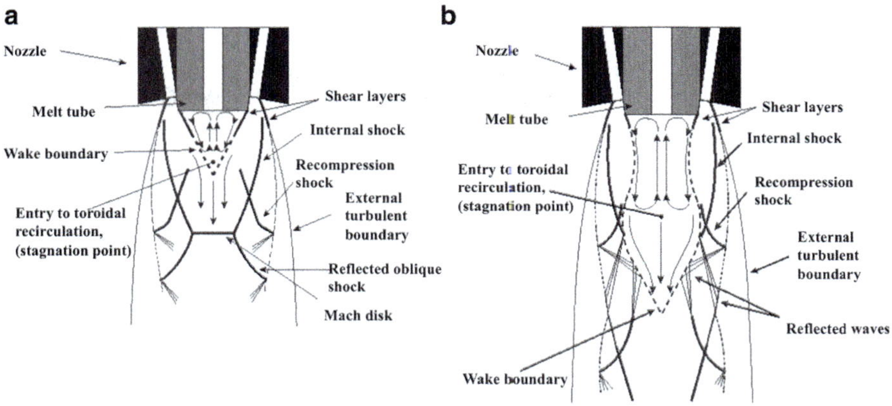

Fig. 3.8 Schematic axi-symmetric gas-only flow fields for Ar or N_2 atomization gas exiting from a close-coupled gas atomization nozzle, adapted from [18]

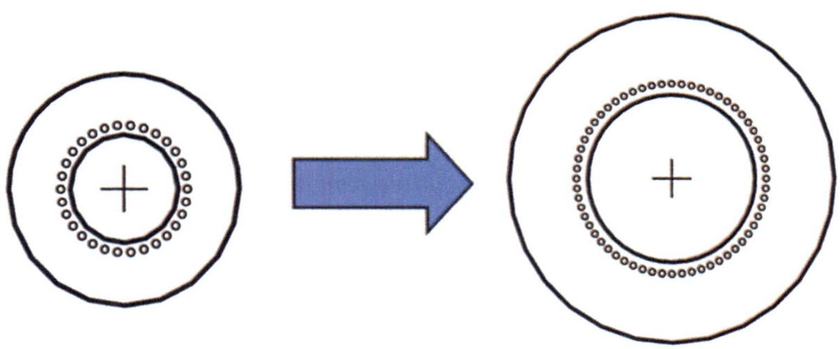

Fig. 3.9 Schematic bottom-view of the enlarged gas atomization nozzle from, (**a**) 30 gas jets with a 10.4 mm diameter central bore to, (**b**) 60 gas jets with a 19.5 mm dia. central bore, adapted from [19]

Thus, attempts were made [1] to perform more controlled close-coupled gas atomization with an expanded nozzle (Fig. 3.9b) that was designed (with a 14° apex angle) to operate only in an open wake condition (Fig. 3.8b) over a wide range of atomization gas manifold pressures. The selection of an open wake gas flow pattern can eliminate complications for analysis of resulting powder sizes from the onset and operation of an additional shock disruption (atomization) mechanism at higher pressures [18], where a wake closure gas flow pattern (see Fig. 3.8a) can dominate. Simple geometric effects from the expanded nozzle central bore can promote melt film thinning across the base of the melt tube tip, driven by the toroidal recirculation in Fig. 3.6, by increasing the melt spreading distance, from the melt orifice center, and by expanding the periphery of the tip [20].

To achieve melt flow stability and further control of filming to maximize control of the initial shear atomization process, a melt tube orifice design was developed that was meant to transform the typically chaotic melt stream flow [19] into stable

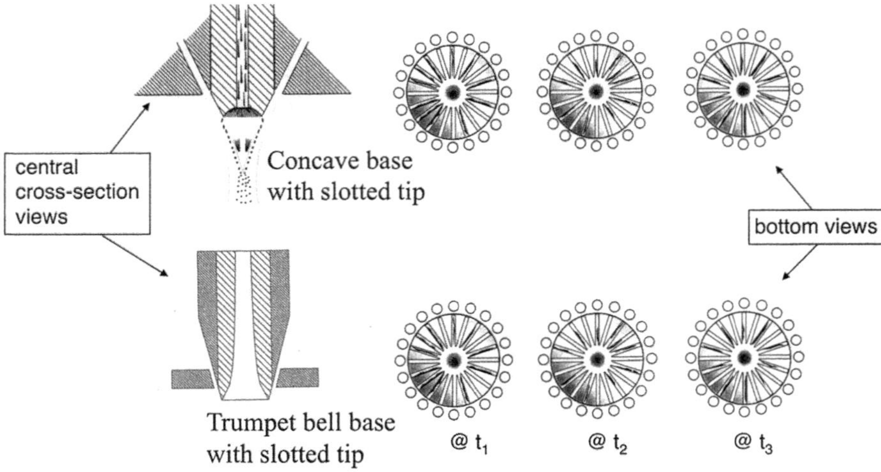

Fig. 3.10 Schematic of two different version of an internally slotted melt pour tube, where the melt loading in the bottom view of the slots is shown for a random sequence of time, adapted from [1]

film segments or thin ligaments (Fig. 3.10). This is accomplished by providing slots or channels on the interior of the melt tube that capture the flowing melt and guide it to the tube periphery without allowing a free stream "wandering" effect, that both overloads and "starves" opposite regions of the gas flow "curtain" with a chaotic frequency. In the first version of this internally slotted melt pour tube; the slots were cut into an axi-symmetric concave recess in the base of the melt tube tip. A later version that appeared to provide further enhancement of the melt distribution uniformity had an internal profile like the opening of a trumpet bell and extended a smaller number (4) of slots up into the tube (not shown in Fig. 3.10). Thus, the intention was to promote more uniform and efficient atomization by using the energetic action of all of the available gas jets, as illustrated in the bottom views of the tips in Fig. 3.10 [19].

From this knowledge base, the challenge was to demonstrate that a specific gas atomization system, i.e., with similar CCGA nozzles and slotted trumpet bell melt pour tubes, could be designed to produce high yields of passivated [21], spherical Al powder over a size range from a moderate average particle diameter (APD) of 40–50 μm to a very coarse APD of 500 μm, without major modifications. Thus, an open wake gas atomization nozzle with a 14° apex angle and an expanded central bore was selected to promote an open wake gas flow pattern for all investigated pressures, permitting dominance of melt film shear as the primary atomization mechanism. The study also utilized an internally slotted pour tube with a trumpet bell orifice to promote melt pre-filming and stability. Using this selected nozzle type and pour tube design, an investigation of the effect of atomization gas supply pressure was attempted for argon gas atomization of spherical Al powders with enhanced control of the size distribution. It should be noted that the same nozzle/

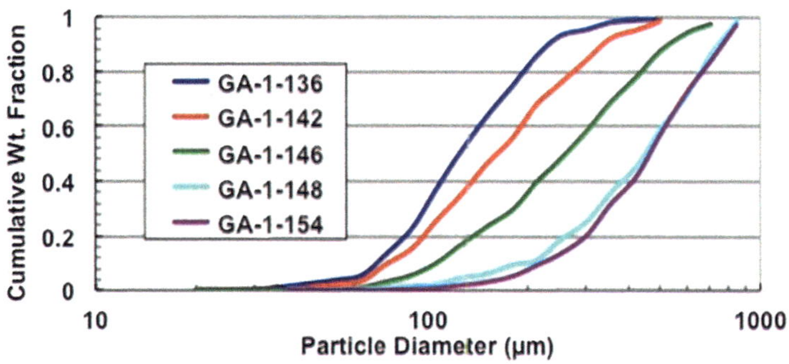

Fig. 3.11 Summary of the cumulative size distribution results from the set of five gas atomization runs that are detailed in Table 3.1

Table 3.1 Summary of atomization results

Run No.	Pressure (kPa/psig)	APD (μm)	s.d. (d_{84}/d_{50})	Melt flow (kg/min.)	G/M
1(GA-1-136)	345/50	138	1.63	2.03	1.42
2(GA-1-142)	207/30	177	1.86	3.75	0.675
3(GA-1-146)	103/15	286	1.76	3.85	0.243
4(GA-1-148)	34/5	449	1.56	1.50	0.963
5(GA-1-154)	69/10	463	1.53	1.50	0.785

pour tube system had been demonstrated previously [19] for high pressure (1.2 MPa or 175 psi) nitrogen gas atomization of spherical Al powder with a moderate APD (46 μm) and a narrow size distribution with standard deviation (d_{84}/d_{50}) of 1.6 that closely matched the fine powder requirements.

Results

Therefore, a series of five gas atomization experiments were conducted over a range of a very low (Ar) atomization gas pressures to develop the capability for close-coupled gas atomization to produce a highly controlled particle size distribution of coarse Al powders. The cumulative size distribution results for all five runs are summarized in Fig. 3.11, above. As shown in Table 3.1, two results from use of the lowest atomization gas pressures produced an average particle diameter of 449–463 μm, very close to the 500 μm target size and with a standard deviation of 1.53–1.56, an extremely narrow size distribution.

The results reveal that the process uniformity needed for this narrow standard deviation appeared to have been achieved by stabilizing uniform melt filming with an expanded discrete jet, close-coupled atomization nozzle and a slotted trumpet bell pour tube.

Discussion

Initial analysis of the size results indicated that decreased atomization gas velocity, acting within the acceleration wave model, was the key controlling variable in this low-pressure CCGA regime. In other words, according to the acceleration wave model (see Eq. (3.1)) for melt break-up of [22], the dominant parameter for control of the mean droplet diameter, termed D_m, is the mismatch in velocity values between the gas and the melt, $\Delta U = U_g - U_L$, where ρ_g is the gas density, ρ_L is the melt density, σ_{LV} is the melt surface tension, μ_L is the melt viscosity, and D_0 is the melt film thickness, defined at the location of contact with the high velocity gas (at the outer edge of the melt pour tube tip) during primary break-up. Note that the Weber number (We) and the Reynolds number (Re) that are contained within Eq. (3.1) also are defined immediately below.

$$\begin{aligned} D_m &= \frac{D_0}{0.027}\left[We_g Re_L\right]^{-0.4} \\ &-\text{for } We_g Re_L > 10^6 \\ We_g &= \frac{\rho_g \Delta U^2 D_0}{\sigma_{LV}} \\ Re_L &= \frac{\rho_L \Delta U D_0}{\mu_L} \end{aligned} \tag{3.1}$$

No consistent influence of simple gas/melt mass flow ratio was detected in the data that is summarized in Table 3.1.

Experimental observation of the atomization spray images indicated that a practical lower limit to atomization gas pressure is about 69 kPa for achieving atomization process uniformity with the atomization method and the other parameters of this study. This is the reason that the APD for the lowest atomization gas pressure was excluded from the linear least squares fit in Fig. 3.12.

3.2.2.4 Pulsatile Shock-Enhanced Disintegration in CCGA

Introduction

Powder processing is an efficient manufacturing method that reduces material waste (e.g., machining scrap) during fabrication of net shape components by the die pressing and sintering approach of conventional powder metallurgy (P/M), as well as during additive manufacturing (AM) of near-final shape objects that are "built" from powders by this very important new P/M approach. However, true overall P/M manufacturing efficiency must also take into account the usable yield of powders that is produced by atomization for the specific process. In an effort to increase the yield of usable powders for a popular AM application (e.g., laser melted/powder bed fusion) that needs a specific powder size range (typically,

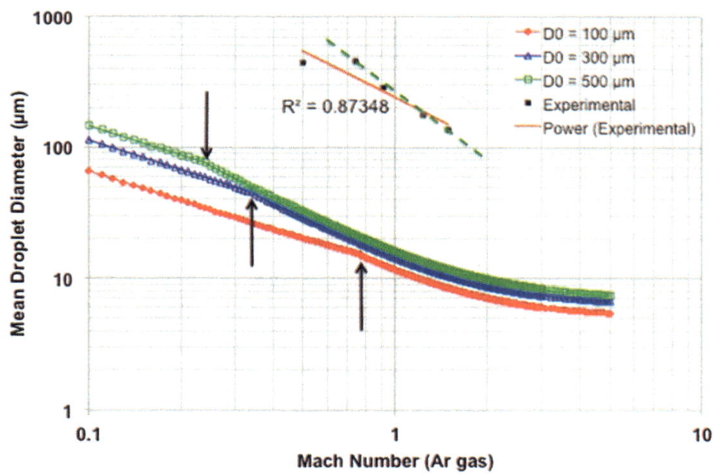

Fig. 3.12 Summary of the predicted size trends of a capillary breakup model at Mach numbers less than that indicated by the arrow on each colored trend line, along with the size trends predicted for the acceleration wave model for Mach numbers greater than the arrowed value, where both model results are shown as a function of the melt film thickness, ranging from 100–500 μm. Also plotted are the experimental results for APD for each experiment from Table 3.1 as *black square* data points with a power law fit (*as a red line*) using all of 5 of the data points and as a linear least squares fit (*as a green dashed line*), neglecting the lowest atomization gas pressure

15–45 μm dia.), an adaptable close-coupled gas atomization nozzle was designed and fabricated. After fabrication, the new "dual manifold" CCGA nozzle was tested initially by gas-only flow analysis to compare the typical aspiration characteristics and gas flow structure as a function of atomization gas supply pressure. This analysis was used to determine the expected atomization performance by analyzing the resulting gas structure, which influences the strength of the near-field recirculation zone and melt pre-filming characteristics prior to primary atomization [23]. After fitting the dual manifold nozzle into the lab-scale CCGA system at Ames Lab, a series of close-coupled atomization trials using a high temperature alloy (Fe–Cr) helped to evaluate the usable powder production performance of the combination nozzle, compared to the typical closed wake nozzle described above. The objective of this section is to highlight key factors that influence the resulting PSD when using an expanded set of CCGA parameters that are modified significantly to eliminate or strengthen the Mach shock disk that indicates closed wake operation.

Dual Manifold Nozzle Design

Compared to the typical high angle (capable of closed wake operation) CCGA nozzle shown in Fig. 3.5a, b, the new dual manifold gas atomization nozzle integrates two types of nozzles into one unit. First, the interior manifold feeds a

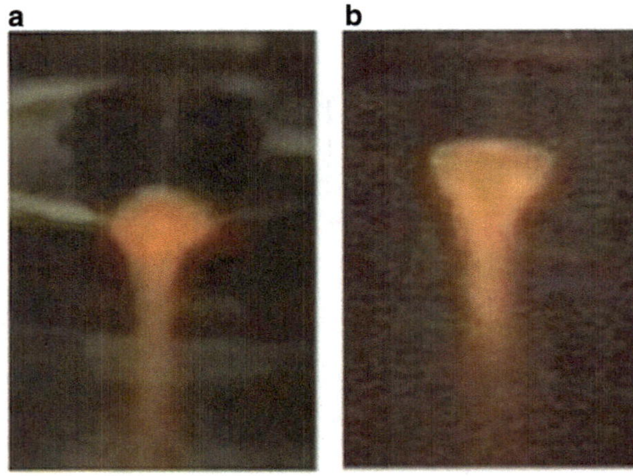

Fig. 3.13 Still images of the atomization spray of molten Al droplets captured from the video recordings of the CCGA runs from Table 3.1, with atomization gas pressures of (**a**) 45 kPa and (**b**) 69 kPa

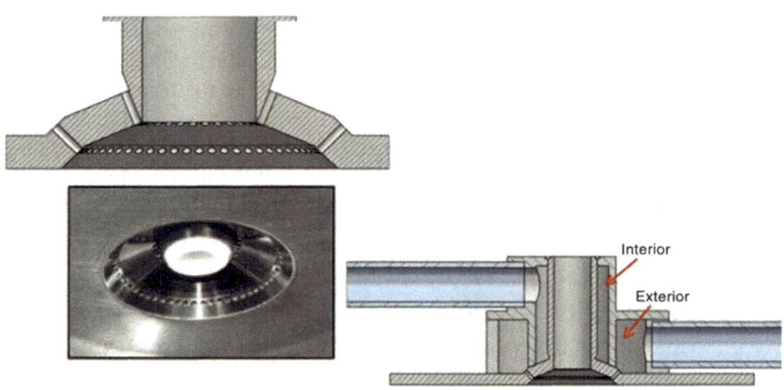

Fig. 3.14 Schematic central cross-sections and photo (*bottom view*) of the dual manifold CCGA nozzle

typical closed-wake (CW) nozzle that contains an interior ring of 30 jets, each with 0.74 mm dia., where the jets have a gas flow apex (high) angle of 45° and are equally spaced around an 11.15 mm annulus, similar to the CCGA nozzle type that produced the aspiration in Fig. 3.5b [24]. However, the dual manifold (DM) CCGA nozzle contains an additional second concentric ring of 60 jets with 0.74 mm dia. and gas flow apex angle of 90° equally spaced around a 21.92 mm annulus in combination with the aforementioned CW design (see Fig. 3.14). The DM-CCGA geometry (in Fig. 3.14) was designed to create an identical gas flow focal point between the two rings of jets, while the exterior ring of jets contains twice the cross-sectional area compared to the interior jets. The nozzle plate for both manifolds was fabricated from a 316 stainless steel plate. Also, it is important to note that the two rings of jets in the DM-CCGA nozzle are hermetically isolated (during operation) and supplied from independent gas manifolds, allowing significant experimental design flexibility (e.g., independent manifold pressures and/or differing atomization gas types). Further information about the DM-CCGA gas nozzle design can be found elsewhere in the literature [23].

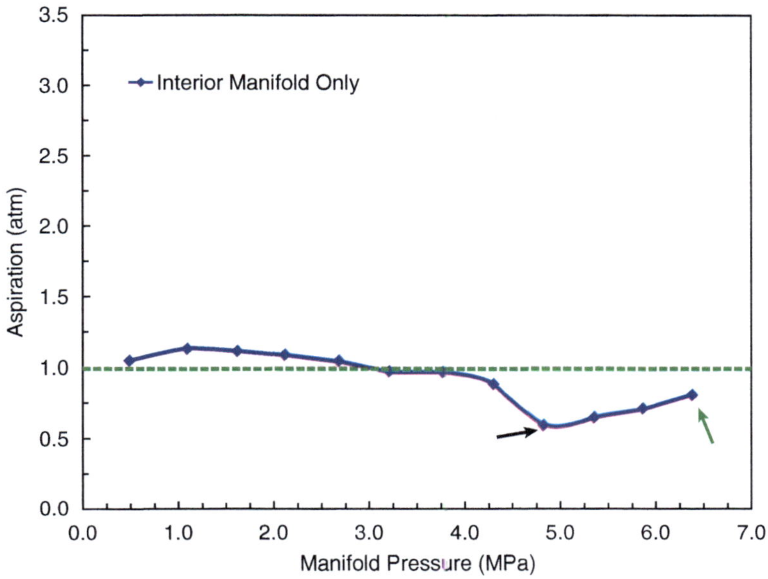

Fig. 3.15 Summary of aspiration results for (Ar) gas-only testing of the interior manifold that feeds the CW nozzle (operating without exterior manifold flow) that shows a characteristic increase in aspiration (suction) at the wake closure pressure (see *black arrow*)

Nozzle Lab Testing

Gas-only analysis using orifice pressure measurements and schlieren images was used to characterize the aspiration effects and gas structure produced by each nozzle using high-purity Ar gas. The schlieren light diffraction images were recorded using a digital camera with an exposure setting of 1/400th of a sec. and an aperture setting of f/5D. A series of matching angle (45°) brass inserts were machined with different extensions (tip protrusion distance), but only the results for one (typical) extension of 2.29 mm (0.090 in.) will be reported for the CW and DM-CCGA nozzles. Each brass insert was attached to a pressure transducer to measure the aspiration pressure at the insert tip. The test results reported in Fig. 3.15 are for the CW nozzle only and look similar to those in Fig. 3.5b. This data also allowed selection (see green arrow) of an experimental manifold pressure of 6.4 MPa (925 psi) for subsequent experiments that compared operation of the single interior (CW) nozzle to the combined dual manifold CCGA nozzle.

The elevated manifold pressure of 6.4 MPa, which is above the wake closure pressure, was selected for these experiments because it had been shown to combine the desirable features of both an enhanced melt disintegration mechanism (from its residual Mach disk shock structure) and a reduced melt flow rate (from reduced suction at the melt orifice, compared to operation at the wake closure pressure). Using the selected pressure (6.4 MPa) to produce an Ar flow pattern from the inner manifold (with a 2.29 mm tip extension), gas-only aspiration measurements were

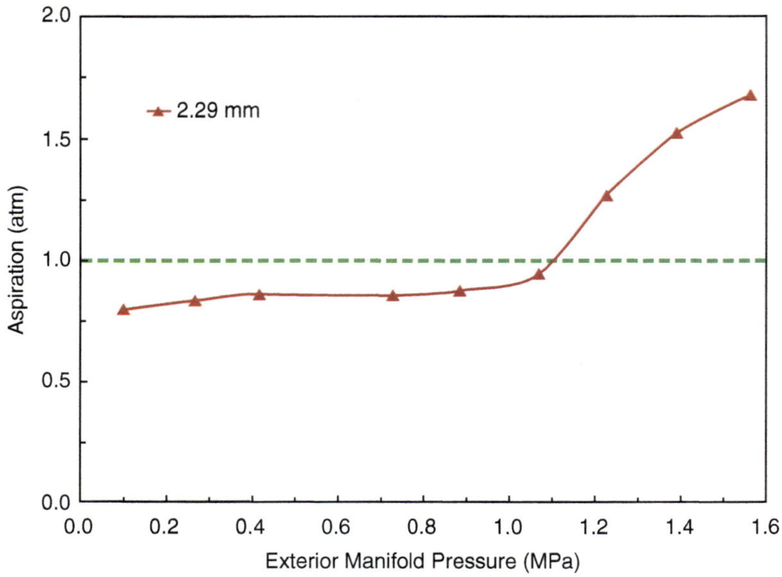

Fig. 3.16 Summary of aspiration results for (Ar) gas-only testing of the combined flow from the interior and exterior manifolds, i.e., where aspiration of the CW nozzle flow (at a fixed pressure of 6.4 MPa) is modified by increasing input gas from the exterior gas jets, showing steady desirable aspiration up to a threshold manifold pressure of about 1.1 MPa

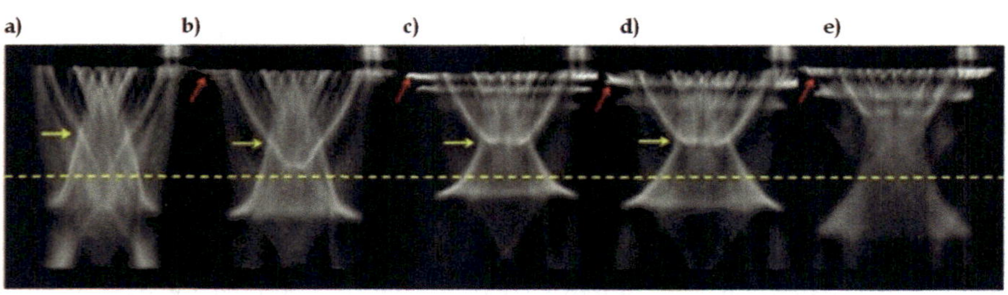

Fig. 3.17 A series of schlieren images observed when the interior manifold pressure was set and held constant at 6.4 MPa and the exterior manifold was set at: (**a**) 0 MPa, (**b**) 0.34 MPa, (**c**) 0.69 MPa, (**d**) 0.97 MPa, and (**e**) 1.52 MPa with a horizontal dashed yellow line indicating the vertical displacement of the Mach disk beyond the original location (**a**) without the influence of gas from the exterior jets, also shown are yellow arrows highlighting the location of the incident and reflective shock node, and red arrows that highlight the expansion waves created from gas exiting the exterior set of jets

performed to characterize the combined flow results that incorporated the exterior gas manifold, as summarized in Fig. 3.16.

In an effort to understand the aspiration results, schlieren (light diffraction) imaging was used to evaluate changes in the gas structure. A series of schlieren images that were captured using a matching 2.29 mm insert extension with a constant interior manifold pressure of 6.4 MPa and varying exterior manifold pressures from 0 to 1.52 MPa are displayed in Fig. 3.17. Interestingly, it appears that the addition of gas from the jets fed by the exterior manifold appeared to "re-close" the original Mach disk region with a low-pressure (0.69 MPa) gas addition. Also, the Mach disk appears to be retained as the exterior manifold pressure is increased above 0.69 MPa (see horizontal dashed yellow line in Fig. 3.17), but shows a shift upward in location. Furthermore, the exterior of the recirculation zone gas pattern (above the Mach disk) appeared much sharper/well defined when gas was flowing from the exterior manifold, comparing Fig. 3.17a (exterior manifold at 0 MPa) to Fig. 3.17b–d. Moreover, the shock node (i.e., intersection between the incident and reflective shocks) was found to be at a constant vertical displacement, and uninfluenced by an increased manifold pressure (see yellow arrows in Fig. 3.17). Also, the new augmented closed wake pattern exhibited a broader Mach disk feature, i.e., it seems that adding sufficient gas into the stagnant zone re-established a larger Mach disk at pressures between about 0.7 MPa and 1.0 MPa.

At manifold pressures greater than 1.1 MPa the aspiration of the DM-CCGA nozzle went positive, i.e., suction was lost to elevated positive pressure. In other words, Fig. 3.16 permitted prediction that an exterior manifold pressure of 0.7 MPa is safely below the positive pressure threshold, but still within the range of enlarged Mach disk formation. Thus, an exterior manifold (Ar) gas pressure of 0.7 MPa (100 psi) was selected for atomization experiments that compared results for Ar atomization of a Fe-16Cr-12Al-0.9 W-0.25Hf-0.20Y at.% alloy by an un-augmented CCGA nozzle to results from the dual manifold-CCGA nozzle. Since the dual manifold nozzle run parameters were selected to maintain aspiration in the closed-wake condition with a broadened and intensified recirculation zone, it was expected that finer powder would be produced.

For the atomization experiment with the dual manifold-CCGA nozzle, the resulting combined gas mass flow rate was measured at 15.7 kg/min (i.e., interior jets: 13.1 kg/min and exterior jets: 2.6 kg/min) and the metal mass flow rate was determined to average 1.15 kg/min, resulting in a high gas-to-metal ratio (GMR) of 13.6. The resulting metal mass flow rate was found to be significantly lower than the predicted value of 11.1 kg/min. (using a modified Bernoulli's equation that combines metallostatic head and aspiration pressure). For the atomization experiment with the conventional CCGA nozzle (inner manifold only), the resulting combined gas mass flow rate was again 13.1 kg/min and the metal mass flow rate averaged higher (with exactly the same melt orifice dia.) at 1.72 kg/min, resulting in a gas-to-metal ratio (GMR) of 7.6. The 50% increase in the resulting metal mass flow rate for the inner manifold nozzle was not expected and an explanation was explored further, given the apparent flickering illumination from the atomization spray during the dual manifold nozzle run, i.e., these observations were both consistent with some low frequency pulsatile atomization mode.

Particle size distribution analysis (by Microtrac) of the resulting as-atomized powders determined an average particle diameter (d_{50}) of 28.6 μm with a standard deviation (d_{84}/d_{50}) of 1.85 for the dual manifold nozzle and an average particle diameter (d_{50}) of 26.5 μm with a standard deviation (d_{84}/d_{50}) of 1.96 for the single manifold nozzle. A direct comparison of the cumulative size distributions of representative powder samples from both runs is shown in Fig. 3.18, showing the close correspondence between both particle size distributions.

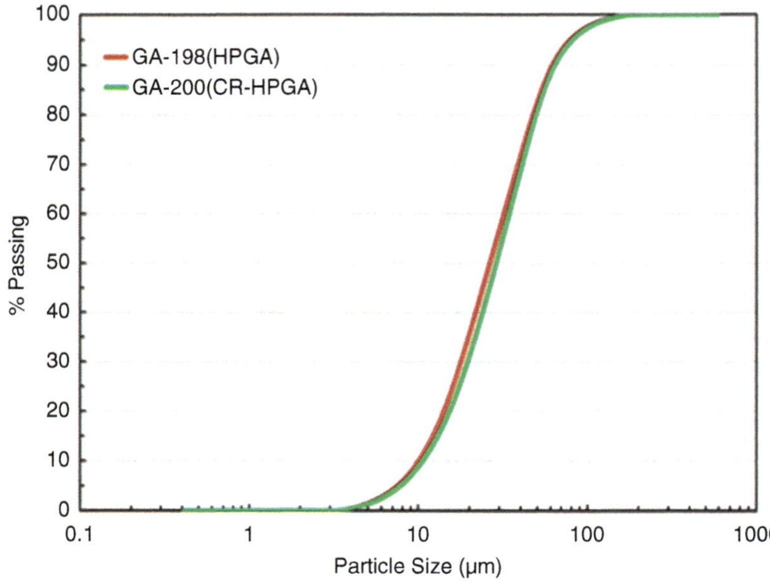

Fig. 3.18 Summary of the cumulative size distribution results for the both types of nozzles

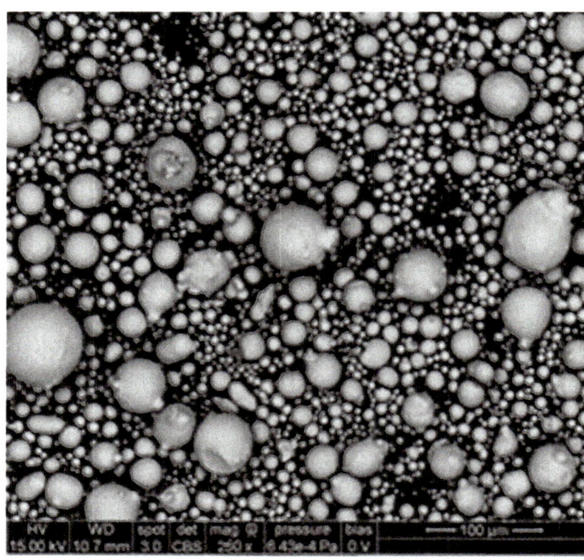

Fig. 3.19 SEM micrograph of the resulting gas atomized powder (dia. < 106 μm) from the single manifold atomization nozzle run

A representative sample of as-atomized powder for the single manifold nozzle is shown in Fig. 3.19. The powders appeared to be quite spherical, with very few surface defects. From examination of the SEM micrographs of powders from both the single and double manifold nozzles, the resulting powder from both nozzles appeared to have a bimodal size distribution, but further analysis will be required to confirm.

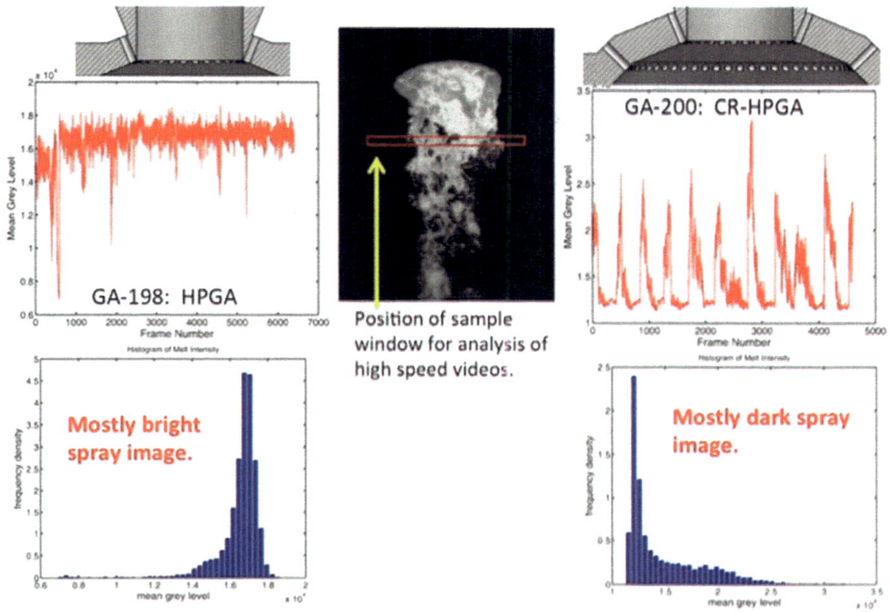

Fig. 3.20 Summary of the high-speed video analysis from atomization with the single manifold nozzle (*left side*) and from atomization with the dual manifold nozzle (*right side*), where an illustration of the analysis method (*center*) highlights the choice of a sample window

To explore the possibility that the abnormal restriction of the melt flow was due to some type of pulsatile gas atomization mechanism (presumed from the flickering of the bright spray), the help of Prof. Andrew Mullis [25, 26] was enlisted for analysis of high speed video sequences from both runs. High-speed video analysis, involving pixel counting within a central window of the spray image (see Fig. 3.20), confirmed the presence of a pulsation effect during the dual manifold atomization trial (see right side of Fig. 3.20). It can be seen from the regular set of bright spikes that the atomization stream pulses between on and off about every 40 ms, i.e., with an apparent frequency of ~11 Hz. Inspection of a series of video frames (not shown) also indicated that the melt stream was briefly interrupted between every pulse, but able to re-establish flow. For the single manifold atomization run, the spray image appeared to have much more continuous melt flow without a strong periodic pattern, which helped to explain the higher melt flow rate than the dual manifold nozzle run.

Thus, using a bit of conjecture, it should be possible to recalibrate the geometry of this novel CR-HPGA nozzle to not only increase the pressure at the stagnation front, but also to potentially decrease the pressure at the stagnation front, while retaining the closed wake condition. This would provide a vast amount of flexibility to adjust the dynamics of the closed-wake pulsatile atomization effect, in order to control this unique phenomenon and gain better insight into how it may be used to "tune" the resulting particle size distribution.

Summary

A new dual manifold gas nozzle (DM-HPGA) with isolated manifolds was demonstrated as an effective method for manipulating the gas structure from a close-coupled gas atomization nozzle. A truncated recirculation zone with a broader Mach disk was achieved with an apparent stronger recirculation effect. The incident and reflective shock node was identified as a threshold marker for continued aspiration when using the DM-HPGA nozzle. High-speed videography of an initial gas atomization trial highlighted a pronounced pulsation effect with a frequency of ~11 Hz. Although there were notable differences (compared to the equivalent single manifold nozzle) in the gas flow induced dynamics of the atomization process, the dual manifold nozzle resulted in just a slightly larger particle size with a slight decrease in particle standard deviation. It was postulated that this type of DM-HPGA nozzle design may provide the unique ability to manipulate the stagnation pressure during closed-wake atomization, where better understanding of its operation may unlock the ability to control the dynamics of droplet spray formation and, hence, powder particle size.

3.3 Hybrid Atomization

3.3.1 Introduction

Conventional gas atomization processes for metal powder production utilize twin fluid atomizers with high-pressure gas and molten metal. Besides the use of Close-Coupled and Free-Fall atomizer types, relevant developments are continuously being made to advance atomizer concepts for improved powder products and more energy and resource efficient processes. The main aims of new atomizer developments are the minimization of (mean) particle size, narrowing of the particle size distribution and the technical processing of complex melt systems for powder applications.

3.3.2 Discussion

In this chapter the development of gas atomization units which use characteristic flow effects of a molten metal stream in relation to the flow of compressed gas for efficient fragmentation of liquids and melts is described. The interaction of gas and liquid melt in the twin-fluid atomization process can be used as a basis for understanding and optimizing melt fragmentation processes. It is known that to increase the efficiency and decrease the resulting particle size of an atomization process, one can increase the specific surface energy before atomization [27]. In the context of twin fluid (gas) atomization, optimal fragmentation must then take place in a region where the velocity difference between the atomization gas and the melt is highest. One method to achieve this would be to first transform the liquid melt into a film, and then atomize it with a high-speed gas flow. This configuration allows the generation of a high velocity difference between the atomization gas and the melt film. The atomization concept based on this configuration is known as hybrid atomization, or otherwise called Pressure-Gas-Atomization, or Prefilming-Hybrid-Atomization.

3.3.3 *Pressure-Swirl Film Formation Plus Gas Jet Disintegration*

3.3.3.1 Introduction

Metal powders are produced for various applications such as the manufacturing of tools and components (sintering), or soldering of metallic parts. Today, there is an increasing demand for large amounts of fine and spherical powders e.g. for metal injection moulding technology. This results in special requirements for the production process of spherical, narrowly size-distributed powders beyond conventional methods [28–30]. Conventional close-coupled or free-fall atomizers are used because of their high through-put, but their gas atomized powders typically show broad size distributions and particles with satellites. The aim of development is an atomizer design for molten metals that produces a narrow particle size distribution of small particle size, and consumes less gas. Pressure-gas-atomization is a disintegration process, which combines a single fluid pressure-swirl nozzle, and an external mixing atomizer to meet these aims. This atomization process became possible after the development of the pressure-swirl-atomizer for molten metals by Sheikhaliev [31]. Originally, the authors called this atomization technique "Centrifugal Hydraulic Atomization". The pressure-swirl technique is commonly used for cold liquids and uses centrifugal forces and cone geometry to transform the liquid into a film. For a long time, it was doubted if this principle could be applied to molten metal because of the much higher operating temperature and material properties required. The advantage of the designed combination of a swirl nozzle and gas atomization is that the melt is first transferred in a liquid film, thus increasing the initial liquid surface prior to the gas atomization process.

3.3.3.2 The Pressure-Swirl Filming/Gas Atomizer Concept

In Fig. 3.21 a schematic of a pressure-swirl nozzle and an image of a water-film highlighted by a laser-light-sheet is illustrated. In the pressure-swirl nozzle molten metal is pressed tangentially into a conical swirl chamber where a swirling liquid flow surrounds a central air core via centrifugal force. As the swirling liquid passes

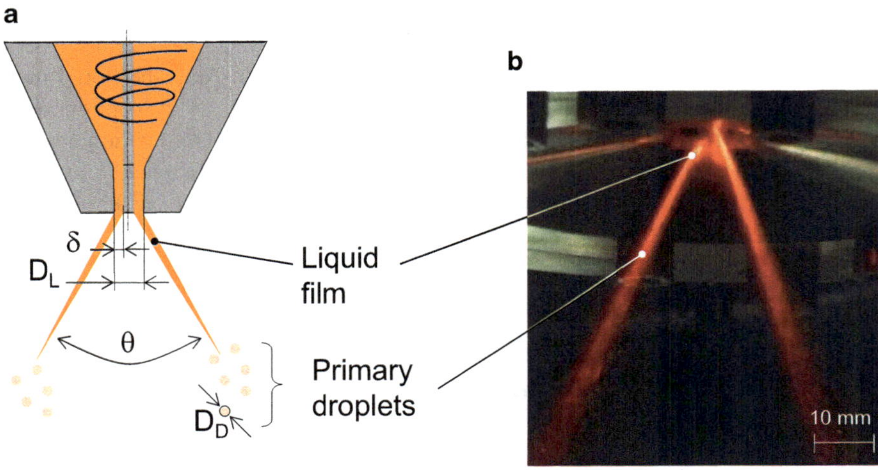

Fig. 3.21 Schematic diagram (**a**) and laser-light sheet highlighted image (**b**) of pressure-swirl atomization

through a cylindrical outlet (D_L) it forms a hollow-cone film (δ). The outer edge of the hollow-cone film subsequently breaks up into droplets (D_D mass median diameter). Droplet size decreases with decreasing sheet thickness, increasing spray angle (θ) and increasing film velocity. Pressure-swirl atomization produces the finest sprays when only pressure is applied to the liquid for atomization.

Figure 3.22 shows a schematic diagram of the pressure-gas atomizer and an image of a water-film highlighted by a laser-light-sheet. It combines the pressure-swirl atomization as a pre-filming step and an external gas atomization to enhance disintegration. This invention was patented in 2002 [33] and initiated intensive scientific investigations [32, 34–38]. Due to the pressure difference between the pressure vessel and spray chamber and centrifugal forces acting on the melt, the molten metal forms a hollow cone film, which is disintegrated into primary droplets by the pressurized gas flow in the vicinity of the pressure-swirl nozzle. These primary droplets are disintegrated further by high velocity nitrogen jets (p_G) issuing from the orifices (D_G) of the ring-gas nozzle. The ring-gas nozzle and the pressure-swirl nozzle are separated by a gap (a). The zone where the gas-flow hits the primary droplets is called the atomization area.

3.3.3.3 Initial Results

Pressure-gas-atomizer experiments have been conducted in a pilot plant where the spray chamber and pressure vessel are the main components with a height of approximately 5 m. The pressure inside the vessel is limited to a value of 1.0 MPa and the volume of the crucible represents approximately 3 dm^3. Before operation the spray chamber is evacuated and refilled with nitrogen to reduce the

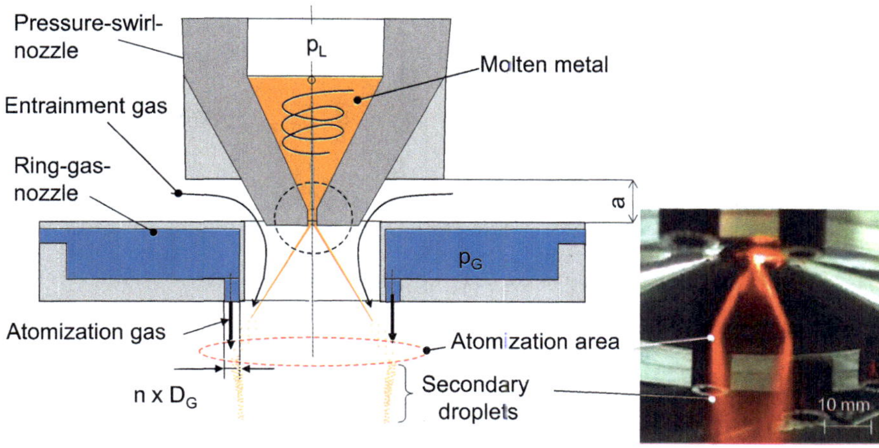

Fig. 3.22 Schematic diagram and laser-light sheet highlighted image of pressure-gas-atomization [32]

oxygen content. Both the crucible and pressure-swirl-atomizer are raised to a temperature of 1200 K by resistance heating. The process is capable of achieving a maximum liquid melt mass flowrates of 200 kg/h, and the ring-gas nozzle can produce gas flowrates up to 300 kg/h. A gas-recirculation system (GR) was installed to improve flow conditions in the disintegration area below the nozzle-system and to avoid recirculation zones inside the spray chamber. Powder particles and gas exit from the bottom of the spray chamber and are fed to a cyclone where the gas and particles are separated. The powder particles remain in the cyclone while most of the gas is recycled by a fan to the top of the spray chamber, and the remaining gas is purged through the exhaust system.

The location at which the high velocity gas flow meets the molten metal film or primary droplets is of vital importance for the disintegration of the melt in pressurized melt/gas atomization. This location (atomization area in Fig. 3.22) is largely effected by the hollow cone angle θ. With a larger hollow cone angle, the atomization area is moved towards the ring-gas nozzle and the gas flow strikes the primary droplets with a greater velocity resulting in more effective atomization.

3.4 Isolated Effects of the Pressure-Swirl Nozzle

It is useful to investigate the atomization effects of the pressure-swirl in isolation to help understand how differences arise from the combined effects of the hybrid process. According to Lefebvre's investigations on the atomization of cold liquids [27], the spray cone passes through several stages as the liquid pressure and flow rate are increased. Figure 3.23 shows the effect of pressure on a pure tin melt spray cone. As the pressure increases the spray cone changes from an onion and tulip

Fig. 3.23 Effect of the pressure p_L on the spray cone angle θ during pressure-atomization of pure tin [32]

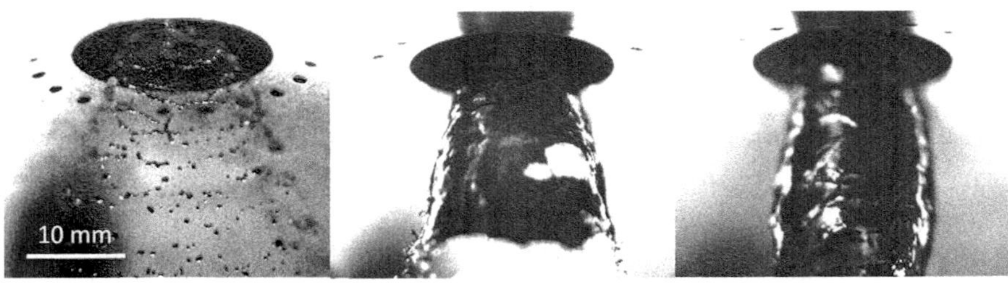

Fig. 3.24 Break-up of liquid metal films: Wave disintegration of Sn (*left*), perforation disintegration of SnCu30 (*center*), and film development of SlSi12 (*right*), adapted from [39]

shape producing fairly large droplets, to a conical sheet producing finer droplets. A pressure difference of about 0.4 MPa is necessary to achieve a fully developed hollow cone with a small cone angle (θ) of about 15°. Higher liquid pressures yield larger cone angles (55°) as demonstrated in the figure.

Figure 3.24 displays the effect of material properties on disintegration mode as three different alloys, pure tin, SnCU30, and AlSi12 are atomized at 0,6 MPa. The image section shows the area below the ring-gas-atomizer (not used) where the issuing of the film from the nozzle exit is hidden by the ring-gas nozzle. Pure tin melt (Fig. 3.24 left side) is shown undergoing wave disintegration as the conical sheet experiences oscillations motivated by the opposing surface tension and aerodynamic forces. In this case the thickness of the sheet varies periodically as a series of contractions and dilations. The edge of the oscillating sheet breaks off to form rings which themselves then break up into primary droplets.

A different disintegration mode known as perforation is observed for SnCu30 (Fig. 3.24, center). In this mode a perforated sheet is formed as surface tension causes holes to appear in the thinning film. As the holes enlarge the liquid between them contracts to form a net of liquid ligaments, or rims of irregular shape which then break up into droplets of varying size [40]. The melt of AlSi12 alloy (Fig. 3.24, right) opens out into a hollow tulip shape maintaining the film structure. The disintegration follows later (outside of the image) through rim formation and perforation. Due to different material properties of AlSi12 (mainly viscosity) the rotation inside the nozzle is slower, and the resulting film is thicker with a smaller cone angle.

3.4.1 *Isolated Effects of the Gas Flow Field*

The gas flow field is formed by 20 discrete under-expanded jets in an annular configuration. Through momentum transfer the jets impart normal and tangential stress on the molten film causing destabilization and disintegration. The gas jets expand from vertical straight bore holes and are bended towards the centre-line of the atomizer. Model experiments with water have been used to investigate the ring-gas nozzle configuration with a focus on the gas outlet angle [35].

Figure 3.25 shows an image of the pressure-gas-atomization process and a gas velocity distribution obtained through numerical simulation. It can be seen that the disintegrated liquid forms an hour-glass-structured spray cone with significant contraction unlike FFA or CCGA. The circumferential symmetry of the flow field is used to calculate the flow in a three-dimensional wedge through the ring-gas-atomizer [41]. It was found that an entrainment zone occurs directly under the ring-gas-atomizer, enclosed by the gas jets.

Fig. 3.26 shows the axial gas velocity distributions along the centre axis, and the radial profiles of the axial velocity at z = 20 mm and z = 40 mm. The entrainment is limited within 20 mm from the pressure-swirl-nozzle and becomes stronger with increasing gas pressure. The axial velocity on the centre axis achieves maximum value around z = 60 mm and then decreases. Secondary atomization mainly occurs in the region between z = 20 and 40 mm. Ideally the molten metal film spreads in

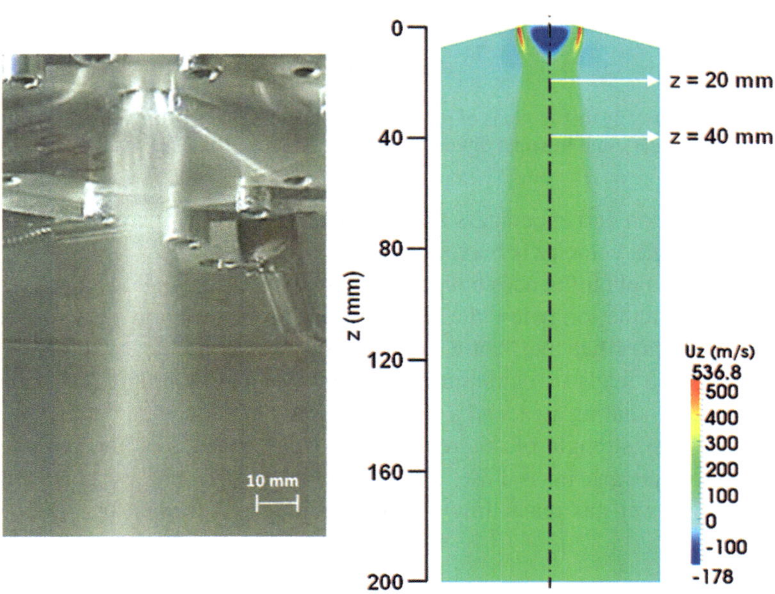

Fig. 3.25 Image of pressure–gas-atomization ($p_L = 0.6$ MPa; $p_G = 1$ MPa) and calculated gas velocity field at atomization pressure $p_0 = 1$ MPa; adapted to [32, 41]

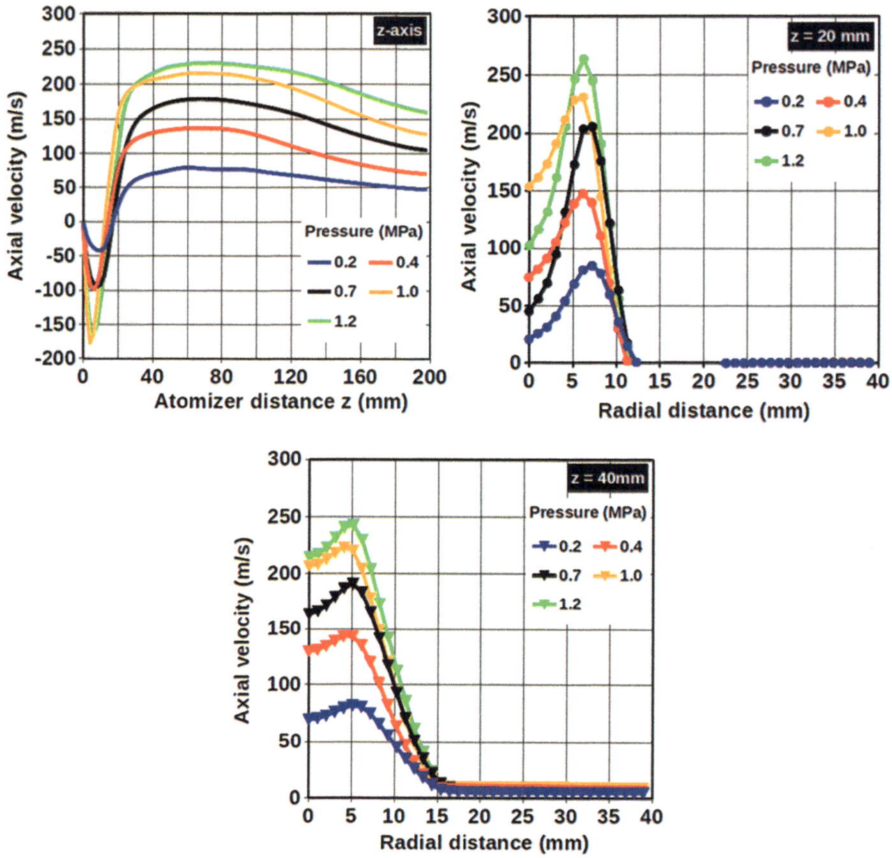

Fig. 3.26 Axial velocity distributions of gas flow field: along center axis (*left*), in radial direction at z = 20 mm (*center*), and in radial direction at z = 40 mm (*right*); [41]

the radial direction and encounters the gas stream 5–10 mm away from the centre axis where the gas velocity is maximized.

The gas flow field influences both the development of the hollow cone angle (θ) and the gas recirculation below the ring-gas nozzle. Because the gas expands as a free jet from each orifice they cause a reduction in pressure close to their exits. This reduced pressure leads to entrainment between the pressure atomizer, ring-gas-nozzles and surrounding area, and amplifies the recirculation zone within the liquid hollow cone. The strength of this recirculation zone depends on the gas pressure, and therefore the gas mass flow. As a result a smaller hollow cone angle is generated by lower gas mass flow and a wider angle is formed by a higher gas mass flow.

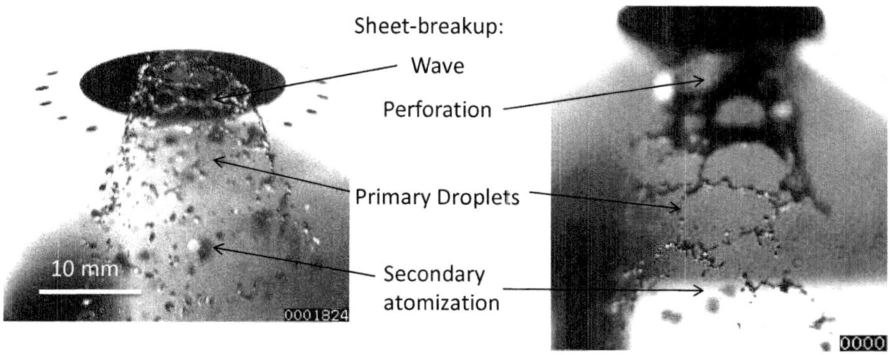

Fig. 3.27 Secondary atomization of primary droplets during pressure-gas atomization adapted from [38]

3.4.2 Effects of Gas Recirculation and Melt Properties

The breakup and disintegration of two molten metals is shown in Fig. 3.27. The left image in the figure displays the wave break-up of a pure Sn film as it fragments to form ligaments, and subsequently primary and secondary droplets. The aerodynamic forces of the gas provide the undulating motion of the Sn film and contribute to its disintegration. The right image displays the breakup of SnCu30 as it first perforates then subsequently disintegrates. The primary break-up of Sn starts earlier compared to SnCu30 due to its material properties. The breakup of the SnCu30-film is also affected by the gas flow. When compared with Fig. 3.24 the addition of gas flow appears to reduce the SnCu30 film to approximately half the length. Primary droplets maintain the main flow direction of the film until the high velocity gas jets cause the disintegration of the droplets. After this secondary atomization, the small individual droplets are not visible due to their high velocity, but leave shadow-like clusters as indicated in the figure. For the SnCu30 melt, the atomization area starts at a greater distance from the ring-gas nozzle than the pure Sn.

3.4.2.1 Improved Results

In Fig. 3.28. the mass median diameter is compared to the gas-to-metal ratio (GMR) for experiments composed of four different alloys, two ring-gas nozzle configurations, and with/without gas recirculation (GR). All results show that the particle size decreases with increasing gas mass flow, indicated by GMR. An increase of the particle size as a result of an increase in copper content (SnCu63) was expected due to the associated change in material properties, however the lower copper content (SnCu30) was observed to have a minor effect on particle diameter compared to the median copper content (SnCu50). The use of the secondary flow via GR allows the molten metal cone to enter the high velocity gas flow closer to the exit of the ring-

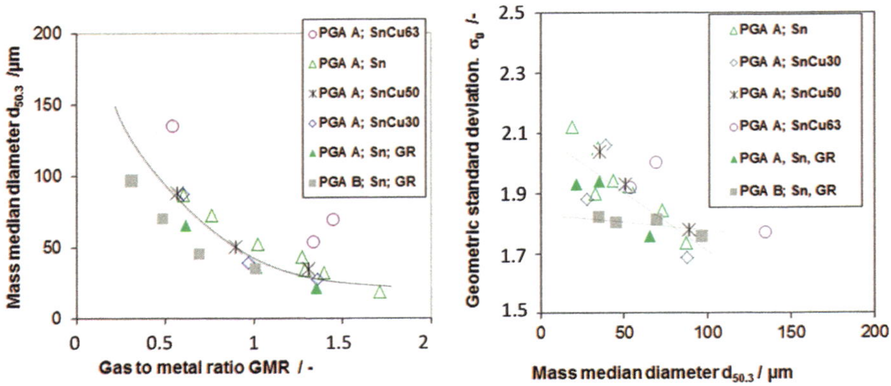

Fig. 3.28 Effect of the GMR on the mass median diameter for different alloys and experimental configurations (*left*), and geometric standard deviation σ_g versus the mass median diameter (*right*); adapted from [36, 37]

gas nozzle. More kinetic energy is available for disintegration in this area thus contributing to a smaller particle size. Atomizer B has half the jet orifice area of atomizer A and produces slightly smaller mass median particle diameters. In conclusion, a mass median diameter of 18 µm for tin particles can be achieved with a distribution between 1 and 75 µm.

Besides mass median diameter, the geometric standard deviation s_g is an important metric in powder production and is defined as the diameter ratio, $d_{84.3}/d_{50.3}$. Figure 3.28 (right) shows the geometric standard deviation compared to the mass median diameter. Conventional gas atomized metal powder has a standard deviation of about 2. Powders produced by the pressure-gas-atomizer, have a geometric standard deviation between 1.6 and 2.1. For Atomizer PGA A, the pure tin and copper alloys were found to be inversely proportional to the mass median diameter. Atomizer PGA B produced powder with a standard deviation of approximately 1.8, and behaves independently of the mass median diameter.

3.4.2.2 Discussion

The surface of gas-atomized powder is generally smooth and spherical, but on a finer scale frequently exhibits a cellular or dendritic morphology, reflecting the solidification mode of the droplets and satellites attached to the particle surface. The flowability of metal powder, a necessary condition for the industrial application, is decreased by satellites and agglomerates. In powder production the solidification time has to be sufficient to allow droplets to form spherical particles, but fast enough to suppress coalescence. To gain an impression of particle sphericity the mean circularity ($C = 4\pi$ projection area/perimeter^2/$C = 1 =$ perfect circle) of the powder is measured by image analysis and illustrated in Fig. 3.29. Circularity of SnCu-powder generated without an optimised flow lie in the range between 0.74

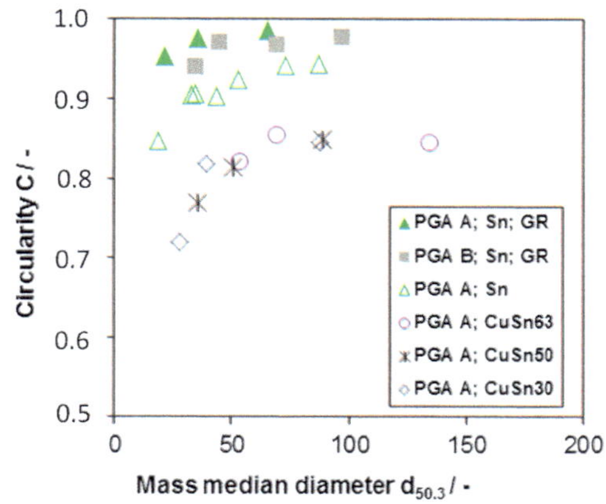

Fig. 3.29 Mean circularity versus mass median diameter; adapted to [36, 37]

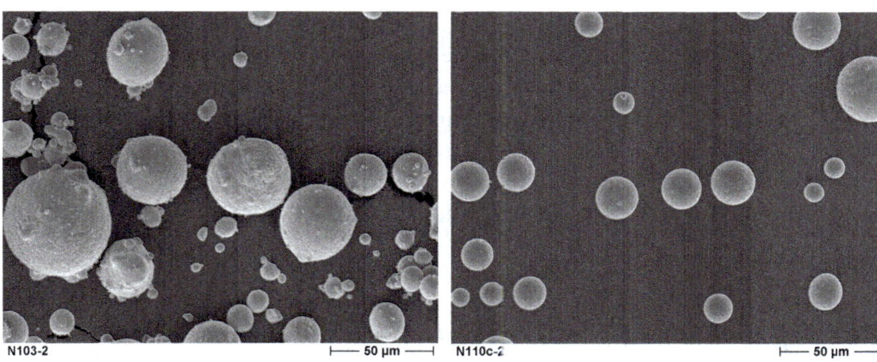

Fig. 3.30 SEM-Image of: SnCu30, $p_G = 1.0$ Mpa, GMR = 1.31, $d_{50.3} = 35$ µm (*left*); Sn; $p_G = 1.0$ MPa; GMR = 1.34; $d_{50.3} = 21$ µm (*right*) [32]

and 0.85. The circularity of particles produced with large superheating and without GR lie in the range of 0.84 and 0.95, and Sn particles produced with GR have a circularity above 0.94, essentially spherical.

This relationship is exemplified by two SEM micrographs of powder produced from pressure-gas-atomization experiments (Fig. 3.30). When GR is not used, the flow conditions near the atomizer are marked by a strong recirculation area. This leads to a high particle concentration in the vicinity of the atomizer and consequently a lot of collisions of larger semi-solid particles and small solid particles, thus resulting in non-spherical particles with satellites (Fig. 3.30, left). By implementing a secondary flow via GR the backflow of small cooled particles is prevented, and the particle concentration is decreased. As such, the small solid particles are prevented from colliding with the larger liquid droplets and the process favors the production of satellite free particles (Fig. 3.30, right).

3.5 Rotary Film Formation Plus Gas Jet Disintegration

3.5.1 Introduction

It is generally a great challenge to atomize highly viscous liquids such as slag or glass melts which have a comparably high viscosity and low surface tension. Gas disintegration of such viscous melts is a difficult task because of the rapid cooling of the ligaments during melt fragmentation that may favor fiber formation instead of a particulate product. Furthermore, decreasing the temperature of the melt within the process increases the melt viscosity that leads to a lower dynamic energy transfer rate between the gas and liquid. Thus, the disintegration time increases and liquid fragments may be transported out of the area where the most effective atomization occurs. If the energy-transfer rate decreases, then a lower efficiency of the atomization process follows and a larger particle size distribution is produced. By using heated atomization gas for the disintegration process, in principal a particulate product instead of a fiber product can be obtained [42, 43]. The efficiency of the atomization process decreases with increasing viscosity, and the minimum particle size is limited by the available energy input [44, 45]. Concepts for atomization of highly viscous liquids or melts have already been developed [46–48]. However current technologies have limitations in regard to producing a sufficiently low particle size and low fiber-to-particle ratio at high melt-mass flow rates.

3.5.2 Rotary Filming/Gas Atomizer Concept

The development of a new atomizer concept is focused on atomizing highly viscous liquids to produce small droplets at high process throughputs. The prefilming hybrid atomizer introduced here is a combination of a single-fluid rotary atomizer and an external mixing twin-fluid atomizer. In the first step the feed material is spread out by a spinning disc via centrifugal forces, so that the initial liquid surface energy increases prior to the second step of gas atomization. Fig. 3.31 (left) shows the schematic of the prefilming hybrid atomizer. The rotary atomizer operates in the sheet formation mode to produce the melt prefilm, while the external mixing twin-fluid atomizer provides the pressurized gas to disintegrate the film once in the atomization zone. The gas and liquid flow field of the prefilming hybrid atomizer is illustrated in Fig. 3.31 (right). During the disintegration, the gas expands through small cylindrical holes and free jets develop by the inner and outer entrainment gas. As a result, from the interaction of the gas jets and system geometry a recirculation gas flow develops below the gas outlet nozzle and rotary disk. This recirculation flow guides the moving film into the atomization zone. The film movement mainly depends on the acting aerodynamic and inertial forces, which are caused by the momentum distribution of the inner entrainment flow, the recirculation gas flow momentum, and the film momentum itself. In the meantime, a sufficiently small gas

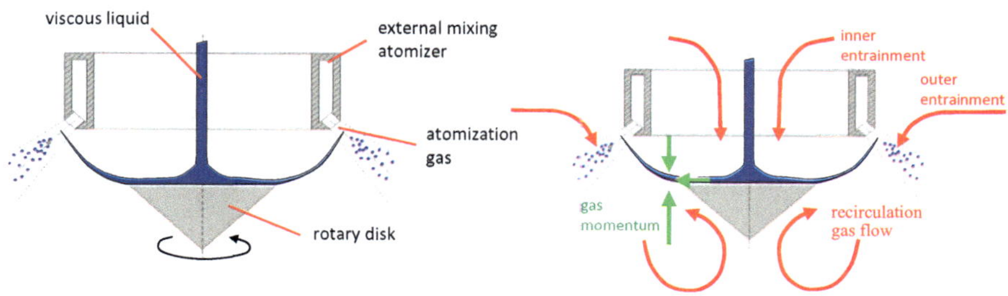

Fig. 3.31 Sketch of the prefilming hybrid atomizer (*left*) and indicated gas flow field with effective momentum transfer [13], with permission

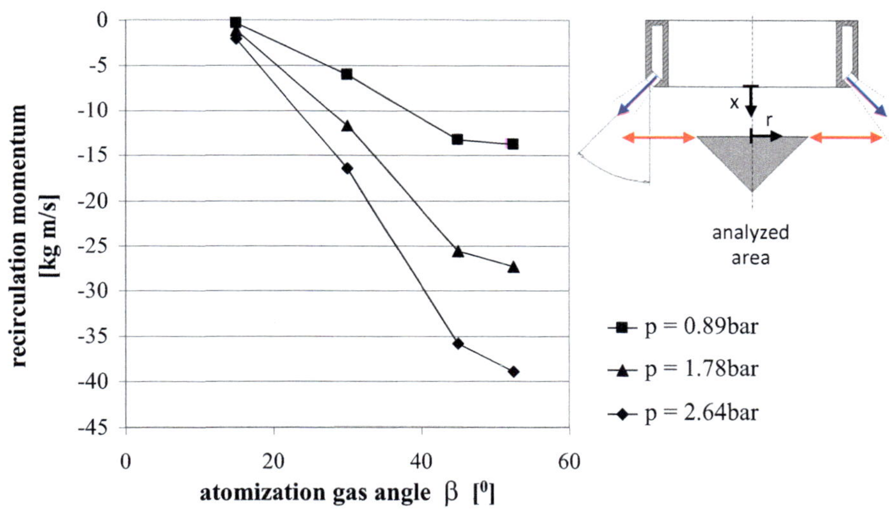

Fig. 3.32 Axial recirculation momentum on a free flowing film [13], with permission

mass flow rate from the inside of the atomizer protects the atomizer against recirculation of the liquid and ensures stable atomization conditions. The aim of the prefilming hybrid atomizer design is to generate a maximum recirculation momentum such that the liquid film is transported close to the gas outlet nozzle, where the most effective atomization occurs.

3.5.3 Numerical Simulation

Numerical simulations are used to derive suitable atomizer geometry for efficient viscous melt atomization. At the highest velocity difference (at the highest gas manifold pressure) between the gas and liquid (see Fig. 3.32), the most efficient atomization area is close to the atomization-gas outlet. Thus, the aim of numerical simulations is to design a gas nozzle where the maximum recirculation momentum

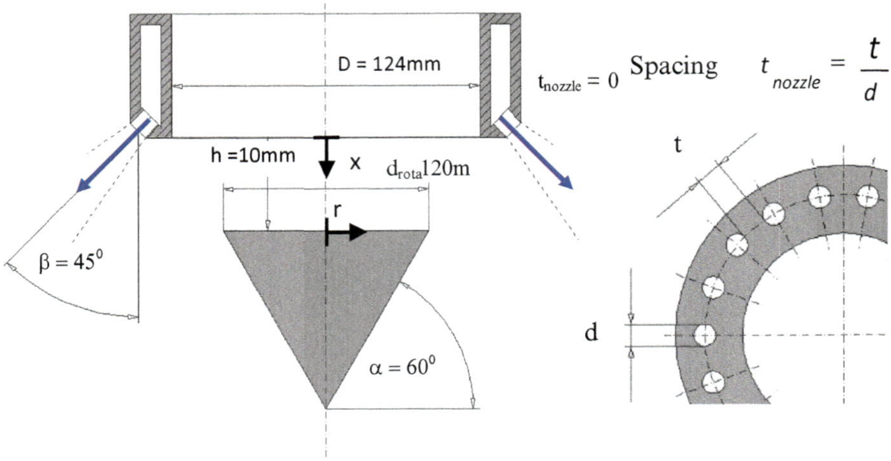

Fig. 3.33 Geometric parameters of the hybrid atomizer design [13], with permission

is available to transport the liquid film upwards towards the outlet of the gas jets. For this, an important geometric parameter, the gas outlet angle, was investigated for its effect on gas velocity and recirculation momentum. Figure 3.32 shows the recirculation momentum compared to the atomization gas angle. It was found that recirculation momentum increases with gas pressure and outlet angle. For outlet angles lower than 15°, the inner entrainment momentum dominates the interaction with the liquid film. Outlet angles larger than 15° affect the recirculation momentum, and optimal recirculation flow conditions begin to arise at 45°.

Figure 3.33 presents the resulting optimum parameter set for the prefilming hybrid atomizer. All geometric parameters of the rotary disk and gas nozzle have been determined through the same simulation method as the outlet angle. As a gas jet spacing of zero was determined optimal, a slit nozzle is used as the gas-flow exit geometry. The atomizer was designed for the processing of mineral melts at a temperature of 1873 K, a viscosity of 1 Pa-s, and a melt mass flow rate of 300 kg/h.

3.5.4 Model Experiments

Experimental results of water and glycerol gas disintegration by hybrid, and conventional free-fall atomization are illustrated in Fig. 3.34. The mass median droplet diameter of the spray, measured by laser diffraction is plotted as a function of air-to-liquid mass-flow ratio (ALR). The following provides a comparison of hybrid and conventional free-fall atomizer efficiencies [8]. The free-fall atomizer (see earlier in this chapter) is a common device for powder production [29] and spray forming applications [49]. For the disintegration of water with low viscosity, the conventional atomizer is more effective (Fig. 3.34 (left)). At a constant liquid-

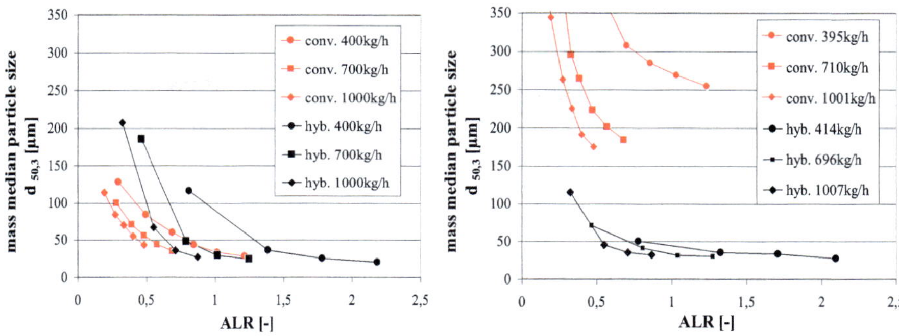

Fig. 3.34 Experimental results for atomizing water by air (*left*) and glycerol by air (*right*) [13], with permission

mass flow rate and identical ALR the hybrid atomizer produces coarser particle sizes and in this case the increase of the specific surface energy via film formation before atomization is not an advantage. A completely different result is observed for the disintegration of liquids with higher viscosity (Fig. 3.34 (right)). At constant ALR, the hybrid atomizer produces 3–5 times finer particles than the conventional atomizer for viscous glycerol atomization. In this case, a mass-median diameter of less than 30 µm was achieved.

3.5.5 *Viscous Melt Atomization Experiments*

A pilot plant for hot-gas atomization has been used to disintegrate viscous melts by hybrid atomization [50]. The main part of the plant is a spray tower (about 5.5 m in height) where the material is melted at the top by an induction heating system. Blast-furnace slag is used as the material to be atomized. The melt temperature before atomization is measured to be 1813–1843 K with a mean temperature of 1826 K. Driven by gravity, the melt flows out through the bottom of the pouring crucible at flow rates up to 300 kg/h. The hybrid atomizer is located directly under the crucible and operates using heated gas. Gas temperatures, and pressures up to 1273 K, and 2 bars (rel.) have been obtained during hybrid atomization runs. Hot gas is produced by means of a discontinuous Cowper heat exchanger, where compressed gas flows through ceramic bulk material that is heated by a propane burner. The desired atomization gas temperature and pressure are then obtained by mixing the heated gas with gas at room temperature. The melt droplets produced from the spray may be quenched and solidified approximately 2 m below the atomization nozzle. The solidified powder (see Fig. 3.35) is collected at the bottom of the spray tower and the fine powder fraction is collected in the cyclone.

Fig. 3.35 shows a SEM micrograph of a mineral melt powder fraction produced by the hybrid atomization system. The particle fraction size ranges from 110 to 350 µm, where mostly spherical particles are seen. By mass, 90% of the product is

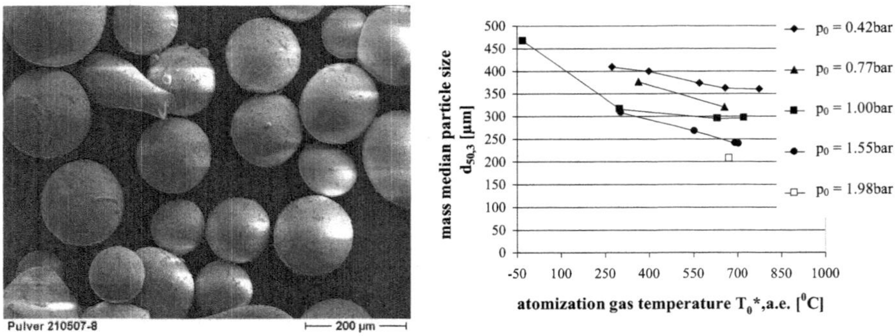

Fig. 3.35 Micrograph of hot-gas atomized mineral melt particles (size fraction 110–350 µm) and results for atomizing mineral melt at an angular disc speed of 1500 L/min. From [13], with permission

(equi-axed) particulate and the remainder is fibrous material. Figure 3.35 (right) shows particle size results for the atomization of mineral slag melt at different atomization gas temperatures and pressures (relative). The gas temperature given in the graph is after expansion (a.e.). The particle size decreases with increasing gas temperature and increasing atomization gas pressure. For this set of experimental conditions, the minimum achieved mass median particle size is 210 µm, and the maximum gas temperature before expansion is 1273 K. Compared with the model experiments and the material properties of glycerol the minimum particle size is rather coarse. This result stems from the maximum atomization gas temperature available in the facility. The maximum achieved temperature is still too low in comparison with the melt temperature. Before and within the disintegration process, the melt is cooled down rapidly, increasing the viscosity and limiting disintegration efficiency.

3.5.6 *Overall Conclusions*

Free-fall gas atomizers were some of the first two fluid atomizer designs to be used for molten metal atomization. In a simple open (unconfined stream) design a melt stream falls from a tundish exit via gravity into the convergence of focused atomization gas jets where it is disintegrated. Thus, free-fall atomizers much less problematic in terms of freezing and clogging when compared to close-coupled gas atomizers, but the arrangement requires that the atomization gas flow travels a relatively long way before hitting the melt. This reduces their velocity and, consequently, the resulting yields of fine powder for a given gas to metal mass flow ratio (GMR).

In close-coupled gas atomization (CCGA) the primary influence of G/M at elevated levels is on control of standard deviation, not on average powder size. Closed-wake conditions in CCGA nozzles are most effective at ultra-fine powder production due to an intense secondary break-up feature. CCGA nozzles operating in open-wake conditions can achieve a broad range of powder size control, especially with effective melt pre-filming.

A new concentric ring (CR) gas die with independent manifolds was demonstrated as a method for manipulating the close-coupled gas structure. High-speed video highlighted a pronounced (visible) pulsation effect during first experimental test of the Dual Manifold-HPGA gas atomization die. Resulting powders contained a slightly improved standard deviation, but a minor increase in average particle size was noted. Further testing will be required to exploit the potential of the concentric ring gas die for well controlled production of fine powders and mixed gas experiments, including high temperature gas/melt surface reactions.

Considerable development has been performed on a type of hybrid atomization approach that involves pressure-swirl melt filming that is linked to a gas atomization nozzle to promote more intense droplet disintegration. For low viscosity metallic melts, this method appears to be very suitable to generate a particle size distribution with a relatively narrow standard deviation (d84/d50) that ranges 1.6–2.1. Also, the particle circularity and lack of attached "satellite" particles on the resulting powders is very good, if an accessible gas recirculation effect is implemented.

The development of a new hybrid atomizer concept is focused on atomizing highly viscous (glassy) liquids to produce small droplets at high process throughputs. This prefilming hybrid atomizer is a combination of a single-fluid rotary disk atomizer and an external mixing twin-fluid atomizer so that molten material is spread out by the spinning disc via centrifugal forces, prior to the second step of heated gas atomization. Special parameters generate a maximum recirculation momentum such that the liquid film is transported close to the atomization (hot) gas outlet, where the most effective atomization occurs. The ability to produce fine spherical powders of the glassy melts depends critically upon the achievable atomization gas temperature.

3.6 List of Symbols

3.6.1 Latin

Symbole	Description
D_0	Melt film thickness (defined at the location of contact with the high velocity gas)
D_m	mean droplet diameter
Re	Reynolds number
U_g	Velocity of the gas
U_l	Velocity of the melt
We	Weber number

3.6.2 Greek

Symbole	Description
ρ_g	Gas density
ρ_L	Melt density
σ_{LV}	Melt surface tension
ΔU	Mismatch in velocity values between the gas and the melt
μ_L	Melt viscosity

References

1. Anderson, I., Terpstra, R., & Figliola, R. (2005). Visualization of enhanced primary atomization for powder size control. In *Advances in powder metallurgy and particulate materials* (pp. 1–17). Princeton, NJ: Metal Powder Industries Federation.
2. McHugh, K. M., Lin, Y., Zhou, Y., Johnson, S., Delplanque, J.-P., & Javernia, E. (2008). Microstructure evolution during spray rolling and heat treatment of 2124 Al. *Materials Science and Engineering, 477*, 26–34.
3. Fritsching, U. (2004). *Spray simulation: Modeling and numerical simulation of sprayforming metals*. Cambridge: Cambridge University Press.
4. Yule, A., & Dunkley, J. (1994). *Atomization of melts*. Oxford: Clarendon Press.
5. Fritsching, U., & Bauckhage, K. (1992). Investigations on the atomization of molten metals: The coaxial jet and the gas flow in the nozzle near field. *PHOENICS Journal of Computational Fluid Dynamics, 1*, 5.
6. Markus, S., Fritsching, U., & Bauckhage, K. (2002). Jet break up of liquid metals in twin fluid atomization. *Materials Science and Engineering: A, 326*, 122–133.
7. Heck, U. (1998). *Zur Zerstäubung in Freifalldüsen*. Düsseldorf: VDI Verlag.
8. Lohner, H., Czisch, C., & Fritsching, U. (2003). Impact of gas nozzle arrangement on the flow field of a twin fluid atomizer with external mixing. In International conference on liquid atomization and spray systems. Sorrento, Italy.
9. Uhlenwinkel, V., Fritsching, U., Bauckhage, K., Urlau U. (1990). Strömungsuntersuchungen im Düsennahbereich einer Zweistoffdüse - Modelluntersuchungen für die Zerstäubung von Metallschmelzen, Chemie Ingenieur Technik – CIT Volume 62, Issue 3, S. 228-229
10. Czisch, C., Lohner, H., & Fritsching, U. (2004). Einfluss der Gasdüsenanordnung auf den Desintegationsvorgang und das Zerstäubungsergebnis bei der Zweistoff-Zerstäubung. *Chemie-Ingenieur-Technik, 76*, 754–757.
11. Heck, U., Fritsching, U., & Bauckhage, K. (2000). Gas-flow effects on twin-fluid atomization of liquid metals. *Atomization and Sprays, 10*, 25–46.
12. Wille, R., & Fernholz, H. (1965). Report on the first European mechanics colloquium on the coanda effect. *Journal of Fluid Mechanics, 23*, 801–819.
13. Czisch, C., & Fritsching, U. (2008). Atomizer design for viscous-melt atomization. *Materials Science and Engineering, 477*(1–2), 21–25.
14. Schwenck, D., Ellendt, N., & Uhlenwinkel, V. (2014). Gas recirculation affects powder quality. In *World confress on powder metallurgy and particulate materials (PM2014)*. Orlando, Florida.
15. Lawley, A. (1992). *Atomization: The production of metal powders* (pp. 102–107). Princeton, NJ: MPIF.
16. Ting, J., Peretti, M. W., & Eisen, W. B. (2000). Control of fine powder production and melt flow rate using gas daynamics. *Advances in Powder Metallurgy and Particulate Materials, 2*, 27–40.

17. Mates, S. P., Ridder, S. D., & Biancaniello, F. S. (2000). Comparison of supersonic length and dynamic pressure characteristics of discrete–jet and annular close–coupled nozzles used to produce fine metal powders. In *Liquid metal atomization: Fundamentals and practice, TMS annual meeting and symposium* (pp. 71–81).
18. Ting, J., & Anderson, I. E. (2004). A computational fluid dynamics (CFD) investigation of the wake closure phenomenon. *Materials Science and Engineering, A379*, 264–276.
19. Anderson, I. E., Terpstra, R. L., Cronin, J. A., & Figliola, R. S. (2006). Verification of melt property and closed wake effects on controlled close-coupled gas atomization processes. In *Advances in powder metallurgy and particulate materials* (pp. 1–16).
20. Unal, A. (1987). Effects of processing variables on particle size in gas atomization of rapidly solidified aluminium powders. *Materials Science and Technology, 3*, 1029–1039.
21. Brandes, E., & Brook, G. (1992). *Smithells metals reference book* (7th ed.). Oxford: Butterworth Heinemann.
22. Ingebo, R. D. (1980). *Atomizing characteristics of swirl blast fuel injectors. NASA Technical Memorandum 79297*. Cleveland, OH: Lewis Research Centre.
23. Rieken, J., Heidloff, A., & Anderson, I. (2013). Moving towards improved ultra-fine powder production for precursor ODS Fe-based alloys, compiled by D. Christopherson & R. M. Gasior, metal powder. *Advances in Powder Metallurgy and Particulate Materials, 2*, 11–22.
24. Anderson, I., Figliola, R., & Morton, H. (1991). Flow mechanisms in high pressure atomization. *Materials Science and Engineering, 148*, 101–114.
25. Mullis, A. M., Adkins, N. J., Aslam, Z., McCarthy, I., & Cochrane, R. F. (2008). Close-coupled gas atomization: High-frame rate analysis of spray-cone geometry. *International Journal of Powder Metallurgy, 44*, 55–64.
26. Mullis, A. M., McCarthy, I., Cochrane, R., & N. J. Adkins. (2016). Investigation of the pulsation phenomenon in close-coupled atomization, Advanced in powder metallurgy and particulate materials. Princeton, NJ: Metal Powder Industries Federation.
27. Lefebvre. (1989). *Atomization and sprays*. New York, NY: Hemisphere.
28. Anderson, I., Terpstra, L., & Rau, S. (2001). SFB-spray forming kolloquium. In *Band 5* (pp. 1–16). Norderstedt: Books on Demand GmbH.
29. Bauckhage, K., & Fritsching, U. (2000). In K. Cooper, I. Anderson, S. Ridder, & F. Biancaniello (Eds.), *Liquid metal atomization: Fundamentals and practice* (pp. 23–36). Warrendale, PA: TMS.
30. Lawley, A. (2000). In I. Anderson & K. P. Cooper (Eds.), *Liquid metal atomization: Fundamentals and practice*. Warrendale, PA: TMS.
31. Dunkley, J., & Sheikhaliev, S. (1995). Single fluid atomization of liquid metals. In *Proceedings of the international conference on powder metallurgy and particulate materials* (Vol. 1, pp. 79–87). Seattle, USA.
32. Achelis, L. (2009). *Kombinierte Drall-Druck-Gaszerstäubung von Metallschmelzen*. Aachen: Shaker Verlag.
33. Uhlenwinkel, V. (2002). Patent Nr. 10237213
34. Lagutkin, S. (2003). *Development of technology and equipment for metal powder production by centrifugal-gas atomization of melt*. Ekaterinburg: Ural Department of Academy of Sciences.
35. Lagutkin, S., Achelis, L., Sheikhaliev, S., Uhlenwinkel, V., & Srivastava, V. (2004). Atomization process for metal powder. *Materials Science and Engineering: A, 383*, 1–6.
36. Achelis, L., & Uhlenwinkel, V. (2007). Characterisation of metal powders generated by a pressure-gas-atomize. *Materials Science and Engineering A, 477*(1–2), 15–20.
37. Achelis, L., Uhlenwinkel, V., Lagutkin, S., & Sheikhaliev, S. (2007). Atomization using a pressure-gas-atomizer. *Materials Science Forum, 534–536*, 13–16.
38. Achelis, L., Sulatycki, K., Uhlenwinkel, V., & Mädler, L. (2010). Spray angle and particle size in the pressure gas atomization of tin and tin-copper alloys. In *Proceeding of the international conference on powder metallurgy*. Florence, Italy.

39. Achelis, L., Uhlenwinkel, V., Sulatycki, K., & Mädler, L. (2010). New approach to generate composite particles. In *Proceeding of the international conference on powder metallurgy and particulate materials* (pp. 1–11). Florida, USA.
40. Fraser, R., & Eisenklam, P. (1953). Research into the performance of atomization of liquids. *Imperial ChemEngSoc, 7*, 52.
41. Li, X. (2014). *Modeling and simulation of the gas-atomization process of metal melts for metal-matrix-composite production*. Aachen: Shaker Verlag.
42. Lohner, H. (2002). *Zerstäuben von Mineralschmelzen mit Heißgas*. PhD. thesis, University Bremen.
43. Czisch, C., Lohner, H., Fritsching, U., Bauckhage, K., & Edlinger, A. (2003). Atomisation process for metal powder. In K. Bauckhage, U. Fritsching, J. Ziesenis, A. Uhlenwinkel, A. Leatham (Eds.), *Proceedings on Spray Deposition and Melt Atomization Conference SDMA 2003*, Bremen, 22–25 June 2003.
44. Strauss, J.T. (1999). Hotter gas increases atomization efficiency. *Metal Powder Report, 11*, 24–28.
45. Dunkley J.J. (2001). In *2001 International Conference on Powder Metallurgy and Particulate Materials PM2TEC 01, 2-29-2-35, 2001*, Metal Powder Industries Federation, Princeton, USA.
46. Fraser, R.P., Dombrowski, N., & Routley, J.H. (1962). *Chemical Engineering Science, 18*, 339–353.
47. Pickering, S.J., Hay, N., Roylance, T.F., & Thomas, G.H. (1985). *Ironmaking and Steelmaking, 12*(1).
48. Campanile, F., & Azzopardi, B.J. (2003). In Cavaliere, A. (Ed.), *CD-ROM Proceedings of International Conference on Liquid Atomization and Spray Systems ICLASS 2003*, Sorrento, Italy, 13-17.07.2003, ILASS-Europe.
49. Fritsching, U., & Bauckhage, K. (2006). Sprayforming of metals. In *Ullmann's encyclopedia of industrial chemistry* (Vol. 7). Weinheim: Wiley VCH.
50. Lohner, H., Czisch, C., Schreckenberg, P., Fritsching, U., & Bauckhage, K. (2005). Atomization of viscous melts. *Atomization and Sprays, 15*(2), 169–180.

Chapter 4
Spray Transport Fundamentals

Xing-gang Li and Udo Fritsching

4.1 Introduction

Each technical production process couples several physical phenomena and sub-processes in a complex way. Therefore, processes (in a first analysis approach) are to be subdivided into individual process steps (e.g. unit operation) for further analysis. In this context a common subdivision of a technical atomization and spray process into modular sub-processes is performed. From the viewpoint of multiphase flow analysis, a typical subdivision for the atomization and spray process in metal melt processing is illustrated in Fig. 4.1. The sketch shows a (relatively rough) subdivision of a general atomization and spray process into five main sub-processes as:

- *media delivery*: the internal liquid flow in the atomizer, the gas flow in transonic condition at the atomizer and the mixing of atomization media and energy, typically liquid and gas,
- *atomization*: the process of fluid disintegration or fragmentation, from the continuous delivery of the fluid or melt and the necessary supporting materials (like gases or additives), to the primary resulting spray structure and droplet/ligament spectrum from the atomization process,
- *spray*: the establishing and spreading of the spray, to be described as a dispersed multiphase flow process with momentum, heat, and mass transfer, and possible

X.-g. Li (✉)
Foundation Institute of Materials Science, Bremen, Germany

General Research Institute for Nonferrous Metals, Beijing, China
e-mail: xing-gangli@163.com

U. Fritsching
Foundation Institute of Materials Science, University of Bremen, Bremen, Germany
e-mail: ufri@iwt.uni-bremen.de

H. Henein et al. (eds.), *Metal Sprays and Spray Deposition*,
DOI 10.1007/978-3-319-52689-8_4

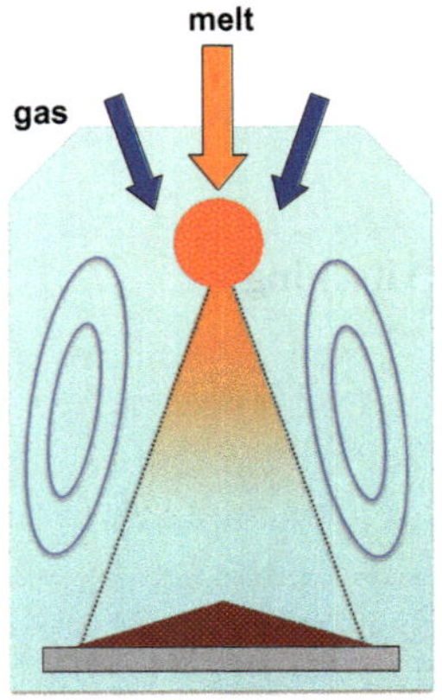

Fig. 4.1 Subdivision of melt atomization and spray process into sub-processes

chemical reactions in all phases, and the exchange between the phases, as well as a possible secondary disintegration process of fluid ligaments or coalescence of droplets,

- *gas recirculation* in spray chamber: the spray flow in the bounded chamber (e.g. spray tower) with gas (and particles) recirculation and spray entrainment,
- *consolidation*: the solidification of the spray droplets to form solid particles in powder production process, or the impact of the spray droplets onto a solid or liquid surface and compaction and possible film formation of the impacting fluid or melt mass as well as the build-up of a remaining layer or preform.

A common viewpoint for analysis of all atomization and spray processes is to be seen from multiphase flow analysis with integral momentum, heat and mass transfer. Spray processes typically involve two-phase flows of liquid droplets and gas, or even three-phase flows of solid, liquid and gas (e.g. within melt atomization or within metal-matrix-composites (MMC) production).

The specific spray evolution and spray transport in melt spray processes to be described here depends on the initial conditions of the spray formation stage. Thus in this chapter (in addition to Chaps. 2 and 3) will discuss special features of the gas atomization.

4.2 Near-Field Gas Flow Dynamics

In most gas atomization cases with external mixing, the central melt jet stream is surrounded by a gas flow from an (slit) jet configuration or a set of discrete gas jets, which are flowing parallel or inclined towards the melt stream. The coaxial atomizer gas usually exits the atomizer at high pressures with high kinetic energy. Two main configurations and types of external mixing twin-fluid atomizers need to be distinguished as they result in different fragmentation characteristics. The first one is the confined or close-coupled atomizer (CCA) and the second is the free-fall atomizer (FFA). The gas flow in the close-coupled atomizer immediately covers the flowing melt jet. Within the confined atomizer the distance between gas exit and melt stream is smaller than in the free-fall arrangement, where the melt jet moves a certain distance in the direction of gravity before the gas flow impinges onto the central melt jet; therefore the close-coupled configuration generally results in higher atomization efficiencies (in terms of smaller particles at identical energy consumption). But the confined atomizer type is more susceptible for freezing problems of the melt at the nozzle tip. This effect can be due to the extensive cooling of the melt by the expanding gas flow, which exits in the close-coupled type near the melt stream and contributes to rapid cooling of the melt at the tip of the melt nozzle. The thermal related freezing problem is most important directly in the initial phase of the spray process when the melt stream is initiated and exits the nozzle for the first time. At that point the nozzle tip is still cool and needs to be heated first either externally or by the hot melt flow.

Analyses of the gas flow field in atomizer configurations have been done computationally and experimentally in external mixing twin-fluid atomizers especially to identify the disintegration potential in terms of the operation conditions (gas pressure, nozzle arrangement...). The geometric configuration and arrangement of the gas jet system as well as the geometry or contour of the individual gas nozzles may be adapted for suitable application of the atomization gas [1–15].

4.2.1 *Close-Coupled Atomizer*

The gas flow behaviour in front of an external mixing twin-fluid atomizer may have an influence on fundamental atomization parameters like the liquid mass flow rate (yield) or the resulting drop size distribution in the spray. The typical concentric gas flow configuration in an atomizer results in a pressure change in the central liquid feed area that causes variations in the pressure ratio between the liquid feed (or reservoir) and the pressure level in front of the liquid exit. The correlation between the gas pressure in the plenum of the atomizer and the orifice pressure (also called aspiration pressure) strongly depends on the arrangements and shapes of the melt nozzle tip [16, 17]. Typical technical atomizers are constructed to give a suction pressure (lower pressure than ambient) at the liquid delivery port rather than overpressure. Also the arrangement and number of discrete jets or the configuration of the gas nozzle as an annular slit nozzle result in different aspiration pressure behavior (see [18]). If the spacing between the individual jets is large enough and the interaction between the jets is small, pressure equalization takes place and in between the discrete jets the aspiration pressure effect is decreased or even completely suppressed. For a slit nozzle, the aspiration pressure effect tends to a maximum.

In the gas-only flow conditions, the atomizing gas pattern flowing across the melt tip moderates itself via dynamic gas expansion and compression to

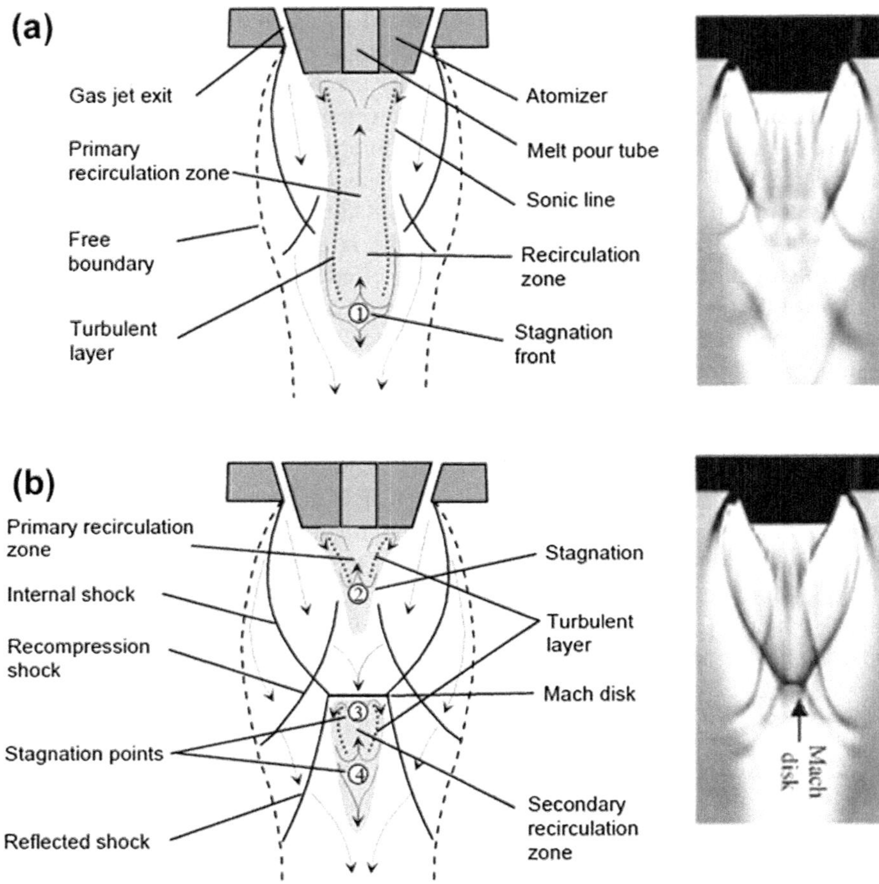

Fig. 4.2 General gas flow structure in gas-only operation of the HPGA (high-pressure gas atomizer) nozzle: (**a**) open-wake condition, the schematic (*left*) showing stagnation front 1 at the wake front, experiment (*right*); (**b**) closed-wake condition, the schematic (*left*) showing the formation of the Mach disk, stagnation front 2, and stagnation point 3 and 4 in the secondary recirculation zone, experiment (*right*) (adapted from [9, 10, 19])

accommodate the pressure between the high velocity gas and the quiescent, ambient surroundings, giving rise to either open-wake or closed-wake conditions, as shown in Fig. 4.2a, b, respectively. The flow pattern at closed-wake condition is characterized by a primary recirculation zone truncated by the sudden appearance of a Mach disk, which is immediately followed by a secondary recirculation zone. Despite of higher-pressured recirculation gas flow in the secondary recirculation zone [11, 12], the Mach disk acts as a barrier and prevents any backward gas flow from the secondary recirculation zone from reaching the primary recirculation zone (thus the so called 'closed-wake' condition for the primary recirculation zone). The wake-closure phenomenon occurs above a critical atomization-gas pressure, termed wake-closure pressure (WCP), at which an abrupt drop in aspiration pressure is seen [9–12]. The wake-closure pressure is highly specific to the nozzle designs and setups. Usually, the stable wake-closure cannot be sustained during melt atomization.

Ting et al. [11, 12] discuss the effect of the aspiration pressure on the flow of gas and liquid at atomization-gas pressures up to 7.6 MPa. The atomization gas flow field results in a fluctuating pressure distribution at the melt outlet. Based on gas dynamic assumptions for the gas flow field in the nozzle vicinity for increasing gas pressures, these authors finally conclude that for increased gas pressures an oscillating flow field of the gas may occur. This effect will result in the temporal occurrence of central orthogonal shock structures (Mach disk) in front of the nozzle. Based on this observation a fragmentation process model is developed which predicts, for gas pressures exceeding a configuration dependent threshold value, a pulsating flow field configuration of the melt flow at the nozzle tip and therefore highly transient atomization behaviour (see chapter on gas atomization).

Flow separation may occur over the outer surface of the melt guide tube (or the liquid nozzle insert) for some conditions associated with atomization pressure and the fundamental design of the atomization nozzle and the melt guide tube, and it is implied that this causes a low pressure area, which may subsequently draw melt up the outer surface of the melt guide tube and even up the inner annulus of the atomization nozzle, as shown in Fig. 4.3 [1]. The wetting of the atomization nozzle by the melt (lick back) can be a cause of nozzle failure. In many cases the nozzle is damaged by the melting of the inner annulus as the melt temperature typically exceeds the intrinsic melting point of the alloy from which the nozzle is constructed. To avoid this, one way is to eliminate or restrict the local flow separation around the tip of the liquid nozzle insert through improving the nozzle design, e.g., by adjusting

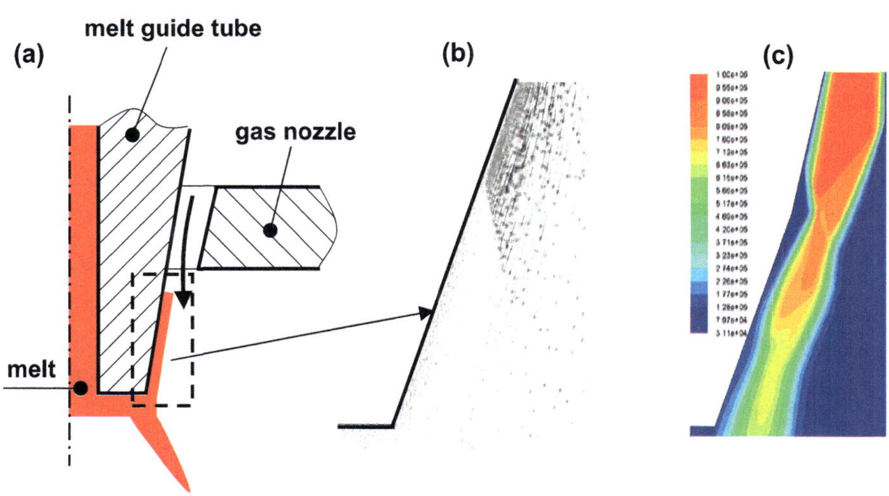

Fig. 4.3 (**a**) Schematic of the wetting of the atomization nozzle by the melt; (**b**) schematic of the gas flow separation by vectoral presentation; (**c**) pressure contour (Pa) on the outer surface of nozzle from numerical simulation (adapted from Aydin and Unal [1], with permission)

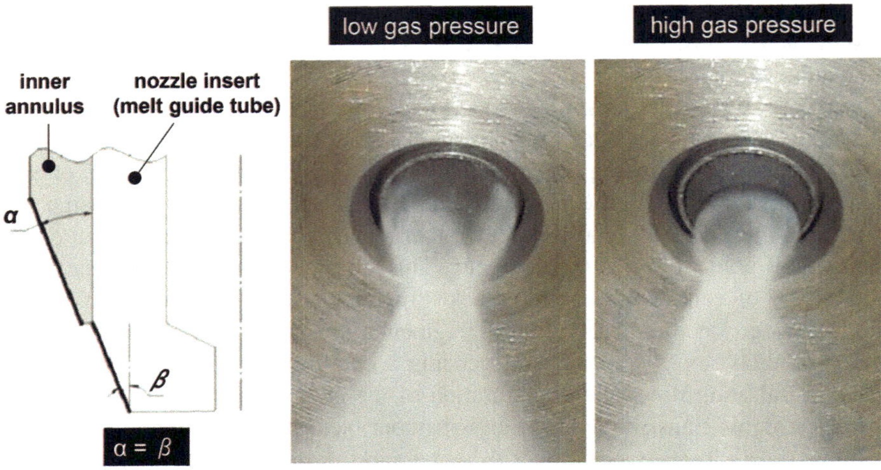

Fig. 4.4 Atomization plumes (water) near the gas- and melt-nozzle exits at two pressure levels (adapted from [20])

the angle matching or the relative placement between the atomization nozzle and the liquid nozzle insert, as shown in Fig. 4.4 [20]. For a specific nozzle design, the flow separation is strongly dependent on the atomizing gas pressure. In this case an extensive hour glass shape of the spray cone is observed that severely influences the overall spray behaviour and movement of the droplets.

Figure 4.5a, b shows the simulated gas flow field (velocity contours and vectors) in an external mixing (close-coupled) atomizer configuration [13]. The gas exit geometry is an annular slit configuration where the gas emerges from a concentric ring slit nozzle. As the high-pressure gas enters the nozzle, the sharp angle with the nozzle causes a detachment of the flow from the wall and a recirculation flow is generated around the corner of the throat. An oblique shock is formed inside the nozzle and reflected by the outer surface of the feeding tube. At the exit, the gas flow expands through a series of Prandtl–Meyer expansion waves and recompression shocks to match the atmospheric pressure inside the atomisation chamber and the high velocity gas continuously overshoots the equilibrium position as the external layer of the jet communicates with the jet core by sound/pressure waves which is slower than the supersonic flow. The gas flow separation at the corner of the melt tube generates a recirculation vortex under the melt exit hole. The recirculation zone is distinguished from the high velocity gas with the sonic lines. The gas flow in the centre of the recirculation zone moves toward the feeding tube and turns outward radially as it moves close to the tip. When the gas flow comes into contact with the sonic boundary, it is pushed inward and flows downstream. Inside the recirculation zone, a turbulent layer separates upstream and downstream flows. At the end of this recirculation zone there is a stagnation front, where the gas velocity falls to around zero.

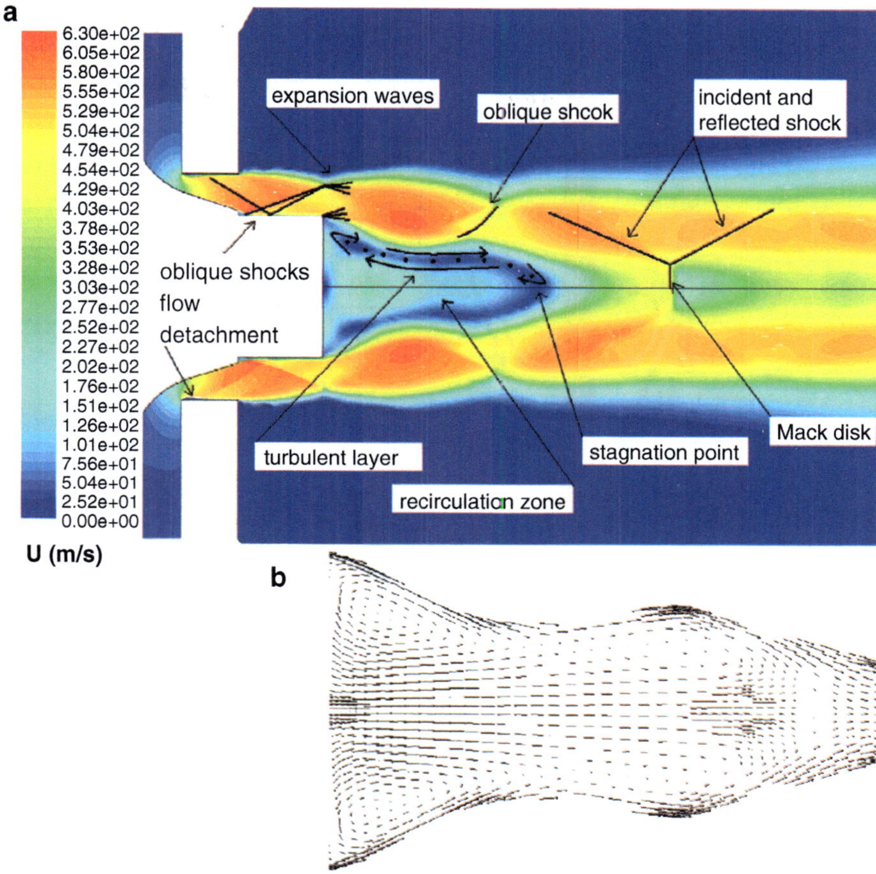

Fig. 4.5 Gas flow field simulation for an external mixing (close-coupled) atomizer: (**a**) velocity contour plot; (**b**) velocity vector depicting flow recirculation under the melt exit (adapted from [13], with permission)

4.2.2 *Free-Fall Atomizer*

Figure 4.6 shows a sketch of an external mixing free-fall atomizer. This atomizer design combines two separate gas nozzle systems, namely, a primary gas nozzle and a secondary gas nozzle, both concentrically surrounding the central jet [21]. The secondary gas nozzle is the main atomization unit, from which the concentric gas jets impinge onto the central liquid jet that is disintegrated due to instabilities from the shearing action of the secondary gas flow and its relative velocity [22]. The atomization area is located underneath the secondary gas nozzle [8, 18].

Figure 4.7 shows the simulated gas flow field (velocity contours) in an external mixing (free-fall) atomizer configuration for different gas exit boundary conditions.

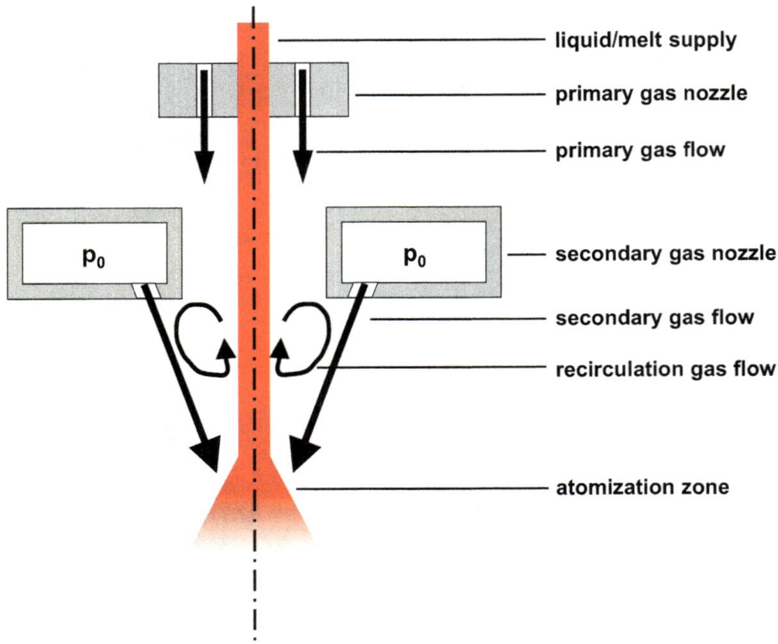

Fig. 4.6 Sketch of a conventional free-fall atomizer

Fig. 4.7 Gas flow field simulation for an external mixing (free-fall) atomizer: (**a**) gas velocity contours: under-expanded exit condition (*left*), ideally expanded exit condition (*right*) (adapted from Heck [8], with permission)

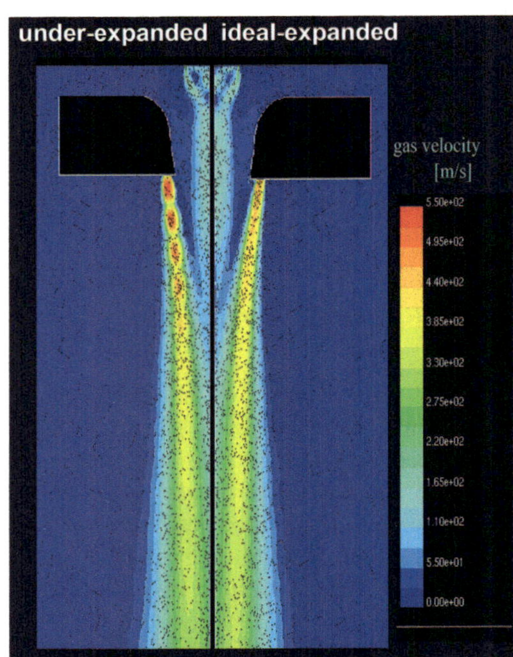

The gas exit geometry is an annular slit configuration where the gas emerges from a concentric ring slit nozzle. The left part of the Fig. 4.7 shows an under-expanded gas flow (left side) and the right side shows the result for the ideal expanded case, both cases have been calculated for identical mass flow rates. In the under-expanded case a shock cell structure in the gas jet is in front of the nozzle.

During operation of the free-fall atomizer, the gas mass flow rates of both gas nozzle systems are controlled by the primary and secondary gas pressures, respectively. An intense recirculation flow may be generated if the pressure ratio between the primary and the secondary gas nozzle is improperly adjusted. The applicable maximum gas pressure of the primary gas nozzle is limited by initial disturbances that may be generated on the liquid jet before reaching the atomization area. Therefore, the secondary gas pressure is limited also because the secondary gas mass flow rate determines the necessary primary gas mass flow rate for prevention of recirculation. This coupling of the two nozzle systems limits the applicability of the free-fall atomizer [23–25].

To improve the operating conditions and to overcome limitations of the free-fall atomizer, a Coanda-flow ring device was developed and installed inside the secondary gas flow [3]. The applied flow device utilizes the Coanda effect [26] and the injector principle to influence the local atomization gas direction. Within the confined flow device the secondary gas flow is deflected in the downward direction, the entrainment mass flow is increased, and, therefore, a primary gas nozzle is not necessary for suppression of gas recirculation.

In Fig. 4.8a, the flow field of the free-fall atomizer without primary gas application is illustrated as streamline distribution (left) and local static pressure distribution (right) at a secondary pressure of 0.5 MPa (rel.). A recirculation flow field generated underneath the secondary gas nozzle is visible. The area with highest pressure is located where the atomization gas streams impinge onto the centre line. The local static pressure exceeds 50 kPa in this area. Fig. 4.8b shows the flow situation of a free-fall atomizer without primary gas application but with the installed Coanda-flow ring device at a secondary gas pressure of 0.5 MPa (rel.). On the left, the flow field is illustrated by its streamline distribution; on the right, the static pressure distribution is displayed. No recirculation flow field is generated in this case. On the right it can be seen that at the entrance of the installed Coanda-flow device an under-pressure area is generated. Depending on the negative pressure ratio between the under-pressure area close to the atomization zone and the environmental gas pressure, the gas flow is deflected downwards. Thus, the mass flow rate increases though the liquid passage of the secondary nozzle suppressing the recirculation completely. The maximum under-pressure within the Coanda-flow device exceeds −60 kPa for the given secondary gas pressure.

Figure 4.9a shows a photograph of the atomization area without the Coanda-flow device at secondary gas pressure 0.5 MPa (rel.) and without primary gas application. Obviously an unstable atomization occurs where droplets are transported from the atomization area upwards towards the atomizer body due to the generated recirculation flow. The instability of the atomization is visible even as three-dimensional structure (asymmetric). When the Coanda-flow device is installed,

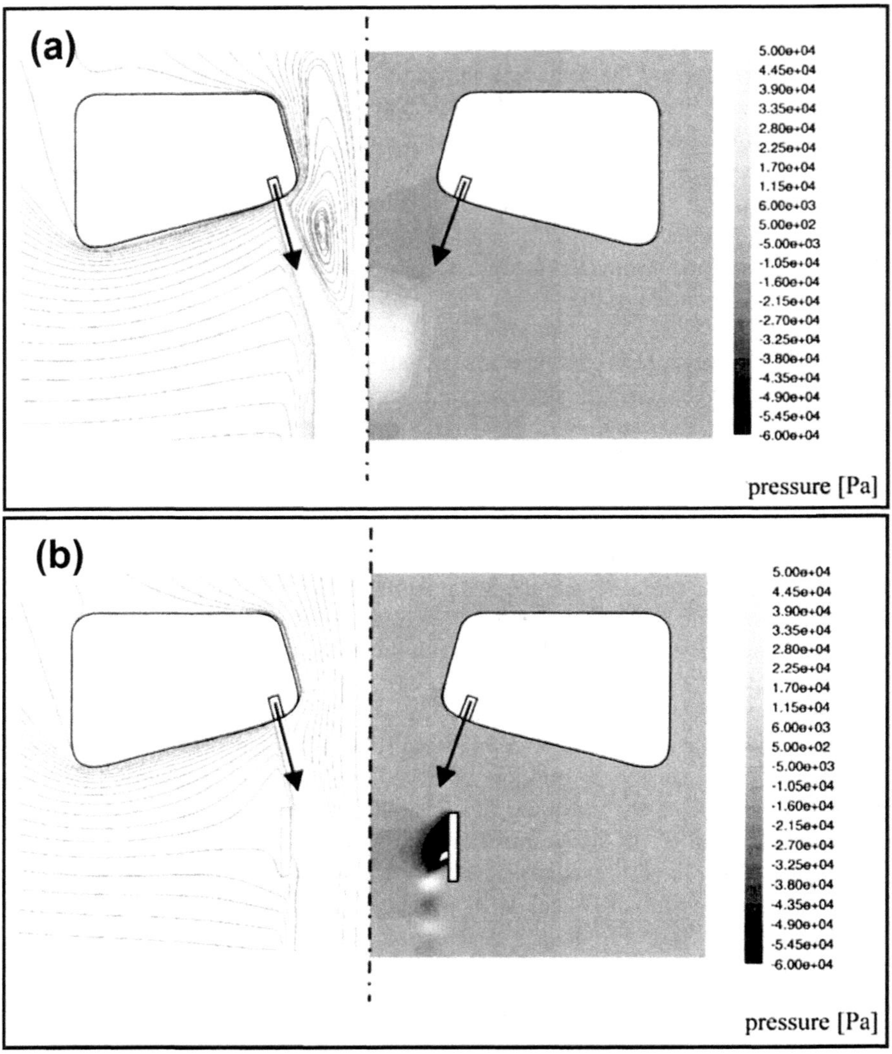

Fig. 4.8 (**a**) Flow field (*left*) of and pressure distribution (*right*) of a conventional free-fall atomizer generated only by the secondary gas flow; (**b**) Flow field (*left*) of and pressure distribution (*right*) of a conventional free-fall atomizer generated only by the secondary gas flow but with installed Coanda-flow device (from [3], with permission)

the atomization at secondary gas pressure 0.5 MPa (rel.) and without primary gas becomes stable, i.e., no major recirculation in the atomization area occurs at all, as shown in Fig. 4.9b. Advantages of the flow-adapted atomizer design with Coanda-flow device are a reduction in gas consumption, as well as the potential of generation of small particles [3, 5].

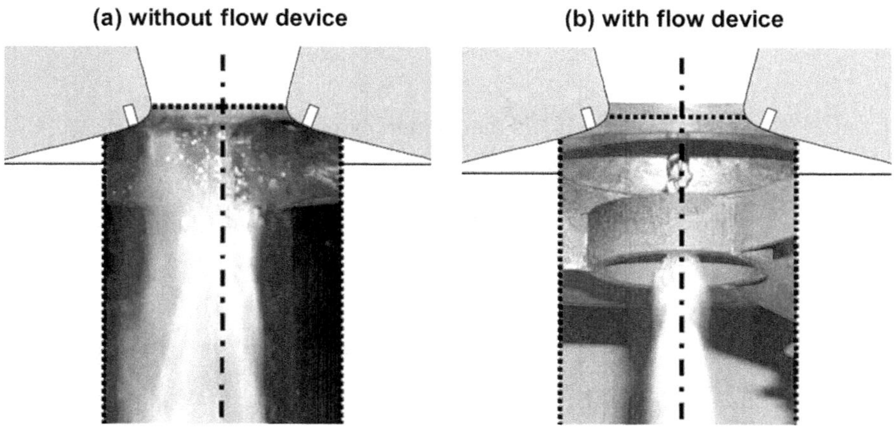

Fig. 4.9 (**a**) Model experiments without Coanda-flow device; (**b**) Model experiments with Coanda-flow device (p_{sec} = 0.5 MPa, no primary gas, liquid mass flow rate 435 kg/h) (from [3], with permission)

4.3 Liquid Metal Jet Disintegration

Within atomization, the bulk fluid (continuous liquid phase) is transformed into a spray system (dispersed phase/droplets). The disintegration process itself is caused either by intrinsic (e.g. potential) or extrinsic (e.g. kinetic) energy, where the liquid, which is typically fed into the process in the form of a liquid jet or sheet, is atomized either due to the kinetic energy contained in the liquid itself, by the interaction of the liquid sheet or jet with a (high-velocity) gas, or by means of mechanical energy delivered externally e.g. by rotating devices. Because the liquid fragmentation process has a major impact on the resulting spray structure (and the spray transport phenomena) the disintegration behaviour and main features of jet break-up will be reflected in this Sect. 4.3.

For jet disintegration processes, the most important characteristic numbers are:

$$\text{the gas (air) to liquid density ratio: } M = \frac{\rho_g}{\rho_l} \tag{4.1}$$

$$\text{the gas (air) to liquid viscosity ratio: } N = \frac{\mu_g}{\mu_l} \tag{4.2}$$

$$\text{the liquid Reynolds number: } \mathrm{Re}_l = \frac{\rho_l u_l d_l}{\mu} \tag{4.3}$$

$$\text{the liquid Weber number: } We_l = \frac{\rho_l u_l^2 d_l}{\sigma} \tag{4.4}$$

$$\text{the aerodynamic (gas) Weber number: } We_g = \frac{\rho_g u_{rel}^2 d_l}{\sigma} \tag{4.5}$$

$$\text{the Ohnesorge number Oh}: \ Oh = \frac{\sqrt{We_l}}{\mathrm{Re}_l} = \frac{\mu_l}{\sqrt{\rho_l \sigma d_l}} \tag{4.6}$$

$$\text{the gas to liquid momentum ratio}: \ I = \frac{\rho_g u_g^2}{\rho_l u_l^2} \tag{4.7}$$

$$\text{and the gas (air) to liquid mass flow ratio}: \ GLR = \frac{\dot{m}_g}{\dot{m}_l} \tag{4.8}$$

4.3.1 Flow Regimes

For a first classification of liquid fragmentation in various configurations, the disintegration behaviour of a liquid jet or sheet is subdivided into different atomization modes depending on the process conditions. As a result of several investigations the disintegration behaviour of a liquid has been globally classified, regimes have been identified and interfaces between the different regimes have been correlated.

4.3.1.1 Twin-Fluid Atomizer

A map of breakup regimes for a liquid jet in a coaxial gas flow was introduced by Farago and Chigier [27]. Hopfinger [28] introduced the momentum ratio between gas and liquid as an additional parameter. The main disintegration modes are illustrated in Fig. 4.10.

1. At small aerodynamic Weber numbers the jet mainly breaks up due to Rayleigh instabilities (axisymmetric and non-axisymmetric Rayleigh mode).
2. If the Weber number is increased, ligaments are formed. The surface tension acts now at smaller scales. Owing to its smaller radius of curvature, the tip of the

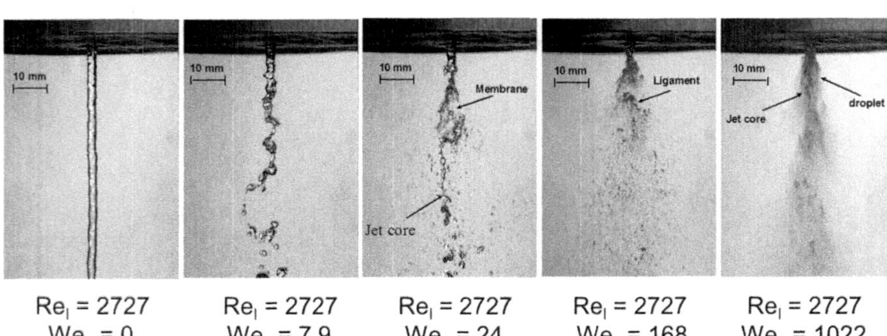

Fig. 4.10 Atomization of a liquid jet in twin-fluid atomization with external mixing: main disintegration modes, $d_l = 3$ mm

ligament recedes and a bulge is formed that may be blown-up by the gas to form a membrane. This breakup regime is called the membrane mode.

3. At further increasing Weber numbers, the size of the ligaments decreases and the breakup occurs in the form of fibers (fiber mode).

Not enough data at present are available to identify the interfaces between the different regimes in a general manner.

The role of an inner gas stream on the stability of annular liquid sheet has been studied in [29–32], as well as applied to gas atomization by Anderson and Figliola [16]. With two coflowing gas streams, Lavergne et al. [31] distinguished three break-up regimes: bubble formation, wind-induced and atomization. They also investigated the instability frequency of the sheet and found that it increases with gas velocity but grows slightly with outer gas swirl. Choi et al. [30] identified three disintegration modes: Rayleigh, bubble break-up and pure-pulsating. Moreover, in Rayleigh mode, cell structures are visible on the surface of the bubbles. In Adzic et al. [29] three major sheet break-up regimes were identified: Kelvin-Helmholtz (K-H) regime, cellular regime and atomization regime. The first regime is divided into three sub-modes: the Rayleigh regime occurs at low inner gas velocity and without significant outer gas velocity; the second sub-mode is shaped like a "Christmas tree" at the injector exit; then, the last sub-mode is characterized by Kelvin-Helmholtz instabilities visible on the sheet surface. In [29–31], there is no swirl in the annular liquid sheet. In [32], swirling conical sheets were investigated and different spray regimes were determined based on different inner air and outer liquid flow momentums. With increasing inner air stream momentum and thereby increasing gas Weber number (We_g), four break-up regimes have been identified: wave-assisted sheet break-up, perforated sheet break-up, segmented sheet break-up, and pulsation spray regime. Some examples are illustrated in Fig. 4.11. The spray exhibits a periodic ejection of liquid whose features are dominantly controlled by the central air jet.

Figure 4.12a shows the lamella breakup of Sn without atomization gas from the ring gas nozzle. The origin of the lamella (marked by two dashed lines) is not visible because it is covered by the ring gas nozzle. The breakup of the Sn lamella begins close to the origin in the covered area. The breakup of Sn-lamella is dominated by wave disintegration where ligaments are the first to form, which subsequently develop into droplets. As shown in Fig. 4.12b, the SnCu30-lamella is considerably longer and the breakup is initiated by sheet fragmentation due to perforation. Even though the surface of the lamella appears rippled, the low amplitudes do not grow with flow distance and the waves do not cause the breakup. Small holes in the lamella grow fast until only ligaments are left. Finally, these ligaments disintegrate and, again, form droplets. The different breakup mechanisms for Sn and SnCu30 are caused by the differences in material properties. External forces are not able to initiate film fragmentation due to wave formation for SuCu30 because the viscosity of the melt is higher and the density of the gas in the vicinity of the lamella is lower because the gas temperatures are higher (due to the higher melting temperature of the melt). Fig. 4.12c and d display the lamella

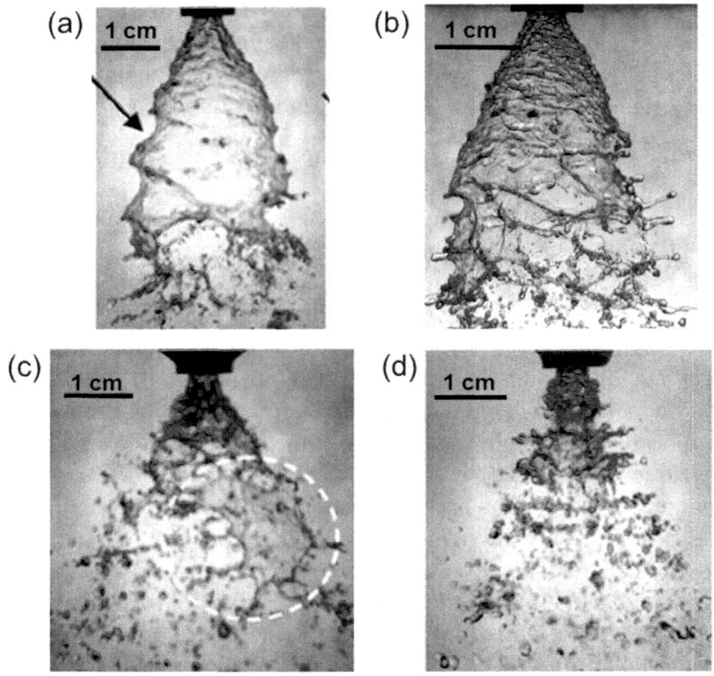

Fig. 4.11 Disintegration modes of a conical liquid sheet (water) from a gas-centred swirl coaxial atomizer: (**a**) wave-assisted sheet breakup, (**b**) perforated sheet breakup, (**c**) pulled-out segmented sheet breakup, (**d**) pulsation spray regime (adapted from [32], with permission)

breakup with the atomization gas from the ring gas nozzle. Since the gas flow from the ring nozzle generates gas entrainment, the breakup of the lamella is affected as well. This effect is particularly visible for the SnCu30 melt (Fig. 4.12d) where there are more holes visible in the lamella and they form closer to the outlet of the nozzle. For the Sn melt (Fig. 4.12c), secondary atomization of the primary fragments has been observed (the marked area).

4.3.2 Numerical Simulation of Liquid Jet Breakup

4.3.2.1 Direct Numerical Simulation

Recent investigations describe the complex behaviour of the liquid jet and the gas/liquid interface, including the primary fragmentation process in atomization, using direct numerical simulation (DNS) of the multiphase flow system (see e.g. [36–46]). The progress in direct numerical simulation of liquid atomization and liquid fragmentation processes as well as in elementary liquid processes as e.g. liquid interface coalescence is tremendous. Appropriate length and time scales

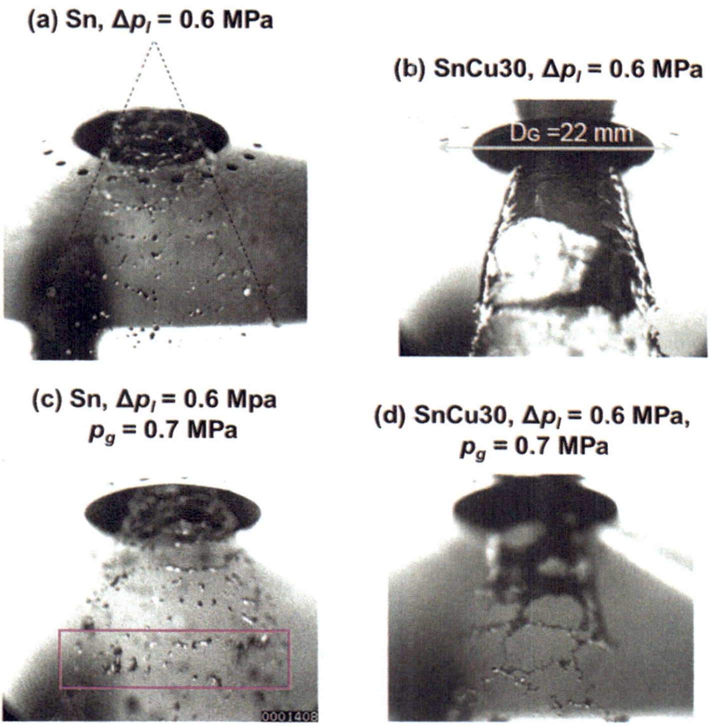

Fig. 4.12 Effect of external gas flow from a ring nozzle on disintegration of swirling conical sheets of liquid metal Sn and its alloy SnCu30 (adapted from [33, 34])

are increasingly resolved. The success of fundamental research in this area is mainly driven by tremendously increasing computer power; however, carrier of the progress is the successful cooperation of related disciplines as physics, chemistry, engineering, and mathematics, respectively.

DNS approaches in liquids are based on the continuum approach, thus the integration of the single-fluid Navier-Stokes equations, identifying the gas/liquid interface at each time. The goal of such a DNS is to resolve all time and length scales solving the governing equations directly. Interface capturing models [37, 47, 48] are usually applied to obtain the gas/liquid interface evolution and topology by solving an advection equation of an indicator (Φ) that can be attributed to the interface

$$\frac{\partial \Phi}{\partial t} + \vec{U} \bullet \nabla \Phi = 0 \tag{4.9}$$

The surface-tension force, which is a singular force active only at the location of the phase interface, can be handled by three different approaches, i.e., the continuum surface force (CSF) method [49], the continuous surface stress (CSS) method [42, 50, 51], and the ghost-fluid method [52–54]. In both CSF and CSS methods the surface tension force is usually spread into a small neighbourhood normal to the phase interface, while a sharp interface can be obtained by use of the ghost-fluid method. Two principle methods within interface capturing scheme are the Volume of Fluid (VOF) method [42, 55] and the Level Set (LS) method [56, 57].

In the VOF method, a volume fraction (f) is defined in each cell as the fraction of the cell containing liquid. If the cell is completely filled with liquid then $f = 1$ and if it is filled with gas then its value should be $f = 0$. At the location of the gas/liquid interface the value of f is between 0 and 1 ($0 < f < 1$). Discontinuity in f propagates according to Eq. (4.9), where Φ is replaced with f. In a physical sense, the equation implies mass conservation of one phase in the mixture. The gas/liquid interface needs to be reconstructed in each cell. Gopala and van Wachem [58] give a useful comparison between different interface sharpening and reconstruction algorithms in a variety of CFD codes. In general, the geometric piecewise linear interface calculation (PLIC) method [59, 60] is commonly used because of its accuracy and applicability for complex flows. In the PLIC method, the interface is approximated by a straight line of approximate inclination in each cell. A typical reconstruction of the interface with a straight line in a cell, which yields an unambiguous solution, is perpendicular to an interface normal vector and delimits a fluid volume matching the given volume fraction f for the cell. In the open source CFD code OpenFOAM, an interface compression scheme has been implemented as interface sharpening methodology, which adds an additional 'artificial' compression term to the LHS of the volume fraction transport equation (see [61]).

In the level set (LS) method, a continuous function F is defined as the signed distance between any point of the domain and the interface. The function F is usually set to zero at the interface, is positive on one side and negative on the other. Solving the advection equation Eq. (4.9), where Φ is replaced by the level set function F, determines the evolution of the interface. A reinitialization algorithm is necessary to keep F as the signed distance to the interface. The major drawback of the LS method is that it does not inherently conserve liquid mass. Since liquid volume errors are proportional in size to the employed grid resolution, grid refinement strategies, such as the refined level-set grid (RLSG) method [62] and structured adaptive mesh refinement [232], can be employed to reduce their influence. An alternative to grid refinement is to augment and correct the LS function by an additional numerical scheme, e.g., coupling LS and VOF (CLSVOF) [63]. The main concept of the CLSVOF method is to benefit from the advantage of both LS and VOF methods: mass loss is limited through the VOF method and a fine description of interface properties is kept with the LS method.

Menard et al. [41], Lebas et al. [64] and Shinjo and Umemura [43] have successfully employed the CLSVOF method in 3D DNS of the primary atomization of a turbulent liquid jet into stagnant air. In [41, 64], the ghost-fluid method is used to model the surface-tension force, while in Shinjo et al. [43], the CSF method is applied. These DNSs have been performed with very fine grid resolution (grid spacing 0.35–2.36 μm). Therefrom more detailed information in liquid atomization process can be revealed. Shinjo and Umemura [43] found that ligament formation occurs both from the liquid jet tip roll-up (Fig. 4.13a) and the liquid core surface (Fig. 4.13b). In Fig. 4.13a, the mushroom-shape tip is created due to the lateral

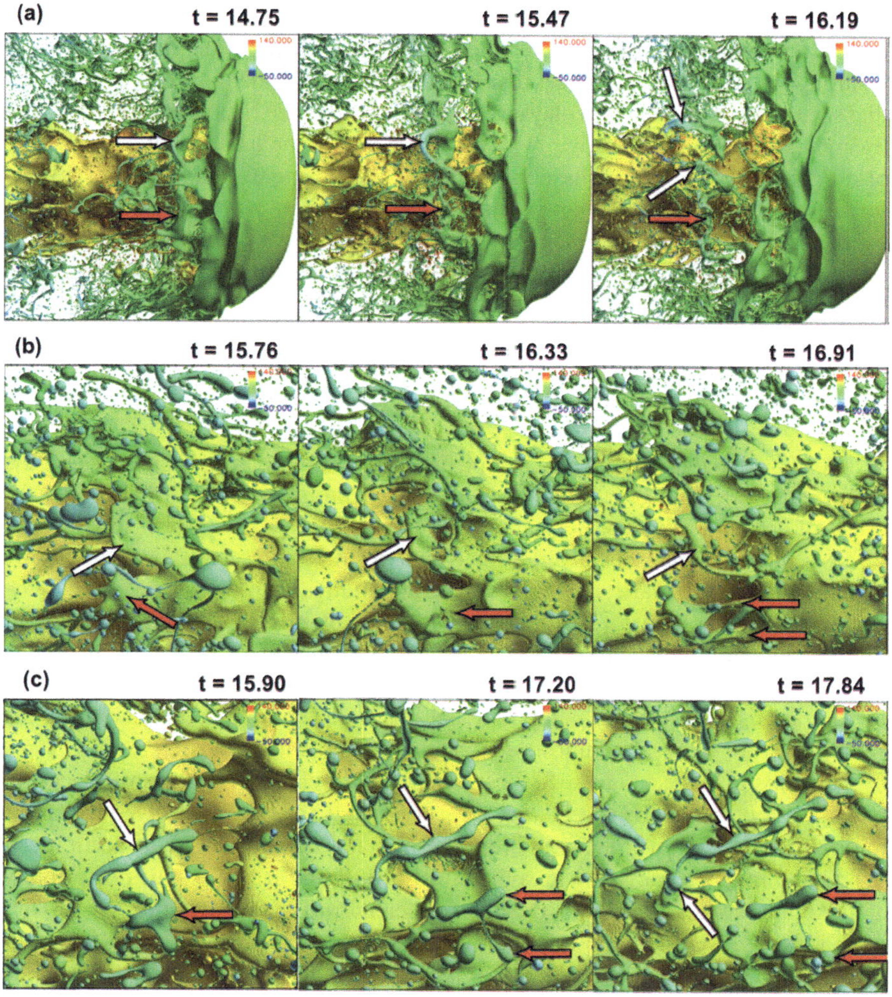

Fig. 4.13 Direct numerical simulation of liquid jet fragmentation: (**a**) Ligament formation from tip edge; (**b**) ligament formation from liquid core; (**c**) droplet formation from ligament. $Re_l = 1470$, $We_l = 14{,}100$; the colour indicates the axial velocity in m/s; the flow is from left to right; non-dimensional time $t = t_{real}/(D/U_l)$ with nozzle diameter D = 0.1 mm and liquid injection velocity $U_l = 100$ m/s. (Adapted from [43], with permission)

liquid spread by impingement against the stagnant gas and roll-up by the initial Rayleigh-Taylor instability; from the mushroom tip edge, quasi-axisymmetric (or ring-like) ligaments (indicated by the red arrow) and streamwise ligaments (indicated by the white arrow) are generated alternately. In Fig. 4.13b, crests are formed on the liquid core due to surface instability and break up into ligaments: in the case indicated by the white arrow, the crest break up into only one ligament; while in the case indicated by the red arrow, the central part of the flat crest soon becomes thinner than the rim, and then a hole is created and grows due to contracting motion by surface tension in the central region of the flat crest, finally resulting in the formation of two ligaments. Shinjo and Umemura [43] pointed out that shear from local vortices plays an important role in the ligament formation, and ligaments are created when $We_g \sim O(1)$, where We_g is the local gas Weber number of a ligament based on the baseline radius of the ligament. In Fig. 4.13c, droplet formation occurs from the ligament tip mostly by the short-wave mode (as indicated by the red arrow), and surface capillary wave propagation is clearly observed (as indicated by the white arrow). Re-collision of ligaments and droplets often happens and enhances next breakup.

4.3.2.2 RANS Approach and LES

In a DNS, the goal is to reduce epistemic uncertainty by resolving all necessary time and length scales inherent in the flow and thus eliminating the need to model the effect of any unresolved scales [37]. In a single-phase turbulent flow, the smallest length scale that must be resolved is the Kolmogorov length scale. Multiphase flows add an additional smallest length scale that requires resolution, the size of the smallest liquid structure. This implies the DNS of primary atomization with a limited domain size, simple geometries and low turbulent Reynolds numbers. Therefore, interface capturing models, usually coupled with turbulence models for the sake of optimizing computational cost, are mainly used to simulate the situations from jet formation to fragmentation near the nozzle exit.

Reynolds-averaged Navier-Stokes (RANS) formulation can be introduced in industrial solvers. In this case, almost any configuration may be resolved but much information concerning the liquid properties remains unknown, and the accuracy of the results may be quite low in some situations. Some RANS studies of primary atomization process based on VOF interface capturing can be found in [15, 66].

In [15], the VOF method together with the Reynolds Stress Model (RSM) was employed to simulate melt jet/sheet formation and fragmentation process from three different twin-fluid atomizer designs, i.e. an annular-slit atomizer, a swirling gas atomizer and an inner gas jet atomizer. It is found that the gas-melt interaction can significantly affect the gas flow dynamics in the near field. The 3D simulation results indicate that the gas swirl atomizer seems to inhibit the atomization process by stabilizing the melt core, while the inner gas jet atomizer enhances the fragmentation of the annular melt sheet, and as a result may improve the powder generation.

In [66], the formation of a swirling conical sheet and its fragmentation near the nozzle exit were numerically investigated based on the VOF approach, the k-ω SST (shear stress transport) turbulence model and the adaptive mesh refinement technique. Fig. 4.14 exhibits the temporal evolution of a swirling conical sheet of melt tin (Sn) from the 3D simulation, which indicates two kinds of instabilities existing on the sheet surface: one grows in the longitudinal direction, while the other in the circumferential direction. The latter wavelength (several millimeters) is one order of magnitude larger than the former (hundred micrometers). The ligament

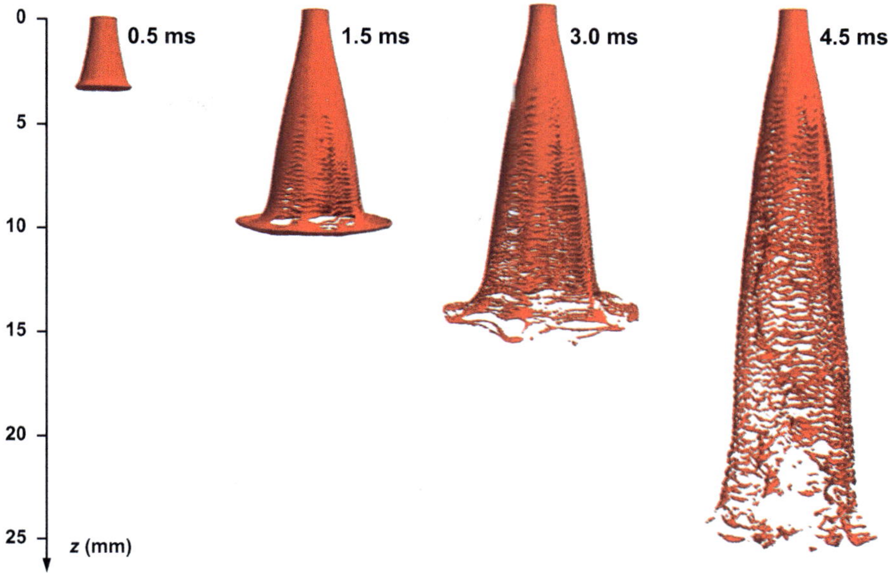

Fig. 4.14 3D simulation: temporal evolution of a swirling conical sheet of melt tin (Sn) from a pressure swirl nozzle, $\Delta p_l = 0.6$ MPa, $D_0 = 1.3$ mm (adapted from [66])

formation can be observed. The inter-connected ligaments form cellular structures (or perforations), and finally disintegrate into droplets. This phenomenon has been confirmed by the experimental observations in [33, 68, 69].

Large eddy simulation (LES) is a promising technique between industrial RANS solvers and academic DNS solvers. It solves the largest scales of the flow while the impact of the smallest scales (or the subgrid scales) is modeled. Zeoli et al. [35] numerically investigated the unsteady features of a hot liquid metal entering an atomization tower by means of a 3D, turbulent, unsteady simulation using the VOF method and the LES turbulence capturing technique. Three modes are predicted by the computational model for close-coupled atomizers, namely nozzle filming, mixed filming and pinch-off, and no-filming (see Fig. 4.15), which are determined by the gas to melt mass flow ratio (GMR). The simulated scenarios of nozzle filming and mixed filming occur at GMRs of 6.6 and 3.3, respectively. When a constant pressure of 12 atm at the melt inlet is applied, the GMR varies between 0.9 and 2.1, creating a non-film working condition. The third mode represents the ideal operating condition of an industrial coaxial gas atomizer. Figure 4.16 depicts the evolution of the melt flow from when the melt enters the domain until when the primary atomization occurs. As the melt stream enters the domain, the strong circulating flow imposes itself on the melt that generates a flattened interface. With the protrusion of liquid extending into the rear stagnation point, the counter-flowing gas forces the metal accumulated at the jet periphery back toward the feeding tube, generating an upside-down mushroom shape. Once the melt extends

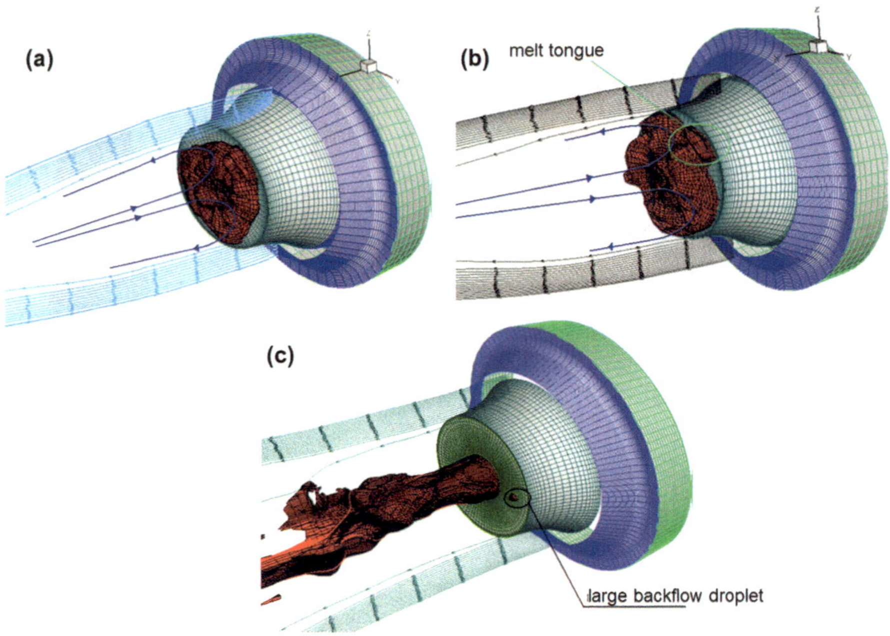

Fig. 4.15 Unsteady melt stream, 3D simulation: (**a**) with melt flow rate of 0.05 kg/s; (**b**) with melt flow rate of 0.10 kg/s; (**c**) with a constant pressure melt boundary condition of 1.2 MPa, corresponding to an unsteady melt flow rate ranging from 0.37 to 0.16 kg/s (adapted from [35], with permission)

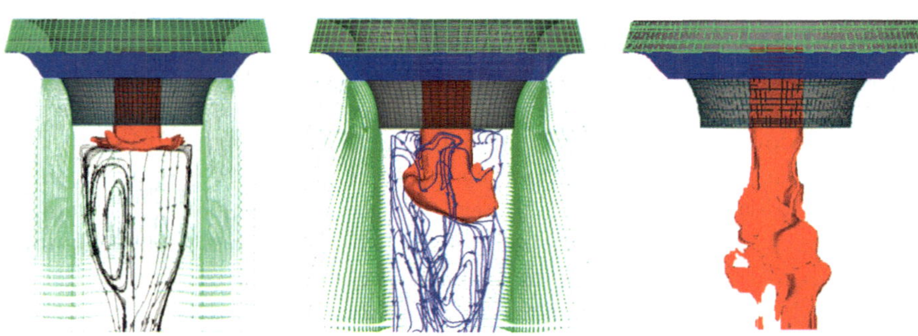

Fig. 4.16 Liquid core evolution and streamlines for the constant pressure melt-inlet boundary condition of 1.2 MPa, 3D simulation (adapted from [35], with permission)

further downstream of the nozzle, the turbulent environment causes several liquid ligaments to form. Initially, the streamlines are positioned under the melt in a well-formed regular distribution, but as the melt moves downstream the streamlines engulf the mushroom creating a highly unsteady condition.

4.3.2.3 Multiscale Model

The primary atomization of a liquid jet into small droplets is a multiscale phenomenon, in which the size of the smallest droplet is often three orders of magnitude smaller than the diameter of the atomizing liquid jet. Direct numerical simulation (DNS) with interface tracking/capturing schemes is very useful in understanding fundamental physics of atomization. However, the range of scales that can be covered by DNS with computing ability today is still often much smaller than practical applications. For example, even with a mesh resolution of 0.35 μm [43], the finest scales of the flow are still not resolved.

To resolve this multiscale challenge in atomization simulation, coupled Eulerian/Lagrangian multiscale models have been proposed in [70–72]. In these multiscale models, interface capturing methods (ICM) (the Eulerian approach) such as VOF, LS and CLSVOF methods describe the large scales of gas-liquid interface during atomization while the Lagrangian particle tracking (LPT) approach is employed to model the droplets smaller than the Eulerian grid size. The purpose of the coupling procedure is to identify separated regions of liquid in gas fulfilling certain criteria, to remove these from the interface capturing representation, and to insert them into a Lagrangian framework, preserving their volume V_d, centre of mass $\mathbf{x_d}$, and momentum $\rho_d \mathbf{U_d}$. The removal criteria can be based on the derivation assumptions of typical Lagrangian spray models, i.e., that individual drops be small and nearly spherical. The size criterion can be expressed in terms of a thresh volume V_{cut} (e.g. the grid volume), in that only those separated liquid structures with volume $V_d < V_{cut}$ are candidates for removal. The shape criterion can be expressed in terms of an eccentricity measure threshold e_{cut}. The multiscale coupling approach links DNS of the liquid core and the dense spray regime to the LPT simulation of the dilute spray regime. Compared with a full DNS, the computational cost will be reduced by this coupling approach, as the unresolved droplets, which are computational expensively tracked in DNS, are only tracked in the LPT frame if these droplets are judged to satisfy Eulerian to Lagrangian transfer criteria.

Figure 4.17 shows the atomizing liquid jet, the remaining tracked phase interface geometry and the transferred Lagrangian drops from multiscale simulation of the primary atomization of a turbulent liquid jet at 20 μs after injection into still compressed air under conditions relevant to diesel engine systems [71]. The injector diameter and the mean injection velocity are $D = 100$ μm and $U = 100$ m/s, respectively. The multiscale simulation has been performed for a liquid Reynolds number $Re_l = 5000$ and a liquid Weber number $We_l = 17{,}000$ at a liquid/gas density ratio $\rho_l/\rho_g = 34$. The refined level set grid (RLSG) method proposed in [62] has been used to track gas/liquid interface during atomization. In Fig. 4.17, roughly 403,000 drops have been transferred into the Lagrangian spray model. Diameters of these transferred drops range from the sub-micron size up to 10.8 μm. The total computer time per time step needed to identify and remove separated structures is less than 1% of the total computer time of a single time step, which demonstrates the high efficiency of the multiscale atomization model and its applicability to realistic problems.

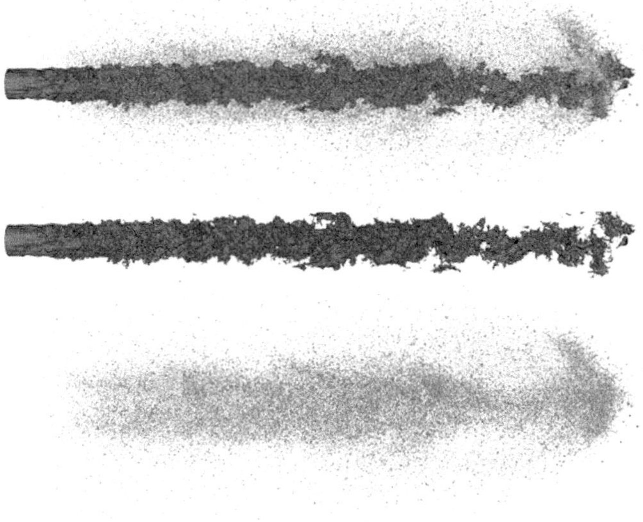

Fig. 4.17 Atomization of turbulent liquid jet 20 μs after start of injection. Liquid core and spray (*top*), tracked phase interface of liquid core (*centre*) and Lagrangian tracked spray (*bottom*). Injection Reynolds number $Re_l = 5000$ and Weber number $We_l = 17{,}000$ (adapted from [71], with permission)

In Tomar et al. [72], a 2D–simulation of atomization of a liquid jet destabilized by a high-speed coaxial gas flow has been performed based on coupled VOF and LPT approach for a high liquid/gas density ratio of 100 and a high momentum ratio ($M = \rho_g U_g^2 / \rho_l U_l^2$) of 16. The probability density functions (PDF) of the droplets are derived in two different zones of the computational domain (near the nozzle inlet and far from it), which indicates two mechanisms of droplets formation, namely, by primary atomization of the jet and another by breakup of bigger fragments of liquid emerging downstream, as shown in Fig. 4.18. The smallest droplets interact with the bigger lumps of liquid downstream leading to a shift in the droplet size distribution to large droplet sizes.

In Grosshans et al. [70], instead of transferring each droplet individually, the statistical parameters of the droplet distributions at a defined layer are extracted from the VOF simulation and applied as starting conditions for the LPT simulation. This layer has to be positioned far enough downstream from the injector, so the spray is diluted enough to be simulated with the LPT approach. In contrast to the direct coupling approach (DCA) in [71, 72], this approach is termed statistical coupling approach (SCA). For statistically stationary sprays, the accuracy of the SCA is only marginally lower than that of the DCA. However, compared with the DCA, the SCA is a computational significantly faster method for two reasons:

- By nature of the numerical methods, the time step and the grid size of the Eulerian simulation by interface capturing schemes are at least one order of

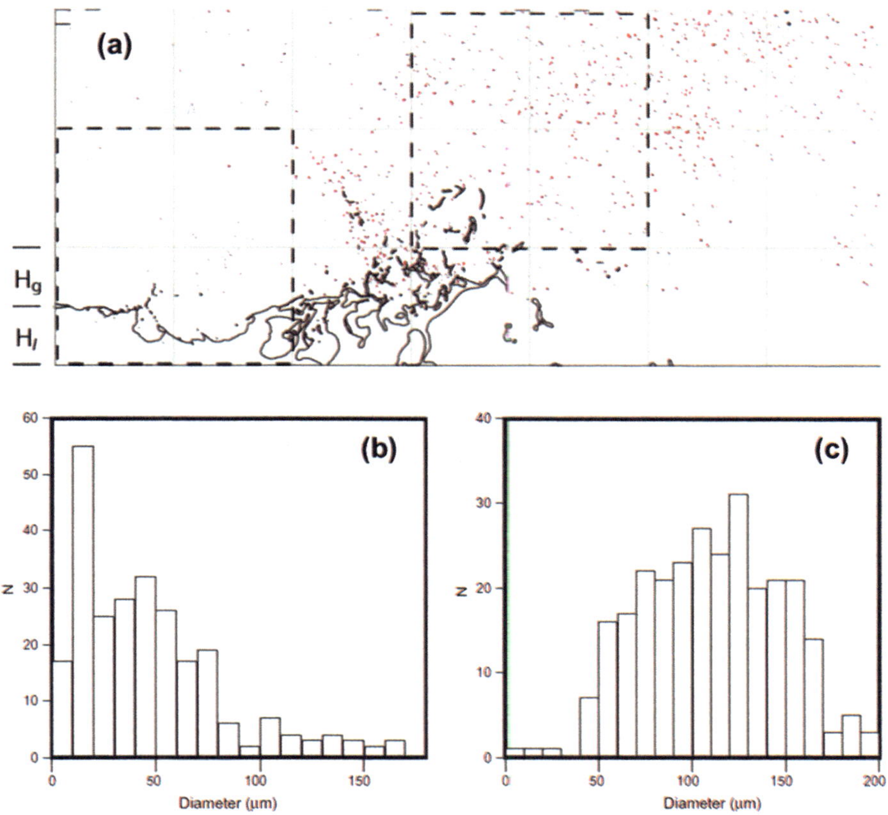

Fig. 4.18 (**a**) Breakup of a liquid jet by a high-speed coaxially flowing gas jet, with a cloud of small droplets, modeled as Lagrangian particles (shown in *red*), formed during the atomization process and spread by the high-speed gas; probability density distributions (pdf) of the diameter of the droplets formed at a location (**b**) near the nozzle inlet and (**c**) further downstream as marked in (**a**) (adapted from [72], with permission)

magnitude smaller than the ones of LPT simulation. Thus, the computational time of a droplet travelling a certain distance in the Eulerian domain is several orders of magnitude larger than in the LPT domain. In the DCA, coupled Eulerian and LPT simulations covering the whole spray development are performed in parallel in a same physical domain, and the computational time during the whole multiscale atomization simulation is determined by the Eulerian simulation. In the SCA, the Eulerian simulation and the LPT simulation can be performed in different physical domains, and the physical domain for the Eulerian simulation is limited within a region only between the nozzle injector and the selected layer.

- In the DCA, each droplet in the particle cloud in the LPT simulation is tracked individually. The large number of droplets in a practical spray cannot be handled on individual bases. In the SCA, the droplets can be tracked in the LPT simulation using the stochastic parcel method. Here, a number of droplets are grouped in one parcel. All the droplets within the parcel are assumed to have the same properties, such as same size, position, velocity, temperature, distortion and composition. Using the stochastic parcel method makes the description of a complete spray process feasible in the SCA.

Based on the principle of the SCA, Li and Fritsching [66] performed integral process modelling and simulation of pressure-gas-atomization (PGA) of molten metal for powder production. The primary disintegration of swirling conical liquid sheets of the molten metal is described by the VOF approach, while the subsequent droplet spray process is simulated through the LPT approach by taking into account the secondary breakup of produced droplets as well as other in-flight spray phenomena. The characteristics of liquid sheet fragmentation such as breakup length and primary droplet size and velocity are derived from the VOF simulation and applied as the initial conditions for the droplet spray process simulation. The coupled atomization and spray simulation obtained a good prediction of the mass median diameter (MMD) of the metal powder produced by the PGA process.

4.4 Secondary Atomization

Secondary breakup in sprays is defined as the disintegration of larger droplets and ligaments into smaller droplets. The breakup of a single droplet in a gas may be caused by relative velocity, turbulence, or shock structure interaction. The main features of binary droplet collisions are presented in Sect. 4.5.2.1.

Numerous studies of dilute sprays as well as of isolated droplets have been performed in order to increase the understanding of secondary atomization mechanism. In-depth reviews on secondary atomization can be found in [74–77]. Shock tube (Faeth and co-workers, 1992–2001) and continuous jet methods (Reitz and co-workers, 1993–2000; [78, 79]) have been employed to study the secondary breakup process. Both methods attempt to subject droplets to a step change in velocity.

Droplet breakup dynamics in flight were numerically investigated in Li and Fritsching [66], based on a pressure-gas-atomization configuration. Melt-tin droplets of different sizes are injected into a developed gas flow field and tracked in a Lagrangian way, as shown in Fig. 4.19 (left). The droplet size (d_d) ranges from 100 to 300 μm. The gas flow field corresponds to an atomization pressure $p_0 = 1.0$ MPa and accordingly a gas mass flow rate 100 kg/h. The injection conditions are supposed to be constant for all the droplets: the initial injection velocity is 10 m/s; and the injection angle is 27.5°. The distributions of Weber number are derived along droplets paths, as shown in Fig. 4.19 (right). For all the tested cases, the maximum Weber number is below $We_g = 50$, and thereby the main breakup mode should be bag or multimode breakup in the atomization zone under the operating condition. For a droplet to disintegrate, its Weber number should

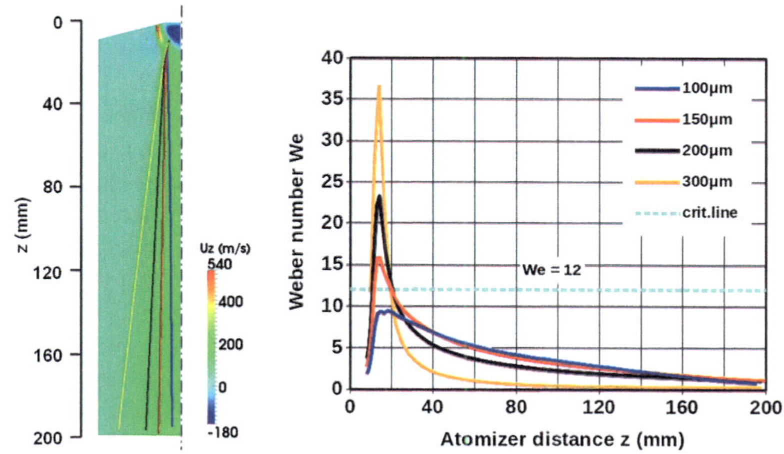

Fig. 4.19 Distributions of Weber number along trajectories of different sized droplets, the gas flow field corresponding to an atomization pressure $p_0 = 1.0$ MPa and the resulting gas mass flow rate 100 kg/h (adapted from [66])

exceed a critical value, usually, $We_{crit} = 11–13$. It can be found that the Weber number, which can meet this breakup criterion, exists only in a very narrow region around the atomizer, which means a very narrow atomization zone. Beyond this region, the aerodynamic forces are too small for droplets to break up, although the droplets may be still in a liquid state.

4.4.1 *Droplet Breakup Models*

At critical conditions, initially the droplet deforms to a disc that finally breaks up into smaller droplets. The total breakup time consists of the deformation time t_{def} and the breakup time $t_{breakup}$. The latter describes the time where no further fragmentation occurs. For computations of the droplet breakup process, models are used that include:

- breakup criteria, if the droplet breaks up or not,
- an estimation of the deformation time t_{def} and the breakup time $t_{breakup}$,
- the resulting droplet size distribution (e.g. in terms of d_{32}),
- the spatial distribution of the resulting droplets and their velocities after breakup.

4.4.1.1 Semi-Empirical Models

In the semi-empirical breakup model, the time it takes to deform and disrupt a droplet is described by the characteristic breakup time

$$t^* = \frac{d_d}{U_{rel}}\sqrt{\frac{\rho_d}{\rho_g}}. \tag{4.10}$$

The initial deformation of a droplet into a disc shape is similar for all breakup modes. The duration of this initial phase is estimated as $t_{def} = 1.6\,t^*$ by Hsiang and Faeth [80, 81]. Thereafter, the droplet will fragment depending on the breakup modes. As proposed in [77], the correlations of $t_{breakup}$ for different intervals of Weber numbers can be formulated as

$$t_{breakup}/t^* = \begin{cases} 6(We-12)^{-0.25}, & 12 \le We < 18 \\ 2.45(We-12)^{0.25}, & 18 \le We < 45 \\ 14.1(We-12)^{-0.25}, & 45 \le We < 351 \\ 0.766(We-12)^{0.25}, & 351 \le We < 2670 \\ 5.5, & We \ge 2670 \end{cases}. \tag{4.11}$$

To account for the viscosity effects observed for Oh > 0.1, a corrected Weber number is used in these relations for the low Oh-number range only [97]:

$$We_{corr} = \frac{We}{1 + 1.077Oh^{1.6}} \tag{4.12}$$

When a droplet breaks up it results in a group of new droplets with a certain size distribution and mean diameter. The Sauter mean diameter d_{32} (SMD) is computed for bag breakup and multimode breakup from the following relation:

$$d_{32}/d_d = 1.5Oh^{0.2}We^{-0.25} \tag{4.13}$$

where d_d is the initial droplet diameter. Sheet stripping breakup is characterized by continuous film stripping leaving a larger core droplet at the end of the breakup process. In the computational model, the maximum stable diameter (d_s) of this droplet is evaluated by

$$d_s = 12\sigma/\rho_g U_{rel}^2 \tag{4.14}$$

The fine fraction of the droplet fragments is distributed as in bag and multimode breakup regimes based on a reduced SMD $d_{32,red}$ which is derived from the Sauter mean diameter d_{32} given by Eq. (4.13) and the maximum stable diameter d_s by Eq. (4.14), as

$$d_{32,red} = \frac{4d_{32}d_s}{5d_s - d_{32}} \tag{4.15}$$

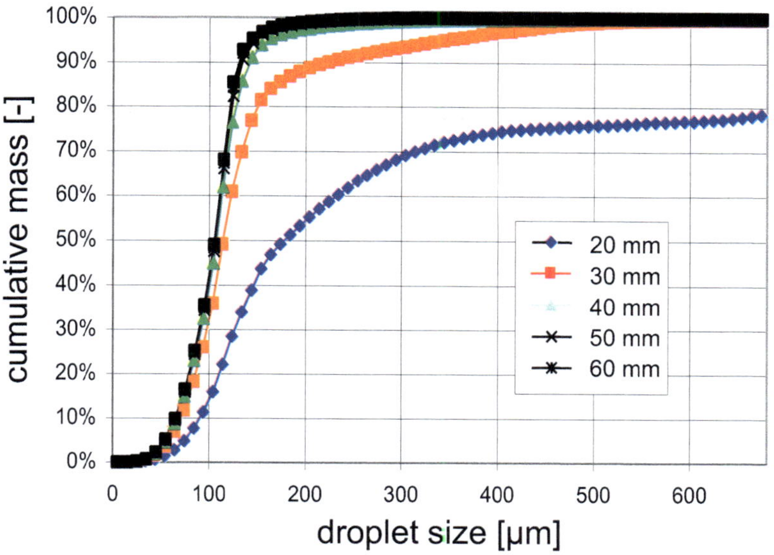

Fig. 4.20 Secondary drop fragmentation in liquid spray: evolution of the drop size distribution versus the distance to the atomizer nozzle (from [82], with permission)

The results based on a semi-empirical breakup model [82] are illustrated in Fig. 4.20. Here, atomization in a twin-fluid gas atomizer configuration is studied. In this direct numerical approach, each particle has been tracked throughout the flow field. Droplet deformation and breakup are considered. If a ligament/droplet breaks up, all daughter droplets created are further tracked. The gas velocity distribution is based on measurements. Figure 4.20 shows a typical cumulative droplet size distribution at increasing distances from the atomizer. Most droplets are created in a short distance behind the region of the primary breakup. Then the droplets are accelerated and spread out due to turbulence.

Other commonly used droplet breakup models include those based on droplet deformation dynamics and those based on the instability growth on the droplet surface. The former include Taylor Analogy Breakup (TAB) model, as well as its extensions such as ETAB and CAB models. The latter include wave breakup model and KH-RT hybrid breakup model.

4.4.1.2 Droplet-Deformation Based Models

The Taylor Analogy Breakup (TAB) model [83] treats the oscillation of a distorting droplet with the analogy of a spring-mass system. The external force is represented by aerodynamic force while the restoring force of the spring by the surface tension (σ). The liquid viscosity (μ_l) represents the damping force. Setting $y = 2x/r$, where x is the displacement of the droplet equator from its spherical (undisturbed) position and r is the undisturbed droplet radius, the governing equation of droplet deformation is

$$\ddot{y} = \frac{C_f}{C_b} \frac{\rho_g}{\rho_l} \frac{U_{rel}^2}{r^2} - \frac{C_k \sigma}{\rho_l r^3} y - \frac{C_d \mu_l}{\rho_l r^2} \dot{y} \tag{4.16}$$

where ρ denotes the density, U_{rel} the relative droplet-gas velocity, and the subscripts g and l denote the gas or liquid properties, respectively. As suggested in [83], the values of the dimensionless constants C_b, C_f, C_k and C_d are 0.5, 1/3, 8 and 5, respectively. It is argued in the TAB model that a necessary condition for droplet breakup is reached when $We_g > We_{crit}$, where the critical Weber number is experimentally determined as $We_{crit} = 6$. The gas Weber number used in TAB, ETAB and CAB model is based on the droplet radius, i.e., $We_g = \rho_g U_{rel}^2 r / \sigma$. For an inviscid liquid, this condition is met when $y(t) > 1$. The breakup time is taken to be the smallest positive root of the deformation equation $y(t) = 1$. By balancing the deformation energy and surface energy before and after breakup, the expression for the Sauter mean radius (r_{32}) of the child droplets after breakup can be derived as

$$\frac{r}{r_{32}} = 1 + \frac{8K}{20} + \frac{\rho_l r^3}{\sigma} \left(\frac{6K - 5}{120} \right) (\dot{y})^2 \tag{4.17}$$

where K is a constant that must be evaluated experimentally by measuring droplet sizes. O'Rourke and Amsden [83] suggest a value of $K = 10/3$. It is assumed that the product droplets are neither distorted nor oscillating, i.e., $y = \dot{y} = 0$.

The Enhanced-TAB (ETAB) model developed in Tanner [84] maintains the droplet deformation dynamics of the TAB model, i.e., droplet breakup occurs when the normalized droplet distortion y exceeds the critical value of 1. However, for each breakup event the ETAB model assumes that the rate of product droplet generation is proportional to the number of the product droplets. From this, the rate of droplet creation, in conjunction of with the mass conservation principle, leads to the basic ETAB law

$$\frac{d\bar{m}(t)}{dt} = -3K_{br}\bar{m}(t) \tag{4.18}$$

where $\bar{m}(t)$ denotes the mean mass of the product droplet distribution. The breakup constant K_{br} depends on the breakup regime (bag or stripping breakup) according to the gas Weber number as follows:

$$K_{br} = \begin{cases} k_1 \omega, & We_g \leq We_{trans} \\ k_2 \omega \sqrt{We_g}, & We_g > We_{trans} \end{cases} \tag{4.19}$$

with $k_1 = 2/9$, $k_2 = 2/9$ and $We_{trans} = 80$ [84]. The droplet oscillation frequency ω is given by

$$\omega^2 = C_k \frac{\sigma}{\rho_l r^3} - \left(\frac{C_d \mu_l}{2\rho_l r^2}\right)^2 \tag{4.20}$$

with $C_k = 8$ and $C_d = 5$ [83]. With an assumption of a uniform product droplet size distribution, Eq. (4.18) becomes

$$r_{new}/r = e^{-K_{br}t} \tag{4.21}$$

where r_{new} and r are the radii of the product and parent droplets, respectively.

The Cascade Atomization and Droplet Breakup (CAB) model proposed by Tanner [85] is a further development of the ETAB model. In the CAB model, the breakup condition is determined by means of the droplet deformation dynamics of the TAB model, and the definition of the rate of droplet creation follows the way in the ETAB model. However, the individual droplet breakup regions, which, in the ETAB model, were restricted to either bag ($We_{crit} < We_g < We_{b,s}$) or stripping breakup ($We_{b,s} < We_g < We_{s,c}$), have been extended to include the catastrophic breakup regime ($We_{s,c} < We_g$) to accommodate the disintegration of the high-speed droplets. Correspondingly, the breakup constant K_{br} for each breakup regime can be expressed as

$$K_{br} = \begin{cases} k_1\omega, & We_{crit} < We_g < We_{b,s} \\ k_2\omega\sqrt{We_g}, & We_{b,s} < We_g < We_{s,c} \\ k_3\omega(We_g)^{3/4}, & We_{s,c} < We_g \end{cases} \tag{4.22}$$

where the droplet oscillation frequency ω is given by Eq. (4.20). In Tanner [85], the regime-dividing Weber numbers are taken to be $We_{b,s} = 80$ and $We_{s,c} = 350$, and the constant $k_1 = 0.05$ is determined, whereas the values for the constants k_2 and k_3 are chosen such that K_{br} is continuous at the regime-dividing Weber numbers, $We_{b,s}$ and $We_{s,c}$. The way to deal with the velocity of the product droplets in the CAB model is same as that in the ETAB model.

4.4.1.3 Wave and KH-RT Hybrid Breakup Model

The Wave breakup model or the Kelvin-Helmholtz (K-H) breakup model, developed by Reitz and Diwarkar [86] and further improved by Reitz [87], assumes that breakup is caused by KH instability on the surface of a cylindrical 'blob' of liquid. Under the assumption that the size of the striped off product droplets are proportional to the length of the fastest growing wave, the wave model postulates that a parent parcel with radius, r, breaks up to form new droplets with radius, r_{new}, as

$$r_{new} = B_0 \Lambda_{KH} \tag{4.23}$$

where Λ_{KH} is the wavelength corresponding to the K-H wave with maximum growth rate (Ω_{KH}) and B_0 is a constant equal to 0.61. Linear analysis of K-H instabilities on a round liquid jet gives the frequency of the fastest-growing wave (Ω_{KH}) and its corresponding wavelength (Λ_{KH}), as follows:

$$\Omega_{KH} = \frac{0.34 + 0.38We_g^{1.5}}{(1+Oh)\left(1+1.4\left(Oh\sqrt{We_g}\right)^{0.6}\right)}\sqrt{\frac{\sigma}{\rho_l r^3}} \tag{4.24}$$

$$\Lambda_{KH} = \frac{9.02r\left(1+0.45\sqrt{Oh}\right)\left(1+0.4\left(Oh\sqrt{We_g}\right)^{0.7}\right)}{\left(1+0.865We_g^{1.67}\right)^{0.6}} \tag{4.25}$$

The definitions of gas Weber number (We_g) and Ohnesorge number (Oh) can be found in Eqs. (4.5) and (4.6), respectively, but with a substitution of droplet diameter with droplet radius in both equations. During breakup, the parent parcel reduces in diameter due to the loss of mass. The rate of change of the radius of the parent parcel is calculated using

$$\frac{dr}{dt} = \frac{r - r_{new}}{\tau_{KH}} \tag{4.26}$$

where τ_{KH} is the breakup time defined by

$$\tau_{KH} = \frac{3.726B_1 r}{\Omega_{KH}\Lambda_{KH}} \tag{4.27}$$

The constant B_1 has been given a variety of values between 1.73 and 60 [88].

In the KH-RT hybrid atomization model [88, 89], the wave model is employed to predict the primary breakup of the intact liquid core of a liquid jet. The secondary breakup of individual droplets is modeled with the wave model in conjunction with the Rayleigh-Taylor (RT) accelerative instability model. The hybrid model is allowed to grow KH instability and RT instability simultaneously, and the fastest growing instability leads to a droplet-breakup event. In the RT model, the frequency of the fastest growing wave is given by

$$\Omega_{RT} = \sqrt{\frac{2}{3\sqrt{3\sigma}}\frac{\left[-g_t\left(\rho_l-\rho_g\right)\right]^{3/2}}{\rho_l+\rho_g}} \tag{4.28}$$

where g_t is the acceleration in the direction of travel and is defined by $g_t = \mathbf{g} \bullet \mathbf{j} + \mathbf{a} \bullet \mathbf{j}$, where **a** is the droplet acceleration, and **j** is the unit vector tangent to the droplet trajectory. The corresponding wavelength is

$$\Lambda_{RT} = 2\pi C_{RT} \sqrt{\frac{3\sigma}{-g_t\left(\rho_l - \rho_g\right)}} \tag{4.29}$$

where C_{RT} is an adjustable constant set equal to 0.1 [89]. The RT waves are only allowed to form on the droplet when the diameter of the droplet is larger than the wavelength of the fastest growing disturbances (i.e. d > Λ_{RT}). The radii of new droplets is given by

$$r_{new} = \Lambda_{RT}/2 \tag{4.30}$$

The breakup time in the RT model is calculated using

$$\tau_{RT} = \frac{C_\tau}{\Omega_{RT}} \tag{4.31}$$

where C_τ is a constant usually equal to unity beyond the breakup length, and is equal to 9.0 within the breakup length.

4.4.1.4 Empirical Extension of Droplet Deformation Model

In many application scenarios, the time scale of deformation and breakup is comparable to or larger than the characteristic scales of the gas flow fluctuations, i.e., the relative velocity varies significantly during the deformation and breakup process [90]. In these cases, the deformation and breakup process and the resulting characteristics of secondary fragments do not depend only on instantaneous local flow situations but on the variation of aerodynamic forces during the disintegration process. As an empirical extension of dynamic droplet deformation model, the breakup model proposed in [91] takes into account the temporal evolution of the aerodynamic loading of the droplet. A critical deformation, derived by the Non-linear Taylor Analogy Breakup (NLTAB) model [92], is used as breakup criterion. The actual disintegration process is described empirically. Fragment sizes resulting from secondary atomization can be predicted by a global statistical atomization product model as proposed in the semi-empirical breakup model, or by the detailed atomization product model, which Bartz et al. [91] obtained by analyzing the experimental data from Chou et al. [93–95]. The detailed atomization product model accounts for the volume fraction of the fine droplets produced by the fragmentation of the bag during bag and bag-plume breakup.

Bartz et al. [96] presented the comparison of droplet trajectories for bag breakup, as illustrated in Fig. 4.21. The trajectory of the droplet in the experiment is plotted as a solid line with circles. The time when the fragmentation of the bag begins, t_1, and a time t_2 during the fragmentation of the rim are marked. The trajectory predicted with the semi-empirical model of Schmehl [97] is displayed as a dotted line. As stated before, in this model the characteristic times are calculated based on

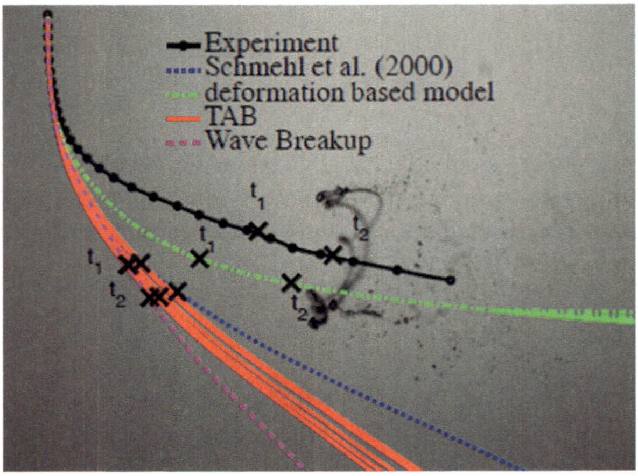

Fig. 4.21 Comparison of trajectories for bag breakup, makers denote droplet position at t_1 (fragmentation of the bag) and t_2 (fragmentation of the rim); high-speed-image at t_2 in the background; $We_{max} = 15$ (adapted from [96])

the instantaneous Weber number at the time step when We_c is first exceeded. The breakup time is overestimated, and fragmentation begins at a location which is not in the field of view. Trajectories calculated with the TAB model are displayed as solid lines. The droplet exceeded the critical deformation at an early stage and then fragments are released. The trajectories deviate from the experimental trajectory. The calculation using the wave breakup model did not predict fragmentation. As stated earlier, the physics behind this model is only valid for high We_g. The droplet trajectory as well as the temporal evolution of the trajectory predicted with the deformation based model showed a better agreement to experimental results than predictions with the other models.

4.5 Spray Behaviour

In terms of a multiphase flow approach to spray analysis, main features of the dispersed spray to be described are:

- the overall spray geometry (e.g. spray angle),
- the local fluxes of droplet mass, momentum and enthalpy (where appropriate) or
- the equivalent droplet concentrations by number or volume,
- the local droplet size and droplet state including velocity and temperature distribution.

The analysis of the spray behaviour is to be divided into two regions, depending on the relevance of droplet/droplet interactions and the possibility and frequency of droplet/droplet collisions. A rough estimate for the relevance of collisional effects for the overall spray behaviour can be done based on the ratio of particle relaxation time τ_p to the time scale between individual particle collisions τ_c. The particle relaxation time is the time that a droplet takes to be accelerated from velocity u_1 to u_2 that is (in the Stokes flow regime) expressed as:

$$\tau_p = \frac{\rho_l d_p^2}{18\mu_g}. \tag{4.32}$$

The time constant between collisions for particles is to be derived from the collision frequency as:

$$\tau_c = \frac{1}{f_c}. \tag{4.33}$$

At low droplet concentrations (dilute flow) the particulate transport mainly is determined by fluid dynamic interactions of the individual particles or droplets with the continuous carrier phase (like drag, lift etc.). At high particle concentrations (dense flow) the influence of particle collisions affects the movement of the droplets and the droplet size distribution. These two regions may be separated in terms of the ratio of the characteristic time scales as

$$dilute: \frac{\tau_p}{\tau_c} < 1; \quad dense: \frac{\tau_p}{\tau_c} > 1. \tag{4.34}$$

In a dense spray the time between droplet collisions is smaller than the droplet relaxation time. Before reaching another steady slip velocity from the droplet gas interaction another collision may occur. Therefore, the droplet movement is determined mainly by collisions and fluid dynamic effects are of less importance. In sprays, the area close to the atomizer typically is a dense spray region, as here the number concentration of droplets is high, while with increasing distance to the atomizer due to the spreading of the spray cone, droplet collisional effects decrease and transition to dilute flow is achieved.

4.5.1 Drop Size Correlations in Sprays

4.5.1.1 Theoretical Analysis

Approaches for general theoretical descriptions of drop size distributions resulting from atomization processes can be based on statistical approaches and tools which derive a probability density function in relation to the process to be analysed. The Maximum Entropy Formalism (MEF) is such a statistical tool that delivers partial information for a specific process (based on a number of compulsatory conditions) by transformation into a suitable distribution function. The Maximum Entropy Formalism is a statistical tool and does not contribute any physical aspects into the analysis of fragmentation processes. Fundamental model developments in the area of applying MEF to the analysis of fragmentation processes of liquids and the derivation of the resulting droplet size distributions in sprays have been done by Sellens [98], Sellens and Brzustowski [99], and Li and Tankin [100]. Their approaches have been further developed for example by Cousin and Dumouchel [101], Kim et al. [102], Li et al. [103], Dumouchel [104], and Hosseinalipour and Karimaei [105]. The fundamental idea of this modelling is the description of a probability density function of the particle size distribution $q_r(d)$ from the Shannon entropy (uncertainty). The Maximum Entropy Formalism allows the prediction of the droplet size probability density function depending on a number of mathematical constraints. These constraints are described based on a number of known conservation properties of the process and their distribution and are equivalent to the moments of different orders of the distribution. Without further limitations, a number of different probability density functions fulfill a given set of constraints.

Another model is the approach of Platzer and Sommerfeld [106]. It originates from a structural effects theory based on the work of Naue and Bärwolf [107] as detailed by Hartmann [108], and provides functions for drop size and volume distribution. In contrast to conventional MEF models, no mean drop diameter needs to be provided as input parameter. The Sauter mean diameter and the drop size distribution are calculated iteratively using only known geometrical information and operating conditions. This is done assuming a structure formation process that is a function of the critical Weber number. Some varying functions are derived, differing in the number of drops in each class and in the function of the critical Weber number describing the number of classes formed. The complete spray model from internal and primary fragmentation flow to the spray structure simulation is described. The two-fluid model including the prediction of the drop size distribution is used for the first part and the Euler/Lagrange method for the second part.

4.5.1.2 Empirical Drop Size Correlations

In melt atomization Lubanska's correlation (for detailed description see Chap. 2) utilizes a liquid Weber number $We_l = \rho_l\, u^2_{max}\, d_l/\sigma_l$ and the inverse of the mass flow rate ratio (1/GMR; GMR = gas to metal mass flow ratio) to describe the drop (and particle) size distribution in the spray [109]. The work of Rao and Mehrotra [110] investigated the influence of nozzle diameter and atomization angle on particle sizes, finding that the mean droplet size decreases with decreasing nozzle diameter and increasing atomization angle. They found a different value of the exponential factor as well as the atomizer constant in Lubanska's correlation. Another modification has been proposed by Rai et al. [111]. These authors studied melt atomization within ultrasonic gas atomizers and also proposed a modification of Lubanska's formula.

A discussion and evaluation of the relevance of Lubanska's formula for the atomization of metal melts in twin-fluid atomization is to be found in [112]. A general overview for a number of empirical correlations for mean and median droplet sizes dependent on operational conditions and nozzle types and geometries for melt atomization within different atomizer configurations can be found in [48, 113, 114].

4.5.2 *Dense Spray*

4.5.2.1 Droplet and Droplet Collisions

When two liquid droplets are interacting during flight, several events may occur. The colliding droplets may: (1) bounce, (2) stably coalesce, (3) temporarily coalesce followed by disruption, or (4) temporarily coalesce followed by fragmentation. For relevance in spray processes, experimental works of water-water and hydrocarbon-hydrocarbon droplet collisions have been studied extensively. Orme [115] reviewed the experimental studies of water and hydrocarbon droplet binary collisions. The basic phenomena of coalescence and separation of droplets during binary collision have been experimentally addressed in some investigations such as by Ashgriz and Poo [116], Qian and Law [117] for water and fuel droplets, Menchaca-Rocha et al. [118] for mercury droplets, Kuschel and Sommerfeld [119] for solution droplets, Kurt et al. [120, 121] for suspension droplets, Gao and Fritsching [122] for melt droplets.

Brenn et al. [123, 124] have investigated the formation of satellite droplets during binary collisions of equal sized droplets. The satellites are formed by contraction or breakup of a ligament which is formed from the mass of the two interacting droplets and may eventually pinch off the remaining portions of the droplets. Stability nomograms of the Weber number (We) and the nondimensional impact parameter (B) have been developed to describe the collision behaviour of the droplets by zones with constant numbers of satellite droplets formed only at sufficiently high Weber number due to either reflexive separation (for small impact parameters) or stretching separation (for intermediate and large impact parameters). In the reflexive separation regime, the formation of one satellite droplet occurs with highest probability. In the stretching separation regime with moderate values of the impact parameter, the satellite droplets are formed at the ends of the ligament by end-pinching mechanism, and the number of satellite droplets increases with the impact parameter, until a maximum is reached. A further increase of the impact parameter leads to the satellite-droplet formation process dominated by the capillary wave growth and thereby a decrease of the number of satellite droplets. An increase of the Weber number generally leads to the formation of more satellite droplets.

Binary collisions of melt droplets have been experimentally investigated in [122] where Carnauba wax as a low melting point material has been used. The droplets from two separate droplet-jets are colored differently, one in red and the other in yellow. The results indicate that the molten wax droplet collision shows similar behaviours as conventional Newtonian droplets. Figure 4.22 illustrates the process of wax droplet collision in the regime of stretching separation. The right parts of all the figures show the images of the dynamic binary collision process, and the left parts show pictures of the solidified particles collected from that side below the collision zone of which the yellow droplet is the main droplet colour. In the collision regime of the stretching separation, depending on the impact parameter

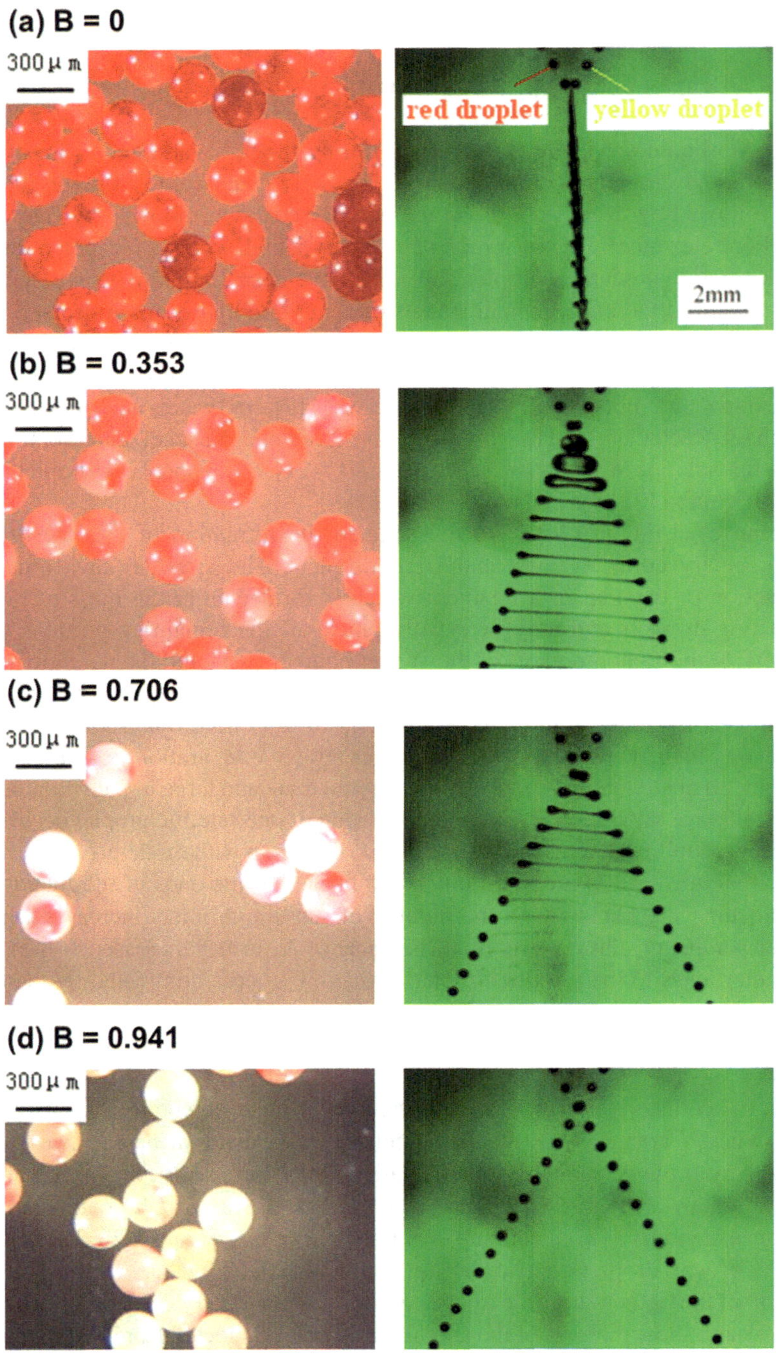

Fig. 4.22 Collision process and solidified particles at various impact parameters (We = 570), (adapted from [122], with permission)

(B), a liquid ligament is formed in between the two separating droplets and after the ligament breakup one or more satellite droplets between the two primary drops are formed. The colour distribution of the collected particles shows the existence of a certain amount of mass transfer in between the two colliding droplets. At smaller values of the impact parameter, this mass transfer is rather large. After the collision, there is a rotation of droplets caused by the angular momentum of the droplets which plays a role in the process of stretching separation and formation of secondary particles. It should be mentioned that binary droplet collisions in this study have been performed only in the liquid-liquid state of both melt droplets. Fritsching and Gao [125] have also studied the droplet-droplet and/or droplet-particle collisions in intersecting wax-melt sprays. Thereby the collisions may lead to coalescence or sticking or aggregation of the particles, depending on the instantaneous solidification states of droplets upon collision.

4.5.2.2 DNS of Droplet Collisions

Direct numerical simulations (DNS) of the behaviour of two droplets during collision have been reported by Nobari and Tryggvason [126, 127], Frohn and Roth [128], Pan and Suga [129], Focke and Bothe [130, 131], Li and Fritsching [132], Focke and Bothe [133, 134], and Kwakkel et al. [135]. Popular methods for DNS are front tracking and front capturing methods. Interface tracking methods are based on the Lagrangian tracking of marker particles that are attached to the interface motion and appear suitable for the interfaces with great regularities and small topological changes. Interface capturing methods capture the interface implicitly by a contour of a particular scalar function, for example, the level set function in the Level Set (LS) method and the volume-fraction advection equation in the Volume of Fluid (VOF) method. One advantage of the interface capturing methods is its ability to represent topological changes, both in 2D or 3D geometry, quite naturally. Especially, the Coupled Level Set and Volume of Fluid (CLSVOF) method combines the sharp interface representation of the LS method with the mass-conserving interface representation of the VOF method. Some principal physical details of drop collisions can be explained by the DNS approach. However, sub-grid models are still necessary for an accurate description of the droplet-collision behaviours occurring in different regimes. These sub-grid models may not be fully predictive, but nevertheless allow studying more details during the entire collision process.

One example is the modeling of retarded coalescence occurring at low Weber numbers. The reason for this phenomenon is that the air is trapped between the approaching droplets and needs a certain time to leave the gap. This effect is also relevant for the collision result bouncing, where the drops do not merge during the complete time of contact. According to Mackay and Mason [136] and Bradley and Stow [137], droplet coalescence may happen when the trapped gas layer thickness reaches a critical value which is within the range of the molecular interaction, typically in the order of magnitude of 0.01 μm, otherwise the droplets will bounce. Resolving this scale in a droplet collision simulation would lead to huge

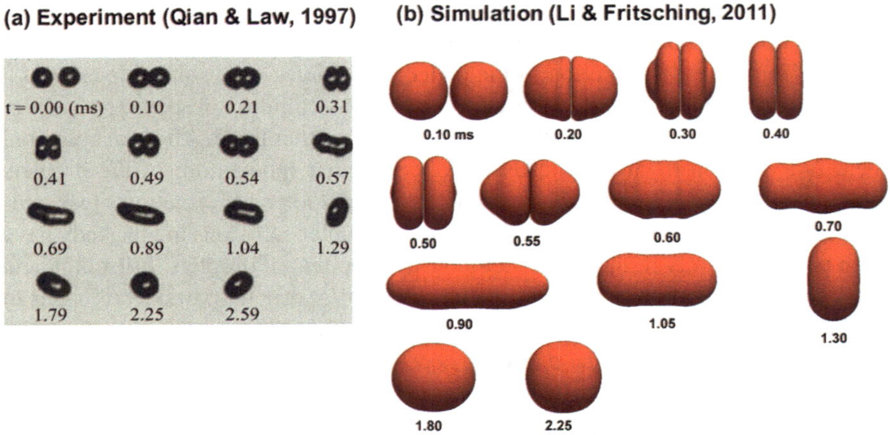

Fig. 4.23 Collision sequences during retarded coalescence process: (**a**) experimental observation, (We, Re, B, D) = (15.2, 139.8, 0.08, 302 μm) (adapted from [117]); (**b**) simulation results, (We, Re, I, D) = (15.2, 139.8, 0, 302 μm) (adapted from [132])

computational effort due to a large number of grid cells or levels of adaptive grid refinement which is currently not feasible in a DNS even with high performance computing techniques.

To appropriately describe the head-on collision process of retarded coalescence or bouncing during VOF simulation, Li and Fritsching [132] have introduced an artificial gas layer between the colliding droplets by use of ghost (or dummy) cells adjacent to the symmetry plane. This artificial gas layer prevents the immediate numerical coalescence when two droplets approach each other. At a critical time when the thickness of the gas layer between the droplets reaches a minimum value, the gas layer can be removed, leading to coalescence, otherwise the two droplets rebound from each other. Figure 4.23 shows collision sequences during retarded coalescence process from the experiment in Qian and Law [117] and the VOF simulation in [132], respectively.

In Kwakkel et al. [135], the coalescence is modelled based on a multiple marker CLSVOF method coupled with a computationally efficient film drainage model in Zhang and Law [138]. The film drainage model predicts if and when two colliding droplets will coalesce. If the contact time between the two colliding droplets exceeds the predicted film drainage time, coalescence is numerically accomplished by merging the marker functions of the two separate droplets.

Experiments show that during the extension of the collision complex an extremely thin fluid lamella encircled by the thick rim often appears in case of large Weber numbers. A lamella rupture is not observed in binary collisions at least for Weber numbers up to 2800 in experiments [134]. In contrast to the physical behaviour, simulations often predict the rupture of the lamella because of the unresolved lamella in the vicinity of the rim. If a simulation is continued with the lamella, the restoring forces acting on the rim are significantly reduced, leading to wrong results. To overcome these problems, lamella stabilization schemes have been developed to keep the lamella from rupturing in [134]. The procedure of the stabilization is to guarantee a minimum thickness of the lamella and subsequently compute the surface tension forces without interactions of the two lamella sides.

From these droplet collision studies, the outcome of a binary collision can be visualized. Typically droplet collision models are implemented into numerical spray codes such as Sommerfeld [139], Georjon and Reitz [140], Aamir and Watkins [141], and Rüger et al. [142].

4.5.2.3 Numerical Models of Droplet Collisions in Sprays

The most common used water-water collision model is based on O'Rourke [143]. This model considers only grazing collisions (bounce) and permanent coalescence. The decision criteria, whether grazing or coalescence collisions occur, is the critical collision angle:

$$\sin^2\varphi_{crit} = \min\left[2.4\frac{f(\gamma = d_{p,1}/d_{p,2})}{We};\ 1\right]. \tag{4.35}$$

In case of a collision angle less than ϕ_{crit}, the droplets coalesce, otherwise a grazing collision occurs. The function f has been fitted by Amsden et al. [144] using experimental results as

$$f(\gamma) = \gamma^3 - 2.4\gamma^2 - 2.7\gamma. \tag{4.36}$$

Podvysodsky and Shraiber [231] proposed an empirical formula for liquid droplet collisions in a wide range of material properties. The mathematical expectation $\phi_{1,2}$ of the ratio between a target droplet's mass change and total mass of "projectile" droplets impacting on the target is:

$$E\left[\frac{\frac{dm_2}{dt}}{m_1}\right] = \varphi_{1,2} = 1 - 0.246\ \mathrm{Re}_{1,2}^{0.407} Lp_2^{-0.096}.\left(\frac{d_2}{d_1}\right)^{-0.278}; \qquad \mathrm{Re}_{1,2} = \frac{u_{1,2}d_1\rho_1}{\mu_1};\quad Lp_2 = \frac{d_2\rho_2\sigma_2}{\mu_2^2} \tag{4.37}$$

The mass of the droplet after the collision is predicted from:

$$m_{2,ac} = m_{2,bc} + \varphi_{1,2}.m_{1,bc} \tag{4.38}$$

The subscripts ac and bc stand for after and before collision respectively. This empirical formula was obtained within the parameter range $30 \leq \mathrm{Re}_{1,2} \leq 6000$; $5 \leq \mathrm{Lp}_2 \leq 300{,}000$ and $1.9 \leq \mathrm{d}_2/\mathrm{d}_1 \leq 12$. The parameter $\phi_{1,2}$ should be in the range $-1 \leq \phi_{1,2} \leq 1$.

Dubrovsky et al. [145] studied collisions between liquid drops and solid particles. According to their experimental results, collision outcome between a fast moving small drop and a large solid particle was always accompanied by liquid break up and certain number of fragments. For small solid particle colliding with a large liquid drop target, they reported four modes of collision outcome; (1) projectile capture by target, (2) shooting through, (3) gas bubble formation, (4) target destruction.

Lagrangian/Eulerian simulation models are most widely used tools to calculate droplet spray behaviour. The gas flow is treated as continuum (Eulerian) and discrete parcels of particles (Lagrangian) are injected into the flow field. Each parcel contains many identical droplets. The Lagrangian step may be performed by simultaneously tracking of all parcels in an integral time step as developed by Amsden et al. [144], or tracking of each complete trajectory of the parcel in serial manner as performed by Crowe et al. [146]. Collision models are incorporated in the main gas-spray flow model. Spray collision models reflect the two approaches: (1) simultaneous tracking of all droplet parcels and search for collision partners, as performed by Amsden et al. [144], or (2) serial tracking of parcel and sampling individual collision partners from statistical information as performed by Sommerfeld [147–149] and Osterle et al. [150]. The second approach significantly reduces the calculation task. However, as the collision partner is sampled from a statistical value, the post-collision properties have to be treated carefully if the information of both partners is required.

In both models the collision partners are the uniform particle cloud contained in each parcel. All parcels presented in a control volume are taken into account in determining the collision probability (all real droplets for O'Rourke model and statistical droplets for Sommerfeld model). Therefore stochastic collision is treated and kinetic theory of gases is applied. The collision cylinder with cross section diameter ($d_{p1} + d_{p2}$) and length ($U_{rel}\Delta t$) where Δt is the time step considered.

The model presented by O'Rourke [143] and implemented in the KIVA computer program [144] tracks all parcels simultaneously and the possibility of collision between each pair of parcels is evaluated by searching all collisional partners that are present in the computational cell. The criterion whether collision takes place or not is determined by the probability of no collision:

$$P_{no\_collision} = e^{-f\Delta t} \tag{4.39}$$

where f is the collision frequency in a collision cylinder:

$$f = \frac{N_{r\_s}}{V}\pi(r_l + r_s)^2|\bar{V}_{r\_l} + \bar{V}_{r\_s}|. \tag{4.40}$$

N_{r_s} is the number of smaller radius droplets, V is the volume of the computational cell, the subscript r_l and r_s describe the properties of larger and smaller droplets respectively. To decide whether a collision takes place, a random number RN is generated from a uniform distribution and compared with $P_{no\_collision}$, if

RN > $P_{no\_collision}$ a collision event is valid. The position of contact point during collision is directly calculated since the collisional partners are both "real" and their trajectories are known. Since the model considers the collision of particles within the same computational cell only, "artefacts" concerning neighbouring particles located in an adjacent cell may arise.

Schmidt and Rutland [151, 152] developed the NTC (No-Time-Counter) collision model that predicts how many collisions should occur in a given cell and then randomly samples collision pairs within the cell. In the NTC model, the collision computation is performed on a special collision mesh that is optimized for both sample size and spatial resolution. The mesh is different every time step to further suppress artefacts.

The Proximity collision model has been developed in [153] and is designed to improve the previous discrepancy in the KIVA code collision model. To counter the "artefacts" problem, possible collision pairs are chosen based on the proximity of the two particles from each other, irrespective of their placement relative to the computational grid. A user-defined collision radius is introduced to define the maximum distance that two particles can be apart from each other to consider for collision. The equations for determining whether collision occurs, and the outcome of the collision, are nearly identical to the KIVA collision model.

Within the models proposed by Osterle and Petijean [150] and Sommerfeld et al. [147, 148] each parcel is tracked in a serial manner and statistical value of particle number density, average and rms velocity-size relationship and other variables are kept and stored in each computational cell. The collision probability assumes a moving droplet while the other droplet (fictitious) is fixed and is calculated within the collision cylinder:

$$P_{collision} = 0.25\frac{N_2}{V}\pi(D_l + D_s)|U_r|\Delta t. \tag{4.41}$$

N_2 is the total particle number density in the computational cell (from the stored particle number in the last Lagrangian step). The particle size of the collision partner (fictitious droplet) is sampled from the local particle size distribution. Rüger et al. [142] validated this collision model in a spray with experimental results. The comparison of measurements and predictions showed very good agreement for the profiles of mean properties of both phases and the local droplet size distributions as well as size-velocity correlations. According to the calculation results, the increase of the integral droplet Sauter mean diameter along the spray distance is mainly due to coalescence, while the impact of droplet evaporation on mean diameter decrease is of minor importance.

In a further extension, Sommerfeld [154] developed an inter-particle collision model taking into account the correlation of the fictitious particle with the velocity of the real particle as a consequence of gas turbulence. The model has been validated with results from large eddy simulations (LES) in the literature. For all test cases considered, good agreement between model calculations with the results obtained by LES has been found.

Pischke et al. [155] developed a hybrid collision algorithm to predict the collision probability: collisions between parcels are described deterministically, i.e., by searching for intersecting trajectories between computational parcels, while droplet collisions stochastically. The collision probability formulation eliminates the influence of a gas-phase mesh or a collision mesh by introduction of a parcel diameter, which is a measurement for the volume represented by a specific parcel. Two non-parametric parcel diameter estimators are derived based on the volume-of-influence approach, an isotropic parcel diameter estimator for homogeneous droplet clouds (sphere-of-influence), and an anisotropic parcel diameter estimator for inhomogeneous droplet clouds of reduced dimensionality (ellipsoid-of-influence). During derivation, the exactness for analytical homogeneous dispersions is preserved. The performance of the anisotropic volume-of-influence approach is demonstrated by synthetic validation cases, indicating that the formulation delivers exact predictions in cases where other existing collision algorithms do not even converge. To improve numerical convergence, a velocity decomposition method is proposed in Pischke et al. [156], splitting the relative velocity between the parcels into a velocity gradient (with little influence on collisions, but high numerical sensitivity) and into velocity fluctuations (with high influence on collisions, and little numerical sensitivity). By removing the spurious contribution of the velocity gradient, the convergence of the collision algorithm is increased from first to second order. Based on the anisotropic volume-of-influence approach and the velocity decomposition method, a Lagrangian collision algorithm has been developed in Pischke [157], which incorporates strategies from Eulerian NTC-algorithms. For efficiency reasons, a dedicated collision mesh is introduced, which separates parcels that are improbable to collide. Overlapping control volumes are introduced to avoid mesh artefacts, which typically occur at collision mesh boundaries, while reducing the number of potential collision partners. With the Lagrangian collision algorithm, highly anisotropic sprays with strong velocity gradients appear numerically solvable.

4.5.2.4 Clustering of Droplets

A typical spray flow behaviour is characterized by highly turbulent flow structures interacting with coherent (periodic) large-scale flow structures. Local accumulation of droplets (clustering) may take place. Figure 4.24 illustrates the typical formation of droplet clusters in a spray [158]. The left side of the figure shows a light sheet image of a water spray (from PIV measurement) and the right side the result of a LES simulation of an identical spray situation using the droplet size distribution that has been correlated from experiments. The formation of strained droplet clusters has been found in strong correlation to large scale (coherent) gas flow vortex patterns.

At what position clusters in the spray are formed mainly depends on the Stokes number of the droplets. In [159] a local Stokes number has been defined to account

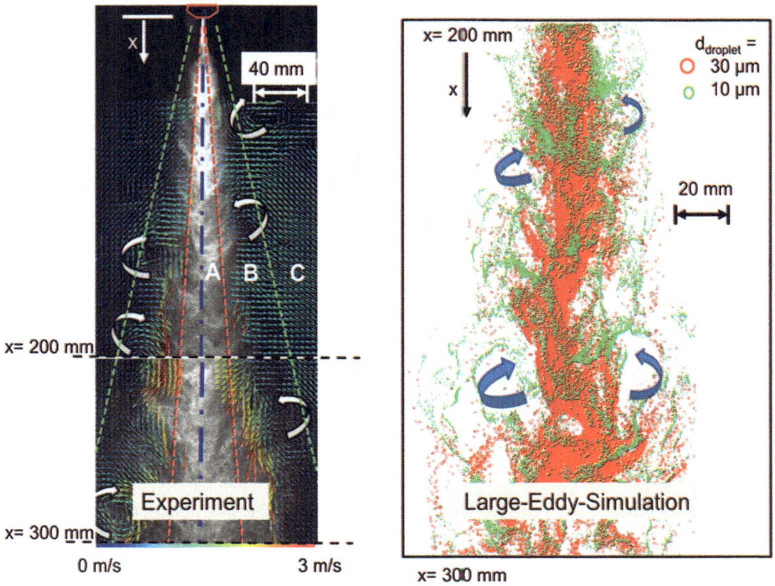

Fig. 4.24 *Left*: Laser light sheet picture from the water spray (PIV); *Right*: Droplet pattern from LES with droplet size distribution of the experiments

for the strong changes in time scale along the axial and the radial direction in the spray flow as:

$$St_{droplet}(x,y) = \frac{\rho_{droplet} \cdot \mathrm{d}^2_{droplet} \cdot U_{gas,\mathrm{x}}(x)}{18 \cdot \eta_{gas} \cdot \mathrm{L}_{ref}(x)} \cdot \frac{1}{1 + 0.15\mathrm{Re}_{droplet}(x,y)^{0.687}}. \quad (4.42)$$

This definition of the local Stokes number in particular depends on the droplet diameter $\mathrm{d_{droplet}}$ and the local fluid time scale $\mathrm{t_{Fluid}} = \mathrm{L_{ref}(x)/U_{gas,x}(x)}$, where $\mathrm{U_{gas,x}}$ is the gas velocity in the axial direction along the spray axis and $\mathrm{L_{ref}(x)}$ is the local half width of the gas velocity profile in the spray flow. The Reynolds number is based on the local relative velocity between the droplets and the gas phase. In order to illustrate the effect of the Stokes number two droplet size classes have been extracted from the simulation of the spray (Fig. 4.24). The droplet size distribution in the numerical simulations is fitted to the experiment. The green circles represent the 10 μm droplets in the flow. At the spray edge ring-like cellular structures are formed. In principle the same effect can be detected for the 30 μm droplets further downstream. The local Stokes number there is of similar order as for the 10 μm droplets further upstream. The intermediate droplets ($\mathrm{d_{droplet}} = 30$ μm) form fir tree like structures, which are concentrated on the spray axis. In the laser light sheet image these structures can be detected more easily, because the light scattering intensity of the droplets is increased with increasing droplet diameter. Clusters consisting of small droplets ($\mathrm{d_{droplet}} = 10$ μm) with local Stokes numbers close to

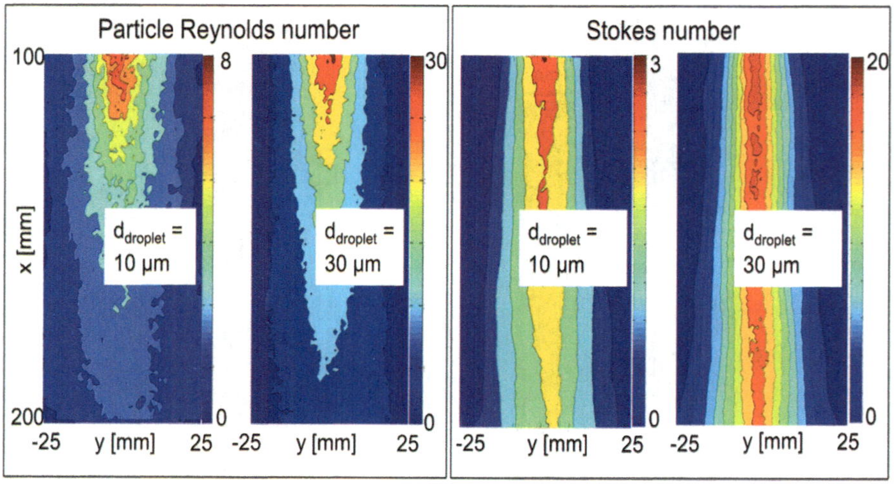

Fig. 4.25 Contour plot of the particle Reynolds number and Stokes number in a spray

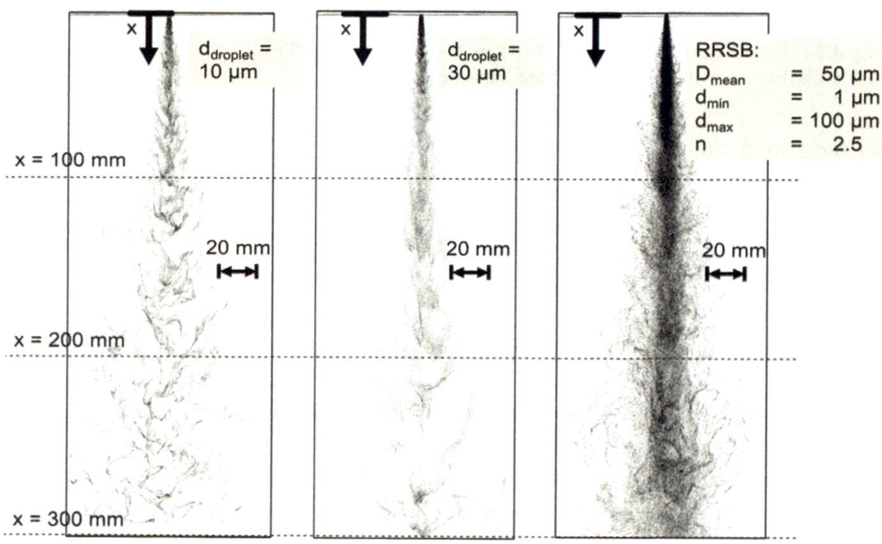

Fig. 4.26 Spray patterns from LES simulation for different droplet sizes

unity can be formed spontaneously at the shear edge as ring-like structures. These structures get entrained into the central region of the spray and merge with the clusters of the intermediate droplets. Some overlapping of the fine structures onto the fronts of the intermediate droplet clusters can be detected, especially in the regions of a lower Stokes number further downstream ($x > 200$ mm).

In Fig. 4.25 the spatial distribution of the Particle Reynolds and Stokes number are evaluated for the single droplet size cases. The particle Reynolds number is much higher than 1 for $x < 100$ mm. The small droplets adapt quickly to the gas flow conditions indicated by $St_{droplet} < 3$ for $x > 150$ mm. The intermediated droplets close to the spray do not adapt to the gas flow conditions in this region of the spray. The Stokes numbers in these areas of pronounced clustering are larger than 10 and thereby an order of magnitude higher than for small droplets. Droplets and particles in sprays do not necessarily require Stokes numbers in the order of unity to form clusters. The case C in Fig. 4.26 depicts the droplet pattern for the polydisperse spray. The structures described before are coexisting in the polydisperse spray. Larger droplets ($d_{droplet} > 50$ μm) can be found on the spray axis without the tendency to form clusters. Small droplets ($d_{droplet} < 10$ μm) can be found equally distributed in the spray, also inside of vortices (tracer-like behavior). The 10 μm droplets show a distinctive cluster pattern in the flow (The case A in Fig. 4.26).

Due to the dynamics of the primary liquid fragmentation process, especially in the dense spray region a clustering of droplets may occur. In most cases the experimental characterization of sprays has been limited to single point statistics of the spray in time or space. Such a description provides as integral information a measure of the mean number density or flux rate of the spray. Progress in measurement techniques like the Phase-Doppler-Anemometry PDA has made temporal correlated droplet measurements and the Particle-Image-Velocimetry has made space correlations information available for sprays. Marx et al. [161–163] developed a multipoint statistical description of a spray. Based on this theoretical approach it is possible to distinguish between steady and unsteady sprays by using the interparticle arrival time τ at a certain position. Steady sprays are defined as those whose interparticle arrival time distribution obeys inhomogeneous Poisson statistics. Unsteady sprays are defined as those whose interparticle arrival time distribution do not obey inhomogeneous Poisson statistics. An example of unsteady

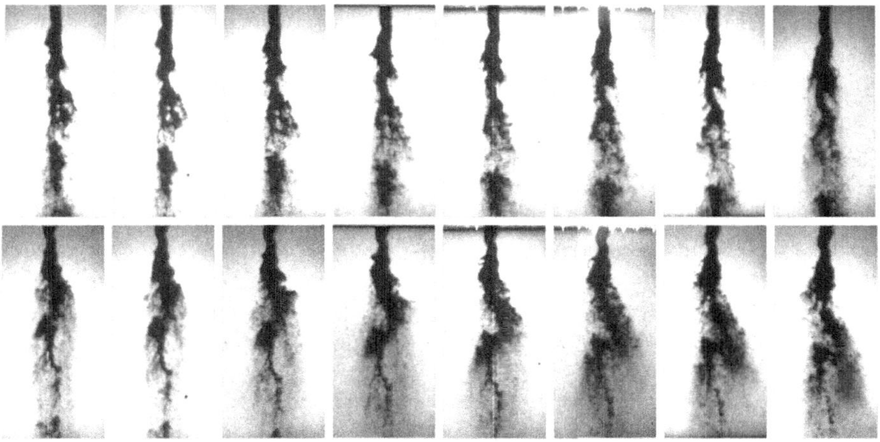

Fig. 4.27 Drop clustering in sprays: high speed photo series of liquid fragmentation process, liquid jet diameter 4 mm, time increment 0.2 ms

spray behaviour with droplet clustering is illustrated in Fig. 4.27 with a high-speed-video sequence of a disintegrating liquid jet. Areas at higher and lower droplet concentrations in the initial spray are to be seen that are attributed to the asymmetric instability and primary fragmentation of the liquid jet.

Heinlein and Fritsching [164] analyzed the spatial particle distribution in the spray of a pressure and twin-fluid atomizer, respectively, by means of point wise and time resolved PDA measurements. Steady and unsteady particle structures are identified by evaluating the interparticle arrival time statistics at a certain position. While for the pressure atomizer, clustering takes place mostly within the central area of the spray, this effect is found for the twin-fluid atomizer mostly in the outside spray area. For both atomizers clustering increases and/or begins only at increasing axial distances. In a droplet cluster, a locally high droplet number density exists which results in relatively short inter arrival times. No significant dependence between the droplet size and the occurrence of droplet clusters has been found; the drop size distribution within a cluster corresponds to the drop size distribution of the entire spray. The presence of clusters can be determined, however, primarily in the lower velocity range, which has to be attributed to the fact that collective groups of droplets have different drag than individual droplets.

Lampa and Fritsching [158] investigated the clustering behaviour of droplets in the spray of a twin-fluid atomizer based on different spray chamber geometries. The PIV-measurements have captured the large scale vortices located at the edge of the spray cone flow. The large eddy simulations (LES) indicate that the gas-droplet interaction within the shear layer is the driving force for the formation of droplet clusters. Besides the aforementioned analysis of local cluster formation in a spray based on pointwise measurements, particle clusters may also be determined quantitatively by evaluating the spatial inhomogeneity of droplet distributions within planar sheets through the spray. These information either are derived from numerical simulation results (filtering the instantaneous spatial drop distribution) or from experimental spray observation with light sheet imaging (as e.g. from PIV measurements). In the latter it is necessary to distinguish the particle clusters from the background noise. Common image processing techniques typically are used for filtering. The result is a binary picture with segregated particle clusters. These particle structures are processed with the Garncarek-Algorithm defined in [165] or by Voronoi-tessallation algorithms [166].

The inhomogeneity index H is the central quantity in the Garncarek algorithm and is defined as the ratio of the distribution of objects in a specific state and the distribution of objects in a random state. The inhomogeneity index increases when there are structures in the spray that show clustering behaviour. The inhomogeneity index is dependent on a certain measurement scale κ and is expressed as

$$H(\kappa, n) = \left(\frac{n(n-1)}{2(\kappa-1)}\right)^{0.5} \left(\frac{1-n-\kappa}{\kappa-1} + \frac{\kappa}{n(\kappa-1)} \sum_{i=1}^{\kappa} n_i^2\right), \qquad (4.43)$$

where n is the the sum of all grey or binary values within the processed picture and κ is the scale on which the inhomogeniety is evaluated. For each interrogation

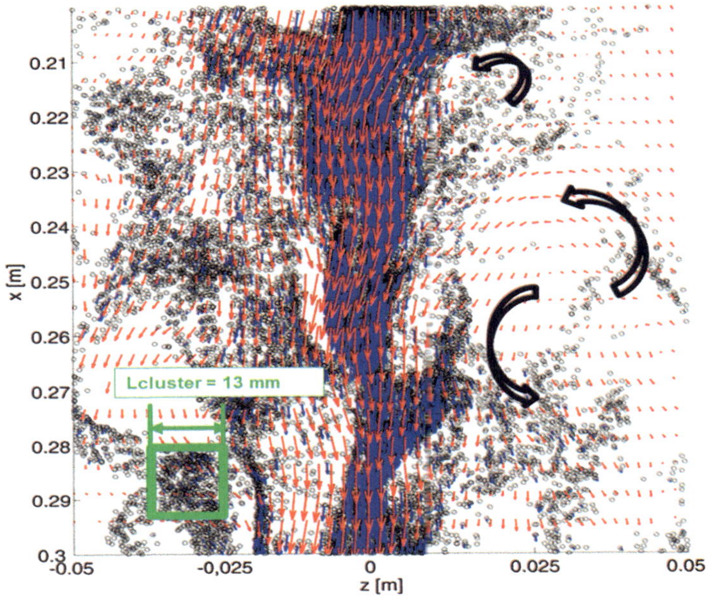

Fig. 4.28 Drop distribution in a spray and identification of a cluster of L = 13 mm

window with the side length $B = L/\sqrt{\kappa}$ the sum of gray values n_i is counted. H is a scalar value for the inhomogeneity in the whole image. For the variation of the box size B, the inhomogeneity index yields a maximum where the characteristic cluster size $B_{Cluster}$ can be found.

An example of cluster size analysis in a spray from an LES simulation result is shown in Fig. 4.28. In this frame of 10 × 10 cm of a central part of the spray a cluster size of L = 13 mm has been identified that is about 400 times the mean droplet diameter. A representative cluster in this image is indicated by the green box.

Some atomizer configurations even show an intrinsic formation of unsteadiness and droplet cluster formation in sprays. An evaluation method of unsteadiness of the spray generated by an internal mixing twin-fluid atomizer (effervescent atomizer) has been presented in [167]. The method of evaluation is based on measurements of pressure fluctuations in the atomizer mixing chamber. Measurements, made under different atomizer operational conditions, showed that the spray unsteadiness depends mainly on the gas-to-liquid ratio (GLR). A decrease in GLR causes the spray to become more unsteady. The relation between atomizer internal bubbly two-phase flow pattern and the spray unsteadiness is elucidated by visualization of the internal two-phase flow using a digital camera and the use of published two-phase flow maps. The findings have been complemented and confronted with the results obtained by the use of the spray unsteadiness evaluation method of [162, 163].

4.5.3 Dilute Spray

The majority of all spray investigations deal with the dilute spray region where mainly gas/droplet interactions determine the spray behaviour. A variety of sophisticated measurement instruments and experimental and numerical techniques are used to investigate details of the gas and droplet behaviour in this spray region. These include:

- non-intrusive laser diagnostic instruments for dispersed multiphase flows, like Phase-Doppler-Anemometry, Diffraction particle sizer, Particle-Image- and Particle-Tracking-Velocimetry, holography, Laser-Induced-Fluorescence, and others and
- numerical models, based on Euler/Lagrange techniques [66, 168], Euler/Euler techniques (Two-Fluid Models), Direct-Numerical- or Large-Eddy-Simulations [158, 159, 169–173], probability density functions [174, 175] among others,

These numerical and experimental techniques have contributed to understanding of multiphase transport processes in sprays. Parameters describing the dilute spray structure identify the behaviour of the droplets in the spray, the interaction with the gas phase, and the macrostructure of the spray development. These results typically are specific for their spray application under investigation (medium, atomizer type, operational parameters etc.).

4.5.3.1 Droplet Movement

Momentum Transfer and Droplet Trajectory

In the analysis of fluidic drops or solid particles in a gaseous atmosphere, the density ratio between gas and particles is typically sufficiently small ($\rho_g/\rho_{d,p} < 10^{-3}$). For molten metal droplets and solid metal particles this density ratio is even smaller. In this special case, the droplet or particle trajectory equation can be simplified as

$$m_{d/p}\frac{du_{d/p}}{dt} = m_{d/p}g + \frac{1}{2}\rho_g\left(u_g - u_{d/p}\right)\left|u_{d/p} - u_g\right|A_{d/p}C_d \qquad (4.44)$$

by only taking into account droplet/particle inertia, gravity and aerodynamic drag, where m is mass, A is projection area normal to the gas flow and C_d is drag coefficient. The drag coefficient C_d in most cases is taken from the analysis of a single spherical solid particle, which can be described in the range of Reynolds numbers as

$$C_d = \begin{cases} \dfrac{24}{Re_g}\left(1 + \dfrac{1}{6}Re_g^{2/3}\right), & Re_g \leq 1000; \\ 0.44, & Re_g > 1000. \end{cases} \tag{4.45}$$

Liquid droplets deviate from solid particles during interaction with gases due to the free mobility of the drop surface, leading to droplet deformation as well as to movement of the drop surface (and the correlated inner circulation of the droplet). In addition, the drop may oscillate.

A time-independent correlation was proposed by Wiegand [176] to approximate aerodynamic deformation effects on the drag coefficients in the sub-critical Weber number range:

$$C_d = 0.28 + \frac{21}{Re_g} + \frac{6}{\sqrt{Re_g}} + We_g\left[0.2319 - 0.1579\log(Re_g) + 0.0471\log^2(Re_g)\right.$$
$$\left. - 0.0042\log^3(Re_g)\right], \quad 5 < Re_g < 2000. \tag{4.46}$$

In Liu et al. [177], the distortion parameter (y), as defined in TAB model (see Sect. 4.4.1.2), was used to modify the drag coefficient to account for the drop deformation as

$$C_{D,y} = C_D(1 + 2.53y) \tag{4.47}$$

where $C_{d,y}$ is the drag coefficient for the deformed drop and C_d is that of the initially spherical drop.

Another geometry-based modeling concept is proposed for the aerodynamic drag force in Schmehl [92]. The drag coefficient of the droplet is given as

$$C_d = C_{d,sphere}f + C_{d,disc}(1 - f), \text{with} f = 1 - E^2 \text{ and } E = 1/Y^3 \tag{4.48}$$

Here E is the droplet aspect ratio and Y is the ratio of the cross-stream diameter to the spherical diameter. E is 1 for the sphere and 0 for the disc. The drop drag coefficient evolves linearly between the two limiting geometries, a sphere ($C_{d,sphere}$) and a disk ($C_{d,disk}$). In addition, for the droplets exposed to high Reynolds numbers (Re_g), a time-dependent correlation was derived by Markus and Fritsching [168] as

$$C_d = \begin{cases} 0.65\,t/t_{def} + 0.44 & for\ 0 \leq t/t_{def} < 1; \\ 1.11 & for\ t/t_{def} > 1. \end{cases} \tag{4.49}$$

In the above correlation, the drag coefficient C_d is linearly increased from 0.44 (sphere) to 1.11 (disc) over the deformation time.

Based on a one-dimensional numerical analysis using the droplet trajectory equation (without coupling effects), a principal investigation for the droplet behaviour in a spray such as the identification of different atomizer nozzle characteristics is possible. Fritsching and Bauckhage [178] analyzed the interaction of droplets and the surrounding gas phase in a water spray as a function of distance from nozzle for single-fluid (pressure) and twin-fluid (gas) atomizers. Results of measurements of droplet velocities and sizes on the spray centreline measured by Phase-Doppler-Anemometry (PDA) have been used for comparison with the one-dimensional model. The two nozzle types and their respective dimensions have been chosen by the criterion of comparable droplet size and velocity spectra in the spray.

For the uncoupled solution of the droplet trajectory an assumption for a velocity distribution of the gas phase is necessary. This boundary condition has been determined from PDA measurements of droplet velocities of the smallest detected droplets in the spray based on the assumption that these droplets act as tracer particles and the local gas velocity equals the measured droplet velocity (zero slip velocity). The initial droplet (starting) velocity value has been prescribed. For the twin-fluid nozzle it has been assumed that the starting velocity of the droplets equals the liquid feed velocity from the nozzle. For the pressure atomizer a constant starting velocity for all droplet size classes has also been assumed. Its value has been obtained by extrapolating the droplet velocities measured for the biggest drop size class detected from the first measurement location in front of the nozzle backwards to a common starting location for the droplets in the vicinity of the nozzle.

In the spray of the twin-fluid atomizer a change of the direction of momentum transfer occurs. This is to be seen in Fig. 4.29. Close to the atomizer the faster moving atomizer gas accelerates all droplets more or less depending on size. The direction of momentum transfer is always from the gas to the slower particles. Therefore, the gas looses a significant amount of kinetic energy within a small distance from the atomizer. At a certain distance from the atomizer the direction of momentum transfer is reversed. At greater nozzle distances all droplets are moving faster than the gas and the direction of momentum transfer is from the droplets to

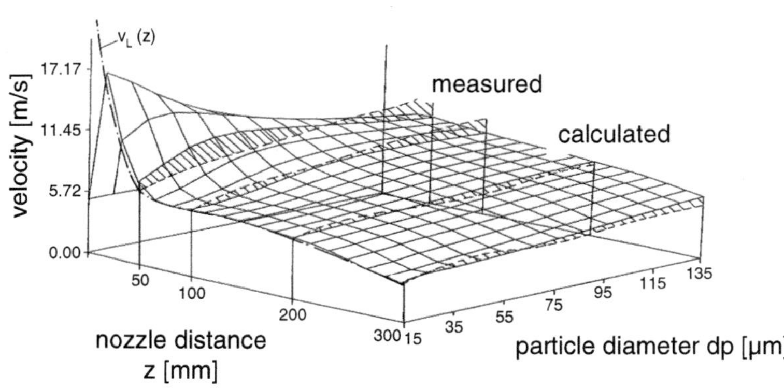

Fig. 4.29 Drop velocity on spray centreline: twin-fluid atomization (from Fritsching and Bauckhage [178], with permission)

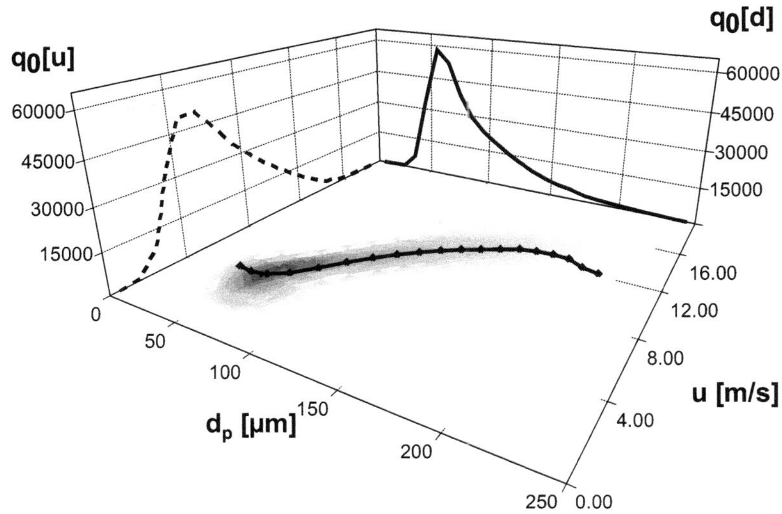

Fig. 4.30 Drop size/velocity correlation in spray (from [179], with permission)

the gas. The droplets in this area accelerate the gas. The point where the momentum transfer changes direction depends on the droplet size. While the smaller droplets already exhibit this change of momentum transfer closer to the atomizer due to their smaller inertia, the bigger particles change the momentum transfer direction in a somewhat greater nozzle distance.

By comparing the characteristics of these nozzle types a common characteristic for technical spray processes can be seen. When the aim of the atomizer application is the achievement of almost identical droplet velocities at a specific distance from the atomizer (e.g. for coating application), this behaviour is achieved much closer to the atomizer for the twin-fluid atomizer. In addition, the reversal of the momentum transfer direction has to be recognized as an important feature for twin-fluid atomizers.

An important property of a droplet spray is the correlation of drop sizes and velocities. This leads to a two-dimensional distribution of drop properties in sprays. Typical behaviour of a drop size/velocity correlation at a single point in the spray of a pressure atomizer is illustrated (from a simultaneous measurement of drop size and velocity by PDA) in Fig. 4.30 [179]. In the ground plane of the plot, the measurement result of several ten thousand individual particles is shown (individual droplet size and velocity), while at the two sides of the plot, the integration in terms of number frequencies of drop sizes and velocities respectively is illustrated.

Turbulent Dispersion of Particles in the Spray

Due to the turbulent structure of the gas flow field, the particles may deviate from the deterministic trajectory, which is governed by the mean gas flow properties.

This effect is called turbulent particle dispersion. A stochastic dispersion model was developed in [139] to model the effect of the gas-fluctuating velocity on the motion of the discrete phases by the turbulence. In the model, the velocity is perturbed in random direction, with a Gaussian random number distribution with variance of $\sqrt{2k/3}$, which characterizes an isotropic turbulence. The interaction time between a particulate and an eddy (τ_I) can be defined as the minimum of the eddy lifetime (τ_E) and the residence time of the particulate in the eddy (τ_r). The eddy lifetime (τ_E) is given as

$$\tau_E = C_E \frac{k}{\varepsilon} \tag{4.50}$$

Where the constant of C_E has a value of 0.3 [139], k is the turbulence energy and ε is the dissipation rate. The residence time of a particulate in an eddy (τ_r) is expressed as

$$\tau_r = \frac{L_e}{\left|\vec{U}_g - \vec{U}_p\right|}, \text{with } L_e = \tau_E \sqrt{2k/3}. \tag{4.51}$$

During the interaction time, a fluctuation velocity is selected by the Gaussian random number generator and added to the mean gas velocity. The modified velocity holds until either the eddy has dissipated or the particulate has left the eddy.

4.5.3.2 Heat Transfer and Droplet Cooling

The heat transfer process across the surface of a moving spherical droplet in a gas flow field is determined by convection, radiation and the release of latent heat (ΔH_f) arising from the solidification. Therefore, the heat balance for an individual particle can be expressed as

$$m_d C_{p,l} \frac{dT_d}{dt} = Nu \pi \lambda_g d_d \left(T_g - T_d\right) - \sigma_s \varepsilon_s \left(T_d^4 - T_w^4\right) \pi d_d^2 + m_d \Delta H_f \frac{df_s}{dt} \tag{4.52}$$

Radiation may occur between the droplet under investigation and the surrounding spray chamber walls as well as between the droplets. Within a typical metal atomization or spray forming application with twin-fluid atomization of the melt, comparison of the heat fluxes resulting from convection and radiation obtains a heat loss due to convection which is two orders of magnitude higher than the heat loss due to radiation. This effect is due to the huge velocity gradients between gas and droplets and the resulting high heat transfer coefficients. Therefore, the heat loss due to radiation from the droplet is usually neglected. In the above heat balance equation, the Nusselt number Nu for convective heat transfer is typically taken from the conventional Ranz and Marshall [180] correlation:

$$Nu = 2 + 0.6\,Re^{0.5}Pr^{0.33} \tag{4.53}$$

An extension of this correlation that takes into account gas turbulence effects during heat and mass transfer from droplets has been derived by Yearling and Gould [181]:

$$Nu = 2 + 0.584\,Re^{0.5}Pr^{0.33}\left(1 + 0.34\sigma_t^{0.843}\right) \tag{4.54}$$

These correlations depend on the local Reynolds number Re, the Prandtl number Pr and the later correlation with the relative turbulence intensity σ_t. The multiplication factor in the brackets of Eq. (4.54) therefore extends the conventional Ranz–Marshall correlation.

Solidification of a melt is to be described by the processes of nucleation and crystal growth. A droplet solidification process model typically is based on equilibrium phase diagrams for slow solidification and time-transfer phase change diagrams and/or experimental solidification investigations for some more realistic (higher) cooling rates. In fact, in spray processes the cooling rate of droplets especially immediately after atomization might be very high (e.g. max. 10^7 K/s). Therefore, the possibility of undercooling prior to nucleation has to be considered. Here a low carbon steel droplet (C = 0.3 w%) is considered as an example. Starting with the initial melt temperature (superheat) T_m, the droplet cools down to liquidus temperature T_l. Depending on the actual cooling rate, the droplet may undercool until it reaches the nucleation temperature T_n before solidification starts. Due to the rapid release of latent heat of fusion during recalescence, the droplet temperature increases until it reaches a local maximum in the cooling curve at T_r. During the following segregated solidification, droplet temperature decreases continuously. At the temperature T_{per} in this specific case, a peritectic transformation takes place at constant droplet temperature. After termination of the peritectic transformation, again segregated solidification occurs until the droplet is completely solidified at T_s. From here on, droplet cooling is in the solid state of the fully solidified particle. The drop cooling behaviour in the various stages of solidification is described based on thermal balances (internal and external). For metal droplets, typically no internal temperature gradient is assumed, as the Biot-number Bi is small for most metal droplets (Bi $\ll$ 1).

Depending on cooling rate and droplet size, the temperature T_n where nucleation occurs can be much lower than the liquidus temperature T_l. The nucleation temperature for continuous cooling is defined as the certain temperature, where the number of nuclei N_n in the droplet volume V_d is identical to one:

$$N_n = V_d \int_{T_l}^{T_n} \frac{J(T)}{\dot{T}}\,dT = 1 \approx \frac{0,01\,J(T_n)\,V_d\,\Delta T_{\mathrm{hom}}}{\dot{T}}. \tag{4.55}$$

Here, J(T) is the nucleation rate and $\dot{T}$ the cooling rate [182, 183]. Hirth [184] has introduced the simplification in the above equation. ΔT_{hom} is the undercooling temperature difference for homogeneous nucleation. The nucleation rate may be expressed as [185]:

$$J(T_n) = K \exp\left(- \frac{16 \pi \sigma_{sl}^2 V_m^2 T_l^2}{3 k T_n \Delta H_{fm}^2 \Delta T_{\text{hom}}^2} \right). \tag{4.56}$$

In the work of Turnbull [186] and Woodruff [187], a correlation has been given between the solid-liquid interfacial energy σ_{sl}, the latent heat of fusion per atom ΔH_{fa} and the atomic volume V_a. In technical processes, heterogeneous nucleation rather than homogeneous nucleation mechanisms limit the degree of undercooling [185]. Only in very small droplets does homogeneous nucleation play an important role during solidification. Based on experimental results for different alloys, Mathur et al. [188, 189] derived a correlation between the actual undercooling and the amount of undercooling necessary for homogeneous nucleation. Simulation of molten metal droplet sprays including solidification submodelling has been done by Lee and Ahn [190], Bergmann [191], Bergmann et al. [23, 73], Pedersen et al. [192], Gjesing et al. [193], and Li and Fritsching [66].

4.5.3.3 Droplet Transport in the Spray

In the spray, a complex two-phase flow situation involving gas and particles/ droplets occurs. The high speed gas looses its momentum by accelerating the droplets. Due to the heat transfer from the droplets, the gas undergoes intensive heating in the zone close to the atomizer. The droplets are accelerated and partly solidified during their flight. Modelling the complete dispersed two-phase flow situation including momentum and heat coupling is used to describe and analyse the spray behaviour [66, 67, 168, 191, 193–197]. As an example modelling, Figs. 4.31 and 4.32 show the behaviour of different droplet size classes on the centreline of a spray of molten steel, atomized with nitrogen [160]. The boundary conditions include an atomizer pre-pressure of $p_2 = 3$ bar abs., a gas to metal flow rate (GMR) of 1.25 and a log-normal droplet size distribution with a mass median diameter of 133 μm. The axial velocity distribution on the spray centreline of gas and particles with different diameters is illustrated in Fig. 4.31. The gas velocity upstream of the atomization area is unaffected by the particles. Close to and in the atomization region the gas velocity decreases due to the momentum transfer associated with acceleration of the droplets. Afterwards, the gas velocity further increases because of the heating of the gas and thermal expansion and, also, due to the entrainment of momentum from the edge of the spray cone. The maximum gas velocity is reached at a nozzle distance of ~330 mm. Afterwards, the gas velocity decreases due to the spreading of the spray cone. Particles are accelerated in the atomization region and further downstream. In the main region of the spray cone, particle velocities for the bigger particles remain approximately constant.

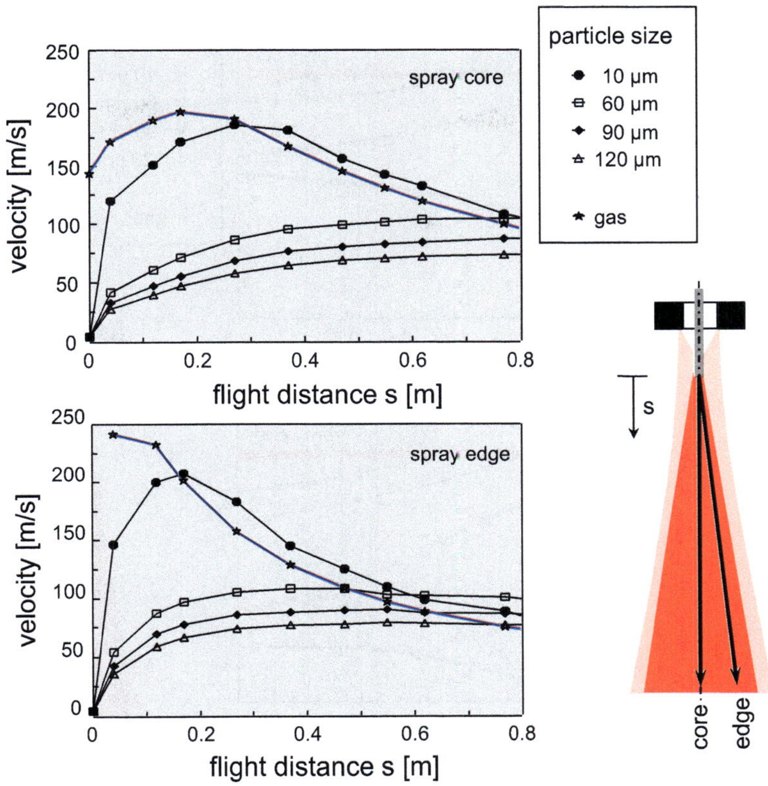

Fig. 4.31 Modelling result for gas and particle velocity behaviour on spray centreline and edge within spray forming of steel (from [160], with permission)

The gas reaches high temperatures in the atomization region up to 1200 K, as can be seen from Fig. 4.32. Further downstream, the local gas temperature on the centreline decreases, but heat transfer from the particles continues, due to the spreading of the spray and mixing with the colder ambient gas. After cooling down to solidification temperature, the phase change occurs inside the particles and the latent heat of the particles is released, whereby the particle temperature remains constant. When the latent heat content of the droplets is removed, the particles further cool down. From these calculations it can be seen, that particles bigger than 210 μm are still in the state of phase change when impacting the substrate/deposit.

Experimental investigations have verified these simulation results. On-line measurement techniques have been used to monitor the spray conditions in spray forming. With a modified Phase-Doppler-Anemometer [198], local droplet sizes and velocities are measured and used for process control [199]. Measurements of individual droplet temperatures (plus droplet sizes and velocities) in metal sprays have been performed by means of a High-Speed-Pyrometer instrument [200–202].

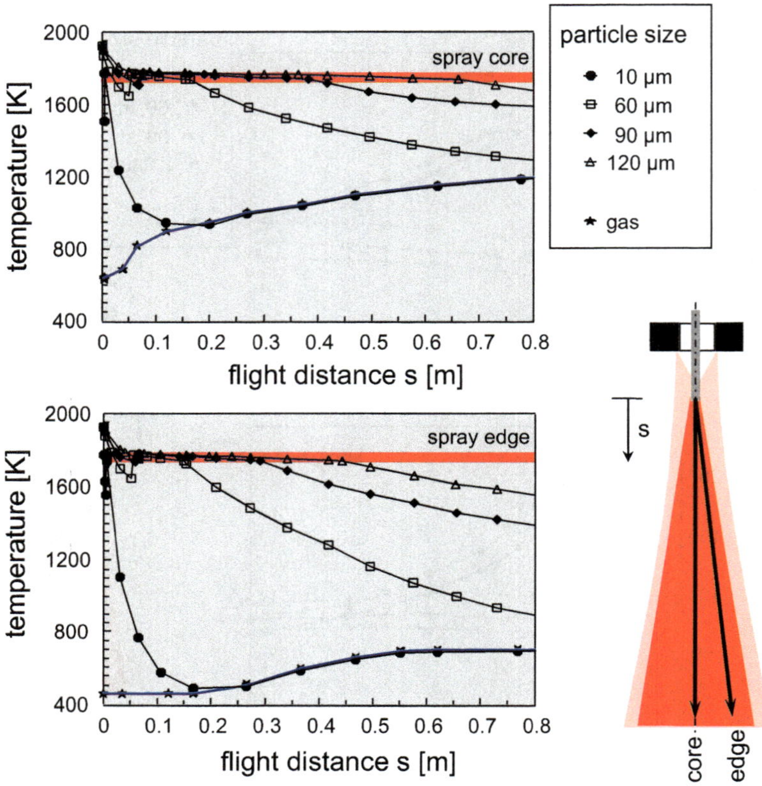

Fig. 4.32 Modelling result for gas and particle temperature behaviour on spray centreline and edge in spray forming of steel (from [160], with permission)

To determine the overall heat and mass fluxes from the spray to the deposit, the droplet data have to be averaged in a suitable manner. Thermal averaging may be performed in two different ways, describing two extreme situations that occur in a molten metal spray [203]. The first averaging method (*the enthalpy method*) describes the thermal status of the particle mass in thermal equilibrium. In this way, the thermal state of a certain particle mass is described after adiabatic equilibration. This means the particle mass has a specific enthalpy which is directly related to its thermal state (temperature and solid fraction).

In an impinging molten metal spray for forming or coating processes, situations may occur where the particle mass should already be fully solidified according to its average properties (when calculated by the enthalpy method), although it still contains some liquid. This may occur when the main spray consists of a large proportion of cold, solidified particles and only a small number of large, fluid, and hot melted particles. These liquid droplets still deliver a certain amount of liquid melt to the mushy layer on the deposit surface, which is not accounted for by the enthalpy method. However, if a *separation averaging method* is used, this difficulty is avoided. Here, the amount of solidified mass or fluid melt mass remaining in the spray is calculated separately.

The enthalpy averaging approach yields the thermal equilibrium condition of the total droplet mass (in terms of a discrete number of droplet size classes i):

$$\overline{h_p} = \frac{1}{\sum_i m_{p,i}} \sum_i \left\{ m_{p,i} \left[\left(c_{pl} \left(T_{p,i} - T_s \right) + \Delta h_f \right) \left(1 - f_{s,i} \right) + c_{ps} \left(\left(T_{p,i} - T_s \right) f_{s,i} + T_s \right) \right] \right\} \tag{4.57}$$

Based on the calculated average specific enthalpy $\overline{h_p}$, the mean spray solid fraction $f_{s,h}$ and the mean temperature T_h are calculated according to the solidification state of the droplet mass. In the separation averaging approach, the non-equilibrium situation in the spray process is recognised. The energy exchange within the total particle mass occurs only via the specific heat content and not via the remaining latent heat content of the particles:

$$f_{s,m} = \frac{1}{\sum_i m_{p,i}} \sum_i \left(m_{p,i} f_{s,i} \right), \tag{4.58}$$

$$T_m = \frac{1}{\left(c_{ps}\, f_{s,m} + c_{pl} \left(1 - f_{s,m} \right) \right) \sum_i m_{p,i}} \sum_i \left(m_{p,i} T_{p,i} \left(c_{ps} f_{s,m} + c_{pl} \left(1 - f_{s,m} \right) \right) \right). \tag{4.59}$$

Differences in the results obtained using the two averaging methods are shown in Fig. 4.33 for the averaged temperatures and solid fractions of the spray along the centreline as a function of the particle flight distance. The results of the enthalpy

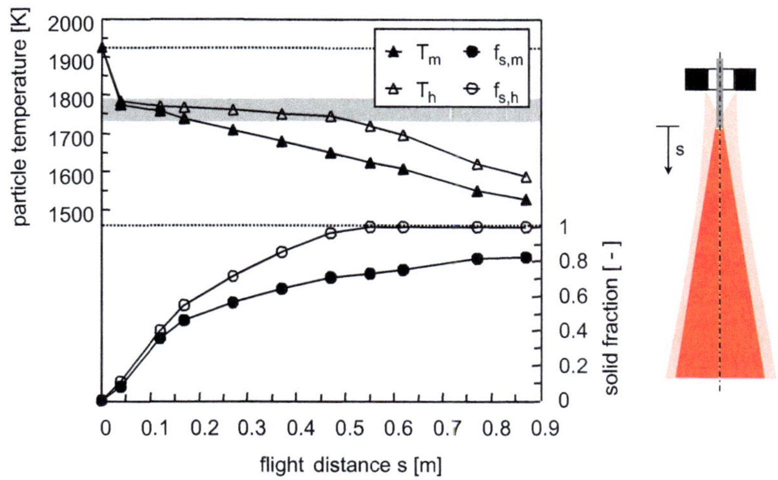

Fig. 4.33 Integral temperature and remaining solid fraction in steel melt spray versus the spray distance, two different averaging methods (from [73], with permission)

method represent the thermal equilibrium with respect to the particle mass, considering all particles from a specific location (point of impingement) together and leaving this mass under adiabatic conditions for an inner compensation process (heat conduction and solidification). This compensation process in spray forming immediately occurs when the sprayed particle mass impinges onto the deposit. In the semi-solidified mixing layer on top of the deposit an equilibration process will take place. Therefore, the enthalpy method properly describes the thermal state of the particle mass *in the top mushy layer* after deposition. In contrast, the separation method describes the instantaneous local thermal state (temperature and solid content) of the particle mass *in the spray*. As can be seen in Fig. 4.33, the instantaneous mean particle temperatures and mean solidification fractions calculated for the whole particle flight distance by the separation model are lower than the values derived by the equilibrium state (enthalpy) method. With respect to the solidification process after the deposition this difference means that the overall particle mass is "undercooled" and may be reheated after the deposition (in the mixing layer) by the latent heat released during solidification.

The mass flux distribution in the spray cone is directly connected to the resulting shape of the sprayed deposit. For varying deposit shapes, either the atomizer and/or the substrate can be moved. For a stationary atomizer, the radial profile of the mass flux distribution in the spray cone can be described as an exponential function. The maximum mass flux is always found on the spray cone centreline and $r_{0,5}$ is the half width of the mass flux profile. The exponential factor has been correlated as $k_1 = 1.3$ [204] and $k_1 = 1.124$ [205] for steel atomization. The half width of the mass flux profile is proportional to the maximum mass flux value as:

$$\dot{m}(r) = \dot{m}_{\max}\, e^{\left(\ln(0.5)\left|\frac{r}{r_{0.5}}\right|^{k_2}\right)}. \tag{4.60}$$

For calculating the mass flux distribution at each position within the spray cone, a single calibrating measurement on the spray centreline in combination with the mass flux correlations is sufficient.

4.5.3.4 Spray Cone Spreading

The spray cone angle determines the coverage and dispersion of a spray in the surrounding environment, i.e. the macro geometry of the spray. Different atomizers yield different spray characteristics as e.g. flat, round or hollow cone. The spray cone angle initially is determined by the liquid exit condition, the main disintegration mode within primary fragmentation (and here the growth rate of asymmetrical disturbances), later on by gas turbulence effects (particle dispersion), interaction with the ambient air (gas entrainment), and finally by gravity.

Within pressure swirl atomization, an increase of spray cone angle with the injection pressure differential (Δp_l) is confirmed by most of experimental

Table 4.1 Analytical and empirical correlations for the spray cone angle α (in SI units)

References	Correlation
[228]	$\sin(\alpha/2) = \frac{(\pi/2)C_D}{K(1+\sqrt{X})}$
[229]	$\alpha = 6K^{-0.15}\left(\frac{\Delta p_l D_0^2 \rho_l}{\mu_l^2}\right)^{0.11}$
[209]	$\alpha = 16.156K^{-0.39}D_0^{1.13}\mu_l^{-0.9}\Delta p_l^{0.39}$
[230]	$\sin(\alpha/2) = \frac{2\sqrt{2}X}{(1+\sqrt{X})\sqrt{1+X}}$

observations based on kinds of fluid viscosities and different nozzle geometries [206–208]. An increase in swirl-ratio $S = \pi (D_s - D_p) D_O/(8A_p)$ leads to an increase of the sheet velocity in tangential direction at the nozzle orifice, resulting in a larger spray angle [208]. The spray cone angle decreases with increasing atomizer constant (K) [209]. An increase in the orifice length/diameter ratio (L_O/D_O) reduces the spray cone angle [208]. An increase in liquid viscosity reduces the spray cone angle [207]. Analytical and empirical correlations for the spray cone angle in pressure-swirl-atomization are summarized in Table 4.1. A way to influence the film formation and the spray and ligament opening angle at the orifice of a pressure swirl atomizer by utilizing the Coanda effect (jet deflection at curved interface) has been illustrated in Kamplade et al. [210]. The liquid film angle deflection and thus the spray by the Coanda orifices mounted to a pressure swirl nozzle has been characterized by geometrical parameters as the Coanda curvature radius R_c/D (versus the orifice diameter) and by the flow- or pressure Reynolds-number. In general at proper conditions it is shown that much higher spray angles than with sharp edged nozzles can be achieved.

For a simplex pressure atomizer with a full spray, the work of Reitz and Bracco [211] concluded that the spray angle can be determined by combining the radial velocity of the fastest growing of the unstable waves in the primary fragmentation process with the axial injection velocity. This hypothesis results in an expression for the spray angle that has been later on simplified by Bracco et al. [212] in the form:

$$\tan\Theta = \frac{2\pi}{\sqrt{3}A}\left(\frac{\rho_g}{\rho_l}\right)^{0.5} \tag{4.61}$$

where A is a function of the orifice length to diameter ratio that must be experimentally determined [213].

The distribution of the sprayed liquid mass in the spray cone is described by the mass flux distribution. Adopting the concept of self-similarity from a single phase jet, the droplet mass flux distribution in a spray may be described (see Fig. 4.34). In full cone sprays, this can be expressed in terms of a general empirical formula as [204]:

$$\frac{\dot{m}}{\dot{m}_{cl}} = e^{-k_1\left(\frac{r}{r_{0.5}}\right)^{k_2}} \tag{4.62}$$

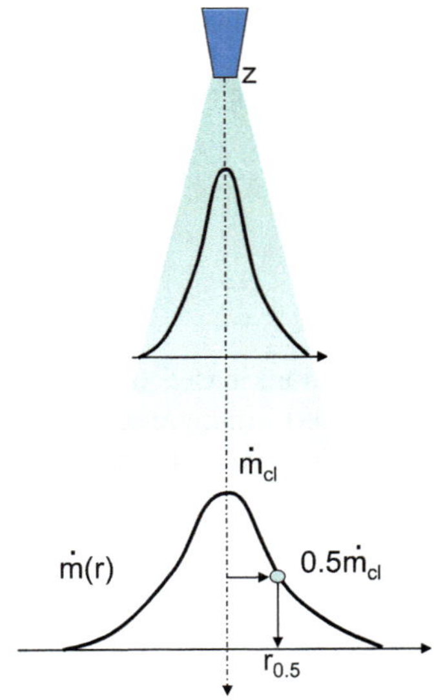

Fig. 4.34 Mass flux distribution in sprays

where $\dot{m}_{cl}$ is the maximum value of the mass flux distribution function on the spray centreline and $r_{0.5}$ is the half width of the mass flux distribution. The proposed empirical constants have been given as $k_1 = \ln(2.0)$ and $1.2 < k_2 < 2.0$. The constant k_2 depends on operational process conditions and liquid properties. The spreading behaviour of the spray cone may be described either by a function of the mass flux on the centreline or by the distribution of the half width of the mass flux distribution versus nozzle distance, where for a Gaussian type distribution:

$$r_{0.5} = \sqrt{k_2 \frac{\dot{M}_l}{\dot{m}_{cl}}}. \tag{4.63}$$

The behaviour of the centreline mass flux has been empirically correlated as Uhlenwinkel [204], Kramer [205]:

$$\dot{m}_{cl} \approx z^{-k_3}. \tag{4.64}$$

For water sprays Uhlenwinkel [204] has found $k_3 \cong 2$ while for metal melt atomization Kramer [205] found $k_3 < 2$. Spray cone spreading (namely the coefficient k_3) depends on atomization parameters. For water atomization the spreading of the spray increases with higher gas flow rates while for metal melt the reverse trend is observed. In this case, the spray cone width narrows when the gas mass flow rate increases with higher pressure [204].

4.5.3.5 Spray/Environment Interaction, Entrainment

The spray system interacts with the environmental gas. Like a single phase jet, it accelerates ambient gas due to friction (entrainment). Thereby, the plume of the dispersed phase (the droplet spray) maintains a narrower width than the complete flowing gas jet.

The spray/gas interaction and gas entrainment may lead in some cases to a change in the spray structure with increasing distance from the atomizer, namely a drop size segregation. Generally, this affect is pronounced in pressure atomizers, as in this case the entrainment effect is strongest, and is typically found in pressure swirl atomization. Here, due to the entrainment of ambient gas into the initially hollow cone spray, droplets are accelerated and transported into the core region of the spray (see sketch in Fig. 4.35b). The primary hollow cone structure is changing towards a solid cone with increasing spray distance. Figure 4.35a illustrates the change as a short exposure view onto the spray (instantaneous structure) from outside in comparison to a light-sheet illumination of the central plane of the spray in Fig. 4.35b. Because of their smaller inertia, smaller droplets are more easily transported into the spray core than bigger drops. Here, a size segregation effect follows, where the droplet size distribution in the spray is changed, typically as shown in the sketch in Fig. 4.35b. As an example, in Fig. 4.35c particle size data are presented as radial distributions of the arithmetic number mean value of the measured droplet size distribution at several planes below the atomizer (as measured by PDA). With increased spray distance the edge of the spray shows increased drop sizes, while in the spray core the smaller droplets are more prevalent.

The total spray/gas entrainment rate depends on the atomizer type. With twin-fluid atomization, entrainment can be estimated based on the overall gas flow rate in circular, turbulent jet and the distance from the nozzle. The ratio of the total gas flow rate to the initial jet flow rate is a linear function:

$$\frac{\dot{Q}_{total}}{\dot{Q}_0} = const. \frac{x}{d_0}. \tag{4.65}$$

The constant is 0.4565 from the analytical solution for a turbulent jet and varies from 0.30 to 0.45 in experimental and simulation studies of single- and two-phase jets and spray flows.

The principal two-phase flow field in a typical powder production chamber employing gas atomization is illustrated in Fig. 4.36 obtained by numerical simulation. The flow is subdivided into three regions [215]: (a) the droplet spray in the core, surrounded by (b) the gas jet and (c) a recirculation zone in the outer area of the spray chamber.

The spray flow field in a bounded environment (spray chamber) yields some specific flow features. Some characteristics are

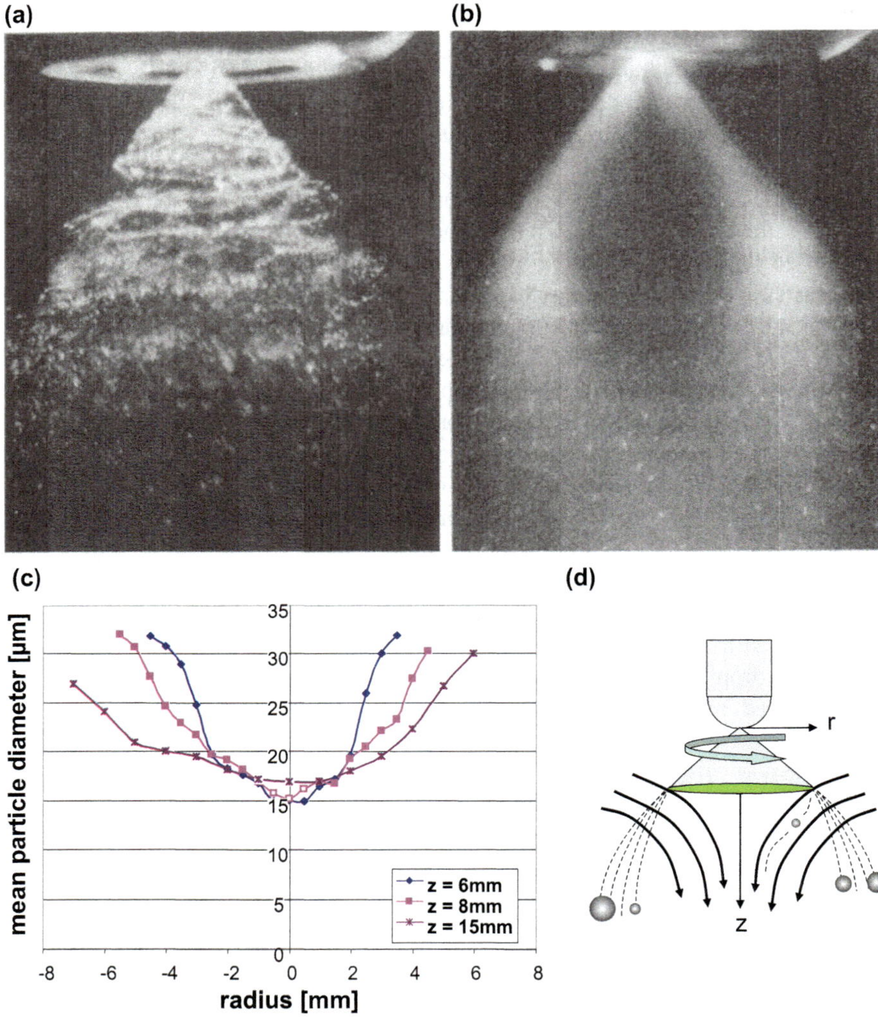

Fig. 4.35 Spray from pressure swirl atomizer: (**a**) short and (**b**) long time exposure, droplet entrainment in hollow cone spray; gas and droplet entrainment in hollow cone spray, (**c**) drop size distribution, (**d**) spray sketch (from Lampe [214], with permission)

- a high velocity core, that expands to the outer chamber walls in a certain distance,
- a recirculation zone, in which the upward flow re-entrains into the main core,
- strong gradients of downstream velocity in radial direction between core zone and recirculation region and
- an outflow region in the lower chamber, where the velocity is more or less in plug flow directed downstream.

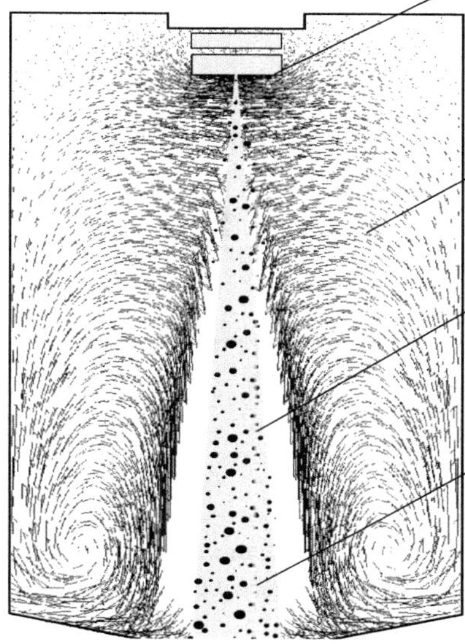

Fig. 4.36 Flow field in a spray chamber for powder production, simulation result

Numerical simulations of bounded spray flows may be carried out on different scales, using numerical approaches like Unsteady-Reynolds-Averaged-Navier-Stokes models (RANS or URANS) and Large-Eddy-Simulation (LES) models with lagrangian particle tracking for the droplet motion [159]. The spray structures in- and outside the central spray cone flow like the recirculation and entrainment zones of gas in between the spray and the chamber wall can be predicted as well as large scale turbulent structures within the spray cone. Recirculation of gas and droplets are investigated for identification of droplet re-entrainment into the spray leading to satellite particle formation in metal powder production.

Recirculation Flow Rate

The dependency on upwards flow rate, recirculation point and velocity gradients have been derived for a gas only case where a gas jet from a cylindrical nozzle emerges into a spray chamber. Three flow rates with inlet flow velocities 50, 100, and 200 m/s for a fixed geometry of a spray tower diameter of 1 m and a nozzle diameter of 4 cm are discussed [216].

The recirculating flow rate (volumetric flor rate that is entrapped in the recirculating vortex) depends linear on the inflow at the nozzle. Figure 4.37 shows the dimensionless profiles at different velocities. The profiles are self-preserving if normalized with the inlet flow rate. The initial slope is similar to the free stream entrainment flow. In the course of spray downstream propagation, the entrainment increases due to radial pressure gradients. Each curve shows a

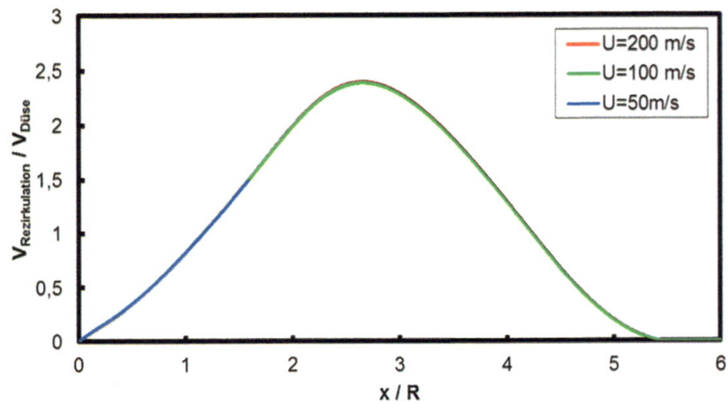

Fig. 4.37 Recirculation flow rates at various velocities at the spray nozzle outlet

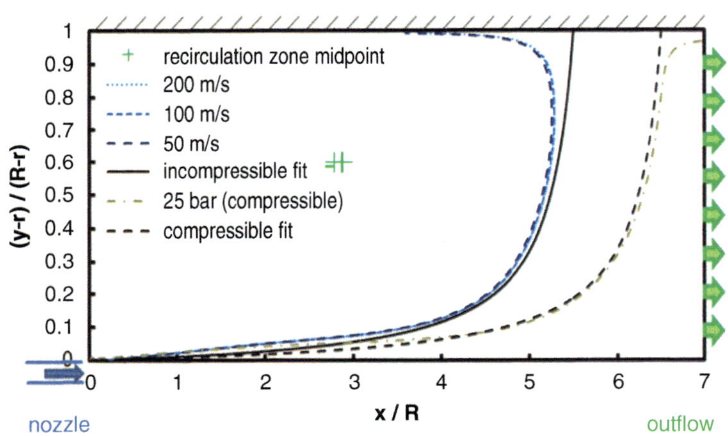

Fig. 4.38 Spray contour using streamlines and reattachment point for incompressible and compressible gas flow

maximum around 0.5 x_R, where x_R is the reattachment length of the spray in the chamber. After the maximum, the recirculating flow decreases until the reattachment point. Due to mass conservation, this value is achieved for all distances longer than the reattachment length.

Reattachment Point

The reattachment point is calculated using a zero gradient boundary condition for the velocity in radial direction at the surrounding wall. Figure 4.38 depicts the spray contour and shows streamlines calculated using Runge-Kutta 4–5 starting slightly below the nozzle radius. The curve should connect with the reattachment point,

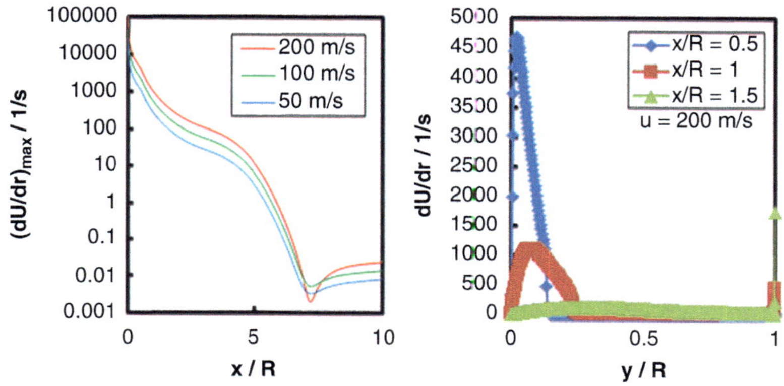

Fig. 4.39 Maximal shear rates for the 3 presented cases (*left*) and radial profiles of shear gradients (*right*)

though due to computational inaccuracies, the curves slip away closer to the wall. The midpoint of the recirculation zone is marked for various flow rates and geometries. The dimensionless reattachment point is 5.5 times the radius R for incompressible flow and 6.5 R for compressible flow. Due to the shock waves expanding from the nozzle, the penetration into the spray chamber is larger for compressible flow calculations.

Shear Gradient

The shear gradients follow a trend as shown in Fig. 4.39 (left): Close to the nozzle, high gradients occur, as a block velocity profile is assumed at the jet exit. Besides this boundary rise in gradient, the near nozzle gradients fall up to a nozzle distance x/R of about 7 R when they reach an inlet flow rate dependent minimum and finally converge to a constant level, which is the gradient equal to a pipe flow. Figure 4.39 (right) shows the gradients development in radial direction. According to the diagram the gradients rise sharply around the outer jet penetration region. This region correlates to the nozzle width at the nozzle outlet and then spreads in radial direction as the jet evolves into the chamber. The evolution causes the velocity to decrease and broadens the area of high gradients. This enlargement causes the maximal gradient to flatten out, until only small gradients remain with are distributed over the entire radius.

Spray Chamber Design

By changing the shape and the size of the spray tower, the effect on the entrainment may be analyzed as done in [159, 216]. Here, a conical part and an orifice have been integrated into the spray chamber model. Variations of the spray tower width at different flow rates of the atomizer are also studied in Fig. 4.40. The flow inside the

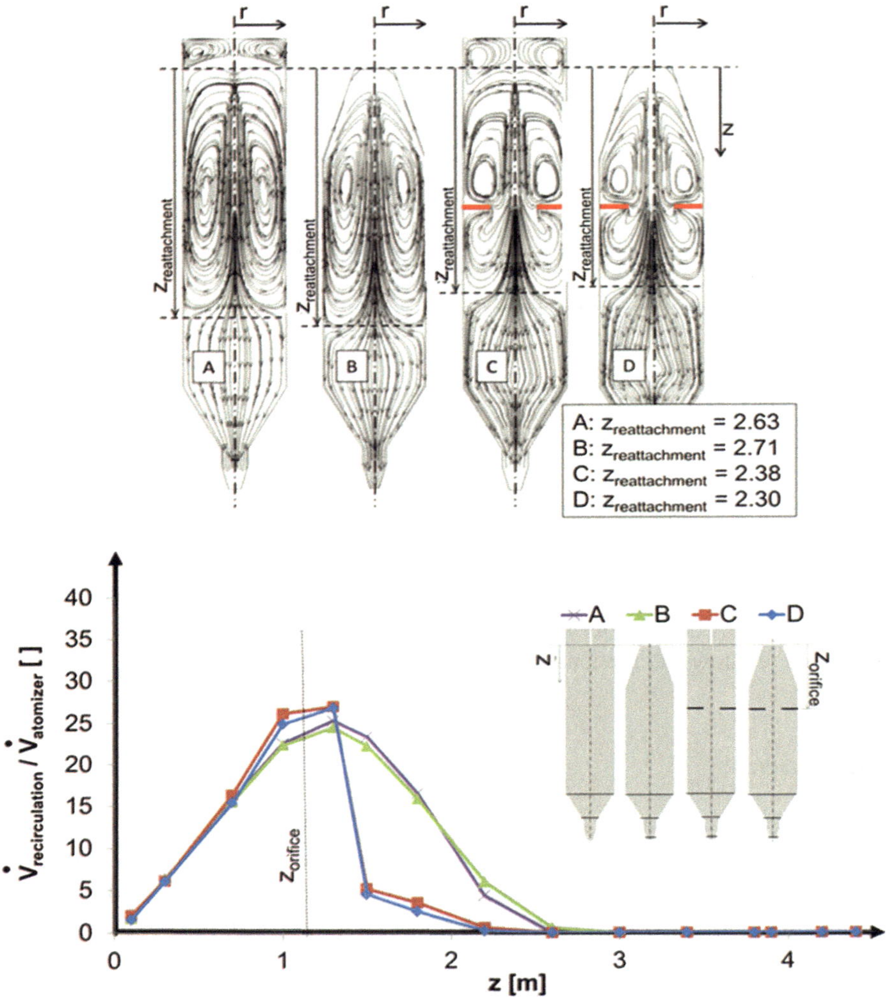

Fig. 4.40 Recirculation gas flow pattern for different tower geometries and entrainment volumetric gas flow dependent on the distance from the atomizer ($V_{atomizer} = 200\ m^3/h$, only gas phase)

spray tower is characterized by the large recirculation zone which is also found in similar studies of concurrent tall-form spray dryers [217]. Through the large recirculation zone the mixing inside the spray tower is enhanced. Air laden with moisture and small particles is recirculated upwards and is entrained into the spray. The result is a relatively homogeneous temperature and species distribution inside the spray tower at $z > 200$ mm.

The recirculation/entrainment rate can be evaluated from the simulations by integrating the velocity profile of the upwards directed gas flow. The resulting recirculating gas flow can be divided by the volumetric flow rate of air through the hot gas nozzle. There is a recirculation maximum in the center of the recirculation zone. From the reattachment point the flow profile is unidirectional. The position of the reattachment could be determined in experiments (Variant A) and agrees with the simulation results ($z_{reattachment} \approx 2.5$ m). The Variant A is without any modifications of the spray chamber design. Variant B is equipped with a conical part in the spray tower to make the recirculation zones close to the nozzle smaller to avoid accumulation of material in these areas. The case number C is investigated to show the change of entrainment flow close to the nozzle in dependence of the spray tower height. The Variant D is a mixture of variant B and C. The variant A shows a flow pattern that is related to concurrent spray drying processes with pneumatic nozzle systems. The flow from the air dispenser is centered and pointing downwards to adapt the direction of the heat flow to the low expansion angle of the spray flow. In some applications the air stream has a swirl component to avoid the uncontrolled deflection of the spray onto the spray chamber walls. In the hot gas atomization process swirl is not necessary because of the strong axial momentum flux from the nozzle.

In variant B the recirculation zones close to the nozzle are minimized by limiting the cross-sectional area. This variant is already used in many spray dryers. The entrainment flow profile is not changed in comparison with variant A. A strong deflection of the gas flow (and thereby also of the particles) in the spray towards the wall is found for variant C showing that hot gas atomization demands tall spray towers. Here, the axial momentum flux of the spray flow is higher than for a conventional concurrent spray drying process. The entrainment flow in front of the orifice is increased because of the strong deflection of the gas, but close to the nozzle the entrainment flow is the same as for variant A and B. Variant D combines the effects of Variant B and C. In the measurements as well as in the simulations the temperature distribution on the centerline is almost the same.

The entrainment flow depends on the flow rate of the atomizer gas and on the tower geometry (Fig. 4.41). The gas flow rate has been varied in this case between 150 and 250 m^3n/h. The analysis of the reattachment points for the different flow rates obtains no significant difference. The entrainment flow rate along the spray axis is identical. An increase of the entrainment flow can be found when the tower radius is increased, while the reattachment point remains almost at the same position. This implies that the retention time of the particles inside the large recirculation zone can only be changed by using spray towers with different radii. Increasing the tower radius by a factor of 2 increases the entrainment/recirculation flow rate by the same factor. Despite the increase in the upwards directed flow the overall velocities are smaller.

Typically within the recirculation zone a large amount of fine particles swirl around in the spray chamber (dust formation). Here preferably the fine particles are to be found due to their small inertia. These particles cool down and solidify in their upward movement and possibly are reentrained into the spray as fine solid particles. As there are a large number of these fines, collisions with spray droplets are likely and result in so-called satellite formation in melt sprayed particles. These satellite formations lower the quality of sprayed powders such as a reduction in free-flow ability. A possibility for satellite formation prevention has been proposed by Dunkley and Telford [218], Achelis and Uhlenwinkel [219], and Achelis

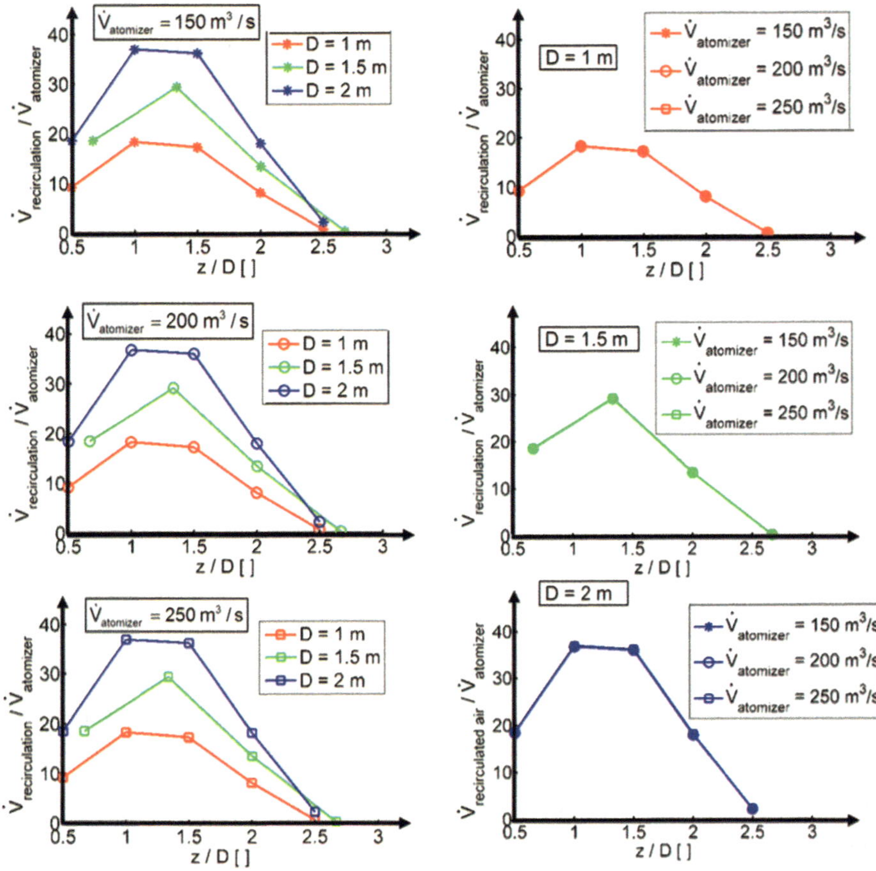

Fig. 4.41 Influence of the spray tower width and volumetric mass flow of the atomizer gas on the entrainment flow conditions

[206]. Here part of the exhaust gas (after particle separation) is re-injected from top into the spray chamber (as illustrated in Fig. 4.42). The recirculation gas flow rate is controlled for replacing the natural jet entrainment flow (conventionally achieved from recirculating gas inside the chamber) into the spray. As result, dust recirculation and satellite droplet formation is reduced. In Fig. 4.43, the spray with anti-satellite device becomes less chaotic as that without anti-satellite device. In this way the number of non-spherical particles is dramatically reduced.

4.5.4 *Three-Phase Spray Process for MMC Production*

Metal-matrix-composite (MMC) particles can be fabricated through mixing ceramic particles and atomized metallic droplets in a spray atomization and co-injection process [65, 67, 183, 220–222]. The process is characterized by a

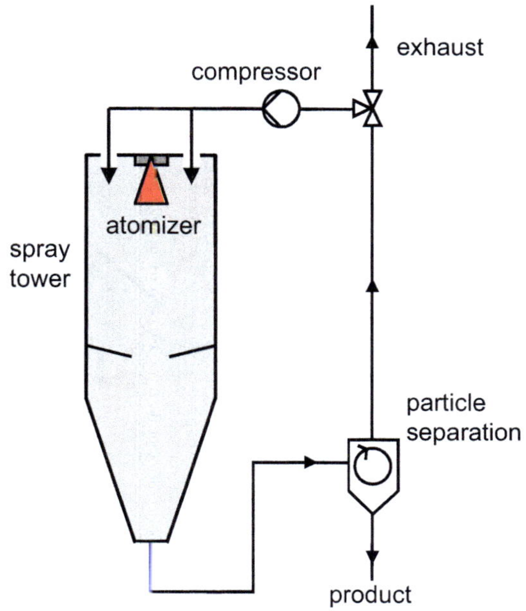

Fig. 4.42 Concept for flow configuration in powder production spray process for decreasing of satellite drop formation

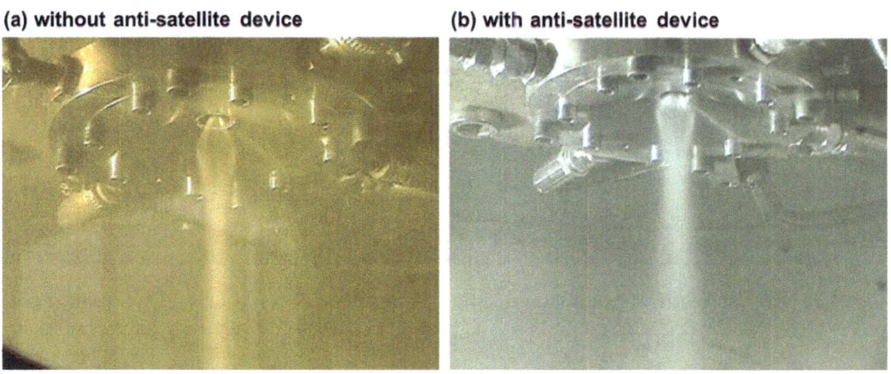

Fig. 4.43 Pressure-gas-atomization process (from [206], with permission)

three-phase spray flow (gas/droplets/particles). Ceramic particles (usually at least one order of magnitude smaller than atomized droplets), conveyed by the atomization gas or via a separate gas-assisted delivery system, are injected into the metallic droplet spray and likely to be incorporated into the droplets or captured by the droplets surface during frequent impingements, forming MMC droplets which are subsequently solidified as MMC particles. This approach will inherently avoid the extreme thermal excursions, with concomitant degradation in interfacial properties and extensive macro-segregation, normally associated with the MMC production in casting processes. By mixing reinforcements with liquid/semi-solid matrix in a spray atomization process, de-mixing problem associated with the

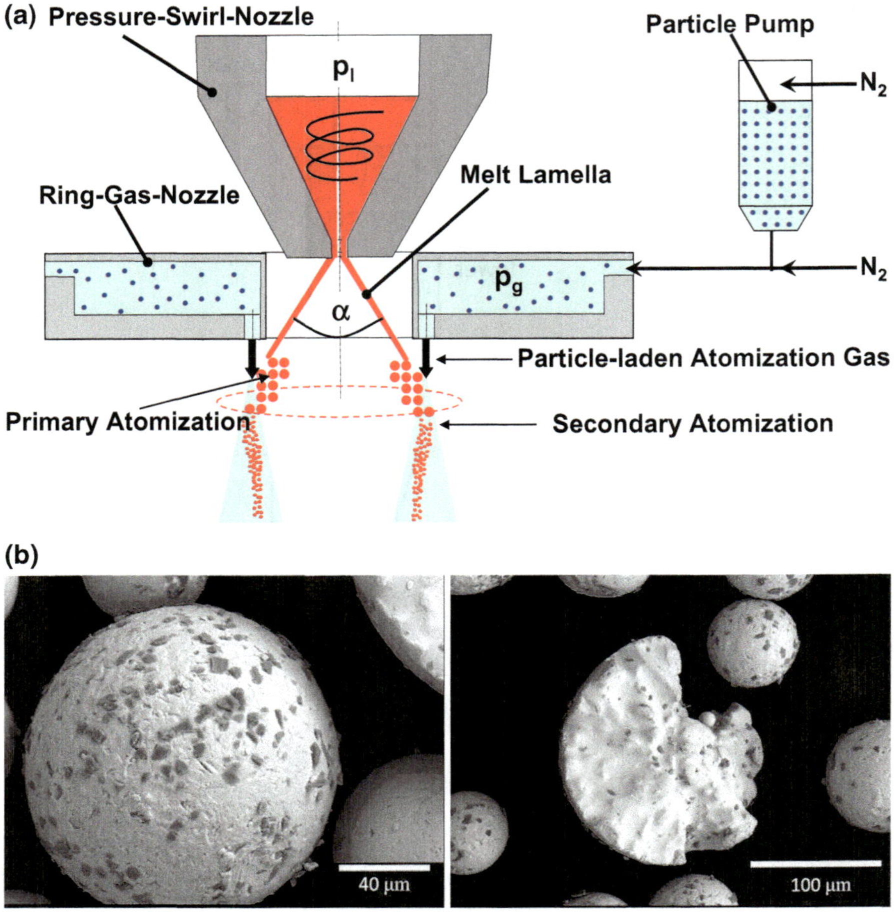

Fig. 4.44 (**a**) Schematic of generation of MMC particles in a pressure-gas-atomization (PGA) process; (**b**) MMC particles: Sn-matrix + SiC-particulates (adapted from [65], with permission)

MMC production in powder metallurgy (PM) process can also be eliminated. A maximum incorporation rate (18 %) of ceramic particles in metal matrix was reported in Eslamian et al. [220], where a prefilming hybrid atomizer was employed for the production of MMC particles.

For MMC-particle production in a spray process a hybrid gas atomization nozzle configuration ideally fits the requirement of gas/particle/liquid dispersion and mixing in the spray flow field. A typical pressure-gas-atomization (PGA) configuration for MMC-particle production is sketched in Fig. 4.44a, where solid particles, continuously supplied by a particle pump and conveyed by the atomization gas, are co-injected and impacted with the liquid droplets in the secondary atomization zone. Figure 4.44b shows some spray-processed MMC particles in a PGA process. The fact that the particulate reinforcements are distributed inside or on the surface of the MMC particles proves the concept of MMC particles. MMC particles can be used in powder processes to generate composite parts or for thermal spray coating applications.

Li [65] presented multi-scale descriptions of particle-droplet interactions in spray processing of MMC particles based on Multiphase Computational Fluid Dynamics (M-CFD) models, in which processes such as liquid atomization and particle-droplet mixing spray (macro-scale), particle-droplet collision (meso-scale), and particle penetration into droplet (micro-scale) are taken into account. Thereby the incorporation efficiency and sticking efficiency of ceramic particles in metal matrix are correlated with the operation conditions and material properties.

4.5.4.1 Particle-Droplet Mixing Behaviour

On the *macro-scale*, the mixing behaviour of ceramic particles (SiC) and metallic droplets (Sn) in a pressure-gas-atomization (PGA) configuration is described by an Eulerian-Lagrangian-Lagrangian (ELL) approach. In the PGA process, ceramic particulates conveyed by the atomization gas are co-injected and impacted with the metallic droplets in the secondary atomization zone. The ELL formulation uses Eulerian description for the continuous gas phase, hence the gas flow through stationary mesh-volumes, while the dispersed phases are tracked by the Lagrangian approach. Figure 4.45 shows the particle-droplet mixing configurations at different atomization gas pressures. Here the TAB model is employed to describe droplet breakup. Increasing atomization gas pressure (i.e. increasing gas-mass-flow rate) causes a spray of tiny droplets which tend to be involved into the particle-laden gas flow, thus leading to a high particle-droplet mixing degree. However, the fast droplet solidification rate at a high atomization gas flow may reduce the particle-

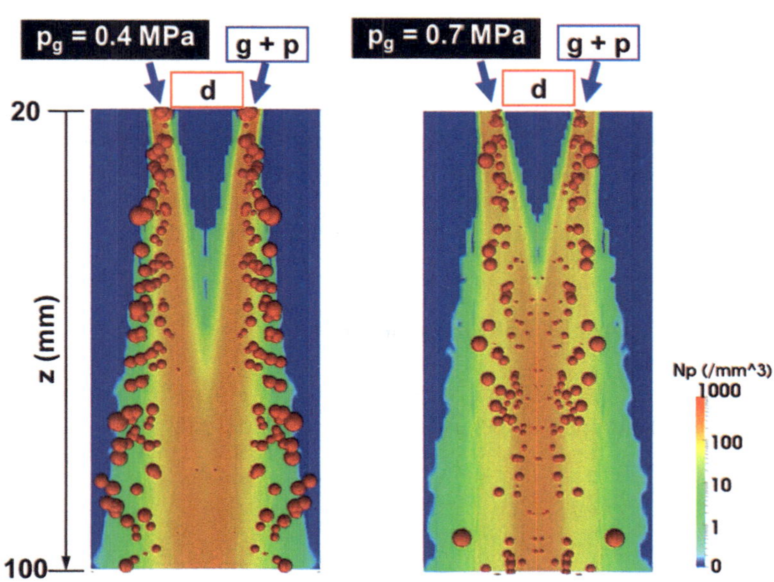

Fig. 4.45 Simulation results: mixing configurations of Sn-droplet sprays and SiC-particle jets at different atomization pressures (p_g), melt mass flow rate 180 kg/h, particle mass flow rate 6.4 kg/h, d—droplets, p—particles, g—gas (adapted from [65])

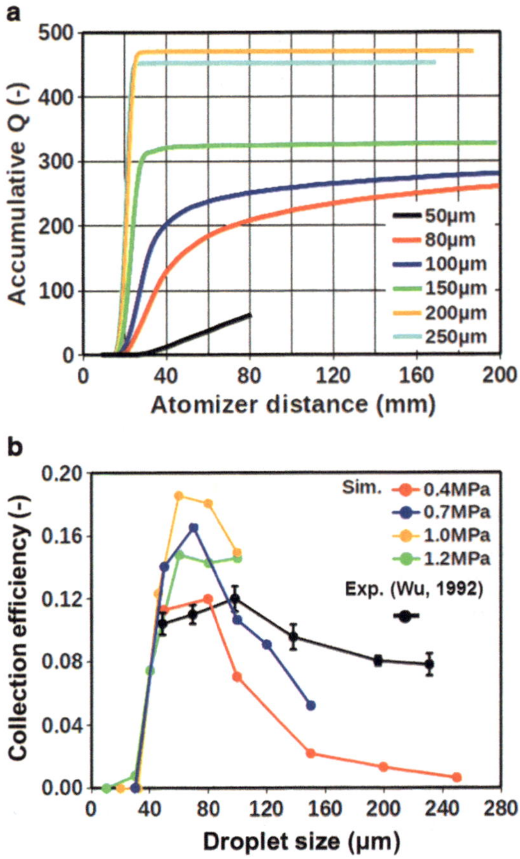

Fig. 4.46 (**a**) Accumulative particle-droplet collision number Q along the droplet flight path vs. droplet size in a gas flow field at atomization pressure $p_g = 0.4$ MPa; (**b**) collection efficiency of SiC-particles by Sn-droplets of different sizes at different atomization-gas pressures (adapted from [65])

droplet mixing quality. A lower atomization gas pressure results in a more radially-expanded droplet spray which tends to intersect the particle-laden gas jets, leading to a poor particle-droplet mixing extent. Therefore, the best particle-droplet mixing configuration is determined at a moderate atomization gas flow.

4.5.4.2 Particle-Droplet Collision Behaviour

On the *meso-scale*, the collision behaviour between a metallic droplet and solid particles along the droplet trajectory is quantitatively described by

$$\begin{aligned} Q &= \int \frac{\pi (d_p + d_d)^2}{4} \bullet \left| \vec{U}_p - \vec{U}_d \right| N_p \bullet \eta dt \\ &\approx \sum \frac{\pi (d_p + d_d)^2}{4} \bullet \left| \vec{U}_p - \vec{U}_d \right| \bullet N_p \bullet \eta \bullet \Delta t \end{aligned} \tag{4.66}$$

Here Q represents the accumulative droplet-particle collision number along the droplet trajectory, d diameter, U velocity, N_p particle number concentration ($/m^3$) in a control volume ΔV, and η is collision efficiency. Δt is a time step for a droplet to move from one position to another. The subscripts p and d denote the physical quantities relative to particle and droplet, respectively. The particle-droplet relative velocity and the particle number concentration along the droplet moving path can be obtained based on a Lagrangian tracking.

The droplets have different trajectories in the computational domain because of their different inertias. Along the droplet trajectory, the particle-droplet relative velocity and the particle number concentration at the droplet location can be extracted from the macro-scale simulation to derive the accumulative particle-droplet collision number. Figure 4.46a shows the accumulative particle-droplet collision numbers along the trajectories of different sized droplets in a gas flow field at $p_g = 0.4$ MPa. The accumulative particle-droplet collision number achieves a maximum value when the droplet flies into a particle-free region or finishes solidification. The collection efficiency of ceramic particles by a metallic droplet is defined as the volume fraction of ceramic particles colliding with the droplet before it achieves a 50% solidification degree (assuming that when the solidification degree in a droplet is over 50%, the particle-droplet collisions will not lead to particle incorporation or sticking in the droplet). The collection efficiency as a function of droplet size and atomization gas pressure is shown in Fig. 4.46b. In general, the collection efficiency first increases with increasing droplet size, reaches a maximum value, and then decreases. The particle-droplet mixing quality at a given operation condition can be described by a characteristic collection efficiency, e.g., the collection efficiency of a droplet with a size of mass median diameter (MMD, $d_{50,3}$) at that operation condition. A maximum characteristic collection efficiency (~16%) is obtained at $p_g = 0.7$ MPa, which implies a best particle-droplet mixing quality under this operation condition.

4.5.4.3 Particle Penetration Behaviour

Force/Energy Balance Approach

Two theoretical approaches for description of the particle penetration behaviour into the liquid droplet have been evaluated. The first approach uses an energy balance to calculate the distance the particle penetrates into the droplet [223], where the initial kinetic energy of the solid particle or the liquid droplet is considered as the driving force for penetration; the change in surface energies and the work done by the viscous drag in the droplet are considered as the forces resisting penetration. The other approach uses a force balance to calculate the penetration [224–226]. The factors taken into account in these physical models include a number of intensive properties such as particle/droplet density (ρ_p/ρ_d), viscosity (μ_l), surface tension (σ) and solid/liquid contact angle (θ), and extensive properties like impact velocity and particle/droplet size (d_p/d_d). Both theoretical approaches assume that the droplet remains spherical during the penetration, without taking into account the information like the cavity formation behind the penetrating particle, the droplet deformation during penetration, and the gas layer rupture between solid particle and liquid surface. However, important conclusions can still be achieved by these theoretical models.

CFD Simulation

The numerical simulation of the penetration behaviour of a solid particle into a liquid droplet is performed based on a multiphase CFD model which describes the multiphase flow situation (gas-particle-droplet) by solving incompressible Navier-Stokes (N-S) equations coupled with Volume of Fluid (VoF) method, Six degrees of Freedom (6-DoF) method and dynamic mesh technique [67]. Two typical outcomes observed in simulations are: (1) the particle partially penetrates into the droplet; (2) the particle completely penetrates into the droplet. The outcomes are categorized with Weber number (We) and Reynolds number (Re) in a regime map. The regime boundary is formulated by

$$\ln(We/We_0) = a(\ln(Re/Re_0))^b, \text{ with} \qquad We = \rho_l U_{rel,0}^2 d_p/\sigma, Re = \rho_l U_{rel,0} d_p/\mu_l \tag{4.67}$$

Here We_0, Re_0, a and b are constants related to solid-liquid contact angle (θ), particle/droplet size ratio (δ), and particle orientation. As shown in Fig. 4.47, the regime boundary shifts to a higher Re-We region with increasing contact angle, and

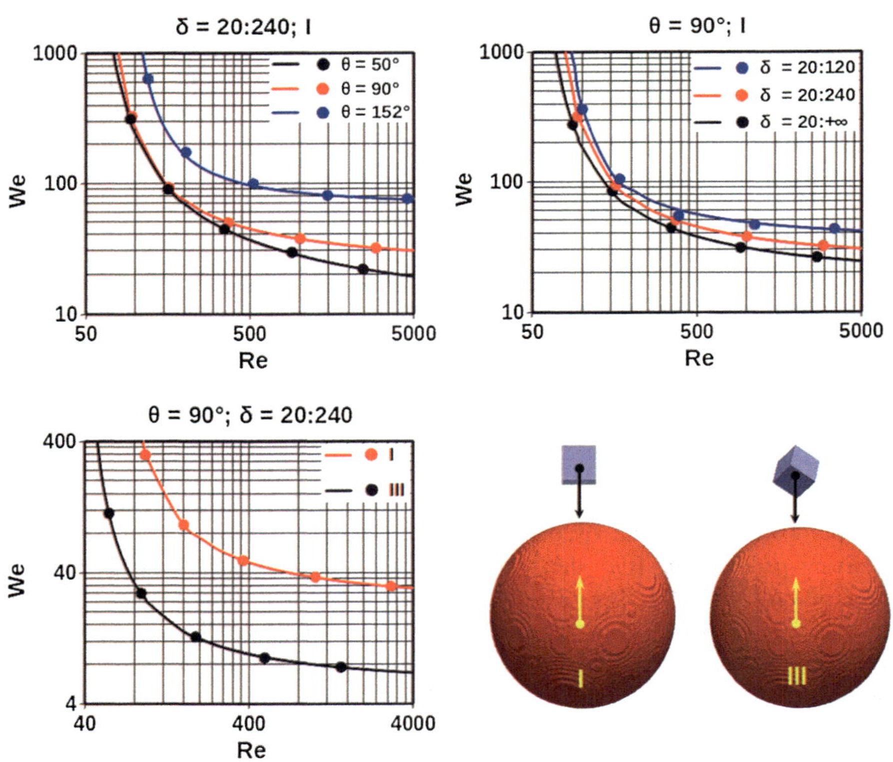

Fig. 4.47 Shift of regime boundary with variation of contact angle (θ), size ratio (δ) and particle orientation in Re-We regime maps, head-on collision (adapted from [66, 67])

slightly to a lower Re-We region with decreasing particle/droplet size ratio (δ). A significant shift of the regime boundary towards the lower Re-We region occurs if the particle-orientation varies from I to III.

4.5.4.4 Incorporation Efficiency and Sticking Efficiency

Based on the derived critical penetration conditions from the micro-scale modelling and simulation, the accumulative collisions derived from the meso-scale modelling and simulation can be thus divided into two groups, as sketched in Fig. 4.48:

1. One group of collisions occurring in the penetration regime is the main source for the incorporation of particulate reinforcements in the interior region of MMC particles. The incorporation efficiency can be roughly described by the volume fraction of these particles to the droplet.
2. The other group of collisions will result in particle partial penetration according to the Re-We regime map. Since most of collisions in this regime occur in the downstream spray flow, it is considered reasonable that most of the solid particles are likely to be captured by the droplet surface because of droplet solidification. The sticking efficiency of particulate reinforcements on the surface of MMC particles can be roughly described by the volume fraction of these particles to the droplet.

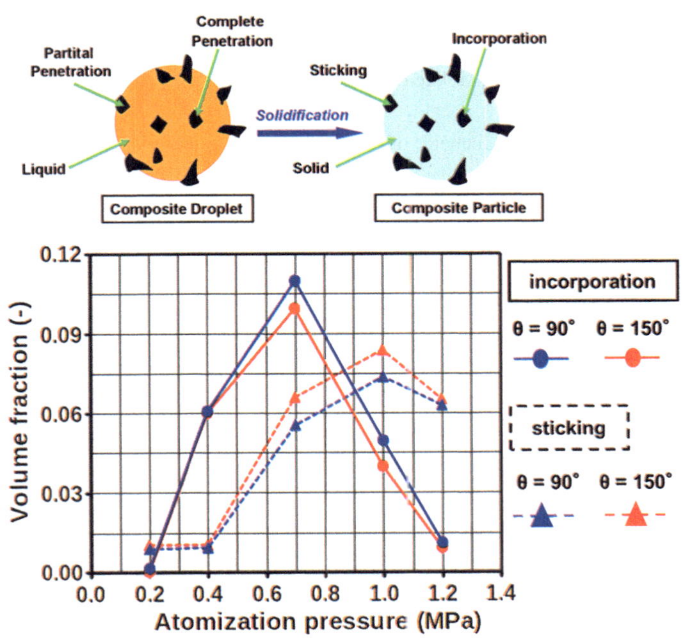

Fig. 4.48 Characteristic particle incorporation efficiency in droplet and sticking efficiency on droplet surface at different atomization process operation conditions (adapted from [65])

Suitable process conditions for optimization of the particle sticking/inclusion ratio may be derived from the M-CFD simulations. Figure 4.48 shows the calculated characteristic incorporation efficiency and sticking efficiency of TiC-particles ($\theta = 90°$) and SiC-particles ($\theta = 152°$) in Tin (Sn) based MMC-particles as a function of atomization gas pressure. It can be found that the incorporation efficiency as well the sticking efficiency first increases with increasing atomization gas pressure, reaches a maximum value, and then decreases; a significant portion of collisions is squandered by droplet surface. The maximum incorporation efficiency occurs at $p_g = 0.7$ MPa, while the maximum sticking efficiency occurs at $p_g = 1.0$ MPa. A larger contact angle results in a smaller particle-incorporation efficiency but a larger particle-sticking efficiency.

4.6 List of Symbols

4.6.1 Latin

Symbols	Description
A	Projection area normal to the gas flow
a	Droplet acceleration
C_d	Drag coefficient
C_{RT}	Adjustable constant (equal to 0.1)
C_τ	Constant usually equal to unity beyond the breakup length
d	(hydraulic) Diameter of the pipe
d	Physical quantitives related to droplet
d_d	Droplet diameter
d_s	Maximum stable diameter of this droplet
f	Volume fraction
f_c	Collision frequency
$f_{s,h}$	Mean spray solid fraction
g	Gravity
GLR	Gas to liquid mass flow ratio
g_t	Acceleration of the wave in the direction of travel
h	Sheet thickness
H	Inhomogeneity index
$\overline{h_p}$	Calculated average specific enthalpy
j	Unit vector tangent to the droplet trajectory
J(T)	Nucleation rate
k	Sheet number related to the sheet thickness
K	Constant that evaluated experimentally by measuring droplet sizes
k	Turbulence energy
k_S	Constant, to be determined by the dimensionality of the specific process under investigation
l	Gas to liquid momentum ratio

(continued)

Symbols	Description
$L_{ref}(x)$	Local half width of the gas velocity profile in the spray flow
M	Gas to liquid density ratio
$\dot{m}_l$	Mass liquid
$\dot{m}_g$	Mass gas
$\bar{m}(t)$	Mean mass of the product droplet
$\dot{m}_{cl}$	Maximum value of the mass flux distribution function on the spray centreline
n	Sum of all grey or binary values within the processed picture
N_2	Total particle number density
N_p	Particle number concentration
Oh	Ohnesorg number
p	Physical quantitives related to particle
$P_{no\_collision}$	Probability of no collision
Q	Accumlative droplet-particle collision numver along the droplet trajectory
$\dot{Q}_0$	Initial jet flow rate
$\dot{Q}_{total}$	Total gas flow rate
r	Radius
$r_{0.5}$	Half width of the mass flux distriution
r_{32}	Sauter mean radius of the child droplets after breakup
Re_l	Liquid Reynolds number
Sha	Shannon entropy
St	Stokes number
t	Time
$\dot{T}$	Cooling rate
T_h	Mean temperature
u_l	Velocity of the fluid
U_{rel}	Relative droplet-gas velocity
V_a	Atomic volume
We_l	Liquid weber number
Y	Distorsion parameter
z	Distance to the nozzle exit

4.6.2 Greek

Symbole	Description
ρ_g	Density liquid
ρ_l	Density gas
μ_g	Gas viscosity
μ_l	Liquid viscosity
σ	Surface tension
ω	Angular velocity
ϕ_{crit}	Critical collision angle
σ_{sl}	Solid-liquid interfacial energy
ΔH_{fa}	Latent heat of fusion

(continued)

Symbole	Description
ΔT_{hom}	Undercooling temperature difference for homogenous nucleation
ε	Dissipation rate
η	Collision efficiency
Λ_{KH}	Wavelength corresponding to the K-H wave with maximum growth rate
τ_c	Time scale between individual particle collisions
τ_E	Eddy lifetime
τ_I	Eddy
τ_{KH}	Breakup time
τ_p	Particle relaxation time
τ_r	Residence time of a particulate in an eddy
Φ	Indicator that can be attributed to the interface?
Ω_{KH}	K-H wave with maximum growth rate
Ω_{RT}	Rayleight–Taylor model: frequency of the fastest growing wave
κ	Scale on which the inhomogeniety is evaluated

References

1. Aydin, O., & Unal, R. (2011). Experimental and numerical modeling of the gas atomization nozzle for gas flow behavior. *Computers and Fluids, 42*(1), 37–43.
2. Czisch, C., Lohner, H., Fritsching, U., Bauckhage, K. (2003). Atomization of Highly Viscous Melts. *Paper presented at 2nd international conference on spray deposition and melt atomization SDMA 2003*, Bremen, Germany, 22–25 June 2003.
3. Czisch, C., & Fritsching, U. (2008). Flow-adapted design option for free-fall atomizers. *Atomization and Sprays, 18*, 511–522.
4. Dielewicz, L.G., van Berg, E., Lampe, M. (1999). Computation of Transsonic Two-Phase Flow in Liquid Metal Jet Atomizers. *Paper presented at ILASS-Europe 99*, Toulouse, France, 5–7 July 1999.
5. Espina, P.I., Piomelli, U. (1998). Numerical simulation of the gas flow in gas metal atomizers. *Paper presented at 1998 ASME fluids engineering division summer meeting: FEDSM'98*, Washington, US, 21–25 June 1998.
6. Espina, P.I., Piomelli, U. (1998). Study of the gas jet in a close-coupled gas metal atomizer. AIAA, paper 98–0959.
7. Espina, P. I. (1999). Numerical simulation of atomization gas flow. In K. Bauckhage & V. Uhlenwinkel (Eds.), *SFB 372: Sprühkompaktieren, Kolloquium, Band 4* (p. 127). Bremen: Universität Bremen.
8. Heck, U. (1998). Zur Zerstäubung in Freifalldüsen. Dissertation, Universität Bremen.
9. Mates, S. P., & Settles, G. S. (2005). A study of liquid metal atomization using close-coupled nozzles, part 1: Gas dynamic behavior. *Atomization and Sprays, 15*, 19–40.
10. Mates, S. P., & Settles, G. S. (2005). A study of liquid metal atomization using close-coupled nozzles, part 2: Atomization behavior. *Atomization and Sprays, 15*, 41–60.
11. Ting, J., Peretti, M. W., & Eisen, W. B. (2002). The effect of wake-closure phenomenon on gas atomization performance. *Materials Science and Engineering A, 326*, 110–121.
12. Ting, J., & Anderson, I. E. (2004). A computational fluid dynamics (CFD) investigation of the wake closure phenomenon. *Materials Science and Engineering A, 379*, 264–276.
13. Zeoli, N., & Gu, S. (2006). Numerical modelling of droplet break-up for gas atomisation. *Computational Materials Science, 38*, 282–292.
14. Zeoli, N., & Gu, S. (2008). Computational validation of an isentropic plug nozzle design for gas atomisation. *Computational Materials Science, 42*, 245–258.

15. Zeoli, N., Tabbara, H., & Gu, S. (2011). CFD modelling of primary breakup during metal powder atomization. *Chemical Engineering Science, 66*, 6498–6504.
16. Anderson, I. E., & Figliola, R. S. (1988). Observations of gas atomization process dynamics. In P. U. Gummeson & D. A. Gustafson (Eds.), *Modern developments in powder metallurgy* (Vol. 20, p. 205). Princeton, NJ: Metal Powder Industries Federation.
17. Mullis, A. M., McCarthy, I. N., & Cochrane, R. F. (2011). High speed imaging of the flow during close-coupled gas atomisation: Effect of melt delivery nozzle geometry. *Journal of Materials Processing Technology, 211*(9), 1471–1477.
18. Lohner, H., Czisch, C., Fritsching, U.. (2003) Impact of gas nozzle arrangement on the flow field of a twin fluid atomizer with external mixing. *Paper presented at 9th international conference on liquid atomization and spray systems ICLASS 2003*, Sorrento, Italy, 13–17 July 2003.
19. Ting, J. (2003). Melt flow rate in melt atomization. *Paper presented at 2nd international conference on spray deposition and melt atomization SDMA 2003*, Bremen, Germany, 22–25 June 2003.
20. Strauss, J.T. (2013). Lick back in close-coupled atomization: A phenomenological study. *Paper presented at 5th international conference on spray deposition and melt atomization, SDMA 2013*, Bremen, Germany, 23–25 September 2013.
21. Fritsching, U., & Bauckhage, K. (1992). Investigations on the atomization of molten metals: The coaxial jet and the gas flow in the nozzle near field. *PHOENICS Journal of Computational Fluid Dynamics, 5*(1), 81–98.
22. Markus, S., Fritsching, U., & Bauckhage, K. (2002). Jet break up of liquid metals. *Materials Science and Engineering A, 326*(1), 122–133.
23. Bergmann, D., Fritsching, U., & Bauckhage, K. (2001). Simulation of molten metal droplet sprays. *Journal of Computational Fluid Dynamics, 9*, 203–211.
24. Fritsching, U. (2004). *Spray simulation: Modeling and numerical simulation of sprayforming metals*. Cambridge, NY: Cambridge University Press.
25. Heck, U., Fritsching, U., & Bauckhage, K. (2000). Gas-flow effects on twin-fluid atomization of liquid metals. *Atomization and Sprays, 10*(1), 25–46.
26. Wille, R., & Fernholz, H. (1965). Report on the first European mechanics colloquium on the Coanda effect. *Journal of Fluid Mechanics, 23*, 801–819.
27. Farago, Z., & Chigier, N. (1992). Morphological classification of disintegration of round liquid jets in a coaxial air stream. *Atomization and Sprays, 2*, 137–153.
28. Hopfinger, E. J. (1998). Liquid jet instability and the breakup process in liquid-liquid agitated atomization in a coaxial gas stream. *Journal of Thermal Engineering Japan, 16*(4), 313–319.
29. Adzic, M., Carvalho, I. S., & Heitor, M. V. (2001). Visualization of the disintegration of an annular liquid sheet in a coaxial air blast injector at low atomizing air velocities. *Optical Diagnostics in Engineering, 5*, 27–38.
30. Choi, C. J., Lee, S. Y., & Song, S. H. (1997). Disintegration of annular liquid sheet with core air flow—mode classification. *International Journal of Fluid Mechanics Research, 24*(1–3), 399–406.
31. Lavergne, G., Trichet, P., Hebrard, P., & Biscos, Y. (1993). Liquid sheet disintegration and atomization process on a simplified airblast atomizer. *Journal of Engineering for Gas Turbines and Power, 115*(3), 461–466.
32. Sivakumar, D., & Kulkarni, V. (2011). Regimes of spray formation in gas-centered swirl coaxial atomizers. *Experiments in Fluids, 51*(3), 587–596.
33. Achelis, L., Sulatycki, K., Uhlenwinkel, V., Mädler, L. (2010). Lamellenzerfall von Metallschmelzen im Düsenbereich eines Druck-Gas-Zerstäubers zur Erzeugung von Kompositpartikeln. *Paper presented at Spray 2010* Heidelberg, Germany, 3–5 May 2010.
34. Uhlenwinkel, V., Achelis, L., Sulatycki, K., Mädler, L. (2010). New approach to generate composite particles. *Paper presented at PMTEC 2010*, Fort Lauderdale, USA, 27–30 June 2010.

35. Zeoli, N., Tabbara, H., & Gu, S. (2012). Three-dimensional simulation of primary break-up in a close-coupled atomizer. *Applied Physics A: Materials Science & Processing, 108*(4), 783–792.
36. Desjardins, O., Moureau, V., & Pitsch, H. (2008). An accurate conservative level set/ghost fluid method for simulating turbulent atomization. *Journal of Computational Physics, 227*, 8395–8416.
37. Gorokhovski, M., & Herrmann, M. (2008). Modeling primary atomization. *Annual Review of Fluid Mechanics, 40*, 343–366.
38. Klein, M., Sadiki, A., Janicka, J. (2001). Influence of the inflow conditions on the direct numerical simulation of primary breakup of liquid jets. *Paper presented at ILASS-Europe 2001*, Zürich, Switzerland, 2–6 September 2001, 475–480.
39. Klein, M., Sadiki, A., Janicka, J. (2002). Untersuchung des Primärzerfalls eines Flüssigkeitsfilms: Vergleich direkte numerische simulation, experiment und lineare theorie. In: *Spray 2002, 7th Workshop über Techniken der Fluidzerstäubung und Untersuchungen von Sprühvorgängen*, TU-Bergakademie, Freiberg, 2002, 63–72.
40. Mayer, W. (1993). Zur koaxialen Flüssigkeitszerstäubung im Hinblick auf die Treibstoffaufbereitung in Raketentriebwerken. Dissertation, Universität Erlangen.
41. Menard, T., Tanguy, S., & Berlemont, A. (2007). Coupling level set/VOF/ghost fluid methods: Validation and application to 3D simulation of the primary break-up of a liquid jet. *International Journal of Multiphase Flow, 33*(5), 510–524.
42. Scardovelli, R., & Zaleski, S. (1999). Direct Numerical Simulation of Free-surface and Interfacial Flow. *Annual Review of Fluid Mechanics, 31*, 567–603.
43. Shinjo, J., & Umemura, A. (2010). Simulation of liquid jet primary breakup: Dynamics of ligament and droplet formation. *International Journal of Multiphase Flow, 36*, 513–532.
44. Zaleski, S., Li, J.. (1997). Direct simulation of spray formation. *Paper presented at 7th international conference on liquid atomization and spray systems ICLASS 1997*, Seoul, Korea, 18–22 August 1997.
45. Zaleski, S., Boeck, T. (2003). Direct numerical simulation of high speed jet atimization. *Paper presented at 9th international conference on liquid atomization spraying systems ICLASS 2003*, Sorrento, Italy, 13–17 July 2003.
46. Zhu, C. (2014). *Numerical Investigation on the Instability and the Primary Breakup of Inelastic Non-Newtonian Liquid Jets*. München: Dissertation, Universität Stuttgart, Verlag Dr Hut.
47. Ashgriz, N. (2011). Numerical techniques for simulating the atomization process. In N. Ashgriz (Ed.), *Handbook of atomization and sprays*. New York, NY: Springer.
48. Fritsching, U., & Li, X. G. (2015). Spray systems. In E. E. Michaelides, J. D. Schwarzkopf, & C. Crowe (Eds.), *Multiphase flow handbook*. Boca Raton, FL: CRC Press.
49. Brackbill, J. U., Kothe, D. B., & Zemach, C. (1992). A continuum method for modelling surface tension. *Journal of Computational Physics, 100*, 335–354.
50. Gueyffier, D., Li, J., Nadim, A., Scardovelli, R., & Zaleski, S. (1999). Volume-of-fluid interface tracking with smoothed surface stress methods for three dimensional flows. *Journal of Computational Physics, 152*, 423–456.
51. Lafaurie, B., Nardonne, C., Scardovelli, R., Zaleski, S., & Zanetti, G. (1994). Modeling merging and fragmentation in multiphase flows with SURFER. *Journal of Computational Physics, 113*, 134–147.
52. Hong, J. M., Shinar, T., Kang, M., & Fedkiw, R. (2007). On boundary condition capturing for multiphase interfaces. *Journal of Scientific Computing, 31*, 99–125.
53. Kang, M., Fedkiw, R., & Liu, X. D. (2000). A boundary condition capturing method for multiphase incompressible flow. *Journal of Scientific Computing, 15*, 323–360.
54. Tanguy, S., & Berlemont, A. (2005). Application of a level set method for simulation of droplet collisions. *International Journal of Multiphase Flow, 31*, 1015–1035.
55. Hirt, C. W., & Nichols, B. D. (1981). Volume of fluid (VOF) method for the dynamics of free boundaries. *Journal of Computational Physics, 39*, 201–225.

56. Osher, S., & Sethian, J. A. (1988). Fronts propagating with curvature-dependent speed: algorithms based on Hamilton-Jacobi formulations. *Journal of Computational Physics, 79*, 12–49.
57. Osher, S., & Fedkiw, R. (2002). Level set methods and dynamic implicit surfaces. In S. S. Antman, J. E. Marsden, & L. Sirovich (Eds.), *Appl Math Sci* (Vol. Vol 153). New York, NY: Springer.
58. Gopala, V. R., & van Wachem, B. G. M. (2008). Volume of fluid methods for immiscible-fluid and free-surface flows. *Chemical Engineering Journal, 141*, 204–221.
59. van Wachem, B. G. M., & Schouten, J. C. (2002). Experimental validation of 3-d Lagrangian VOF model: Bubble shape and rise velocity. *AIChE Journal, 48*(12), 2744–2753.
60. Zaleski, S., Li, J., Succi, S., Scardovelli, R., Zanetti, G.. (1995). Direct numerical simulation of flows with interfaces. *Paper presented at 2nd international conference on multiphase flow*, Kyoto, Japan, April 1995.
61. Wardle, K. E., & Weller, H. G. (2013). Hybrid multiphase CFD solver for coupled dispersed/segregated flows in liquid-liquid extraction. *International Journal of Chemical Engineering, 2013*, 1–13.
62. Herrmann, M. (2008). A balanced force refined level set grid method for two-phase flows on unstructured flow solver grids. *Journal of Computational Physics, 227*(4), 2674–2706.
63. Van der Pijl, S. P., Segal, A., Vuik, C., & Wesseling, P. (2005). A mass-conserving level-set method for modeling of multiphase flows. *International Journal for Numerical Methods in Fluids, 47*, 339–361.
64. Lebas, R., Menard, T., Beau, P. A., Berlemont, A., & Demoulin, F. X. (2009). Numerical simulation of primary break-up and atomization: DNS and modelling study. *International Journal of Multiphase Flow, 35*, 247–260.
65. Li, X. G. (2014). *Modeling and simulation of the gas-atomization process of metal melts for metal-matrix-composite production*. Aachen: Dissertation, Universität Bremen, Shaker-Verlag.
66. Li, X.G., & Fritsching, U. (2017). Process Modeling Pressure-Swirl-Gas-Atomization for Metal Powder Production. *Journal of Materials Processing Technology, 239*, 1–17.
67. Li, X. G., & Fritsching, U. (2014). Numerical investigation of solid particle penetration into liquid droplet. *Materialwissenschaft und Werkstofftechnik, 45*(8), 666–682.
68. Park, J., Huh, K. Y., Li, X. G., & Renksizbulut, M. (2004). Experimental investigation on cellular breakup of a planar liquid sheet from an air-blast nozzle. *Physics of Fluids, 16*(3), 625–632.
69. Wahono, S., Honnery, D., Soria, J., & Ghojel, J. (2008). High-speed visualisation of primary break-up of an annular liquid sheet. *Experiments in Fluids, 44*(3), 451–459.
70. Grosshans, H., Szasz, R. Z., & Fuchs, L. (2014). Development of an efficient statistical volumes of fluid-Lagrangian particle tracking coupling method. *International Journal for Numerical Methods in Fluids, 74*, 898–918.
71. Herrmann, M. (2010). A parallel Eulerian interface tracking/Lagrangian point particle multi-scale coupling procedure. *Journal of Computational Physics, 229*, 745–759.
72. Tomar, G., Fuster, D., Zaleski, S., & Popinet, S. (2010). Multiscale simulations of primary atomization. *Computers and Fluids, 39*, 1864–1874
73. Bergmann, D., Fritsching, U., & Bauckhage, K. (2000). A mathematical model for cooling and rapid solidification of molten metal droplets. *International Journal of Thermal Sciences, 39*, 53–62.
74. Chryssakis, C., & Assanis, D. N. (2008). A unified fuel spray breakup model for internal combustion engine applications. *Atomization and Sprays, 18*, 1–52.
75. Faeth, G. M., Hsiang, L. P., & Wu, P. K. (1995). Structure and breakup properties of spray. *International Journal of Multiphase Flow, 21*, 99–127.
76. Guildenbecher, D. R., Lopez-Rivera, C., & Sojka, P. E. (2009). Secondary atomization. *Experiments in Fluids, 46*(3), 371–402.

77. Pilch, M., & Erdmann, C. A. (1987). Use of breakup time data and velocity history data to predict the maximum size of stable fragments for acceleration-induced breakup of a liquid drop. *International Journal of Multiphase Flow, 13*, 741–757.
78. Cao, X. K., Sun, Z. G., Li, W. F., Liu, H. F., & Yu, Z. H. (2007). A new breakup regime for liquid drops identified in a continuous and uniform air jet flow. *Physics of Fluids, 19*(5), 057103.
79. Zhao, H., Liu, H. F., Li, W. F., & Xu, J. L. (2010). Morphological classification of low viscosity drop bag breakup in a continuous air jet stream. *Physics of Fluids, 22*(11), 114103.
80. Hsiang, L. P., & Faeth, G. M. (1992). Near-limit drop deformation and secondary breakup. *International Journal of Multiphase Flow, 18*(5), 635–652.
81. Hsiang, L. P., & Faeth, G. M. (1993). Drop Properties after secondary breakup. *International Journal of Multiphase Flow, 19*(5), 721–735.
82. Markus, S., Fritsching, U. (2003). Spray forming with multiple atomization. *Paper presented at 2nd international conference on spray deposition and melt atomization SDMA 2003*, Bremen, Germany, 22–25 June 2003.
83. O'Rourke, P.J., Amsden, A.A. (1987). The TAB method for numerical calculation of spray droplet breakup. Los Alamos National Laboratory Report LA-UR-87-2105.
84. Tanner, F.X. (1997). Liquid jet atomization and droplet breakup modeling of non-evaporating diesel fuel sprays. SAE paper 970050.
85. Tanner, F. X. (2004). Development and validation of a cascade atomization and drop breakup model for high-velocity dense sprays. *Atomization and Sprays, 14*(3), 211–242.
86. Reitz, R.D., Diwarkar, R. (1987). Structure of high pressure fuel sprays. SAE paper 870598.
87. Reitz, R. D. (1987). Modeling atomization processes in high-pressure vaporizing sprays. *Atomisation and Spray Technology, 3*, 309–337.
88. Patterson, M.A., Reitz, R.D. (1998). Modeling the effects of fuel spray characteristics on diesel engine combustion and emissions. SAE paper 980131.
89. Beale, J. C., & Reitz, R. D. (1999). Modeling spray atomization with the Kelvin-Helmholtz/ Rayleigh-Taylor hybrid model. *Atomization and Sprays, 9*, 623–650.
90. Schmehl, R. (2004). *Tropfendeformation und Nachzerfall bei der technischen Gemischaufbereitung*. Germany: Dissertation, Universität Karlsruhe.
91. Bartz, F.O., Schmehl, R., Koch, R., Bauer, H.J. (2010). An extension of dynamic droplet deformation models to secondary atomization. *Paper presented at ILASS-Europe 2010*, Brno, Czech, 6–8 September 2010.
92. Schmehl, R. (2002). Advanced modeling of droplet deformation and breakup for CFD analysis of mixture preparation. *Paper presented at ILASS-Europe 2002*, Zaragoza, Spain, 9–11 September 2002.
93. Chou, W. H., Hsiang, L. P., & Faeth, G. M. (1997). Temporal properties of drop breakup in the shear breakup regime. *International Journal of Multiphase Flow, 23*(4), 651–669.
94. Chou, W. H., & Faeth, G. M. (1998). Temporal properties of drop breakup in the bag breakup regime. *International Journal of Multiphase Flow, 24*(6), 889–912.
95. Dai, Z., & Faeth, G. M. (2001). Temporal properties of secondary drop breakup in the multimode breakup regime. *International Journal of Multiphase Flow, 27*(2), 217–236.
96. Bartz, F.O., Guildenbecher, D.R., Schmehl, R., Koch, R., Bauer, H.J., Sojka, P.E. (2011). Model comparison for single droplet fragmentation under varying accelerations. *Paper presented at ILASS-Europe 2011*, Estoril, Portugal, 5–7 September 2011.
97. Schmehl, R. (2000). CFD analysis of fuel atomization, secondary droplet breakup and spray dispersion in the premix duct of a LPP combustor. *Paper presented at 8th international conference on liquid atomization and spray systems ICLASS 2000*, Pasadena, CA, 16–20 July 2000.
98. Sellens, R. W. (1989). Prediction of the drop size and velocity distribution in a spray based on the maximum entropy formalism. *Particle and Particle Systems Characterization, 6*, 17–23.
99. Sellens, R. W., & Brzustowski, T. A. (1985). A prediction of the drop size distribution in a spray from first principles. *Atomization and Spray Technology, 1*, 89–102.

100. Li, X., & Tankin, R. S. (1987). Droplet size distribution: A derivation of Nukyama-Tanasawa type distribution function. *Combustion Science and Technology, 56*, 65–76.
101. Cousin, J., & Dumouchel, C. (1996). Effect of viscosity on the linear instability of a liquid sheet. *Atomization and Sprays, 6*, 563–576.
102. Kim, W. T., Mitra, S. K., Li, X., Prociw, L. A., & Hu, T. C. J. (2003). A predictive model for the initial droplet size and velocity distributions in sprays and comparison with experiments. *Particle and Particle Systems Characterization, 20*, 135–149.
103. Li, X., Li, M., & Fu, H. (2005). Modeling the initial droplet size distribution in sprays based on the maximization of entropy generation. *Atomization and Sprays, 15*, 295–321.
104. Dumouchel, C. (2006). A new formulation of the maximum entropy formalism to model liquid spray dropsize distribution. *Particle and Particle Systems Characterization, 23*, 468–479.
105. Hosseinalipour, S. M., & Karimaei, H. (2016). A new model based on coupling of MEP/CFD/ILIA for predition of primary atomization. *The Canadian Journal of Chemical Engineering, 94*, 792–799.
106. Platzer, E., Sommerfeld, M. (2003). Modelling of turbulent atomisation with an Euler/Euler approach including the drop size prediction. *Paper presented at 9th international conference on liquid atomization spray systems ICLASS 2003*, Sorrento, Italy, 13–17 July 2003.
107. Naue, G., & Bärwolf, G. (1992). *Transportprozesse in Fluiden*. Leipzig: Deutscher Verlag für Grundstoffindustrie.
108. Hartmann, D. (1993). Theoretische Untersuchungen zur Tropfenbildung bei Dispergierprozessen. Dissertation Martin-Luther-Universität Halle.
109. Lubanska, H. (1970). Correlation of spray ring data for gas atomization of liquid droplets. *Journal of Metals, 2*, 45–49.
110. Rao, K. P., & Mehrotra, S. P. (1980). Effect of process variables on atomization of metals and alloys. In H. H. Hausner, H. W. Antes, & G. D. Smith (Eds.), *Modern developments in powder metallurgy: Principles and processes* (pp. 113–130). Princeton, NJ: MPIF and APMI International.
111. Rai, G., Lavernia, E. J., & Grant, N. J. (1985). Factors influencing the powder size and distribution in ultrasonic gas atomization. *Journal of Metals, 37*(8), 22–26.
112. Bauckhage, K., & Fritsching, U. (2000). Production of metal powders by gas atomization. In K. P. Cooper, I. E. Anderson, S. D. Ridder, & F. S. Biancanello (Eds.), *Liquid metal atomization: Fundamentals and practice* (pp. 23–36). Warrendale, PA: TMS.
113. Liu, H. (2000). *Science and engineering of droplets: Fundamentals and applications*. Norwich, CT: William Andrew Publishing.
114. Yule, A. J., & Dunkley, J. J. (1994). *Atomization of melts*. Oxford: Clarendon Press.
115. Orme, M. (1997). Experiments on droplet collision, bounce coalescence and disruption. *Progress in Energy and Combustion Science, 23*, 65–79.
116. Ashgriz, N., & Poo, J. Y. (1990). Coalescence and separation in binary collision of liquid drops. *Journal of Fluid Mechanics, 221*, 183–204.
117. Qian, J., & Law, C. K. (1997). Regimes of coalescence and separation in droplet collision. *Journal of Fluid Mechanics, 331*, 59–80.
118. Menchaca-Rocha, A., Huidobro, F., Martinez-Davalos, A., Michaelian, K., Perez, A., Rodriguez, V., & Carjan, N. (1997). Coalescence and fragmentation of colliding mercury drops. *Journal of Fluid Mechanics, 346*, 291–318.
119. Kuschel, M., & Sommerfeld, M. (2013). Investigation of droplet collisions for solutions with different solids content. *Experiments in Fluids, 54*(1440), 1–17.
120. Kurt, O., Fritsching, U., Schulte, G. (2007). Binary collisions of droplets with fluid and suspension particles. *Paper presented at ILASS-Europe 2007*, Mugla, Turkey, 10–13 September 2007.
121. Kurt, O., Fritsching, U., Schulte, G. (2008). Secondary droplet formation during binary suspension droplet collisions. *Paper presented at ILASS-Europe 2008*, Como Lake, Italy, 8–10 September 2008.

122. Gao, S., & Fritsching, U. (2010). Study of binary in-flight melt droplet collisions. *Materialwissenschaft und Werkstofftechnik, 41*(7), 547–554.
123. Brenn, G., Valkovska, D., & Danov, K. D. (2001). The formation of satellite droplets by unstable binary drop collisions. *Physics of Fluids, 13*(9), 2463–2477.
124. Brenn, G., & Kolobaric, V. (2006). Satellite droplet formation by unstable binary drop collisions. *Physics of Fluids, 18*, 087101-1–087101-18.
125. Fritsching, U., & Gao, S. (2010). Droplet-particle collisions in intersecting melt sprays. *Atomization and Sprays, 20*(1), 31–40.
126. Nobari, M. R. H., & Tryggvason, G. (1996). Numerical simulations of three-dimensional drop collisions. *AIAA Journal, 34*, 750–755.
127. Nobari, M. R. H., Jan, Y. J., & Tryggvason, G. (1996). Head-on collision of drops—A numerical investigation. *Physics of Fluids, 8*, 29–42.
128. Frohn, A., & Roth, N. (2000). *Dynamics of droplets*. Berlin: Springer.
129. Pan, Y., & Suga, K. (2005). Numerical simulation of binary liquid droplet collision. *Physics of Fluids, 17*, 082105.
130. Nikolopoulos, N., Nikas, K. S., & Bergeles, G. (2009). A numerical investigation of central binary collision of droplets. *Computers and Fluids, 38*, 1191–1202.
131. Nikolopoulos, N., Theodorakakos, A., & Bergeles, G. (2009). Off-centre binary collision of droplets: A numerical investigation. *International Journal of Heat and Mass Transfer, 52*, 4160–4174.
132. Li, X.G., & Fritsching, U. (2011). Numerical investigation of binary droplet collisions in all relevant collision regimes. *J. Comput. Multiphase Flows, 3*(4), 207–224.
133. Focke, C., & Bothe, D. (2011). Computational analysis of binary collisions of shear thinning droplets. *Journal of Non-Newtonian Fluid Mechanics, 166*, 799–810.
134. Focke, C., & Bothe, D. (2012). Direct numerical simulation of binary off-center collisions of shear thinning droplets at high Weber numbers. *Physics of Fluids, 24*, 073105-1–073105-18.
135. Kwakkel, M., Breugem, W. P., & Boersma, B. J. (2013). Extension of a CLSVOF method for droplet-laden flows with a coalescence/breakup model. *Journal of Computational Physics, 253*, 166–188.
136. Mackay, G. D., & Mason, S. G. (1963). The gravity approach and coalescence of fluid drops at liquid interfaces. *Canadian Journal of Chemical Engineering, 41*, 203–212.
137. Bradley, S. G., & Stow, C. D. (1978). Collision between liquid drops. *Philosophical Transactions of the Royal Society of London, Series A, 287*, 635–675.
138. Zhang, P., & Law, C. K. (2011). An analysis of head-on droplet collision with large deformation in gaseous medium. *Physics of Fluids, 23*, 042102-1–042102-22.
139. Sommerfeld, M. (1996). *Modellierung und numerische Berechnung von partikelbeladenen turbulenten Strömungen mit Hilfe des Euler/Lagrange Verfahrens*. Aachen: Verlag Shaker.
140. Georjon, T. L., & Reitz, R. D. (1999). A drop-shattering collision model for multidimensional spray computations. *Atomization and Sprays, 9*, 231–254.
141. Aamir, M.A., Watkins, A.P. (1999). Dense propane spray analysis with a modified collision model. *Paper presented at ILASS-Europe 99*, Toulouse, France, 5–7 July 1999.
142. Rüger, M., Hohmann, S., Sommerfeld, M., & Kohnen, G. (2000). Euler/Lagrange calculations of turbulent spray: the effect of droplet collisions and coalescence. *Atomization and Sprays, 10*(1), 47–82.
143. O'Rourke PJ (1981). Collective drop effects on vaporizing liquid sprays. Dissertation, Los Alamos National Laboratory, New Mexico.
144. Amsden, A.A., O'Rourke, P.J., Butler, T.D. (1989). KIVA-II: A computer program for chemically reactive flows with sprays. Los Alamos National Laboratory Report LA-11560-MS.
145. Dubrovsky, V. V., Podvysotsky, A. M., & Shraiber, A. A. (1992). Particle interaction in three-phase polydispersed flows. *International Journal of Multiphase Flow, 18*(3), 337–352.
146. Crowe, C. T., Sharma, M. P., & Stock, D. E. (1977). The particle-source-in-cell method for gas droplet flow. *Journal of Fluids Engineering, 99*, 325–332.

147. Sommerfeld, M., & Zivkovic, G. (1992). Resent advances in the numerical simulation of pneumatic conveying through pipe systems. In H. Hirch et al. (Eds.), *Computational methods in applied science*. Brussels: First European Computational Fluid Dynamics.
148. Sommerfeld, M., Kohnen, G., Rüger, M. (1993). Some open question and inconsistencies of Lagrangian particle dispersion models. *Paper presented at the 9th symposium on turbulent shear flows*, Kyoto, Japan, 16–18 August 1993.
149. Sommerfeld, M. (1995). The importance of inter-particle collisions in horizontal gas-solid channel flows. In D. E. Stock et al. (Eds.), *ASME fluids engineering conference, FED-288* (pp. 333–345). Hilton Head, SC: ASME.
150. Osterle, B., & Petijean, A. (1993). Simulation of particle-to-particle interaction in gas-solid flows. *International Journal of Multiphase Flow, 9*(19), 199–211.
151. Schmidt, D. P., & Rutland, C. J. (2000). A new droplet collision algorithm. *Journal of Computational Physics, 164*, 62–80.
152. Schmidt, D. P., & Rutland, C. J. (2004). Reducing grid dependency in droplet collision modeling. *Journal of Engineering for Gas Turbines and Power, 126*, 227–233.
153. Bauman SD (2001). A spray model for an adaptive mesh refinement code. Dissertation, Madison University Wiconsin.
154. Sommerfeld, M. (2001). Validation of a stochastic Lagrangian modelling approach for inter-particle collisions in homogeneous isotropic turbulence. *International Journal of Multiphase Flow, 27*, 1829–1858.
155. Pischke, P., Cordes, D., & Kneer, R. (2012). A collision algorithm for anisotropic disperse flows based on ellipsoidal parcel representations. *International Journal of Multiphase Flow, 38*, 1–16.
156. Pischke, P., Cordes, D., & Kneer, R. (2012). The velocity decomposition method for second-order accuracy in stochastic parcel simulations. *International Journal of Multiphase Flow, 47*, 160–170.
157. Pischke, P. (2014). Modeling of collisional transport processes in spray dynamics. Dissertation, RWTH Aachen Univeristy.
158. Lampa, A., & Fritsching, U. (2013). Large eddy simulation of the spray formation in confinements. *International Journal of Heat and Fluid Flow, 43*, 26–34.
159. Lampa, A., & Fritsching, U. (2011). Spray Structure analysis in atomization processes in enclosures for powder production. *Atomization and Sprays, 21*(9), 737–752.
160. Bergmann, D., Fritsching, U., Crowe, C.T. (1995). Multiphase flows in the spray forming process. *Paper presented at 2nd international conference on multiphase flow*, Kyoto, Japan, 3–7 April 1995.
161. Marx, K. D., Edwards, C. F., & Chin, W. K. (1994). Limitations of the ideal Phase-Doppler system: Extension to spatially and temporally inhomogeneous particle flows. *Atomization and Sprays, 4*, 1–40.
162. Edwards, C. F., & Marx, K. D. (1995). Multipoint statistical structure of the ideal spray, Part I: Fundamental concepts and the realization density. *Atomization and Sprays, 5*, 435–455.
163. Edwards, C. F., & Marx, K. D. (1995). Multipoint statistical structure of the ideal spray, Part II: Evaluating steadiness using the interparticle time distribution. *Atomization and Sprays, 5*, 457–505.
164. Heinlein, J., & Fritsching, U. (2006). Droplet clustering in sprays. *Experiments in Fluids, 40*, 464–472.
165. Czainski, A. (1994). Quantitive characterization of inhomogeneity in thin metallic films using Garncarek's method. *Journal of Physics D: Applied Physics, 27*, 616–622.
166. Yang, R. Y., Zou, R. P., & Yu, A. B. (2002). Voronoi tessellation of the packing of fine uniform spheres. *Physical Review E, 65*, 041302.
167. Jedelsky, J., & Jicha, M. (2008). Unsteadiness in effervescent sprays: A new evaluation method and the influence of operational conditions. *Atomization and Sprays, 18*(1), 49–83.
168. Markus, S., & Fritsching, U. (2006). Discrete breakup modeling for melt sprays. *International Journal of Powder Metallurgy, 42*(4), 23–32.

169. Apte, S. V., Gorokhovski, M. A., & Moin, P. (2003). LES of atomizing spray with stochastic modeling of secondary breakup. *International Journal of Multiphase Flow, 29*, 1503–1522.
170. Bellan, J. (2000). Perspectives on large eddy simulations for sprays: Issues and solutions. *Atomization and Sprays, 10*, 409–425.
171. Irannejad, A., & Jaberi, F. (2014). Large eddy simulation of turbulent spray breakup and evaporation. *International Journal of Multiphase Flow, 61*, 108–128.
172. Jones, W. P., & Lettieri, C. (2010). Large eddy simulation of spray atomization with stochastic modeling of breakup. *Physics of Fluids, 22*, 115106-1–115106-12.
173. Lampa, A., Fritsching, U. (2014). Impact of droplet clustering on heat transfer in spray processes. *Proceedings ILASS 2014—26th European conference on liquid atomization and spray systems*, Bremen, Germany, 8–10 Sep. 2014.
174. Gorokhovski, M. A., & Saveliev, V. L. (2003). Analyses of Kolmogorovs model of breakup and its application into Lagrangian computation of liquid sprays under air-blast atomization. *Physics of Fluids, 15*, 184–192.
175. Lasheras, J. C., Eastwood, C., Martinez-Bazan, C., & Montanes, J. L. (2002). A review of statistical models for the breakup of an immiscible fluid immersed into a fully developed turbulent flow. *International Journal of Multiphase Flow, 28*, 247–278.
176. Wiegand, H. (1987). *Die Einwirkung eines ebenen Strömungsfelds auf frei bewegliche Tropfen und ihren Widerstandsbeiwert im Reynoldszahlenbereich von 50 bis 2000. Fortschrittberichte VDI 7(120)*. Düsseldorf: VDI-Verlag.
177. Liu, A.B., Mather, D., Reitz, R.D.. (1993) Modeling the effect of drop drag and breakup on fuel sprays. SAE Technical Paper 930072.
178. Fritsching, U., & Bauckhage, K. (1987). Die Bewegung von Tropfen im Sprühkegel einer Ein- und einer Zweistoffdüse. *Chemie Ingenieur Technik, 59*(9), 744–745.
179. Schulte, G. (1995). *Zweidimensionale Verteilunen von Partikeleigenschaften*. Aachen: Shaker Verlag.
180. Ranz, W. E., & Marshall, W. R. (1952). Evaporation from Drops-I and II. *Chemical Engineering Progress, 48*, 141–173.
181. Yearling, P. R., & Gould, R. D. (1995). Convective heat and mass transfer from single evaporating water, methanol and ethanol droplets. *ASME FED, 223*, 33–38.
182. Lavernia, E. J., & Wu, Y. (1996). *Spray Atomization and Deposition*. Chichester: J Wiley & Sons.
183. Lavernia, E. J. (1996). Spray atomization and deposition of metal matrix composites. In K. Bauckhage & V. Uhlenwinkel (Eds.), *Kolloquium des SFB 372, Sprühkompaktieren* (Vol. 1, pp. 63–122). Bremen: Universität Bremen.
184. Hirth, J. P. (1978). Nucleation, undercooling and homogeneous structures in rapidly solidified powders. *Metallurgical Transactions A, 9*(3), 401–404.
185. Libera, M., Olsen, G. B., & van der Sande, J. B. (1991). Heterogeneous nucleation of solidification in atomized liquid metal droplets. *Materials Science and Engineering A, 132*, 107–118.
186. Turnbull, D. (1950). Formation of crystal nuclei in liquid metals. *Journal of Applied Physics, 21*, 1022–1028.
187. Woodruff, D. P. (1973). *The solid-liquid interface*. Cambridge: Cambridge University Press.
188. Mathur, P., Apelian, D., & Lawley, A. (1989). Analysis of the spray deposition process. *Acta Metallurgica, 37*(2), 429–443.
189. Mathur, P., Annavarapu, S., Apelian, D., & Lawley, A. (1989). Process control, modeling and applications of spray casting. *Journal of Metals, 10*, 23–28.
190. Lee, E., & Ahn, S. (1994). Solidification progress and heat transfer analysis of gas atomized alloy droplets during spray forming. *Acta Metallurgica et Materialia, 42*(9), 3231–3243.
191. Bergmann D (2000). Modellierung des Sprühkompaktierprozesses für Kupfer- und Stahlwerkstoffe. Dissertation, Universität Bremen.
192. Pedersen, T.P., Hattel, J.H., Pryds, N.H., Pedersen, A.S., Buchholz, M., Uhlenwinkel, V. (2000). A new integrated numerical model for spray atomization and deposition: Comparison between numerical results and experiments. *Paper presented at SDMA 2000*, Bremen, Germany, 26–28 June 2000 (pp. 813–824).

193. Gjesing, R., Hattel, J., & Fritsching, U. (2009). Coupled atomization and spray modelling in the spray forming process using open foam. *Engineering Applications of Computational Fluid Mechanics, 3*(4), 471–486.
194. Fritsching, U. (1995). Modelling the spray cone behaviour in the metal spray forming process: Momentum and thermal coupling in two-phase flow. *PHOENICS Journal of Computational Fluid Dynamics, 8*(1), 68–90.
195. Grant, P. S., Cantor, B., & Katgerman, L. (1993). Modelling of droplet dynamic and thermal histories during spray forming-I: Individual droplet behaviour. *Acta Metallurgica et Materialia, 41*(11), 3097–3108.
196. Grant, P. S., Cantor, B., & Katgerman, L. (1993). Modelling of droplet dynamic and thermal histories during spray forming-II: Effect of process parameters. *Acta Metallurgica et Materialia, 41*(11), 3109–3118.
197. Pedersen, T.B. (2003). Spray forming-a new integrated numerical model. Dissertation, Technical University of Denmark.
198. Ziesenis, J. (2003). Weiterentwicklung der PDA-Meßtechnik zur on-line Prozeßkontrolle beim Sprühkompaktieren. Dissertation, Universität Bremen.
199. Bauckhage, K. (1998). Use of the phase-doppler-anemometry for the analysis and the control of the spray forming process. *Paper presented at PM^{2}TEC'98*, Las Vegas, USA, 31 May–4 June, 1998.
200. Delshad Khatibi, P., Ilbagi, A., Beinker, D., & Henein, H. (2011). In-situ characterization of droplets during free fall in the drop tube-impulse system. *Journal of Physics: Conference Series, 327*, 012014.
201. Delshad Khatibi, P., & Henein, H. (2014). The robustness of the two-colour assumption in pyrometry of solidifying AISI D2 alloy droplets. *Materialwissenschaft und Werkstofftechnik, 45*(8), 736–743.
202. Krauss, M., Bergmann, D., & Fritsching, U. (2002). In-situ particle temperature, velocity and size measurements in the spray forming process. *Materials Science and Engineering A, 326* (1), 154–164.
203. Bergmann, D., Fritsching, U., Bauckhage, K. (1999). Averaging thermal conditions in molten metal sprays. *Paper presented at TMS-Annual Meeting, EPD Congress*, San Diego, USA, February 28–March 4.
204. Uhlenwinkel, V. (1992). Zum Ausbreitungsverhalten der Partikeln bei der Sprühkompaktierung von Metallen. Dissertation, Universität Bremen.
205. Kramer, C. (1997). Die Kompaktierungsrate beim Sprühkompaktieren von Gauß-förmigen Deposits. Dissertation, Universität Bremen.
206. Achelis L (2009). Drall-Druck-Gas-Zerstäubung von Metallschmelzen, Dissertation, Universität Bremen, Shaker-Verlag, Aachen.
207. Mulhem, B., Khoja, G., Fritsching, U., Schulte, G. (2006). Break-up of hollow cone and flat suspension lamellae of pressure atomizers. *Paper presented at ICLASS 2006*, Kyoto, Japan, 27 August–1 September 2006.
208. Musemic, E., Gaspar, M., Weichert, F., Müller, H., Walzel, P. (2010). Experimental examination of the liquid sheet disintegration process using combined photography and fibre based measuring techniques. *Paper presented at ILASS-Europe 2010*, Brno, Czech Republic, 6—8 September 2010.
209. Ballester, J., & Dopazo, C. (1994). Discharge coefficient and spray angle measurements for small pressure-swirl nozzles. *Atomization and Sprays, 4*(3), 351–367.
210. Kamplade, J., Musemic, E., Walzel, P. (2013). Investigation on pressure swirl nozzles with Coanda deflection outlets. *Paper presented at ILASS-Europe 2013*, Chania, Greece, 1—4 September 2013.
211. Reitz, R.D., Bracco, F.V. (1979). On the dependence of spray angle and other spray parameters on nozzle design and operating conditions. SAE paper 790494.
212. Bracco, F.V., Chehroudi, B., Chen, S.H., Onuma, Y. (1985). On the intact core of full cone sprays. SAE Trans 94 paper *850126*.

213. Lefebvre, A. H. (1989). *Atomization and sprays*. New York, NY: Hemisphere.
214. Lampe, K. (1994). Experimentelle Untersuchung und Modellierung der Mehrphasenströmung im düsennahen Bereich einer Öl-Brenner-Düse. Dissertation, Universität Bremen
215. Conelly, S., Coombs, J. S., & Medwell, J. O. (1986). Flow characteristics of metal particles in atomised sprays. *Metal Powder Report, 41*, 9.
216. Lampa, A., Sander, S., Schwenck, D., Fritsching, U. (2016). Recirculation, entrainment and cluster formation in bounded sprays. *ICMF-2016—9th international conference on multiphase flow*, May 22nd–27th 2016, Firenze, Italy.
217. Harvie, D. J. E., Langrish, T. A. G., & Fletcher, D. F. (2002). A computational fluid dynamics study of a tall-form spray dryer. *Food and Bioproducts Processing, 80*(3), 163–175.
218. Dunkley, J.J., Telford, B. (2002). Control of satellite particles in gas atomization. *Paper presented at World Congress on Powder Metallurgy and Particulate Materials PM2TEC 2002*, Orlando, FL, USA, 16–21 June 2002.
219. Achelis, L., & Uhlenwinkel, V. (2008). Characterisation of metal powders generated by a pressure-gas-atomizer. *Materials Science and Engineering A, 477*, 15–20.
220. Eslamian, M., Rak, J., & Ashgriz, N. (2008). Preparation of aluminum/silicon carbide metal matrix composites using centrifugal atomization. *Powder Technology, 184*, 11–20.
221. Li, B., & Lavernia, E. J. (2000). Particulate penetration into solid droplets. *Metallurgical and Materials Transactions A: Physical Metallurgy and Materials Science, 31*(2), 387–396.
222. Wu, Y., & Lavernia, E. J. (1992). Interaction mechanisms between ceramic particles and atomized metallic droplets. *Metallurgical Transactions A, 23*(10), 2923–2937.
223. Zhang, J., Wu, Y., & Lavernia, E. J. (1994). Kinetics of ceramic particulate penetration into spray atomized metallic droplet at variable penetration depth. *Acta Metallurgica et Materialia, 42*(9), 2955–2971.
224. Hoeven, M.J. (2008). Particle-droplet collisions in spray drying. Dissertation, University of Queensland.
225. Majagi, S. I., Ranganathan, K., Lawley, A., & Apelian, D. (1992). Spray forming of metal matrix composites. In E. J. Lavernia & M. Gungor (Eds.), *Microstructural design by solidification processing* (pp. 139–149). Warrendale: The Minerals Metals & Materials Society.
226. Wu, Y., Zhang, J., & Lavernia, E. J. (1994). Modeling of the incorporation of ceramic particulates in metallic droplets during spray atomization and co-injection. *Metallurgical and Materials Transactions B: Process Metallurgy and Materials Processing Science, 25*(1), 135–147.
227. Fritsching, U., & Lampa, A. (2015). Droplet clustering in spray processes. *ICLASS 2015, 13th triennial international conference on liquid atomization and spray systems*, Tainan, Taiwan, August 23–27, 2015.
228. Giffen, E., & Muraszew, A. (1953). *The Atomization of Liquid Fuels*. New York: John Wiley and Sons.
229. Rizk, N.K., & Lefebvre, A.H. (1985). Internal flow characteristics of simplex swirl atomizers. *J. Propul. Power, 1*(3), 193–199.
230. Rivas, J.R.R., Pimenta, A.P., & Rivas, G.A.R. (2014). Development of a mathematical model and 3D numerical simulation of the internal flow in a conical swirl atomizer. *Atomization and Sprays, 24*(2), 97–114.
231. Podvysotsky, A.M., & Shraiber, A.A. (1994). Coalescence and breakup of drops in two-phase flows. *International Journal of Multiphase Flow,* 10, 195–209.
232. Nourgaliev, R.R., & Theofanous, T.G. (2007). High-fidelity interface tracking in compressible flows: Unlimited anchored adaptive level set. *Journal of Computational Physics, 224*(2), 836–866.

Chapter 5
Spray Impingement Fundamentals

Sanjeev Chandra and Javad Mostaghimi

5.1 Introduction

Chapter 5 will review the dynamics of both single droplets and sprays of molten metals landing on solid surfaces. Droplets landing on a solid substrate flatten out and solidify; the splats may either be disc shaped or fragmented, depending on impact conditions. Multiple droplets impacting on a surface fuse with each other to form a solid layer. The dynamics of single droplet impact and solidification are discussed. The thermal contact resistance between the droplet and substrate on the solidification rate is important in determining the solidification rate and the splat shape. An overview is given of numerical models to simulate droplet impact and solidification. The impact and coalescence of multiple droplets in a spray to form a solid layer is described. Monte Carlo and smoothed particle hydrodynamics methods can be used to simulate droplet impact and coalescence in a spray and predict properties such as coating thickness and porosity.

A fundamental step in all metal spray forming and coating processes is the impact of molten droplets on a solid surface. In an ideal spray deposition process impacting droplets hit a solid surface, flatten out and solidify, forming cylindrical disks known as splats. Subsequent molten droplets land on those already frozen on the surface, raising the temperature at the interfaces between droplets sufficiently to cause remelting, so that they coalesce to form a dense, uniform deposit that is free of pores. The droplets lose heat very rapidly to the substrate while they are spreading out since there is a very large area, relative to the droplet volume, for heat transfer. This produces very high cooling rates, producing a fine-grained microstructure in the solidified material.

S. Chandra (✉) • J. Mostaghimi
Department of Mechanical and Industrial Engineering, University of Toronto, Toronto, ON, Canada
e-mail: chandra@mie.utoronto.ca; mostag@mie.utoronto.ca

H. Henein et al. (eds.), *Metal Sprays and Spray Deposition*,
DOI 10.1007/978-3-319-52689-8_5

An actual spray deposition process may be quite different from this idealized scenario, depending on spray parameters such as the size, velocity and temperature of droplets, and the thermophysical properties of the droplets and substrate. Impacting droplets may not form circular disks but break-up instead, leaving small fragments on the surface. Droplets landing on these particles may not completely fill voids around them, leaving pores in the solidified layer and reducing its strength.

The mechanism that leads to splashing and fragmentation of impacting droplets is therefore an important concern in spray processes, and an understanding of the parameters that control it can help avoiding conditions that lead to porosity in the spray deposit. When a droplet lands on a substrate, fluid instabilities perturb the edges of the spreading liquid sheet. Surface roughness, or any debris on the surface amplifies these instabilities. If droplet freezing starts before spreading is complete solid protuberances in the flowing liquid act as obstructions and cause droplet splashing. The rate of heat transfer from a droplet to the substrate under it is a very important factor in determining how rapidly it solidifies, and depends on the thermal properties of the substrate, its temperature, and the condition of the droplet-substrate interface. Contaminants or trapped air bubbles can significantly reduce the rate of heat transfer and therefore affect the splat shape and microstructure.

Many experimental studies have been devoted to studying the impact dynamics of molten metal droplets and sprays landing on a surface. Direct observations can be difficult because of the short length (tens of microns) and time scales (a few milliseconds), and high temperatures (several hundred degrees Kelvin) involved. Numerical models offer another way of investigating droplet dynamics, but simultaneously solving mass, momentum and energy conservation equations while simulating rapid deformation of free liquid surfaces, and incorporating realistic boundary conditions presents its own challenges.

When a droplet collides with a surface, there are three phases involved: liquid (the droplet), solid (the substrate) and gas (the surrounding atmosphere). A droplet is described by two impact parameters, diameter (D_o) and impact velocity (V_o), and three physical properties: liquid density (ρ), viscosity (μ), and liquid-gas surface tension (σ). Combining these into non-dimensional groupings we obtain the Reynolds number ($Re = \rho V_o D_o/\mu$) and Weber number ($We = \rho V_o^2 D_o/\sigma$). The Weber number is a ratio of the inertial forces, which drive splashing, to surface forces that hold the droplet intact. Similarly, the Reynolds number is a ratio of the droplet inertia to viscous forces that damp out motion. Droplets are more likely to break-up during impact when Re and We are large.

If the droplet is freezing as it impacts, the rate of solidification depends on the phase change properties of the droplet, which are the melting temperature (T_m) and latent heat of fusion (H_f), and the rate of heat transfer from the droplet to the surface which depend on the specific heat (c) and thermal conductivity (k) of the droplet and the substrate temperature (T_s). Combining these properties into dimensionless groups gives the Stefan number ($Ste = c(T_m - T_s)/H_f$) and Peclet number ($Pe = \rho c V_o D_o/k$).

When molten metal comes suddenly in contact with a rough, solid surface, air is trapped in crevices at the liquid solid interface, creating a temperature difference between the molten metal and the substrate, whose value depends on surface finish, contact pressure and material properties. To quantify the magnitude of this effect the thermal contact resistance (R_c) is defined as the temperature difference between the droplet (T_d) and substrate (T_w) divided by the heat flux (q'') between the two.

$$R_c = \frac{T_d - T_w}{q''} \tag{5.1}$$

The thermal contact resistance can be non-dimensionalized by defining a Biot number ($Bi = D_o/R_c k$).

At a minimum the impact dynamics of a molten metal droplet impacting and freezing on a solid surface are characterized by these dimensionless numbers: *Re*, *We*, *Ste*, *Pe* and *Bi*. In addition the properties of the substrate may also affect droplet impact. Both the average substrate roughness (R_a) and the wettability of the surface, as quantified by the liquid-solid contact angle (θ), can also influence droplet impact.

5.2 Impact of Molten Metal Droplets on Surfaces

5.2.1 Photographing Droplet Impact

Several studies have been carried out in which millimeter sized droplets were dropped onto solid substrates from heights ranging from a few centimeters to several meters and the collision photographed. Fukanuma and Ohmori [1] photographed the impact of tin and zinc droplets on a solid plate and found under for the conditions of their experiment the time for solidification was much longer than that for droplet deformation so that freezing had no influence on the extent of droplet spread. Inada and Yang [2] observed droplet-substrate contact during impact of lead droplets on a quartz plate using holographic interferometry. Pasandideh-Fard et al. [3] photographed the impact of tin droplets on a stainless steel substrate and used a thermocouple to measure substrate temperature variations during impact.

Aziz and Chandra [4] photographed the impact of 2.7 mm diameter molten tin droplets with onto a heated stainless steel plate. The height from which droplets were released was adjusted to vary droplet impact from 1 to 4 m/s and tests were done in an inert atmosphere to minimize oxidation. A single-shot photographic technique was used to capture droplet impact. Some of their images are seen in Fig. 5.1 which shows molten tin droplets landing on a 25 °C stainless steel substrate with impact velocities of 1 m/s (Fig. 5.1a), 2 m/s (Fig. 5.1b), and 4 m/s (Fig. 5.1c). Each row in Fig. 5.1 represents the same dimensionless time ($t^* = tV_o/D_o$) while the real time (t) from the instant of impact is also given next to each frame. At a low

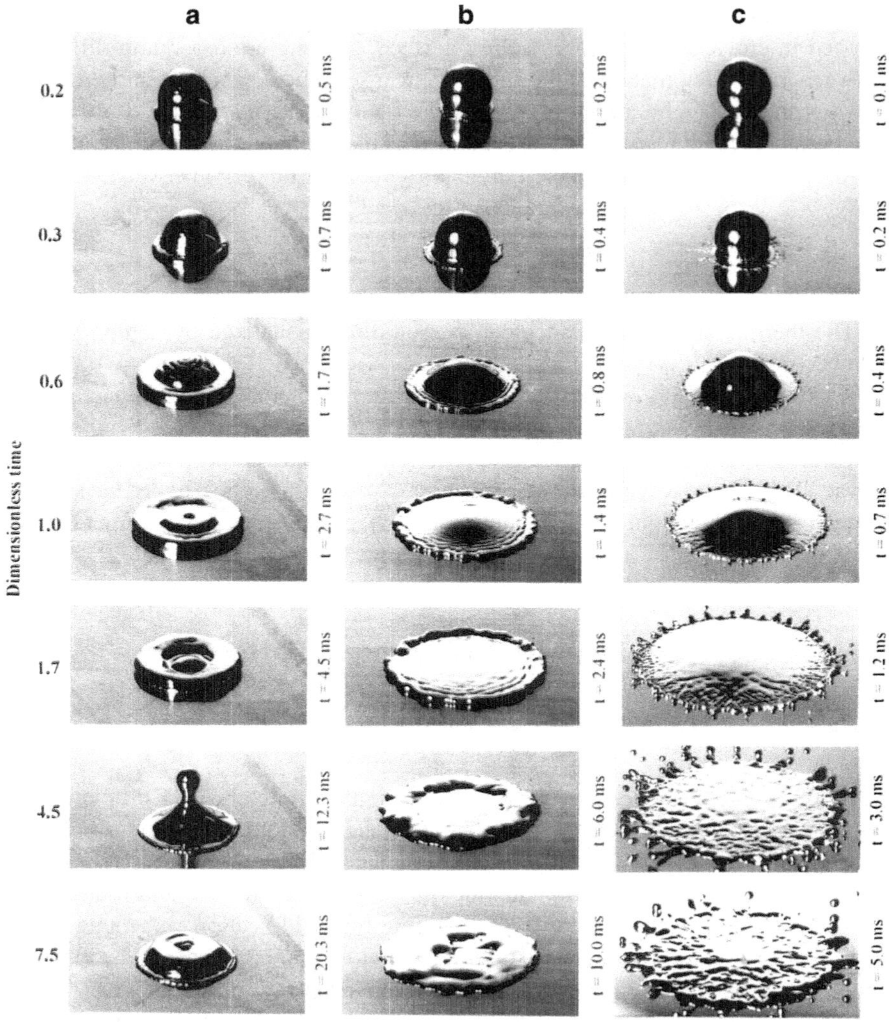

Fig. 5.1 Impact of molten tin droplets on a stainless steel surface at temperature 25 °C with velocity (**a**) 1 m/s, (**b**) 2 m/s and (**c**) 4 m/s [4]

impact velocity $V_o = 1$ m/s (Fig. 5.1a), the droplet reached its maximum spread a little after $t^* = 1.0$. The molten layer was pulled back by surface tension, and recoiled above the surface ($t^* = 4.5$). Increasing the impact velocity to 2 m/s (Fig. 5.1b) increased the splat diameter and reduced the splat thickness. The recoil of the droplet was also greatly diminished, so that there was only a small flow of liquid back from the edges of the splat towards its center. At the highest velocity, 4 m/s (Fig. 5.1c), fingers were seen forming around the edges of the impacting drop very early during impact ($t^* = 0.6$) which grew larger until their tips detached to form small satellite droplets ($t^* = 4.5$). The growth of the fingers was stopped by the droplet solidifying so that the final splat shape was reached by approximately $t^* = 4.5$, with little change after that time. On a surface that was maintained at 240 °C, above the melting point of tin (232 °C), there was extensive splashing in droplets impacting at a velocity of 4 m/s, so that they shattered upon impact [4].

Studies of large molten metal droplet landing at low velocity give insight into the dynamics of spreading; however, they do not seem adequately simulate the splashing of plasma particles. The impact Weber and Reynolds number of such droplets is much lower than those in typical spray forming applications ($We \sim 10^2$ and $Re \sim 10^3$ in experiments, compared to Re and $We \sim 10^3$–10^4 in applications). For low We and Re droplet solidification suppresses splashing, since the impacting liquid does not have enough momentum to jet over the solidified layer near the edges of droplets and splash.

Droplet size also affects heat transfer, since it alters the relative resistance to heat conduction of the droplet itself, relative to the thermal contact resistance between the particle and substrate. The ratio between the two is given by the Biot number ($Bi = D_o/(R_c k)$). For the 2.7 mm diameter tin droplets of Fig. 5.1 $Bi \sim 10^2$ assuming $R_c = 10^{-6}$ m^{2}K/W, and thermal contact resistance can be neglected; for a spray particle with diameter two orders of magnitude smaller, $Bi \sim 1$, and contact resistance controls heat transfer from the particle to the substrate.

To observe droplet impact at high We Mehdizadeh et al. [5] built an apparatus in which molten tin droplets impinged on a steel plate mounted on the rim of a rotating flywheel, giving impact velocities of up to 40 m/s and $We \sim 10^3$. Dhiman and Chandra [6] used the same apparatus to photograph impact of tin droplets on solid plates for a range of impact velocities (10–30 m/s), substrate temperature (25–200 °C) and substrate materials (stainless steel, aluminium and glass). Droplet Reynolds number ranged from 2.2×10^4 to 6.5×10^4 and Weber number from 8.0×10^2 to 7.2×10^3. Figure 5.2 shows images of 0.6 mm diameter tin droplets impacting on a mirror-polished stainless steel substrate with 20 m/s velocity. Each column shows successive stages of droplet impact on a substrate at initial temperature (T_w) varying from 25 to 200 °C (indicated at the top of the column). The first picture in each sequence shows a droplet prior to impact, and the last shows the final splat shape. Droplets hitting a cold substrate ($T_w = 25$–150 °C) splashed extensively, producing small satellite droplets and leaving a splat with irregular edges. The final splat surface was rough along the periphery, showing the region where it first solidified very rapidly; the center was smoother, marking the area where surface tension forces had enough time to smoothen the surface before the onset of solidification. The extent of splashing decreased and eventually disappeared as substrate temperature was increased. No splashing was visible on a surface at 180 °C. Solidification did not start until fairly late during spreading; localized freezing at several spots acted to obstruct spreading of the splat and produced an irregular shaped splat even though there was no splashing. At $T_w = 200$ °C solidification was sufficiently delayed that droplets spread to form thin discs. Freezing around the droplet periphery during spreading on a substrate at low temperature obstructs liquid flow and triggers splashing. When substrate temperature is increased, freezing is slowed down and the droplet spreads

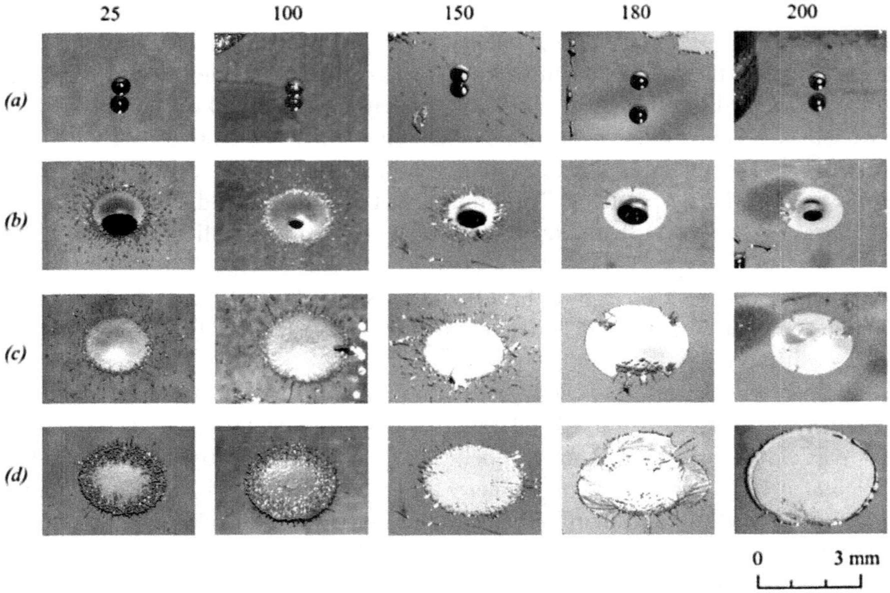

Fig. 5.2 Impact of 0.6 mm diameter tin droplets with 20 m/s velocity on a stainless steel substrate at varying temperature. Each image in a column shows successive stages of impact. The initial surface temperature in °C is indicated at the top of the column [6]

in the form of a thin liquid sheet without any splashing. The transition temperature, though difficult to identify exactly, lies between $T_w = 150$ and 180 °C.

5.2.2 *Droplet Impact Dynamics*

A molten metal droplet impacting on a solid surface spreads out into a thin sheet that expands radially until it reaches its maximum diameter (D_{max}) There has been many attempts to derive analytical expressions for the extent of maximum spread, typically non-dimensionalized by the initial diameter to give the spread factor $\xi_{max} = D_{max}/D_o$ [7–9]. Most of the analytical models use an energy balance in which the droplet energy before and after impact are equated, accounting for the energy dissipation during impact. Before impact a droplet possesses both kinetic and surface energy. After impact, when the droplet is at its maximum extension, the kinetic energy is zero and the surface energy is a function of the splat diameter and the advancing liquid-solid contact angle (θ_a). Pasandideh-Fard et al. [9] estimated the work done in deforming the droplet against viscosity from a simple model of stagnation point flow in a droplet landing on a surface. The effect of solidification in restricting droplet spread was modelled by assuming that all the kinetic energy stored in the solidified layer is lost [4]. An energy balance gave an expression for the maximum spread factor:

$$\xi_{max} = \frac{D_{max}}{D_o} = \sqrt{\frac{We + 12}{\frac{3}{8}Wes^* + 3(1 - \cos\theta_a) + 4\frac{We}{\sqrt{Re}}}} \tag{5.2}$$

where s^* is the dimensionless solid layer thickness ($s^* = s/D_o$).

Liquid-solid contact angles during spreading and recoil of tin droplets on a stainless steel were measured from enlarged photographs by Aziz and Chandra [4] and the advancing contact angle was found to be almost constant at $\theta_a = 140°$. The growth in thickness of the solidified layer (s^*) can be calculated using an approximate analytical solution developed by Poirier and Poirier [10]. The model assumes that heat transfer is by one-dimensional conduction; there is no thermal contact resistance at the droplet-substrate interface; the temperature drop across the solid layer is negligible; the substrate is semi-infinite in extent and has constant thermal properties. The dimensionless solidification thickness was expressed as a function of the Stefan number, Peclet number and $\gamma = k\rho c$:

$$s^* = \frac{2}{\sqrt{\pi}} Ste \sqrt{\frac{t^* \gamma_w}{Pe \gamma_d}} \tag{5.3}$$

where the subscripts w and d refer to substrate and droplet properties respectively. Substituting Eq. (5.9) into (5.8) gives the maximum spread of a droplet that is solidifying during impact:

$$\xi_{max} = \sqrt{\frac{We + 12}{WeSte\sqrt{\frac{3\gamma_w}{2\pi Pe \gamma_d}} + 3(1 - \cos\theta_a) + 4\left(\frac{We}{\sqrt{Re}}\right)}} \tag{5.4}$$

Predictions of ξ_{max} from Eq. (5.4) were found to agree reasonably well with experimental data.

5.2.3 *Thermal Contact Resistance*

Wang and Matthys [11] reviewed much of the available literature on measurements of thermal contact resistance (or its reciprocal, the thermal contact conductance) and compiled an exhaustive list of data available in the literature. Values of contact resistance show a large range of values, from 10^{-3} to 10^{-6} m^2K/W, and depend on the physical properties of liquid metal and substrate, surface roughness and liquid velocity.

A widely used method of measuring contact resistance has been to release a drop of molten metal onto a solid surface, measure transient substrate or droplet temperature, and fit the data to a numerical or analytical heat transfer model to calculate contact resistance [11–15]. The response time of the temperature measurement technique is important, since solidification is very rapid and it may not be clear if solid-solid or solid contact resistance is being measured. Liu, Wang and Matthys [12] and Wang and Matthys [11] used an optical pyrometer to measure the cooling rate of nickel and copper droplets falling on copper plates. Contact resistance was initially low (~10^{-6} m^2K/W) and then increased rapidly ($R_c > 10^{-4}$ m^2K/W) as the metal solidified. Loulou, Artyukhin and Bardon [13] used thermocouples with response time of 0.1 ms to measure substrate and droplet temperature for impacting tin, lead and zinc drops and calculated $R_c \sim 10^{-4}$ m^2K/W at times much after the droplet had solidified. Aziz and Chandra [14] used a thermocouple with 10 μs response time to record substrate temperature variation under tin droplets and used an analytical heat transfer model to calculate $R_c \sim 10^{-6}$ m^2K/W. Wang and Qiu [15] measured heat transfer to a copper block on whose surface was sputtered a thin Constantan film that acted as a thermocouple with 5 μs response time. When the block was dropped onto the surface of a pool of molten solder from heights of a few centimeters, contact resistances of approximately 10^{-5} m^2K/W were recorded, decreasing with greater impact velocity.

Heichal and Chandra [16] fabricated fast temperature sensors that had a response time of 40 ns and used them to measure substrate temperature variation under molten metal droplets (aluminum alloy 380 and bismuth) landing on solid plates (steel and brass). Thermal contact resistance between and was measured experimentally. The diameter of the droplets was kept constant (4 mm) while varying droplet impact velocity (1–3 m/s), substrate temperature (25–300 °C) and surface roughness (0.06–5.0 μm). Thermal contact resistance during the first few milliseconds of impact was obtained by matching measured surface temperature variation with an analytical solution of the 1-D transient heat conduction equation and found to range from 10^{-7} to 3×10^{-6} m^2K/W, increasing with surface roughness and decreasing with rising impact velocity. The thermal contact resistance could be calculated from the following equation

$$BiR_a^* = \frac{WeR_a^*}{WeR_a^* + \pi/2} \tag{5.5}$$

where Ra^* is the normalized surface roughness ($Ra^* = R_a/D_o$).

McDonald, Moreau and Chandra [17] plasma-sprayed molybdenum and yttria-stabilized zirconia particles (38–63 μm diameters) onto glass and Inconel plates held at either room temperature or 400 °C. Samples of Inconel were also preheated for 3 h, and then air-cooled to room temperature before spraying. Photographs of the splats were captured by using a fast charge-coupled device (CCD) camera. A rapid two-color pyrometer was used to collect thermal radiation from the particles during flight and spreading to follow the evolution of their temperature. The temperature evolution was used to determine the cooling rate of the spreading particles. It was found that particles on the heated or preheated surfaces had cooling rates that were significantly larger than those on surfaces at room temperature, suggesting that the thermal contact resistance was smaller on the heated and preheated surfaces. An analytical heat conduction model was developed to predict the thermal contact resistance at the interface of the plasma-sprayed particles and the surfaces. The analysis showed that the thermal contact resistance between the heated or preheated surfaces and the splats were more than an order of magnitude smaller than that on non-heated surfaces held at room temperature.

Thermal contact resistance between an impacting particle and a non-heated solid substrate has been attributed to the presence of volatile compounds on the surface, which evaporate under the hot splat and form a gaseous barrier between the two surfaces [17–20]. On heated surfaces, these adsorbates/condensates are almost completely vaporized, improving splat–substrate contact and greatly reducing the thermal contact resistance at the splat–substrate interface.

5.2.4 Surface Roughness Effects

Figure 5.3 shows the effect of increasing surface roughness on droplet impact [23]. Each column in the figure shows the impact of a 2.2 mm diameter tin droplet impacting with 4 m/s velocity on surfaces of different average roughness, having R_a 0.07, 0.56, and 3.45 μm respectively. The time after impact is indicated on the left side of the images. On the smoothest surface (Fig. 5.3a) small fingers were observed around the periphery of the drop immediately after impact with some of these detaching to form satellite droplets. Increasing the roughness of the stainless steel substrate to $R_a = 0.56$ μm produced significant changes in droplet spreading (see Fig. 5.3b) Instead of thin fingers there were large, triangular projections around the periphery of the drop early during spreading ($t = 0.3$ ms) which then broke loose ($t = 0.6$ ms) and continued to travel outwards, leaving behind a solidified circular splat ($t = 7.9$ ms). Increasing the roughness even further to $R_a = 3.45$ μm produced further changes in the droplet shape during spreading (Fig. 5.3c). Again there were triangular projections around the drop ($t = 0.3$ ms), but these did not detach ($t = 1.1$ ms). In this case solidification of the droplet was much slower, so that it remained liquid and surface tension forces pulled back the edge of the droplet ($t = 7.9$ ms). The final splat had a distinctive star-like shape.

On a smooth surface the thermal contact resistance between the droplet and surface is low because little air is trapped in surface cavities. Therefore solidification is rapid, starting before the droplet has fully spread. Increasing the surface roughness raises contact resistance, and lets the droplet spread to a greater extent before it freezes. Therefore droplets spread further on a rough surface than on a smooth surface when the substrate temperature was low enough to cause freezing. On a hot surface, where there was no solidification, surface roughness had little effect on droplet spread [23].

Shinoda et al. [24] plasma sprayed molten zirconia particles onto quartz substrates in which dimpled patterns had been etched and found that air entrapped in the dimples increased thermal contact resistance and promoted fragmentation of splats.

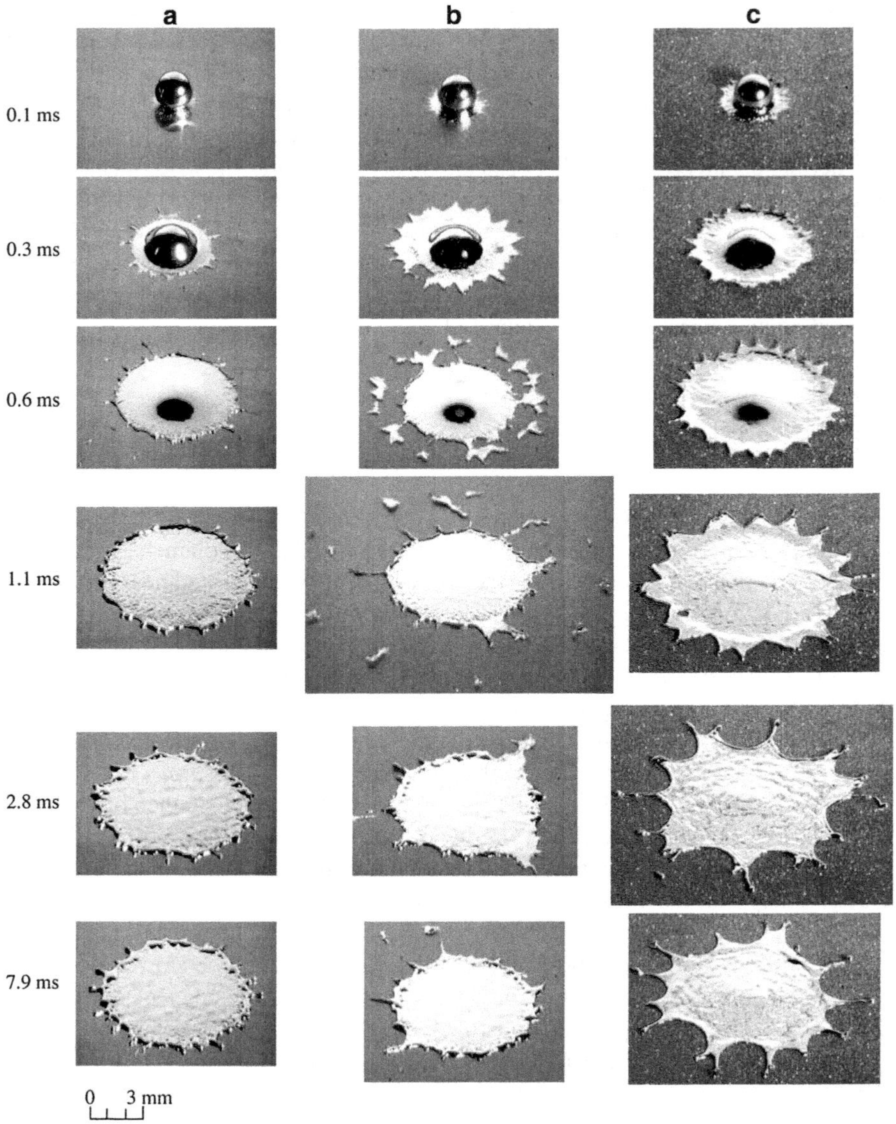

Fig. 5.3 The impact of 2.2 mm diameter molten tin droplets with 4.0 m/s velocity on a stainless steel plate at a temperature of 240 °C with surface roughness R_a (**a**) 0.07 μm, (**b**) 0.56 μm and (**c**) 3.45 μm [23]

5.2.5 Droplet Splashing and Fragmentation

When a droplet impacts a solid surface it spreads into a thin circular sheet. If the impact velocity is sufficiently large fluid instabilities create undulations around the edge of the spreading sheet that grow larger and form fingers. The fingers detach and form satellite droplets, a process that is commonly known as "splashing".

Figure 5.4 shows photographs of successive stages during the impact of a 2.7 mm diameter molten tin droplet impacting with a velocity of 4 m/s on a stainless steel plate [3]. Both drop and plate are at a temperature of 240 °C, above the melting point of tin (232 °C) so that impact is isothermal. The drop, initially spherical, begins to deform very rapidly upon contact and a thin liquid sheet begins to spread radially under it. The liquid-solid contact line edge of this sheet becomes unstable as it advances and a periodic disturbance is visible around it. Once the droplet reaches its maximum extension surface tension, which is very strong in molten metals, begins to pull it back. Because the molten metal does not wet the steel substrate well the fingers grow longer and break-up into smaller satellite droplets. The remaining liquid bounces off the substrate.

As the droplet approaches the substrate the gas between them has to be expelled and its density and viscosity determine how rapidly this occurs [25, 26]. The gas film trapped at the liquid-solid interface forms a bubble [27]. Then as the edges of the droplet spread out they face resistance from the surrounding atmosphere that has to be pushed back. Xu, Zhang and Nagel [28] demonstrated that lowering the pressure of the surrounding atmosphere suppresses splashing in droplets of water and organic liquids.

Some of the difficulty in predicting when splashing will occur can be attributed to uncertainties about surface wettability and the effect of the surrounding atmosphere. However, there is a certain ambiguity about the concept of "splashing" itself. Several different break-up modes are grouped under the same term, even though the mechanism of each may be quite different. Rioboo et al. [29] identified three different types of splashing. Immediately after impact, as the liquid sheet under the droplet spreads out, its edge becomes unstable and fingers around the edge begin to break off and form small droplets. This has been termed "prompt splash" and occurs when the edge of the lamella is still in contact with the surface.

The second type of splashing has been termed "corona" splashing: the liquid lamella lifts off the surface, the edge becomes unstable so that fingers grow at regular spaced intervals and the tips of these break off in the crown-like shape characteristic of splashing drops. Many studies have been devoted to predicting when corona splashes will occur. Mundo, Sommerfeld and Tropea [30] found that droplets splashed only if the so-called "splash parameter" $K = We^{1/2}Re^{1/4}$ exceeds a critical value $K = 57.7$. Cossali, Coghe and Marengo [31] developed an empirical correlation between K, R_a and the liquid lamella thickness h.

The air film trapped under the impacting droplet plays an important role in creating instabilities. Xu, Zhang, Nagel [28] demonstrated that if the pressure in the

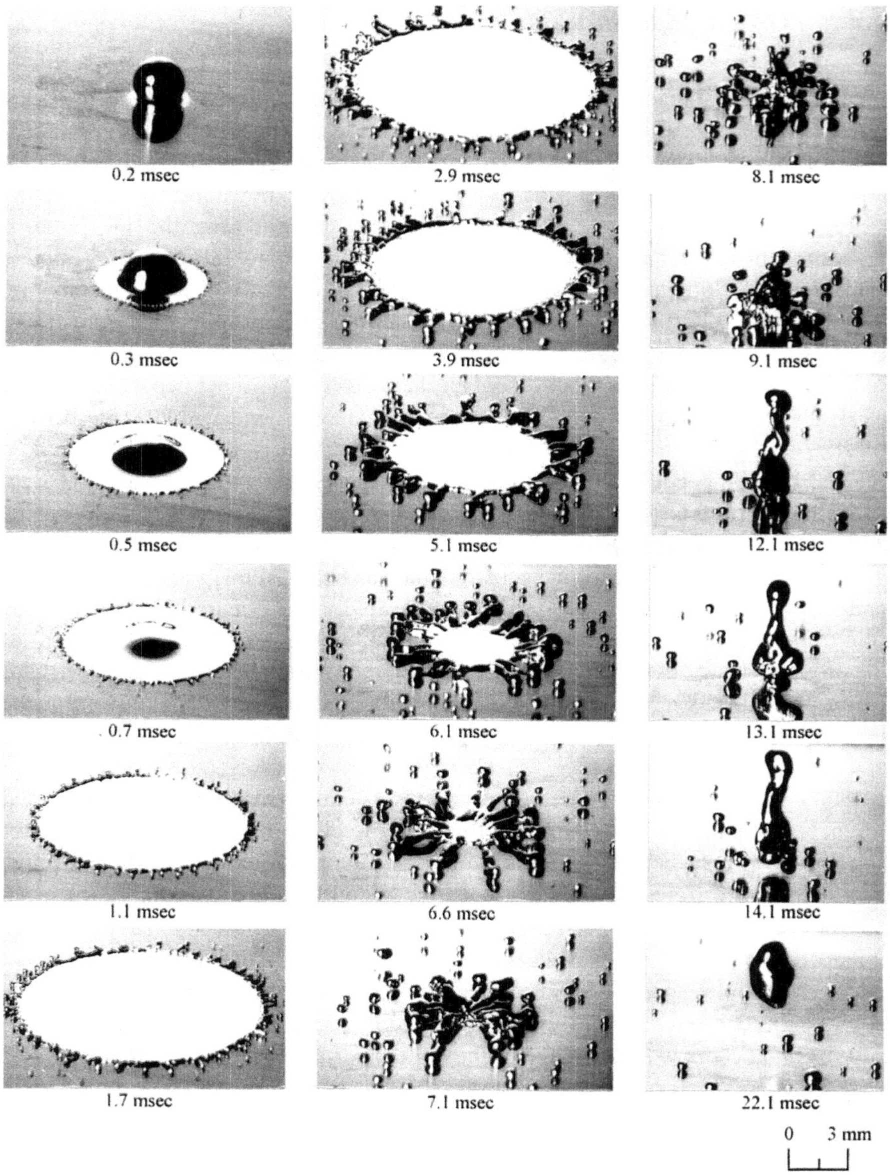

Fig. 5.4 Splashing of a 2.7 mm diameter molten tin droplet during impact with velocity 4 m/s on a stainless steel surface at temperature 240 °C. The droplet and substrate are both above the melting point of tin (232 °C) so there is no freezing [3]

atmosphere surrounding an impacting drop is reduced corona splashes are suppressed. Prompt splashing, however, persists even in the absence of surrounding gas [32].

The third type of splashing is known as "receding break-up", in which the droplet remains intact until it has spread to its maximum extent and then, as surface tension forces pull it back, the fingers formed due to instabilities around its periphery grow longer and begin to breakup into smaller droplets. If the liquid-solid contact angle is small, less than 90°, neighboring fingers along the edges of the spreading liquid sheet tend to merge with each other and disappear. However, if the contact angle is large, as is the case with droplets of molten metal, the cylindrical fingers become unstable and disintegrate. Figure 5.4 shows this type of splashing.

Apart from these three mechanisms, there are two others that can cause break-up of impacting droplets. If a droplet impacts on a substrate that is cold enough to cause freezing, the solid layer formed at the liquid-substrate interface acts as a barrier. The spreading liquid hits the solid mass obstructing its path, jets upwards and disintegrates. This is known as freezing-induced splashing [6] and whether it occurs depends on the rate of heat transfer between the droplet and substrate, which is controlled by the substrate temperature, substrate thermal properties, and the thermal contact resistance at the liquid-solid interface.

There is yet one more mechanism that leads to droplet fragmentation, when impact velocities are very high so that the liquid film becomes very thin and air bubbles trapped under it break through. These punctures in the liquid grow larger and can eventually lead to complete disintegration of the droplet [21, 22].

5.2.6 Splat Shapes

The properties of solid deposits produced by molten metal sprays depend on the shape of solid splats formed by the impact of individual droplets. If the splats fragment upon impact they do not coalesce well with each other and the deposits formed tend to be porous and have low strength. If a dense deposit is desired splat break-up must be avoided; round, disk-shaped splats should be formed by impacting droplets flattening out and solidifying.

There are two, entirely different, mechanisms by which a particle can fragment during impact. If the thermal contact resistance under the splat is very low, and cooling is very rapid, it begins to solidify as it spreads. The solid layer obstructs and destabilizes the flow of liquid, leading to fingers being formed around its edges. At the other extreme, if contact resistance is very high, the particle remains liquid and spreads into a very tin film that ruptures internally. In this case the splat is also fragmented, but its shape is different, appearing as a small central core surrounded by a ring. Disk shaped splats are formed if the value of thermal contact resistance lies between these two extremes, so solidification starts after the particle has already flattened out and does not obstruct the liquid flowing outwards, but is still sufficiently rapid to prevent the splat from spreading so thin that it ruptures internally.

Dhiman et al. [33] proposed a single parameter to estimate the importance of freezing during solidification and predict the likelihood of splat break-up.

When a molten droplet lands on a solid surface it spreads into a thin splat of uniform thickness h. If the substrate is at a temperature lower than the melting point of the droplet a solid layer of thickness s grows in it during the time it takes to reach its maximum spread. The solidification parameter is defined as the ratio of the solid layer thickness to splat thickness ($\Theta = s/h$). Dhiman et al. [33] developed an analytical expression to calculate the value of Θ as a function of the dimensionless parameters Re, We, Ste, Pe and Bi. The magnitude of Θ can be used to predict what the final shape of the splat will be, and what the mechanism of break-up, if it occurs, is. Three outcomes are possible during spreading:

(1) A very thin solid layer ($\Theta << 1$), has no effect on spreading. The splat spreads into a thin sheet liquid, ruptures internally and fragments, producing a small central splat surrounded by a ring of debris,
(2) If solid layer growth is significant ($\Theta \sim 0.1$–0.3), it will restrain the splat from spreading too far and becoming thin enough to rupture, producing a disk-type shape and
(3) If solidification is very rapid ($\Theta \sim 1$), the solid layer obstructs the outward spreading liquid and produces a splat with fingers radiating out from its periphery.

Comparison with experimental photographs [33] showed the value of the solidification parameter gave a reasonably accurate method of predicting the shape of the final splat. Figure 5.5 shows photographs of splats taken both during and after impact, illustrating the three different modes of splat impact. In Fig. 5.5a, for molybdenum and nickel particles landing on substrates at room temperature, thermal contact resistance was high ($\sim 10^{-5}$ m^{2}K/W) and $\Theta \sim 0.01$. The splats spread into a thin film that ruptured internally and fragmented. The final splats all showed a central portion at the point of impact that adhered strongly to the substrate, surrounded by a ring.

Raising the substrate temperature reduced the thermal contact resistance by an order of magnitude, since it evaporated adsorbed contaminants on the surface. Figure 5.5b shows impact of zirconia and nickel particles on surfaces heated to 400 °C which had $R_c \sim 10^{-6}$ m^{2}K/W and correspondingly $\Theta \sim 0.1$. Solidification occurred near the end of droplet flattening, when the spreading liquid did not have enough momentum to jet over the solid rim and instead came to rest forming a circular splat with smooth edges.

If Θ was increased further ($\Theta \sim 0.4$, see Fig. 5.5c), solid layer growth was sufficiently rapid to obstruct flow of liquid early during spreading. The liquid had enough momentum that it jetted outward, producing fingers radiating out from the central splat. For nickel particles spreading on a steel substrate oxidized by heating to 640 °C (Fig. 5.5c) the splat was intact, and smooth at the centre, where solidification was slow. The edges, which solidified very rapidly, have a rough surface since surface tension did not have time to level irregularities before solidification occurred. Molybdenum splats on a glass surface heated to 400 °C also have fingers radiating out.

	Splat Impact Mode	Pictures during impact	Splat images
a)	**Fragmentation** (Slow solidification) Θ~0.01	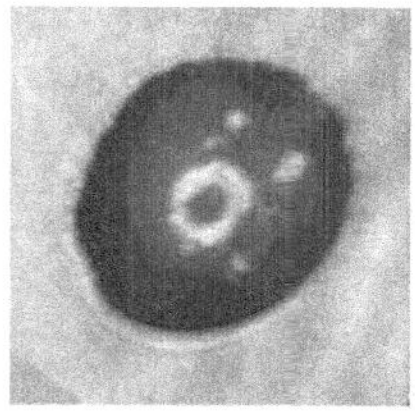 Mo on glass, room temp	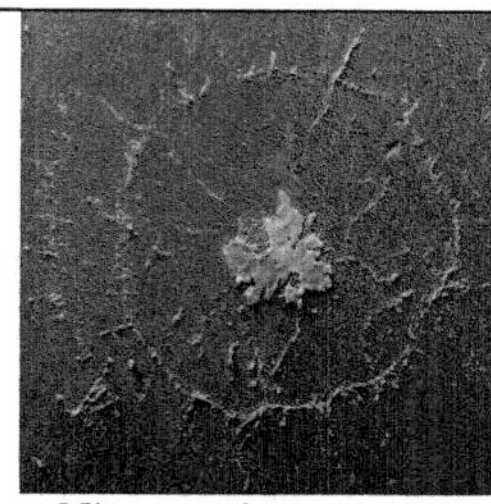 Ni on steel, room temp
b)	**Disk Splats** (Intermediate rate of solidification) Θ~0.1	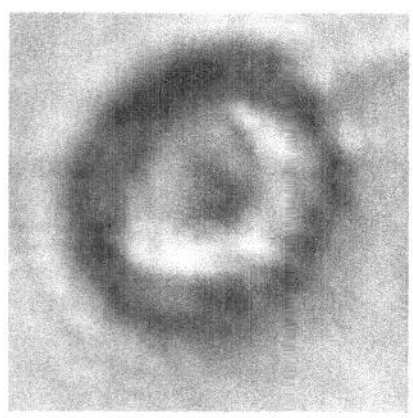 ZrO_2 on glass, 400°C	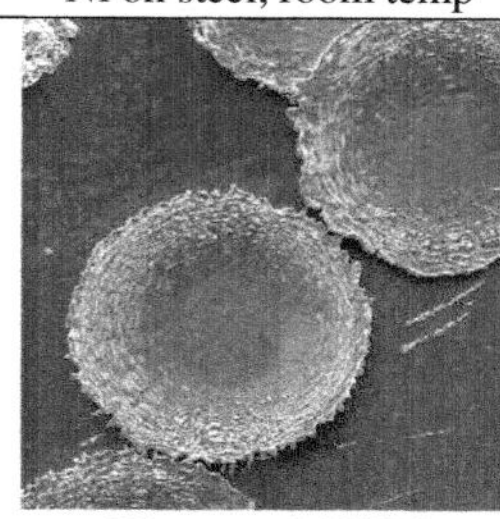 Ni on steel 400°C
c)	**Freezing induced break-up** (Rapid solidification) Θ~0.4	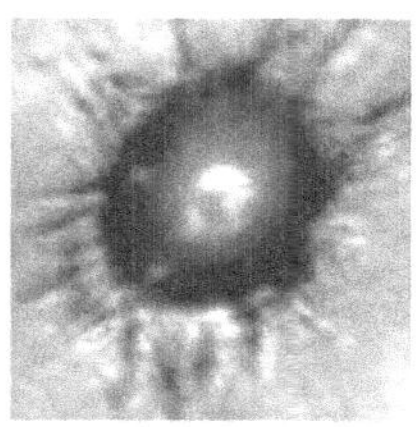 Mo on glass, 400°C	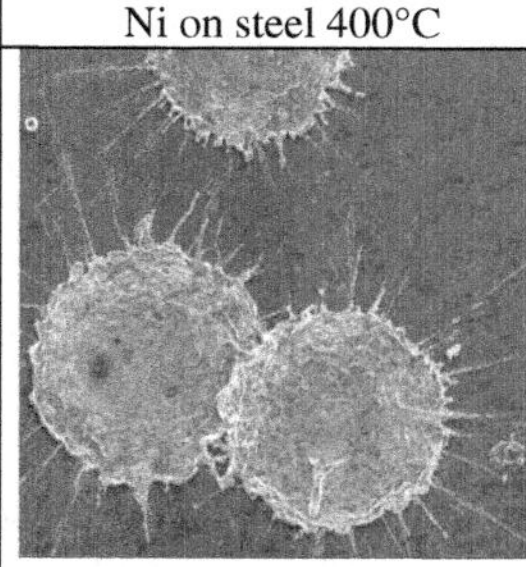 Ni on steel 640°C

Fig. 5.5 Photographs during and after impact for splats (**a**) Fragmenting during impact, (**b**) forming disk splats and (**c**) undergoing freezing-induced break-up [33]

5.2.7 Residual Stresses and Splat Deformation During Solidification

Curling up at the edge of splats is frequently observed in thermal spray coatings and is one of the main sources of coating porosity. After the droplet impact on the surface, it spreads across the surface and solidifies. Then, as it cools down to room temperature, it shrinks. If a portion of the bottom is bonded to the substrate it cannot shrink, while the upper surface of the splat is free to contract, so that stresses are created in the splat. To relieve these stresses the unbonded portion of the splat, along its periphery, curls up.

The degree of curling up is affected by several factors such as stresses generated by mismatch of thermal expansion coefficient at the coating interface, surface tension of the liquid splat, surface roughness and remelting. Fukanuma [34] proposed a physical and mathematical model for the production of porosity by considering deformation of a molten particle during thermal spray coating processes. He observed that most pores exist at the periphery of splats, starting at about 0.6 times the spat radius (R) from its center. Cirolini et al. [35] developed a model for the deposition of a thermal barrier plasma-sprayed coating assuming that curling was caused by the temperature drop across the splat when the solidification front just reached the top.

Xue et al. [36] studied the curling up of splats formed by impact of molten metal drops both experimentally and numerically. Splat curling up was assumed to be entirely due to shrinkage of splats as they cool from a high initial temperature to the substrate temperature. A simple analytical model was used to estimate the magnitude of the curl-up angle. Splat curl-up angle was found to increase with the area of the splat bonded to the substrate and decrease with higher substrate temperature. Increasing thermal expansion coefficient or Young's modulus of the splat increased curl-up angle.

Figure 5.6 shows two cross-sections through two splats formed by plasma spraying nickel powders onto a stainless steel 303 substrate initially at 400 °C [36]. In Fig. 5.6a the splat starts curling up at about 0.38R, and the curl-up angle was measured to be 5.5°. In Fig. 5.6b, the splat starts curling up at about 0.49R, and the curl-up angle was measured to be 6.5°. A greater radius of bonding of the splat to the substrate results in a larger curl-up angle.

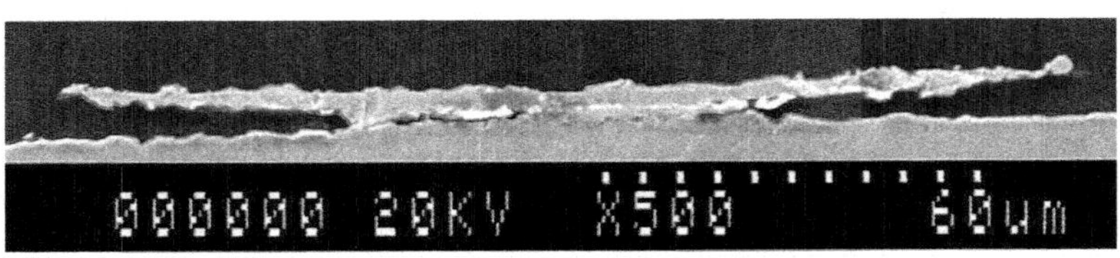

(a)

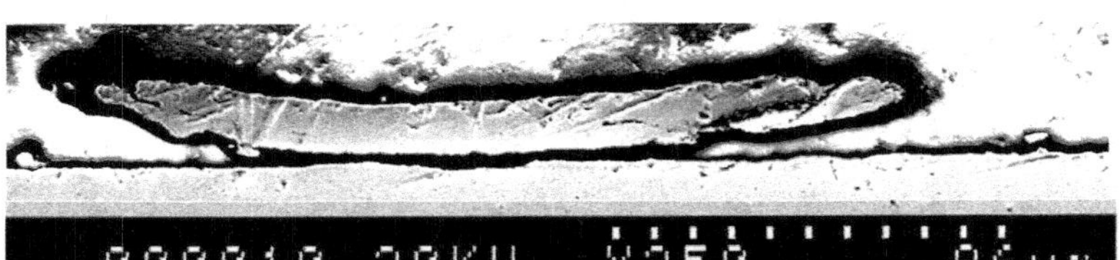

(b)

Fig. 5.6 Cross-sections through the centers of nickel splats on a stainless steel 303 substrate with initial surface temperature 400 °C (**a**) Bonded fraction = 38%R, (**b**) Bonded fraction = 49%R [36]

5.3 Numerical Modeling of Droplet Impact

5.3.1 Governing Equations

For impact of a droplet on a surface, we assume that the flow is incompressible and laminar. The equations governing the conservation of mass and momentum in an Eulerian frame of reference are then given as:

$$\frac{\partial \rho}{\partial t} + \nabla \cdot (\rho \mathbf{V}) = 0 \tag{5.6}$$

$$\frac{\partial (\rho V)}{\partial t} + \nabla \cdot (\rho V V) = -\nabla p + \nabla \cdot \overline{\overline{\tau}} + \boldsymbol{F}_b + \boldsymbol{F}_{ST} \tag{5.7}$$

where $\mathbf{V}$ is the velocity vector, ρ is the fluid density, p is the pressure, $\overline{\overline{\tau}}$ is the shear stress tensor, $\mathbf{F}_b$ represents body forces such as gravity acting on the fluid elements, $\mathbf{F}_{ST}$ is the surface tension force which acts only near the fluids interface [37]. Assuming the fluids are Newtonian, the shear stress tensor is expressed as

$$\overline{\overline{\tau}} = \mu\left(\nabla V + \nabla V^T\right) \tag{5.8}$$

where μ is the dynamic viscosity. Using the one-field approach and having several fluids in the domain, each with velocity field $\mathbf{V}_k$, we may assume that all fluids move with the local center of mass velocity $\mathbf{V}$ [38]

$$\mathbf{V_k} = \mathbf{V} \tag{5.9}$$

5.3.2 Interface Tracking

There are a number of methods described in the literature to resolve the interface between two immiscible and incompressible fluids. These include the volume-of-fluid (VOF) method [39], level-set (LS) method [40], combined level-set volume-of-fluid (CLSVOF) method, height function (HF) method [41], and volume-of-fluid with advecting normal (VOF-AN) method [42]. The results presented here will be based on an improved VOF method [43].

One of the most common and robust approaches for tracking interfaces is the volume-of-fluid (VOF) approach. In this method a scalar function *f*, is defined to mark the space where each fluid resides. In the case of two immiscible fluids, the values are assigned zero in one fluid and unity in the second one. Since all fluids are assumed to be incompressible, *f* is passively advected with the flow and, thus, it satisfies the advection equation:

$$\frac{\partial f}{\partial t} + \mathbf{V} \cdot \nabla f = 0 \tag{5.10}$$

where,

$$f(\vec{r}) = \begin{cases} 1, & \vec{r} \in fluid\ 1 \\ 0, & \vec{r} \in fluid\ 2 \end{cases} \tag{5.11}$$

$\vec{r}$ is the position vector. The interface normal and curvature can be calculated from the VOF data by:

$$\widehat{n} = \frac{\nabla f}{|\nabla f|}$$
$$\kappa = -\nabla \cdot \widehat{n}$$

The numerically discretized form of f is the fraction of a numerical control volume occupied by fluid 1, i.e.,

$$F = \frac{1}{V} \int_V f\, dv \tag{5.12}$$

Thus $F = 1$ in cells that are fully occupied by fluid 1 and $F = 0$ for cells that are filled with fluid 2. Thus $0 < F < 1$ for those cells that contain both fluids. Equation (5.12) is numerically solved using the Young's algorithm [43].

Based on the volume fraction of each phase, mixture properties in the interface cells are defined as

$$\begin{aligned} \rho &= f\rho_d + (1-f)\rho_b \\ \mu &= f\mu_d + (1-f)\mu_b \\ \kappa^{-1} &= f/\kappa_d + (1-f)/\kappa_b \end{aligned} \tag{5.13}$$

where d and b refer to dispersed and bulk phases, respectively.

The surface tension force, $\boldsymbol{F}_{ST}$, which is non-zero only at the interface, can be expressed as

$$\boldsymbol{F}_{ST} = \sigma(T)K\nabla f + \nabla_{\|}\sigma(T)\,|\nabla f| \cong \sigma(T)K\nabla f \tag{5.14}$$

where K is the local interface curvature, T is the interface temperature, and $\nabla_{\|}$ the tangential surface derivative. The first term on the right hand side corresponds to the temperature-dependent normal surface tension component, while the second term corresponds to the Marangoni force. The Marangoni convection force is negligible during droplet impact. Thus (5.7) becomes:

$$\frac{\partial(\rho \boldsymbol{V})}{\partial t} + \nabla \cdot (\rho \boldsymbol{V}\boldsymbol{V}) = -\nabla p + \nabla \cdot \mu\left(\nabla \boldsymbol{V} + \nabla \boldsymbol{V}^{\boldsymbol{T}}\right) + \sigma(T)K\nabla f + \boldsymbol{F}_b \tag{5.15}$$

5.3.3 Heat Transfer and Solidification

We assume that solidification occurs at melting temperature and we neglect viscous dissipation. Densities of liquid and solid phases are assumed to be constant and equal to each other. The energy equation is then written as:

$$\frac{\partial h}{\partial t} + (\mathbf{V} \cdot \nabla)h = \frac{1}{\rho} \nabla \cdot (k \nabla T) \tag{5.16}$$

Energy equation has two dependent variables; these are temperature T and enthalpy h. we employed the method of Cao et al. [44] and transformed the energy equation in terms of enthalpy alone. The main advantage of this method is that it solves the energy equation for both phases simultaneously. The transformed energy equation is as follows [44]:

$$\frac{\partial h}{\partial t} + (\mathbf{V} \cdot \nabla)h = \frac{1}{\rho} \nabla^2(\beta\, h) + \frac{1}{\rho} \nabla^2 \phi \tag{5.17}$$

where in the solid phase:

$$h \leq 0; \quad \beta = \frac{k_s}{C_s}, \ \varphi = 0 \tag{5.18a}$$

at the liquid-solid interface:

$$0 < h < H_f; \quad \beta = 0, \ \phi = 0 \tag{5.18b}$$

and in the liquid phase:

$$h \geq H_f; \quad \beta = \frac{k_l}{C_l}, \ \varphi = -\frac{H_f k_l}{C_l} \tag{5.18c}$$

where ϕ is a new source term, and H_f is the latent heat of fusion. Subscripts l and s refer to liquid and solid properties, respectively. The energy equation has now only one dependent variable, the enthalpy, h. The relationship between temperature and enthalpy is given by

$$T = T_m + \frac{1}{k}(\beta h + \phi) \tag{5.19}$$

where T_m is the melting point of the droplet. Heat transfer within the substrate is by conduction only. The governing equation is

$$\rho_w C_w \frac{\partial T_w}{\partial t} = \nabla \cdot (k_w \nabla T_w) \tag{5.20}$$

where subscript w indicates the substrate. At the free surface, we used an adiabatic boundary condition. Note that, initially, the dominant heat loss from the droplet is due to heat conduction to the substrate, and later on, conduction and convection to the solidified layer. Estimates of heat loss by convection from the droplet surface to the surrounding gas showed that it is three orders of magnitude lower than heat conduction to the substrate. Therefore, the adiabatic condition at the free surface is reasonable. This condition can however, be easily modified to a convective, radiative, or mixed boundary condition.

5.3.4 Initial and Boundary Conditions

Initial conditions, i.e., droplet size, impact velocity, substrate and droplet temperatures, liquid-substrate contact angle, and thermal contact resistance are given along with the thermophysical properties of the droplet and substrate. Heat conduction within the substrate is accounted for.

The incomplete contact between the drop and the substrate results in a temperature discontinuity across the contact surface. The effect can be incorporated in the model via definition of the thermal contact resistance, R_c (see Eq. (5.1)). Values of R_c are provided as an input to the model. Although in principle R_c could vary with time and/or position on the interface, we assumed it to be a constant. In practice, R_c typically varies between 10^{-6} and 10^{-7} m^2 K/W.

Computation of velocity field has to account for the presence of a moving, irregularly shaped solidification front on which the relevant boundary conditions are applied. We treat the solidified regions by a modified version of the fixed velocity method. In this approach, a liquid volume fraction Θ is defined such that $\Theta = 1$ for a cell completely filled with liquid; $\Theta = 0$ for a cell filled with solid; and, $0 < \Theta < 1$ for a cell containing a portion of the solidification front. Normal and tangential velocities on the faces of cells containing only solidified material are set to zero. The modified continuity and momentum equations are then given by [45]:

$$\nabla \cdot (\Theta \mathbf{V}) = 0 \tag{5.21}$$

$$\frac{\partial (\Theta \mathbf{V})}{\partial t} + (\Theta \mathbf{V} \cdot \nabla)\mathbf{V} = \frac{-\Theta}{\rho} \nabla p + \Theta \upsilon \nabla^2 \mathbf{V} + \frac{\Theta}{\rho} \mathbf{F}_b \tag{5.22}$$

$$\frac{\partial f}{\partial t} + (\Theta \mathbf{V} \cdot \nabla) f = 0 \tag{5.23}$$

The modified Navier-Stokes, volume of fluid, and energy equations are solved on an Eulerian, rectangular, staggered mesh in a 3D Cartesian coordinate system. Details of the computational procedure are described by Bussmann et al. [43] and Pasandideh-Fard et al. [9].

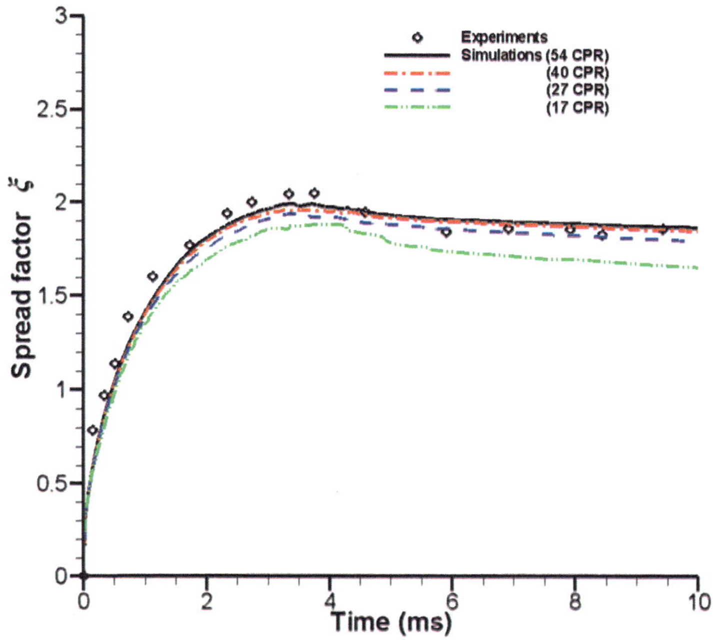

Fig. 5.7 Spread factor versus time for a tin droplet impacting on a stainless steel substrate. Impact conditions: impact velocity 1 m/s, initial droplet temperature 513 K, substrate temperature 298 K, droplet diameter 2.7 mm, melting point of tin 505 K. Experimental points from Aziz and Chandra [4]

5.3.5 *Simulations of Droplet Impact*

To validate the numerical model, results are compared to experimental measurements of Aziz and Chandra [4]. Figure 5.7 shows the spread factor ratio versus time for the impact of a 2.7 mm tin droplet impacting on a stainless steel substrate at 1 m/s and 513 K. The substrate temperature is 298 K and the melting point of tin is 505 K [46]. As shown in the figure, the predictions are in excellent agreement with the experimental results as the resolution of the numerical calculations is increased to 27 cell-per-radius (cpr) or higher. Figure 5.8 shows a comparison between experimental measurements and numerical simulations of the impact of a tin droplet on a previously deposited and solidified tine splat [47]. The comparison between the predictions and experiments is again excellent for such a relatively complicated situation.

5.3.5.1 Effect of Solidification on Break-up

Figure 5.9 shows the different stages of the normal impact of a 60 μm nickel droplet on a smooth stainless steel substrate at 290 °C [48]. The impact speed is 73 m/s and

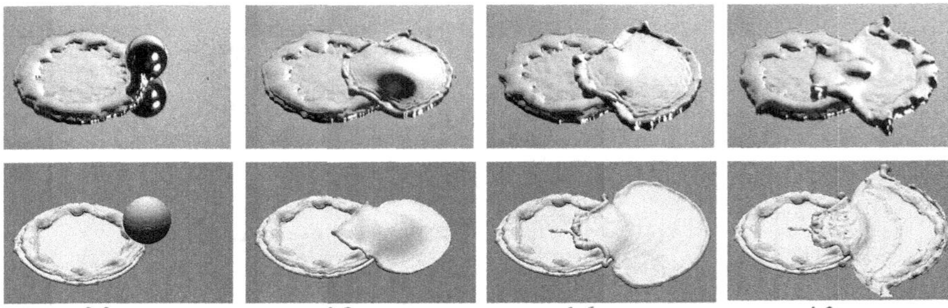

Fig. 5.8 Comparison of photographs and computer generated images of a 2.2 mm diameter tin droplet landing with a velocity of 2.5 m/s at a point 3.0 mm from the center of a solidified splat [47]

the initial droplet temperature is 1600 °C. Thermal contact resistance is assumed to be 10^{-7} m^2K/W. This case corresponds to $Re = 7892$, $We = 1419$, $Ste = 1.67$, and $Pr = 0.043$; hence, $Ste/Pr = 38.3$, which indicates the effect of solidification on droplet spreading is important. As droplet starts spreading, instabilities around the rim appears. These instabilities result in generation of a number of fingers as well as breakup of the finger tips into smaller drops (Fig. 5.9). Examination of the numerical results shows that, for this impact conditions, these instabilities occur due to solidification. This is demonstrated in Fig. 5.10. As the thermal contact resistance is increased by an order of magnitude (Fig. 5.10c), solidification occurs at a slower rate and the splat assumes a circular disc shape. The effect of substrate temperature has been found to be of great importance in affecting the dynamics of the impact on metallic substrates [33].

5.3.5.2 Effect of Surface Roughness on Impact Dynamics

Raessi et al. [49] and Parizi et al. [50] studied the effect of the roughness on the dynamics of the droplet impact, silicon substrates were etched and patterned with cubes of 1, 2, and 3 μm. The distance between the cubes were the same as their height. Figure 5.11 shows good agreement between the final splat shape of a nickel droplet impacting on the 1 μm rough surface and the numerical predictions. For these patterned surfaces, as the roughness increases, the final splat shape is no longer circular (Fig. 5.12). This effect is particularly important for the case of 3 μm roughness. The effect is due to the fact that solidification rate depends on the direction of the spreading droplet. As shown in Fig. 5.12b, the calculation of the splat shape in the absence of solidification results in disc like splat. Figure 5.13a, b show the contact area of the liquid droplet with the surface of the substrate. As show, the contact is maximum at an angle of 45°. Increased contact results in a faster rate of solidification, hence, the spreading is arrested in the directions with high contact.

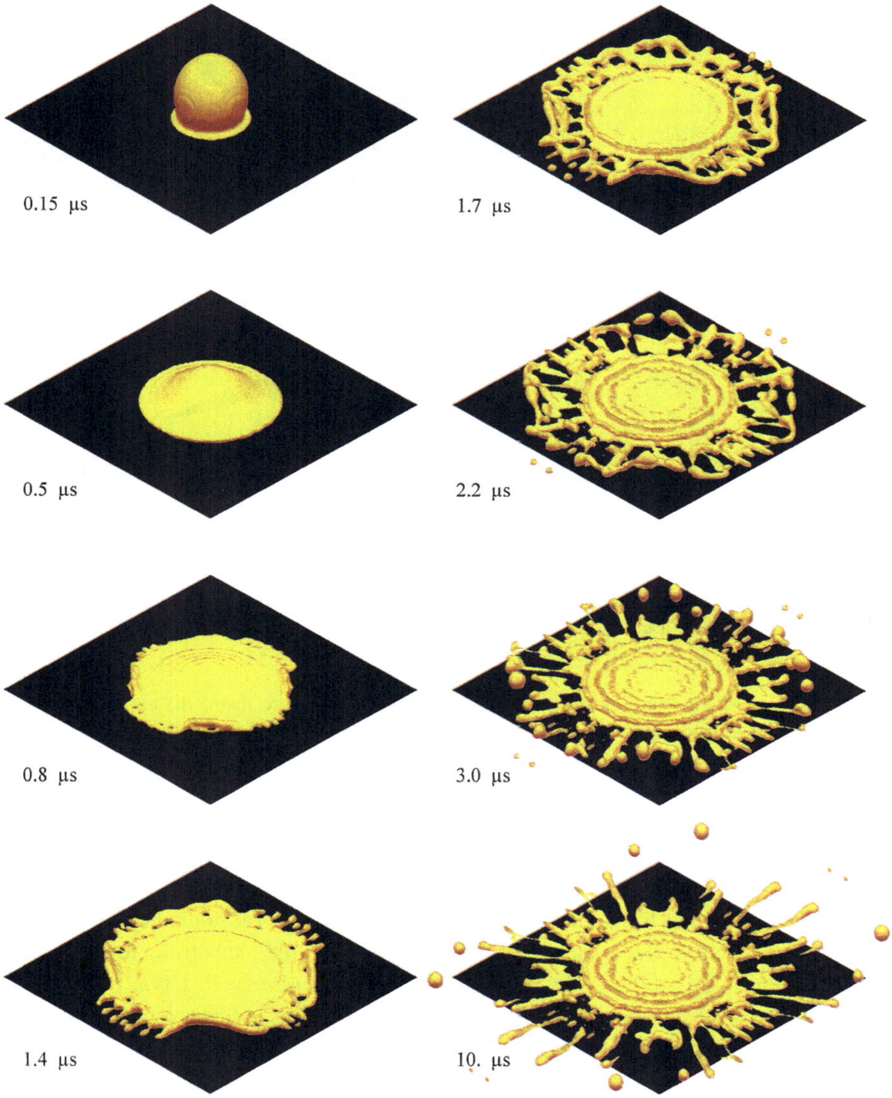

Fig. 5.9 Simulations showing the impact of a 60 μm diameter molten nickel particle at 1600 °C landing with a velocity of 73 m/s on a stainless steel plate initially at a temperature of 290 °C. The contact resistance at the substrate surface was assumed to be 10^{-7} m^2K/W [48]

5.3.5.3 Impact of Partially Molten Droplets

Wu et al. [51] and Alavi et al. [52] studied the impact of partially molten zirconia and partially molten nickel droplets, respectively. The droplets are melted on the outer layer and have a solid core. This situation often occurs in thermal spray

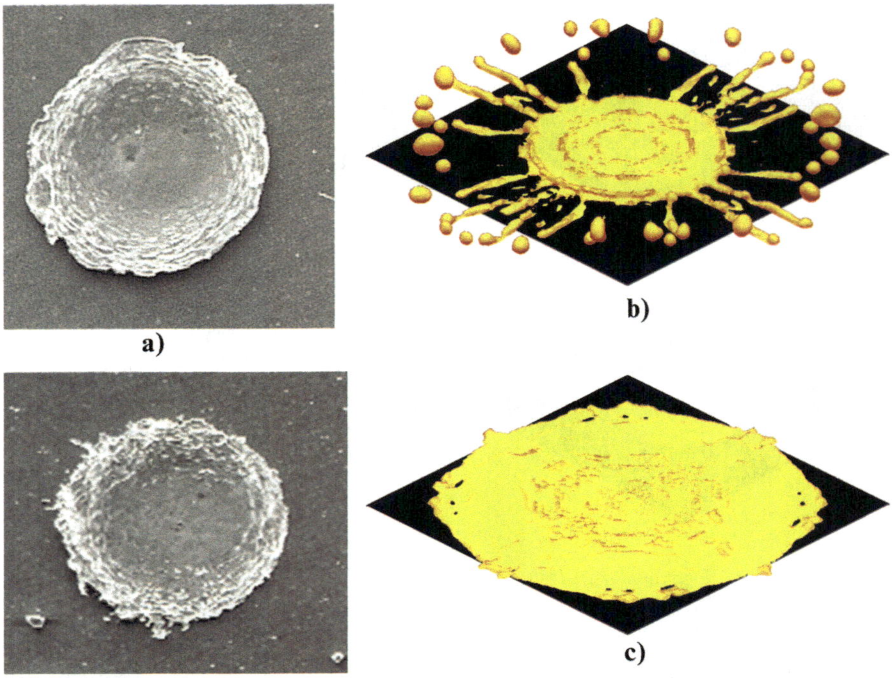

Fig. 5.10 Nickel splat shapes on a steel plate initially at 400 °C from (**a**) experiments, (**b**) numerical model assuming a contact resistance of 10^{-7} m^{2}K/W and (**c**) numerical model assuming a contact resistance of 10^{-6} m^{2}K/W, [48]

coating process when a particle is not heated sufficiently and does not fully melt. Insufficient heat transfer may be due to the trajectory of the particle as well as its big size.

Figure 5.14 shows the dynamics of the impact of a fully molten nickel droplet on a smooth, stainless steel substrate. Because of the high substrate temperature, solidification rate is rather slow and little splashing is observed. Upon impact, the drop starts spreading and solidifying on. Some splashing is observed after around 1 μs after the impact. The final splat is shaped as a flat disk with raised rims. As the streamlines illustrate, vortices are generated in the gas flow during the particle impingement. These vortices influence the amount of the material detached from the particle during splashing. It may be noted that at 1 μs, in addition to the main vortex flow, another circulation is observed which is caused by the movement of the splashed droplet.

Figure 5.15 shows spreading of a partially molten nickel droplet. Compared to the fully molten case, the presence of the hard core results in a reduction in spreading and less splashing. Furthermore, as we might have expected, because of the unmelted core, there is a bump in the center of the final splat. Alavi et al. [52] show that increase in the impact speed will have no effect on the size of bump, increases splashing, and decreases the thickness of the final splat. A larger unmelted core promotes splashing.

Fig. 5.11 Effect of surface roughness on spreading of a nickel droplet on a silicon substrate [49]

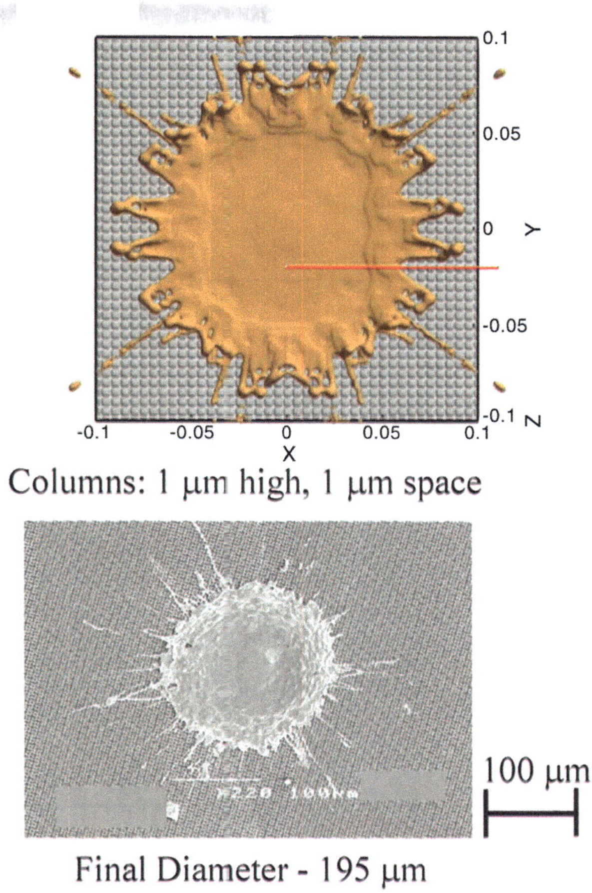

5.4 Spray Impingement

5.4.1 Coalescence of Droplets Impacting on a Surface

Spray deposition involves the impingement of a very large number of droplets that first land on a bare substrate, and then, as the deposit grows thicker, on previously accumulated splats. The growing mass of metal on the substrate loses heat by conduction to the substrate and by convection and radiation to the surrounding atmosphere. If incoming droplets add energy to the deposit faster than it is lost, the temperature of the metal will increase and the spray will land in a layer of molten metal. If heat transfer to the surroundings is sufficiently fast to allow droplets to cool down and freeze after impact, they will form solid splats. For a molten droplet to fuse with a solid deposit it must have enough energy to cause remelting in the material under its impact point, which then solidifies again.

Amon et al. [53] did experiments in which a feedstock wire located directly over the substrate was melted using a plasma welding torch. Droplets typically 1–10 mm

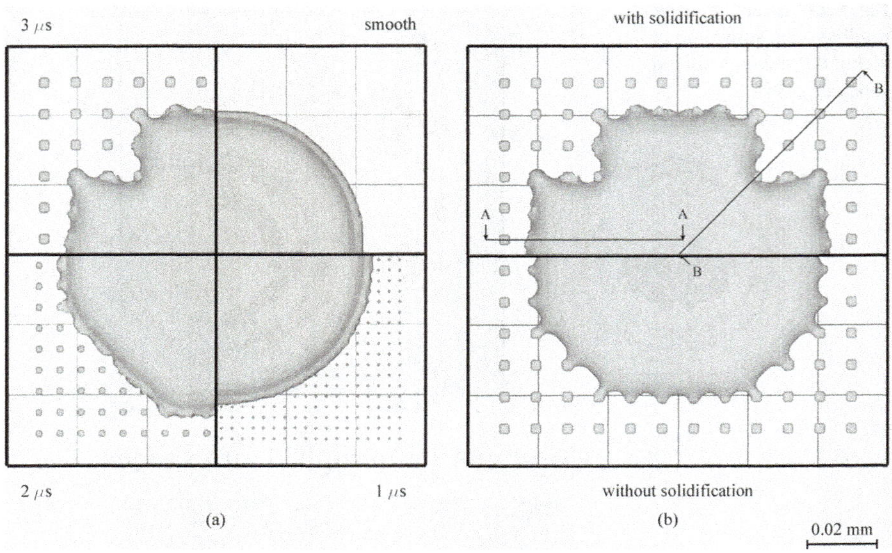

Fig. 5.12 Comparison between (**a**) the shape of alumina splats on different surface conditions in the presence of solidification and (**b**) splat shape on a substrate with 3 μm roughness and an alumina droplet on the same substrate and the corresponding time but without solidification. 40 μm diameter alumina droplets at 2055 °C impacting with a velocity of 65 m/s onto alumina substrates initially at 25 °C and at different surface roughness [49]

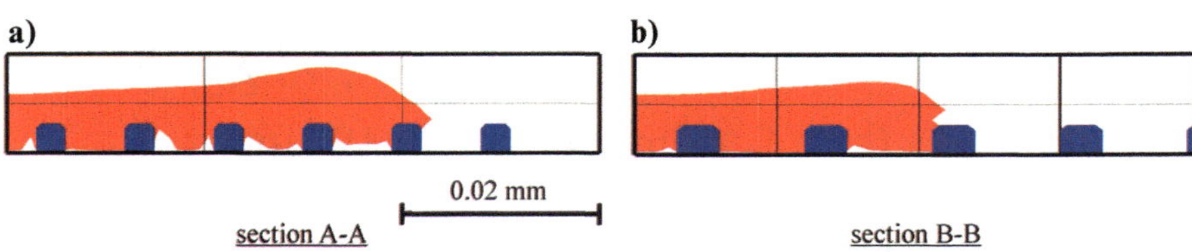

Fig. 5.13 Cross section of the alumina splat on a substrate with 3 μm roughness in the directions shown in Fig. 5.6b. The cubes on the substrate and the splat are shown in *blue* and *red*, respectively [50]

in diameter fell from the tip of the wire and onto a substrate where they fused with each other. Orme and Huang [54] formed deposits by directing onto a substrate a stream of approximately 100 μm diameter droplets formed by vibrating a molten metal jet issuing out of a fine nozzle. Fang et al. [55, 56] used a pneumatic droplet generator to deposit molten tin droplets on top of each other to form vertical columns. They used an energy balance around the tip of a column, equating the rate at which energy was carried in by impacting droplets to the rate of heat conduction through the column, to calculate the maximum frequency at which droplets could be deposited while still solidifying after impact.

Ghafouri-Azar et al. [57] studied the coalescence of 2.2 mm diameter tin droplets deposited in lines on a substrate, each offset by a small amount from the other. Figure 5.16 shows splats formed by depositing four tin drops along a straight

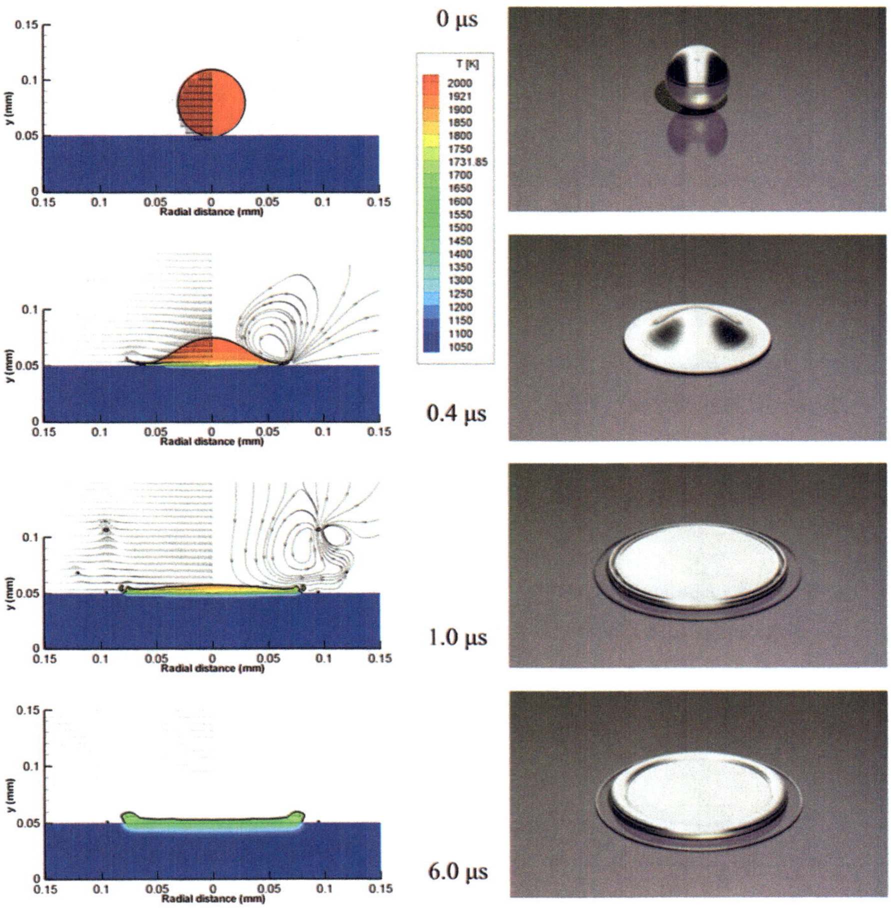

Fig. 5.14 Fully Molten Nickel Particle: 60 μm; Initial Temperature 1921 K; Impact Velocity 100 m/s; Substrate Temperature: 1050 K; Thermal contact resistance 10^{-6} m^2 K/W. The *thin black line* inside the particle shows the solidification front [52]

line, with the center of each drop offset by 2.0, 3.0 and 3.0 mm respectively from that of the previous one They used numerical simulations to predict the shapes of splats formed by interacting droplets and to calculate where sufficient remelting occurred for spalts to fuse with each other.

5.4.2 Spray Deposition

A metal spray consists of a large number of droplets that are transported by a by high velocity gas jet. The droplets have a size distribution that is determined by how the spray was generated, typically either by atomizing a jet of molten metal or by

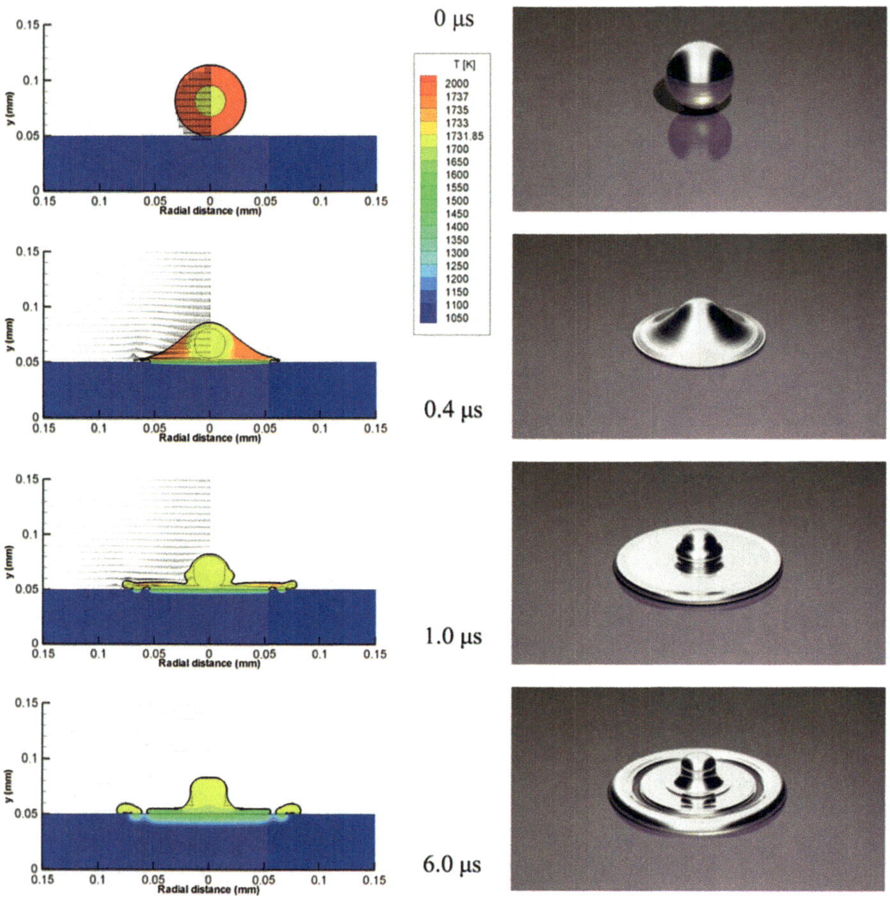

Fig. 5.15 Semi-Molten Nickel Particle impact; 64 μm Dia.; Solid Core Dia. 28.7 μm; Initial Temperature: 1737 K; Impact Velocity: 100 m/s; Substrate Temperature: 1050 K; Thermal contact resistance 10^{-6} m^2 K/W. The *thin black line* inside the particle shows the solidification front [53]

melting powders injected into a high temperature gas. Drag forces accelerate the droplets so that their impact velocity is a function of both their size and the gas velocity. Similarly, heat transfer due to convection and radiation between the droplets and the surrounding environment determines droplet temperature at the moment of impact.

When droplets land on a surface that is at a temperature below the melting point of the droplet material they should, in principle, begin to solidify. In reality a large degree of undercooling may be required to initiate phase change [58]. At the molten metal deposit formed by coalescence of droplets accumulates it loses heat both to the substrate and the surrounding gas. If the rate of heat loss is less than that at which heat is added to the deposit by the impinging droplets a thin liquid layer is maintained at the surface of the deposit [59]. The thickness of this liquid layer can

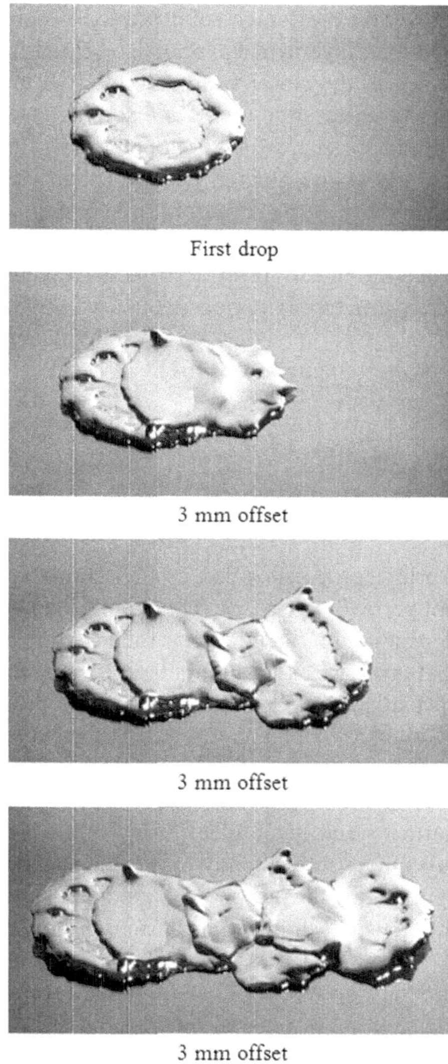

Fig. 5.16 Splats formed by depositing four 2.2 mm diameter tin drops along a straight line, with the center of each drop offset by 2.0, 3.0 and 3.0 mm respectively from that of the previous one [57]

be controlled by adjusting the mass flux deposited by the spray or by varying the heat transfer to the substrate. This may be necessary in spray forming applications where it is important to minimize run-off of the liquid metal. In coating applications, where the layers are very thin and the mass deposition rates relatively low, impinging droplets typically land on solidified substrates.

As the solid deposit grows the temperature distribution in it can be calculated by solving the heat conduction equation with appropriate boundary conditions, which include the rate of heat removal through the substrate and the rate of heat addition with the spray [59, 60]. If an alloy is being sprayed it does not have a unique freezing point, but a liquidus temperature at which solidification starts and a solidus temperature at which solidification is complete. The sprayed deposit then consists of a solid layer and a partially solidified "mushy" layer, in which the liquid fraction increases the closer you get to the free surface.

The microstructure of the solidified layer depends on its thermal history. The bottom surface, which is in contact with the substrate, undergoes rapid quenching and has a fine grain structure. Since it takes time for a mushy layer to form on the surface of the deposit once spraying commences, droplets that land initially on the solid layer may not weld completely, leading to the formation of porosity. Once the mushy layer is formed droplets landing in coalesce completely and the microstructure is homogeneous with the boundaries of individual droplets no longer visible. Near the edges of the spray, where heat transfer is the highest, the deposit tends to remain porous [61].

5.4.3 *Porosity Formation*

During spray deposition molten droplets fuse together to form a solid layer that is not perfectly dense, but contain pores and cracks that may or may not be desirable, depending on the function of the coating. In general low porosity is desirable since that increases the strength of the material and makes it impervious. In some specialized applications porosity may be desirable, such as in thermal barrier coatings, where the insulating properties are improved by the presence of air pockets. Several different mechanisms have been identified that can create porosity: curling up of splats due to thermal stresses, entrapment of gas under impacting particles and incomplete filling of cavities in the already deposited coating. Protuberances may already exist on a rough substrate, or they may be created during spraying by the presence of unmelted particles in the spray, or as a result of satellite droplets detaching from impacting droplets and solidifying on the surface.

Pores formed by gas entrapment are typically very small, and found at the interface between splats in thermal spray coatings. Based on transmission electron microscopy of plasma-sprayed coatings, McPherson and Shafer [61] showed that the interfaces between lamellae consist of regions of perfect contact alternating with gaps of 0.01–0.1 μm which probably arise from absorbed or entrapped gas between impinging droplets and previously solidified layers.

Splat curl up is cause by residual stresses in the splat as it cools and shrinks after being sprayed on the surface. The bottom surface of the splat is attached to the substrate and cannot shrink, while the upper surface is free to contract. The resulting stresses are relieved either by the edges of the splat curling up. Fukanuma [34] developed a model for porosity formation during thermal spray coating processes by considering deformation of a molten particle, and showed that most of the porosity is near the periphery of the splat, starting at a distance from its center of about 0.6 times the spat radius (R). Porosity was sensitive to particle velocity, ambient gas pressure, particle diameter, and molten material viscosity.

Xue et al. [62] studied impact of molten droplets on a rough surface where liquid is driven into the crevices between asperities on the surface by liquid pressure while surface tension restrains it from completely filling gaps, leaving voids. An analytical model was developed to calculate the volume of these voids that predicted, to within an order-of-magnitude, the volume of voids measured from experiments.

Porosity in spray formed deposits has been found to decrease with increasing melt flow rate, since the surface of the layer remains liquid and there are no unmelted particles to create pores. However, beyond a certain critical flow rate the porosity increases again since the liquid layer becomes deeper and the amount of air entrapped under impacting droplets increases [63].

5.5 Modeling Spray Impingement

Thermal spray coatings are formed by the impact and deposition of millions of molten and semi-molten droplets on a substrate. While we can accurately model impact of many droplets on a surface, it is not yet computationally possible to deposit millions of droplets using computational fluid dynamics and predict the microstructure of the coating. In order to develop such a model, based on Monte Carlo approach, Ghafouri-Azar et al. [64, 65] developed a 3-dimensional model of coating formation. The model was further developed by Xue et al. [62, 66] and Parizi et al. [67].

One of the promising modelling approaches for detail simulation of coating formation is smoothed particle hydrodynamics approach, first introduced by Gingold and Monaghan [68] and Lucy [69] in 1977. The method is Lagrangian and can handle many droplet impact events simultaneously in a very efficient manner.

Section 5.1 will summarize the stochastic approach to predicting microstructure of coatings while in Sect. 5.2 the basic SPH approach is described.

5.5.1 Monte-Carlo Simulations

Ghafouri-Azar et al. [64, 65] proposed a 3-dimensional, stochastic model of thermal spray coating formation that can predict coating porosity, roughness as a function of spray parameters. The model assigns values of droplet size, velocity and temperature T, dispersion angle and azimuthal angle to molten droplets on the substrate by generating random values of these properties, assuming that their properties follow known distributions with user-specified means and standard deviations [66] that can be obtained for specific experiments by diagnostic instruments. Once the impact conditions of the individual droplet are selected from these distributions, the splat size is calculated by a simple analytical expression proposed by Aziz and Chandra [4] (Eq. (5.4)).

Interaction of an impacting droplet over a previously deposited splat was considered in the following manner. When a droplet lands overlapping a previously deposited splat, it will not spread into a disk-shaped splat but will assume a shape that depends on its distance from the center of the splat under it. Based on experimental results, and some detailed simulations of sequential droplet impact using a three-dimensional model, Ghafouri-Azar et al. [57] developed four possible scenarios for the second splat shape formed by two-droplet interactions. To select one of these scenarios, the distances between the droplet impact point and the center points of all previously deposited splats were evaluated. The smallest distance was then used to determine splat shape according to the rules which were established by approximating detailed numerical simulations of droplet interactions on a substrate and observations of interacting plasma sprayed splats collected on a surface during experiments. The surface area of non-circular splats was assumed to be the same as it would have been had the splats remained circular.

According to Xue et al. [36, 62] porosity is formed because of the incomplete filling of the interstices on previously deposited splats, since surface tension prevents molten material from entering small gaps. The model assumes that the impacting molten droplet is in contact with a series of uniform hemispherical asperities on the surface along the splat radius. In order to calculate the volume of the voids created between the hemispheres and the liquid layer, the equilibrium profile of the liquid meniscus is calculated using a method in which the total potential and surface energies of the system are minimized. Knowing the shape of the liquid meniscus and the profile of the asperity, one can use some geometrical expressions and then integrate the gap area over the total length of the splat to calculate the volume of the incompletely filled voids. In addition to curl up and incomplete filling of interstices, a third phenomenon may result in the formation of porosity. The small, satellite droplets which are formed when droplets splash and are settled on the surface also promote the formation of porosity. Based on this assumption, the number of satellite droplets, their sizes, and locations are approximated using the theories presented in reference [66]. These satellite droplets, in turn, create surface roughness and the incomplete filling of the coating layer will create more porosity.

Figures 5.17 and 5.18 show the result of the stochastic model for an yttria-stabilized zirconia plasma sprayed coating [66]. The calculations were based on the following spray parameters: average particle diameter and standard deviation of 25 and 5 μm, respectively; Average particle temperature and standard deviation of 3000 K and 50 K, respectively; average impact speed and standard deviation of 100 m/s and 10 m/s, respectively, dispersion angle and standard deviation of 3.0 and 0.58°, respectively and a uniform distribution in the azimuthal direction. The gun standoff distance was 0.12 m, and the average powder feed rate for the cases studied was 0.35 g/s. The gun moved constantly back and forth along the length of the substrate with a speed of 0.6 m/s. The model correctly predicts the effect of different operating parameters on coating porosity, roughness and thickness [66].

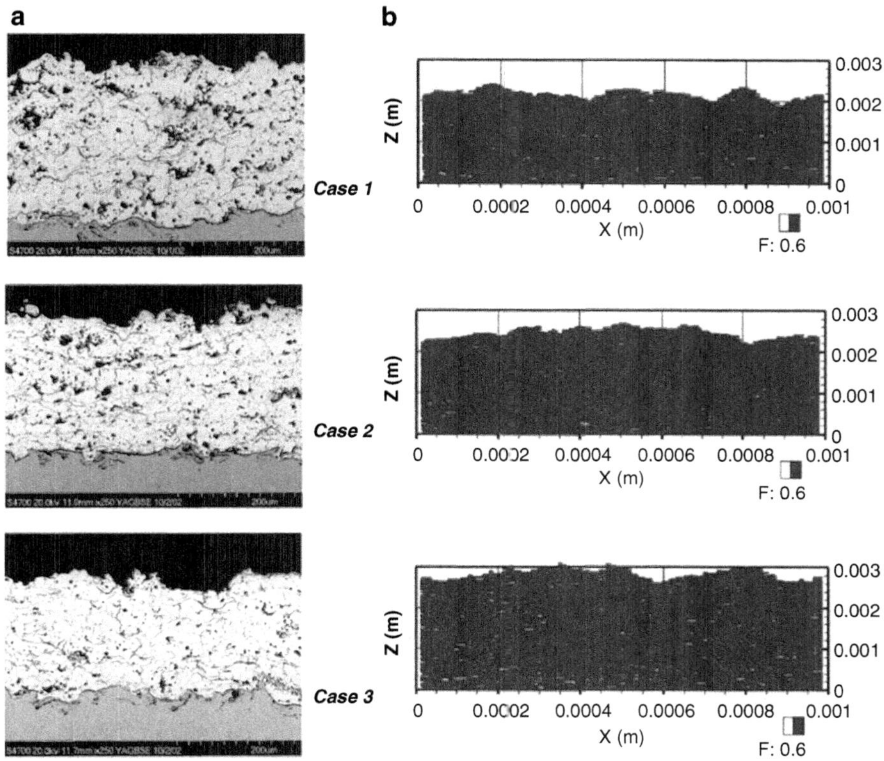

Fig. 5.17 Cross-sections of three YSZ coating cases from (**a**) experiments, (**b**) simulations [66]

Stochastic models are very useful in predicting microstructure of coatings as a function of operating conditions and show the dependence of porosity, roughness and thickness on different parameters.

5.5.2 *Smoothed Particle Hydrodynamics*

Smooth particle hydrodynamics or SPH was introduced and developed by Gingold and Monaghan [68] and Lucy [69] in 1977. In SPH, computational domain is discretized using fluid particles. Each particle has density and mass to represent a lump of fluid moving around with the velocity of the fluid at that location in a Lagrangian manner. Properties of these particles are smoothed over a distance known as the smoothing length. This means that the properties of a particle of interest can be calculated from its neighboring particles. The contribution of neighbors is weighted using a kernel function that mostly depends on the distance between neighboring particles.

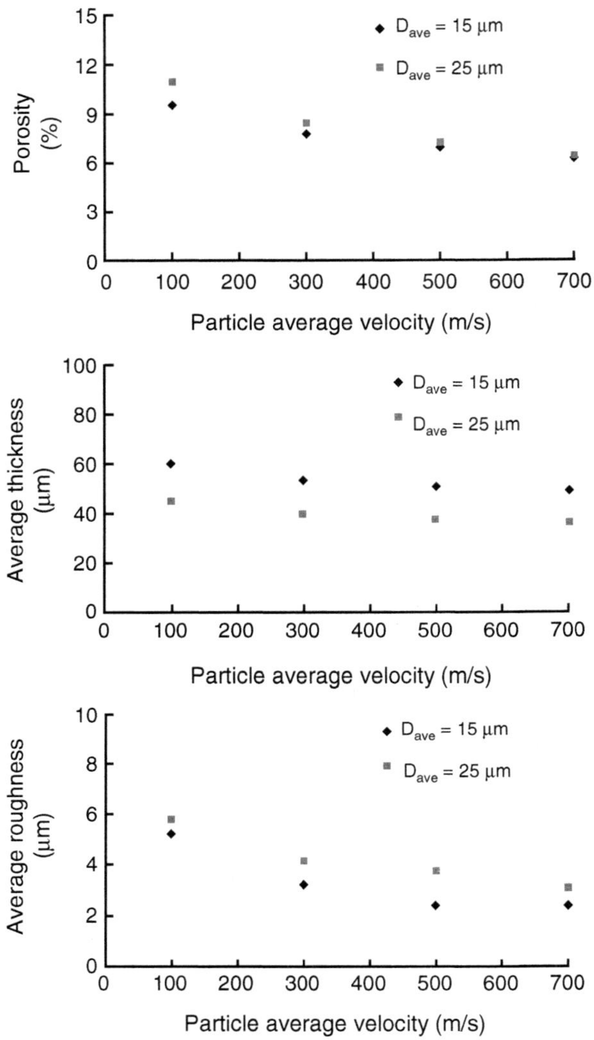

Fig. 5.18 Variations of coating porosity, average thickness and average roughness with particle size and impact speed [66]

Since its inception, SPH has been extensively used in simulating different physical phenomena in fields like astrophysics, fluid sciences, oceanography, ballistics, etc. One of the major subjects studied in SPH is interfacial flows. Practical studies like tsunami simulations [70], simulation of floating bodies like ships [71] and multiphase studies [72] are among them. In multiphase flows, numerical study of droplets has been of interest to many researchers due to applications in fields like spray coating and inkjet printing. Recently, Farrokhpanah et al. [73] studied droplet impact on a surface and proposed and implemented a model for contact angle.

SPH solves Navier-Stokes equations in a Lagrangian framework. In this frame, Eq. (5.6, 5.7) for an isothermal case are written:

$$\frac{D\rho}{Dt} = -\rho\nabla \cdot \boldsymbol{V} \tag{5.24}$$

$$\frac{D\boldsymbol{V}}{Dt} = \frac{1}{\rho}\left[-\nabla p + \nabla \cdot \overline{\overline{\tau}} + \boldsymbol{F}_b + \boldsymbol{F}_{ST}\right] \tag{5.25}$$

where $\frac{D}{Dt} = \frac{\partial}{\partial t} + V.\nabla$ is the substantial derivative, $\boldsymbol{F}_b$ represents external body forces such as gravity. The surface tension force, $\boldsymbol{F}_{st}$, is approximated based on the Continuum Surface Force (CSF) model of Brackbill et al. [37].

The continuity and momentum equations are closed by the equation of state, which calculates pressure using density in the form of (see Monaghan [74])

$$P = P_0\left(\frac{\rho}{\rho_0}\right)^{\gamma} + b \tag{5.26}$$

where γ=7 and 1.4 for liquid and gas phases, respectively, b is a background pressure, and P_0 represents a reference pressure adjusted to keep maximum density deviations from ρ_0 in the order of O (1%).

In SPH, the local values of dependent variables are interpolated by an integral interpolant. For example, quantity A, which is a function of spatial coordinate system, may be exactly expressed as:

$$A(\boldsymbol{r}) = \int A(\boldsymbol{r}')\delta(\boldsymbol{r} - \boldsymbol{r}')d\boldsymbol{r}' \tag{5.27}$$

where $\boldsymbol{r}$ is spatial coordinates, d$\boldsymbol{r}'$ is the differential volume element, and δ is the Dirac delta function. The above may be approximated by a kernel, W, as follows

$$A(\boldsymbol{r}) = \int A(\boldsymbol{r}')\, W(\boldsymbol{r} - \boldsymbol{r}', h)d\boldsymbol{r}' \tag{5.28}$$

The kernel is defined by:

$$\lim_{h\to 0} W(\boldsymbol{r} - \boldsymbol{r}', \boldsymbol{h}) = \boldsymbol{\delta}(\boldsymbol{r} - \boldsymbol{r}') \tag{5.29}$$

For a particle with mass m_i and density ρ_i at position $\boldsymbol{r}_i$, the integral may be approximated by:

$$A_i = \sum_{j=1}^{N} m_j \frac{A_j}{\rho_j} W(\boldsymbol{r}_i - \boldsymbol{r}_j, h) \tag{5.30}$$

where N is the total number of particles in the domain. In practice, the summation is limited to a limited number of particles which are in the neighborhood of particle i since W rapidly approaches zero with distance from particle i. The most commonly used kernel is the cubic spline, which, in one dimension, has the following form:

$$W(x,h)/h = \begin{cases} \frac{1}{6}\left[(2-q)^3 - 4(1-q)^3\right]\}, & for\ 0 \le q \le 1 \\ \frac{1}{6}(2-q)^3, & for\ 1 \le q \le 2 \\ 0 & for\ q > 2 \end{cases} \tag{5.31}$$

where $q = |x|/h$. For a detailed description of the application of SPH technique to fluid equations the reader is referred to the review article by Monaghan [74].

Das and Das [75, 76] and Farrokhpanah et al. [73] presented a method for applying contact angle on a horizontal surface during the impact of a drop using SPH. The model is capable of accurately applying contact angle to a stationary and a moving contact line. In the method, the prescribed value of contact angle is used to adjust the interface profile near the triple phase point. This is done by adjusting the surface normal close to the contact line and interpolating the drop profile into the boundaries. Farrokhpanah et al. [73] developed a parallel, GPU (Graphic Processing Unit) compatible SPH solver to capture interface evolution during droplet impact. To improve stability and performance of the solver, a customized reduction algorithm is used on the shared memory of GPU. Speedup using a variety of different memory management algorithms on GPU-CPU were studied. The algorithm was validated using the Rayleigh-Taylor instability test.

Figure 5.19 shows the SPH results for the case of a two-dimensional impact of a water droplet [73]. A constant contact angle is imposed during the impact of the droplet with a radius of 250 μm. The chosen values of constant contact angles for each case are 50, 70, 90, 100, 110, 130, 145, 160, and 175°. Droplet is impacting the surface from a distance of 375 μm at a velocity of 1 m/s under gravitational acceleration of 9.8 m/s^2. The calculated Reynolds and Weber numbers are 440 and 6.86, respectively. The computational domain is a square with sides of 3×375 μm filled with 10,000 particles. 776 particles sweep the surface of the drop (approximately 23 particles per radius) and the rest of particles form the surrounding air. Results are benchmarked against an identical case simulated using a VOF solver. For contact angles larger than or equal to 90°, drops spread on the solid surface and after reaching their maximum spread, they recoil. Figure 5.20 shows the maximum diameter of the water droplet after impact as a function of the contact angle. As contact angles become larger and larger than 90° there is a strong recoil following maximum spread. Farrokhpanah et al. also implemented an empirical relationship suggested by Kistler [77] for the dynamic contact angle. The predicted results of the spread factor were in excellent agreement with experimental measurements of Šikalo et al. [78].

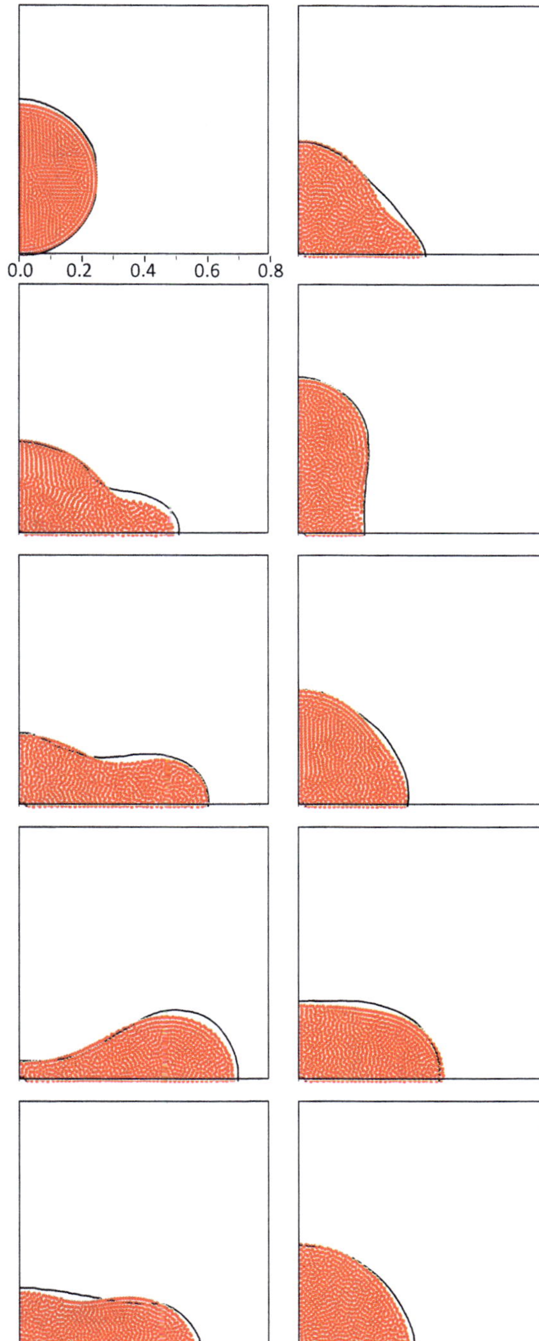

Fig. 5.19 Droplet impact on a solid surface for contact angle of 90°, comparing SPH (*red filled circle*) and VOF (*solid line*) results [73]

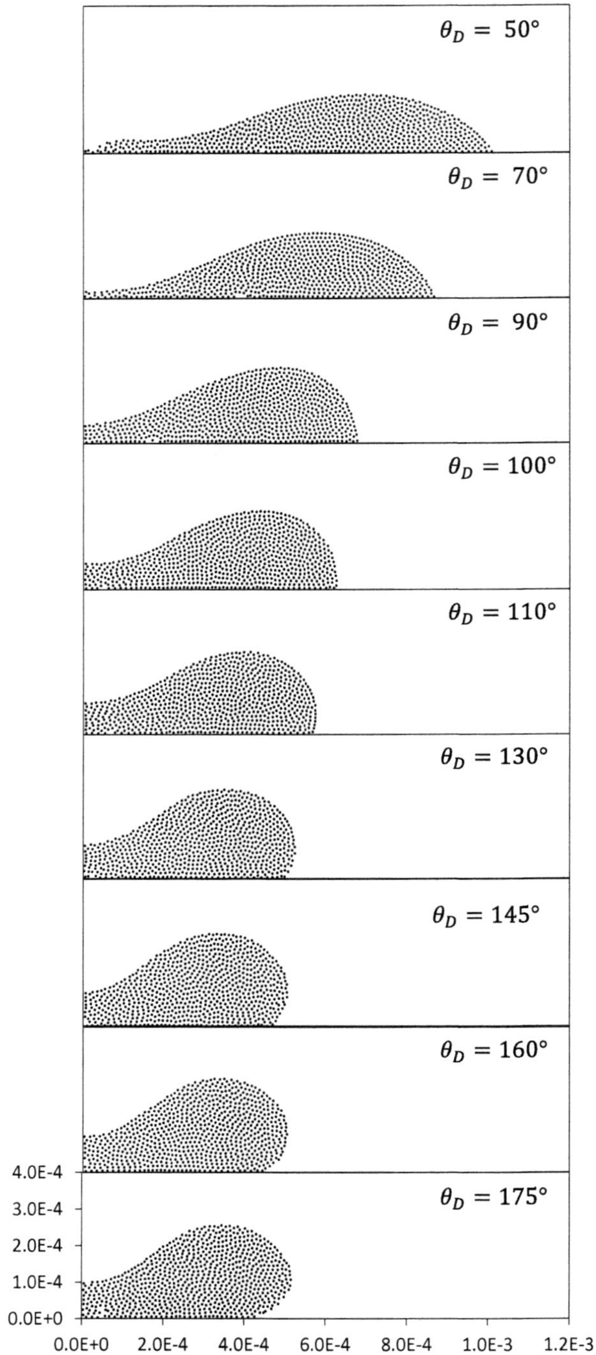

Fig. 5.20 Impacted drops shown at their maximum expanded diameter for various constant contact angles imposed during impact [73]

5.6 Summary

The impact of molten droplets on a solid surface is a basic step in all metal spray forming and coating processes. The microstructure and shape of splats formed by droplets that have flattened out and solidified determines the properties of the layer formed by coalescence of splats. Splat shapes depends on the properties of both droplets (diameter, velocity and temperature) and the substrate (roughness, surface contaminants). If droplets fragment during impact and leaves smaller particles on the surface, the solidified deposit may contain pores and have low strength.

The dynamics of impact of a single droplet have been studied extensively and found to depend on dimensionless Reynolds and Weber numbers that are a function of droplet diameter and impact velocity, and physical properties such as the liquid density, viscosity and surface tension. The solidification rate depends on the Stefan, Peclet and Biot numbers, which include the effect of the thermal contact resistance between the droplet and substrate. Surface roughness affects both droplet spreading and heat transfer.

Numerical models are a powerful tool for investigating the impact of single and multiple droplets. A number of techniques such as volume of fluid (VOF) and level set (LS) have been developed to track the motion of interfaces between two fluids. Comparison with experiments has shown that simulations can accurately predict the motion of impacting droplets and interactions between droplets.

When a large number of molten metal droplets are sprayed onto a surface they coalesce to form a solid deposit. The microstructure of the material depends on is thermal history, which is a function of the rate at which droplets are sprayed on the substrate and their temperature. If solidification is too rapid pores may form in the material due to air being trapped, curl-up of splats because of thermal stresses, or incomplete filling of crevices in the previously deposited material. Coating formation by the deposition of a large number of droplets has been simulated using Monte Carlo models that can predict properties such as porosity, roughness and thickness. Smoothed particle hydrodynamics models offer a computationally efficient way of modeling interfacial flows, including atomization, droplet impact and solidification.

5.7 List of Symbols

5.7.1 *Latin*

Symbole	description
Bi	Biot number
b	Bulk phase
c	Specific heat of the droplet
D_{max}	Diameter max

(continued)

Symbole	description
D_o	Diameter
d	Dispered phase
f	Fraction of a numerical control volume
$\mathbf{F}_{\mathrm{ST}}$	Surface tension force (only near fluids interface)
$\mathbf{F}_{\mathrm{b}}$	Body forces (e.g. gravity)
h	Liquid lamella thickness
h	Enthalpy
H_f	Heat of fusion
k	Thermal conductivity
$\widehat{\mathrm{n}}$	Interface normal
p	Pressure
Pe	Peclet number
q''	Heat flux
r	Splat radius
R_a	Substrate roughness
Ra^*	Normalized surface roughness
R_{c}	Thermal contact resistance
Re	Reynolds number
s^*	Dimensionless solid layer thickness
Ste	Stefan-number
T_d	Temperature droplet
T_m	Melting temperature
T_w	Temperature substrate
T_w	Initial temperature
$\mathbf{V}$	Velocity vector
V_o	Impact velocity
We	Weber number

5.7.2 Greek

Symbole	description
ξ_{max}	Spread factor
$\overline{\overline{\tau}}$	Shear stress tensor
$\gamma = k\rho c$	Product of thermal conductivity, density and specific heat
Θ	Liquid volume fraction
θ	Liquid-solid contact angle
θ_{a}	Advancing liquid-solid contact angle
μ	Viscosity
σ	Surface tension
κ	Curvature
ρ	Fluid density

References

1. Fukanuma H., Ohmori A. (1994). Behavior of molten droplets impinging on flat surfaces. *Proceedings of the 7th National Thermal Spray Conference*, 563–568.
2. Inada, S., & Yang, W.-J. (1994). Solidification of molten metal droplets impinging on a cold surface. *Experimental Heat Transfer, 7*, 93–100.
3. Pasandideh-Fard, M., Bhola, R., Chandra, S., & Mostaghimi, J. (1998). Deposition of tin droplets on a steel plate: Simulations and experiments. *International Journal of Heat and Mass Transfer, 41*, 2929–2945.
4. Aziz, S. D., & Chandra, S. (2000). Impact, recoil and splashing of molten metal droplets. *International Journal of Heat and Mass Transfer, 43*, 2841–2857.
5. Mehdizadeh, N. Z., Raessi, M., Chandra, S., & Mostaghimi, J. (2004). Effect of substrate temperature on splashing of molten tin droplets. *ASME Journal of Heat Transfer, 126*, 445–452.
6. Dhiman, R., & Chandra, S. (2005). Freezing-induced splashing during impact of molten metal droplets with high Weber numbers. *International Journal of Heat & Mass Transfer, 48*, 5625–5638.
7. Madejski, J. (1976). Solidification of droplets on a cold surface. *International Journal of Heat & Mass Transfer, 19*, 1009–1013.
8. Bennet, T., & Poulikakos, D. (1994). Heat transfer aspects of splat-quench solidification: Modelling and experiment. *Journal of Materials Science, 29*, 2025–2039.
9. Pasandideh-Fard, M., Qiao, Y. M., Chandra, S., & Mostaghimi, J. (1996). Capillary effects during droplet impact on a solid surface. *Physics of Fluids, 8*, 650–659.
10. Poirier, D. R., & Poirier, E. J. (1994). *Heat transfer fundamentals for metal casting* (2nd ed.pp. 41–42). Warrendale, PA: Minerals, Metals and Materials Society.
11. Wang, G. X., & Matthys, E. F. (1996). On the heat transfer at the interface between a solidifying metal and a solid substrate. In E. F. Matthys & W. G. Truckner (Eds.), *Metal spinning, strip casting and slab casting*. Warrendale, PA: Minerals, Metals and Materials Society.
12. Liu, W., Wang, G. X., & Matthys, E. F. (1995). Thermal analysis and measurements for a molten metal drop impacting on a substrate: cooling, solidification and heat transfer coefficient. *International Journal of Heat and Mass Transfer, 38*, 1387–1395.
13. Loulou, T., Artyukhin, E. A., & Bardon, J. P. (1999). Estimation of thermal contact resistance during the first stages of metal solidification process: II- experimental set-up and results. *International Journal of Heat and Mass Transfer, 42*, 2119–2127.
14. Aziz, S. D., & Chandra, S. (2000). Impact recoil and splashing of molten metal droplets. *The International Journal of Heat and Mass Transfer, 43*, 2841–2857.
15. Wang, W., & Qiu, H. H. (2002). Interfacial thermal conductance in rapid contact solidification process. *International Journal of Heat and Mass Transfer, 45*, 2043–2053.
16. Heichal, Y., & Chandra, S. (2005). Predicting thermal contact resistance between molten metal droplets and a solid surface. *Journal of Heat Transfer, 127*, 1269–1275.
17. McDonald, A., Moreau, C., & Chandra, S. (2007). Thermal contact resistance between plasma-sprayed particles and flat surfaces. *International Journal of Heat and Mass Transfer, 50*, 1737–1749.
18. Pershin, V., Lufita, M., Chandra, S., & Mostaghimi, J. (2003). Effect of substrate temperature on adhesion strength of plasma-sprayed nickel coatings. *Journal of Thermal Spray Technology, 12*, 370–376.
19. Cedelle, J., Vardelle, M., & Fauchais, P. (2006). Influence of stainless steel substrate preheating on surface topography and on millimetre and micrometer sized splat formation. *Surface and Coatings Technology, 201*, 1373–1382.
20. Li, C.-J., & Li, J.-L. (2004). Evaporated-gas-induced splashing model for splat formation during plasma spraying. *Surface and Coatings Technology, 184*, 13–23.

21. Mehdizadeh, N. Z., Lamontagne, M., Moreau, C., Chandra, S., & Mostaghimi, J. (2005). Photographing impact of molten molybdenum particles in a plasma spray. *Journal of Thermal Spray Technology, 14*, 354–361.
22. McDonald, A., Lamontagne, M., Moreau, C., & Chandra, S. (2006). Impact of plasma-sprayed metal particles on hot and cold glass surfaces. *Thin Solid Films, 514*, 212–222.
23. Shakeri, S., & Chandra, S. (2002). Splashing of molten tin droplets on a rough steel surface. *International Journal of Heat and Mass Transfer, 24*, 4561–4575.
24. Shinoda, K., Raessi, M., Mostaghimi, J., Yoshida, T., & Murakami, H. (2009). Effect of substrate concave pattern on splat formation of yttria-stabilized zirconia in atmospheric plasma spraying. *Journal of Thermal Spray Technology, 18*, 609–618.
25. Mani, M., Mandre, S., & Brenner, M. P. (2009). Precursors to splashing of liquid droplets on a solid surface. *Physical Review Letters, 102*, 134502.
26. Mani, M., Mandre, S., & Brenner, M. P. (2010). Events before droplet splashing on a solid surface. *Journal of Fluid Mechanics, 647*, 163–185.
27. Mehdi-Nejad, V., Mostaghimi, J., & Chandra, S. (2003). Air bubble entrapment under an impacting droplet. *Physics of Fluids, 15*, 173–183.
28. Xu, L., Zhang, W. W., & Nagel, S. R. (2005). Drop splashing on a dry smooth surface. *Physical Review Letters, 94*, 184505.
29. Rioboo, R., Tropea, C., & Marengo, M. (2001). Outcomes from a drop impact on solid surfaces. *Atomization and Sprays, 11*, 155–165.
30. Mundo, C., Sommerfeld, M., & Tropea, C. (1995). Droplet-wall collisions: experimental studies of the deformation and breakup process. *International Journal of Multiphase Flow, 21*, 151–173.
31. Cossali, G. E., Coghe, A., & Marengo, M. (1997). The impact of a single drop on a wetted solid surface. *Experiments in Fluids, 22*, 463–472.
32. Xu, L., Barcos, L., & Nagel, S. R. (2007). Splashing of liquids: Interplay of surface roughness with surrounding gas. *Physical Review E, 76*, 066311.
33. Dhiman, R., McDonald, A., & Chandra, S. (2007). Predicting splat morphology in a thermal spray process. *Surface and Coatings Technology, 201*, 7789–8801.
34. Fukanuma, H. (1994). A porosity formation and flattening model of an impinging molten particle in thermal spray coatings. *Journal of Thermal Spray Technology, 3*, 33–44.
35. Cirolini, S., Harding, J. H., & Jacucci, G. (1991). Computer simulation of plasma-sprayed coatings – I. Coating deposition model. *Surface and Coatings Technology, 48*, 137–145.
36. Xue, M., Chandra, S., & Mostaghimi, J. (2006). Investigation of splat curling up in thermal spray coatings. *Journal of Thermal Spray Technology, 15*, 531–536.
37. Brackbill, J., Kothe, D., & Zemach, C. (1992). A continuum method for modeling surface tension. *Journal of Computational Physics, 100*, 335–354.
38. Kothe, D. B. (1998). Perspective on Eulerian finite volume methods for incompressible interfacial flows. In H. C. Kuhlmann & H. J. Rath (Eds.), *Free surface flows* (pp. 267–331). New York: Springer.
39. Hirt, C. W., & Nichols, B. D. (1981). Volume of fluid (VOF) method for the dynamics of free boundaries. *Journal of Computational Physics, 39*, 201–225.
40. Osher, S., & Fedkiw, R. (2001). Level set methods: An overview and some recent results. *Journal of Computational Physics, 169*, 463–502.
41. Afkhami, S., & Bussmann, M. (2008). Height functions for applying contact angles to 3D VOF simulations. *International Journal of Numerical Methods in Fluids, 61*, 827–847.
42. Raessi, M., Mostaghimi, J., & Bussmann, M. (2007). Advecting normal vectors: A new method for calculating interface normal and curvatures when modeling two-phase flows. *Journal of Computational Physics, 226*, 774–797.
43. Bussmann, M., Mostaghimi, J., & Chandra, S. (1999). On a three-dimensional volume tracking model of droplet impact. *Physics of Fluids, 11*, 1406–1417.

44. Cao, Y., Faghri, A., & Chang, W. S. (1989). A numerical analysis of Stefan problems for generalized multi-dimensional phase-change structures using the enthalpy transforming model. *International Journal of Heat and Mass Transfer, 32*, 1289–1298.
45. M. Pasandideh-Fard (1998). *Droplet impact and solidification in a thermal spray process*. Ph. D. thesis, Department of Mechanical and Industrial Engineering, University of Toronto, Toronto, Canada.
46. S. Alavi, M. Pasandideh-Fard, J. Mostaghimi (2012) *Simulation of fluid flow and heat transfer including phase change during the impact of semi-molten particles*. ASME 2012 Heat Transfer Summer Conference, Rio Grande, Puerto Rico, USA, July 8–12.
47. Ghafouri-Azar, R., Mostaghimi, J., & Chandra, S. (2004). Numerical study of solidification of a droplet over a deposited frozen splat. *International Journal of Computational Fluid Dynamics, 18*, 133–138.
48. Pasandideh-Fard, M., Pershin, V., Chandra, S., & Mostaghimi, J. (2002). Splat shapes in a thermal spray coating process: Simulations and experiments. *Journal of Thermal Spray Technology, 11*, 206–217.
49. M. Raessi, J. Mostaghimi, M. Bussmann (2005). Droplet impact during the plasma spray coating process – Effect of surface roughness on splat shapes. *Proceedings of 17th international symposium on plasma chemistry, Toronto, Ontario, Canada*, 916–917.
50. Parizi, H. B., Rosenzweig, L., Mostaghimi, J., Chandra, S., Coyle, T. W., Salimi, H., Pershin, L., McDonald, A., & Moreau, C. (2007). Numerical simulation of droplet impact on patterned surfaces. *Journal of Thermal Spray Technology, 16*, 713–721.
51. Wu, T. C. M., Bussmann, M., & Mostaghimi, J. (2009). The impact of partially molten YSZ particle. *Journal of Thermal Spray Technology, 18*, 957–964.
52. Alavi, S., Pasandideh-Fard, M., & Mostaghimi, J. (2012). Simulation of semi-molten particle impacts including heat transfer and phase change. *Journal of Thermal Spray Technology, 21*, 1278–1293.
53. Amon, C. H., Schmaltz, K. S., & Prinz, F. B. (1996). Numerical and experimental investigation of interface bonding via substrate remelting of an impinging molten metal droplet. *ASME Journal of Heat Transfer, 118*, 164–172.
54. Orme, M., & Huang, C. (1997). Phase change manipulation for droplet-based solid freeform fabrication. *Transactions of the ASME, 119*, 818–823.
55. Fang, M., Chandra, S., & Park, C. B. (2007). Experiments on remelting and solidification of molten metal droplets deposited in vertical columns. *ASME Journal of Manufacturing Science and Engineering, 129*, 311–318.
56. Fang, M., Chandra, S., & Park, C. B. (2009). Heat Transfer during deposition of molten aluminum alloy droplets to build vertical columns. *Journal of Heat Transfer, 131* .paper 112101
57. Ghafouri-Azar, R., Shakeri, S., Chandra, S., & Mostaghimi, J. (2003). Interactions between molten metal droplets impinging on a solid surface. *International Journal of Heat and Mass Transfer, 46*, 1395–1407.
58. Bergmann, D., Fritsching, U., & Bauckhage, K. (2000). A mathematical model for cooling and rapid solidification of molten metal droplets. *International Journal of Thermal Sciences, 39*, 53–62.
59. Mathur, P., Apelian, D., & Lawley, A. (1989). Analysis of the spray deposition process. *Acta Metallurgica, 31*, 429–443.
60. Bergmann, D., & Fritsching, U. (2004). Sequential thermal modelling of the spray-forming process. *International Journal of Thermal Sciences, 43*, 403–415.
61. McPherson, R., & Shafer, B. V. (1982). Interlamellar contact within plasma-sprayed coatings. *Thin Solid Films, 97*, 201–204.
62. Xue, M., Chandra, S., Mostaghimi, J., & Salimijazi, H. R. (2007). Formation of pores in thermal spray coatings due to incomplete filling of crevices in patterned surfaces. *Plasma Chemistry and Plasma Processing, 27*, 647–657.

63. Cai, W. D., & Lavernia, E. J. (1997). Modeling of porosity during spray forming. *Materials Science and Engineering, A226–228*, 8–12.
64. Ghafouri-Azar, R., Mostaghimi, J., Chandra, S., & Charmchi, M. (2003). A stochastic model to simulate the formation of a thermal spray coating. *Journal of Thermal Spray Technology, 12*, 54–69.
65. Ghafouri-Azar, R., Mostaghimi, J., & Chandra, S. (2006). Development of residual stresses in thermal spray coatings. *Computational Materials Science, 35*, 13–26.
66. Xue, M., Chandra, S., Mostaghimi, J., & Moreau, C. (2008). A stochastic model to predict the microstructure of plasma sprayed zirconia coatings. *Modelling and Simulations in Material Science and Engineering, 16*, 065006.
67. Parizi, H. B., Mostaghimi, J., Pershin, L., & Jazi, H. S. (2010). Analysis of the microstructure of thermal spray coatings: A modeling approach. *Journal of Thermal Spray Technology, 19*, 736–744.
68. Gingold, R. A., & Monaghan, J. J. (1977). Smoothed particle hydrodynamics: Theory and appications to non-spherical stars. *Monthly Notices of the Royal Astronomical Society, 181*, 375–389.
69. Lucy, L. B. (1977). A numerical approach to the testing of the fission hypothesis. *Astronomical Journal, 82*, 1013–1024.
70. Liu, P. L. F., Yeh, H., & Costas, S. (Eds.). (2008). *Advances in coastal and ocean engineering: Advanced numerical models for simulating tsunami waves and runup*. Singapore: World Scientific Publishing 10.
71. B. Cartwright, P. H. L. Groenenboom, D. Mcguckin (2004). Examples of ship motions and wash predictions by smoothed particle hydrodynamics, *9th international symposium on the practical design of ships and other floating structures, Germany*.
72. Hu, X., & Adams, N. (2006). A multi-phase SPH method for macroscopic and mesoscopic flows. *Journal of Computational Physics, 213*, 844–861.
73. Farrokhpanah, A., Samareh, B., & Mostaghimi, J. (2015). Applying contact angle to a two dimensional multiphase smoothed particle hydrodynamics model. *Journal of Fluids Engineering, 137*, 041303–041301.
74. Monaghan, J. (2012). Smoothed particle hydrodynamics and its diverse applications. *Annual Review of Fluid Mechanics, 44*, 323–346.
75. Das, A. K., & Das, P. K. (2009). Simulation of drop movement over an inclined surface using smooth particle hydrodynamics. *Langmuir, 25*, 11459–11466.
76. Das, A. K., & Das, P. K. (2010). Equilibrium shape and contact angle of sessile drops of different volumes – Computation by SPH and its further improvement by DI. *Chemical Engineering Science, 65*, 4027–4037.
77. Kistler, S. F. (1993). The hydrodynamics of wetting. In J. C. Berg (Ed.), *Wettability* (pp. 311–429). New York: Marcel Dekke.
78. Šikalo, Š., Wilhelm, H. D., Roisman, I. V., Jakirli, S., & Tropea, C. (2005). Dynamic contact angle of spreading droplets: Experiments and simulations. *Physics of Fluids, 17*, 062103.

Chapter 6
In-Situ, Real Time Diagnostics in the Spray Forming Process

Pooya Delshad Khatibi, Hani Henein, and Udo Fritsching

The structure and material properties of spray formed products depend directly on the thermal state of the semi-solid droplets before their impact, of the substrate and of the already deposited layer. Monitoring specific droplet properties as i.e. droplet temperature, velocity and size as well as mass and enthalpy fluxes provide a unique tool for optimizing the material properties as well as controlling spraying conditions during deposition (as sketched in Fig. 6.1).

In-situ, real time diagnostics of metal melt atomization processes and in the spray forming process is a challenging task due to the hostile environment in such high temperature spray processes. However, such data is required in order to drive proper process understanding, validate process models leading to improved process control. Such efforts yield improved product quality. In the spray forming process, proper process control is necessary to achieve the desired product shape and microstructure, while minimizing overspray. In-situ real time diagnostic have been reported in measuring melt mass flow rates, in flight droplet/particle characteristics such as velocity, size and temperature distributions.

During the most commonly used form of the spray forming process, a continuous molten metal stream is atomized by impinging high speed inert gas jets with a very high velocity (up to the sound velocity). In the generated spray cone the broad spectrum of the generated droplet size distribution (app. 5–500 μm) depends on the type of the used material, the gas to metal mass flow ratio (GMR), and the velocity of the inert gas in the atomization area, the droplets reach the substrate with a velocity in a range of app. 70–140 m/s.

P. Delshad Khatibi (✉) • H. Henein
Advanced Materials and Processing Lab, University of Alberta, Edmonton, AB, Canada, T6G 2G6
e-mail: delshadk@ualberta.ca; hhenein@ualberta.ca

U. Fritsching
Foundation Institute of Materials Science, University of Bremen, Bremen, Germany
e-mail: ufri@iwt.uni-bremen.de

H. Henein et al. (eds.), *Metal Sprays and Spray Deposition*,
DOI 10.1007/978-3-319-52689-8_6

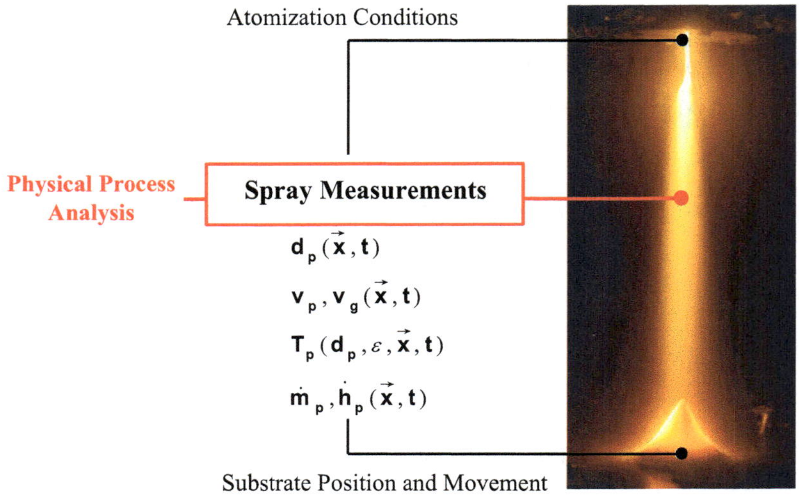

Fig. 6.1 Spray diagnostics for spray forming process control (from [1], reproduced with permission)

In a molten metal droplet spray from a gas atomization process, smaller droplets (5–50 μm) cool very fast and may completely solidify after a short flight distance to solid metal powder particles [2, 3]. Larger droplets (50–100 μm) contain a higher amount of latent heat and thermal energy, thus still fly during the state of phase change (semi solid) or even still in a completely liquid state (>100 μm) [4]. The impacting droplets, in different thermal states and sizes, form a partly solidified layer on top of the sprayed product. In the rest of this chapter, the term droplet(s) or particle(s) will be used interchangeably to mean droplets that are liquid, semi-solid or fully solid. When a distinction is required this will be made clear.

The reproducibility of products obtained by spray forming processes strongly depends on the ability to monitor and control key variables of the process. For this reason there is a tremendous interest for on-line measurements of particle temperature, velocity, diameter, mass flux and specific enthalpy in the spray cone to detect any changes that could affect the quality of the deposited material. Those measurements help to control the atomizing conditions as well as the substrate position and movement. Since spray formed products are essentially an agglomeration of particles projected on the substrate in a molten or partially molten state, the diameter, velocity and especially the temperature of those particles are variables that best characterize the spraying conditions.

The measurement of in-situ droplet characteristics (e.g. size, concentration, velocity) have relied on any of the following principles:

- Optical sensing (high-speed imaging, shadowgraphy, particle image velocimetry)
- Thermal radiation sensing (pyrometry, infrared imaging)

- Light scattering sensing (laser and phase Doppler anemometry, diffraction analysis, spectroscopy).

Some of the measuring instruments that have been used in this context:

- Shadowgraphy (Particle Master from LaVision GmbH in Göttingen, Germany) to measure droplet velocity and diameter, Shadowgraph is a high-magnification shadow imaging system, which is used for visualizing particles.
- High-Speed-Pyrometry DPV-2000 (Tecnar Automation Ltée, St. Hubert Quebec, Canada) for droplet radiation measurements aiming to measure droplet velocities, sizes and temperatures. The DPV-2000 is a high-speed two colors pyrometer that measures the effective radiant energy of the particles.
- Phase-Doppler-Anemometry PDA (University Bremen, Bremen, Germany as well as DANTEC—Kopenhagen, Denmark) for measurement of droplet size and velocity utilizing the analysis of light scattering from moving particles.

Principles of operation and limitations associated with the measurement instruments will be addressed. Emissivity behavior of the falling droplets is also investigated using a thermal model of droplet cooling coupled with the temperature of primary phase undercooling. Estimation of droplet temperature is discussed followed by analysis of the precision of in-situ, real time measurements.

6.1 Principles of In-situ Particle Diagnostics and Sensors

Solidification of alloys is a complex phenomenon arising in many modern experimental techniques and industrial technologies related to casting and surface processing. The variation of different solidification conditions (such as undercooling or cooling rate) provides the possibility to control the morphology and size of crystal structures, which substantially influences physical and chemical properties of alloys.

It is well known that SDAS (secondary dendrite arm spacing) is strongly influenced by cooling rate, as shown in [5], where SDAS is the secondary dendrite arm spacing, $\frac{\Delta T}{t_{sL}}$ is the solidification cooling rate composed of a solidification temperature range ΔT and a solidification time t_{SL}, and B and n are experimentally-determined constants with units of μm (K/s) and dimensionless, respectively.

$$SDAS = B\left(\frac{\Delta T}{t_{SL}}\right)^{-n} \tag{6.1}$$

In addition to the cooling rate, undercooling of the liquid prior to grain nucleation or the formation of second and subsequent phases has a significant effect on the final microstructure of metallic alloys. For example, Behulova et al. [6] showed that in metal powder generated by rapid solidification, a variation of morphological features correlated with a variation of particle sizes. Thus the objective of many efforts to make in-situ measurements is ultimately to identify the thermal history of the droplets. Such efforts are complicated by the very complex nature of atomization of molten metals. While the focus of these measurements is aimed to understand the gas atomization process, several measurements have been reported using a single fluid technique.

The complexities of gas atomization involve coupled transport phenomena of heat, momentum and mass transfer. The process is further complicated by the chemistry of the alloy to be atomized as well as its reactivity. In gas atomization, gas jets are directed onto a melt stream falling into the atomizing chamber. The break-up of the stream results in the generation of a multitude of droplets that flow in a high velocity gas stream exchanging heat with the gas and undergoing spheroidization of each droplet when an inert gas is used. The droplets continue to flow with the gas stream, losing heat and approaching the deposit in spray forming. Typically the droplets undercool below the liquidus of the primary phase. This is followed, at some distance away from the gas nozzle, by the nucleation of the primary phase in each droplet. As a droplet size distribution is generated, different droplets undergo nucleation at different distances away from the gas nozzle. Furthermore, the concentration of droplets in the cross section of the spray varies as the droplets travel towards the deposit. This simple overview of the atomization process provides the context for the in-situ measurements to be described in this chapter.

A comprehensive overview of the possible instruments that could be considered for in-situ measurements has been presented by Jiang et al. [7]. In this chapter, we present those instruments that have been used by researchers for in-situ measurements in atomizers as well as the results reported. These instruments include:

- conductivity of the melt for melt flow measurements,
- single particle counter technique (PCSV)
- ensemble particle counter (EPSV)
- phase-doppler-anemometry (PDA)
- shadowgraphy
- Infra-red pyrometry

6.2 Melt Flowrate

The gas to metal ratio is a critical parameter in gas atomization as it has a direct relation to the median droplet size generated. The gas flowrate into an atomizer may be readily measured either using flowmeters or inferred from the measurement of the supplied gas pressure. The measurement of the melt is somewhat more complex. This is due to melt reactivity and high temperature.

Le et al. developed a conductivity probe to measure the flowrate of molten metal in a draining crucible during atomization of zinc [8]. Using graphite rods or varying

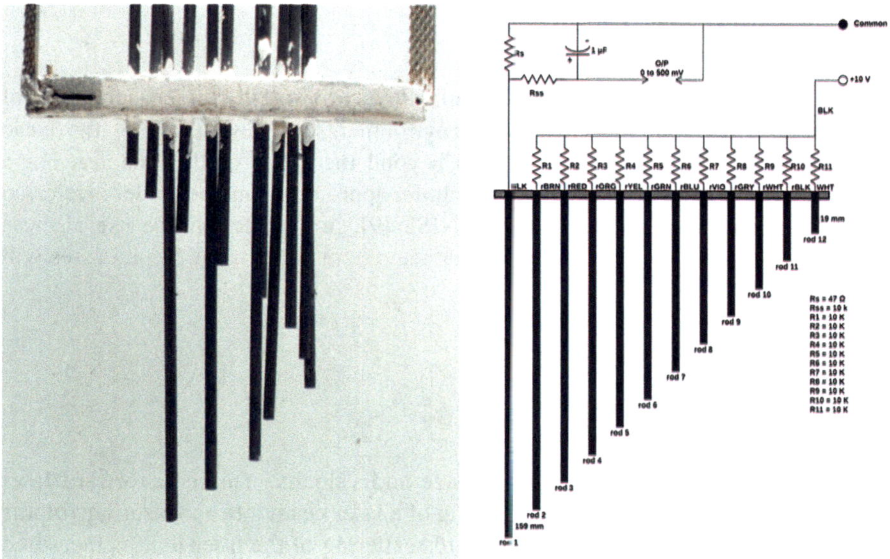

Fig. 6.2 Graphite rods of varying lengths to be immersed into molten metal (**a**) photograph of rods, (**b**) schematic of circuit [8]

lengths and immersed into a bath of molten zinc, the height of the melt is recorded as it recedes from the crucible. An image of such a set-up and a schematic of the circuit is shown in Fig. 6.2a, b. The principle of the technique is based on a measurement of the changing resistivity of the melt as the height in the crucible decreases. Careful placement of the graphite rods at known heights provides a direct conversion of the resistivity of the melt to position of the melt surface. Using this technique, Le et al. characterized the discharge coefficient of the delivery tube transporting the melt from the crucible to the atomizing chamber. The role of gas jets on accelerating the melt flow was quantitatively characterized. It was also shown that the inside roughness of the delivery tube results in an increase in flow resistance for the melt, yielding erratic melt flowrate. Such a sensor must be modified for steel melts due to the dissolution of graphite in steel. By using ceramic rods and a circuit that measures the capacitance of the melt rather than its resistance the flowrate of steel may be measured.

In the work described by Le et al. [8], ten graphite rods were used. This was necessitated by the fact that the melt was fed into the atomizer directly from the induction melt crucible. In operations where the melt crucible is poured into a tundish which in turn feeds melt into the delivery tube, fewer rods would be required. In this instance, efforts would be directed to maintain a constant head of melt in the crucible.

6.3 Laser Based Techniques

A number of laser based sensors are available to carry out in-situ measurements of droplets and are adaptable to hostile environments. A description of the basic principles of each of the techniques is beyond the scope of this chapter. For a good description of Mie scattering, Fraunhofer approximation, the reader is referred to an excellent overview published by NIST [9]. In this section, the use of these techniques as they have been used to characterize melt atomization processes will be described.

6.3.1 PCSV-P

PCSV-P stands for particle counting, size and velocity. This is a forward laser scattering technique based on the focusing of a laser beam into a measuring volume through which droplets flow. The forward scattering of the laser light is measured by a detector which after passing through an algorithm provides the size and velocity of each droplet going through the measuring volume. The PCSV is designed to fit into a probe and can be place inside a water cooled jacket. Because the operator fixes the location of the measuring volume, the droplet characteristics of concentration, size and velocity are determined with respect to a specific location in the spray. Some of the PCSV characteristics used for evaluating the atomization of zinc are shown in Table 6.1. The unit is composed of two laser beams each having a different measuring volume and therefore size range capability.

To carry out the in-situ measurements of droplet size, two ports were machined into an atomizing shell 0.91 and 2.4 m downstream of the nozzle. Zinc atomization was carried out using a Coanda nozzle [10] with air and argon as the atomizing gases. Repeat measurements with the PCSV were taken during an atomization experiment and compared to another with higher atomizing gas pressure. The results are shown in Fig. 6.3 [11]. The four measurements taken at 862 kPa run gas pressure are very reproducible. Furthermore, the expected trend of smaller particle size with higher atomizing gas pressure is readily apparent. The challenge with a PCSV system is that the location of the measuring volume must be carefully selected. A comparison with in-situ measurements taken using laser diffraction technique will now be compared with the PCSV Mie scattering technique.

Table 6.1 PCSV characteristics

	Small beam	Large beam
Laser type	He-Ne (divided into two lasers	
Waist size (mm)	20	300
Particle size range (mm)	0.2–5	5–200
Velocity range (m/s)	0.1 to supersonic	
Operating temperature	Up to 1400 °C	
Particle concentration	10^7 particles/cm^3	100 ppm

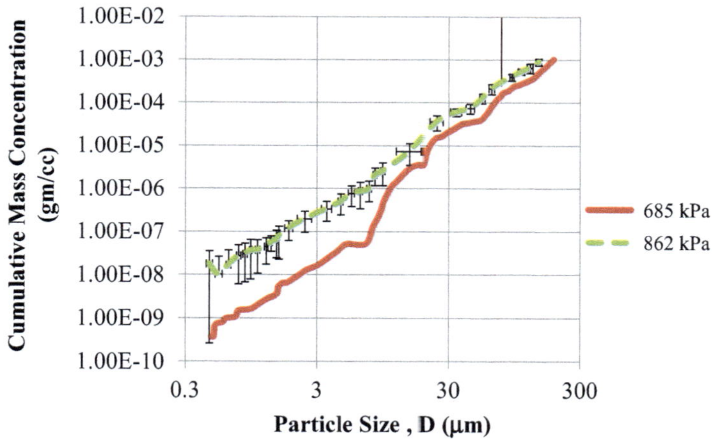

Fig. 6.3 Size distribution of zinc powder measured in-situ for 685 kPa (100 psi) and 862 kPa (125 psi) measured 0.91 m below the gas nozzle [11]

Table 6.2 Run conditions for comparing PCSV and EPCS

P_{N2} (kPa)	$\dot{G}/\dot{M}^a$ (avg)	T_{Zn} (K)
690	0.75	797
862	1.45	885
1034	1.45	897
862	1.17	885
1034	1.23	887

[a]Ratio of average gas to metal mass flowrates

6.3.2 *EPSV*

Laser diffraction is a commonly used technique for laboratory measurements of samples of powders. A version that has adapted for in-situ measurements is an ensemble particle sizing instrument (EPCS) [12]. A He-Ne laser is collimated and directed through a stream of powders flowing in a gas. When the laser intercepts a particle, it is diffracted at an angle in proportion to the size of the particle. Only one diffraction is assumed for each interaction with the laser beam. All the diffractions are recorded on a ring detector and the corresponding equivalent spherical size is reported. The size range that can be measured varies from 1 to 500 μm with particle loadings as high as 1 kg/m^3. Sampling intervals may be selected as desired. For the work to be reported a 5 s interval was used. This enables numerous measurements of size can be made during atomization. Thus, unlike the PCSV, it lends itself more easily as a process monitoring tool. The disadvantage of the EPCS system is that measurements must be made when the powders and gas are at low temperatures in order to reduce the effects of beam steering [13].

A series of atomization experiments were carried out with Coanda gas nozzles and zinc with both the EPCS and PCSV probes inserted into the tower exit. The melt throughput was measured using the graphite probes described earlier. The atomizing run conditions are shown in Table 6.2. The powder size results are shown

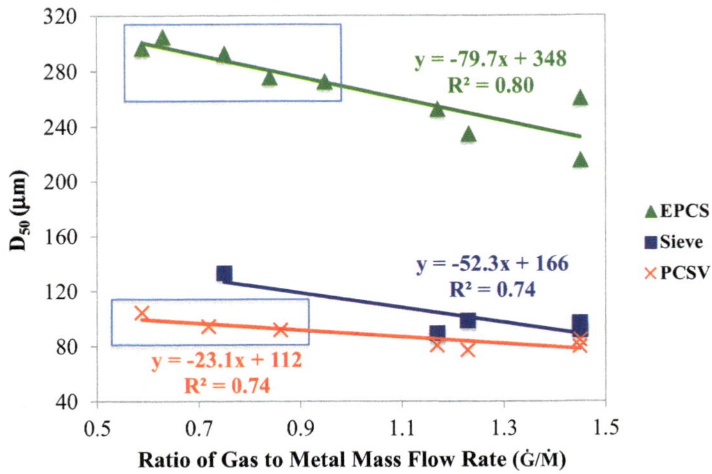

Fig. 6.4 $\dot{G}/\dot{M}$ versus median particle size for the atomization of zinc. A comparison of the in-situ measurements of powder size using EPCS and PSCV instruments compared with a sieve analysis of the powder atomized

in Fig. 6.4. The results from the EPCS and PCSV are compared with sieve analysis results of representative samples taken following the completion of each experiment (Sieve + Cyclone). The boxed data represent measurements taken at different times during the same atomization run. It is evident form the results in Fig. 6.3 that as the mass gas to metal ratio increases the median particle size also decreases. The EPCS seems to follow a similar slope of the collected sample. While the PCSV does not relay the dependence between the $\dot{G}/\dot{M}$ and the median powder size. Given the fast and repeated set of data that may be collected with the EPCS, this instrument can be considered for use in industry. In fact, similar instruments are found in use in plant operations in other industries such as food, pharmaceuticals and chemicals.

6.3.3 Light Scattering Sensing: Phase-Doppler-Anemometry

Phase-Doppler-Anemometry (PDA) is a laser based instrument for detecting droplets/particles size and velocity in sprays [14]. The PDA principle has been further developed to characterize even sprays of molten metal droplets and metal particle properties in flight. Applying this optical measurement technique to the spray-forming process (Fig. 6.5), the pointwise size and the velocity of the droplets within the spray can be determined [15, 16]. In addition, process quantities such as particle number flux and mass flux can be evaluated considering the PDA data. By extending PDA to an on-line measurement system, process analysis even very close to real time is possible, thus on-line PDA is employed to monitor and control the spray forming process. However, the operator training required to correctly use

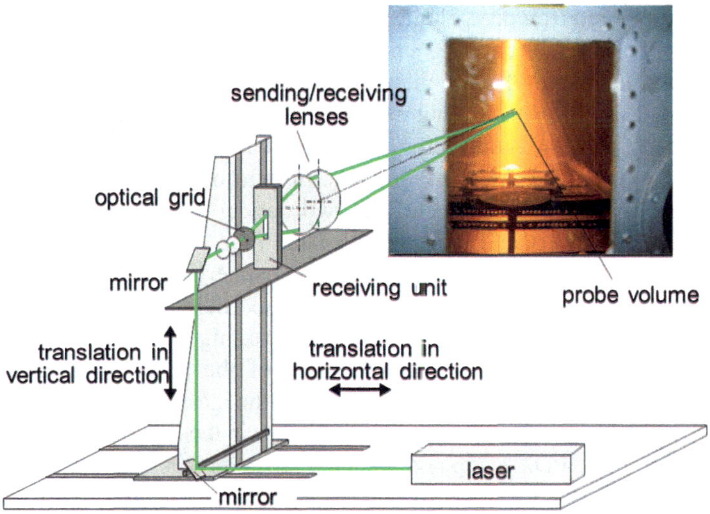

Fig. 6.5 PDA applied at a spray forming facility with moveable measurement point location (adapted from [15])

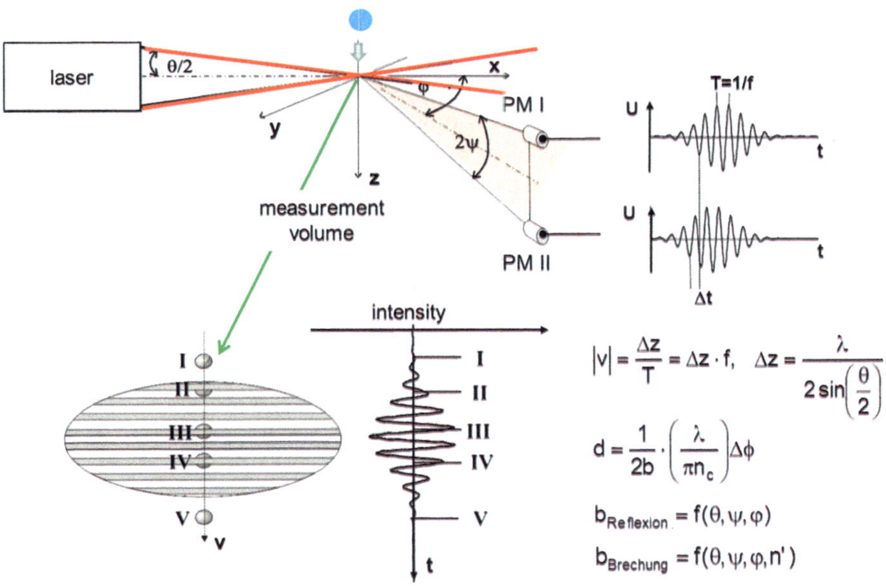

Fig. 6.6 Arrangement of a Phase-Doppler–Anemometer PDA in forward scattering arrangement (e.g. for water droplets) (adapted from [16])

the PDA is far greater than that for the EPCS system as it is a more complex instrument.

A PDA device basically consists of two laser beams and two optical detectors (Fig. 6.6, left). The two laser beams intersect, defining the PDA measuring volume (probe volume or interference volume), the structure of which may be characterized by parallel planes of alternating light intensity (fringe pattern) as a result of interference phenomena of the coherent electromagnetic waves. When a particle passes through the measuring volume the scattered light creates a corresponding signal voltage in the photodetectors (Doppler bursts) as illustrated in Fig. 6.6 (right).The frequency of the PDA bursts is proportional to the particle velocity. Owing to the spatial arrangement of the two optical detectors, a temporal delay between the PDA bursts can be observed. This time shift or phase difference is directly proportional to the particle diameter (for spherical droplets). Furthermore, as a measurement system counting individual particles, PDA gives additional information concerning the number of particles passing through the probe volume per unit time. This permits the determination of the particle number flux and (considering the particle diameter and its density) the mass flux. The specific optical properties of the (solidifying) droplets and particles in the liquid metal spray require a modification of the PDA setup and the burst signal processing. For application in this arrangement, typically reflective scattered light is detected by using a back-scatter arrangement of the laser/detector configuration. Owing to the high cooling rates of the droplets in flight, mainly the smaller particles may be partially or completely solidified after flight distances of about 0.10 m. At smaller distances to the atomizer, PDA measurements in the spray typically are not possible owing to the high particle concentration, thus leading to crossover signals of multiple particles in the measurement volume [17].

Metal droplet solidification in flight leads to rough particle surface structures (Fig. 6.7) and as a result the particles produce some diffuse scattered (reflected) light. In addition, when the laser is passing through the dust laden spray chamber, the laser light is scattered by these particles. This reduces the intensity and increases the noise within the scattered light. Therefore, the detected PDA signals (bursts) carefully need to be evaluated.

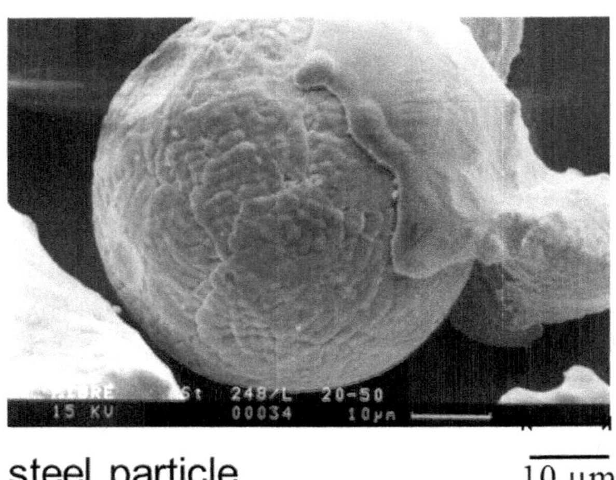

Fig. 6.7 Steel particle from gas atomization, surface solidification structure (adapted from [16])

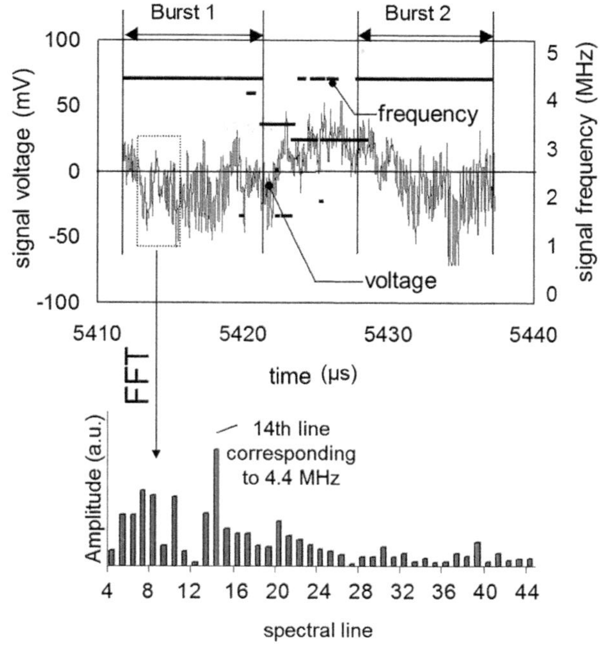

Fig. 6.8 FFT of PDA signals (adapted from [16])

As to be seen in Fig. 6.8, a Fourier-transform data processing may be applied to the PDA signals, a typical response of which is summarized in Fig. 6.8b [18, 19]. If the initial signal includes a dominating periodic portion, the resulting line spectrum contains a characteristic maximum peak (14th line in Fig. 6.8). The horizontal position of this peak is proportional to the frequency of the most powerful periodic signal fraction. To apply Fourier transformation to PDA signal analysis, the continuously detected scattered light signal is converted into frequency step by step. In contrast to a particle creating a constant predominating frequency versus time while passing through the probe volume (horizontal lines in Fig. 6.8), in the event of signal noise (the section between the two bursts in Fig. 6.8a) a steady variation of the predominating frequency can be observed. The length of a constant frequency band is a distinguishing mark for a signal formed by a single particle moving through the probe volume, thus Fourier transformation of PDA signals can be used to separate the bursts from signal noise. In this way, even bursts with a signal power much lower than the noise power can be detected and evaluated by considering the introduced criterion (low signal to noise ratio SNR).

A typical PDA-based analysis of the spray demonstrates the significant influence of the atomization gas pressure on the mean value of the particle diameter $d_{1,0}$ and velocity $v_{1,0}$ (Fig. 6.9). The higher the kinetic energy of the atomization gas brought to the disintegration process, the greater is the new melt surface area formed, i.e. the mean particle diameter decreases. As the kinetic energy of the atomization gas is increased by raising its pressure, the gas velocity is also increased and the droplets generated from atomization are accelerated to higher velocities.

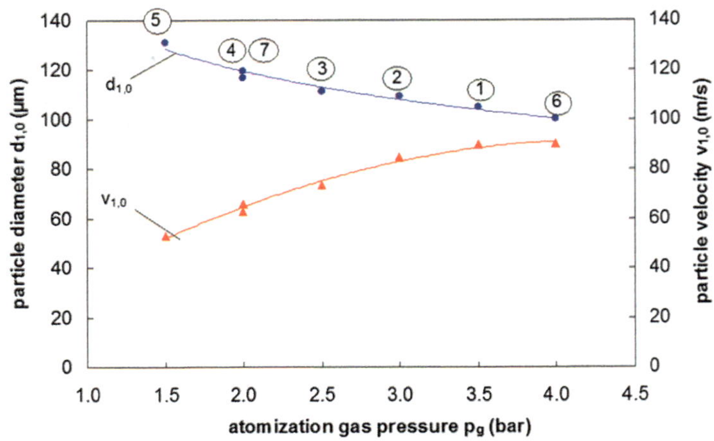

Fig. 6.9 PDA results in copper spray (adapted from [16])

The parameters, particle diameter and velocity, are essential to the heat transfer processes in the molten metal spray. These parameters define the thermal conditions in the process and thus affect the characteristics of the spray formed material such as its microstructure, grain size and chemical homogeneity. To monitor and control the spray forming process, the local mean particle diameter and velocity may be utilized as control parameters. Other components of a PDA processor are an analog-to-digital converter (ADC) and a digital signal processor (DSP) (Fig. 6.10). Both are adapted to a computer board. For signal processing the analog voltage delivered by the optical detectors is converted into a machine-usable digital signal by the ADC. Subsequently, the transformed signal is transmitted to the DSP where the bursts are separated from signal noise considering the Fourier-based algorithm. The detected bursts are then evaluated with reference to the particle diameter, the particle velocity and the mass flux. The resulting diameter, velocity and mass flux distributions provide information about the state of atomization and spray [15, 20].

Furthermore, the on-line PDA data can be used as input for the atomization process control loop (Fig. 6.11). The automatic gas control loop consists of the PDA measurement unit, the regulating valve and the control unit. This arrangement allows the automatic adjustment of the mean particle velocity or diameter by varying the atomization gas pressure. For illustration, sudden variations in the atomizing gas pressure was imposed in a sample spray run in order to examine the temporal behavior of the atomization process. Figure 6.12 summarizes the variations of the on-line measured mean value of the particle diameter $d_{1,0}$ (or D_{50}) and droplet velocity as well as the corresponding atomization pressure P_g during the experiment. In the case of particle diameter, both the actual and the desired value are shown, i.e. using the closed control loop a definite particle diameter $d_{1.0}$ set is proposed and the required adjustment of the current particle diameter $d_{1.0}$ is done by varying the atomization gas pressure is observed.

Fig. 6.10 PDA processor (adapted from [16])

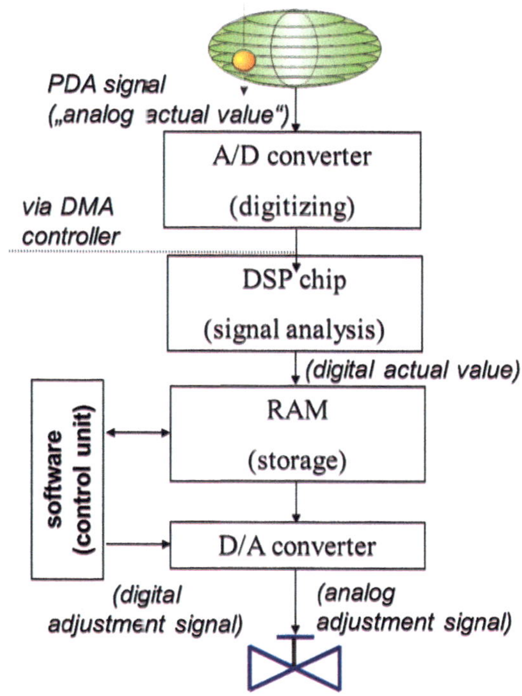

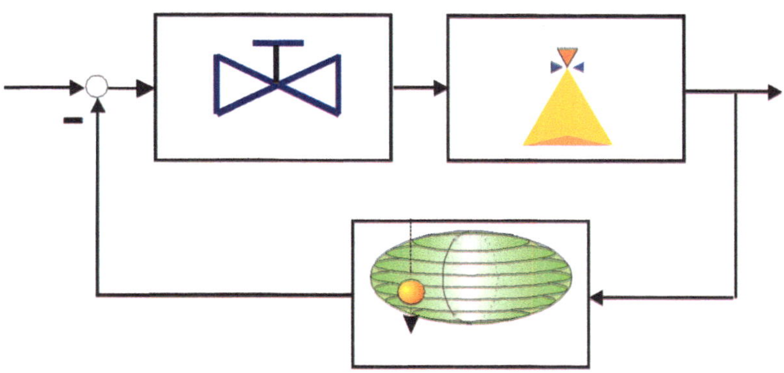

Fig. 6.11 Spray forming control cycle (adapted from [16])

The graphs agree with typical correlations of particle diameter and gas pressure. Therefore, with increasing gas pressure an immediate decrease in the mean particle diameter is reported by the PDA along with an increase in droplet velocity. By contrast, more time is needed to adjust the actual particle diameter to the set diameter, e.g. in Fig. 6.12 about 50 s is needed to achieve the necessary modification of the current particle diameter to the set value [17].

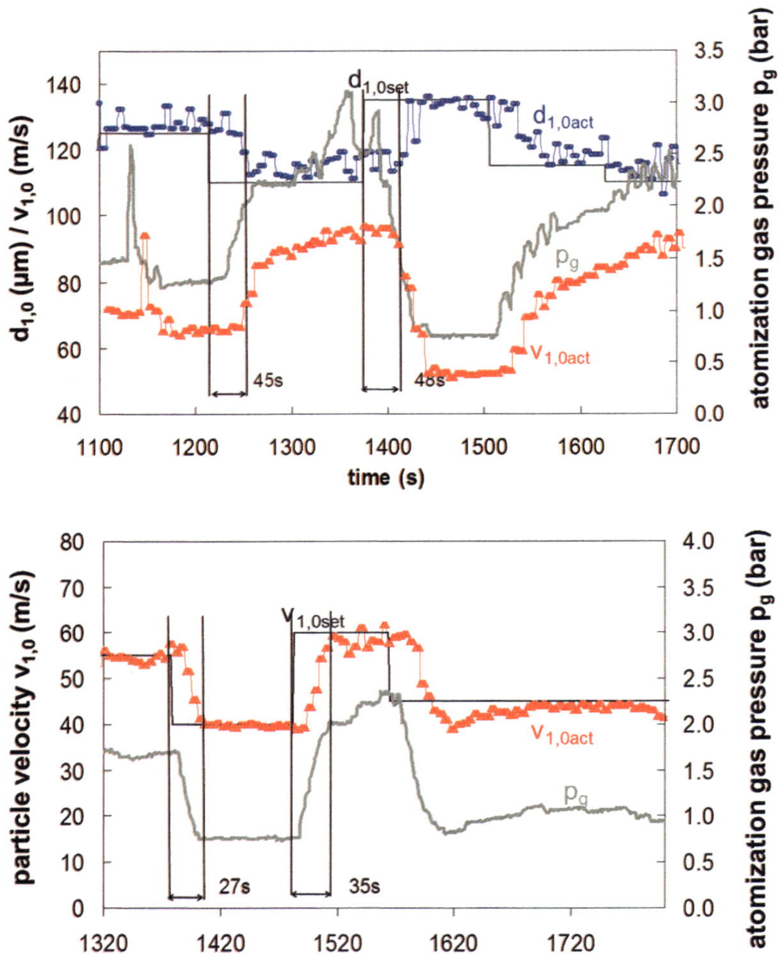

Fig. 6.12 Control of spray forming process (here: atomizer gas pressure variation) (adapted from [16])

Similar results were obtained for the particle velocity (Fig. 6.12). However, in comparison with the average value of the particle diameter, the mean particle velocity shows a greater dependency on the atomization gas pressure and the curve for adjusting velocity in Fig. 6.7b shows a stronger response. As a result, the time for automatic adjustment after changing the set value is reduced to approximately 30 s [17].

For the evaluation of mass flux, the number of particles passing through the measurement volume is counted and weighted with the individual particle size. Therefore, the particle number is sorted into the particle size class i and di is the

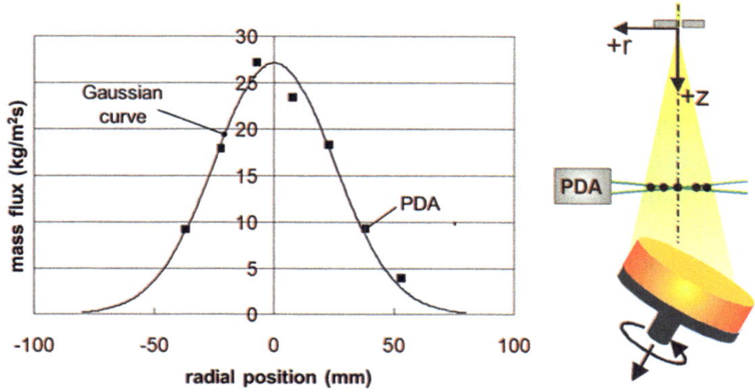

Fig. 6.13 PDA measured mass flux distribution in metal spray and comparison to analytical Gaussian distribution function (adapted from [16])

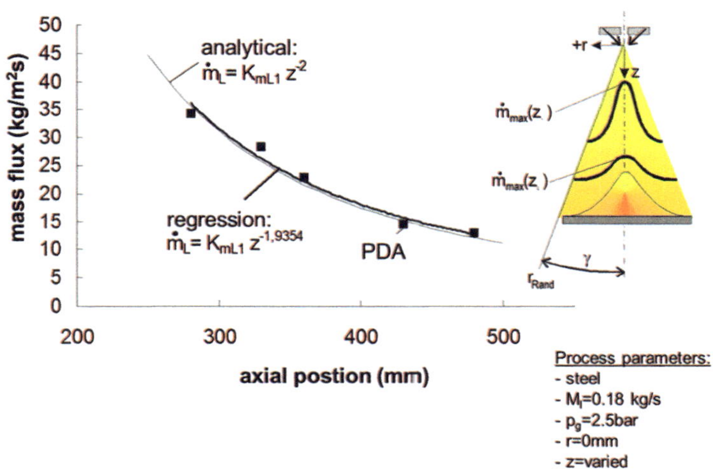

Fig. 6.14 Mass flux on steel spray centerline (adapted from [16])

mean diameter of the relevant size class. Dividing the total particle mass M_p by the cross-section of the probe volume A_{pv} and considering the measurement period t_m yields the PDA-measured mass flux. An example is shown in Fig. 6.13 in terms of the radial mass flux distribution in a steel spray at a distance of about 320 mm from the atomizer. The measurements are compared to a typical Gaussian distribution that has been found to fit the mass flux profile in sprays [21, 22]. In addition, Fig. 6.14 illustrates the behavior of the centerline mass flux with increasing distance from the atomizer. An empirical power law is regressed to the data and provides an excellent fit to the PDA measurements.

6.4 Optical Based Techniques

The Particle Master (LaVision GmbH in Göttingen, Germany) is a high-magnification shadow imaging system which is used for visualizing particles (Fig. 6.15). It is based on high resolution imaging with pulsed backlight illumination. This technique is independent of the shape and material of the particles and allows the investigation of sizes down to the micron scale.

In order to illuminate the falling droplets, a double pulse laser can be used. By coupling the light source with a double frame camera, it is also possible to investigate the velocity of the particles. Data from the shadowgraph will be used to get particle size, particle position, particle shape and velocity during free fall. The position of the shadowgraph can be controlled with a translation stage in all three axes.

A two-step segmentation algorithm is applied on the images taken by the shadowgraph. The first segmentation is to find the location of the particles which is called the bounding box. In this step, shadow picture of the droplet is subtracted from a reference image (background atmosphere image) and an inverted shadow image is then achieved (Fig. 6.16). For the first segmentation a user defined global threshold is used to detect the particles (Fig. 6.16). The global threshold is relative to the difference between maximum and minimum intensity of the inverted image. The droplets images which are above the global threshold are considered for the next step (#1, #2, #3 and #4 in Fig. 6.17). If the peak intensity is below the global threshold, that peak will be ignored for the further calculations.

In the second step of the segmentation, areas of the image which are above the global threshold are selected (Fig. 6.18 for particle #3). High level and low level

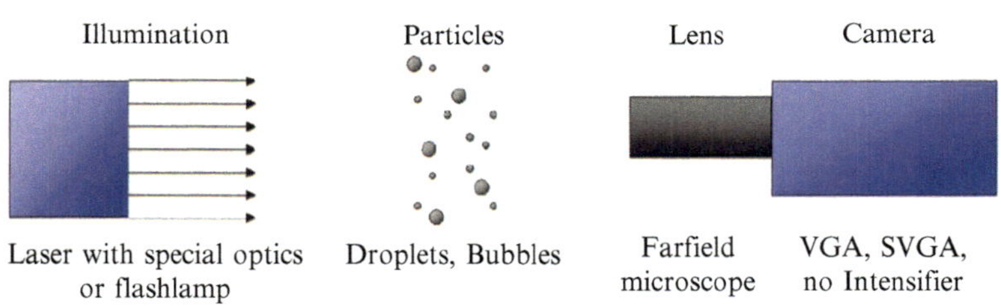

Fig. 6.15 Schematic view of a shadowgraph system [23] (courtesy: LaVision GmbH, Göttingen, Germany)

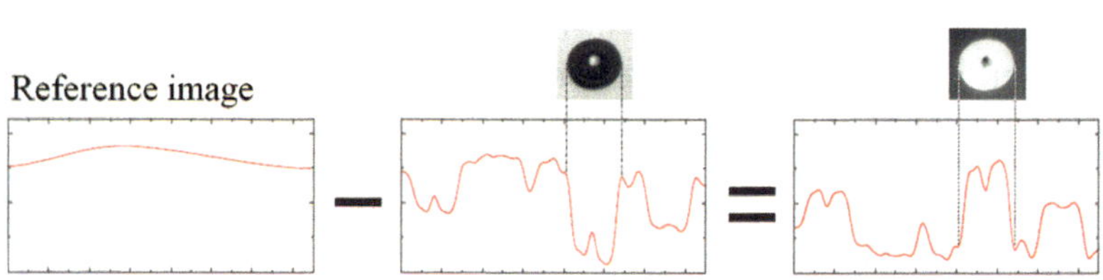

Fig. 6.16 Inverting the shadow image using a reference image [23] (courtesy: LaVision GmbH, Göttingen, Germany)

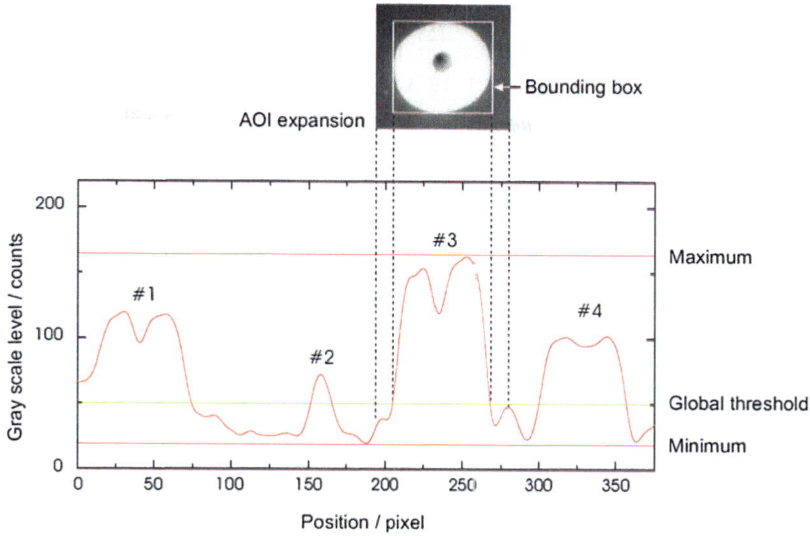

Fig. 6.17 Thresholds for the first segmentation [23] (courtesy: LaVision GmbH, Göttingen, Germany)

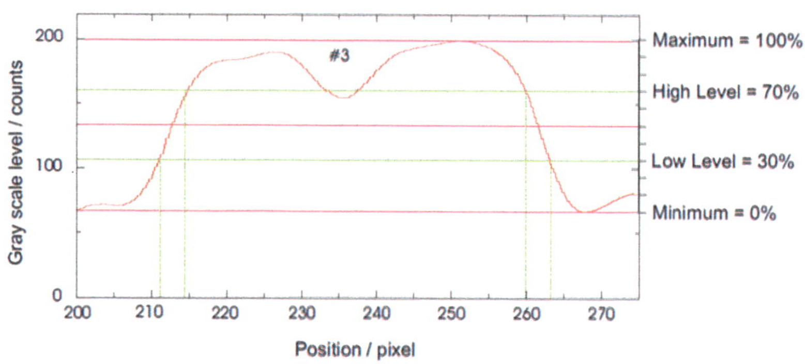

Fig. 6.18 Thresholds for the second segmentation [23] (courtesy: LaVision GmbH, Göttingen, Germany)

grey counts can be determined by user based on the image focus needed for the calculations (Fig. 6.18). If high values of "high level" counts and low values of "low level" counts are selected, less-focused droplets will be detected by the shadowgraph which results in errors in particle size calculations. Global threshold, high level and low level values of 60% and 40% were chosen respectively for the work described in this chapter [23].

In order to calculate the velocity of the droplets, the shadowgraph software first applies sizing algorithm to each frame of the source images and stores information about position and size of each droplet. Pictures taken from both frames then analyzed by the software and based on the location of the droplets in both pictures and time between each frame. The velocity of the droplet is then calculated. The droplets in two pictures are only accepted if the diameter deviation is within the given range (±5% diameter deviation in this work) [23]. The application of the shadowgraph to atomization of molten metals will be discussed in Sect. 6.4 above.

6.5 Pyrometery

6.5.1 *Description of the DPV Pyrometer Spray Monitoring System*

The high-speed-pyrometry system (DPV-2000 from Tecnar/Canada) has been specifically designed for thermal spray applications but a version has been tailored for the spray forming process. The monitoring system as illustrated in Fig. 6.19 consists of three main components: (a) the sensing head located near the spray cone collecting the thermal radiation emitted from the hot particles, (b) the detection cabinet containing the optical components and photodetectors and (c) a computer equipped with the required digitizing and computing boards (a high precision oscilloscope with a detection algorithm including adjustable rejection criteria for signal processing). A calibration module is used with two different calibration units: the first unit is for temperature calibration and the second is for diameter calibration, or for trial runs respectively. Both methods are based on a tungsten lamp as calibration standard that has been evaluated using a high precision pyrometer [24–26]. The lamps can be adjusted for four representative temperatures in a position in front of a rotating pinholed disk, representing an in-flight particle with a corresponding velocity.

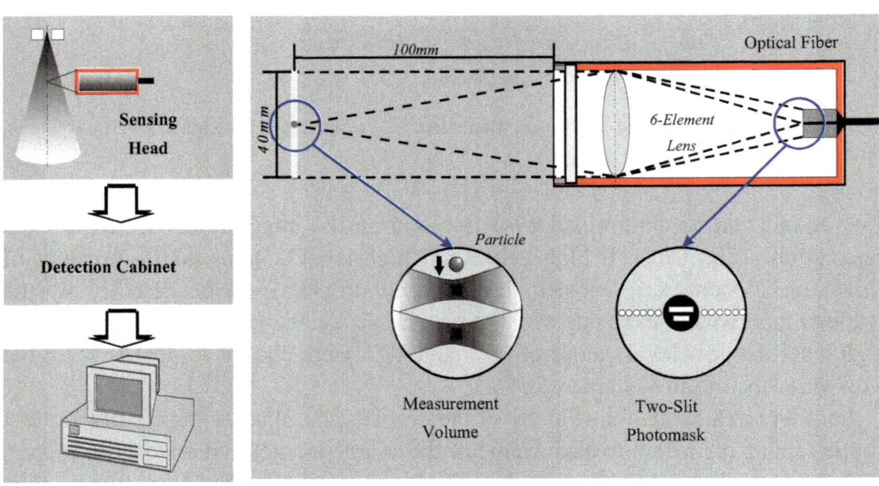

Fig. 6.19 System components and alignment with details of the sensing head (from [1], reprint with permission)

6.5.1.1 The Sensing Head

As illustrated in Fig. 6.19, the sensing head is aimed perpendicular to the spray cone (typical distance 100 mm) providing continuous monitoring of the spray process. Thermal radiation emitted by in flight particles is collected by this tubular shaped passive optic. Individual particles are detected as they pass though the sensors measurement volume, a 0.35 mm^3 region centered around the sensors focal point. Light from particles travelling in the focal plane is collected by a 6-element, 1.8 cm diameter lens (magnification factor 2.83) specially designed to minimize chromatic and spherical aberrations for wavelength ranging from $\lambda = 700$ to 1000 nm.

The collected light is imaged on the end of the optical fiber bundle composed of two distinct arrangements of fibers. The first arrangement consists of a group of 48, 200 μm core optical fibers whose ends are aligned along a straight line forming a 40 mm long linear array. As discussed in the next section, the light collected by this group of fibers is detected by a linear CCD line sensor providing on-line monitoring of the width and position of the particle jet relatively to the spray axis. The second arrangement consists of a 400 μm core fiber located in the middle of the array, whose end is covered by a two-slit photomask. Only the light impinging on the two transparent slits engraved on the opaque mask can reach the end of the optical fiber and thus be transmitted to the detection cabinet. Each slit width and the area between the slits has been designed that particles up to 600 μm in diameter will have their image totally included (*magnification factor x area between the slits*). The applied photomask consists of two different slit lengths for higher precision. They prevent the detection of particles not travelling straight through the measurement volume.

6.5.1.2 The Detection Cabinet

The collected radiation is transmitted through the aforementioned optical fiber bundle to a detection cabinet (see Fig. 6.20). The detection system for the central fiber (400 μm) is essentially a high-speed, two color pyrometer. Light from this fiber is focused through a dichronic mirror onto two photo-detectors, D_1 and D_2. During measurements in thermal spraying conditions (e.g. plasma guns) it is necessary to use an interference filter in front of each detector, F_1 and F_2 respectively. The bandpass of both filters is about $\lambda = \pm\ 25$ nm and their center wavelengths are for D_1: $\lambda_1 = 787$ nm and for D_2: $\lambda_2 = 995$ nm. The detectors are silicon avalanche photodiodes with integrated pre-amplification. The data is statistically compiled, displayed and saved.

The line array generated by the 48 smaller (200 μm) fibers is imaged onto a CCD line array. The determined intensity profile is also digitized and displayed from the control module along with the offset of the intensity peak from the central fiber. A beam splitter allows illumination of the fibers by a laser diode focused on the line array through a cylindrical lens. This enables visualization of the sensors field of

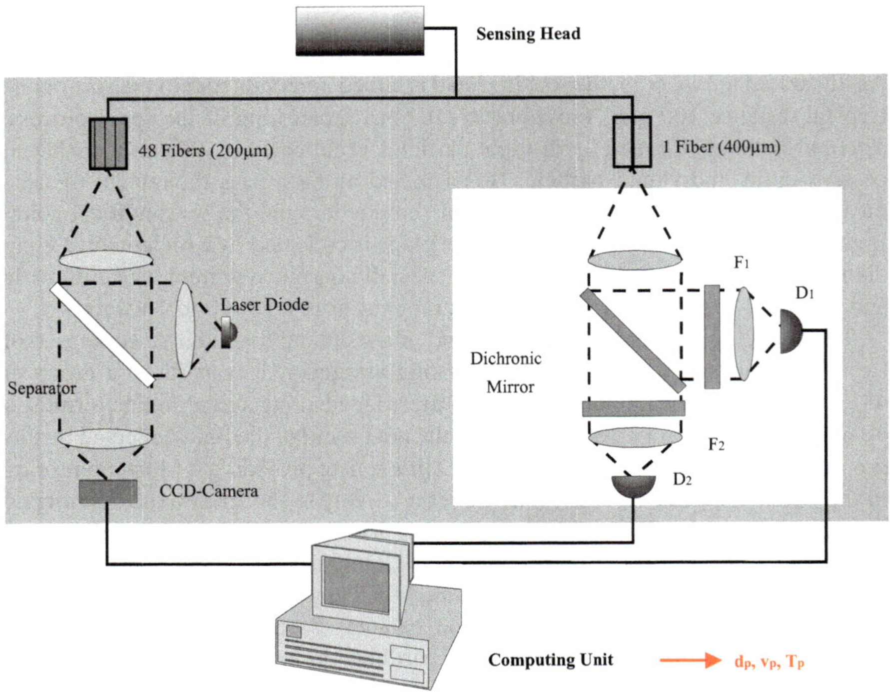

Fig. 6.20 Description of the detection module head (from [1], reprint with permission)

view for positioning the sensor head relative to the spray plume. After detection the signals are transmitted to the computer for analysis followed by the calculation of particle temperature, velocity and diameter.

6.5.2 Signal Processing and Calculation of the Parameters

Light from a particle travelling in the sensors focal plane enters the central fiber only when at least part of its image is located within one of the masks two slits. The characteristic signature of a particle crossing the sensors focal point is a two peak signal representing 100%, 0% and again 100% of its radiation within the corresponding detector bandwidth. It is caused by the two-slit photomask. The signal is transmitted to the computer where the input range versus the capture depth is recorded. A characteristic ideal signal from detector D1 and D2 respectively is shown in Fig. 6.21. The shape of the measured two-peak signal depends on and changes with the particle velocity, temperature, diameter and trajectory.

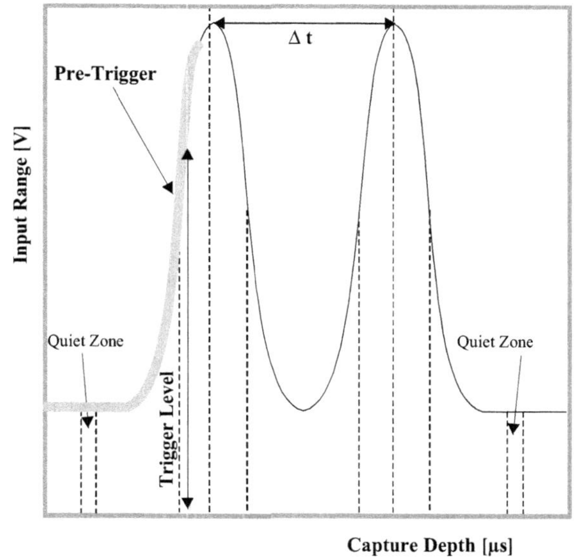

Fig. 6.21 Pattern of an ideal particle signal measured with the high-speed-pyrometer head (from [1], reprint with permission)

6.5.2.1 Velocity

The particle velocity, v_p, is computed from measuring the time interval between the two peaks and is the most simple and precise particle parameter carried out by the measurement system:

$$v_p = \frac{s}{\Delta t} \cdot \textit{optical magnification of the lens} \tag{6.2}$$

where s represents the center to center distance between the slits and t the time of flight between the two slits. The time of flight is measured by a high-speed, high precision scope board. The relative error of velocity measurements is on the order of 0.5% [26].

6.5.2.2 Temperature

The particle temperature, T_p, is obtained from the ratio of signal intensities from each detector according to Planck's law and assuming that emissivity of particle surface is the same at both wavelengths ($\varepsilon(\lambda_1) = \varepsilon(\lambda_2)$):

$$T_p = \frac{c_2(\lambda_1 - \lambda_2)}{\lambda_1 \cdot \lambda_2} \cdot \left(\frac{1}{\ln R + 5 \cdot \ln \frac{\lambda_1}{\lambda_2}} \right) \tag{6.3}$$

where c_2 is the second radiation constant and R is defined as the energy ratio of $E(\lambda_1)/E(\lambda_2)$. Both energies are calculated through the time integral of the two peak signal (area under the curve) normalized by the velocity. As long as one stays within the temperature measurement range of the system (900–4000 °C) and goes through a conscientious temperature calibration, the accuracy on T_p is 3% or better for thermal spray systems [26, 27].

6.5.2.3 Diameter

Taking Planck's Law into consideration, the melted particles are spherical or at least very close to be, one can derive the diameter of a particle d_p from,

$$d_p = \sqrt{\frac{E(\lambda_i)}{c_3 \cdot \varepsilon(\lambda_i)}} = \sqrt{\frac{E(\lambda_i)}{d_c}} \tag{6.4}$$

where i represents 1 or 2 and c_3 is the third radiation constant. Since $\varepsilon(\lambda_i)$ is extremely difficult to define, the calculation system includes a coefficient d_c, which includes both c_3 and $\varepsilon(\lambda_i)$. For a new material with a thermal history emissivity $\varepsilon(\lambda_i)$, and thus d_c, is unknown at the beginning of the experiment. In order to obtain precise d_p values, one must follow, for each different material, a calibration procedure that corrects the non-calibrated diameter measurement with the adjustment of d_c after a comparison with the real existing diameter determined from another classification system (e.g. sieving). When this calibration is performed, the precision on d_p is between 7 and 15% for thermal spray systems [26].

In order to increase the measurement precision, a detection algorithm identifies valid particle signatures from which temperature, velocity and diameter are calculated. The following describes some signal rejection criteria. Particles travelling outside the focal plane are not totally eclipsed by the two-slit mask and corresponding signatures do not have a deep valley between the peaks. The detection algorithm uses adjustable rejection criteria for distorted or incomplete signals. An alternative data evaluation of signals in post-processing mode will be described below.

6.5.3 DPV Application to Gas Atomized Metal Sprays

As an example of a DPV measurement in a metal spray, three different melt materials are compared in spray forming experiments (low carbon steel Ck35, tool steel X20Cr13, copper alloy CuNi). These materials were atomized at the same atomization pressure of 0.5 MPa. The results are shown in Fig. 6.22 and exhibit more or less the same measured trend of particle temperature versus particle diameter. The influence in using different materials atomized at a constant spray pressure is evidently demonstrated. The examined materials of low carbon Ck35 and tool steel X20Cr13 have nearly the same melting point. Thus a comparable

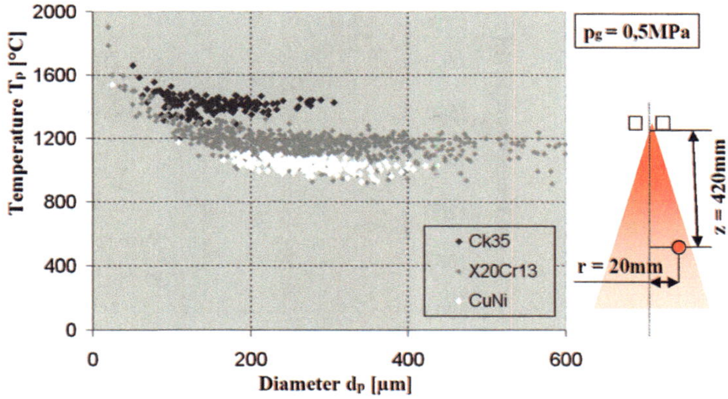

Fig. 6.22 Spray forming of different materials (Ck35, X20Cr13, CuNi) at a constant atomization gas pressure: particle temperature versus particle diameter head (from [1], reprint with permission)

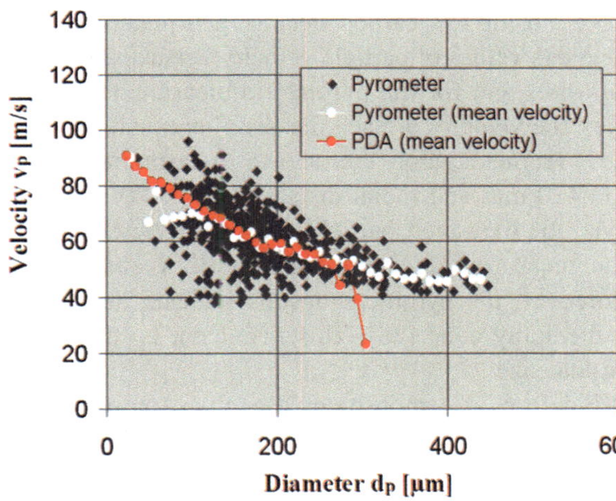

Fig. 6.23 Pyrometer results in comparison to PDA measurements: particle velocity versus particle diameter, $\Delta T = 150$ K, $p_0 = 0.35$ MPa, $z = 420$ mm, $r = 20$ mm, $d_0 = 4$ mm, head (from [1], reprint with permission)

temperature range for the change of phase from liquid to solid is expected. The lower particle temperature measured by the high-speed-pyrometer of tool steel is caused by a smaller radiation emissivity gradient of the material in general. Due to the use of a larger tundish outlet diameter for the spray run using tool steel ($d_0 = 5$ mm) in comparison to low carbon steel ($d_0 = 4$ mm) and the thereby increased metal mass flow rate, larger particle sizes in the spray result and are detected. The measurement results of copper alloy CuNi are located in the region of the lowest possible measurable temperature range. Although, the material has a higher emissivity gradient than steel only the bigger particles with a higher amount of thermal energy can be detected by the measuring system, because of its lower melting point.

Results obtained using the high-speed-pyrometer are compared to results determined in the same spray run by PDA (see Sect. 6.3.3 in this chapter). Figure 6.23

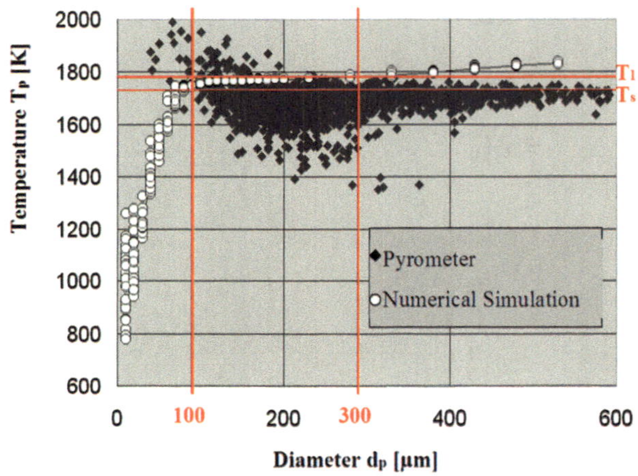

Fig. 6.24 Pyrometer results in comparison to numerical simulations: particle temperature versus particle diameter, $\Delta T = 150$ K, $p_0 = 0.35$ MPa, $z = 420$ mm, $r = 20$ mm, $d_0 = 4$ mm head (from [1], reprint with permission)

illustrates the measured particle velocity versus particle size in a spray forming process using low carbon steel as melt material. The continuous lines characterize the mean values of particle velocity versus particle size, for PDA measurements as dark dots and for the pyrometric measurements in white dots. The dark squares mark the velocity of each measured and validated particle received by employing the pyrometer. For the shown measurement, location at an axial distance $z = 420$ mm and radial distance of $r = 20$ mm from the centerline of the spray cone, the measured particle velocity decreases with increasing particle diameter. The mean values determined by PDA and the pyrometer are nearly similar. Moreover, the pyrometer results indicate the turbulent behavior of the droplets in the spray cone [28]. This is evident by the range of velocities for a constant particle size.

Figure 6.24 demonstrates a comparison of pyrometric measured particle temperature versus particle diameter with calculated results from a numerical simulation at a constant measurement position ($z = 420$ mm, $r = 20$ mm), atomizing gas pressure and material (low carbon steel). It is obvious to note that the measurements carried out by the high-speed-pyrometer show a smaller number of small particles than the calculated values. This phenomenon leads to a measurement error caused by the low emissivity of small particles ($\sim d_p < 100$ µm). Measured and calculated values exhibit the same state of phase change. It seems that even particles in a totally liquid state $\gg$300 µm) with a lower emissivity gradient than semi solid particles are assembled from the measurement system into the incorrect particle class or bin (<100 µm, >1800 K).

Problems concerning temperature range during measurements of hot particles in the spray forming process need to be addressed. The pyrometric system has been originally developed for monitoring particles in thermal spray processes, not in spray forming. Here, typically much higher particle temperatures are to be found and they are typically in the liquid state. Figure 6.25 demonstrates a typical plot of particle temperature versus particle diameter in spray forming on the left and shows

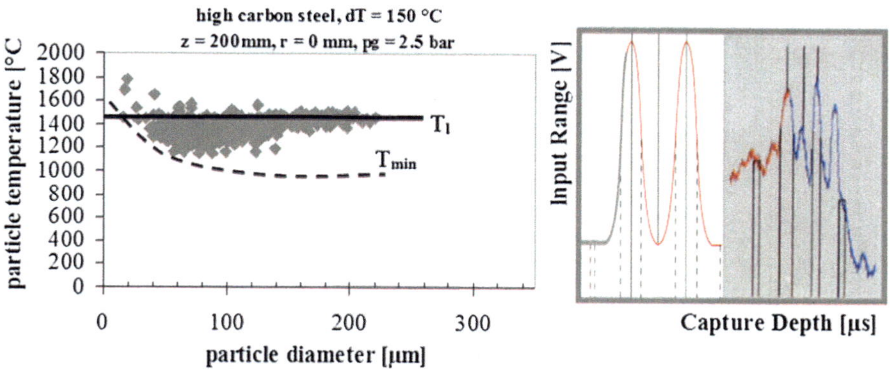

Fig. 6.25 Pyrometric problems within signal analysis (adapted from [29])

on the right an ideal signal in comparison to a real detected signal from the spray forming process.

Most of the particles in spray forming are in the state of phase change. A necessary lower detection limit caused by the sensitivity of the detectors in the measurement system leads to a minimum measurable temperature T_{min}. This limit in spray forming typically leads to a reduction in the detection of particles smaller than 50 μm. Also, when comparing an ideal theoretical signal with a measured one, the measurement system obviously has to deal with a lot of background noise (emitted light from the spray) from a relatively high particle concentration (much higher than in thermal spray processes). The problems in data evaluation within spray forming have been tackled by means of an off-line data evaluation procedure instead of the conventional on-line evaluation. Therefore, an extra PCI board has been installed for signal processing.

The first step within off-line monitoring of particle signals is the continuous sampling and storage of the detector signals without any rejection criteria from the computing system. The second step is to derive suitable data analysis procedures that best determine the temperature, velocity and size of particles. Concerning the background noise, it is necessary to take it into account for the calculation criteria, for example the set of the zero allowance, to avoid mistakes.

Figures 6.26 and 6.27 show typical results from the conventional internal and the new external data evaluation of the high-speed pyrometer. The signals have been measured and calculated in a spray forming experiment with steel, an atomizing gas pressure of $P_g = 3.5$ bars, a melt superheat of $\Delta T = 150$ °C and a measurement position in an axial distance of $z = 280$ mm and a radial distance of $r = 30$ mm. In comparison to the conventionally evaluated signals from the pyrometer, a greater number of particles at the upper and lower edges of the distribution have been detected as can be seen in Fig. 6.28 on the right.

The pyrometric internal detection rate in spray forming is typically about 25 particles per second. The PCI board detects 4000 particles per second and after signal evaluation still 1500 calculated particles per second remained. The

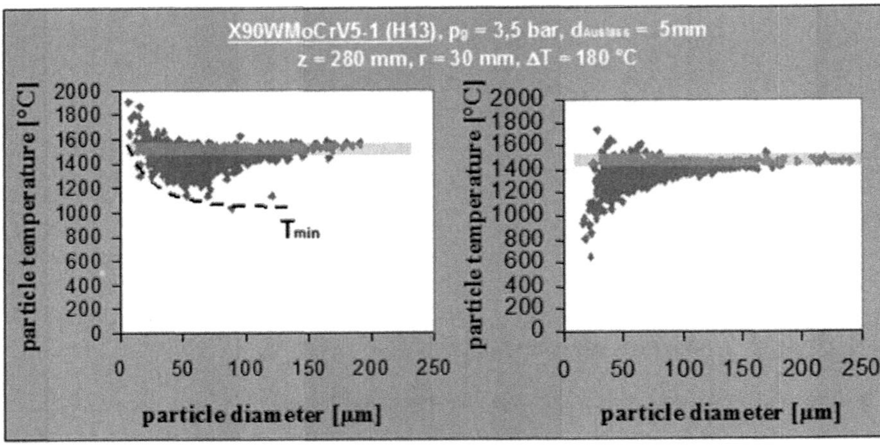

Fig. 6.26 Result of pyrometric measurement in the spray forming process with internal (*left*) and external (*right*) data evaluation (grey bar area = solidification temperature range) (adapted from [29])

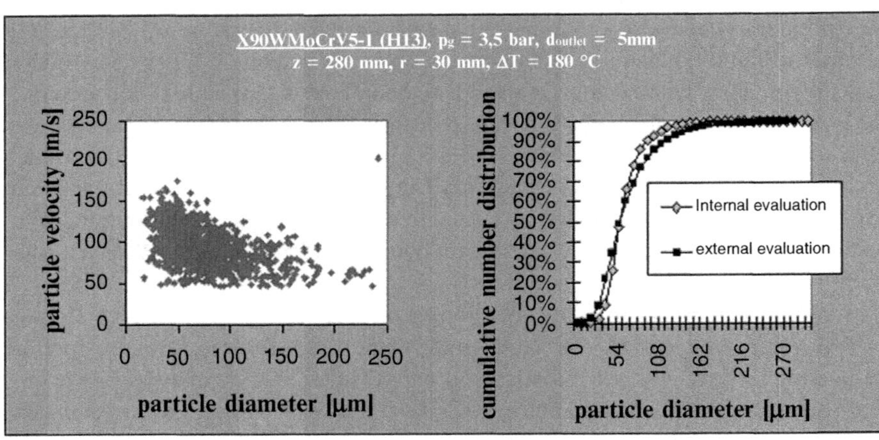

Fig. 6.27 Calculated signals from the external and internal data evaluation (adapted from [29])

adaptation of the analyzing criteria to the spray forming process without changing any components of the system (for example detector type) improves the amount of detected particles and affect the detection limits in a positive direction.

Two typical general process parameters to influence the spray properties in spray forming are: the atomizer gas pressure and the melt superheat (in the crucible/tundish prior to atomization). The input of these process variables in the spray properties is shown in Figs. 6.28 and 6.29 at a certain point in the spray at a distance of 200 mm below the atomizer on the spray centerline. In both cases high carbon steel is spray formed. The cumulative distribution of droplet/particle properties at this specific location is given. The effective gas atomization pressure has a minor

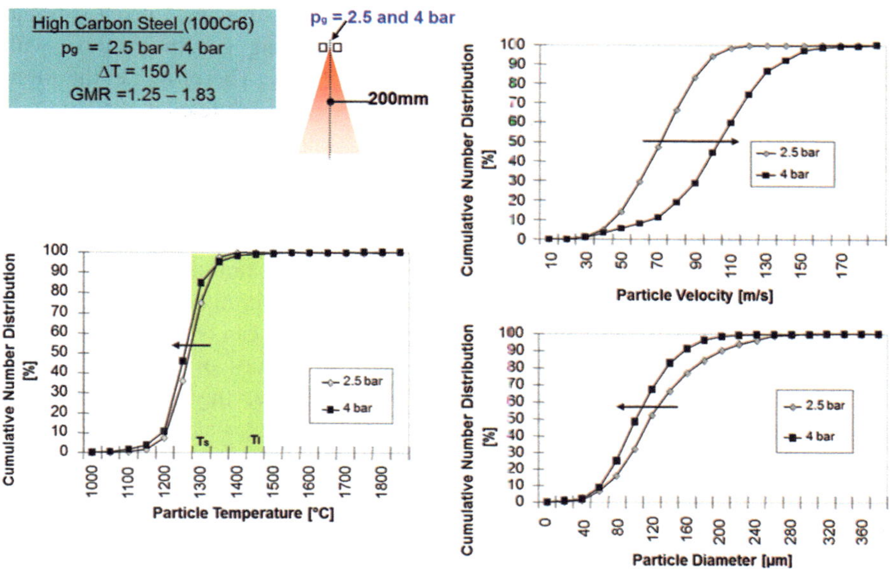

Fig. 6.28 Droplet/particle temperature, velocity as well as size distribution at a certain distance below the atomizer with changing the atomization gas pressure

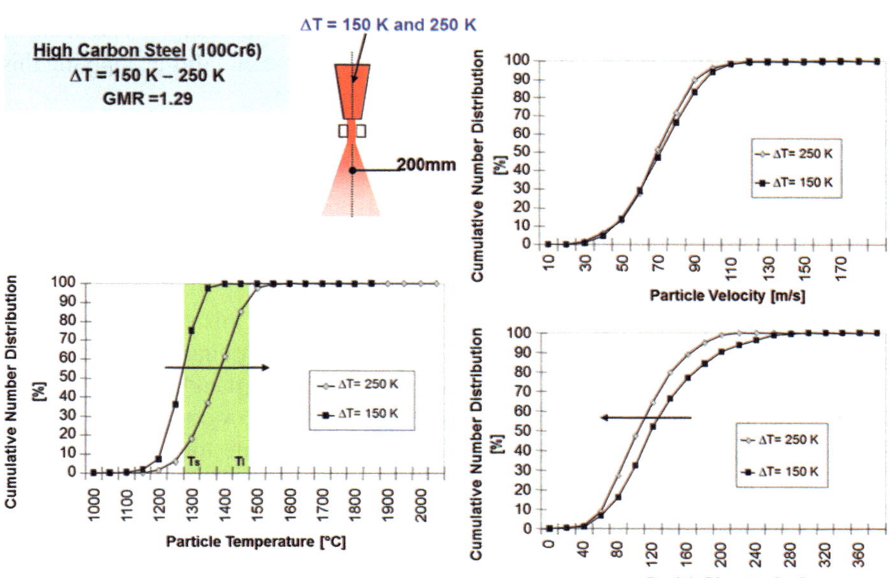

Fig. 6.29 Droplet/particle temperature, velocity as well as size distribution at a certain distance below the atomizer with changing the melt superheat

impact on the particle temperature distribution while the strongest impact is found for the particle velocity. Here, especially the increasing gas momentum with increasing gas pressure leads to a higher particle velocity. Therefore i.e. the flight time of individual droplets in the spray (residence time of particles in the spray) is mainly affected the atomization pressure.

The increase in melt superheat has the strongest impact on the particle temperature, as seen in Fig. 6.28. Here only slight changes in particle velocity as well as particle size are found, but the higher melt superheat directly reflects the higher particle temperature in the spray. The droplets created in atomization are slightly smaller with increasing melt superheat due to the change in material properties of the melt stream here (mainly melt viscosity). In general, it can be concluded to this point that the DPV-2000 provides good reasonable values of droplet size and general trends in droplet temperature. A closure look at the latter variable is warrented. However, because of the chaotic nature of the gas atomization process a more controlled atomization system is desired. The following section will thus address efforts to make droplet tempature measurements in a single fluid atomization system.

6.5.4 DPV-2000 Hypotheses

There are several hypotheses which are considered in DPV-2000 measurements [30]. Some of them can be considered as the principles in calculations and some of them should be explored in order to find out if they are meaningful in applying this instrument to molten metal atomization (IA) application:

- Molten droplets are considered as gray body emitters, i.e. $\varepsilon\ (\lambda_1) = \varepsilon\ (\lambda_2)$,
- Molten droplets should be spherical (or close to spherical); which in any atomization system almost all of the droplets are spherical when atomized in an inert gas environment,
- Regardless of the size, all droplets have similar $\varepsilon\ (\lambda_i)$.

These hypotheses will be evaluated in the remainder of this chapter using free fall droplets in a single fluid atomization system, Impulse Atomization (IA).

Figure 6.30 shows droplets of copper and D2 tool steel at 4 and 10 cm below the IA crucible, respectively. These images were taken by shadowgraph. It can be seen from this figure that almost all droplets are spherical at the minimum distances (4 cm in D2 and 10 cm in copper) from crucible where DPV-2000 has been located.

6.5.5 DPV-2000 Diameter Measurement

Using Planck's law and considering that all droplets are spherical, the diameter of a particle was expressed in Eq. (6.5), d_c is given by:

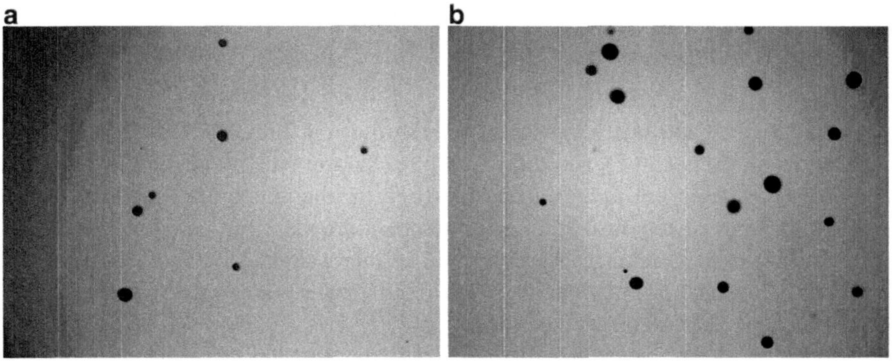

Fig. 6.30 Shadowgraph images of droplets during free fall, (**a**) molten D2 tool steel at 4 cm below the melt crucible and (**b**) molten copper at 10 cm below the melt crucible

$$d_c = \frac{C_1.\varepsilon(\lambda_i).\lambda_i^{-5}}{e^{C_2/\lambda_i T} - 1} \tag{6.5}$$

As shown in Eq. (6.5), the diameter coefficient is dependent on the temperature of the droplet. Since at each distance, the temperature of droplet of different sizes is different, one should calibrate DPV-2000 device at each measuring distance. After calibration of DPV-2000 at each measuring distance, it has been reported by DPV-2000 manufacturer (TECNAR) [30] that the precision of particle size measurement is between 7 and 15% and that it is greatly dependent on the validity of the initial hypotheses, especially sphericity of molten droplets.

In this case the shadowgraph has been used to calibrate the DPV-2000 diameter measurement. As described earlier, shadowgraph technique is independent of the shape and material of the particles. The measured particle size using shadowgraph was used to calibrate diameter measurement of DPV-2000.

During atomization, the shadowgraph was continuously collecting droplet size and velocity data at 5 Hz frequency. As an example, during atomization of pure copper at 15 cm, a total number of 5200 droplets were detected by shadowgraph during the entire atomization time (approximately 2 min) with $D_{50} = 580$ μm. For the same atomization run, DPV-2000 collected 295 droplets with D_{50} of 652 μm.

According to DPV-2000 reference manual [31], the "new d_c" (diameter coefficient) (after calibration) can be calculated using the following equation:

$$New\ d_c = Current\ d_c \times \left[\frac{Current\ Diameter}{Target\ Diameter}\right]^2 \tag{6.6}$$

where "New d_c" is the diameter coefficient after calibration, "Current d_c" is the diameter coefficient which is used before the calibration, "current diameter" is D_{50} measured by DPV-2000 before calibration and "Target Diameter" is D_{50} measured by shadowgraph. Before calibration, "Current d_c" value was 6197. After conducting the calibration, the "New d_c" value for 15 cm would be 7845.

6.5.6 DPV in Single Fluid Atomization and Powder Characterization

In order to evaluate the temperature measuring capabilities of the DPV-2000, validation of the results are required. While, calorimeter probes could be used, they are considered not to be sensitive enough for the temperature validation. No work has been reported on another direct temperature measuring method. Hence, an indirect approach is required as described in this section. The evaluation of the temperatures where primary nucleation starts in droplet solidification or the droplet temperature at the end of recalescence.

Nucleation of a phase occurs after some undercooling has taken place. This is the temperature of the droplet below its equilibrium liquidus temperature (for the primary phase) where the energies of the formation of a solid are favourable. Since, the droplet temperature and its liquid phase is in the metastable state, solid forms very rapidly from the nucleant releasing a large amount of enthalpy of transformation into the droplet. Initially, this release of heat occurs at a rate greater than the rate of heat absorbed by the gas surrounding the droplet. This results in an increase in the droplet temperature. This phenomenon is named recalescence.

D2 tool steel (manufactured by Böhler-Uddeholm) with 1.55% C, 11.8% Cr, 0.40% Mn, 0.80% Mo, 0.80% V and Fe in balance (all in wt.%) was melted in an induction field of a drop tube IA technique to produce D2 steel powder. Melting and eutectic temperatures of this alloy are 1394 °C and 1270 °C, respectively. In the IA system, molten metal is pushed through orifices, forming ligaments, which eventually breakdown and spheroidize into droplets in an inert atmosphere (see Chapter 2). After heating the D2 steel to a temperature of 1600 °C inside an alumina crucible using an induction furnace, atomization was then started. During atomization, the liquid steel is pushed through 37 orifices with the diameter of 300 μm on a nozzle plate at the bottom of the crucible via a vibrating plunger. Falling droplets were cooled in nitrogen atmosphere having a maximum oxygen content of 8 ppm and they lose their heat to the surrounding stagnant gas.

A 3D translation stage was designed, constructed and installed in the drop tube to allow for measurements of radiant energy using a DPV-2000 (Tecnar Automation Ltée) and velocity and droplet size in flight using a shadowgraph (Particle Master from LaVision GmbH in Gottingen, Germany). In the shadowgraph, a pulsed laser combined with the diffuser optics (as the light source) illuminated in-flight droplets 5 times per second. Backlight of droplets inside the measurement volume of $6 \times 6 \times 6$ mm was captured by a high resolution imaging system. It is possible to investigate droplet sizes down to 5 μm using the shadowgraph device. DPV-2000 can detect droplets with <1000 μm diameter. Shadowgraph device is also capable to image particles from 10 to 1000 μm falling at velocities ranging from 0.01 to 10 cm/s. In the particle size algorithm, the appropriate values are set for detection thresholds.

Using the 3D stage, the droplet size, velocity and radiant energy can be measured at different locations below the nozzle plate. In this study, these measurements were

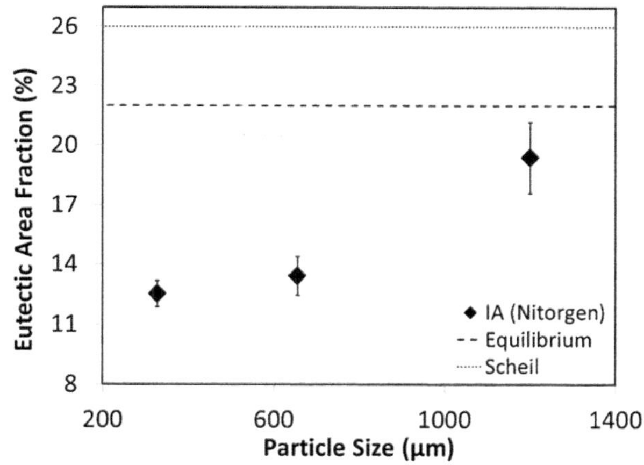

Fig. 6.31 Eutectic area fraction as a function of particle size for IA D2 tool steel

taken at distances of 4, 18 and 28 cm below the nozzle plate. The focal point of both shadowgraph and DPV-2000 was set to an imaginary vertical line going through the center of the nozzle plate; therefore, any unfocused droplets would be rejected by the criteria set by the operator in the respective software of each instrument.

Subsequent to the atomization processes, the IA D2 tool steel powder was washed using toluene and methanol, and then sieved into different size ranges based on MPIF Standard 05 [32]. Size ranges of 300–355 μm, 600–710 μm and 1000–1400 μm were chosen from the IA powder for further analysis. For each of the different powder size fractions, specimens were mounted in epoxy resin, ground, polished, and then carbon coated in preparation for metallography. Scanning electron microscopy (SEM) was used to the area fraction of eutectic. The error bars reported in the results represents the standard deviation from the mean based on the length and area measurements.

Figure 6.31 shows the effect of particle size on the eutectic fraction in IA particles based on image analysis. First, this graph shows that the eutectic area fraction is smaller for the particles that were more rapidly solidified. Second, for the IA particles, the error bars increase with increasing particle size. This indicates that, similar to Al-Cu and Al-Fe systems [33, 34], each D2 tool steel particle is not homogeneous in microstructure.

Using the liquidus and eutectic temperatures of D2, equilibrium and Scheil eutectic fractions are also provided in Fig. 6.31, for comparison purposes. As can be seen, the measured values of eutectic area fraction for D2 tool steel are considerably less than what is predicted by the two limiting theoretical cases.

Using the results shown in Fig. 6.31 and the phase diagram of D2, it is possible to estimate the nucleation temperature of the eutectic. This is accomplished by extending the liquidus and solidus lines of the phase diagram (using Thermocalc v5.02.30 and the TCFE6 database) until the calculated eutectic equals that reported in the figure. The temperature corresponding to this condition is taken to be the nucleation temperature of the eutectic and the difference between this temperature and the eutectic temperature is the eutectic undercooling (see Fig. 6.32). This

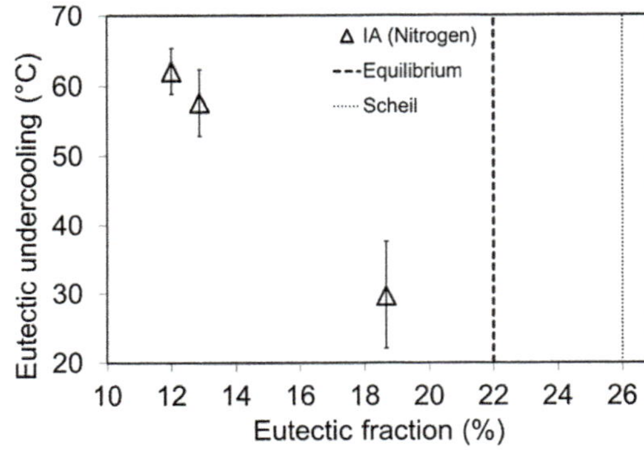

Fig. 6.32 Effect of eutectic undercooling on the amount of eutectic in IA D2 tool steel

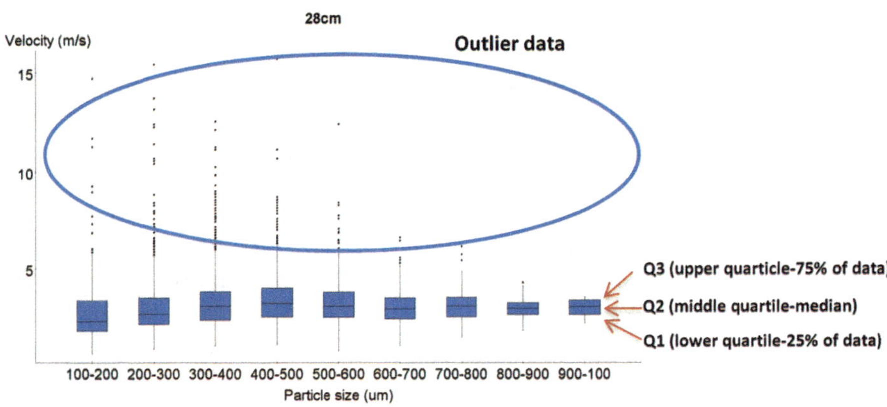

Fig. 6.33 Velocity as a function of droplet size of falling droplets collected by shadowgraph at 28 cm falling distance

approach has also been reported elsewhere for Al-Cu [35]. As can be seen in Figure 6.32, larger undercoolings are found to occur when the eutectic fraction is small.

6.5.6.1 Velocity, Size and Radiant Energy Measurements

During atomization, the shadowgraph was continuously collecting droplet size and velocity data at 5 Hz at different falling heights. Using statistical analysis, one can plot Q1, Q2 and Q3[1] for velocity data for different range of particle sizes in the form of a boxplot. Figure 6.33 shows a plot of velocity versus particle size of the droplets at 28 cm below the nozzle plate. A total number of 5255 droplets were measured during the entire atomization time of approximately 2 min at this height. Same set

[1]Q1 is the lower quartile, Q2 is the middle quartile (median) and Q3 is the upper quartile.

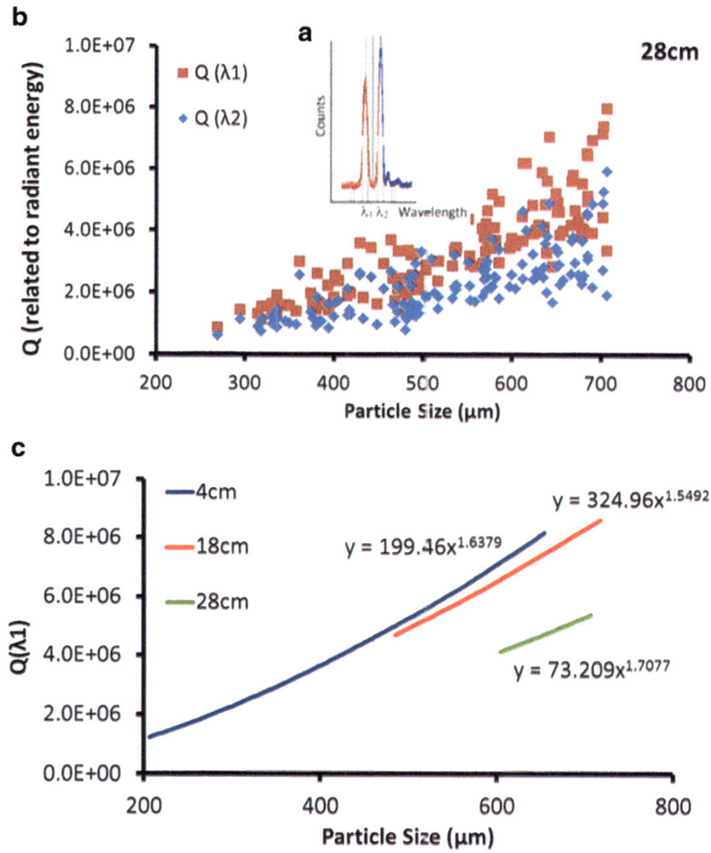

Fig. 6.34 (**a**) The signal counts measured at two different wavelengths for a single droplet and (**b**) range of droplets at 28 cm below the nozzle plate, (**c**) power curve fitted radiant energy vs. droplet diameter curve measured at λ_1 at different heights for the droplets that are fully liquid, including those that are undercooled

of data has been collected for falling droplets at 4 and 18 cm below the nozzle plate. This figure shows that at each particle size range, median values for velocity are almost in the same range and the average equals to approximately 3.2 m/s. This average velocity value for droplets falling at 4 and 18 cm heights are approximately 2.3 and 2.6 m/s, respectively.

While the shadowgraph was collecting the velocity and diameter of falling droplets at the center of the plume, the DPV-2000 was measuring the radiant energy of the droplets from the same location. Figure 6.34a shows the signal counts of each droplet that DPV-2000 measured at two wavelengths (at 28 cm). The area under the curves shown represents Q, which is proportional to the radiant energy, and as such it is a dimensionless quantity. The radiant energies measured for different droplet sizes at two different wavelengths are shown in Fig. 6.34b. Since the as-collected data from DPV-2000 are scattered and make it difficult to work with data, a power regression was determined on DPV-2000 output data. Figure 6.34c shows power curve fitted radiant energy at wavelength 1 ($Q(\lambda_1)$) versus droplet diameter curve measured at λ_1 at different heights for the droplets that are fully liquid (including the undercooled droplets).

6.5.6.2 Droplet Temperature Calculations

Various models of droplet cooling and temperature distribution as well as solidifation may be utilized for comparison and interpretation of experimental findings [36–41]. A mathematical model was also utilized to predict the temperature of falling D2 steel molten droplets at different heights [42]. Briefly, this model calculates the droplet velocity and temperature as a function of time and distance traveled. The model uses the initial velocity of the droplets and the forces of gravity and drag to find the subsequent trajectory of the droplets. It also considers the fraction solidified of the droplets using equilibrium calculation in order to determine the temperature of the droplets and the latent heat generation during solidification. It should be noted that this model cannot calculate eutectic fraction as a result of microsegregation at the end of the solidification. However, for the current objectives of this chapter, the model is primarily used to predict the thermal history of superheated and undercooled liquid droplets.

Based on the methodology described in [43], using the dendrite coarsening model [44] and eutectic undercooling (already described), it is possible to calculate solidification temperature range. With solidification temperature range, and the eutectic undercooling temperatures derived from the eutectic fractions, one can determine the nucleation temperature of the primary phase and its corresponding undercooling. Figure 6.35 shows the relationship between particle size and primary phase undercooling. It can be seen that primary phase undercooling is decreasing as the particle size is increasing.

The primary undercooling values will be used in the droplet thermal model as part of the analysis of online measurements of the falling droplets. Using the thermal model, the velocity measured by shadowgraph, and the primary

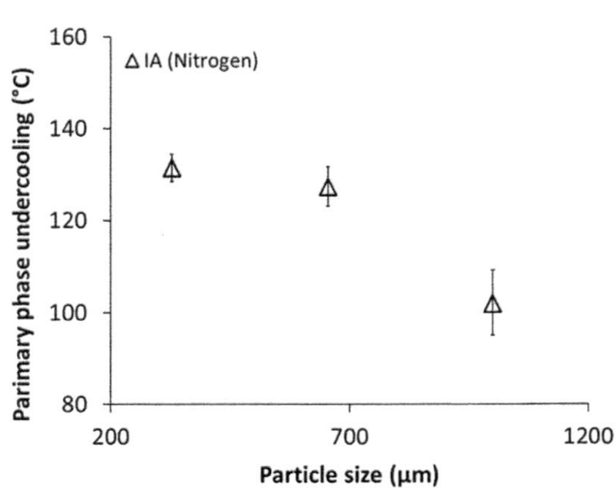

Fig. 6.35 Effect of particle size on the primary phase undercooling

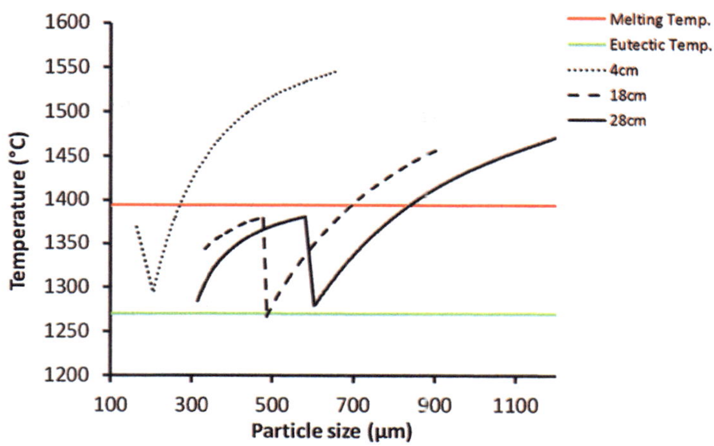

Fig. 6.36 Temperature of the falling droplets at different heights

undercooling temperatures, one can calculate thermal history of different size droplets at different falling heights (Fig. 6.36) [42]. Figure 6.36 shows that the droplets experience primary undercooling before they start solidifying. After solidification starts, due to the generation of latent heat, there is an increase in the temperature of the droplet which is followed by the decrease in the temperature, due to the heat loss from the droplets. Using the graph in this figure, one can identify the droplet sizes that are fully undercooled at the measuring heights (206, 485 and 604 μm droplets sizes at 4, 18 and 28 cm measuring heights, respectively). These droplets will be the limit of the region where further analysis will be developed in the following section.

6.5.6.3 Validity of Two Color Pyrometry Assumption

Emissivity of a material depends on different factors such as temperature and wavelength [45] as well as geometry, oxidation condition, surface roughness, etc. In the current study, since the temperature of the droplets is changing with droplet size (Fig. 6.36), the radiation relation is getting more complicated (more than one variable present in the equation). In order to normalize the effect of droplet size on the equation and simplification, the equation can be written as:

$$\frac{Q(\lambda_i)}{d^2} = K_i\, C_1 \alpha(\lambda_i) \lambda_i^{-5} \tag{6.7}$$

where

$$K_i = \frac{1}{e^{C_2/\lambda_i T} - 1} \tag{6.8}$$

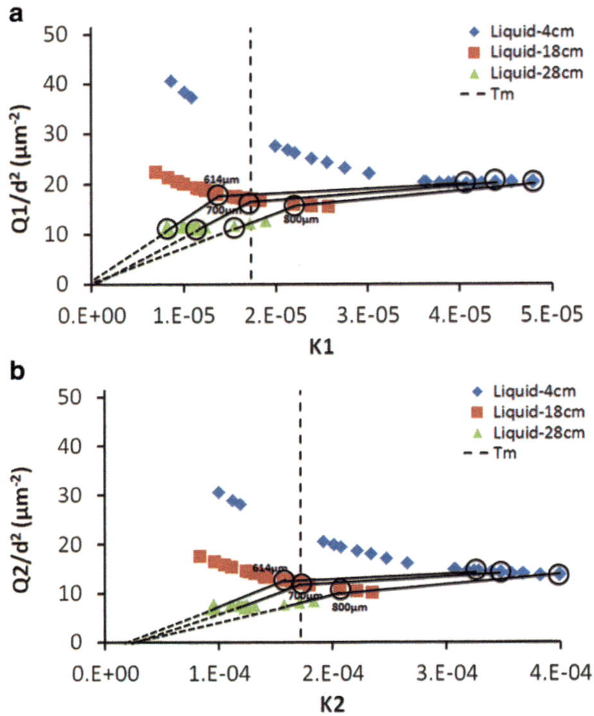

Fig. 6.37 $Q(\lambda_i)/d^2$ versus K_i graph at different heights for the liquid droplets at different wavelengths (**a**) λ_1 and (**b**) λ_2

Since in Eqs. (6.7) and (6.8), C_1, C_2 and λ_i are constants, one can plot $Q(\lambda_i)/d^2$ versus K_i at different heights where the slope of the graph is related to $\alpha(\lambda_i)$ at that height for a constant droplet size. Figure shows $Q(\lambda_1)/d^2$ versus K_1 values for the liquid droplets at different heights during free fall. It also shows the K_1 value related to the liquidus temperature of D2 steel (dashed vertical line). All the droplets left of the line are experiencing primary phase undercooling. Therefore, they are all in the liquid state. Also, in Fig. 6.37, there are solid lines connecting $Q(\lambda_i)/d^2$ values at different heights for the same droplet size. Again, according to Eq. (6.7), $Q(\lambda_i)/d^2$ versus K_i slope (for the specific particle size) should be a linear line passing zero points of the coordinate system. This behaviour is visible between 18 and 28 cm data points. But, when it goes to 4 cm data, it does not follow the trend. This needs further analysis, but it can be related to the larger variation of emissivity from 4 to 18 cm compared to that from 18 to 28 cm for the same particle size. Alternately, this may be due to the DPV-2000 sensor being affected by the radiation from the bottom of the melt crucible due to its close proximity to it. Therefore, only the results at 18 and 28 cm will be further discussed. Also, $Q(\lambda_i)/d^2$ versus K_i slope (for the specific particle size) is changing by changing the particle size. This can show that the emissivity is also changing with particle size, since they have different temperatures and different surface areas. An advantage of plotting $Q(\lambda_i)/d^2$ versus K_i is that direct effect of emissivity can be reduced in the calculation to only the slope of the plot and thus will not affect the subsequent analysis and discussion.

Also, by comparing $Q(\lambda_i)/d^2$ versus K_i at two different wavelengths (λ_1 and λ_2) for a specific particle size shown in Figure 6.37a, b, one can see that the slope of the solid lines connecting $Q(\lambda_i)/d^2$ versus values at different height is different in different wavelengths (λ_1 and λ_2). As an example, this slope for 800 μm particle in λ_1 and λ_2 based calculations is about $73 \times 10^4\ \mu m^{-2}$ and $5 \times 10^4\ \mu m^{-2}$, respectively. As discussed before, $Q(\lambda_i)/d^2$ versus K_i slope is representing the emissivity of the droplets. These results show that emissivity is also different at different wavelength and therefore, grey body assumption ($\varepsilon(\lambda_1) = \varepsilon(\lambda_2)$ at two close wavelength) is not valid in the condition that DPV-2000 is used. Therefore, DPV-2000 should be considered as one color pyrometer for further analysis of the temperature history of droplets.

6.5.6.4 Droplet Size and Temperature Estimation at the Solidification Point

Following the start of solidification, the temperature of the droplets increases as a result of the release of the latent heat of solidification. Consequently, a discontinuous change in K values occurs in the $Q(\lambda)/d^2$ versus K plot. $Q(\lambda_1)/d^2$ versus K_1 plot for semi-solid droplets at different heights is shown in Fig. 6.38. It should be noted that the $Q(\lambda_1)$ values used in this plot are the experimental data that are averaged for each particle size range which lie inside the measurement accuracy of DPV-2000. The fraction of droplets solidified at different heights is also plotted in Fig. 6.38. It can be observed that toward the lower temperatures (smaller K_1 values), the fraction solidified is increasing.

In order to monitor the behavior of droplets in the liquid (fully undercooled) and semi-solid state (before and after the start of solidification), Fig. 6.37 was superimposed with Fig. 6.38 as shown in Fig. 6.39. It should be noted that in this figure, the error bars have been removed for the better presentation and to easily identify the average trend of the results. The error bars will be further discussed in the following section. The linear extension of the semi-solid data as well as its

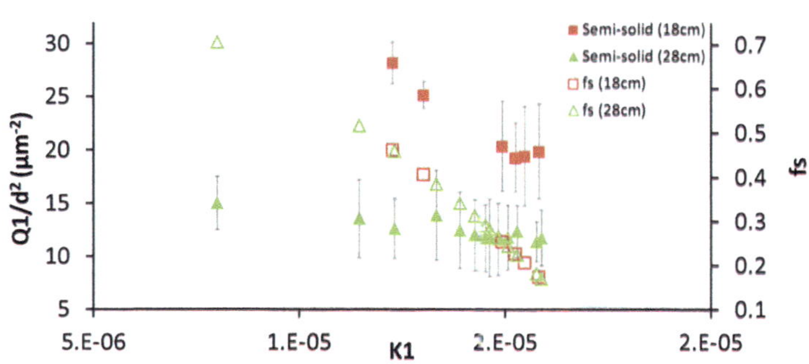

Fig. 6.38 $Q(\lambda_1)/d^2$ and fraction solidified versus K_1 graphs at different heights for the semi-solid droplets

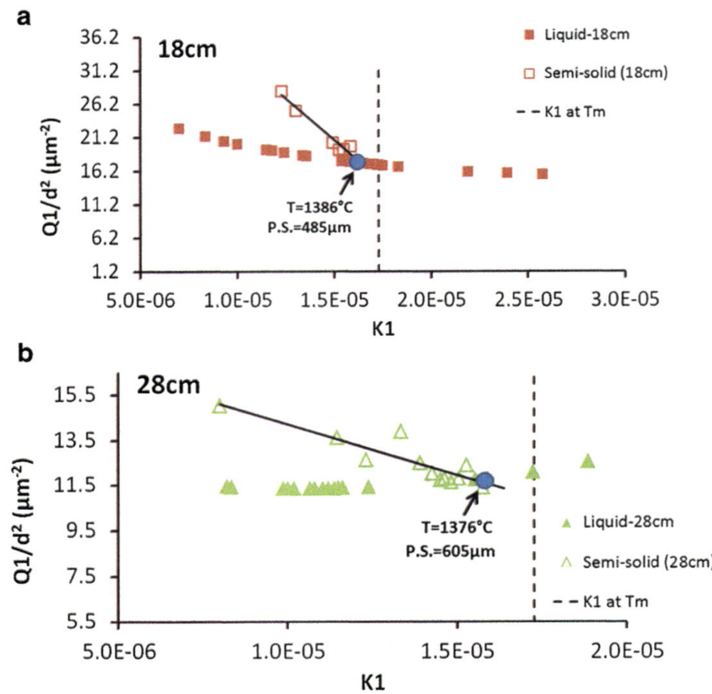

Fig. 6.39 $Q(\lambda_1)/d^2$ versus K_1 graphs at different heights for both liquid and semi-solid droplets at (**a**) 18 cm and (**b**) 28 cm heights

intersection with liquid data is also shown in Fig. 6.39. At 18 and 28 cm distances below the melt crucible, the intersection points correspond to 1386 and 1376 °C temperatures, respectively. It seems that the intersection points are attributed to be close to when recalescence ends. As a matter of fact, the finite solid fraction and temperature rise are related to the droplet size at which the solidification has just started, which corresponds to the end of recalescence. Consequently, the temperature at the intersection points can be equivalent to the maximum temperature that droplet achieves following recalescence.

In order to confirm the values of droplet size and temperature at which solidification has just started (achieved from Fig. 6.38), these values can be used in the thermal heat flow model (at 18 and 28 cm heights). The 485 and 605 μm droplets and their respective initial velocities were used in the thermal model and the height in which the solidification begins was determined by the model. It should be noted that the experimentally estimated primary phase undercooling values were used in the model. The model results showed that for 485 and 605 μm droplets, the solidification was starting at 18.5 and 29.5 cm heights, respectively. This information shows that the results from DPV-2000 are within 0.5 cm (or within 2%) and 1.5 cm (or within 5%) of the model results for 485 μm droplet (at 18 cm) and 605 μm (at 28 cm), respectively.

Using the temperature at the end of recalescence, one can calculate fraction of primary phase formed after recalescence using Scheil method. Using measured fraction of total primary phase (from SEM analysis) and calculated fraction of primary phase after recalescence (from Scheil equation), fraction of primary phase formed during recalescence can be calculated. The advantage of this method is to understand the temperature behaviour of droplets and recalescence fraction, without using assumptions and complexity of a microsegregation model.

6.5.6.5 Precision of In-situ, Real Time Diagnostics

DPV-2000 collects data from falling droplets depending on the several variables such as temperature, size, velocity and trajectory of the droplets. If any of these parameters do not meet DPV-2000 requirements (i.e. very slow droplets, not-vertically falling droplets, low temperature, etc.), DPV-2000 does not collect any information from the droplets or measures it with low precision. For example, only droplets with temperature between 1000 and 4000 °C can be detected by DPV-2000. Temperatures close to the detection limits would be measured by DPV-2000 but with low precision [30]. These cause some errors attributed to DPV-2000 calculations (error bars in figure for semi-solid droplets). These errors can be attributed in part to instrument and measurement errors.

On the other hand, nucleation is also a random phenomenon. Nucleation occurs spontaneously and randomly, and as has been shown earlier, primary phase nucleation requires undercooling of the liquid. A droplet is solidifying when nucleation requirements are met (i.e. chemical potential balance in the system) which are called solidification requirements.

In order to collect the solidification information using DPV-2000, both instrument and measurement errors, and solidification requirements should take place simultaneously. Therefore, collecting information from a droplet (with the qualified properties for DPV-2000) at the precise time or position when it starts to solidify has a very low probability. Also, the time duration of recalescence from its start to its completion is much smaller than the measurement capability of DPV-2000. For example, using thermal model, for 450 μm droplet, the duration of recalescence was calculated to be about 10^{-5} s. Also, at 28 cm height, the time of flight of the same size droplet through the two sensors in DPV-2000 is about 3×10^{-6} s. Therefore, DPV-2000 cannot detect droplet's recalescence during its free fall in front of DPV-2000. This further underscores that the temperatures shown in Fig. 6.39 represent the droplet temperature following recalescence for the respective droplet sizes.

6.6 Summary

The atomization process of molten metals is a very hostile process involving high temperature, inert hot gases, metallic droplets that are potentially pyrophoric. We have come a long way in successfully making in situ real time measurements of droplet size, velocity. Droplet temperature can be estimated with the aid of microstructural post mortem data. These results have been obtained due to the development of molten metal flow sensor, and the use of off the shelf instruments such as PCSV-P, EPCS, PDA, Shadowgraph and DPV 2000. Such data will be very indispensable for use in validating models of molten metal sprays in terms of identifying the size, size distribution, velocity and fraction of solid of droplets. Thus, enabling more effective design and location of the substrate in spray deposition processes.

In industrial operation, it can be considered to install level sensors for the height of molten metal in the tundish to ensure constant molten metal flowrate. In addition, gas flowmeters are indispensable to monitor gas consumption and determine the gas to metal mass flowrate. This is a critical variable to maintain a stable size distribution. Finally, the EPCS has proven itself to be a valuable instrument for real time powder size and distribution measurement in real time in other industrial sectors. It is reasonable to expect to see them used in metal powder production systems in the near future.

6.7 List of symbols

6.7.1 *Latin*

Symbole	Description
B	Experimentally-determined constants (μm)
c_2	Second radiation constant
d_c	Coefficient, includes both c_3 and $\varepsilon(\lambda_i)$
d_p	Diameter of particle
E	Energy
n	Experimentally-determined constants (no dimension)
R	Energy ratio of $E(\lambda_1)/E(\lambda_2)$
s	Center to center distance between the slits
t	Time
v_p	Particle velocity

6.7.2 *Greek*

Symbole	Description
ε	Emissivity of particle surface
λ	Wave length
$\varepsilon(\lambda_i)$	Thermal history emissivity

References

1. Krauss, M., Bergmann, D., Fritsching, U., & Bauckhage, K. (2002). In-situ particle temperature, velocity and size measurement in the spray forming process. *Materials Science and Engineering A, 326*, 154–164.
2. Bergmann, D., Fritsching, U., & Bauckhage, K. (2000). A mathematical model for cooling and rapid solidification of molten metal droplets. *International Journal of Thermal Sciences, 39*, 53–62.
3. Bergmann, D., & Fritsching, U. (2004). Sequential thermal modelling of the spray-forming process. *International Journal of Thermal Sciences, 43*, 403–415.
4. Bauckhage, K., Bergmann, D., Tillwick, J. 1999. Die massen- und enthalpiebilanzierung des sprühkegels als kopplung für die modellvorstellung des materialaufbaues in der mix-schicht, Universität Bremen, SFB 372 Kolloquium, No. 4 (pp. 139–170).
5. Helmersson, G., & Burgdorf, K. (1996). Effects of process parameters on microstructure of gas atomized powder. *Scandinavian Journal of Metallurgy, 25*, 51–58.
6. Behulova, M., Moravcik, R., Kusy, M., Caplovic, L., Grgac, P., & Stancek, L. (2001). Influence of atomisation on solidification microstructures in the rapidly solidified powder of the Cr–Mo–V tool steel. *Materials Science and Engineering A, 304-306*, 540–543.
7. Jiand, G., Henein, H., & Siegel, M. W. (1990). Overview: Intelligent sensors for atomization. *The International Journal of Powder Metallurgy, 26*(3), 253–268.
8. Le, T., Stefaniuk, R., Henein, H., & Huôt, J.-Y. (1999). Measurement and analysis of melt flowrate in gas atomization. *International Journal of Powder Metallurgy, 35*(1), 51–59.
9. Jillavenkatesa, A., Dapkunas, S. J., & Lum, L.-S. H. (2001). *Particle size characterization.* Washington, DC: NIST Special Publication 960-1.
10. Seaton, C., Henein, H., & Glatz, M. (1987). The atomization of liquid metals using the Coanda effect. *Powder Metallurgy, 30*, 37.
11. Henein, H., Meyer, P. L., Holve, D. J., & Kuhni, M. A. (1992). On-line measurement of particle size distribution in zinc atomization. *International Journal of Powder Metallurgy, 28*, 149–159.
12. Boyko, C. M., Le, T. H., & Henein, H. (1993). Ensemble and single particle laser probe sizing results for gas atomized zinc powders. *Particle and Particle Systems Characterization, 10*, 266–270.
13. Yolken, Y. T., & Birnbaum, G. (1993). Intelligent processing of materials: Technical activities. In *Materials science and engineering laboratory*. Gaithersburg, MD: NAS-NRC Assessment Panel.
14. Bauckhage, K., Flögel, H.-H., Fritsching, U., & Hiller, R. (1988). The phase-Doppler-difference method, a new laser-Doppler technique for simultaneous size and velocity measurements, Part 2: Optical particle characteristics as a base for the new diagnostic technique. *Particle and Particle Systems Characterization, 5*, 66–71.
15. Ziesenis, J. 2002. Weiterentwicklung der PDA-Meßtechnik zur on-line Prozeßkontrolle beim Sprühkompaktieren, Dissertation, Universität Bremen.
16. Ziesenis, J., & Bauckhage, K. (2003). Spray forming: Controlling the atomization result with regard to particle properties. *Particle and Particle Systems Characterization, 20*, 290–297.
17. Ziesenis, J., & Bauckhage, K. (2002). Absorption and scattering of light by highly concentrated two-phase flows. *Particle and Particle Systems Characterization, 19*, 195–202.
18. Tillwick, J., Uhlenwinkel, V., & Bauckhage, K. (1999). Analysis of the spray forming process using backscattering phase-Doppler anemometry. *International Journal of Heat and Fluid Flow, 20*(5), 530–537.
19. Wriedt, T., Bauckhage, K., & Schöne, A. (1989). Application of Fourier analysis to phase-Doppler-signals generated by rough metal particles. *IEEE Transactions on Instrumentation and Measurement, 38*(5), 984–990.

20. K. Bauckhage, P. Schreckenberg. 1989. Control of powder-metal production. A new application of phase-Doppler-anemometry. In *Proceedings of international conference on mechanics of two-phase flows* (pp. 1–6). Taipei, Taiwan.
21. V. Uhlenwinkel, M. Buchholz, C. Kramer, & K. Bauckhage. 1998. Characterization of a free fall atomiser and its influence to the spray forming process. In *PM world congress: Thermal spraying/spray forming* (pp. 537–542). Granada.
22. Uhlenwinkel, V., Buchholz, M., & Bauckhage, K. 1999. Particle mass flux in the spray cone of a free fall atomizer. In *Proceedings of TMS meeting*. San Diego.
23. ParticlMaster Manual, LaVision GmbH, Göttingen, Germany
24. Blain, J., Nadeau, F., Pouliot, L.. 1997. An integrated infrared sensor for on-line monitoring of thermally sprayed particles, Canada (pp. 1–6).
25. Moreau, C., Gougeon, P., Lamontagne, M., Lacasse, V., Vaudreuil, G., & Cielo, P. (1994). On-line control of the spraying process by monitoring the temperature, velocity and trajectory of in flight particles. In *Thermal spray conference* (pp. 431–437). Boston, MA: ASM.
26. Pouliot, L., Blain, J., & Nadeau, F. (1998). *DPV 2000 reference manual: In flight particle sensor for thermal spraying systems*. Québec: Tecnar Automation.
27. Eckert, R. G., & Goldstein, R. J. (1976). *Measurements in heat transfer, University of Minnesota, School of Mechanical and Aerospace Engineering*. London: Hemisphere.
28. Ziesenis J., Tillwick, J., Krauss, M., & Uhlenwinkel, V. 2000. Analysis of molten metal atomization process using modified phase-Doppler-anemometry. In *Proceedings of TMS*. Nashville, USA, March 12–16.
29. Krauss, M., Hua, Y., Cui, C., Fritsching, U. 2003. Diagnostic of the mushy zone by in-line measurements and numerical simulations. In *Proceedings of Spray Deposition and Melt Atomization SDMA*. Bremen, June 22–25.
30. Tecnar Automation Ltée, DPV-2000 calculation principles, 2011.
31. TecnarAutomation Ltée, DPV-2000 reference manual, Saint-Bruno, Québec.
32. Standard test methods for metal powders and powder metallurgy products. Metal Powder Industries Federation. Princeton, NJ, 1993.
33. Henein, H., Buchoud, V., Schmidt, R., Watt, C., Malakov, D., Gandin, C.-A., et al. (2010). Droplet solidification of impulse atomized Al-0.61Fe and Al-1.9Fe. *Canadian Metallurgical Quarterly, 49*, 275–292.
34. Prasad, A., Henein, H., Maire, E., & Gandin, C.-A. (2006). Understanding the rapid solidification of Al-4.3Cu and Al-17Cu using x-ray tomography. *Metallurgical and Materials Transactions A, 37*, 249–287.
35. A.-A. Bogno, P. Delshad Khatibi, H. Henein, C.-A. Gandin. 2013. Quantification of primary dendritic and secondary eutectic undercoolings of rapidly solidified Al-Cu droplets. In *Materials science and technology symposium: Light metals for transportation*. Motreal.
36. Cui, C., Fritsching, U., Schulz, A., & Li, Q. (2005). Mathematical modeling of spray forming process of tubular preforms. *Acta Materialia, 53*, 2775–2784.
37. Fritsching, U. (2004). *Spray simulation: Modeling and numerical simulation of sprayforming metals*. Cambridge, UK: Cambridge University Press.
38. Freyberg, A. V., Henein, H., Uhlenwinkel, V., & Buchholz, M. (2003). Droplet solidification and gas-droplet thermal coupling in the atomization of a Cu-6Sn alloy. *Metallurgical and Materials Transactions B, 33b*, 243–253.
39. Prasad, A., & Henein, H. (2008). Droplet cooling in atomization sprays. *Journal of Materials Science, 43*, 5930–5941.
40. Prasad, A., Mosbah, S., Henein, H., & Gandin, C.-A. (2009). A solidification model for atomization. *ISIJ International, 49*, 992–999.
41. Zeoli, S. K. N., & Gu, S. (2008). Numerical modelling of metal droplet cooling and solidification. *International Journal of Heat and Mass Transfer, 51*, 4121–4131.
42. Delshad Khatibi, P., Ilbagi, A., Beinker, D., & Henein, H. (2011). In-situ characterization of droplets during free fall in the drop tube-impulse system. *Journal of Physics Conference Series, 327*, 012014.

43. Delshad Khatibi, P., Henein, H., & Phillion, A. B. (2013). Microstructural investigation of D2 tool steel during rapid solidification. *Powder Metallurgy, 57*, 70.
44. Kurz, W., & Fisher, D. J. (1998). *Fundamentals of solidification*. Enfield, NH: Enfield Publishing & Distribution Company.
45. Holman, J. P. (1997). *Heat transfer* (8th ed.). McGraw Hill.

Chapter 7
Microstructural Evolution in Spray Forming

Patrick S. Grant, Guilherme Zepon, Nils Ellendt, and Volker Uhlenwinkel

7.1 Introduction

Spray forming is a casting process in which the molten metal is directly converted to a solid bulk with unique characteristics. When processed under optimum conditions, spray formed materials typically present microstructures composed of refined polygonal (non-dendritic) grains, uniformly distributed with low levels of micro- and macro-segregation. This set of characteristics is achieved regardless of the alloy system, making spray forming an attractive process to produce alloys where processing by conventional casting techniques is complicated. This chapter is dedicated to presenting the mechanisms that take place when the atomized droplets arrive at the deposit surface, and how the spray-formed microstructures evolve during deposition. It will be seen that spray forming is a *self grain-refining casting process* and cannot be considered a rapid solidification technique. Section 7.6 will address the main differences between the microstructural evolution in spray forming and other spray deposition or "thermal spray" processes. These processes

P.S. Grant (✉)
Department of Materials, University of Oxford, Oxford OX1 3PH, UK
e-mail: patrick.grant@materials.ox.ac.uk

G. Zepon
Department of Materials Engineering (DEMa), Federal University of São Carlos (UFSCar), São Carlos-SP, Brazil
e-mail: zepon@ufscar.br

N. Ellendt
University of Bremen, Bremen, Germany
e-mail: ellendt@iwt.uni-bremen.de

V. Uhlenwinkel
Foundation Institute of Materials Science, University of Bremen, Bremen, Germany
e-mail: uhl@iwt.uni-bremen.de

H. Henein et al. (eds.), *Metal Sprays and Spray Deposition*,
DOI 10.1007/978-3-319-52689-8_7

include plasma spraying, high velocity oxy-fuel, wire arc spraying, detonation gun spraying, etc. In this way, it will show why spray forming is such a unique process. This chapter is also dedicated to presenting how the porosity—an intrinsic feature of spray-formed microstructures—is generated and how the processing parameters affect its type, size and distribution. Furthermore, the generation of other defects related to the solidification and/or to the cooling of the spray formed product after deposition—such as residual stresses and hot cracks—and their influence on the product quality and material properties will be presented. Finally, this chapter will also discuss the effect of the atomization gas (Ar, N_2 or He) on the final product quality in terms of porosity and chemical composition of steels, superalloys, and copper alloys.

7.2 Spray Forming as a Bulk Process

Although having some common features with powder metallurgy and rapid solidification, spray forming is distinctly a casting process. It is operated in industry at a melt flow rate of up to 20 kg/min equivalent for steels per atomizer to produce large billets of up to 400 mm in diameter and weighing up to several tonnes. This is done directly from the melt, as shown in Fig. 7.1.

Despite the comparatively large billet sizes that can be produced, rather than containing large columnar grains normally associated with such large cast billets, the spray formed billet microstructure comprises polygonal (non-dendritic) grains in the range of 20–50 μm, and a comparatively narrow dispersion of grain diameters as shown in Fig. 7.2 [1]. Critically, the grain size distribution is almost insensitive to position in the billet and macro-segregation levels across the billet radius or along the axial length—a key consideration for advanced Al, Ni, Fe and other

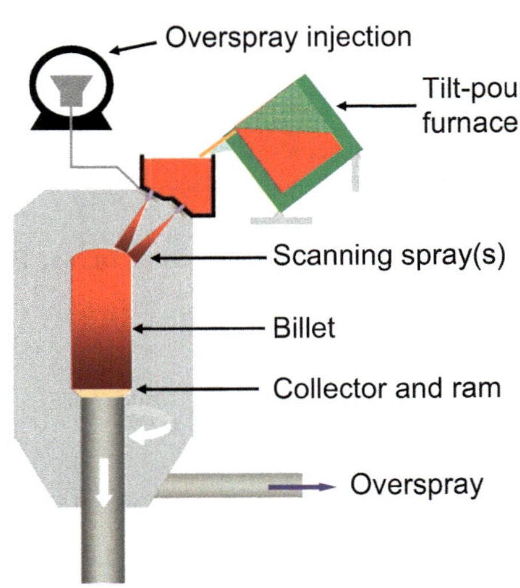

Fig. 7.1 Schematic diagram of a typical spray forming arrangement for the manufacture of large diameter billets, including the re-injection of overspray

Fig. 7.2 A 350 mm spray formed Al alloy billet (~30 kg) and a typical polygonal spray formed grain structure with low porosity

multi-element alloys—are reduced by up to an order of magnitude when compared with billets of similar size produced by conventional continuous or batch casting routes [2]. It is these two principal characteristics of microstructural refinement and low levels of elemental macro-segregation that makes spray forming an attractive process for the casting of complex composition alloys.

This section focuses primarily on the microstructure and nature of the phase transformations taking place in the spray formed billet, and their link to heat and mass flows during manufacture. A detailed description is provided on how the heat and mass flows at the point of deposition—when the droplet spray meets the evolving spray formed billet surface—interact in order to produce the unusual but highly desirable polygonal spray formed microstructure, which is an area that continues to be misunderstood. The process is shown to have solidification features unique to spray forming that lead to a potent intrinsic grain refining effect, necessarily involving a significant amount of re-melting. This re-melting is also essential in promoting the homogenization of billet elemental distributions.

However, the evolution of porosity is also an intrinsic feature of spray forming, and both the tendency for porosity and different types of microstructure that can arise when spray forming conditions are not carefully controlled are explained in terms of the thermal and solid fraction distributions on the spray forming billet surface. Spray forming allows the inter-mixing of a "cold" spray of pre-solidified overspray powders from previous spray forming production runs as shown in Fig. 7.1 to increase yield, and a similar approach can be used to form composites such as metal matrix composites and metal-metal composites [3–5]. Furthermore, two "hot" sprays can also be intermixed either at the micro-scale by co-deposition to make in-situ complex composition alloys, or can be deposited sequentially in discrete layers to make unusual and potentially attractive laminated or layered materials [6]. However, these are beyond the scope of this Chapter.

7.3 The Evolution of the Spray Formed Grain Size and Shape–Thermal Conditions

The spray forming process can be seen as sequential steps of melting and atomization, droplet and spray dynamics, deposition of the droplets at the evolving deposition surface, and final solidification and cooling. Because the first two steps are considered comprehensively elsewhere in this book, here only the last two steps are considered, with a particular focus on how they relate to final billet microstructure.

The first step is to identify the boundary conditions at the evolving surface of a large spray formed billet, and for simplicity, it is assumed initially that this surface is at a pseudo-steady state. This means that regardless of position in space or time in the process, the surface has the same shape and the same heat and mass flow conditions. Indeed, it is often the objective of the spray forming plant operator to ensure that this condition persists for the maximum fraction of the overall production run, as it should yield a billet with parallel sides and consistent microstructure [7]. The pseudo-steady state ignores the transients at the beginning of deposition when the spray first intersects the substrate and the billet begins to form, and the final transient when the melt supply to the atomizer becomes exhausted. In both transients, heat and mass flow rates are evolving, and the microstructure (grain size, porosity, etc.) in the final billet shows spatial variations. It is generally preferred to minimize the fraction of material deposited under transient conditions as these are usually "scalped" or machined away, undermining the process yield.

Returning to the pseudo-steady state at the billet surface, other sections in this book have shown that the incoming spray comprises a distribution of droplet diameters and consequently a distribution of thermal and solid fraction conditions. It can be envisaged that the smallest droplets in the depositing spray are cold and fully solid, the intermediate sized droplets (around the mass median diameter) are of intermediate temperature and likely to be partially solidified/mushy, and that the largest, slowly cooling droplets may be fully liquid. If the different solid fraction conditions of the droplets are integrated over the usually log-normal droplet diameter distribution, almost all researchers agree that, on average, the spray is in the mushy condition, with approximately 50% of the alloy latent heat removed in the short period of time from the creation of the droplets by atomization to the point of their deposition. Of course, the fraction of solid, mushy and liquid droplets—and therefore the overall, average spray solid fraction—can be manipulated by changes in the process parameters. For example, simply increasing the distance between the point of atomization and deposition will tend to increase the fraction of solid particles and the overall spray solid fraction. It is then obvious that if conditions are chosen where the solid fraction is high, powder is being manufactured and there is insufficient liquid fraction to bind the solid particles together at deposition; if the liquid fraction is excessive, there is little benefit of atomization and essentially only melt dispensing is taking place.

Between these extremes, the concept of an optimum liquid fraction emerges: enough liquid fraction to bind (strictly to partially re-melt, as described later) the solid component of the spray, but sufficient solid fraction that a significant proportion of the alloy latent heat has been removed in the spray. This concept of optimum liquid fraction on the surface of the growing billet is critical in understanding how to produce billets of low porosity with microstructural refinement [1, 2, 8–10]. However, a second key concept is required to understand how spray forming works: the extensive literature on the microstructure of atomized powders shows that the solid component of the depositing spray will be in the form of fine-scale dendrites or cells, but an optimized spray formed microstructure shows no evidence of dendrites or cells (Fig. 7.2). How do the solid and liquid components of the depositing spray interact at the microstructural scale to yield a microstructure with no trace of the droplet dendrite structure? This requires a more detailed understanding of the solidification conditions in spray forming, immediately before and after deposition of the droplets.

Droplet diameters in spray forming are usually considerably coarser (mass mean diameter $d_{50} \sim 100$ μm) than those used in powder production ($d_{50} \sim 40$ μm) and a typical log-normal distribution of droplet diameters produced by atomization in spray forming is shown in Fig. 7.3 for $d_{50} \sim 100$ μm, on both a mass (or volume) and number basis.

As described earlier, Fig. 7.3 shows how at the point of deposition the droplet spray will comprise a mixture of solid, mushy and liquid droplets. By mass or volume, the spray comprises a minority of solid powders, with most droplets either mushy or entirely liquid; on a number basis, the depositing spray is predominantly solid, and this solid component of the spray will be in the form of fine-scale dendrites. This fine-scale dendritic structure arises from a high rate of heat extraction and/or an undercooling before solidification is initiated by nucleation so that the metastable liquid then rapidly transforms to solid (and where external heat extraction is largely irrelevant). In both cases, solid/liquid interface velocities can

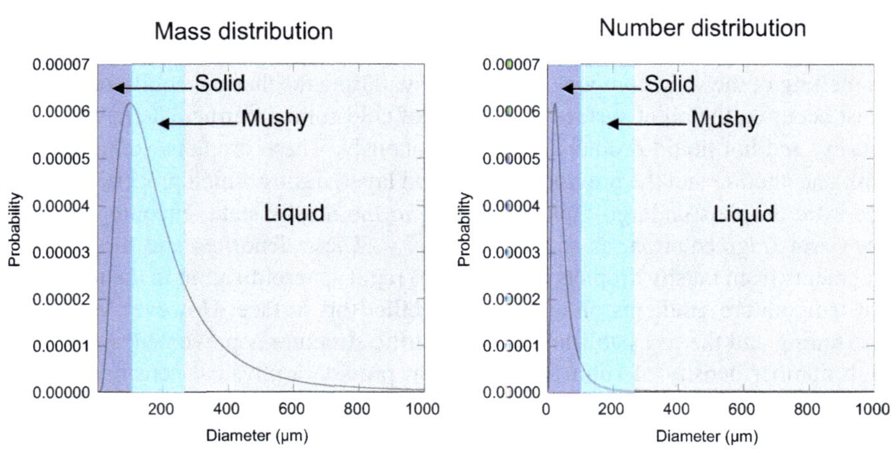

Fig. 7.3 A typical droplet diameter log-normal distribution produced by atomization during spray forming, and typical proportions of solid, mushy and liquid droplets at the point of deposition

be in the mm/s range, and diffusion distances are necessarily highly restricted, leading to the microstructural refinement well-known in powder atomized materials. Since undercooling is sensitive to droplet diameter (via the cooling rate) and droplets in spray forming are larger than used in powder metallurgy, a significant volume fraction of undercooled droplets in spray forming should not be expected. Instead, and especially in commercial alloys, heterogeneous nucleation of solidification close to the alloy liquidus under the modest cooling rates experienced by most droplets in spray forming should be expected. Once the droplets are deposited and a mushy region at the top surface is formed, there will be no undercooled liquid whatsoever since as explained earlier, under any set of sensible conditions, the top surface equilibrated or average solid fraction is always greater than zero, and the surface mushy layer will contain thousands of nucleation sites/embryonic grains arising from the solid component of the spray.

From the description above, we can see that spray forming, while having some similarities with powder production, strictly, spray forming is certainly *not* a rapid solidification process: only about half of the alloy latent heat is removed in the spray stage and under conditions where fast solid/liquid interface velocities can be expected; the remaining half of the solidification occurs in the billet, which due to its much larger physical dimensions (diameters of up to 400 mm or more) cannot cool particularly rapidly (1–10 K/s) and solid/liquid interfaces are similar to conventional casting (0.01–0.1 mm/s).

7.4 The Evolution of the Spray Formed Grain Size and Shape: Polygonal Grains

Returning to the critical question of why the spray formed microstructure contains no remnant of the fine-scale dendrites or cells, only polygonal, equiaxed grains (Fig. 7.2), it can now be understood that this can only occur if there is significant re-melting of the solid component of the spray, during the thermal equilibration that must occur on the billet surface as millions of cold solid, intermediate temperature mushy, and hot liquid droplets arrive continuously. These droplets exchange heat with one another and the previously deposited layer, during which process the cold, dendritic droplets undergo rapid re-heating to the mushy state, encouraging dendrite arm fragmentation, as shown in Fig. 7.4. These dendrites and the dendrite fragments from mushy droplets then undergo rapid spheroidization in the relatively flat temperature gradients on and near the billet top surface. However, excessive coarsening and the re-establishment of dendritic structure is prevented by the very high number density of solid fragments that provide a physical constraint to the development of large, extended grain morphologies [2, 11].

The partial re-melting and spheroidization behaviour of the solid component of the spray is exploited in the re-cycling of powder overspray from earlier spray forming productions runs. Here, cold overspray powder, which is carefully

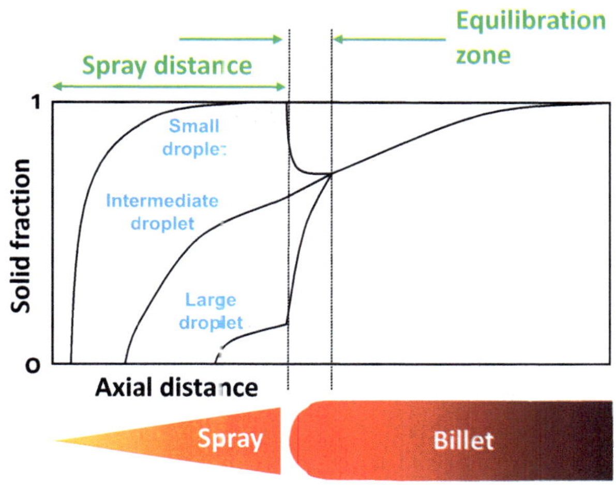

Fig. 7.4 A schematic description of how solid fraction varies in spray forming for different sized droplets and for the evolving billet

maintained under inert atmosphere and free of contamination, is re-injected back into the spray forming process throughout deposition, using a dedicated powder feed system and as shown in Fig. 7.1. The mass feed rate of powder is carefully controlled and the cold powders are injected close to the point of atomization. Some powders are incorporated directly into the larger droplets formed by atomization; others co-deposit with the hotter atomized droplets at the billet surface. When conditions are suitably optimized—the overall liquid fraction of the top surface is correctly controlled—no remnant of the individual, re-injected cold powder dendritic microstructure can be found in the spray formed microstructure, even when significant fractions of cold powder are introduced. As well as providing a useful contribution to overall process yield and therefore the economic viability of spray forming, this behaviour provides proof of the copious re-melting that must take place during spray forming under optimum conditions.

Thus, the presence of significant liquid in the spray formed billet is both a desirable and essential requirement of spray forming. Manipulation of the atomization parameters or geometrical arrangement of the substrate to increase cooling rates in pursuit of traditional rapid solidification benefits are likely counterproductive: the dendritic structure of the solid component will persist rather than partially re-melt, and porosity will increase due to insufficient liquid fraction required for the integration and homogenization of the solid component of the spray [12]. Operating under optimum conditions, spray forming is neither a powder metallurgy nor a rapid solidification process but a *self grain-refining casting process* in which the atomization step is used only to initiate solidification in millions of locations in the melt, before "re-assembling" the droplets back together into a bulk billet in a process that must involve partial re-melting [2]. The mushy, highly grain-refined billet then cools and solidifies slowly, under entirely conventional conditions, as shown in Fig. 7.4.

7.5 The Evolution of the Spray Formed Grain Size and Shape: Micro and Macro-Segregation

The partial re-melting and fragmentation of the solid and mushy component of the spray also provides beneficial chemical micro-homogenization effects. The first to re-melt, inter-dendritic regions of the powders during thermal equilibration are generally those regions enriched in solute atoms (for elements with a partition coefficient with the liquid of less than unity). On re-melting, the solute atoms are released, mixed and homogenized by rapid liquid-state diffusion into the more dilute liquid from the hotter, larger droplets; the more cycles of re-melting as hot droplets continue to arrive, the greater the homogenization effect.

The homogenization effect is shown schematically in the portion of the equilibrium phase diagram for a binary alloy in Fig. 7.5 where droplets A and B with relatively high and low temperatures T_1 and T_2 respectively impact on a mushy billet top surface of intermediate temperature T_{sur}. As indicated by the vertical arrows, droplet A is quenched to T_{sur} by the deposition whereas droplet T_2 is re-heated to T_{sur}. The time for the thermal equilibration $t^T{}_{eq}$ can be estimated from:

$$t_{eq}^{T} \sim \frac{x^2}{\alpha} \tag{7.1}$$

where x is the thermal equilibration distance and α is the thermal diffusivity. For x of up to several hundreds of microns for the largest depositing droplets and

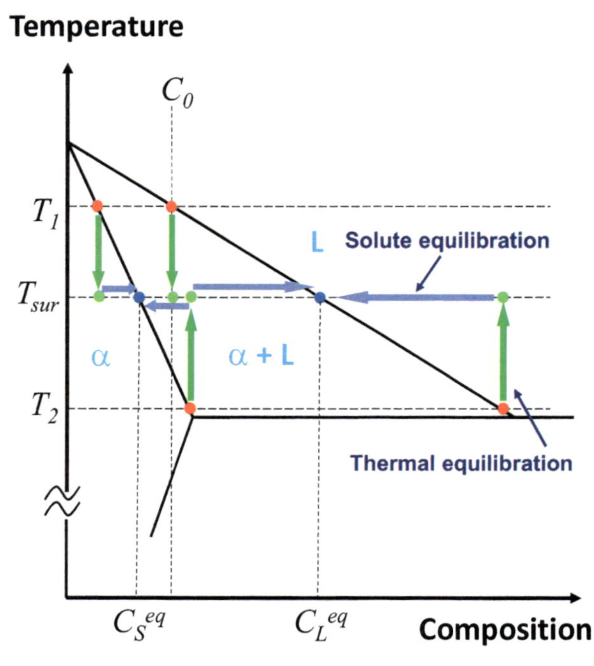

Fig. 7.5 A section of a binary phase diagram showing the thermal and solute equilibration processes that take place as droplets of different temperatures deposit onto a mushy billet top surface

assuming a typical value of $\alpha = 10^{-5}\ \mathrm{m^2\ s^{-1}}$ for metals, then $t^T{}_{eq} < 0.1$ s. Figure 7.5 shows that as well as thermal equilibration, there is a driving force for equilibration of the solid and liquid solute concentrations indicated by the horizontal arrows, where the relatively solute enriched liquid of the colder droplets is mixed into the more dilute liquid associated with the hotter billet top surface. The timescale for solute equilibration $t^C{}_{eq}$ can be estimated similarly from:

$$t_{eq}^{C} \sim \frac{x^2}{D} \tag{7.2}$$

where D is the diffusion coefficient of solute atoms in the liquid. Assuming $D = 10^{-8}\ \mathrm{m^2\ s^{-1}}$ and over the same length scale, the equilibration time for solute atoms is typically the order of seconds. It is the re-heating and mixing effect that leads to homogenization and low levels of "coring" or micro-segregation.

Figure 7.6 shows the maximum to minimum variation in the concentration of elements Cr, Fe, Nb, Mo, Ti and Al in three Ni superalloy IN718 dendritic overspray powders of diameter 25, 100 and 120 μm as well as similar data from a spray formed IN718 ring manufactured at the same time. For all the powder diameters, Nb showed the greatest tendency for microsegregation [2]. However, all the overspray particles showed greater Nb microsegregation than the as spray formed microstructure, despite the much slower cooling rate of the spray formed ring. Therefore, although fast solidification speeds in small droplets can restrict elemental segregation distances in powders, there is only one solidification event and the resulting micro-segregation pattern is quickly frozen. In spray forming,

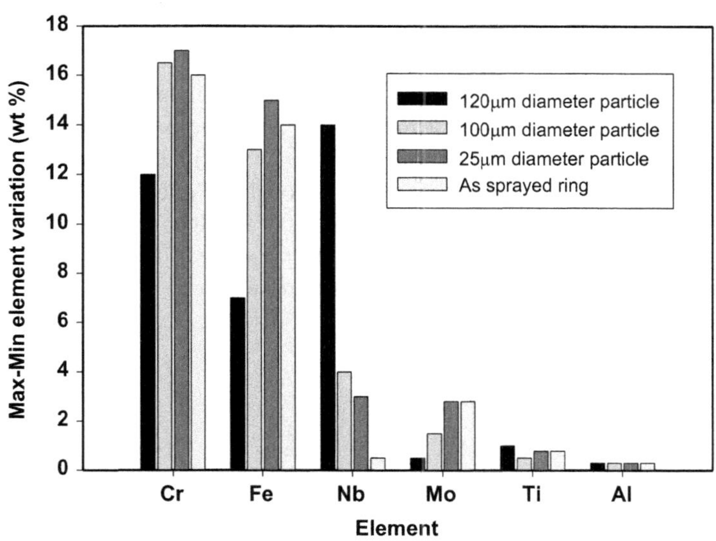

Fig. 7.6 The differences in micro-segregation behavior of Ni superalloy IN718 dendritic powders and the polygonal grains in a bulk spray formed ring (from [2] with permission)

although final billet grain sizes are coarser than in powders they show reduced micro-segregation because of the repeated re-melting cycles that provide a homogenization effect.

Additional benefits also arise in spray forming in terms of macro-segregation i.e. chemical variations over the length of the billet itself. It is well-known that macro-segregation in large billets, such as aluminium billets produced by direct chill (DC) casting, occurs due to (1) shrinkage induced flow of solute-rich liquid over significant distances in the final stages of solidification, (2) gravity effects such as either the floating or sedimentation of equiaxed grains with densities different to that of the liquid, and (3) convection of solute-rich liquid due to liquid density differences from point to point in the billet, usually due to differences in temperature causing differences in the liquid solute concentration and thus liquid density.

In the case of shrinkage induced flow, this tends to be reduced in spray forming because, as previously described, typically more than half the alloy latent heat has already been removed in the spray phase of the process so that as the billet is formed, it already has a comparatively flat composition gradient from billet centre to billet edge. Solidification and shrinkage of the minority fraction of residual liquid may induce some long range movement of the typically solute-rich liquid, but because the starting point is an already substantially solidified billet, the final magnitude of macro-segregation is always reduced when compared with billets of similar dimensions produced conventionally [13, 14]. Grain sedimentation or floating effects (freckling) are completely suppressed by spray forming since the billet solid fraction is too high everywhere across the billet diameter (there is no centre line liquid sump as in DC casting) for any long range motion of individual grains. Similarly, inter-connected liquid in the billet is at relatively low fraction and in a tortuous network around grains that makes long-range convective flow difficult.

In summary, while partial re-melting in spray forming solidification is a critical aspect in the formation of the polygonal and spatially invariant grain diameter, it also produces significant micro- and macro-segregation benefits. These features combine to make spray forming a particularly attractive bulk casting process for those alloys that require significant concentrations of multiple alloying elements such as Ni superalloys and some types of steels. Although the microstructural refinement and excellent chemical homogeneity of the billets can allow spray formed materials to compete with powder metallurgy routes otherwise used for these types of alloys, the solidification path and the final microstructures are quite distinct.

7.6 Layering/Non-Layering

As previously described, a polygonal, equiaxed structure that is the same from place to place, even in large billets, is a characteristic feature of spray formed materials. This integral, coherent structure where no remnant of the droplet-by-droplet nature of the way the billet formed is evident is in stark contrast to other spray deposition or "thermal spray" processes. The key difference between spray forming and these processes, such as plasma spraying, high velocity oxy-fuel, wire arc spraying, detonation gun spraying, etc., is the very much higher mass flow rates involved in spray forming. In thermal spraying processes, individual droplet impacts generally have time to solidify before another droplet arrives at the same point. Although there is some heat exchange between deposited and depositing droplets, the arriving droplet generally does not have sufficient "superheat" (excess specific heat content above the liquidus temperature) to effect significant re-melting of the underlying, previously deposited material so that (1) no significant liquid fraction is present on the evolving top surface for more than a few milliseconds, and (2) each droplet impact or "splat" retains its discrete structure in the final coating or billet. In spray forming, the mass flux of droplets arriving at the substrate, and the associated heat flux per unit area, is much greater so that not only is transient liquid present, but the entire top surface is a mixture of solid and liquid, and this mushy zone extends deep into the underlying billet (see Fig. 7.4). Under these preferred conditions, any layered structure is eliminated [12].

Figure 7.7 shows schematically the thermal history of a small volume of material incorporated into a billet top surface with an average bulk temperature T_{sur} in the alloy mushy zone during spray forming, under two different conditions A and B, and for an alloy with a solidus temperature T_S. Under condition A, the top surface receives material as it passes under the droplet spray due to spray scanning and/or rotation of the substrate, the temperature rapidly rises, and reaches and briefly exceeds the average top surface temperature. The material then begins to cool as it exits the droplet spray and the temperature falls below the solidus temperature T_S. On the next pass under the spray, after an inter-pass time interval Δt, the material is re-heated above T_S as further droplets arrive on the top surface. Under these conditions some inter-mixing between the material and the newly deposited material occurs, and some elemental homogenization and microstructural integration can be expected. Under condition A, Fig. 7.7 shows two re-melting cycles before the material becomes sufficiently embedded into the bulk of the billet and distant from the top surface that pulses of heat associated with continuing layers of deposition are insufficient to effect further re-melting cycles.

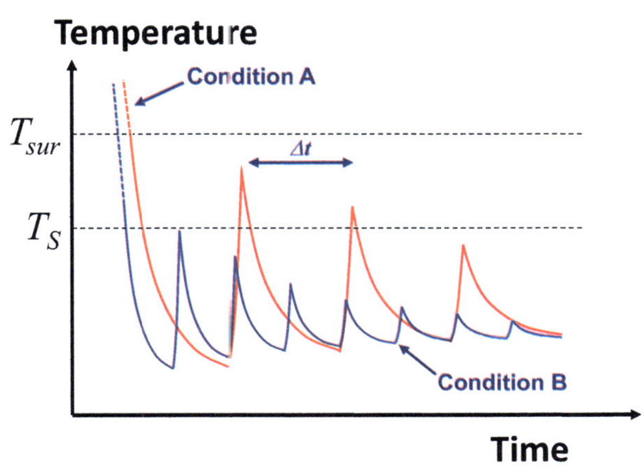

Fig. 7.7 The thermal profiles a small volume of material incorporated into a billet during spray forming under two different conditions

Figure 7.7 shows that under condition B, the point on the surface travels under the spray twice as frequently (Δt is halved), for example because the atomizer scan speed or the substrate rotation speed has doubled. The depositing spray thermal condition is unchanged and so the underlying average temperature of the top surface T_{sur} is also unchanged because the top surface temperature is controlled only by a macro-scale heat rate balance over longer time scales than the inter-pass time. However, at shorter timescales and at the microstructural scale, and considering the thermal history experienced by local regions of deposited material, Fig. 7.7 shows that under condition B there are now no cycles of re-melting. It can be expected therefore that the final microstructure under these conditions will retain some character of each deposited layer, there will be limited homogenization, and the formation of equiaxed grains is not favoured.

Changes in the spray forming geometrical arrangement can affect the balance of heat flux arrival at the preform and the rate of heat extraction from the preform. For example, when large area substrates are used in spray forming, such as in the production of hollow tubes, the heat flux per unit area of substrate due to the arriving spray is much lower than in the production of billets, heat is extracted more efficiently to the substrate and the gas flow, and the inter-pass time becomes extended. Under these conditions, an undesirable layered structure becomes more likely and each revolution of the substrate might be distinct (as a thin layer of pancake or splat grains) in the final as-sprayed microstructure.

A more severe problem may arise if a hollow tube is manufactured by multiple *axial* passes of the spray as the substrate is rotated and moved axially in and out of the spray. In this arrangements it becomes almost impossible to prevent "inter-layer" porosity and microstructural heterogeneity in the form of a region of layered grains retained in the final microstructure. Here, the inter-pass time interval is too extended and the average top surface temperature below the solidus temperature so there is insufficient re-heating to effect inter-layer re-melting and coherence of the microstructure. Moreover, over long inter-pass intervals, limited surface oxidation might occur so that inter-layer cohesion is impossible. Therefore, best practice to avoid layering is always to try to choose a deposition strategy that prevents deposited material from significant cooling until deposition is complete in that region. In the case of large diameter rings and tubes, or even thick sprayed strips on a flat substrate, substrate pre-heating [15] and even continuous heating of the substrate during deposition using an induction coil internally positioned along the axis of the rotating substrate can be effective in reducing inter-layer banding (and in the case of strip or 3D shapes, in the control of residual stresses [16]).

7.7 Porosity

Porosity is one of the most important criteria of spray-formed materials in regards to quality, since porosity can reduce considerably their mechanical, thermal and electrical properties [17, 18]. While controlled porosity levels can lead to improved properties in certain applications (i.e. insulators), metallurgic products usually require high mechanical properties such as tensile, impact and fatigue strength, which are sensitive to porosity. Thus, in applications that require high mechanical properties, porosity must be eliminated through further thermomechanical processing. Porosity is not a characteristic unique to spray-formed products, but also is present in products made from different metallurgical routes such as powder metallurgy and conventional casting, for example, and the porosity level obtained from these processes can only in rare cases reach zero. However, spray-formed products with porosity levels in a range of 0.2–3.0% can be achieved by optimization of the process parameters [15, 19, 20–22]. During the last years several efforts were made in order to understand and to model the formation of pores in the spray forming process as well as to predict the porosity levels in spray forming products [21, 23–25].

The porosity in spray formed materials is dependent on several factors such as processing parameters, thermodynamic properties of the material and of the atomization gas, and the solidification and surface conditions during the spray deposition process. This huge number of variables makes the porosity a very complex scenario to simulate. Three main types of porosity are described in the literature based on their formation mechanisms. The first type is known as interstitial porosity and takes place when, upon impingement, solidified individual droplets overlap each other, forming interstices. If the average fraction of liquid in the droplets is too low and insufficient to fill the interstices in the deposition zone, the porosity is formed. As this sort of porosity occurs when cold spray conditions are present, this porosity is also named "cold porosity". The second type is the gaseous porosity and is caused by a high liquid fraction at the deposit surface as a consequence of hot spray conditions. The gaseous porosity can result from gas rejection due to the difference of solubility between solid and liquid phases and from mechanical gas entrapments. Impinging droplets also entrap gas into the deposit, which is not dissolved and cannot leave the deposit due to solidification. The last type is the solidification shrinkage which results from the different densities of the liquid and solid phases. As this type of porosity is also related to a high fraction of liquid during the deposition, both the gaseous porosity and the solidification porosity are usually referred to as "hot porosity". Cold porosity usually takes place in the vicinity of the substrate due to the efficient heat dissipation that is experienced by the material during the first stages of the deposition process (see Fig. 7.8). Figure 7.8c shows an example of interconnected irregular pores characteristic of cold porosity. Gaseous porosity is usually spherical, with a range size from 100 to 250 μm, Fig. 7.8b, and can be present along the whole deposit depending on the spray forming conditions. The smallest pores that are present in spray-formed materials may be attributed, in some cases, to solidification shrinkage.

In the late 1990's Cai, W. and Lavernia, E.J. [23–25] proposed a simplified model to describe the porosity which considers the fraction of liquid in the deposition zone as the most important parameter to define the porosity formation. The model assumes that deposition occurs in two stages. First, the solidified droplets impinge on the deposition surface forming a random, dense particle

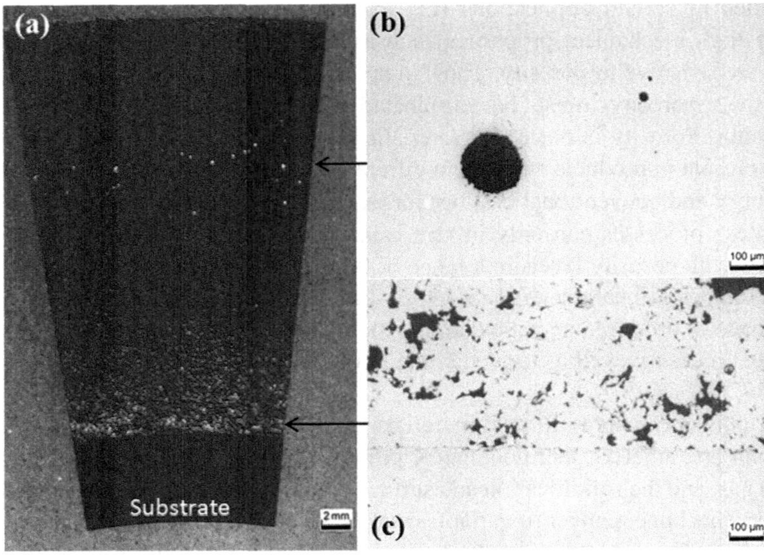

Fig. 7.8 (**a**) Cross-section of a spray-formed 100Cr6 steel ring, (**b**) gaseous porosity and (**c**) cold porosity in the vicinity of the substrate

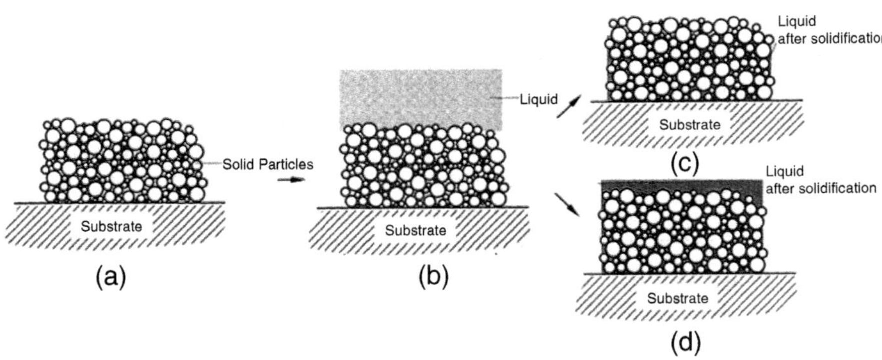

Fig. 7.9 Schematic diagram showing the two-stage mechanism of forming a deposited materials proposed by Cai, W. and Lavernia, E.J (**a**) solid particles forming a random dense particle packing structure, (**b**) liquid droplets impinging on the particle packing structure, (**c**) and (**d**) liquid flowing into the particle packing structure and subsequently solidifying (from [25] with permission)

packing structure. During the second stage, liquid droplets impinge on the resultant particle packing structure, flow into the particle packing structure and finally solidify. A schematic illustration of the model is presented in Fig. 7.9. Porosity is then estimated considering if the fraction of liquid during the second stage is sufficient to fill the voids of the particle packing structure. Based on the model's assumptions, two cases are considered. When the volume of liquid, once solidified, is smaller than that of the voids in the particle packing structure (Fig. 7.9c), interstitial porosity (or cold porosity) is dominant, and the porosity level is calculated based on the volume of the unfilled voids after solidification of the liquid fraction. On the other hand, when the volume of liquid, once solidified, is higher than that of the voids (Fig. 7.9d), the hot porosity is dominant. In this model, it is assumed that the source of hot porosity is mainly due to solidification shrinkage and the gaseous porosity is neglected. The hot porosity is calculated by a simple volumetric shrinkage factor during the solidification. In this simplified model when the interstitial porosity is predominant the solidification porosity is neglected, and vice-versa, which in practice is not true since these two types of porosity can and probably will coexist simultaneously. Mathematical models to describe the average fraction of solid, gas atomized droplet size distribution, droplet dynamics, droplet thermal history (droplet undercooling, liquid-phase cooling, nucleation, recalescence and solidification) and particle packing density were applied to describe the influence of the process parameters in the porosity level of spray formed deposits. The mathematical results showed that the porosity values vary with the main process parameters such as atomization gas pressure, melt flow rate and deposition distance with "V"-shaped behavior (Fig. 7.10), with higher porosity

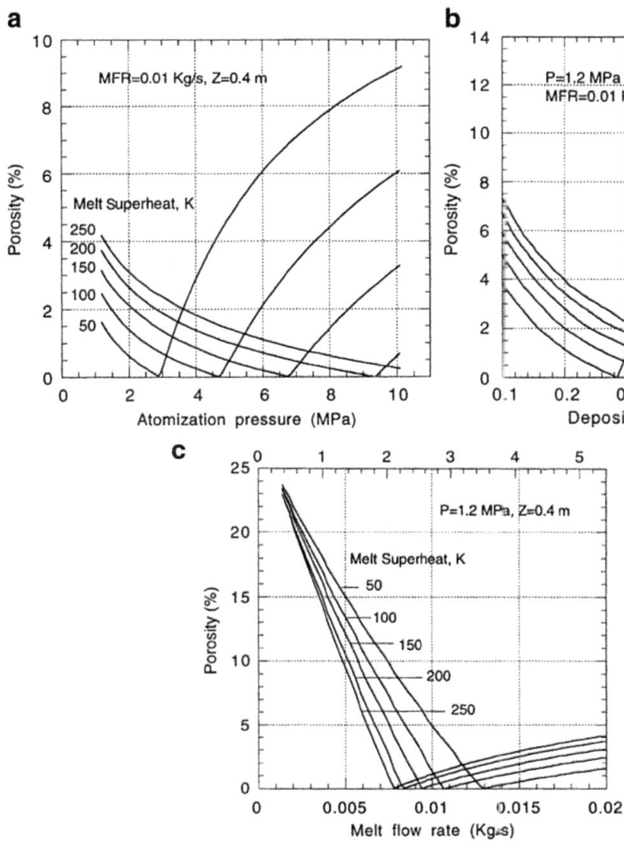

Fig. 7.10 Calculated porosity as function of (**a**) Atomization gas pressure, (**b**) Spray distance and (**c**) Melt flow rate (from [25] with permission)

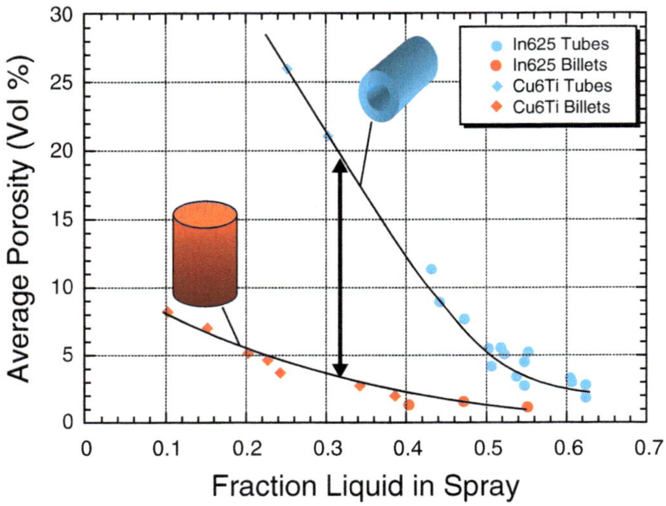

Fig. 7.11 Average porosity versus liquid fraction in the spray during spray forming of a billet and tube of Inconel 625 and Cu-6Ti alloy (adapted from [26])

when cold process conditions are applied (for instance, high atomization gas pressures, low melt flow rates and high spray distances). Although the V-behavior of porosity is observed in practical cases, an experimental validation of the model is still pending. Furthermore, the complexity of the models used, the amount of parameters and properties needed and the exclusion of other factors that also influence the porosity such as substrate conditions and flow field around the spray cone, make this model difficult to be used in practical situations. For instance, it was shown that for a pair of same materials with same fractions of liquid considerable different porosity levels is achieved when a billet or a tube is spray formed (Fig. 7.11). Such difference comes from the different substrate conditions found in both cases [26].

A more practical approach for predicting the porosity levels of spray formed materials was proposed by Uhlenwinkel, V. and Ellendt, N. [21] in which the porosity is related to the deposit surface temperature during spray forming. The authors proposed that independent of the process parameters, for the same material, an identical porosity level will be achieved if the deposit surface temperature is the same. Figure 7.12 presents an example of porosity level as function of deposition surface temperature with data collected from several superalloy ring spray forming runs. Similar to the model presented before, the porosity values also vary with the deposit surface temperature by a V-sharped behavior, with higher porosity values found when low deposit surface temperatures are observed. Meyer et al. [15] showed that the same relation between the porosity level and the deposit surface temperature can be observed in spray-formed sheets and different materials such as Cu alloys (Al-bronze and Sn-bronze) and nitriding steel. Moreover, by using the concept of dimensionless enthalpy of deposit surface (h^*_{surf}), Meyer et al. showed

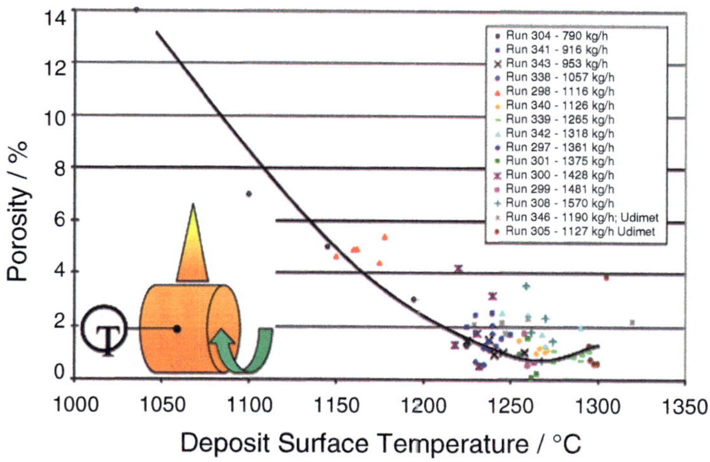

Fig. 7.12 Porosity versus deposit surface temperature of spray formed superalloy rings [21]

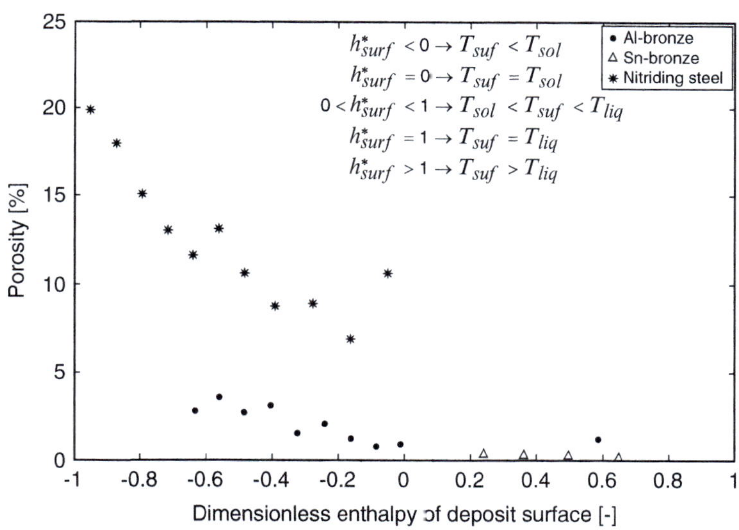

Fig. 7.13 Effect of the dimensionless enthalpy of deposit surface on porosity of Al-bronze, Sn-bronze and nitriding steel (adapted from [15])

that the optimum porosity level can be reached when deposit surface temperatures close to the alloy solidus temperature are present, independent of the material and the process parameters (Fig. 7.13). The authors also showed that cold porosity in the vicinity of the substrate can be considerably reduced by pre-heating the substrate at temperatures close to the alloy solidus temperature. However, the magnitude of the minimum porosity level is strongly dependent of the material. Figure 7.14 shows the as spray-formed porosity levels (expressed as relative density) of different classes of alloys such as Al-alloys, Ni-alloys, Cu-alloys and steel reached under

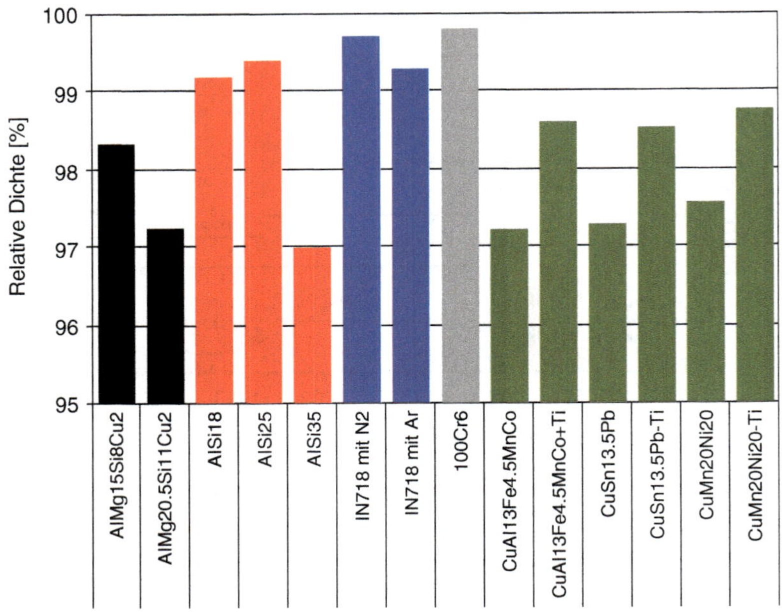

Fig. 7.14 Relative densities of different as-sprayed alloys under optimized process conditions (adapted from [21, 22])

optimized process conditions. As can be seen, relative densities higher than 99% can be reached for some materials such as 100Cr6 steel, Inconel 718 and Al-Si alloys [21, 22].

Since the spreading of the individual droplets on the deposit surface is highly dependent on their Reynolds and Weber number, process windows based on thermal surface or droplet conditions can shift if different atomizers with different droplet size and velocity distributions are used. The use of a pressure-gas-atomizer for the coating of steel tubes with a CuSn15.5 alloy at low mass flow rates led to very low core porosities <0.5% [27]. For this type of atomizer and the reduced mass flow rate, very short deposition distances between 64 and 104 mm gave the best results. At such a low deposition distance, the droplet velocity is higher than for conventional atomizers at deposition distances in the range of 400–700 mm, while the droplet diameter is similar. Under such conditions, the additional amount of kinetic energy promotes densification during droplet impact.

As a summary, four major influence parameters on the formation of porosity can be extracted. The thermal state of the droplet and the deposit surface define the time scale available for droplet spreading until the droplet has fully solidified whereas the droplet diameter and impact velocity define the time which is needed in terms of droplet deformation. If similar time scales can be achieved, this will result in a low porosity level. If the solidification process takes too long, droplet spreading will be incomplete and cold porosity will occur. If the droplets are too hot, no droplet spreading will occur because the droplets can impact into the mushy zone. In this case, gas will be entrapped, resulting in hot porosity.

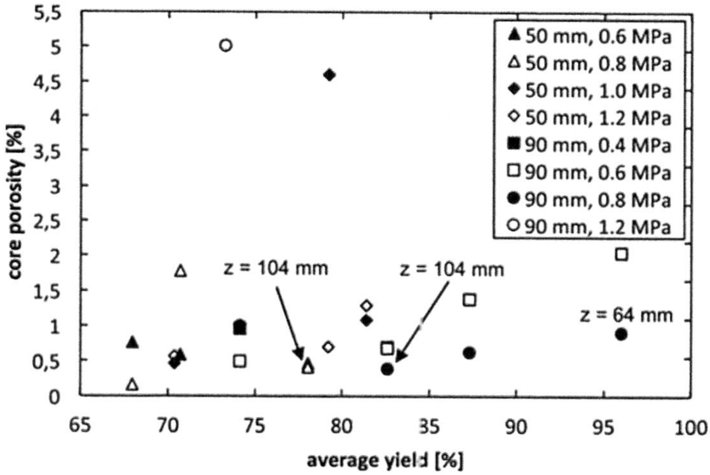

Fig. 7.15 Core porosity of tubes (diameter: 50 and 90 mm) spray-formed using a pressure-gas-atomizer at different atomization pressures

7.8 Residual Stress and Hot Cracks

Thermal gradients within spray formed deposits, during the deposition process and during the cooling after the spraying period, can vary enormously depending on the material properties of the spray formed alloy and the process parameters used [28–31]. Whereas the material properties are fixed for each alloy, the process parameters such as melt superheat, melt flow, gas-to-metal flow ratio (GMR), spray distance, substrate geometry and substrate material can be varied to achieve the desirable process variables such as enthalpy and mass flux distribution in spray, heat flux from the surface of the deposit to the substrate and to the surrounding gas environment [28–32]. The thermal history of the spray formed deposit can be optimized by selecting appropriate process parameters in order to achieve the desired metallurgical properties. This section will discuss the generation of residual stresses and hot cracks in spray formed products, which are defects intimately related to the thermal history of the spray formed billet.

Most manufacturing and fabrication processes induce some level of residual stress in materials. The distribution and magnitude of these residual stresses can vary drastically from one manufacturing process to another. The presence of residual stresses may have a significant effect on the product quality and the material properties. Shape distortion, reduction of the fatigue resistance and accelerated stress corrosion are typical undesirable effects caused by residual stresses. Moreover, residual stresses may initiate and promote premature damage, such as microcracking. Residual stresses in spray formed material may arise from two primary sources: (1) the billet thermal gradients during cooling from the deposition temperature to ambient temperature may produce non-uniformity of elastic and plastic strain, which results in thermal residual stresses in the deposit, and

(2) liquid-solid and/or solid-solid phase transformations during the cooling process can generate residual stresses due to the variation of volumetric change of the spray formed material that can be different in each position of the deposit depending on the phase transition time and fraction [33–36].

Ho and Lavernia [33–35] published a series of works presenting results of both numerical simulation and experimental results on residual stresses in spray formed deposits of different materials such as Ni_3Al, 6061 Al-alloy and a SiC-Al metal matrix composite (MMC). Figure 7.16 presents the results of the calculated and measured von Mises' equivalent stresses along the center of a Gaussian-shaped Ni_3Al deposited on a water-cooled copper substrate [33]. The fully grown preform, as shown in Fig. 7.16a, at the end of deposition was modeled by using the finite element method (FEM) to solve the coupled temperature-displacement problem simultaneously. As boundary conditions, the initial temperature at the top surface was assumed to be the solidus temperature of the alloy (1658 K) and the initial temperature at the bottom surface was 1150 K based on previous measurements. The temperatures inside the preform were interpolated. The heat coefficients at the top and bottom surfaces were estimated as 21 and 550 W m^{-2} K^{-1}, respectively, and the ambient temperature was assumed to be 300 K. Figure 7.16b shows that

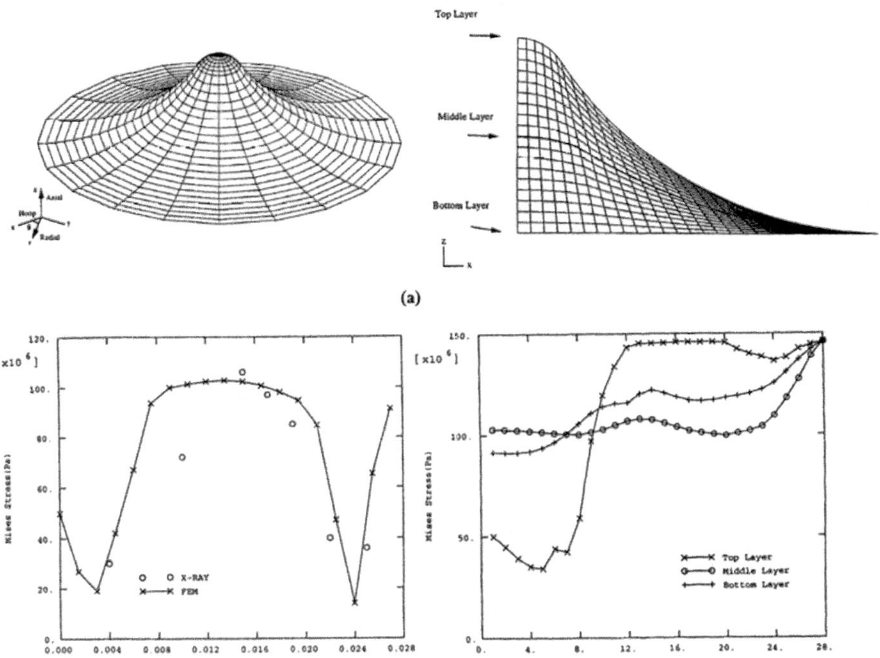

Fig. 7.16 (**a**) Geometrical shape and two-dimensional element mesh used in numerical simulation of the Ni_3Al preform, (**b**) The von Mises' equivalent stresses estimated by FEM and measured by X-ray diffraction at the center of the Ni_3Al preform and (**c**) The von Mises' equivalent stress along the top, middle and bottom layers in the Ni_3Al preform as function of node number [33]

even with the simplicity of the applied model, the FEM results are in a good agreement with stresses measured by X-ray diffraction at the center line of the preform. The results showed that at the center region the distribution of equivalent stresses is relatively homogenous at approximately 100 MPa. The yield stress of Ni_3Al is relatively low at 300K (depending on grain size) and plastic deformation or possibly cracks may then develop in this region. Fig. 7.16c shows that the equivalent stresses along the radial direction at the top, middle and bottom layers have higher values for the region closer to the outer edge of the preform and, thus, plastic deformation or cracks may be present in that region as well [33]. Such results show that residual stresses can easily arise in the spray forming process depending on the cooling conditions.

In a similar work simulating the thermal history and the residual stresses of a Gaussian-shaped deposit of 6061 Al-alloy, it was shown that the residual stresses can vary considerably if the magnitude of the conductive heat transfer coefficient between the substrate and the deposited material is varied [35]. Physically, a variation of conductive heat transfer is easily achieved by changing the substrate material. Figure 7.17a–c presents the von Mises equivalent stresses in an A2 tool steel spray-formed Gaussian-shaped deposit (similar to that shown in Fig. 7.16a) calculated by numerical simulation along the top, middle and bottom layers

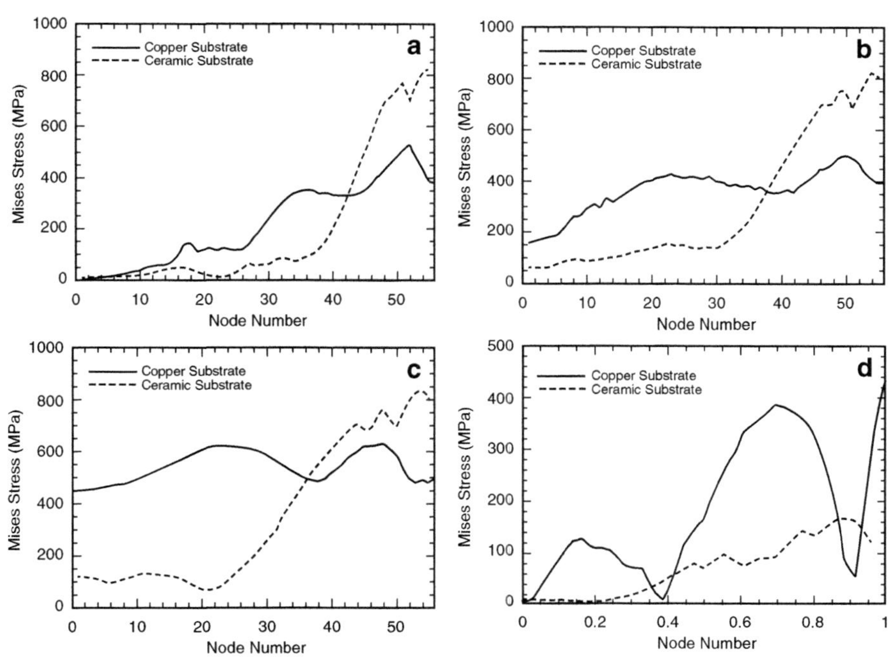

Fig. 7.17 The von Mises equivalent stresses in spray-formed A2 tool steel as function of node number along (**a**) the top layer, (**b**) the middle layer, (**c**) the bottom layer and (**d**) along the center line (in this case as function relative distance from the top to the bottom of the deposit) [36]

considering two different substrate materials: copper and ceramic [36]. It can be seen that: (1) in all cases, the lowest stress levels occur at the center positions and the stresses increase toward the outer edge of the deposit, (2) there is a transition point in the radial distance of the deposit, regardless the layer position, where the stresses in the deposit on the ceramic substrate becomes higher than in the case of copper substrate, and (3) in both cases, but more clearly in the copper substrate case, the equivalent stresses increase from the top to the bottom of the deposit. The last point can also be described as function of the thickness of the deposit. In this way, it can be seen that the stresses increase with a decrease in the deposit thickness. Despite the thickness of the different layers, the variation in magnitude and distribution of the residual stresses can be addressed to the different heat extraction capacity of each substrate. For a ceramic substrate, due to the limited heat extraction capacity, the heat is lost primarily by convection and radiation on the upper surface. Thus, the cooling rates are higher in the surface region than in the center region of the deposit, which results in higher stresses in the regions next to the surface. On the other hand, when the copper substrate is considered, as the heat transfer coefficient between the deposit and the substrate is high, the dissipation of heat from the deposit surface by convection and radiation is relative low when compared with the heat dissipation promoted by conduction to the substrate, especially for the bottom part of the deposit. In consequence of effective heat conduction toward the substrate, the magnitude of the stress along the radial directions tends to be more uniform. Furthermore, once a more pronounced gradient temperature is formed, a higher variation in the residual stresses is expected to appear along the axial direction. This is shown in Fig. 7.17d which shows the von Mises equivalent stress along the center line of the deposit for both cases may be copper and ceramic substrates [36]. For the copper substrate, the stresses vary strongly along the axial direction. At the bottom surface, the steeper temperature gradient causes the highest stress. At a distance away from the bottom part of the deposit, the cooling effect of the substrate diminishes and so do the stresses. Further up, the cooling rate increases again because of the narrowing of the Gaussian deposit. Hence, the temperature gradient along the axial direction changes throughout, resulting in a significant variation of the residual stresses. In the case of the ceramic substrate, the residual stresses along the axial direction are maintained relatively low when compared with the radial stresses. It is important to point out that the simulations presented so far did not consider the effect of any phase transitions on the residual stresses. Especially for spray-formed tool steels, such as A2, D2 and H13, the austenite to martensite phase transformation can significantly affect the residual stresses in the deposit due to the variation of yield stress and also due to the volume expansion that accompanies the formation of martensite. The effect of phase transformations on the magnitude and distribution of residual stresses in spray formed materials is lacking in the literature.

In spray formed MMCs, residual stresses are developed as a result of the mismatch of the thermal expansion coefficient between reinforcement and matrix. The presence of high levels of residual stresses in the reinforcement-matrix

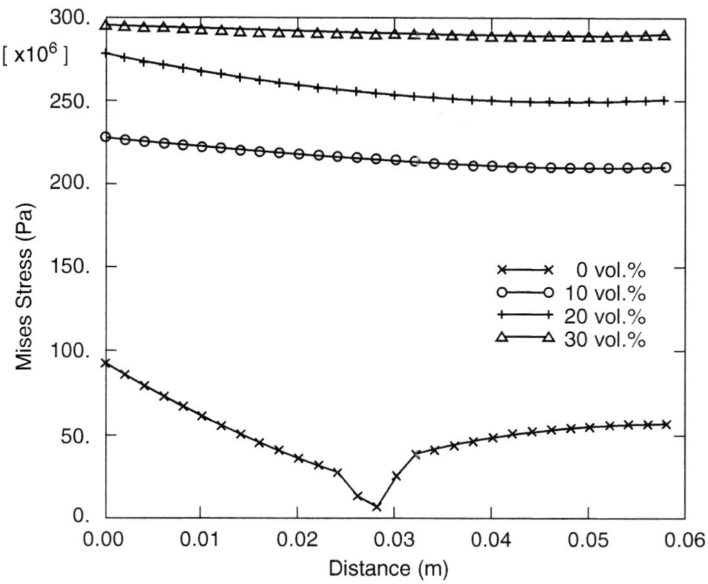

Fig. 7.18 The von Mises' stresses in the matrix at the interface matrix/reinforcement as function of the distance from the bottom of the co-deposited deposit of A357Al-SiC MMC calculated by FEM (from [34] with permission)

interface may reduce the desired properties of the composite. For instance, Ho and Lavernia [34] showed, using numerical simulation, that addition of spherical SiC reinforcement particles in A357 Al-alloy matrix significantly increase the residual stress in the matrix next to the reinforcement-matrix interface. As can be seen in Fig. 7.18, an increase in the volume fraction of reinforcement is also accompanied by increasing the residual stresses. The authors also demonstrated that the residual stresses in the matrix at the reinforcement-matrix were influenced by the thermal gradient in the same way, as presented in the previous examples. Although residual stress is an important aspect of the metallurgical quality of spray formed products only a few publications about this topic has been reported in recent years.

Hot cracking is another type of defect caused by the thermal gradients and is also related to residual stresses within spray formed billets and preforms [37–39]. Hot cracks are usually near the billet top or "crown" and cannot be seen from the outside and are only revealed by ultrasonic examination or by sectioning the billet. Hot cracks are developed when "hot spots" within the deposit during the solidification of the material are present [39]. Depending on the material properties and the spray conditions applied, the remaining liquid within the preform in the final stages of solidification can be totally enclosed by already solidified material as can be seen in Fig. 7.19. In this situation, the continuing shrinkage of the hot spot during its solidification cannot be fed by the movement of residual liquid or accommodation of the solidified grains and residual stresses are generated. If the negative

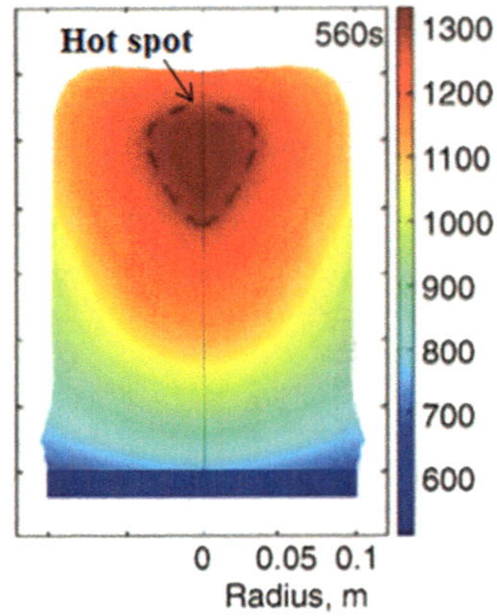

Fig. 7.19 Thermal profiles in spray-formed 100Cr6 steel billet simulated considering cold spray forming conditions and showing the formation of a hot spot within the deposit. The dashed line denotes the location of zero liquid fraction (from [28] with permission)

hydrostatic stresses due to shrinkage or tensile stresses due to movement of the solid are high enough a crack can be initiated. Materials with low thermal conductivity coefficient, for example 100Cr6 steel (30 W m^{-1} K^{-1}), are more susceptible to formation of hot spots since the temperature distribution within the deposit tends to be more non-uniform [28]. However, even when materials with high thermal conductivity, which tend to develop a more uniform temperature distribution, are spray formed, hot spots can be generated if non-appropriate spray conditions are applied [29]. For instance, Fig. 7.20a shows a picture of different crack orientations near to the crown of a CuCrZr spray-formed billet, which has thermal conductivity in the range of 170–340 W m^{-1} K^{-1} [39]. The crack widths extended 10 mm and the crack lengths reached more than 100 mm. Fisher et al. [39] evaluated a series of spray formed billets and showed that cracks usually appear in the upper quarter of the billet. Moreover, they showed that there is a dependency between the crack position and its orientation. Usually transverse cracks are close to the symmetry axis, while longitudinal and especially oblique cracks more often appear at the outer radial positions.

Fisher et al. [39] demonstrated that the temperature difference between the core and the surface of the billet can be used as an indicator for the probability of hot cracks. In other words, this temperature difference is a simplified representation of the thermal gradient within the billet. Figure 7.21 presents the results of the difference between the core and the surface (at 88% of the total height) of CuCrZr billets calculated by numerical simulation considering three different spray conditions: (A) standard conditions; (B) cold spray condition, where a lower temperature of the impacting particles relative to the standard condition was considered; and (C) low metal flow rate condition, where the metal flow rate considered was 66% of standard conditions [39]. Both conditions (B) and (C) represent cold spray conditions and led to similar results. Fig. 7.21 shows that the the standard condition billet showed a higher temperature difference until approximately 200 seconds after the

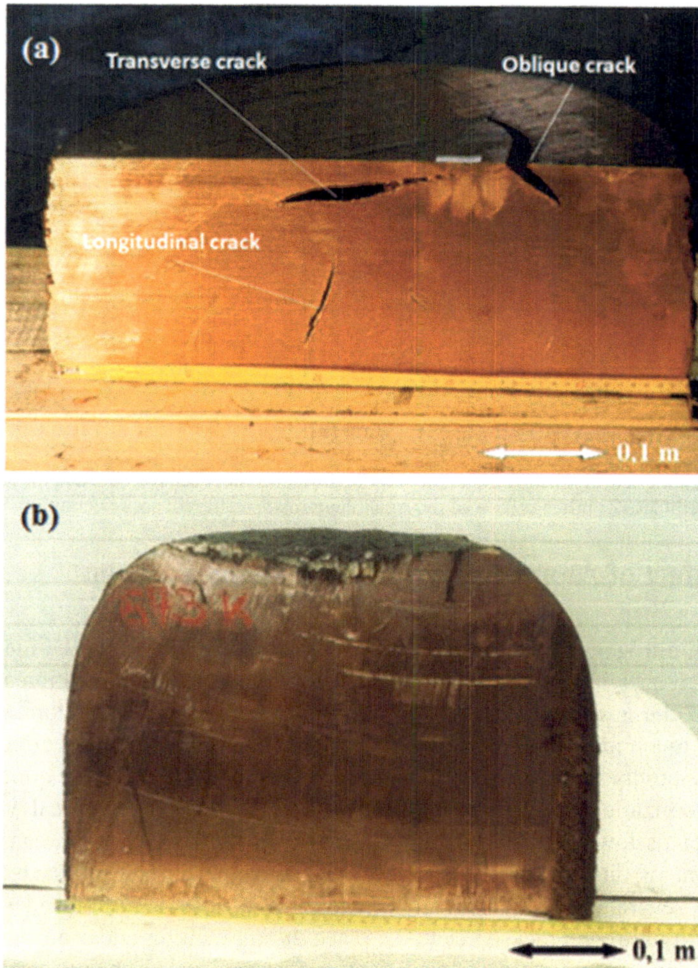

Fig. 7.20 (**a**) Different crack orientations near to the head of a spray-formed CuCrZr billet, (**b**) cross section of the crown region of a spray-formed CuCrZr billet using a lower melt flow rate than the spray conditions of the billet shown in (**a**) [39]

end of the spraying. On the other hand, the cold spray conditions (B) and (C) had much lower temperature differences between the core and the surface and, therefore, a reduction or even prevention of crack formation was expected. Fig. 7.20b shows the cross section of the crown of a CuCrZr billet spray formed with a lower mass flow rate than that shown in Fig. 7.20a. In this case the presence of large cracks was totally eliminated, consistent with the numerical modelling.

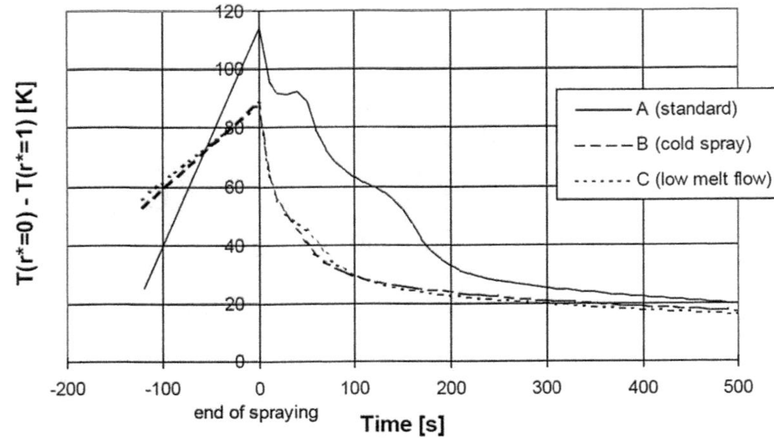

Fig. 7.21 Numerical results of radial temperature difference between the core and the surface of a spray-formed CuCrZr billet at 88% of the total height using different spray conditions [39]

7.9 Effect of Atomization Gas Nitrogen and Argon

Typically, nitrogen or argon are used in spray forming to protect the melt and to avoid reactions with oxygen. Helium has been used only for atomization and spray forming if the gas consumption is low in the case of single fluid atomization, as helium is much more expensive. Helium has a much higher cooling effect due its high conductivity whereas the cooling effect of nitrogen and argon are quite similar. In gas atomization, nitrogen is the first choice from an economical viewpoint because of its low price compared with argon. The use of nitrogen can increase the content of nitrogen in the spray formed alloy perform, because nitrogen is soluble in several elements and can react to form nitrides or carbonitrides.

Generally nitrogen pick-up can take place during melting under nitrogen atmosphere or during atomization. If an induction furnace is used, the nitrogen pick-up during melting is unavoidable; similar problems generally do not arise with the use of argon. Nitrogen pick-up can occur during atomization because the total surface, $A_{total,}$ of a given volume of melt in direct contact with the nitrogen-only atmosphere becomes very large (e.g. 60 m^2/l assuming all droplets are 100 μm in diameter). Even if the average flight time before impact is typically 10 ms, the nitrogen pick-up can nevertheless be significant.

7.9.1 *Steels*

Nitrogen is not an inert with respect to molten steels. It is used as an alloying element to improve mechanical and corrosion properties for special steels. The solubility of nitrogen in pure iron and steel melts, and in the solid state plays an important role in controlling material properties and has been studied intensively [40]. Pure iron exhibits a nitrogen solubility of 470 ppm (0.047 wt%) at its melting point under atmospheric pressure. The nitrogen solubility is also affected by the concentration of the other alloying elements. While C, Si, Ni and Co reduce the solubility, Mn, Ta, Cr, Nb and V increase solubility. An empirical correlation to calculate the solubility of nitrogen in steel has been formulated [41, 42]:

$$[N] = \sqrt{p}\, 10^{k} \tag{7.3}$$

with the nitrogen concentration [N] given in wt%, the pressure p given in bar and the exponent k which depends on the absolute temperature T in K and the wt% of the alloying elements according to:

$$\begin{aligned} k = & -\frac{293}{T} - 1.16 \\ & - \left(\left(\frac{3757}{T} - 0.81\right)(0.072[C] + 0.051[Si] - 0.015[Mn] - 0.039[Cr] \right. \\ & - 0.0103[Mo] + 0.0093[Ni] - 0.095[V] - 0.0056[W] - 0.059[Nb] \\ & -0.031[Ta] - 0.35[O] + 0.044[N]) \\ & + 0.5\left(\frac{5132}{T} - 1.48\right)(0.0215[C]^2 + 0.000005[Mn]^2 + 0.00058[Cr]^2 \\ & +0.00249[V]^2 + 0.00068[Nb]^2) \\ & \left. + 0.167\left(\frac{8124}{T} - 3.06\right)\left(-0.0000068[Cr]^3 - 0.00000401[V]^3\right)\right) \end{aligned} \tag{7.4}$$

Nitrogen is cheap compared with other alloying elements and stabilizes austenite in austenitic steels and reduces the tendency of ferrite generation at high temperatures, and martensite generation at low temperatures [40]. Furthermore, nitrogen increases the strength and the corrosion resistance of austenitic steels.

Nitrogen pick-up and its influence on the microstructure of spray formed high alloyed steels was investigated in [42]. The alloys, typically tool steels, were melted on a laboratory scale induction furnace (150 kg), placed in a vessel which was purged with argon to keep the nitrogen level low, or with nitrogen to increase the nitrogen level. The nitrogen content was 2–6 times higher in the spray formed material if the alloy was melted under nitrogen. The subsequent carbide precipitates were small and evenly distributed within the preforms but the toughness was higher for the alloys with less nitrogen for the steel grades X40CrMoV5-1, HS6-5-2C, X153CrMoV12.

Large spherical pores formed in spray formed low alloy steels that were melted under and atomized with nitrogen. In those steels the solubility in the melt was much higher than in the ferrite that formed first during solidification. When the solubility limit was exceeded, the nitrogen was released and large nitrogen pores were nucleated and grew, and could not move to the surface of the billet due to high solid fraction. This problem can be avoided by using argon during melting to prevent the nitrogen pick-up.

7.9.2 Superalloys

Both argon and nitrogen have been used successfully as the atomizing gas in spray forming of Ni-based superalloy billets [43, 44] and rings [45]. Typically, these alloys are melted under argon to avoid oxygen and nitrogen pickup. Argon atomization is used when no oxygen or nitrogen pickup is desired and when a small amount of argon filled porosity can be tolerated. Welding of argon atomized superalloys (e.g. Rene 41) can result in some thermally induced porosity in the welding zone if the porosity in the as-sprayed material is too high [46].

The content of nitrogen is limited for some applications (e.g. turbine engine discs) to about 100 ppm to achieve excellent properties in the final product. The nitrogen dissolves into the alloy and frequently precipitates as sub-micrometre titanium and niobium carbonitrides. These precipitates have been observed to have an effect on key mechanical properties [44]. Nitrogen atomization leads to a significant pickup of nitrogen, for example in alloy IN718. Approximately 500 ppm nitrogen was measured in a spray formed IN718 ring compared to 50–70 ppm during argon atomization, which was the same as measured in the feedstock [45]. The porosity level during argon atomization was optimized for spray formed rings to less than 1.0 vol.% in the as-sprayed condition (see Fig. 7.22) [45].

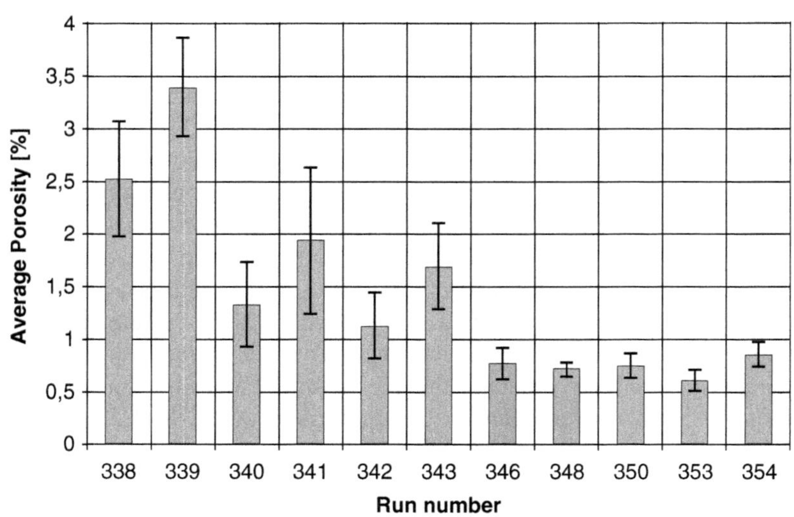

Fig. 7.22 Porosity in spray formed superalloy rings (IN718) using argon for atomization where porosity was measured by image analysis of the unetched microstructure [45]

7.9.3 Copper Alloys

Nitrogen is not soluble in copper and does not react with copper. Therefore, argon is not necessary for the spray forming of copper alloys. The porosity in as-sprayed copper alloys has been reduced successfully by adding small amounts (0.25 wt%) of titanium. Any entrapped nitrogen diffuses within the alloy and fine-scale titanium nitrides are formed with the overall effect of helping to reduce the porosity.

7.10 List of Symbols

7.10.1 Latin

Symbol	Description
D	Diffusion coefficient of solute atoms in the liquid
p	Pressure
[N]	Nitrogen content (in%)
T	Temperature
T	Temperature in Kelvin
t^{C}_{eq}	Timescale (solute equilibration)
T_{sur}	Surfacte temperature
t^{T}_{eq}	Time (thermal equilibration)
x	Thermal equilibration distance

7.10.2 Greek

Symbole	Description
α	Thermal diffusivity

References

1. Grant, P. S. (1995). Spray forming. *Progress in Materials Science, 39*, 497–545.
2. Grant, P. S., et al. (2007). *Metallurgical and Materials Transactions A, 38a*, 1520–1529.
3. Grant, P. S., Chang, I. T. H., & Cantor, B. (1996). Spray forming of Al/Sic metal-matrix composites. *Journal of Microscopy, 177*, 337–346.
4. Schneider, A., Uhlenwinkel, V., Harig, H., & Bauckhage, K. (2004). Overspray injection in spray forming of CuSn13.5 billets. *Materials Science and Engineering: A, 383*, 114–121.
5. Tang, Y. P., Tan, D. Q., Li, W. X., Pan, Z. J., Liu, L., & Hu, W. B. (2007). Preparation of Al-Fe-V-Si alloy by spray co-deposition with added its over-sprayed powders. *Journal of Alloys and Compounds, 439*, 103–108.

6. Cui, C., Schulz, A., & Uhlenwinkel, V. (2014). Materials characterization and mechanical properties of graded tool steels processed by a new co-spray forming technique. *Materialwissenschaft und Werkstofftechnik, 45*, 652–665.
7. Frigaard, I. A. (1995). The Dynamics of spray-formed billets. *Siam Journal of Applied Mathematics, 55*, 1161–1203.
8. Mathur, P., Apelian, D., & Lawley, A. (1989). Analysis of the spray deposition process. *Acta Metallurgica, 37*, 429–443.
9. Annavarapu, S., Apelian, D., & Lawley, A. (1988). Processing effects in spray casting of steel strip. *Metallurgical Transactions A, 19*, 3077–3086.
10. Lavernia, E. J., & Grant, N. J. (1988). Spray deposition of metals—A review. *Materials Science and Engineering, 98*, 381–394.
11. Liang, X., Earthman, J. C., & Lavernia, E. J. (1992). On the mechanism of grain formation during spray atomization and deposition. *Acta Metallurgica et Materialia, 40*, 3003–3016.
12. Cantor, B., Baik, K. H., & Grant, P. S. (1997). Development of microstructure in spray formed alloys. *Progress in Material Science, 42*, 373–392.
13. Mingard, K. P., Alexander, P. W., Langridge, S. J., Tomlinson, G. A., & Cantor, B. (1998). Direct measurement of sprayform temperatures and the effect of liquid fraction on microstructure. *Acta Materialia, 46*, 3511–3521.
14. Mingard, K. P., Cantor, B., Palmer, I. G., Hughes, I. R., Alexander, P. W., Willis, T. C., et al. (2000). Macro-segregation in aluminium alloy sprayformed billets. *Acta Materialia, 48*, 2435–2449.
15. Meyer, C., Ellendt, N., Madler, L., Muller, H. R., Reimer, F., & Uhlenwinkel, V. (2014). Spray forming of high density sheets. *Materialwissenschaft und Werkstofftechnik, 45*, 642–651.
16. Jones, P. D. A., Duncan, S. R., Rayment, T., & Grant, P. S. (2003). Control of temperature profile for a spray deposition process. *IEEE Transactions on Control Systems Technology, 11*, 656–667.
17. Payne, R. D., Moran, A. L., & Cammarata, R. C. (1993). Relating porosity and mechanical-properties in spray formed tubulars. *Scripta Metallurgica Et Materialia, 29*, 907–912.
18. Srivastava, V. C., Schneider, A., Uhlenwinkel, V., & Bauckhage, K. (2004). Effect of porosity and reinforcement content on the electrical conductivity of spray formed 2014-Al alloy plus SiCp composites. *Journal of Material Science, 39*, 6821–6825.
19. Ellendt, N. (2010). *Einfluss der Prozessparameter auf Porosität und Mikrostruktur sprühkompaktierter übereutektischer Al-Mg2Si-Legierungen*. Dissertation, University Bremen, Bremen.
20. Ellendt, N., Stelling, O., Uhlenwinkel, V., von Hehl, A., & Krug, P. (2010). Influence of spray forming process parameters on the microstructure and porosity of Mg2Si rich aluminum alloys. *Materialwissenschaft und Werkstofftechnik, 41*, 532–540.
21. Uhlenwinkel, V., & Ellendt, N. (2007). Porosity in spray-formed materials. *Materials Science Forum, 534–536*, 429–432.
22. Muller, H. R., Ohla, K., Zauter, R., & Ebner, M. (2004). Effect of reactive elements on porosity in spray-formed copper-alloy Billets. *Materials Science and Engineering: A, 383*, 78–86.
23. Cai, W. D., & Lavernia, E. J. (1997). Modeling of porosity during spray forming. *Materials Science and Engineering: A, 226*, 8–12.
24. Cai, W. D., & Lavernia, E. J. (1998). Modeling of porosity during spray forming: Part II. Effects of atomization gas chemistry and alloy compositions. *Metallurgical and Materials Transactions B, 29*, 1097–1106.
25. Cai, W. D., & Lavernia, E. J. (1998). Modeling of porosity during spray forming: Part I. Effects of processing parameters. *Metallurgical and Materials Transactions B, 29*, 1085–1096.
26. Doherty, R. D., Annavarapu, S., Cai, C., & Warner, L. K. (1997). Modelling based studies for control and microstructure development in spray forming. In K. Bauckhage & V. Uhlenwinkel (Eds.), *SFB 372: Spruehkompaktieren, Kollowquium Band 2* (p. 45). Bremen: University Bremen
27. Ellendt, N., Uhlenwinkel, V., & Madler, L. (2014). High yield spray forming of small diameter tubes using pressure-gas-atomization. *Materialwissenschaft und Werkstofftechnik, 45*, 699–707.

28. Cui, C., Fritsching, U., Schulz, A., Tinscher, R., Bauckhage, K., & Mayr, P. (2005). Spray forming of homogeneous 100Cr6 bearing steel billets. *Journal of Materials Processing Technology, 168*, 496–504.
29. Meyer, O., Fritsching, U., & Bauckhage, K. (2003). Numerical investigation of alternative process conditions for influencing the thermal history of spray deposited billets. *International Journal of Thermal Sciences, 42*, 153–168.
30. Mi, J., & Grant, P. S. (2008). Modelling the shape and thermal dynamics of Ni superalloy rings during spray forming part 1: Shape modelling—Droplet deposition, splashing and redeposition. *Acta Materialia, 56*, 1588–1596.
31. Mi, J., & Grant, P. S. (2008). Modelling the shape and thermal dynamics of Ni superalloy rings during spray forming. Part 2: Thermal modelling—Heat flow and solidification. *Acta Materialia, 56*, 1597–1608.
32. Cui, C. S., Fritsching, U., Schulz, A., & Li, Q. C. (2005). Mathematical modeling of spray forming process of tubular preforms—Part 2. Heat transfer. *Acta Materialia, 53*, 2775–2784.
33. Ho, S., & Lavernia, E. J. (1996). Thermal residual stresses in spray atomized and deposited Ni3Al. *Scripta Materialia, 34*, 527–536.
34. Ho, S., & Lavernia, E. J. (1996). The effect of ceramic reinforcement on residual stresses during spray atomization and co-deposition of metal matrix composites. *Scripta Materialia, 34*, 1911–1918.
35. Ho, S., & Lavernia, E. J. (1997). The effect of heat transfer coefficient on thermal residual stresses in spray deposited materials. *Scripta Materialia, 36*, 283–290.
36. Hu, H. M., Lavernia, E. J., Lee, Z. H., & White, D. R. (1999). Residual stresses in spray-formed A2 tool steel. *Journal of Material Research, 14*, 4521–4530.
37. Bauckhage, K., Schulz, A., & Uhlenwinkel, V. (2003). Characteristic features and specific qualifications of the sprayforming process to be generalized. *Materialwissenschaft und Werkstofftechnik, 34*, 6–12.
38. Fischer, J. E., Uhlenwinkel, V., Schroeder, R., Jordan, N., Hansmann, S., Muller, H. R. (1999). Reduction of crack probability during spray forming of billets. In *EPD Congress, San Diego, USA*, February 28–March 4 (pp. 117–128).
39. Fischer, J. F., Uhlenwinkel, V., Schroder, R., Jordan, N., Hansmann, S., & Muller, H. R. (1999). Thermal cracking in large diameter spray formed billets. *International Journal of Powder Metallurgy, 35*, 27–34.
40. Rechsteiner, A. A. (1994). *Metallkundliche und metallurgische Grundlagen zur Entwicklung stickstoffreicher, zäher, hochfester austenitischer Stähle*. Dissertation, ETH Zürich, Zürich.
41. Medovar, B., Saenko, V. Y., Grigerenko, G., Pomarin, Y. M., & Kumysh, V. (1996). *Arc-slag remelting of steels and alloys* (p. 36). Berlin: Cambridge International Publishing.
42. Schulz, A., Uhlenwinkel, V., Bertrand, C., Kohlman, R., Kulmburg, A., Oldewurtel, A., et al. (2003). *Nitrogen pick-up during spray forming of high alloyed steels and its influence on microstructure and properties of the final products*. Paper presented at 2nd international conference on spray deposition and melt atomization, SDMA2003, Bremen, Germany, 22–25, June 2003, pp 4–31.
43. Benz M. G., Sawyer T. F., Carter W. T., Zabala RJ, Dupree P. L. (1993). Nitrogen in spray formed superalloys. Second international conference on spray forming. p. 171–181.
44. Benz, M. G., Sawyer, T. F., Carter, W. T., Zabala R. J., & Dupree, P. L. (1994). Nitrogen in spray formed superalloys. *Powder Metallurgy, 37*, 213–218.
45. Uhlenwinkel, V., Ellendt, N., Walter, M., & Tockner, J. (2006). Spray forming and post processing of superalloy rings. *Continuous Casting, 249*.
46. Huron, E. S. (1993). Properties of spray formed superalloy rings. In *Proceedings of the Second International Conference on Spray Forming*, 1993 (pp. 155–164). Swansea, UK: Woodhead Publishing.

Chapter 8
Processing Aspects in Spray Forming

Guilherme Zepon, Nils Ellendt, Volker Uhlenwinkel, and Hani Henein

8.1 Process Description

8.1.1 Single Fluid Atomization

Single fluid atomization techniques are used for both metal powder production and spray forming. Compared to gas atomization there are significant differences. Single fluid atomization has less inert gas consumption. Median particle sizes are larger, the size distribution is significantly smaller, and particle velocities are lower. Due to the low particle velocities upon impact (low kinetic energy), it is not evident how well the droplets will spread over the deposit. This issue will be discussed further in section 8.4.

Here, three different single fluid atomization methods are discussed in conjunction with spray forming such as:

G. Zepon
Department of Materials Engineering (DEMa), Federal University of São Carlos (UFSCar), São Carlos-SP, Brazil
e-mail: zepon@ufscar.br

N. Ellendt
University of Bremen, Bremen, Germany
e-mail: ellendt@iwt.uni-bremen.de

V. Uhlenwinkel (✉)
Foundation Institute of Materials Science, University of Bremen, Bremen, Germany
e-mail: uhl@iwt.uni-bremen.de

H. Henein
Advanced Materials and Processing Lab, University of Alberta, Edmonton, AB, Canada, T6G 2R3
e-mail: hhenein@ualberta.ca

H. Henein et al. (eds.), *Metal Sprays and Spray Deposition*,
DOI 10.1007/978-3-319-52689-8_8

- Impulse atomization
- Centrifugal atomization
- Uniform droplet spray

Typically, all processes operate under inert gas to avoid oxidation. Since inert gas consumption is low, more expensive gasses like helium can be used to increase cooling rates.

8.1.1.1 Impulse Atomization

The principle of Impulse Atomization is described in Chap. 2. Mean particle diameter can be changed by varying the nozzle diameter. The melt flow is changed by varying the number of nozzles. Reported particle sizes (mass median diameter) are in the range of 200–800 μm and are therefore much larger than gas atomized particles, which are in the range of 10–100 μm. However, the size distribution is significantly smaller for Impulse Atomized particles, with a geometric standard deviation typically between 1.4–1.6. Normally, the particle velocities at the nozzle exit are lower than 10 m/s. Several alloys have been processed in the laboratory using Impulse Atomization in the spray forming mode. These will be discussed in this section as well as in the last section of this chapter dealing with the microstructural evolution of a spray-formed billet.

Small Cu-6wt%Sn billets with diameters of 50 mm and heights up to 60 mm were spray-formed using the Impulse Atomization technique [1]. The droplets were deposited onto a rotating and a non-rotating low carbon steel plate. Because the mass flux in the cross section of the spray is relatively uniform compared to a gas atomized spray, the deposit need not be rotated. The melt temperature (1150–1350 °C) and the melt flow (120–367 kg/h) were varied by changing the number of nozzles, the diameter of nozzles, and the distance between the nozzle plate and the substrate (250 and 400 mm). The yield (weight of the deposit divided by the melted feedstock) was measured to be between 81–88% and the average porosity ranged between 0.5–3.2%. The pore size was similar to a spray-formed material generated by gas atomization using a free fall atomizer. As-sprayed grain sizes were significantly larger compared to the gas atomized material (Fig. 8.1). The average solid fraction of the droplets before impingement was calculated to be between 28–50%. This result was used to calculate both the temperature history and the cooling rate in the deposit.

8.1.1.2 Centrifugal Spray Deposition (CSD)

Centrifugal spray deposition (see Chap. 2 as well) was invented by Singer [2]. He showed that it is possible to spray form aluminum strips. However, the centrifugal spray deposition (CSD) technique is more suitable to produce rings (Fig. 8.2). In this process a molten metal stream is falling onto a rotating disc (water cooled

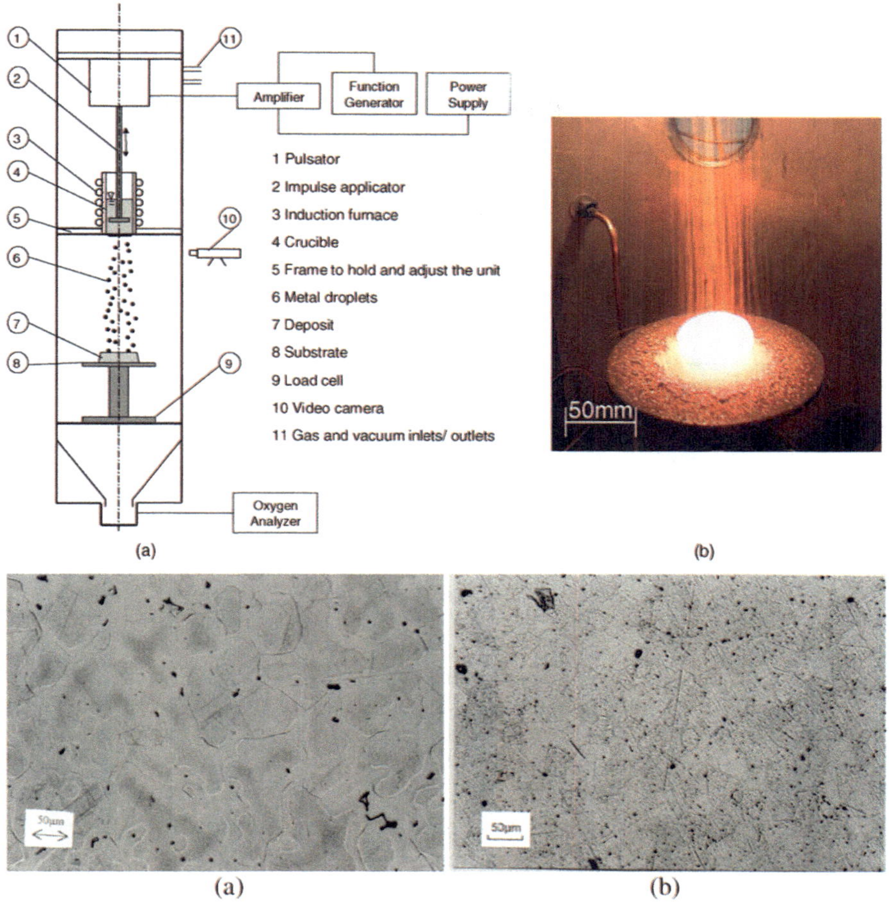

Fig. 8.1 Schematic of Impulse Atomization unit (*upper (a)*), image of spray deposition process (*upper (b)*), microstructure of a spray-formed billet using Impulse Atomization (**a**), and using Gas Atomization (**b**) (from [1] with permission)

copper). The molten metal spreads out radially and is atomized. Different modes of atomization have been observed. After a certain flight distance the particles deposit on a ring-shaped substrate. The substrate is moving up and down to achieve a uniform thickness of the deposit over the entire substrate height. This process can operate under any pressure conditions lower than atmospheric. Lower pressures result in less gas in the pores which simplifies the densification of the material during post-processing.

It has been demonstrated that Ni-based superalloys (IN 718 [3] and Ti-based materials (Ti-48Al-2Mn-2Nb in at% [4]) can be produced with this process (Fig. 8.2). The melting capacity was between 20–50 kg, but the melt flow rate was not reported. Although a mean particle size was not measured, the rotating speed was known to be approximately 3000 rpm. As discussed in Chap. 2, a rotating

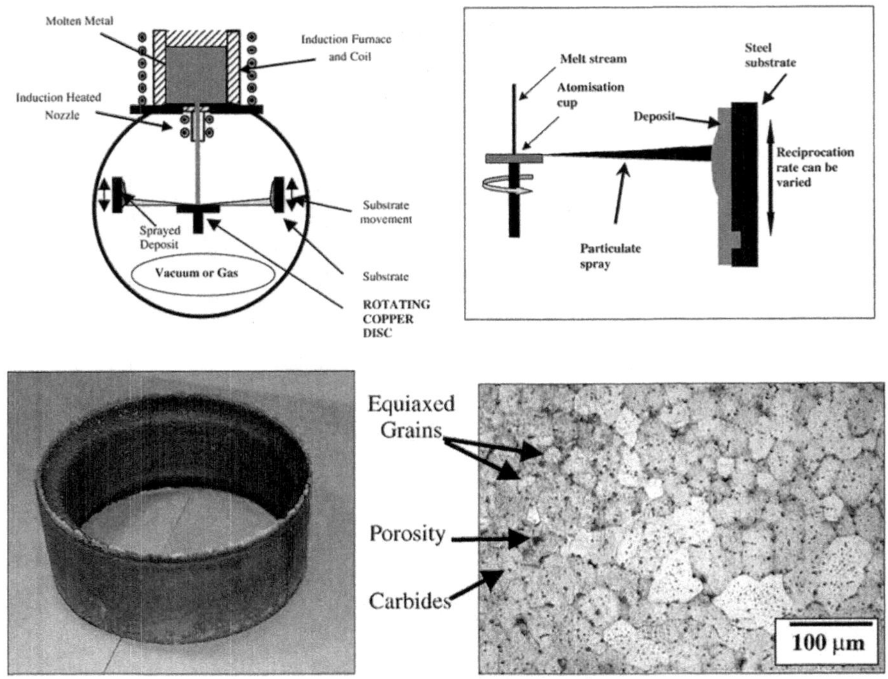

Fig. 8.2 Schematic of Centrifugal Spray Deposition (CSD) (*top left*), process details (*top right*), spray-formed IN718 ring with 340 mm diameter (*bottom left*) and typical microstructure (*bottom right*) (from [3] with permission)

speed of 3000 rpm is relatively low and therefore can only result in particle diameters of several hundred micrometers. These droplets do not lose much heat and most droplets are expected to be fully liquid before they hit the substrate/deposit. The substrate diameter was 350–400 mm and the height was 125 mm for the spray-formed rings.

In the case of superalloy rings, it has been demonstrated that after post processing (HIP, rolling, and heat treatment) aeronautical specifications were achieved. It was found that the yield (deposit weight/atomized material) of this process step is approximately 95%.

8.1.1.3 Uniform Droplet Spray Forming

The Uniform Droplet Spray forming process (UDS) was developed as a basic research tool [5]. The process is exhibited in Fig. 8.3. The melt in the crucible is pressurized and generates a molten metal stream through the nozzle at the bottom of the crucible. A vibrating plunger with a frequency adjusted to the nozzle diameter generates a periodic disturbance that leads to a break-up of the metal stream. Subsequently, a chain of uniform droplets is generated. This phenomena is

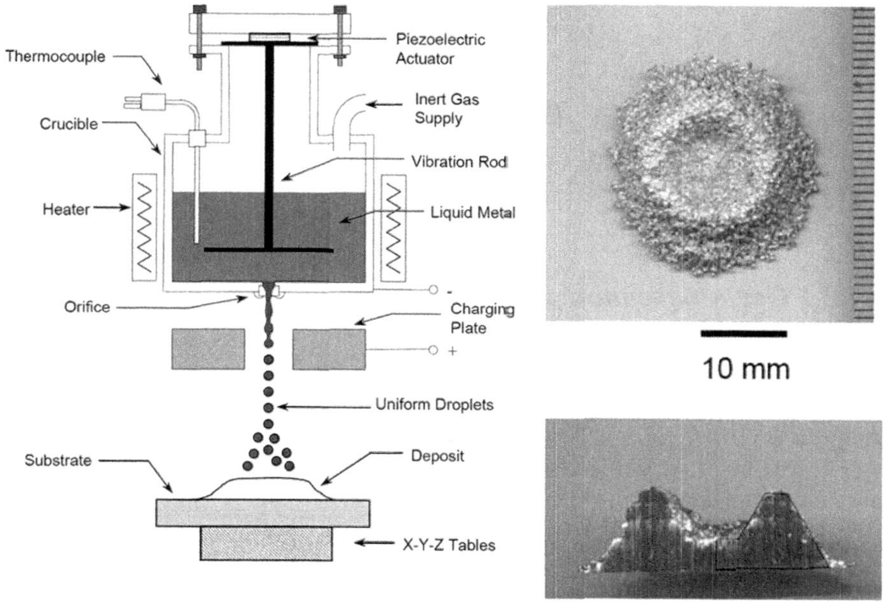

Fig. 8.3 Schematic of uniform droplet spray apparatus (*left*), top view and cross section of the ZnSn20 deposit (*top* and *bottom right* respectively), adapted from [5]

commonly referred to as Rayleigh break-up [6]. Depending on the nozzle diameter and the frequency, it is possible to produce uniform droplets between 75–800 μm in diameter. The initial droplet velocities are below 10 m/s and the droplet diameters are 1.89 times the nozzle diameter. The droplets pass a charging plate, after which the electrical repulsion of the charged droplets creates a diverging spray whose diameter depends on the amount of voltage applied to the plate. The droplets deposit onto a preheated substrate and solidify.

Deposits were generated using a Zn-20wt%Sn alloy with a liquidus and *solidus* temperature of 380 °C and 199 °C respectively. The deposit is typically about 20 mm in diameter (see Fig. 8.3) and up to 10 mm thick. The deposition process lasts about 10 s and the droplet rate was typically 4300 Hz. The melt temperature was approximately 440 °C. A liquid fraction of 76–91% before impact was calculated. At the deposit surface the maximum liquid fraction was between 25–37% (calculated from the pyrometer temperature measurement). The substrate temperature varied between 120–175 °C. During the different experiments an initial cooling rate between 6–27 K/s was measured (from the measured temperature history in the deposit).

The average grain size λ (measured using the linear intercept method) was correlated to the initial cooling rate **ε** in the deposit, the solidification time t_f before eutectic solidification, and the initial liquid fraction l_f in the deposit:

$$\lambda\ [\mu m] = 27.1\ \varepsilon^{-0.52}\ [K/s] \tag{8.1a}$$

$$\lambda\ [\mu m] = 1.9\ t_f^{\ 0.43}\ [s] \tag{8.1b}$$

$$\lambda\ [\mu m] = 56.4 f_l^{\ 4.0} \tag{8.1c}$$

The dimensions are given in brackets.

8.1.2 Gas Atomization

Three different gas atomization methods are used for spray forming. These are the free-fall-atomizer (FFA), the close-coupled atomizer (CCA) and the pressure-gas–atomizer (PGA) [7, 8].

The free-fall-atomizer has been used widely for spray forming (Fig. 8.4) and has the following features:

- Robust process which can run for a long period with constant melt flow
- High melt mass flow and production rates
- The secondary nozzle can be moved periodically (scanning atomizer) to widen the spray cone in one direction
- Wide spray cone
- Mass median diameter of the droplets typically between 80–200 μm

This atomizer is mainly used for industrial applications. Various setups discussed here can be found in Fig. 8.5. Billets, tubes, rings, and sheets are possible preforms that can be achieved because of the scanning secondary nozzle of the free-

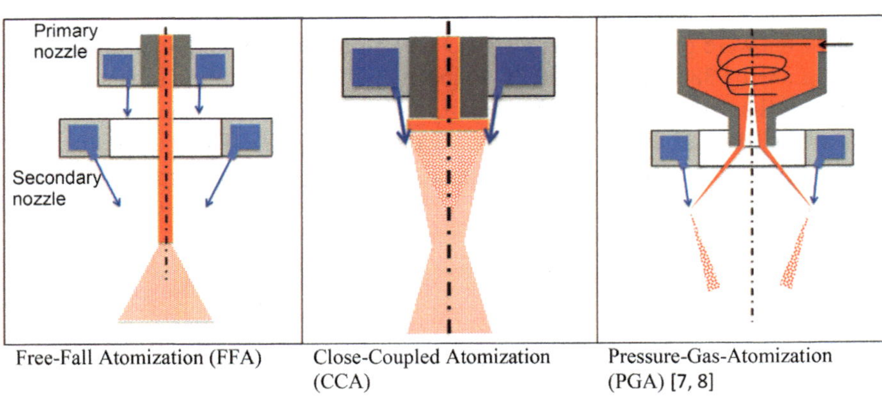

Fig. 8.4 Schematic of different gas atomization techniques which have been used for spray forming

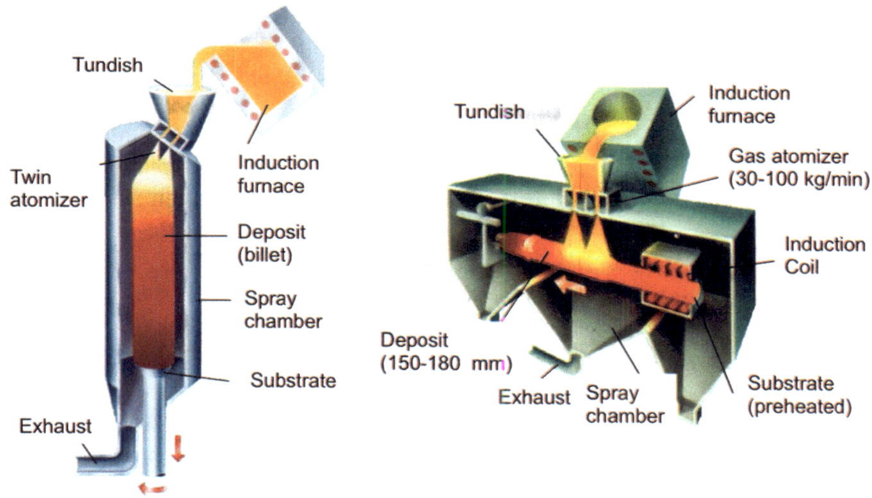

tool steel billet,

400 mm diameter, approx. 4000 kg /Danspray/

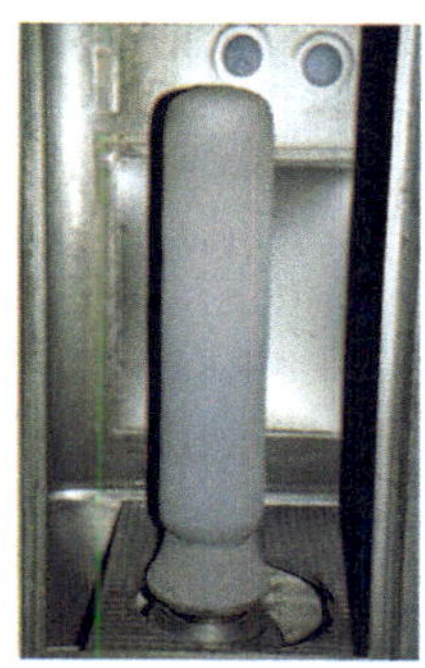

aluminum alloy billet,

300mm diameter /PEAK/

Parameter	unit	
Vessel purge gas	--	N_2
Atomizer gas	--	N_2
Melt temperature	°C	1204
Melt superheat	K	125
Nozzle diameter	mm	2 x 6.0
Melt flow rate	kg/s	0.58
Gas flow rate	kg/s	0.34-0.38
GMR	--	0.58-0.66
Initial spraying distance	mm	373
Scan angle	°	± 6
Scan frequency	Hz	15.2
Substrate rotation	Hz	2.45
Substrate material	--	AISI 4140
Substrate preheating	°C	850 -1100

Fig. 8.5 Spray forming of a billet and a ring with a twin free-fall-atomizer [9] (*upper*), industrial scaled spray-formed billets (*middle*), spray-formed Cu-Al-alloy ring and process parameters (*lower*)

fall-atomizer. A round substrate is rotating at a distance of about 300–500 mm from the atomizer where the spray hits the substrate. By scanning the secondary nozzle up to ±10° the diameter of the billet can be adjusted. The substrate must be withdrawn in conjunction with the growth rate of the billet to keep the deposit surface at the same distance to the atomizer. Typical billet diameter ranges from 120–450 mm. Large diameters are only possible if a twin atomizing system is applied. Production rates between several hundred to 2000 kg/h are possible with just one atomizer.

Tubes can be spray-formed without the scanning mode. The rotation of the substrate and movement through the spray cone is sufficient to generate a tube. The substrate must be preheated to achieve a better bond between substrate and deposit.

Sheets are produced with a linear movement of the substrate and a scanning of the secondary nozzle perpendicular to the direction of the substrate. Rings are normally produced by using a rotating substrate and one or two scanning atomizers [10]. The yield (deposit weight/feedstock weight) depends on the composition, process parameter, and the substrate or deposit size.

Both close-coupled and pressure gas atomization (CCA and PGA respectively) are similar concerning the particle size and velocity but different compared to free-fall atomization. The mass median diameters are significantly smaller and the mean particle velocities are much higher for CCA and PGA. On the other hand, the mass flow is much lower, as is the spray angle. Therefore, the spray distance between substrate and atomizer can be reduced to 100 mm or less. Scanning of CCA and PGA is not possible. Thus far, PGA cannot be used for high melting point alloys. CCA is very robust and often used under industrial conditions. This atomizer can also be used for Gaussian shaped and tubular deposits to produce small amounts of sample for research and development applications. High yields and low porosities even for small diameter tubes have been achieved with CCA, which makes this atomizer interesting for future developments. The reason for the high yield is not clear. It is suspected that high impact particle velocities increase the yield. With the PGA yields between 70–95% were achieved. For the 90 mm diameter substrate (D = 90 mm) the highest distance results in a yield of 75% and increases with less distance from the atomizer (Fig. 8.6). So far the maximum yield for a 50 mm diameter substrate was 90%.

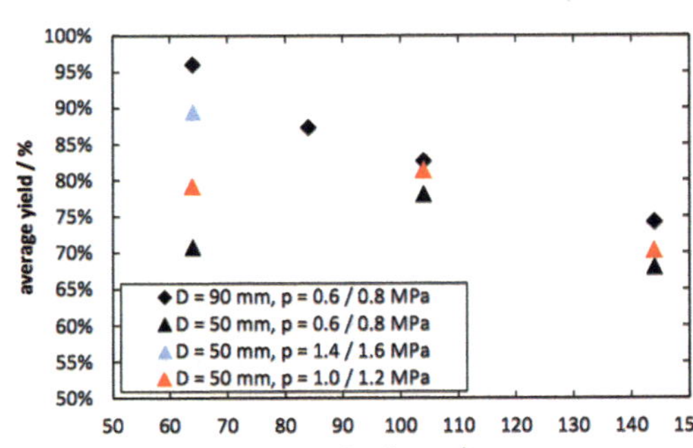

Fig. 8.6 Spray-formed Cu-Sn10 segments with different substrate diameters (25, 50, 90 mm) using PGA (pressure-gas atomization) (*left*). Yield versus spray distance for different substrate diameters D and gas pressure p (*right*) [11]

8.1.3 Metal Matrix Composite

Composites with a metallic matrix and a second material embedded in it are called metal matrix composites (MMC). MMC are generated to improve the material properties such as wear resistance, strength, and thermal conductivity. Typical matrix materials are Al-, Cu-, Mg-, Ni-, Ti- or Fe-alloys. Particles with a globular shape are often used for the second material, or sometimes short or long fibers are used. Generally, oxides (Al_2O_3, MgO, ZrO_2), carbides (SiC, TaC, TiC, WC, B_4C), nitrides (Si_3N_4, TiN, TaN, ZrN), borides (TaB_2, ZrB_2, TiB_2, WB), and pure elements (C, Si) have been used as a secondary phase in different processes. A MMC can be achieved in one process step by spray forming with particle injection. In the early 90s, the first publications concerning spray forming of MMC were introduced [12]. Most papers on spray-formed MMC focus on microstructure, material properties, and post-processing [13–18]. Here, the key aspect is the process and the distribution of the particular phase in the matrix [19–22].

The particulate phase is injected via the primary nozzle or an additional nozzle (injection nozzle, see Fig. 8.7). In both cases the particles are fed by a particle feeder into the gas pipe and are pneumatically transported to the nozzle. Two different MMC's were extensively investigated. The first was a Cu-Ni2-Si matrix with graphite and the second was an Al-Cu4-Si-Mg matrix with SiC particles.

It must be mentioned that the wettability between molten Cu-Ni2-Si and graphite is low and therefore will not promote the generation of a composite during spray forming. All results are taken from spray-formed billets with typical diameters of 120 mm. The main process parameters are: feedstock at 30 kg, superheat of the melt at 345 K, melt flow at 630 kg/h, and gas flow at 815 kg/h. The mass median diameter of the graphite particles was varied between 18–70 μm and the particle to melt ratio (PMR = particle mass flow/melt mass flow) was varied from 0–0.045. The volume fraction of the graphite was measured at different radial positions of the billet as shown in Fig. 8.8. It is clear that the volume fraction of graphite depends on the radial position in the billet and on the mass median diameter of the injected particles. Small mass median diameter particles (18 and 24 μm) achieve the highest values. Nevertheless, the values are typically far below 3 vol% and the particle

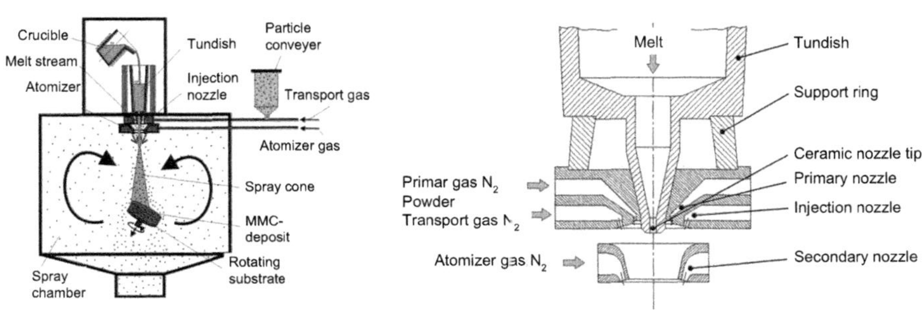

Fig. 8.7 Sketch of the spray forming plant with particle injection system (*left*), Free-fall atomizer with particle injection via additional nozzle (*right*)

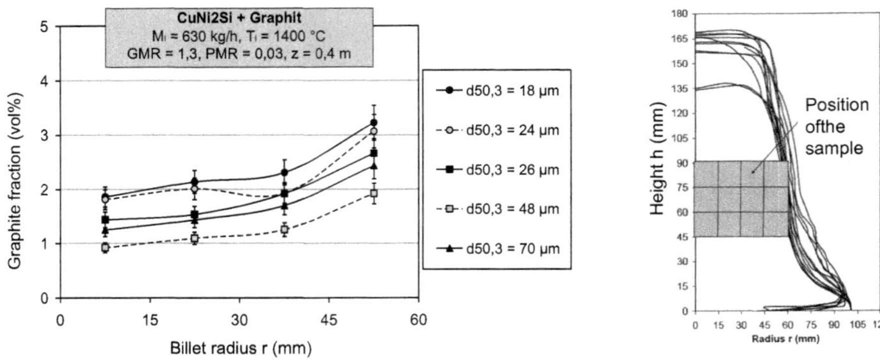

Fig. 8.8 Vol% of graphite in the Cu-Ni2-Si billet versus the radial position and for different mass median particle diameters (*left*), and billet contour and sample position (*right*) [21]

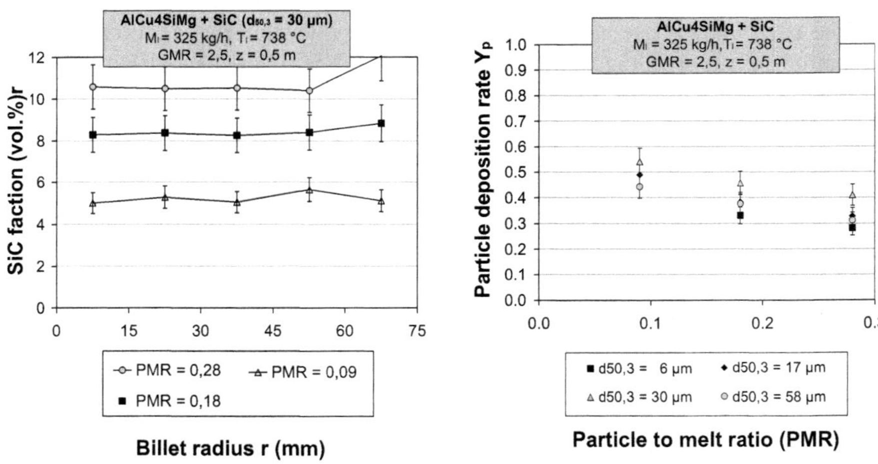

Fig. 8.9 Vol% of SiC_p in the Al-Cu4-Si-Mg billet versus the radial position for different PMR (*left*) and particle yield versus PMR with different mass median particle diameter (*right*) [21]

yield is always less than 15%. The particle yield is defined as the amount of particles in the deposit divided by the amount of injected particles.

Another composite material with better wettability is Al-Cu4-Si-Mg/SiC_p [21, 23, 24]. Here, the feedstock was 45 kg of molten metal with a superheat of 100 K. The melt and gas flow were 325 and 815 kg/h, respectively. The powder flow varied from 0 kg/h up to a maximum of 90 kg/h (PMR = 0.28) and the billet diameter was approximately 150 mm. The results (Fig. 8.9) clearly show a homogenous distribution of the SiC_p content versus the radial position and how the content increases with the PMR. However, the particle yield decreases with increasing PMR. Furthermore, the particles size affects the particle yield as well. The highest values are achieved with a mass median particle diameter of 30 µm. Both smaller and larger mass median particle diameters result in a lower yield.

Figure 8.10 displays the microstructure of the Al-Cu4-Si-Mg/SiCp. On the left side, the unetched microstructure indicates the homogenous distribution of the particles. The upper picture represents the MMC generated with the largest particles (mass median 58 μm) and the pictures below display the composites with smaller particles (30, 17, and 6 μm). The right side illustrates the etched microstructure with a higher magnification. It is clearly visible that the SiC_p particles are mainly located at the grain boundaries.

The co-injection system has also been used successfully to inject overspray [19]. This has been implemented in industrial production of aluminium alloys to increase the yield of the spray forming of billets. Overspray injection significantly reduces the processing cost due to its higher yield and lower gas consumption.

8.1.4 Co-Spray Forming

Co-Spray Forming (CSF) was developed to generate multi-layer or graded materials [25–27]. There are several applications showing the interest and demand of materials with at least two layers such as sliding bearings, heat pipes for power plants, and drilling pipes in the oil industry. The principle plant design is shown in Fig. 8.11. Two different alloys are melted in separate crucibles and poured into the tundishes to generate two spray cones simultaneously. This unique technique was realized at the University of Bremen. The whole melt apparatus was placed in a double wall and water cooled vacuum chamber. One crucible had a steel capacity of 35 kg and the second had a capacity of 80 kg. Both were equipped with a direct inductive heating system. The vacuum was applied once before melting to reduce the oxygen in the vacuum vessel, and once more at the end to reduce impurities of the melt. Figure 8.11 provides a view into the spray chamber during co-deposition of two steels. The distance between the two atomizers was fixed at 340 mm. The maximum dimensions of a sheet were 1000 × 250 mm, and the maximum length of a tube was 800 mm. Both flat and tubular substrates can be preheated by induction heaters. Both atomizers were scanning to generate a bi-metal sheet. Scanning was not necessary to produce a bi-metal tube. The temperature in the substrate was measured with thermocouples during the process. Furthermore, the deposit surface temperature was monitored with a scanning pyrometer. The co-spray forming process has been used to generate different bi-metal materials (Steel A/Steel B, Copper alloy/ Steel, cobalt alloy/steel).

Overspray of both alloys was circulating in the spray chamber. This caused some particles to become embedded in the matrix of the other alloy (Fig. 8.12). The figure below shows an example of a copper alloy matrix with steel particles embedded. The steel particles are spherical and solidified before they hit the deposit surface and are not re-melted because the initial temperature of the Cu alloy is far below the solidus temperature of the steel. It is worth mentioning that the porosity level in this as-sprayed condition was very low. When this kind of composition is undesired, both spray cones must be separated as it affects the material properties of the matrix.

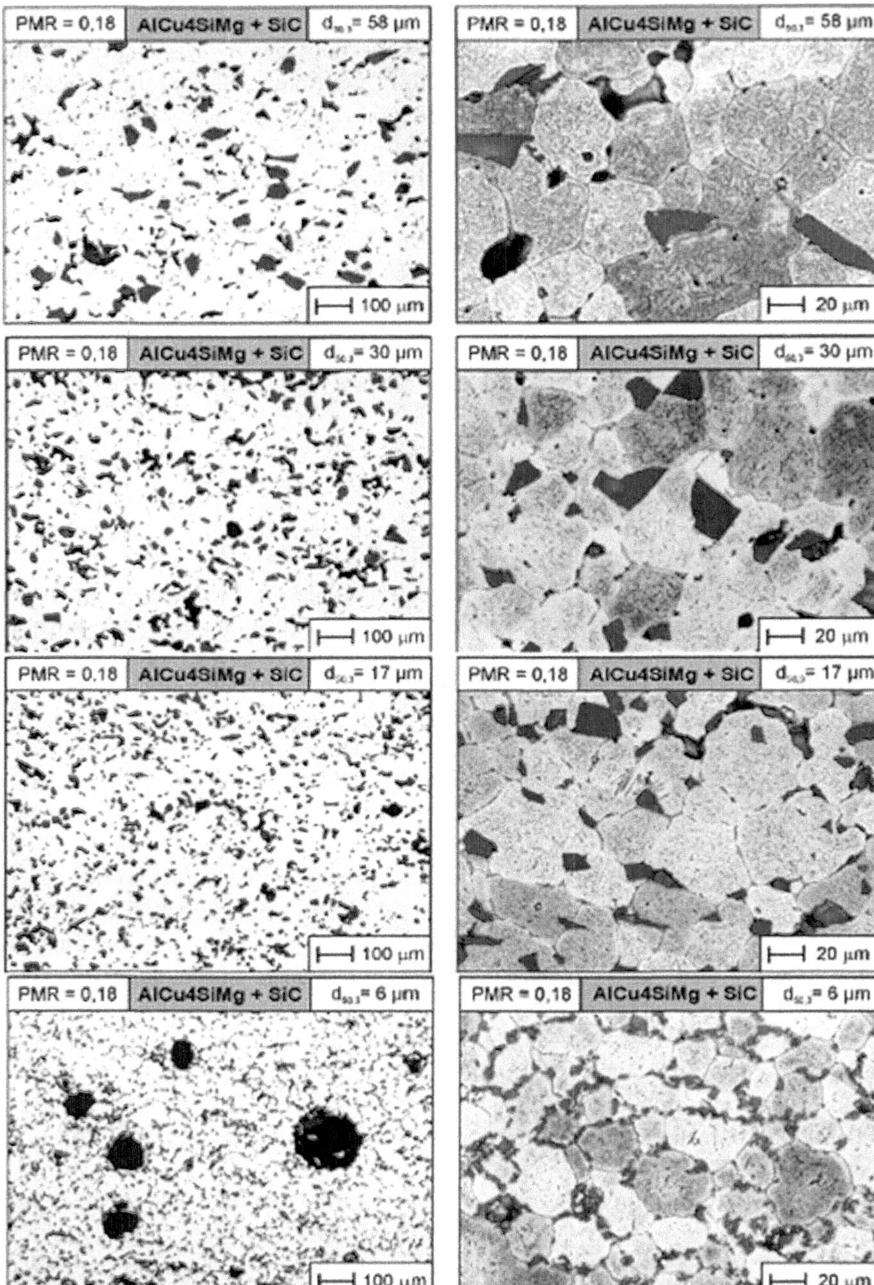

Fig. 8.10 Microstructure of spray-formed MMC (Al-Cu4-Si-Mg/SiC_p) unetched (*left*), etched (*right*), mass median particle diameter varied (6, 17, 30, and 58 µm) [21]

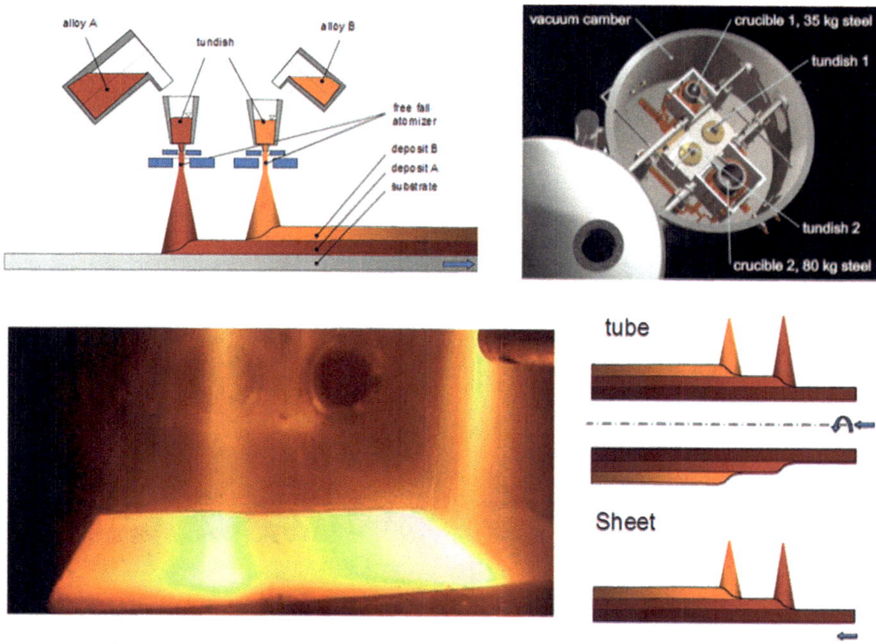

Fig. 8.11 Schematic of the co-spray forming process (*upper left*), view into the vacuum chamber with two separate crucibles and tundishes (*upper right*, with courtesy of Megatherm), view into the spray chamber at the University Bremen (*lower*)

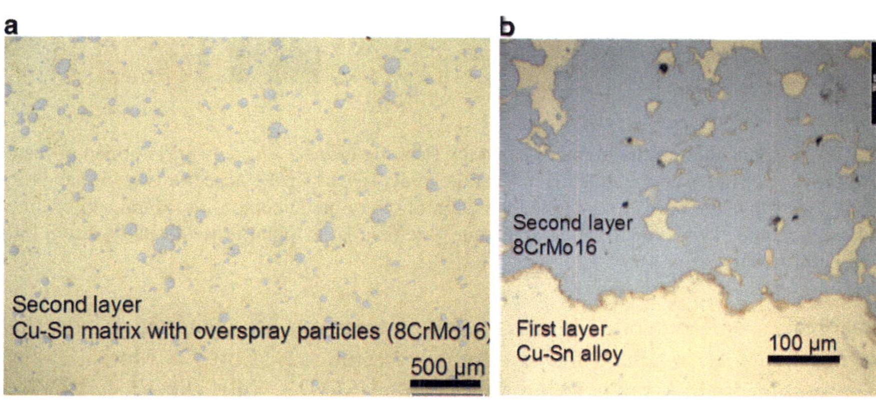

Fig. 8.12 (**a**) Co-spray forming, second layer with Cu-based matrix (Cu-Sn alloy) and steel particles embedded. (**b**) Second layer with steel based matrix and Cu-Sn particles embedded

Thus far, the co-spray process has been used to generate bi-metal tubes [25, 28] and graded sheets [29]. Roller bearings are manufactured from hot bars sheared into small billets and subsequently transported to the first forming stage (Fig. 8.13) [25, 30]. The shear blades require high wear resistance at high temperatures. These

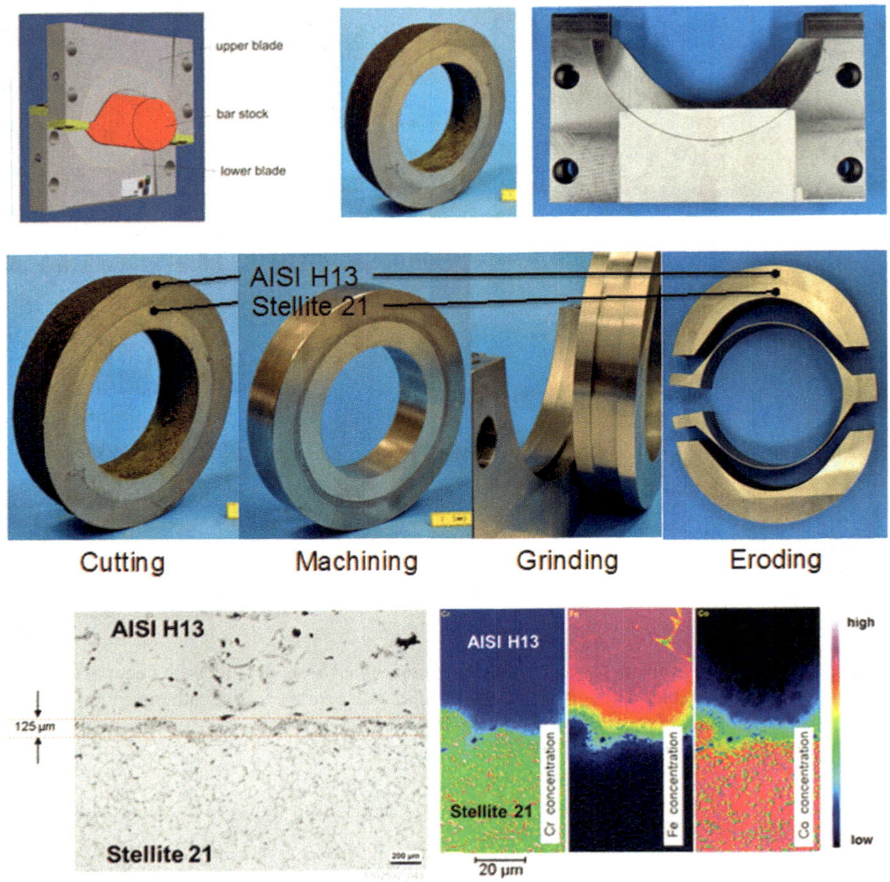

Fig. 8.13 Application of the tool: hot bar stock shearing (*upper, left*), co-spray-formed bi-metal ring and metal composite shearing blade insert in holder (*upper right*), steps from the bi-metal ring to the shearing blade (*middle*), interface structure of the bi-metal composite (Murakami etching, *bottom left*), and element distribution at the interface (EPMA mapping) (*bottom right*) (from [30] with permission)

shear blades were generated by the co-spray forming process and tested in industrial environments. Here, a hard-facing alloy layer (Stellite21, Co-Cr28-Mo-Ni, 2.498) was combined with a hot working steel body (AISI H13, X40CrMoV5-1, 1.2344). Both alloys were spray-formed in a single step. First a 10 mm thick Stellite 21 layer was sprayed onto a low carbon steel tube followed by a 15 mm thick AISI H13 layer on top. While the Stellite 21 layer showed negligible porosity, the H13 layer showed increased porosity in the vicinity of the interface. The interface itself does not show any defects, which leads to a strong bonding between the two materials as shown in Fig. 8.13. The figure displays the steps from the cut ring segment to the final component.

It was proved by electron beam microprobe that there is a metallurgical bonding between the AISI H13 and the Stellite21. The elements Co and Fe show a gradient in a diffusion zone of approximately 10 μm (Fig. 8.13). Machining and heat treatment was applied without detachments between the two materials. Tensile tests were carried out and the fracture was typically not within the interface in the AISIH13 material where the porosity was higher. The newly developed spray-formed shearing blade inserts were successfully tested for more than 1000 cuts and showed less wear than conventional deposit-welded shear blades [25].

In oil production, pipes are used as drilling risers and casings. Those parts are often subjected to severe wear and corrosion inducing conditions. Therefore, a wear and corrosion resistant bi-metallic pipe on the basis of a supermartensitic stainless steel (SMSS) was generated by co-spray forming [28]. The first layer of the bi-metallic tube was a SMSS with an addition of boron (0.3–1.0 wt.%) to improve the wear resistance. First experiments showed that the layer does not stick to the low carbon steel substrate (diameter 140 mm) surface. This problem was solved by preheating the substrate with an inductive heating system to a temperature above 800 °C (see Fig. 8.14). The first layer (SMSS + B) was approximately 8 mm thick and showed a high relative density of more than 99%. The second layer (SMSS) possessed a low relative density in the vicinity of the interface. About 70% of the second layer (approx. 25 mm thick) had a high relative density as well.

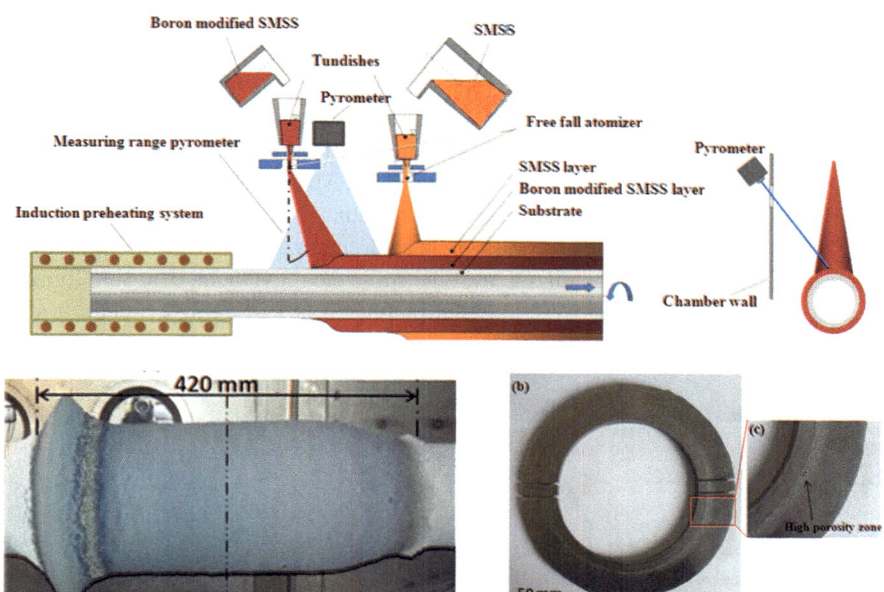

Fig. 8.14 Co-spray forming with inductive preheating of the substrate (*above*), as-sprayed product with a first layer of SMSS + B and a second layer SMSS (*below*), adapted from [28]

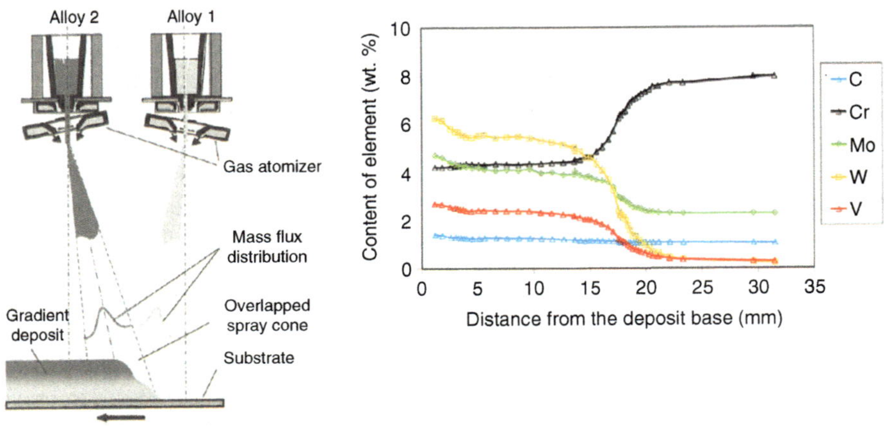

Fig. 8.15 Co-spray forming of tool steels, first layer (HS6-5-3C) second layer (HS6-5-2C) or X110CrMoV8-2 [31, 32] (*left*), main element distribution of the graded deposit after soft annealing (*right*) (from [29] with permission)

Another application for bi-metallic co-spray forming was developed to meet the requirements of micro rotary swaging. The tools should be selected with consideration to abrasive and adhesive wear, compressive strength, and toughness [29, 31, 32]. The co-spray-formed material properties can be achieved by selecting two different steels which are sprayed in such a way that a gradient within the interface is produced. A gradient is necessary to reduce the possibility of critical stresses at the interface during heat treatment and in the rotary swaging process. Here, the co-spray forming was used with HS6-5-3C (M3:2) steel as the first layer and with X110CrMoV8-2 or HS6-5-2C (M2) steel as the second layer. The total thickness was up to 100 mm to allow hot rolling for densification. The gradient was achieved by tilting both spray cones as shown in Fig. 8.15. This leads to an overlapping of materials and results in a graded zone of 5–10 mm.

The first layer was sprayed onto a preheated ceramic sheet and both atomizers scanned perpendicular to the direction of the substrate movement. The new forming tools were successfully applied in the micro plunge rotary swaging of stainless steel wires [29].

8.2 Spray Cone

8.2.1 Mass Flux Distribution

8.2.1.1 Measuring of Local Mass Flux in a Molten Metal Spray

The mass flux distribution and the local sticking efficiency in the spray cone are mandatory to calculate the deposit geometry. Mass flux distributions have been

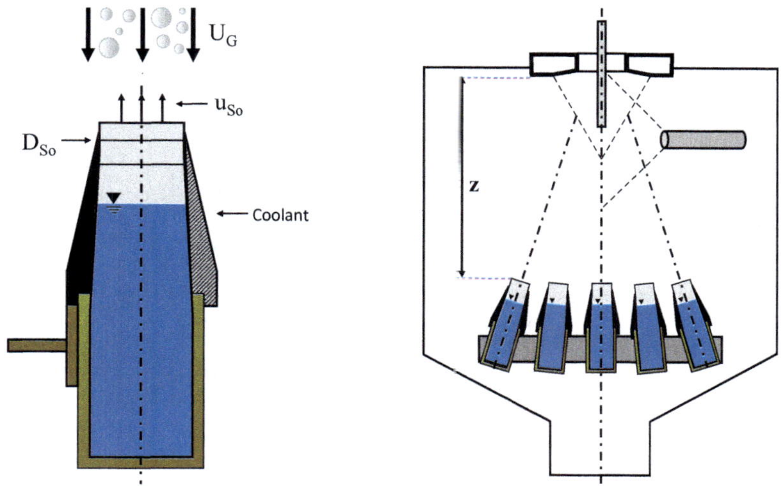

Fig. 8.16 Sketch of a sampling tube (*left*) and sampling tubes placed in a spray chamber (*right*)

measured for free-fall-atomizers using different alloys. Those results were used to formulate empirical correlations. The local particle mass flux $\dot{m}(r,z)$ in the spray cone is defined as

$$\dot{m}(r,z) = \frac{M}{A\Delta t} \tag{8.2}$$

with the particle mass M collected by a sampling tube with an opening area A during an opening time Δt.

Different types of sampling tubes have been used. Figure 8.16 shows an example of a sampling tube with an opening diameter between 8–20 mm. The tip of the orifice has a sharp edge to avoid particle deposition on the orifice tip. The tube was filled with a coolant to quench the particle temperature and to avoid agglomeration of particles. Typically, the opening time was less than 5 s to limit the temperature of the coolant. These sampling tubes were placed at different radial positions in the spray cone and the opening area was adjusted perpendicular to the anticipated particle flow direction.

A Phase-Doppler-Anemometer was also used to measure the local particle mass flux. This system can measure the diameter and velocity of up to 100,000 particles per second passing a measuring area of a few square millimeters. Typically, it is assumed that the mass flux distribution is symmetrical and a two-dimensional description (radial and axial) is sufficient.

8.2.1.2 Free-Fall-Atomizer

The spray cones of free fall atomizers have been investigated intensively. Figure 8.17 displays the spray cone of a free-fall atomizer and it is clear to see that the

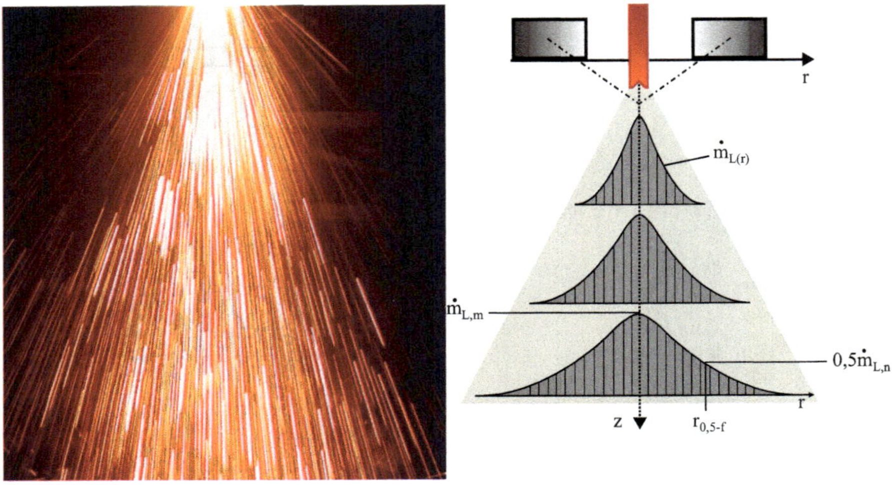

Fig. 8.17 Spray cone of a free-fall-atomizer (*left*) and schematic of particle mass flux profiles in a spray cone (*right*)

particles follow straight lines. The figure on the right shows schematically the mass flux profiles $\dot{m}_L\,(r)$ in the spray cone at different distances from the atomizer. The particle mass flow $\dot{M}_L$ at each distance z remains constant, but the maximum mass flux $\dot{m}_{L,m}$ decreases and the half radius $r_{0,5\text{-}f}$ increases with the distance from the atomizer. If these mass flux profiles are self-similar like two-phase free jets, then the descriptions of the profiles are easier. This assumption means that all dimensionless profiles can be described by one equation. Therefore, the particle mass flux must be normalized by its maximum on the centerline and the half radius.

Figure 8.18 shows examples of self-similar profiles. The curve indicated with $k_2 = 2.0$ is the Gaussian equation, and the other curve with $k_2 = 1.4$ is often used to describe the normalized mass flux profile in a two-phase free jet. On the left side, the results of a normalized mass flux profile are presented. Here, the distance z from the atomizer was varied from 600–1000 mm and the melt was a low carbon steel. A best fit of the results is displayed and indicated with $k_2 = 1.3$. If this fit is assumed, the particle mass flux $\dot{m}_{L,m}$ can be described by two values: the half radius and the maximum mass flux on the centerline. Both values are interconnected by the melt flow (see Fig. 8.18 (lower right)). Therefore, the half radius can be calculated if the mass flow is available. Finally, there is only the mass flux on the centerline which must be identified to calculate the mass flux profile. The mass flux on the centerline was measured for different alloys and gas mass flows. The results are presented in Fig. 8.18 (lower left). The mass flux increases with the gas mass flow. This means the spray cone becomes smaller. Additionally, the mass median diameter of the droplet decreases because a higher gas mass flow is applied. This general behavior holds for all three melts. Furthermore, the results show the highest values for tin, medium values for copper, and the lowest values for steel. It is assumed that there is

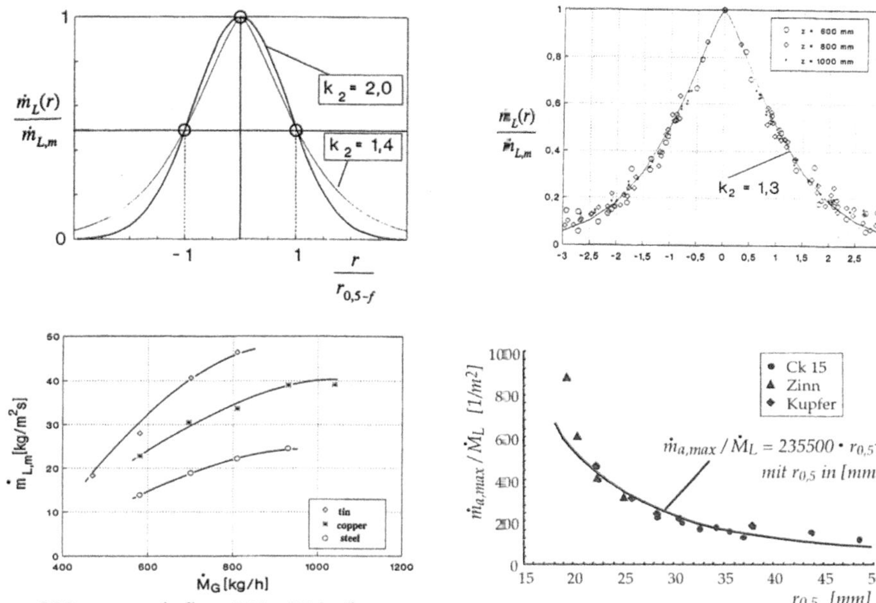

z = 800 mm, melt flow 790 ±70 kg/h

Fig. 8.18 Normalized particle mass flux profile (*upper left*), experimental data of normalized particle mass flux profiles of low carbon steel (*upper right*), particle mass flux on the centerline versus gas mass flow for different melts (*lower left*), adapted from [33], correlation between particle mass flux on the centerline/melt flow (and the half radius ($r_{0,5}$) for different melts (*lower right*), adapted from [34]

a correlation between the mass flux on the centerline and the inverse of the mass median diameter. That is, $\dot m_{L,m} \sim 1/d_{50,3}$.

8.2.1.3 Empirical Correlation

To describe the mass flux profiles it is assumed that the dimensionless profiles are self-similar and can be described by the following equation [35]:

$$\frac{\dot m_L}{\dot m_{L,m}} = e^{\left(\ln 0{,}5\left[r/r_{0,5}\right]\right)^{k2}} \tag{8.3}$$

Here, $\dot m_{L,m}$ is the particle mass flux on the centerline and $r_{0,5}$ is the half radius. The exponent k_2 is typically 1.3–1.4.

The empirical correlation to calculate the mass flux on the centerline is given as follows [35]:

$$\dot{m}_{L,m}\left[kg/m^2s\right] = A\, z[m]^{-1,75}\dot{M}_G[kg/s]^{1,65}\left(0.6 + \frac{\dot{M}_L}{\dot{M}_G}\right)^{1,12} \tag{8.4}$$

Here A is a factor that depends on the design of the atomizer and the alloy, z is the distance from the atomizer (in m), and $\dot{M}_G$ and $\dot{M}_L$ are the gas and the melt flows (both in kg/h) respectively.

The half radius $r_{0,5}$ only depends on the melt flow and the mass flux on the center line. This equation is derived from the mass balance [35]:

$$r_{0,5} = \left(0.151\frac{\dot{M}_L}{\dot{m}_{L,m}}\right)^{0,5} \tag{8.5}$$

A practical way to use these equations for other melts is to determine the melt flux on the centerline at an arbitrary distance (not too close to the atomizer for practical reasons). With the measured mass flux, the factor A in Eq. (8.4) can be determined for a new alloy or a different atomizer design.

8.2.2 *Heat Transfer and Enthalpy Distribution*

In the spray forming process, the melt is atomized into droplets to increase its surface area and allows rapid extraction of its heat before the droplets are deposited to form a product. As the microstructure development is coupled to the thermal history of the droplets and deposit, the rapid cooling step of the droplets has a major influence on the properties of the as-sprayed product. The thermal state of the droplets during their impingement on the substrate or deposit surface also has a great impact on their spreading behavior. A high solid fraction can lead to incomplete spreading and interparticular porosity ("cold porosity"), while a high liquid fraction can lead to gas entrapment ("hot porosity"). Thus, knowledge of the droplet cooling behavior for different process windows is necessary to identify optimized process windows.

8.2.2.1 Caloric Probes

The caloric probe measurement technique is an invasive offline technique. The idea of a caloric probe (Fig. 8.19) is to catch particles in an adequate quenchant during a defined time interval which will lead to an increase of the quenchant temperature to be measured. After the experiment, the collected mass in each caloric probe is measured using the temperature rise to the equilibrium temperature of particles and quenchant, and the amount of enthalpy released from the particles can be calculated. If this enthalpy is added to the particle enthalpy at the equilibrium temperature, the particle enthalpy at collection time is known. The design of

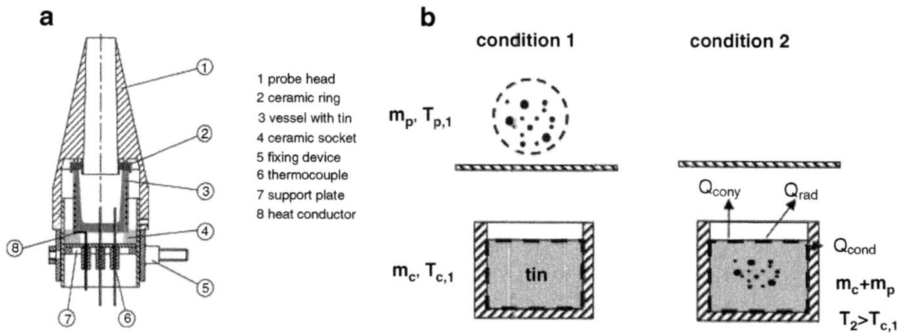

Fig. 8.19 (**a**) Sketch of caloric probe, (**b**) Energy balance before (condition 1) and after (condition 2) collection of particles from the molten metal spray. The balance boundary is the *broken line* (from [36] with permission)

the sampling tubes is similar to the one used for the measurement of the local mass flux with the exception that, instead of water or oil, a quenchant that does not vaporize must be employed. One possible quenchant suggested by Buchholz et al. [36] is tin, which is held in a liquid state at the time of droplet collection. From the measurement of the collected mass in the caloric probes, the mass flux can also be determined [37] so that mass flux collecting tubes can be replaced and particle enthalpy can be delivered at the same time. However, it should be noted that the use of caloric probes requires more measurement and preheating instrumentation than the simple water- or oil-filled mass flux collecting tubes.

8.2.2.2 Distribution of Specific Enthalpy in the Spray Cone

The time-consuming preparation and evaluation of the experiments require a model for the distribution of the enthalpy in the spray cone whose constants can be determined using only a few measurements, as is the case for the distribution of mass flux. While the spatial integral of mass flux at a given spray cone (which equals the mass flow) is constant, the spatial integration of the product of mass flux and specific enthalpy (which equals the total particle heat flow rate) is decreasing with the spray cone distance due to particle cooling. The model suggested by Ellendt et al. [38] consists of three equations: one equation for the axial decrease of total mass flux, one equation for the axial decrease of the enthalpy peak value on the center axis, and one equation which describes the self-similar radial distribution of specific enthalpy. In this model, it is assumed that the mass median particle diameter $d_{50,3}$ derived by Lubanska [39] is a measure for the convective heat loss, the surface to volume ratio of the particles, and for the initial load of the spray so that the droplet cooling is governed by this parameter.

The total heat flow rate $\dot{H}_{total}|_z$ for a nozzle distance z is given by

$$\dot{H}_{total}|_z = 2\pi \int_0^\infty \dot{m}(r,z) \cdot h(r,z)\mathrm{d}r, \tag{8.6}$$

where $\dot{m}(r,z)$ and $h(r,z)$ are the spatial distributions of mass flux and specific heat respectively. If the total heat flow rate is divided by the melt mass flow, one obtains the average specific enthalpy $\bar{h}(z)$. This value can also be expressed by the melt enthalpy h_{melt}, the equilibrium enthalpy h_∞, and an exponential decay that depends on the axial distance z and the droplet diameter $d_{50,3}$:

$$\bar{h}(z) = \frac{\dot{H}_{total}|_z}{\dot{M}_L} = h_\infty + (h_{melt} - h_\infty) \cdot \exp\left(a_1 \cdot z^{b_1} \cdot d_{50,3}^{c_1}\right), \tag{8.7}$$

or in non-dimensional form as

$$\bar{h}^*(z) = \exp\left(a_1 \cdot z^{b_1} \cdot d_{50,3}^{c_1}\right). \tag{8.8}$$

Similar equations are set for the axial decrease of enthalpy in the center of the spray:

$$\begin{aligned} h_0(z) &= h_\infty + (h_{melt} - h_\infty) \cdot \exp\left(a_2 \cdot z^{b_2} \cdot d_{50,3}^{c_2}\right); \\ h_0^*(z) &= \exp\left(a_2 \cdot z^{b_2} \cdot d_{50,3}^{c_2}\right). \end{aligned} \tag{8.9}$$

Figure 8.20 shows the axial decay of average and peak specific enthalpy for the atomization of pure copper and the model constants for copper, two bronze alloys, and two steels as determined by Ellendt et al. [38]. As expected, the specific enthalpy in the spray cone center shows higher values than the average enthalpy, which means there is also a radial decay.

For the radial distribution of specific enthalpy at the nozzle distance z, it is assumed that the type of distribution is similar to the radial distribution of mass flux:

$$\begin{aligned} \bar{h}(r)|_z &= h_\infty + (h_0(z) - h_\infty)\exp\left(-\ln(2) \cdot \left(\frac{r}{r_{0.5,h}}\right)^{k_h}\right); \\ \bar{h}^*(r)|_z &= \exp\left(-\ln(2) \cdot \left(\frac{r}{r_{0.5,h}}\right)^{k_h}\right). \end{aligned} \tag{8.10}$$

The half radius of the distribution of specific enthalpy $r_{0.5,h}$ is different to that of the mass flux distribution $r_{0.5}$. However, their ratio $r_{0.5}/r_{0.5,h}$ can be calculated by solving the integral equation for the total heat flow rate. An exponent of $k_h = 1.8$ was determined, for which the auxiliary value

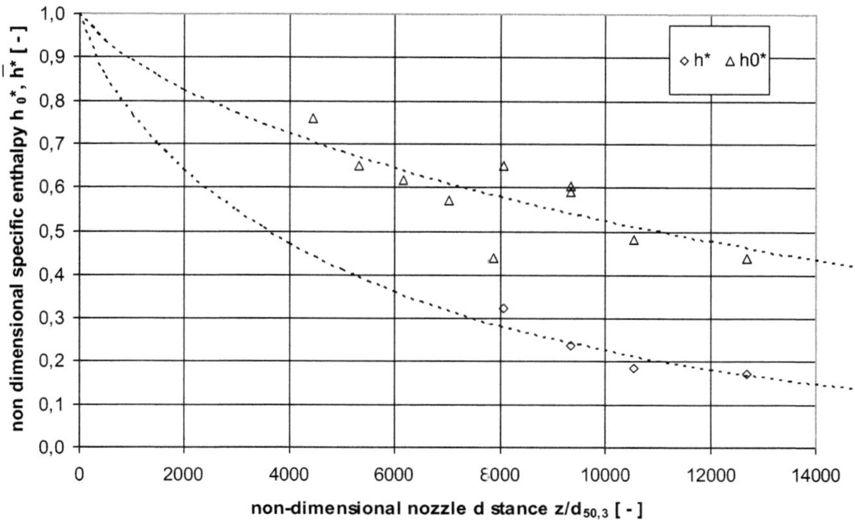

alloy	a_1 [-]	a_2 [-]	h_∞ [J/kg]
Cu	$-1.5.10^{-3}$	$-0.65.10^{-3}$	40
CuSn6	$-1.7.10^{-3}$	$-0.57.10^{-3}$	80
CuSn13.5	$-1.1.10^{-3}$	$-0.7.10^{-3}$	40
C35	$-1.15.10^{-3}$	$-0.19.10^{-2}$	80
AISI52100	$-0.7.10^{-3}$	$-0.43.10^{-3}$	80

All alloys:
$b_1=b_2=0.75$ [-]
$c_1=c_2=-0.75$ [-]
$k_h=1.8$ [-]

Fig. 8.20 Measured data of non-dimensional maximum specific enthalpy (h_0*) and average specific enthalpy ($\bar{h}^*$) on the center line during the atomization of copper (*left*) and model constants (*right*), adapted from [38]

$$\xi = \frac{\dot{H}_{total}|_z - \dot{M}_L \cdot h_\infty}{2\pi \cdot \dot{m}_0(z) \cdot r_{0.5}^2 \cdot (h_0 - h_\infty)} \tag{8.11}$$

can be used to approximate the ratio of half radii as

$$\frac{r_{0.5,h}}{r_{0.5}} = \frac{\xi}{a + b \cdot \xi + c \cdot \xi^2}$$
$$(a = 0.1958130894,\ b = 0.7666652641,\ c = -0.7909785624). \tag{8.12}$$

Figure 8.21 shows the radial distribution of specific enthalpy for pure copper, a tin bronze and a steel.

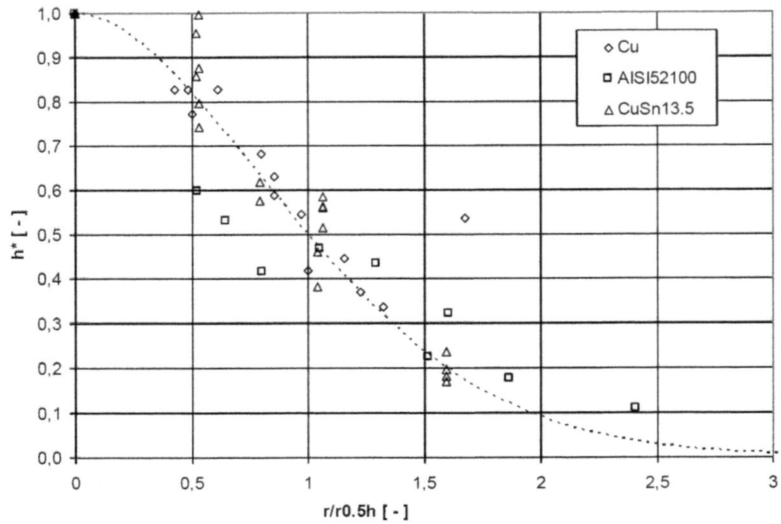

Fig. 8.21 Radial distribution of specific enthalpy for the atomization of Cu, bronze CuSn13.5, and the steel AISI52100, adapted from [38]

8.3 Deposition

8.3.1 Overspray, Yield, and Sticking Efficiency

A major disadvantage of the spray forming process is the unavoidable generation of overspray powder, which can reduce the process yield considerably. In spray forming, yield can be defined as the ratio of feedstock and final deposit mass. The generation of overspray can be divided into three parts. The first one is called geometric overspray and results from a smaller extension of the substrate than the spray cone. Geometric overspray can be easily reduced by the appropriate design of the spray nozzle and the appropriate selection of process parameters such as spray distance, which determines the spray cone dimensions. This sort of overspray can be more problematic in spray forming of small preforms such as small diameter tubes, in which controlling the spray cone dimensions can be more complicated. In Fig. 8.22, the calculated processing yield is shown for a conventional free-fall-atomizer operating with a fixed melt superheat, atomization gas pressure, and melt flow rate [11]. Assuming that spray distances lower than 0.4 m are not allowed due to hot spray conditions, one can see that tubes with a diameter of 25 mm cannot be produced with yield of more than 40%. However, Ellendt et al. [11] reported spray forming of small diameter tubes (90 mm) with up to 96% of yield and less than 1% of porosity by using pressure-gas-atomization, which reduces the melt flow rates. This makes shorter spray distances and smaller substrates diameters possible.

The second source of overspray is aerodynamic overspray caused by the deflection of the particles around the deposit. Aerodynamic overspray is also influenced

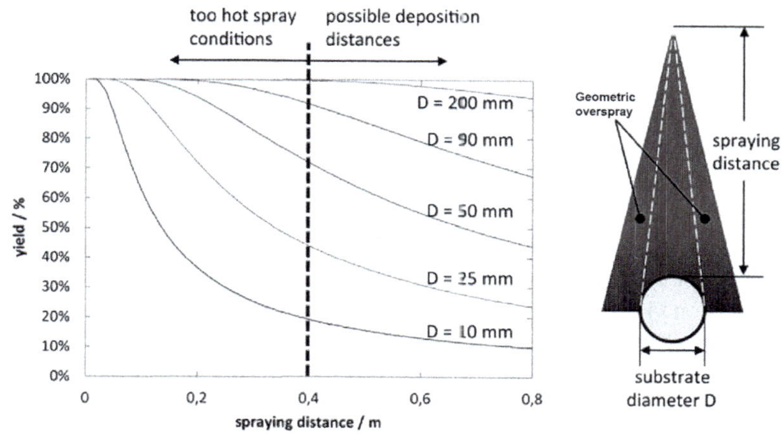

Fig. 8.22 Calculated spray forming yield based on the geometric overspray of small diameter tubes (from [11] with permission)

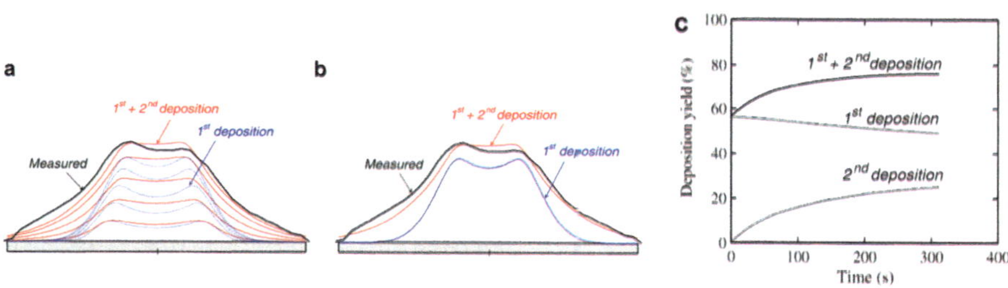

Fig. 8.23 The simulated (**a**) dynamic shapes at every 80 deposition cycles. (**b**) The final shape of the Ni superalloy ring using (1) the primary deposition only model and (2) the primary deposition plus redeposition model, with the measured ring final shape superimposed. (**c**) The calculated mass deposition yields resulting from primary deposition only, redeposition only and primary deposition plus redeposition (from [40] with permission)

by the geometry of the deposit, but represents only a small part of the total overspray. The third overspray source is due to particles bouncing off the surface. During the deposition stage, the droplets hit a cold or pre-heated substrate followed by the surface of the growing deposit. Different effects can be observed depending on the impact conditions: (1) the droplets may change their shape and stick onto the surface; (2) the droplets may bounce off; and (3) the primary droplet can be separated into several secondary droplets which splash back. In a study of spray forming Ni superalloy rings, Mi and Grant [40] showed that the droplet redeposition after splashing play an important role in the deposit shape evolution and in the final process yield. The authors developed a numerical model to simulate the dynamic shape evolution of the Ni superalloy rings during spray forming, which calculate separately (1) the droplet primary deposition, (2) splashing, and (3) the droplet redeposition. Figure 8.23a, b show the comparison between the measured final ring shape with the numerical results when only the primary deposition is considered, and when both primary deposition and redeposition are considered. One can see that when only the primary deposition model is applied the numerical results fail to predict the final measured shape. This leads to a smaller overall area with a more pronounced double-peak profile and a lower thickness of the peripheral regions. On the other hand, when the redeposition model is also considered, the numerical results can closely predict the final measured shape. Moreover, as shown in Fig. 8.23c, the results showed that in the case of the spray-formed Ni superalloy rings the droplet redeposition was calculated to contribute more than 20% to the final deposition yield. This significant contribution from the droplet redeposition on the evolution of the final shape of spray-formed deposits must also have a strong effect on the thermal history and, consequently, on the final microstructure of spray-formed products [41].

Differing from yield, the sticking efficiency (SE) is defined as the ratio of the sticking mass flux to the impacting mass flux, and it does not consider the geometric overspray. Thus, SE is a measure for the effectiveness of the compaction process. The SE is a complex physical phenomenon which occurs on a microscopic scale when high velocity droplets (solid, partially solid, and liquid) impact the rough surface during deposition. This made the value of SE difficult to be estimated based on fundamental analyses or models, which led investigators to study the SE through measurements of macroscopic variables such as liquid or solid fraction in the spray cone or surface temperature of the deposit [42–46]. Mathur et al. [42] proposed a model to calculate the SE assuming that the total SE results from the product of two main components. The first is the geometric component (SE_θ), which depends on the angle θ of incidence between the spray direction and surface normal. The model states that SE_θ increases continuously from $\theta = 90°$ (deposit surface parallel to the spray direction) to unity at $\theta = 0°$ (deposit surface perpendicular to the spray direction). The second is the thermal component (SE_T) which depends on the values of liquid fraction in the spray (f_l) and the liquid fraction in the deposit surface (F_l). The authors assume that liquid droplets always adhere to the deposit surface, but solid droplets stick only to partially liquid or fully liquid surfaces, and bounce off from solid surfaces. According to the model, an optimum sticking efficiency is achieved when partially liquid and fully liquid droplets impinge partially liquid deposit surfaces, since both the solid and liquid in the spray can adhere to this surface. The following equations quantify the model described above:

$$SE = SE_\theta SE_T = SE_\theta(n_s f_s + n_L f_L)\varepsilon \tag{8.13}$$

$$SE_\theta = \cos\theta \ \ or \ \ SE_\theta = cos^2\theta \tag{8.14}$$

$$n_s \approx 1 - 0.75(1 - F_L) \quad \text{and} \quad n_L \approx 0.98, \tag{8.15}$$

where n_s and n_L are the sticking coefficients of the solid and liquid from the spray respectively, f_s and f_L are the fraction of solid and liquid in the spray respectively, F_L is the fraction of liquid on the preform surface, and ε is a parameter which varies

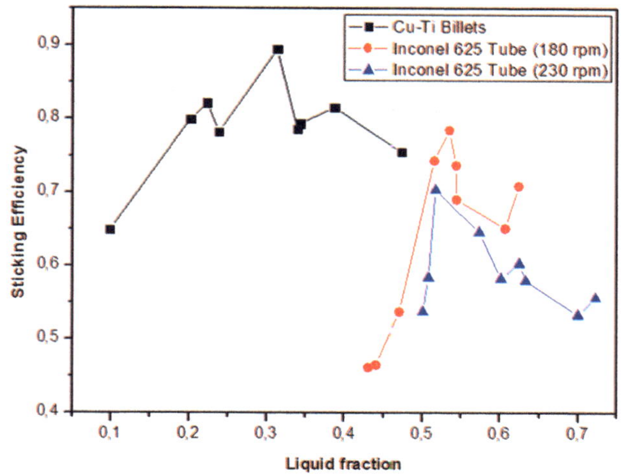

Fig. 8.24 Measured values of SE for Cu-6wt%Ti billets and Inconel 625 tubes as function of liquid fraction in the spray, adapted from [43]

from 0–1. As can be seen, it is assumed that on the solid surface ($F_L = 0$) 75% of the solid bounces off ($n_s = 0.25$), and that 98% of the liquid in the spray adheres to the preform. Although Warner et al. [43] has shown that for different alloys a maximum value of SE indeed takes place at intermediate values of the liquid fraction of the spray (see Fig. 8.24), an experimental proof of the whole model is still pending in the literature. As can be seen in Fig. 8.24 the liquid fraction of the spray in which the optimum SE value is achieved is very dependent on the alloy chemical composition. On the other hand, the results obtained from the spray forming of Inconel 625 tubes show that for the same alloy, the liquid fraction (~0.52) of the spray is the same where SE is maximized. But the SE values are still dependent on other process parameters, such as the substrate velocity, for example.

Kramer et al. [44] studied the effect of the main process parameters on the sticking efficiency of a Gaussian-shaped carbon steel deposit by using several measurement techniques. The SE at a specific position on the surface of the deposits $\{x,r\}$ and certain spray time $\{t\}$ is defined below:

$$SE(r,x,t) = \frac{_s(r,x,t)}{_p(r,x)} \tag{8.16}$$

where $\dot{m}_s$ is the sticking mass flux at x and r, and t and $\dot{m}_p$ are the impacting mass flux at x and r (which was considered constant with the spray time). The SE was calculated for different spray conditions. The sticking mass flux was measured by video analysis, and the impacting mass flux through collecting probe and sieving.

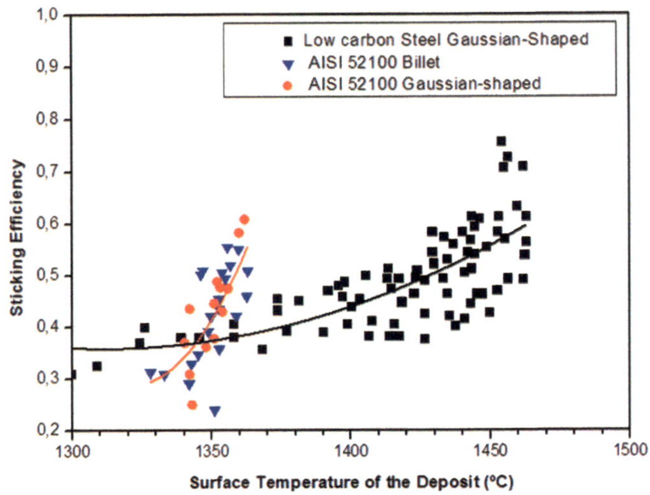

Fig. 8.25 SE as function of the surface temperature of deposit of low carbon steel and 100Cr6 (AISI 52100) bearing still, adapted from [44, 45]

The authors showed that the most important and critical parameter that control the SE is the surface temperature of the deposit, as can be seen in Fig. 8.25. Other parameters such as impact angle of the droplets, for instance, barely affect the SE. Buchholz, M. et al. [45] reported the same dependency of the SE with the surface temperature of the deposit for 100Cr6 bearing steel (AISI 52100). Furthermore, the authors showed that the same behavior of the SE as a function of the surface temperature of the deposit is observed independent of whether a billet or a Gaussian-shaped deposit is spray-formed (see Fig. 8.25). Kramer et al. [46] suggested the following empirical formula to describe the SE as function of surface temperature of the deposit:

$$SE(T_{surf}) = C_1 + C_2 \left(\frac{T_{surf}}{T_m} \right)^{C_3} \tag{8.17}$$

where T_{surf} and T_m are the are the surface and *liquidus* temperature of the deposit respectively. C_1, C_2 and C_3 are experimentally defined constants. Even though the dependency of the SE with the main process parameters has been described since early 1990's, an accurate model to predict the SE in spray forming process is still lacking in the literature. However, both models described above are frequently used in integrated models to predict the shape and thermal history of spray-formed products [40, 41, 47–50].

8.3.2 Deposit Temperature and Cooling Rate (or Thermal Evolution of Deposit)

Figure 8.26 illustrates the thermal transition of the three different regions within the spray forming process: (1) the melt flow in the tundish; (2) particle cooling in the spray; and (3) cooling of the deposit. Initially, the melt is cooled while flowing from the top of the tundish to the exit of the tundish in the atomization nozzle [51]. In the spray flow, the melt droplets are cooled mainly by convective heat transfer to the surrounding atomizer gas flow. As the droplets are accelerated up to 50–100 m/s, the mean residence time of the particles in the spray is considerably short, approximately 10^{-1} s [51–55]. During flight, the melt droplets are cooled down at high cooling rates (10^2–10^5 K/s), and the solid fraction in the spray increases [51–54, 56]. Finally, the deposit is built up through deposition of the impacting droplets/particles from the spray. The impacting particles deliver mass and heat to the deposit. Due to thermal diffusion and latent heat inside the deposition zone, the temperature of the deposit material increases suddenly just after the compaction and the material comes into equilibrium. The metallurgical quality of spray-formed deposits, e.g., porosity, segregation, microstructure, residual stresses, and presence or absence of cracks, is always determined by the cooling and solidification behavior of the deposit. The thermal history of the deposit is enormously influenced by the material properties of the alloy and process parameters such as: enthalpy and mass flux distribution in spray, heat flux from the surface of the deposit to the substrate and surrounding gas environment, and geometry of the preform [41, 47, 51, 57, 58]. This section will present how such parameters influence the temperature history of spray-formed deposits and their effects on the final metallurgical quality of the spray-formed products.

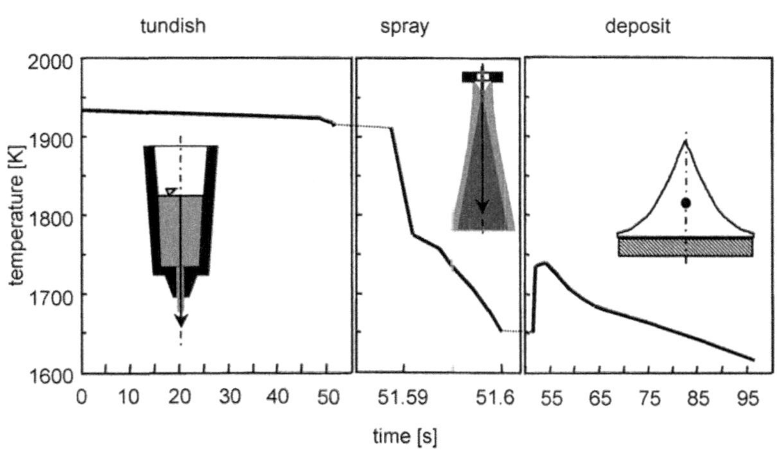

Fig. 8.26 Sequential thermal history in the three steps process of the spray forming process (from [51] with permission)

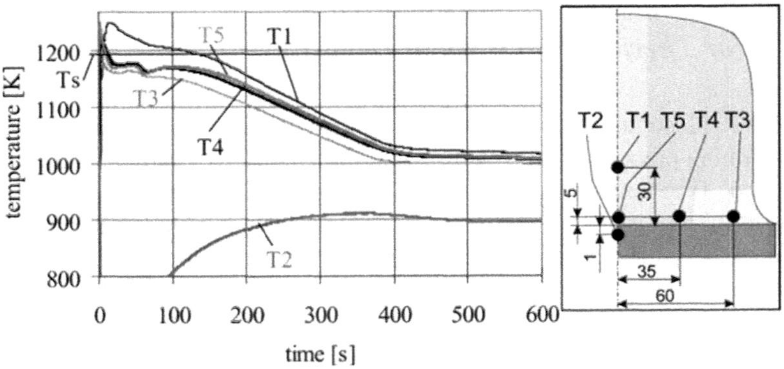

Fig. 8.27 Measured temperatures (*left*) and locations of the measurement points within the substrate and billet (*right*) in a spray forming process of a CuSn6 alloy (from [57] with permission)

Table 8.1 Process parameters of the spray forming of CuSn6 billet [57]

Melt superheat [K]	250
Gas to metal ration (GMR)	0.72
Melt flow rate [kg/s]	0.32
Spray distance [mm]	500
Spray time [s]	360
Initial temperature of the substrate [K]	291

Figure 8.27a presents temperature measurements during the spray forming of a CuSn6 billet with processing parameters listed in Table 8.1 [57]. The measurements were performed by thermocouples positioned at various radial distances from the rotational axis at different heights above and within the substrate as shown in Fig. 8.27b. It can be seen that the measured maximum temperature within the billet is above the *solidus* temperature ($T_s = 1189$ K) of the alloy. In the first stages of the deposition a lot of heat is transferred from the lowest part of the billet into the substrate. This is clearly seen by the strong increase of the substrate temperature (T1) and the drop of the temperature within the bottom of the billet (T3–T5) during the first few seconds. The increasing temperature within the substrate leads to a strong decrease of the cooling rate in the billet. One can see that the temperature measured by the sensor T1, which is only 30 mm above the substrate, is kept above the *solidus* temperature for a considerably longer time (approximately 140 s), while the temperature at the bottom of the billet (T3–T5) drops under the *solidus* temperature in less than 20 s. With increasing time and growth of the billet, the heat flux from the billet to the substrate decreases and the loss of heat from the surface increases. Comparison between the sensors T1 and T5 show that the temperature decreases within the billet in the direction of the substrate plate. However, because of the high thermal conductivity of cooper alloy (153 W/m/K) the radial temperature gradient is relatively low. Such temperature gradients can be clearly seen in Fig. 8.28, which shows the overall temperature and the local liquid fraction distribution in the billet at different times calculated by numerical

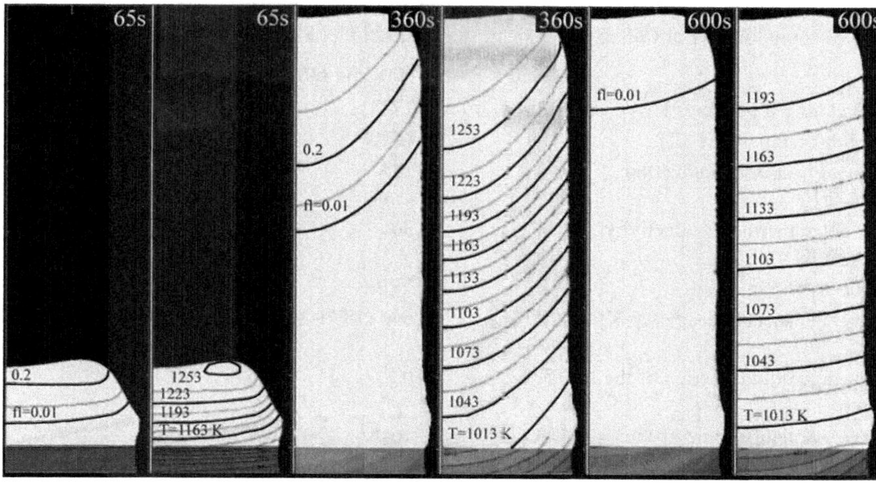

Fig. 8.28 Numerical simulation results showing the overall temperature and local liquid fraction distribution at different times of a CuSn6 billet spray-formed using standard spray conditions showed in Table 8.2. The total time of the spraying process is 360 s (from [57] with permission)

simulation. The thermal and physical properties of the alloy as well as the standard boundary conditions used for the numerical simulation of the spray forming of the CuSn6 billet are presented in Table 8.2 [57]. It is important to notice that the standard condition in the numerical simulation was set up considering an average liquid fraction of the spray constant equals to 0.5. The liquid fraction within the billet was calculated as a function of temperature based on the CuSn equilibrium phase diagram. Figure 8.28 shows that during the whole spraying period (up to 360 s) the calculated liquid fraction is high. After the spraying period the billet cools down slowly and the residual liquid in the mushy zone solidifies. The low gradient temperature is evident from top to bottom of the billet, as well as along the radius due to the high thermal conductivity of the copper alloy. This is especially valid along the radius after the spraying period since the convective heat transfer is low. Therefore, a more constant radial temperature distribution is expected.

The temperature gradient of a billet is strongly dependent on the materials' properties. The calculated overall temperature of a spray-formed 100Cr6 (AISI52100) bearing steel billet at different times, using similar standard process boundary conditions of the CuSn6 spray-formed billet (see Table 8.2), is presented in Fig. 8.29a [58]. Although there are similarities in the process conditions, one can see that due to the low thermal conductivity of 100Cr6 steel (30 W/m/K), the temperature difference from the top of the billet to the bottom as well as along the radius is relatively high. Even when the convective heat transfer coefficient decreases at the end of the spraying period, a non-uniform radial temperature distribution is observed. One can see that after spraying (>360 s), the billet cools down slowly and the residual liquid is enclosed by the totally solidified material. In this case, if shrinkage is suppressed, residual stress may arise and initiate hot cracks.

Table 8.2 Material properties and standard boundary conditions used for the numerical simulation of spray forming of CuSn6 and 100Cr6 steel billets [57, 58]

	CuSn6	100Cr6 (AISI 52100)
Liquidus temperature [K]	1325	1724
Solidus temperature [K]	1189	1570
Latent heat of solidification [kJ/kg]	200	287
Average thermal conductivity [W/m/K]	153	30
Density [kg/m^3]	8484	7810
Average specific heat [J/kg/K]	478 (Cu, 1023–1301 K)	640 (1570–1800 K), 724 (1570–1400 K)
Average liquid fraction of the spray	0.5	0.5
Average temperature of the impinging spray [K]	1295	1648
Convective heat transfer coefficient (billet surface) during spray [W/m^2/K]	280	$hg = h_{max}\exp\left[-1.65\left(\frac{D}{D_{max}}\right) + 0.85\left(\frac{D}{D_{max}}\right)^2\right]$ where D = distance to the top surface of the billet and Dmax = reference distance = 400 mm
Convective heat transfer coefficient (billet surface) after spray [W/m^2/K]	10	10
Temperature of ambient air and spray chamber [K]	523	523
Emissivity of the billet surface	0.18	0.5
Coefficient heat transfer between billet and substrate [W/m^2/K]	1000	1000
Initial temperature of the substrate [K]	303	303

In both cases, for the CuSn6 alloy and the 100Cr6 steel, higher temperature gradients are located at the base of the billets at the beginning of the process. At the top of the billets the temperature gradients are lower due to the high enthalpy in the form of latent heat of solidification contained in the mushy zone. In both cases, it is shown that cooling rates of 10^2–10^3 K/s can be found in the very beginning of the process and decreases to 10^0–10^1 in the following cooling process.

Whereas the materials properties are fixed, the process parameters can be changed to optimize the thermal history of the spray-formed deposits and, consequently, their final metallurgical quality. Three main parameters can be changed: (1) the specific enthalpy of the impacting spray; (2) the heat loss during the spray process; and (3) the heat loss after the spraying period. The specific enthalpy of the spray can be easily changed by varying the melt superheat and/or the gas to metal ratio (GMR), which results in a change of the average liquid fraction of the impacting spray. Figure 8.29b, c presents the temperature profiles of the 100Cr6

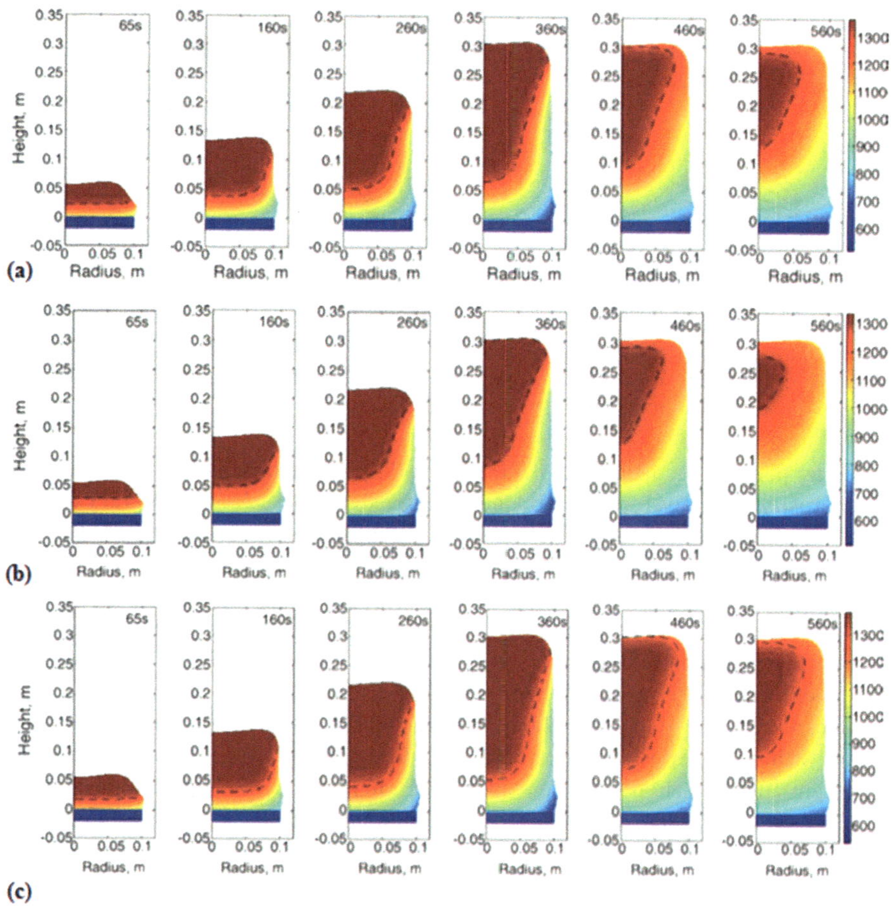

Fig. 8.29 Numerical simulation results showing the overall temperature at different times of a 100Cr6 steel billet (AISI 52100) spray-formed using (**a**) standard (Table 8.2.); (**b**) cold (liquid fraction in the spray = 0.3); and (**c**) hot (liquid fraction in the spray = 0.6) spray conditions. The total time of the spraying process is 360 s. The *dashed lines* denote the location of zero liquid fractions (from [58] with permission)

steel billet in cold and hot spray forming conditions respectively [58]. The overall temperatures in both cases were calculated using the same process parameters showed in Table 8.2 with changes to the average temperature and liquid fraction of the impinging spray. For the hot and cold conditions these are 1623 K and 0.3, and 1673 K and 0.6 respectively. The temperature distribution in the billet in the cold spray conditions (Fig. 8.29b) is similar to the standard spray condition. However, the residual liquid within the billet is largely reduced at the end of spraying. On the other hand, when hot spray conditions are applied (Fig. 8.29c) the billet experiences a slow cooling and solidification process. In this case, the calculated liquid fraction in the hot billet is considerably high during the whole

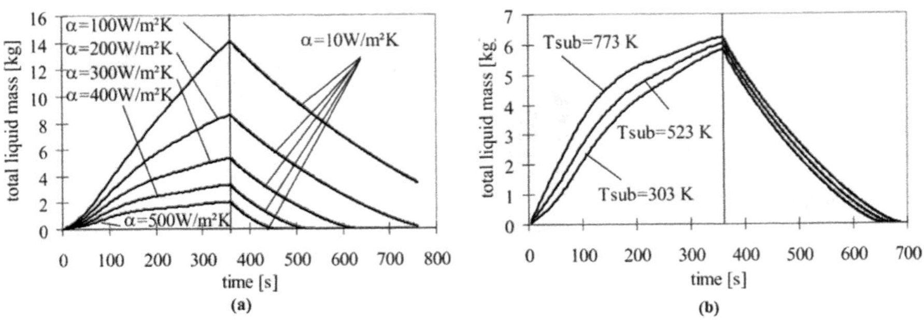

Fig. 8.30 Computed total liquid mass within the CuSn6 billet versus time for (**a**) different heat transfer coefficients at the billet surface during spraying and (**b**) different initial substrate temperatures (from [57] with permission)

spraying period. The influence of the residual liquid fraction within the deposit on the porosity of spray-formed products is extensively described in literature. Low residual liquid fraction may generate interstitial porosity—also called cold porosity—where the amount of residual liquid is not enough to fill up the cavities between the overlapped solid particles. For hot spray conditions, where high residual liquid fraction is present, gas entrapment is probable to occur and generate the so-called hot porosity. Furthermore, if extreme hot spray conditions are applied, hot cracks due to the rising of residual stresses during the final stages of solidification are also expected to occur. Thus, the thermal history and the amount of residual liquid within the deposit must be optimized for each alloy by controlling the input spray enthalpy to achieve the desired product quality.

Heating loss during the spray process is influenced by the temperature of the ambient gas and the initial temperature of the substrate. The effect of changing the temperature of the ambient gas can be represented numerically by the changing of the convective heat transfer (α) at the billet surface. Figure 8.30a illustrate the simulation results of spray forming the CuSn6 billet by varying the convective heat transfer coefficient at the billet surface during the spraying period. The computations for convective heat transfer coefficients of 400 W/m^2/K and 500 W/m^2/K show a relatively low change of the total liquid mass within the billet during the spray time. A decrease of the convective heat transfer coefficient leads to a reduction of the heat flux to the environment, increasing the total enthalpy within the billet. Such results show that the total residual liquid within the billet can be controlled by varying the convective heat transfer coefficient. Physically, an almost identical effect on the thermal history can be obtained by changing the temperature of the ambient gas, since the heat flux ($\dot{q}_{\alpha}$) from the surface of the billet to the gas convection is expressed by [57]:

$$\dot{q}_{\alpha} = \alpha(T_s - T_g) \tag{8.18}$$

where T_s is the local temperature of the surface and T_g is the temperature of the ambient gas. The effect of the initial temperature of the substrate is shown in

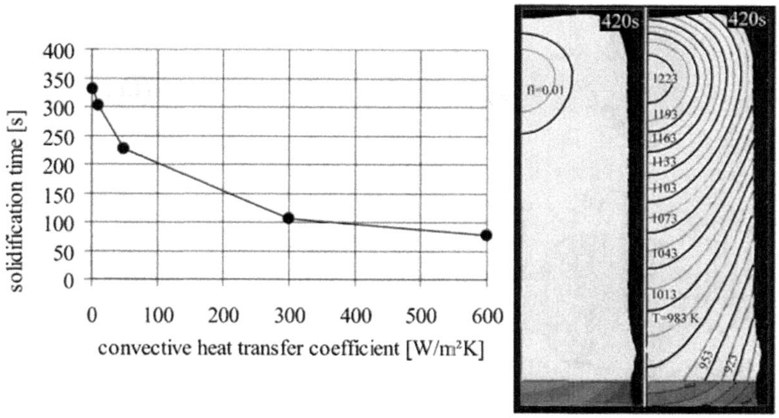

(a) (b)

Fig. 8.31 (**a**) Solidification time of the remaining liquid within the CuSn6 billet versus convective heat transfer coefficient after the spraying period. (**b**) Temperature profiles and local liquid fraction within the CuSn6 billet calculated with a high cooling rate after the spraying period ($\alpha = 600$ W/m^2/K) (from [57] with permission)

Fig. 8.30b. The variation of the initial substrate temperature (which can be easily controlled by preheating systems) influences the heat flux at the bottom of the billet in the initial stages of the deposition process. This correlation is mainly important for the production of spray-formed plates, strips, and tubes which are relatively thin.

During cooling after the spraying period, the convective heat transfer and the environment temperature may be easily controlled. For instance, by maintaining the gas flux on the deposit surface after the end of the melt flux, the convective heat transfer at the surface of the deposit can be considerably increased. Figure 8.31a shows the solidification time of the remaining liquid within the CuSn6 billet in the cooling period for different convective heat transfer coefficients. Besides the decrease of the solidification time with the increase of the convective heat transfer, the temperature field and the liquid fraction distribution within the billet are also considerably changed by varying the cooling conditions. Figure 8.31b shows the calculated temperature profiles and local liquid fraction in the CuSn6 billet 60 s after the end of the spraying period using a convective heat transfer of 600 W/m^2/K^1. One can see that the thermal gradients in radial directions are considerably changed when compared with the standard spray condition showed in Fig. 8.28. Due to the high heat loss across the billet surface, the liquid at the upper portion rapidly solidifies and the mushy zone area is enclosed. A hotspot is thus generated. In this case the temperature profiles of the CuSn6 billet is similar to the 100Cr6 steel billets processed in cold spray conditions shown in Fig. 8.29b.

The geometry of the preform also has a great influence on the thermal history of the spray-formed deposits. Figure 8.32a–c show the surface distribution of a 100Cr6 steel tubular preform at different spray times calculated using the same

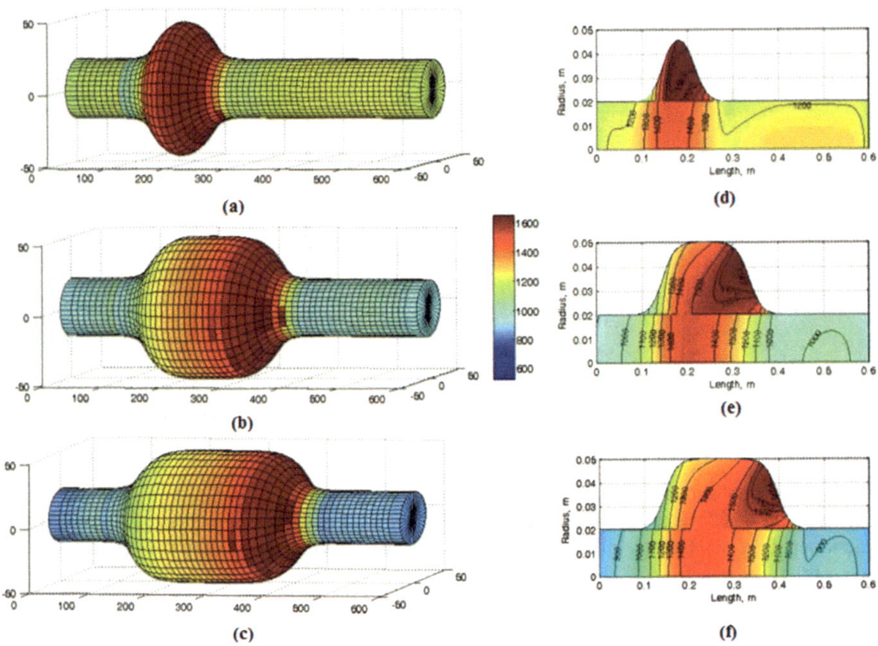

Fig. 8.32 Surface temperature distribution and temperature distribution through the longitudinal section of the tubular preform of 100Cr6 steel (AISI 52100) under standard spray condition at spraying time of: (a) and (d) 30 s; (b) and (e) 90 s; (c) and (f) 120 s (from [47] with permission)

standard conditions showed in Table 8.2 (but with initial temperature of the substrate at 1373 K). In this case the mandrel substrate rotates counter-clockwise at a frequency of 2.5 Hz and translates to the left at a speed of 2 mm/s, collecting the impinging droplets on its surface and forming a tubular deposit [47]. The highest temperature is always positioned below the spray cone and moves to the right as the deposit grows. Once the atomizing gas is spraying toward the preform during the deposition period, the mandrel cools fast and the temperature of the early depositing material decreases. Due to the fast rotation of the mandrel the temperature variation along the circumference of the deposit and the mandrel is very small. The temperature distributions across the deposit and the mandrel at different times are shown in Fig. 8.32d–f. It can be seen that during the whole process the area where the residual liquid within the deposit is present (indicated by the isothermal curve of 1570 K) is considerably smaller when compared to the 100Cr6 steel billet illustrated in Fig. 8.29a. In the case of the billet, the enthalpy of the impacting spray is uniformly distributed on the surface of the deposit during the entire deposition period, generating a temperature gradient from the top to the bottom of the billet. On the other hand, in the case of the tubular preform, the input enthalpy of the impacting spray acts only in the region under the deposition zone. Moreover, the previously deposited layer will cool down through heat transfer to the gas environment and the substrate, generating a temperature gradient from the centerline of the spray cone towards the left side of the deposit (translation direction). One can see that after depositing for a period, the deposition material has a strong influence on the substrate; however, it is limited in the area below the deposit. The regions of the mandrel not covered by the deposit are more influenced by the environmental gas cooling. As can be seen in Fig. 8.33a, b, the amount of residual liquid within the tubular deposit can also be controlled by changing the input enthalpy of the impinging spray in the same way shown for the billet.

As can be seen thus far, the parameters of the spray forming process can be conscientiously varied to achieve the desired metallurgical characteristics of the products. For instance, Fig. 8.34 shows the simulated heat flows and liquid fraction

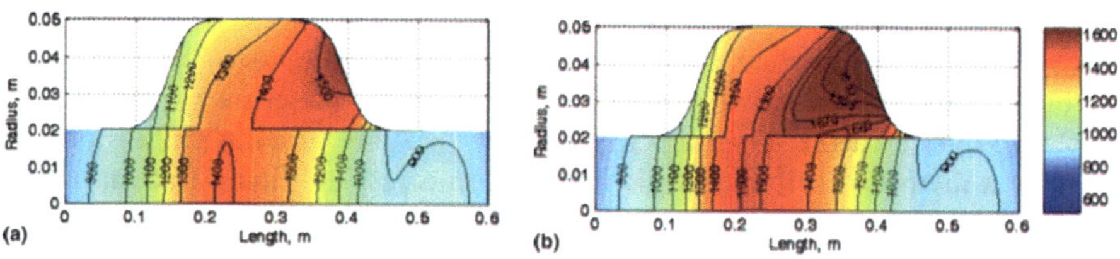

Fig. 8.33 Effect of the enthalpy input from the spray on the thermal profiles of a 100Cr6 steel (AISI 52100) tubular preform at the deposition time of 120 s with average liquid fraction in the spray of (**a**) 0.3 and (**b**) 0.7 (from [47] with permission)

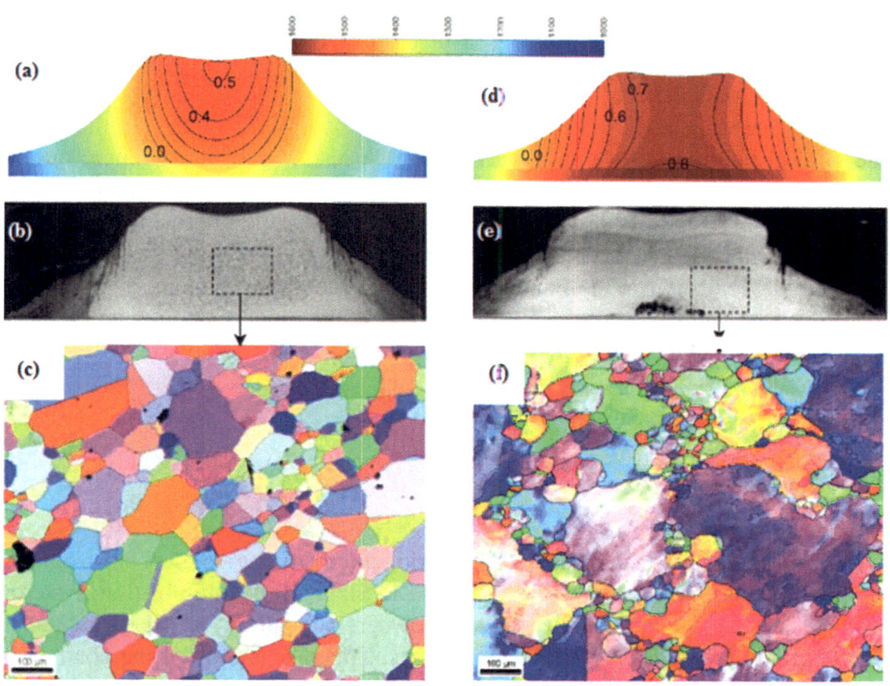

Fig. 8.34 (**a**) and (**d**) Simulated heat flow and liquid fraction contours at the end of the spraying period; (**b**) and (**e**) Macrostructure of the etched ring cross-section; and (**c**) and (**f**) EBSD orientations maps for the marked areas showing the as-spray-formed microstructures of the IN718 rings 1 and 2, respectively (from [41] with permission)

Table 8.3 Processing parameters of two Ar spray-formed IN718 alloy rings [41]

Ring	Gas flow (kg/s)	Melt flow (kg/s)	GMR	Pouring temperature (K)	Substrate heating	Spray forming time (s)	Ring surface temperature[b] (K)
1	0.34	0.35	0.97	1772	Off[a]	390	1523
2	0.35	0.32	1.09	1767	On[a]	380	1562

[a]The heater was operated at 50–60 kW to pre-heat the substrate, but turned off immediately when spray forming started
[b]Preform surface average temperature during steady-state spray forming measured using two-wavelength pyrometer

contours, the macrostructure, and the microstructure of two Ar spray-formed IN718 superalloy rings processed with the two distinct sets of parameters (presented in Table 8.3) [41]. The set of parameters used in the spray forming process of ring 1 led to a thermal distribution which resulted in a residual liquid fraction in the central region of the preform in the range of 0.4–0.5 (Fig. 8.34a). The resulting macrostructure of this central region was relatively uniform (as shown in Fig. 8.34b), and comprised of equiaxed grains typical of spray-formed materials with an average size of 42 μm (Fig. 8.34c) and an average residual porosity of 1.5%. However, in the areas toward the preform edges there were strongly chilled zones that led to a higher micro porosity level. On the other hand, ring 2 was spray-formed with a set of parameters that led to hot deposition conditions due to heating the substrate. Figure 8.34d shows that the residual liquid fraction at the end of the spraying period in the central region of the deposit was greater than 0.7. Due to the continuous heating of the substrate, the lower part of the cross section at the end of the spraying period has a liquid fraction of 0.8, resulting in a macropore due to shrinkage as shown in Fig. 8.34e. The EBSD orientation map of the marked area close to the macropore shown in Fig. 8.34f reveals that the microstructure in this region is composed of abnormally coarsened grains with finer grains between them. The comparison between the Fig. 8.34a–f is evidence of the differences of thermal histories and, consequently, the differences of the final metallurgical quality achievable through changing the process parameters.

8.4 Microstructure Evolution

A rigorous quantitative analysis of the resulting microstructure of deposits from gas atomization has not yet been accomplished. Single fluid atomization systems provide an ideal opportunity to generate powders and spray formed ingots under highly controlled and reproducible conditions. This discussion will be conducted with reference to previous work carried out for Impulse Atomization. Several reported studies provide insight into the mechanism by which spray formed ingots result in such finely distributed precipitated phases [59, 60].

Figure 8.31 shows that the solidification time of a deposit in gas atomized spray forming of a CuSn6 alloy has a range from several hundred seconds to as low as 20 s (near the substrate). This solidification time is not sufficiently quick enough to lead to a rapid solidification structure. It is in fact concluded in the previous section that the ingot is not fully solidified even after spraying ends (360 s), Thus, complete solidification of the deposited droplets must occur within the ingot upon cooling. Given the size of the deposited ingot, it is difficult to imagine how such a relatively large ingot size can contain phases with fine scale secondary phases as reported in this book. This is in contrast to the coarseness of the structure and of these secondary phases that occur when a liquid alloy is poured into an ingot mold with the same dimensions as the spray deposited ingot. Some mechanism, other than heat and momentum transfer, must be taking place to create such a fine structure. This section will address this issue.

The atomization of CuSn6 was introduced in Sect. 8.1.1.1. Figure 8.1 clearly shows the microstructure in the Impulse Atomized spray formed sample to be much larger than the microstructure in gas atomized spray formed sample. Hardness and porosity measurements were made in both ingots. The results are shown in Fig. 8.35 [59]. The dashed lines in the figure represent the range of values obtained from gas atomized spray formed ingots, while the data points are for various ingots atomized and spray formed using Impulse Atomization. By varying the distance between the atomizing nozzle and the substrate, different fraction solids of the median droplets arrived at the deposit.

It is clear from Fig. 8.35 that the range in porosity in samples generated from both processes fit within the same range between 1–5%. Interestingly, the hardness values are lower for the gas atomized spray formed samples than for the Impulse

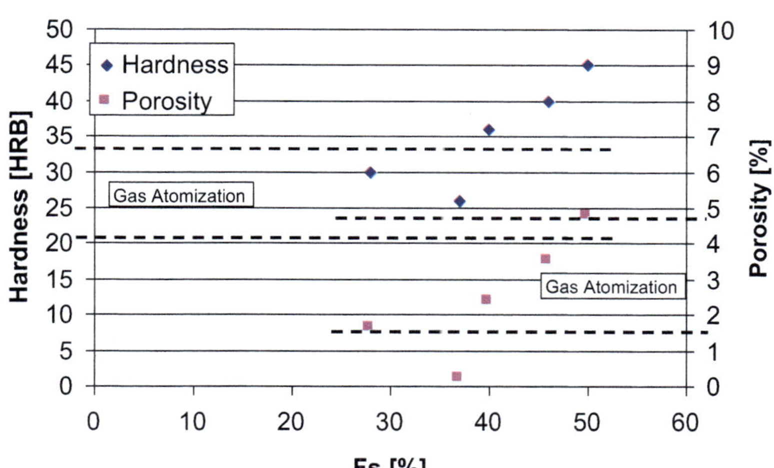

Fig. 8.35 Hardness and porosity versus fraction solid of droplets arriving at the deposit for Impulse Atomized CuSn6 compared with results from gas atomized spray formed ingot of the same alloy

Table 8.4 IA run conditions for Al-0.61 wt%Fe alloy powder [60]

Run number	Gas	D_{50} (μm)	Geo σ (−)
030108I	N2	392	1.37
030108 K	He	435	1.42

Atomized spray formed samples with hardness values going up to 45 HRB. The highest hardness value for the gas atomized spray formed sample is about 33 HRB. The gas atomized samples possess a smaller grain size than the Impulse Atomized samples, which is contrary to the Hall-Petch relation which relates yield stress (by analogy hardness) to the inverse of the square root of grain size. Hence, some other feature of the microstructure is controlling the hardness of the ingots and not the grain size.

Experiments are reported for Impulse Atomization of Al-0.61wt%Fe in both powder and spray formed strip [59, 61, 62]. Table 8.4 provides the run conditions for atomizing Al-0.61wt%Fe powder [61, 62]. An FESEM micrograph of a powder particle 925 mm in diameter from Run # 030108I is shown in Fig. 8.36a. It is clear that a dendritic structure is not visible. However, the cells observed in 2D are in fact secondary dendrite arms when observed in 3D [63–66]. Figure 8.36b shows the measured cell space for powders atomized in N_2 and He. A eutectic structure is visible in the inter-dendritic region. Measurements of the volume fraction of eutectic for a range of particle sizes are shown in Fig. 8.37. Also shown in Fig. 8.37 are the volume fractions of eutectic under equilibrium and Scheil solidification conditions. It can be seen that the amount of eutectic found in the powder particles is far below that calculated for either Scheil or equilibrium. This led to the conclusion that the eutectic nucleated below its equilibrium eutectic temperature considering that each droplet having a single nucleation point for the eutectic.

Figure 8.38 presents the schematic of the Impulse Atomization unit adapted for the spray forming of the Al-0.61 wt% Fe strip. The experimental conditions are presented in Table 8.5. An image of the 5 mm thick strip of Al-0.61wt%Fe produced is shown in Fig. 8.39a, and a micrograph of the microstructure in the middle of the strip is shown in Fig. 8.39b. The result of the measurements of cell spacing in the strip is shown in Table 8.6. When compared with the cell spacing in powders (Fig. 8.36b), it is clear that the cell sizes in the strip is considerably larger than those in the powder. This is simply due to the much slower cooling rate in the strip compared to the powder. This was observed also in the CuSn6 experiments discussed earlier.

Comparing the eutectic fractions in Fig. 8.37 and in Table 8.6, it appears that the eutectic fraction in the strip is slightly lower than that in the powders. If the nucleation of the eutectic intermetallic in the strip occurred at a single nucleation site, the eutectic fraction would clearly have been much closer to either the equilibrium, or more likely the Scheil values shown in Fig. 8.37. Instead, we observe that the eutectic in the strip is considerably less than the Scheil or equilibrium processes and only slightly less than that in the powder. This leads to the only plausible conclusion that multiple nucleation sites of the AlFe intermetallic must have occurred in the strip.

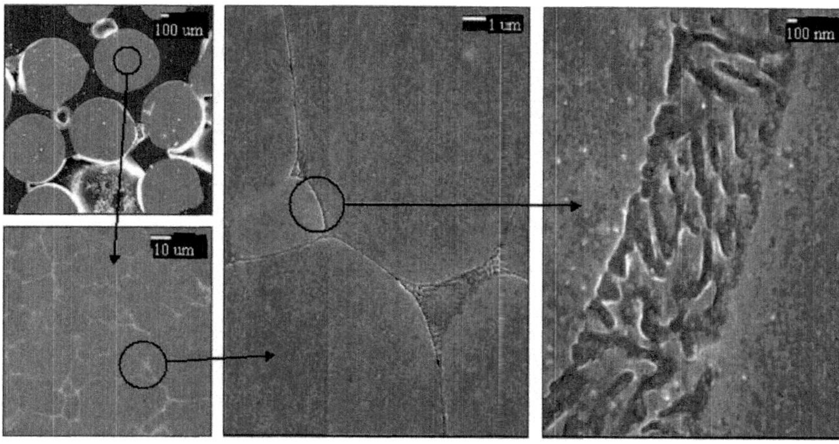

(a)

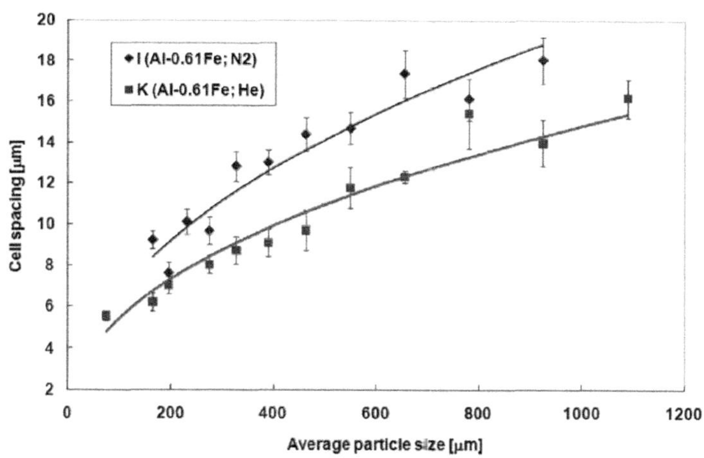

(b)

Fig. 8.36 (**a**) FESEM micrograph of N2 atomized, 925 μm diameter powder particle of Al-0.61 wt% Fe showing the finess of the microstructure as well as its interdendritic region; (**b**) cell spacins versus particle size for the Al-0.61wt%Fe powders, adapted from [60]

The same observation was made when comparing the resultant microstructure of D2 Tool Steel for powders and spray formed ingots; both were produced using Impulse Atomization [60, 67, 68]. D2 Tool Steel possesses approximately 1.5 wt % C, 0.4 wt% Mn, 0.3 wt% Si, 11.8 wt% Cr, 0.8 wt% Mo, and 0.8 wt% V. Its equilibrium structure is ferrite with a eutectic structure of ferrite and M_7C_3. During solidification from the liquid, the alloy first forms austenite followed by a eutectic reaction of austenite and M_7C_3 lamela. Using X-ray diffraction, it was determined that the structure of the powders at room temperature consisted of austenite and the austenite M_7C_3 eutectic. Thus, a structure formed at 1200 °C remained stable in powder form at room temperature [60, 68]. Only upon reheating the powders did the metastable austenite transform into ferrite [67].

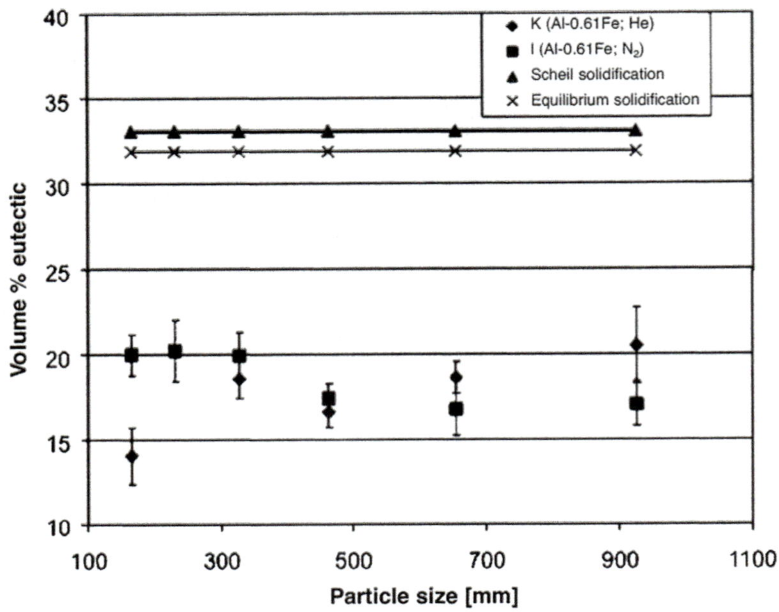

Fig. 8.37 Eutectic fractions in Al-0.61wt%Fe droplets compared with the fractions for equilibrium and Scheil solidifications

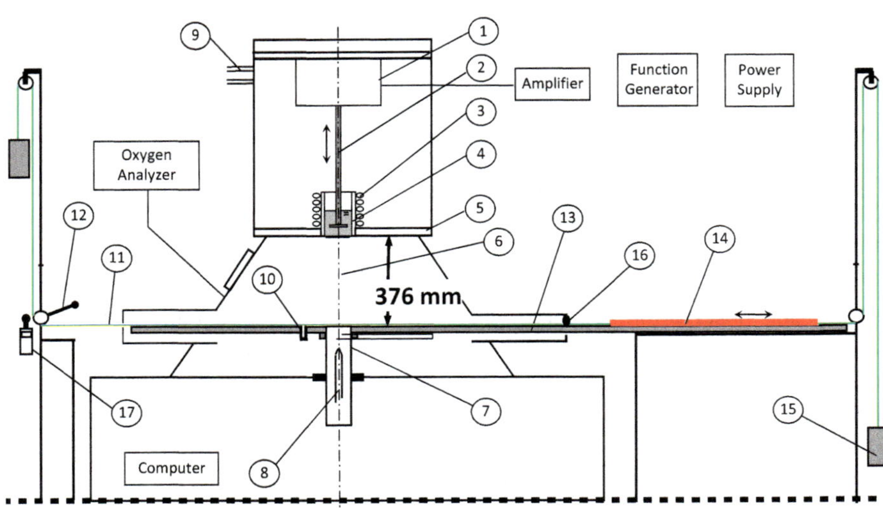

Fig. 8.38 Schematic of the Impulse Atomization of Al-0.61wt%Fe spray formed strip. The components are: (1) pulsator, (2) impulse applicator, (3) induction furnace coil, (4) crucible, (5) frame to hold and adjust the unit, (6) metal spray, (7) support pipe, (8) cooling nozzle, (9) gas inlets/oulets, (10) type K thermocouple, (11) transportation cable, (12) crank, (13) sliding plate, (14) substrate, (15) weight, (16) sealing, and (17) digital tacometer

Table 8.5 Run conditions for the impulse atomization of spray formed Al-0.61 wt% Fe strip

Run number	090503–02	082603–03	082603–05
Run time (s)	109	71	44
Deposit mass (g)	198	136	73
Mass flow (g/s)	1.82	1.91	1.66
Substrate feed (mm/s)	5.78	8.07	7.69
# Layers	1	1	1
Deposite length (mm)	630	575	335

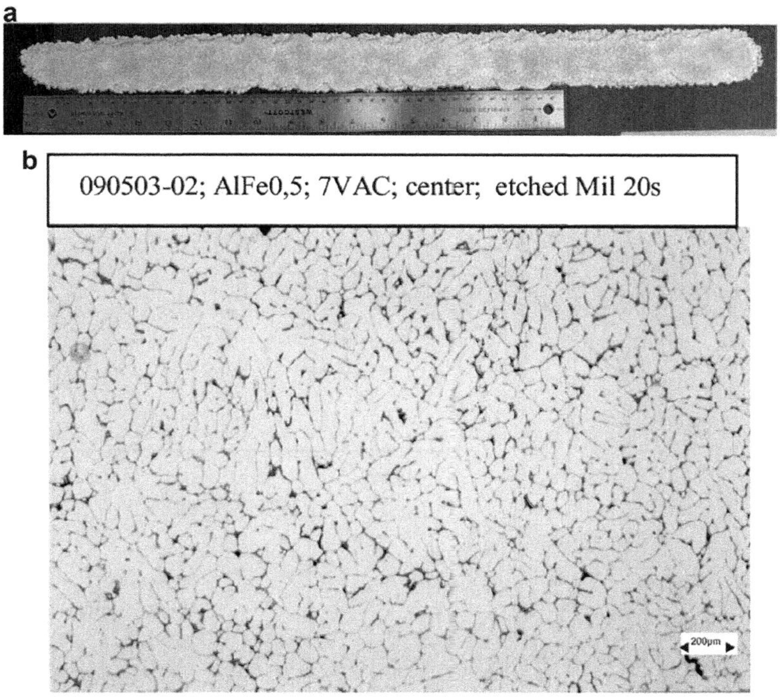

Fig. 8.39 (**a**) Image of a typical 5 mm thick strip of Al-0.61wt%Fe; (**b**) micrograph of the microstructure of the mid-thickness of the spray formed strip [59]

Table 8.6 Cell spacing and eutectic fraction in spray formed strip

Run number	Units	090503–02	082603–03	082603–05
Average cell spacing	µm	44.2	42.9	41.2
Standard deviation	µm	6.2	6.7	2.6
Percent eutectic		15.7	13.7	15.7
Standard deviation		3.1	2.6	3.1

Further confirmation of the metastability of the powder microstructure was reported by carrying out quantitative measurements of the eutectic fraction in the atomized D2 powders. The results are shown in Fig. 8.40. The eutectic fraction for D2 solidified under equilibrium or Scheil solidification conditions are also plotted in Fig. 8.40. Similar to what was observed with Al-0.61wt%Fe, the eutectic fraction in the D2 powders are far lower than that for either equilibrium or Scheil conditions. This clearly points to eutectic undercooling to have taken place [68].

Small ingots of D2 were spray deposited using Impulse Atomization. The microstructure of the spray deposited ingots were analyzed using X-ray diffraction and by quantitative measurements of the eutectic fraction [60]. Figure 8.41 shows the X-ray diffraction results from different spray formed ingot samples described next.

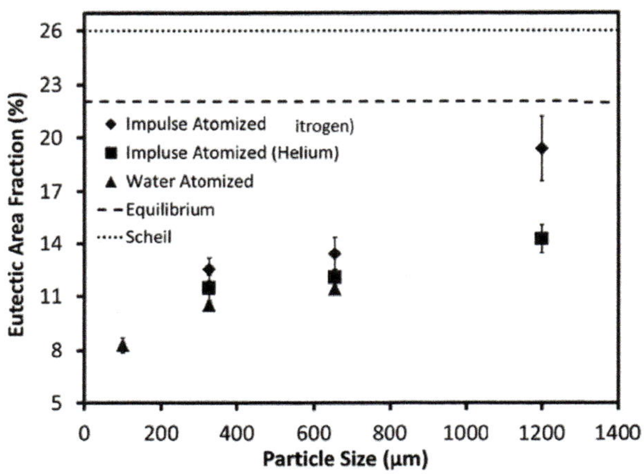

Fig. 8.40 Eutectic fraction as a function of particle size for D2 powders compared with eutectic fraction under equilibrium and Scheil solidification conditions

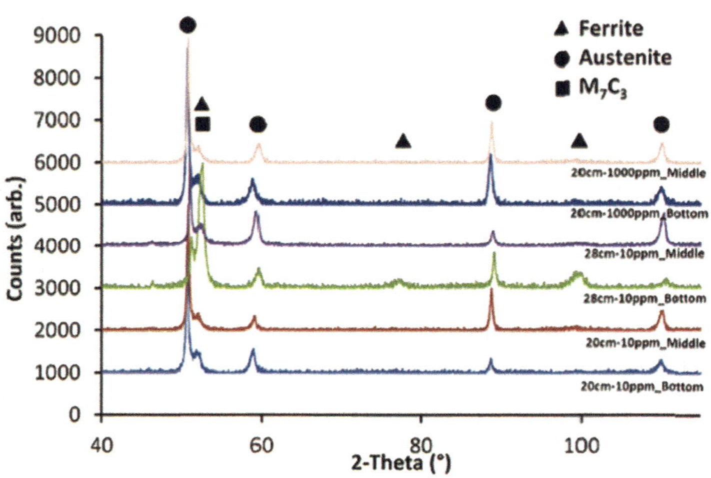

Fig. 8.41 X-ray diffraction spectra of several samples of spray formed D2 Tool Steel spray formed using Impulse Atomization

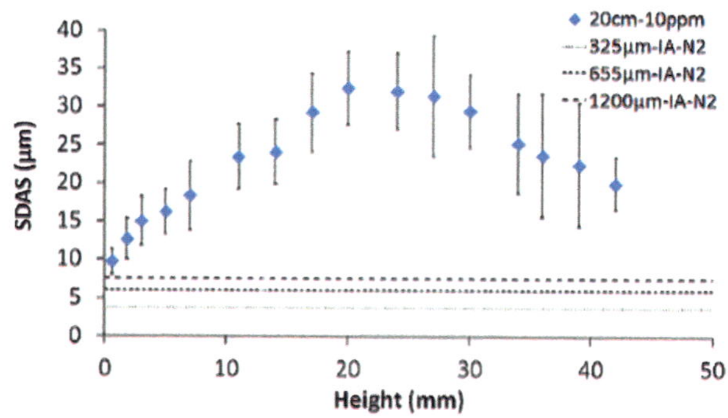

Fig. 8.42 Secondary dendrite arm space (cell spacing) versus height for the spray formed ingots compared with the SDAS of atomized powders for D2 Tool Steel [60]

Figure 8.41 shows the X-ray diffraction results from three different spray formed samples. Two samples were spray formed by collecting the semi-solid droplets 20 cm away from the atomizing nozzle. For these two runs, one was carried out with 10 ppm oxygen in the spray chamber and the other with 1000 ppm oxygen in the spray chamber. The third run was carried out by capturing the semi-solid droplets 28 cm from the atomizing nozzle and with an oxygen content of 10 ppm. For each experiment, two samples were investigated from each spray formed ingot. A sample is taken from the bottom of the ingot close to the substrate, and another from the middle of the spray formed ingot. From the X-ray results shown in Fig. 8.41, it is clear that in all samples austenite is the dominant phase present. In the ingot bottom samples, traces of ferrite are observed. It is postulated that the ferrite formed as a result of self-heat treating of the steel during spray forming. The first set of semi-solid droplets falling onto the substrate will solidify rapidly, most likely forming austenite. As more semi-solid droplets land on the first deposited layers, the heat evolved from the solidification and cooling of these solidifying droplets heat treat the lower layers on the deposit. Thus, some transformation from austenite to ferrite can occur as seen in the X-ray spectra shown in Fig. 8.41.

As with the Al-0.61wt%Fe samples, the cell spacing of powders and spray deposited ingots (SDAS) was measured and compared. A representative result is shown in Fig. 8.42. The SDAS for powders are shown for three different powder sizes and ranges between 4–6 µm. By contrast the SDAS for the spray formed ingots are considerably larger even for the region of the ingot close to the substrate. This clearly indicates that the cooling rate in the spray formed ingot was much slower than for the powders [60].

The eutectic fraction for the same sample was measured and is shown in Fig. 8.40. The results shown in the figure are representative of all three run conditions, the X-ray spectras of which were shown above. The eutectic fraction for the spray ingot is shown with the data points in Fig. 8.40. This result is similar to the eutectic fraction measured in powders with diameters of 655 and 325 µm. It is well below the eutectic fraction calculated if solidification occurred under either equilibrium or Scheil conditions. Thus, multiple nucleation events must have occurred for the eutectic in different regions of the deposit.

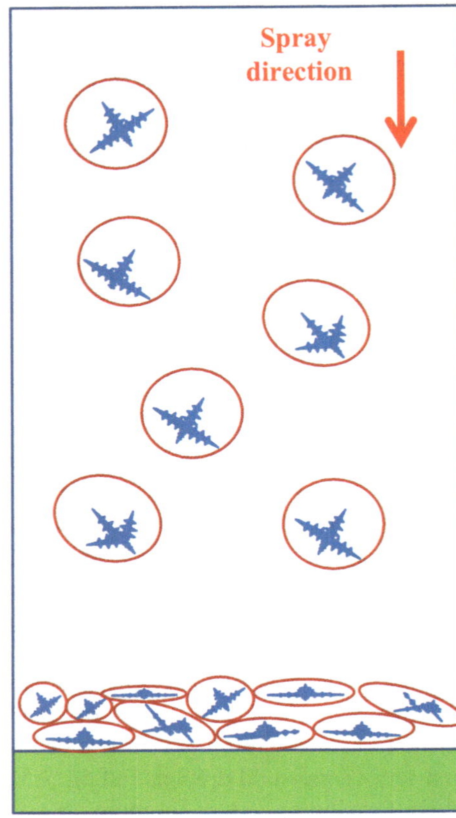

Fig. 8.43 Schematic depicting semi-solid droplets falling onto a substrate/deposit and flattening with their oxide coating intact

The solidification of an alloy in the spray formed ingot has been proposed to occur as follows. When droplets are atomized and are falling in the spray, a thin oxide layer forms and coats each droplet. Within each droplet the primary phase nucleates and starts to grow. The semi-solid droplets reach the substrate/deposit, and those without the oxide layer completely formed will break. The unbroken droplets will flatten as they form the deposit due to their low momentum and the liquid still present within. This scenario is schematically shown in Fig. 8.43. It is then in this condition that the second phase must nucleate (M_7C_3 for D2 and an AlFe intermetallic for Al-0.61wt%Fe) in each flattened droplet as it fully solidifies in the deposit. Thus, in each of these flattened droplets a nucleation event must occur of the second phase. Thus undercooling of the eutectic will be prominent in each of these flattened droplets, leading to eutectic undercooling similar to that occurring in atomized droplets. This mechanism has been termed the slushy balloon model.

The microstruture of a spray formed D2 ingot where the droplets were collected 28 cm from the nozzle plate and with 10 ppm oxygen in the atomizing tower is shown in Fig. 8.44. The layered structure in the deposit is clear evidence of the

Fig. 8.44 Etched deposit of spray formed D2 Tool Steel atomized by Impulse Atomization where the semi-solid droplets are collected 28 cm from the atomizing nozzle and the atmosphere had 10 ppm oxygen. The *arrows* point to droplets that landed onto the deposit fully solid

flattened droplets that formed upon landing. Equally evident is the fine structure of the eutectic observed. The arrows point to two droplets that landed in the deposit fully solid. Their shapes are clearly delineated due to the presence of an oxide layer on the droplets. Thus, the presence of metastable phases, the metastable level of eutectic in the deposit, and the layered structure in the deposit are all confirmation of the slushy balloon model.

It has been suggested that this mechanism occurs in Impulse Atomization only because the droplets land on the deposit in such a gentle state. In gas atomization, the droplets have far greater momentum, therefore the oxide likely breaks on impact with the deposit. This explanation does not hold, however. For if that were to occur, the alloy will solidify with the liquids of all the droplets blended together. The deposit will then solidify in the same manner as a cast ingot, with one nucleation event for the eutectic. Instead all structures observed from gas atomized spray formed ingots present a finely distributed structure of non-primary phase solids. This leads to the conclusion that this slushy balloon model is also present in spray formed gas atomized ingots. A quantitative analysis remains to be carried out—as has been done with Impulse Atomization—to present further evidence that this mechanism is active in gas atomized spray formed ingots.

8.5 List of Symbols

Latin

Symbole	Description
A	Factor based on the design of the atomizer and the alloy
C_1, C_2 and C_3	Experimentally defined constants
d	Droplet diameter
f_L	Fraction of liquid in the spray
F_L	Fraction liquid on the preform surface
f_s	Fraction of solid in the spray
$\dot{H}_{total}\|_z$	Total heat flow rate for nozzle distance z
h_0*	Maximum specific enthalpy
$\bar{h}(z)$	Average specific enthalpy
h_{melt}	Melt enthalpy
h_∞	Equilibrium enthalpy
$\dot{M}_G$	Gas flow
$\dot{M}_L$	Melt flow
$\dot{m}_{L,m}$	Particle mass flux (on the center line)
n_L	Sticking coefficient of the liquid
n_s	Sticking coefficient of the solid
r	Radius
SE	Sticking efficiency
SE_T	Sticking efficiency (thermal component)
SE_θ	Sticking efficiency (geometric component)
t_f	Solidification time
T_m	Liquidus temperature
T_{surf}	Surface Temperature
z	Distance from the atomizer
ξ	Auxiliary value

Greek

Symbole	Description
λ	Average grain size
ε	Initial cooling rate

References

1. Ellendt, N., Schmidt, R., Knabe, J., Henein, H., & Uhlenwinkel, V. (2004). Spray deposition using impulse atomization technique. *Materials Science and Engineering A, 383*(1), 107–113.
2. Singer, A., & Kisakurek, S. (1976). Centrifugal spray deposition of aluminium strip. *Metals Technology, 3*(1), 565–570.
3. Barratt, M., Dowson, A., & Jacobs, M. (2004). The microstructure and properties of IN718 rings produced by centrifugal spray deposition. *Materials Science and Engineering A, 383*(1), 69–77.
4. Chen, W., Song, X. P., Qian, K. W., & Gu, H. C. (1998). The lamellar microstructure and fracture behavior of γ-based TiAl alloy produced by centrifugal spray deposition. *Materials Science and Engineering A, 247*(1), 126–134.
5. Cherng, J.-P. J. (2002). *The effects of deposit thermal history on microstructure produced by uniform droplet spray forming*. Cambridge, MA: Massachusetts Institute of Technology.
6. Strutt, J. W., & Rayleigh, L. (1878). On the instability of jets. *Proceedings of the London Mathematical Society*, 10(4).
7. Uhlenwinkel, V., Achelis, L., Sheikhalliev, S., & Lagutkine, S. (2003). A new technique for molten metal atomization. In *ICLASS 2003, Sorrento, Italy*, 13–17 July, 2003. ILASS-Europe.
8. Lagutkin, S., Achelis, L., Sheikhaliev, S., Uhlenwinkel, V., & Srivatava, V. (2004). Atomization process for metal powder. *Materials Science and Engineering A, 383*(1), 1–6.
9. Leatham, A. (1999). Spray forming: Alloys, products and markets. *Metal Powder Report, 54*(5), 28–37.
10. Meyer, C., Ellendt, N., Mädler, L., Carter, W. T., & Uhlenwinkel, V. (2013). Spray forming of rings with a twin atomizer. In SDMA 5th International Conference on Spray Deposition and Melt Atomization, 23–25 September, Berlin, Germany.
11. Ellendt, N., Uhlenwinkel, V., & Mädler, L. (2014). High yield spray forming of small diameter tubes using pressure-gas-atomization. *Materialwissenschaft und Werkstofftechnik, 45*(8), 699–707.
12. Singer, A. (1991). Metal matrix composites made by spray forming. *Materials Science and Engineering A, 135*, 13–17.
13. Cui, C., Schulz, A., Uhlenwinkel, V., & Zoch, H.-W. (2011). Spray-formed stainless steel matrix composites with co-injected carbide particles. *Metallurgical and Materials Transactions A, 42*(8), 2442–2455.
14. Cui, C., Schulz, A., Uhlenwinkel, V., Zobel, F., & Zoch, H.-W. (2010). Spray forming of stainless steel matrix composites with injection of hard particulates. *Materialwissenschaft und Werkstofftechnik, 41*(7), 524–531.
15. Srivastava, V., Schneider, A., Uhlenwinkel, V., & Bauckhage, K. (2005). Spray processing of 2014-Al + SiC P composites and their property evaluation. *Materials Science and Engineering A, 412*(1), 19–26.
16. Zhao, Y., Grant, P., & Cantor, B. (1993). The microstructure of spray-formed Ti-6Al-4 V/SiCf metal-matrix composites. *Journal of Microscopy, 169*(2), 263–267.
17. Lawley, A., & Apelian, D. (1994). Spray forming of metal matrix composites. *Powder Metallurgy, 37*(2), 123–128.
18. Cui, C., & Schulz, A. (2013). Modeling and simulation of spray forming of clad deposits with graded interface using two scanning gas atomizers. *Metallurgical and Materials Transactions B, 44*(4), 1030–1040.
19. Schneider, A., Uhlenwinkel, V., Harig, H., & Bauckhage, K. (2004). Overspray injection in spray forming of CuSn13. 5 billets. *Materials Science and Engineering A, 383*(1), 114–121.
20. Schneider, A., Uhlenwinkel, V., Wriedt, T., Harig, H., & Bauckhage, K. (2000). *Study on the distribution of Al2O3 and graphite particles in spray formed copper (CuNi2Si) MMC*. Paper presented at SDMA 2000, Bremen, Germany, 26–28 June 2000.
21. Schneider, A. (2005). *Sprühkompaktieren mit Injektion von Feststoffpartikeln*. Dissertation, University Bremen, Bremen, Germany.

22. Uhlenwinkel, V., Schneider, A., Wriedt, T., Harig, H., & Bauckhage, K. (2002). Effect of particle injection during spray forming of Cu-Sn billets. *Advances in Powder Metallurgy and Particulate Materials, 4*, 4–192.
23. Schneider, A., Srivastava, V., Uhlenwinkel, V., & Bauckhage, K. (2004). Spray forming of 2014-Al alloy based composites with injection of SiC particulates. *Zeitschrift für Metallkunde, 95*(9), 763–768.
24. Zhang, J., Sun, Z. Q., Chen, G. L., Liu, X. J., Cui, H., & Duan, X. J. (1997). Microstructure and properties of spray-deposited 2014 + 15 vol pct SiC particulate-reinforced metal matrix composite. *Metallurgical and Materials Transactions A, 28*(5), 1261–1269.
25. Rentsch, R., Grohmann, O., Schulz, A., & Uhlenwinkel, V. (2015). Application of a composite hot shearing tool manufactured by co-spray forming. *Materials Science Forum, 825–826*, 771–778.
26. Meyer, C., Ellendt, N., & Uhlenwinkel, V. (2011). *Spray forming of tubular multilayer materials*. Paper presented at European congress on advanced materials and processes, Montpellier, France, 12–15 September 2011.
27. Uhlenwinkel, V., Achelis, L., Sulatycki, K., Uhlenwinkel, V., & Mädler, L. (2010). *Spray forming of multilayer materials*. Paper presented at PMTEC 2010, Fort Lauderdale, USA, 27–30 June 2010.
28. Zepon, G. (2016). *Spray forming of wear and corrosion resistant bimetallic pipes: From the alloy design to the semi-industrial process* (p. 169). São Carlos, Brazil: Federal University of Sao Carlos.
29. Cui, C., Schulz, A., Steinbacher, M., Moumi, E., Kuhfuss, B., & Böhmermann, F. (2015). Development of micro rotary swaging tools of graded tool steel via co-spray forming. *Manufacturing Review, 2*, 22.
30. Grohmann, O., Meyer, C., Schulz, A., Uhlenwinkel, V., Heinzel, C., & Brinksmeier, E. (2014). Analyse eines durch das Co-Spray-Verfahren hergestellten Werkzeuges zur Warmumformung. *HTM Journal of Heat Treatment and Materials, 69*(4), 235–240.
31. Cui, C., Schulz, A., & Uhlenwinkel, V. (2013). Co-spray forming of gradient deposits from two sprays of different tool steels using scanning gas atomizers. *Steel Research International, 84*(11), 1075–1084.
32. Cui, C., Schulz, A., & Uhlenwinkel, V. (2014). Materials characterization and mechanical properties of graded tool steels processed by a new co-spray forming technique. *Materialwissenschaft und Werkstofftechnik, 45*(8), 652–665.
33. Uhlenwinkel, V., & Bauckhage K. (1993). Mass flux profile and local particle size in the spray cone during spray forming of steel, copper and tin. In: *Proceedings of the 2nd International Conference on Spray Forming (ICSF II), Swansea, 884 UK*, September 13–15, 1993.
34. Kramer, C. (1997). *Die Kompaktierungsrate beim Sprüuhkompaktieren von Gauß-Förmigen deposits*. Dissertation, University Bremen, Bremen, Germany.
35. Uhlenwinkel, V. (1992). *Zum Ausbreitungsverhalten der Partikeln bei der Sprühkompaktierung von Metallschmelzen*. Düsseldorf: VDI-Verlag.
36. Buchholz, M., Uhlenwinkel, V., v Freyberg, A., & Bauckhage, K. (2002). Specific enthalpy measurement in molten metal spray. *Materials Science and Engineering A, 326*(1), 165–175.
37. Minisandram, R.S., Forbes-Jones R.M., Suthar, P.K., Carter, W.T., Uhlenwinkel V., & Ellendt N. (2006). Characterization of spray in nucleated casting of superalloys. In: *Proceedings of the 3rd International Conference on Spray Deposition and Melt Atomization (SDMA 2006) and 6th International on Spray Forming (ICSF VI), Bremen, Germany*, September 4–6, 2006.
38. Ellendt, N., & Uhlenwinkel, V. (2006). Experimental investigation and modeling of the specific enthalpy distribution in a spray cone. In *Materials science forum* (Vol. 534, pp. 417–420). Aedermannsdorf: Trans Tech Publications.
39. Lubanska, H. (1970). Correlation of spray ring data for gas atomization of liquid metals. *Journal of Metals, 22*, 45. doi:10.1007/BF03355938.

40. Mi, J., & Grant, P. S. (2008). Modelling the shape and thermal dynamics of Ni superalloy rings during spray forming. Part 1: Shape modelling, droplet deposition, splashing and redeposition. *Acta Materialia, 56*(7), 1588–1596.
41. Mi, J., & Grant, P. S. (2008). Modelling the shape and thermal dynamics of Ni superalloy rings during spray forming. Part 2: Thermal modelling, heat flow and solidification. *Acta Materialia, 56*(7), 1597–1608.
42. Mathur, P., Annavarapu, S., Apelian, D., & Lawley, A. (1991). Spray casting - An integral model for process understanding and control. *Materials Science and Engineering A, 142*(2), 261–276.
43. Warner, L., Cai, C., Annavarapu, S., & Doherty, R. (1996). Modelling microstructural development in spray forming: Experimental verification. In: *Proceedings of the 3rd International Conference on Spray Forming (ICSF III), Cardiff, UK*, September 9–11, 1996.
44. Kramer, C., Bauckhage, K., & Uhlenwinkel, V. (1996). The sticking efficiency at the spray forming of metals. In: *Proceedings of the 3rd International Conference on Spray Forming (ICSF III), Cardiff, UK*, September 9–11, 1996.
45. Buchholz, M., Uhlenwinkel, V., & Ellendt, N. (1999). The effect of deposit temperature on the sticking efficiency during spray forming In: *Proceedings of the 4th International Conference on Spray Forming (ICSF IV), Baltimore, Maryland, USA*, September 13–15, 1999.
46. Kramer, C., Uhlenwinkel, V., & Bauckhage, K. (1998). Spray forming: The sticking efficiency and its dependency on affecting parameters. In E. Lavernia (Ed.), *Solidification 1998, Indianapolis, IN* (pp. 401–413).
47. Cui, C., Fritsching, U., Schulz, A., & Li, Q. (2005). Mathematical modeling of spray forming process of tubular preforms - Part 2. Heat transfer. *Acta Materialia, 53*(9), 2775–2784.
48. Cui, C., Fritsching, U., Schulz, A., & Li, Q. (2005). Mathematical modeling of spray forming process of tubular preforms – Part 1. Shape evolution. *Acta Materialia, 53*(9), 2765–2774.
49. Mi, J., Grant, P. S., Fritsching, U., Belkessam, O., Garmendia, I., & Landaberea, A. (2008). Multiphysics modelling of the spray forming process. *Materials Science and Engineering A, 477*(1–2), 2–8.
50. Zhang, G., Li, Z., Zhang, Y., Mi, J., & Grant, P. S. (2010). Modeling the deposition dynamics of a twin-atomizer spray forming system. *Metallurgical and Materials Transactions B, 41*(2), 303–307.
51. Bergmann, D., & Fritsching, U. (2004). Sequential thermal modelling of the spray-forming process. *International Journal of Thermal Sciences, 43*(4), 403–415.
52. Grant, P. S., Cantor, B., & Katgerman, L. (1993). Modeling of droplet dynamic and thermal histories during spray forming. 1. Individual droplet behavior. *Acta Metallurgica et Materialia, 41*(11), 3097–3108.
53. Bergmann, D., Fritsching, U., & Bauckhage, K. (2000). A mathematical model for cooling and rapid solidification of molten metal droplets. *International Journal of Thermal Sciences, 39*(1), 53–62.
54. Krauss, M., Bergmann, D., Fritsching, U., & Bauckhage, K. (2002). In-situ particle temperature, velocity and size measurements in the spray forming process. *Materials Science and Engineering A, 326*(1), 154–164.
55. Pariona, M. M., Bolfarnini, C., dos Santos, R. J., & Kiminami, C. S. (2000). Application of mathematical simulation and the factorial design method to the optimization of the atomization stage in the spray forming of a Cu-6% Zn alloy. *Journal of Materials Processing Technology, 102*(1–3), 221–229.
56. Fritsching, U., & Krauss, M. (2004). Particle temperature measurements in the atomization of molten metals. *Chemie Ingenieur Technik, 76*(6), 787–790.
57. Meyer, O., Fritsching, U., & Bauckhage, K. (2003). Numerical investigation of alternative process conditions for influencing the thermal history of spray deposited billets. *International Journal of Thermal Sciences, 42*(2), 153–168.

58. Cui, C., Fritsching, U., Schulz, A., Tinscher, R., Bauckhage, K., & Mayr, P. (2005). Spray forming of homogeneous 100Cr6 bearing steel billets. *Journal of Materials Processing Technology, 168*(3), 496–504.
59. Henein, H. (2010). Why is spray forming a rapid solidification process? *Materialwissenschaft und Werkstofftechnik, 41*(7), 555–561.
60. Delshad-Khatibi, P. (2014). *Microstructural investigation of D2 tool steel during rapid solidification*. PhD Thesis, University of Alberta, Edmonton, AB, Canada.
61. Henein, H., Buchoud, V., Schmidt, R., Watt, C., Malakov, D., Gandin, C.-A., et al. (2010). Droplet solidification of impulse atomized Al-0.61Fe and Al-1.9Fe. *Canadian Metallurgical Quarterly, 49*(3), 275–292.
62. Chen, J., et al. (2011). Microstructure evolution of atomized Al-0.61 wt pct Fe and Al-1.90 wt pct Fe alloys. *Metallurgical and Materials Transactions B, 42*(3), 557–567.
63. Bedel, M., Reinhart, G., Bogno, A.-A., Gandin, C.-A., Jacomet, S., Boller, E., et al. (2015). Characterization of dendrite morphologies in rapidly solidified Al-4.5wt.%Cu droplets. *Acta Materalia, 89*, 234–246.
64. Prasad, A., Henein, H., & Conlon, K. (2006). Quantification of rapid solidification events in Al-Cu powders. *Metallurgical and Materials Transactions A, 37*(5), 1589–1596.
65. Prasad, A., Henein, H., Baire, E., & Gandin, C.-A. (2006). Understanding the rapid solidification of Al-4.3Cu and Al-17Cu using X-ray tomography. *Metallurgical and Materials Transactions A, 37A*(1), 249–257.
66. Prasad, A., Henein, H., Maire, E., & Gandin, C.-A. (2004). X-ray tomography study of atomized Al-Cu droplets. *Canadian Metallurgical Quarterly, 43*(2), 273–282.
67. Delshad-Khatibi, P., Henein, H., & Phillion, A. B. (2016). Microstructure and mechanical characterization of rapidly solidified Cr-C tool steel: Annealing effects. *Advanced Powder Technology, 27*(5), 2076–2083.
68. Delshad-Khatibi, P., Phillion, A. B., & Henein, H. (2014). Microstructural investigation of D2 tool steel during rapid solidification. *Powder Metallurgy, 57*(1), 70–78.

Chapter 9
Characterization of as-Spray-Formed Products

Alwin Schulz and Chengsong Cui

9.1 Introduction

Spray-formed materials are characterized by fine and homogeneous microstructures and typically show much better hot workability than conventional materials. On the other hand, porosity is commonly present in spray-formed materials, and in the worst case cracks and significant segregation may appear. Over the past two decades, a considerable amount of research has been devoted towards characterization of spray-formed materials. A variety of techniques have been used to investigate them with regard to material homogeneity, porosity, grain structure, phases, solid solubility, inclusions and hot workability. For example, segregation has been investigated by means of SOES/GDOES on the macro scale and by means of electron microprobe analysis on the micro scale. Porosity has been measured and studied by means of the Archimedes' method, optical microscopy and image analysis techniques. Microstructures have been examined and evaluated by means of optical microscopy, XRD, SEM + EDX, etc. Size, shape, and distribution of precipitates/carbides have been quantitatively evaluated by means of image analysis. Deformation behavior of spray-formed materials has been studied by means of compression tests. In addition, soundness has been analyzed by color penetration tests. This chapter will focus on the characterization of spray-formed materials in the last two decades. The most effective and widely used techniques for the material characterization will be described, and the representative results of the investigations will be presented.

A. Schulz (✉) • C. Cui
Foundation Institute of Materials Science, Bremen, Germany
e-mail: aschulz@iwt.uni-bremen.de; cscui@iwt.uni-bremen.de

H. Henein et al. (eds.), *Metal Sprays and Spray Deposition*,
DOI 10.1007/978-3-319-52689-8_9

9.2 Porosity

Porosity is an important, but undesirable characteristic of spray-formed materials. The formation of porosity is greatly influenced by the thermal conditions of the impinging droplets and the spray-formed deposits, as reported by a number of investigators [1–8]. According to its morphologies and mechanisms of formation, porosity in spray-formed materials is generally classified into three types: interstitial porosity, gas porosity, and shrinkage.

1) Interstitial porosity, also called cold porosity, is usually observed under cold spray condition when the liquid fraction of the impinging droplets is low. In the deposition stage, solid particles impact on one another and overlap, resulting in interconnected, nonspherical pores if there is insufficient liquid to fill the interparticle or intersplat pores. Accordingly, interstitial pores are present primarily at former droplet boundaries and at grain boundaries, and exhibit a highly irregular morphology.
2) Gas porosity, often called hot porosity, is caused by gas entrapment, which is associated with the splatting process of droplets during impact [9]. The gas entrapment is dependent on the fraction of liquid, the viscosity, and the thickness of the mushy layer. Small gas pores are formed at low fraction of liquid in the mushy layer, and large spherical gas pores may be induced at high fraction of liquid due to the coalescing of trapped gas bubbles. Within the thin mushy zone, large bubbles cannot form since small bubbles may migrate a short distance to the deposit surface prior to solidification. Dissolved gas in the molten metal may also cause gas porosity since it can have the tendency to nucleate, grow, coalesce, and escape during solidification if there is a drastic reduction in gas solubility as the metal solidifies.
3) Solidification shrinkage can lead to the formation of pores as a result of the large differences in density between the liquid and the solid phases. Cai and Lavernia [10] suggested that solidification shrinkage is significant only when an excess of liquid is present in the spray upon impingement. However, this type of porosity is not so distinguished as the other types of porosity.

To summarize, porosity of spray-formed materials has been investigated extensively in the last few decades. The amount, size, distribution and morphology of porosity have been investigated based on various methods: Archimedes' method, metallography, and image analysis.

9.2.1 *Relative Density and Porosity Distribution*

Immersion method based on Archimedes' principle has been frequently used to measure the density of spray-formed materials. Based on the experimentally measured density ρ_e and the theoretical density ρ_t of the material, the relative density $\Delta\rho$ can be calculated by

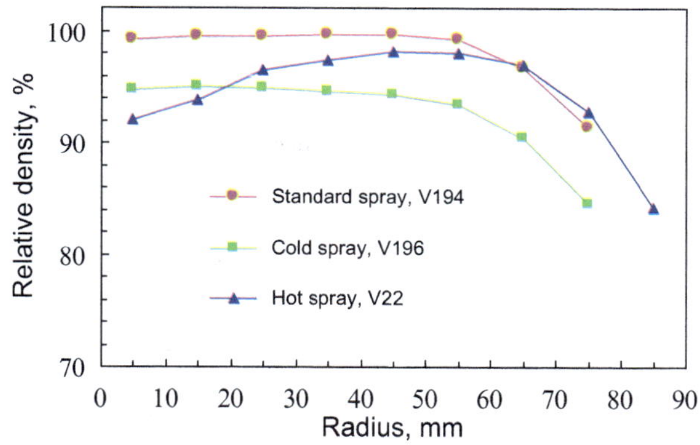

Fig. 9.1 Relative density distributions of spray-formed 100Cr6 (SAE52100) billets determined by immersion method based on Archimedes' principle, reprinted with permission from [11]

$$\Delta\rho = \frac{\rho_e}{\rho_t} \times 100\% \tag{9.1}$$

and the amount of porosity can be calculated by

$$P = \left(1 - \frac{\rho_e}{\rho_t}\right) \times 100\% \tag{9.2}$$

By measuring the density of small samples taken at different positions of spray-formed deposits, the distribution of porosity in the deposits can be determined. Figure 9.1 shows typical radial density distributions of spray-formed 100Cr6 (SAE52100) billets prepared under different spray conditions: standard spray, cold spray, and hot spray [11]. The radial density distributions of the billets are evidently associated with their cooling and solidification conditions during spray forming. A high density level, more than 99% of the theoretical density, is determined in the central area of the standard billet. The peripheral area of the billet is characterized by an increasing amount of porosity. The density of the cold sprayed billet decreases by about 5% both in the central area and in the peripheral area. The density of the hot sprayed billet exhibits a quite different distribution along the radius, i.e., low density both in the center and at the periphery.

The porosity of spray-formed deposits can also be measured by means of macro-etching. The porous areas appear darker than the dense areas when exposed to certain chemicals. Figure 9.2 shows the transverse sections of spray-formed 20MnCr5 (SAE5120) billets prepared under different thermal conditions. The samples were immersed in the etchant 1000 mL H_2O + 700 mL HCl + 300 mL HNO_3 at ambient temperature for 10 min [12]. The etched surfaces show different porosity profiles, which are very sensitive to the thermal conditions of the deposits during spray forming. Under the cold spray condition, the etched surface is dark and shows a structure similar to tree rings. This indicates that the rotating deposit surface might be more or less completely solidified before it re-entered the spay cone and the deposit surface experienced an insufficient re-melting process as it received newly deposited droplets. If the spray contains higher enthalpy, the deposit

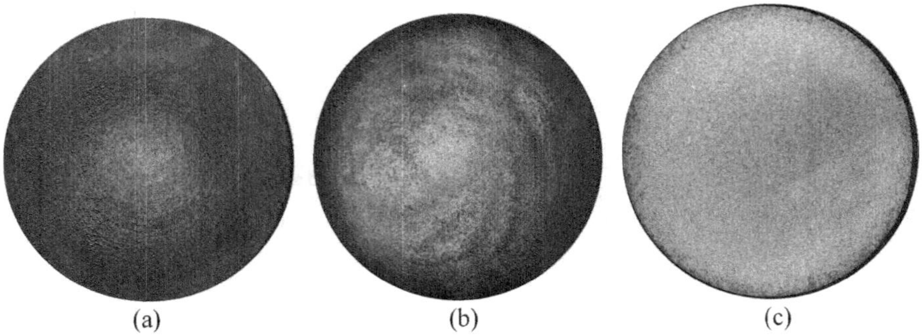

Fig. 9.2 Macrostructures of spray-formed 20MnCr5 (SAE52100) billets under different thermal conditions: (**a**) cold spray, (**b**) intermediate spray, and (**c**) relatively hot spray (disc diameter approx. 100 mm). The samples were etched with 1000 mL H_2O + 700 mL HCl + 300 mL HNO_3 for 10 min at ambient temperature [12]

surface temperature is higher and the mushy zone of the deposit becomes large, accordingly the porosity in the deposit is reduced. In the case of relatively hot spray, the bright macro-etched section indicates that the porosity has been significantly reduced.

The porosity of spray-formed deposits can be quantitatively measured by means of image analysis techniques too. For example, Achelis et al. [13] used the image analysis software Leica QWin to measure the porosity of two flat deposits of Al-Si alloy on metallographic sections. One deposit was spray formed on a relatively cold copper substrate, and another deposit was spray formed on a relatively hot aluminum substrate. The measurement results are presented in Fig. 9.3. A porous zone (3.5 mm thick) appears at the bottom of the deposit on the cold copper substrate (Fig. 9.3a). The average porosity in this zone is 7.4 $\pm$ 3.4%, and the average porosity of the whole deposit is 4.8 $\pm$ 2.0%. In contrast, the porosity zone in the deposit on the hot aluminum substrate is reduced to a mean value of 2.1 $\pm$ 1.5% (Fig. 9.3b). The average porosity of the whole deposit is 1.2 $\pm$ 0.9%. It is evident that the substrate temperature and the thermal conditions of the impinging droplets are important factors for the formation of porosity. A droplet cools by losing its heat to the substrate during impingement. Cold substrates hinder the spreading of the semi-solid droplets due to rapid solidification during impingement. Successive droplets cannot fill up the interstices between the solid particles and thus the formation of cold porosity is promoted. The results show that a reduction of porosity near the substrate surface can be achieved by increasing the substrate temperature from 100 to 400 °C.

Comparison of porosity measurement by Archimedes' principle and digital image analysis has also been made by Matthaei-Schulz et al. [14]. They measured the radial distribution of porosity in a spray-formed 100Cr6 billet, as shown in Fig. 9.4. It is obvious that digital image analysis gives more information about porosity distribution. In general, both measurement results are in good agreement. There is a significant deviation only at high porosity level (above 10%) because the immersion liquid can penetrate the open pores and cause significant measurement errors.

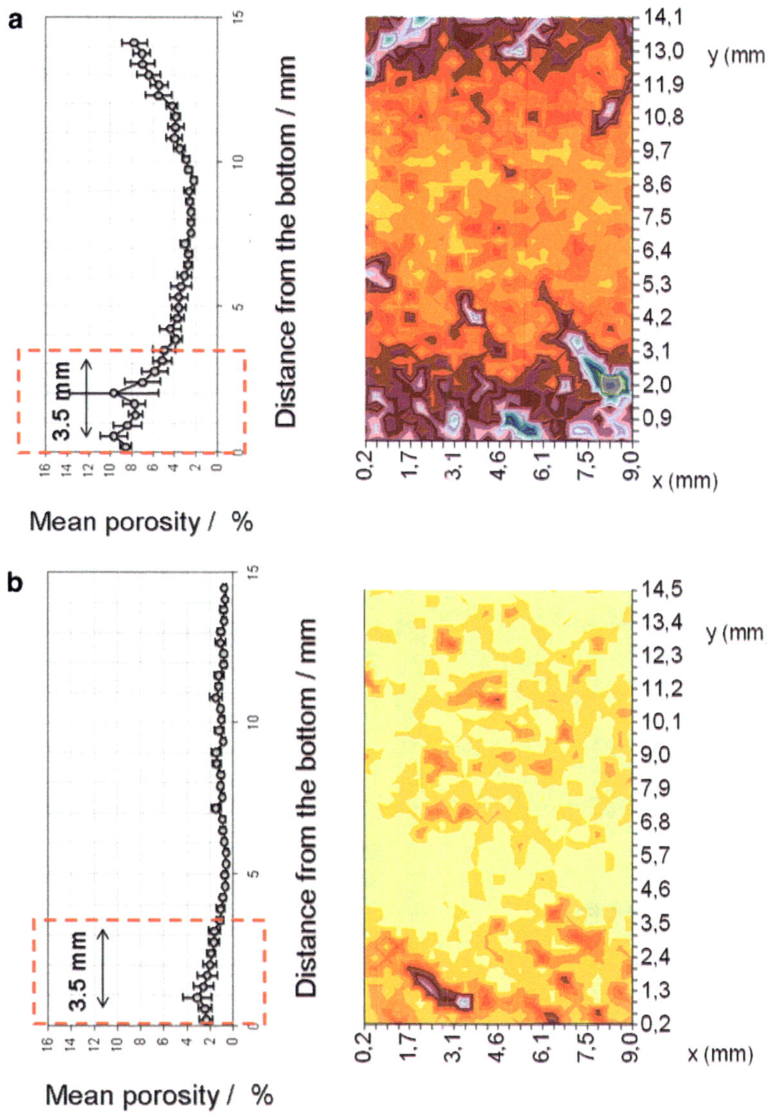

Fig. 9.3 Porosity distribution in as-spray-formed flat deposits of Al-Si alloy determined by means of image analysis technique: (**a**) substrate: T_s = 100 °C, material: Cu, thickness: 30 mm; (**b**) substrate: T_s = 400 °C, material: Al, thickness: 30 mm [13]

The penetrant dye test (also called Met-L-Check test) has also been used to determine the porosity in the surface regions of spray-formed deposits, although the test is primarily used for detecting cracks in machined parts. First, the deposit surface (after peeling) was cleaned and sprayed with a red penetrant. The penetrant was held for at least 10 min and then was wiped. If there were deep cracks or pores

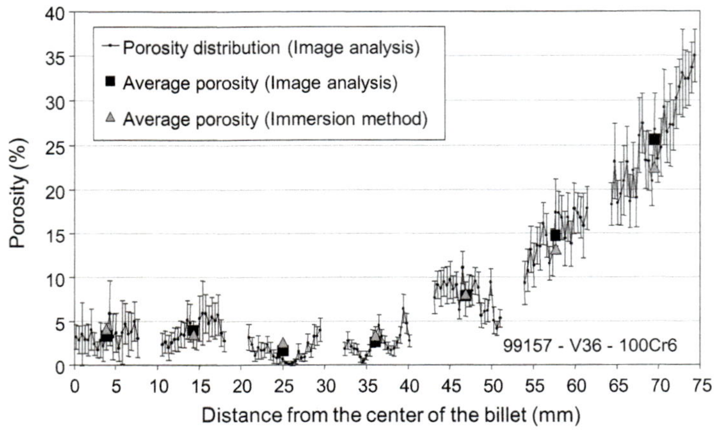

Fig. 9.4 Comparison of porosity measurement by Archimedes' principle and digital image analysis on a spray-formed 100Cr6 (SAE52100) billet [14]

Fig. 9.5 Spray-formed MMC billets after machining and Met-L-Check test (*left*: X46Cr13+TiC (overall porosity 2.1%), *right*: X2CrMo17-2-2+TiC (overall porosity 5.1%) [15]

in the surface regions, the penetrant inside these spaces would remain on the surface. Then, a white developer was sprayed to form a thin and even coating on the surface and held for 10 min. The residual red penetrant was able to diffuse into the white coating and indicate the residence of pores or cracks (as seen in Fig. 9.5).

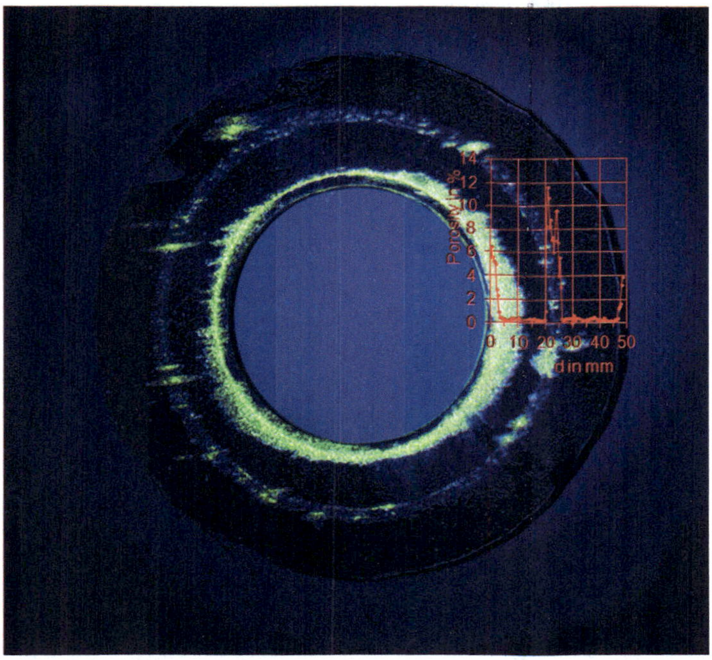

Fig. 9.6 Spray-formed bimetallic tube of H13 and Stellite 21 after fluorescent penetrant infiltration under UV-A light (the diagram shows the result of quantitative image analysis on metallographic sections from the same sample) [16]

Therefore, this method can be used to check whether the surface porous zone of billets is sufficiently removed prior to hot working.

An alternative qualitative test for visualization of pores is fluorescent inspection, a procedure similar to that of Met-L-Check test. In the case of fluorescent inspection, fluorescent penetrant is applied to the surface and allowed time to seep into pores or cracks in the material. After removal of the penetrant on the outer surface, the residual penetrant in the defects can be revealed by ultraviolet radiation with an intensity appropriate to the intent of the inspection operation. This visualization must take place in a dark room to ensure good contrast between the glow emitted by the penetrant in the porous areas and the unlit surface of the material. Figure 9.6 shows a section of a spray-formed bimetallic tube of Stellite 21 (inner ring) and hot-work steel H13 (outer ring) after fluorescent inspection (the substrate is remained inside the deposit). In comparison, the measurement results of quantitative image analysis of the deposit are combined with the fluorescent inspection result. It shows that both measurements are in a good agreement. In addition, it is also noticed that in case of large amounts of big pores, the penetrant may escape from the pores or the gap between the substrate and the Stellite 21 ring and cause measurement errors.

To avoid such uncertainties and to detect flaws close below the surface of ferromagnetic materials the magnetic particle test can be used according to

Fig. 9.7 Magnetic particle test on a spray-formed bimetallic cutting tool after cutting several AISI1045 rods of 120 mm diameter (inner part: Stellite 21, outer part: H13) [16]

ASTM E1444. In this test, a suspension of small fluorescent magnetic particles (1–12 μm) is applied to the sample surface and a strong magnetic field is applied to the sample. In ferromagnetic materials, the electromagnetic field runs parallel to the sample surface. Cracks or large pores at or close to the sample surface disturb the electromagnetic field and allow the magnetic flux to leak. If an area of flux leakage is present, the fluorescent magnetic particles will be attracted to this area, resulting in a concentration of magnetic particles. This structure becomes visible under ultraviolet light. For example, cracks and pores in a spray-formed bimetallic cutter of Stellite 21 and H13 after service are clearly revealed by the use of magnetic particle test, as shown in Fig. 9.7. A sharp line separates the ferromagnetic H13 (outer part) from the paramagnetic Stellite 21 (inner part), indicating a porous interface between the two parts. It is also seen that magnetic particles can penetrate the surface via open pores and indicate them in the ferromagnetic material and in the paramagnetic material as well.

9.2.2 Porosity Morphology

The porosity morphology can be observed clearly on the polished microsections of spray-formed deposits. As already mentioned, it is greatly influenced by the thermal conditions of the deposits during spray forming. Under intermediate thermal conditions, only isolated small pores can be found in the deposits (Fig. 9.8a). On the other hand, large amounts of interstitial porosity can be found throughout the cold sprayed deposit (Fig. 9.8b). Under high magnification observation, many separate primary particles are seen inside the irregular shaped pores, indicating that these particles have fully solidified before deposition and have not experienced re-melting and mixing processes with surrounding material due to local cold conditions. Large spherical pores become dominant in the hot sprayed deposit (Fig. 9.8c). The spherical pores primarily originate from gas entrapment, associated with the high volume of liquid in the mushy zone of the deposit as well as in the impinging droplets [11].

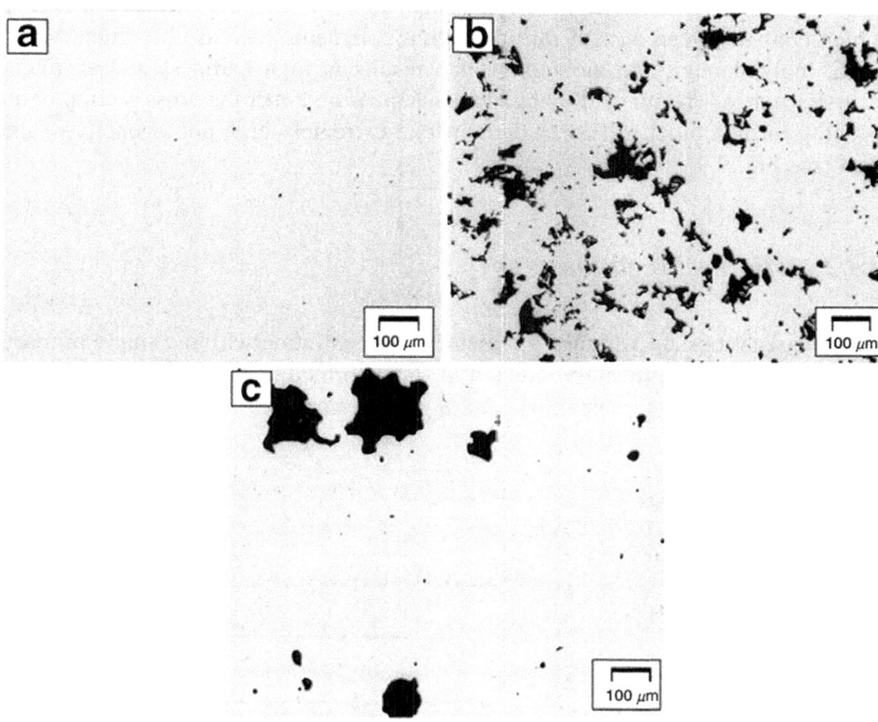

Fig. 9.8 Optical micrographs of porosity in spray-formed billets prepared under different spray conditions: (**a**) intermediate spray, (**b**) cold spray and (**c**) hot spray, reprinted with permission from [11]

9.3 Material Homogeneity and Cleanliness

9.3.1 Macrosegregation

Since most droplets in the spray cone are partially solidified before deposition and cool rapidly during impingement, macro-solute-redistribution that frequently takes place in castings can be eliminated in the mushy zone of the spray-formed deposits. Accordingly, the spray-formed deposits are typically free of macro-segregation. Schruff et al. [17] made a systematic investigation of large tool steel billets spray formed at the former Dan Spray A/S in Taastrup (Denmark) and made comparisons with ESR (electro slag remelting) steels and PM (powder metallurgy) steels. The billet dimensions were around 500 mm in diameter and 2.5 m in length, with a weight of approximately 4 tons. The studies on the spray-formed cold-work tool steel X153CrMoV12 (AISI D2) revealed a very high homogeneity, a fine and uniform carbide distribution, and a high wear resistance in combination with a high ductility. These positive results led to the development of a new spray-formed tool steel with trade name ESP23. The composition of this steel is characterized by a high content of carbon and carbide forming elements such as chromium, vanadium, molybdenum, and niobium, which results in high hardness and excellent wear resistance. The main alloying elements measured over the cross section of an as-spray-formed billet of ESP23 demonstrate extremely high homogeneity of the steel (see Fig. 9.9).

9.3.2 Microsegregation

Microsegregations are variations in element concentration within a single primary crystal. They are significantly reduced in spray-formed materials due to the large amounts of reheating, remelting, mixing, and equilibration occurring in the

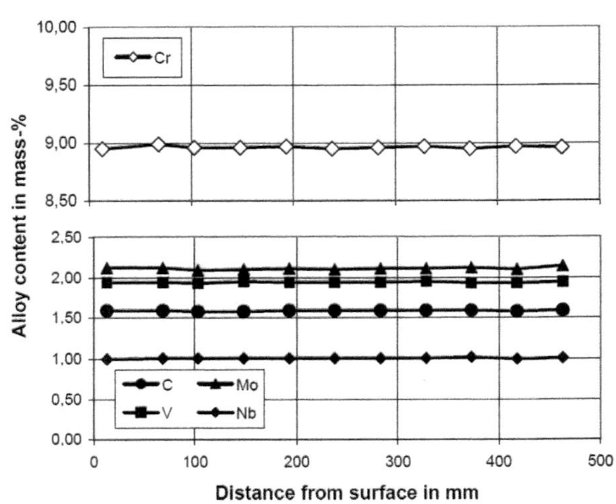

Fig. 9.9 Distributions of the main alloying elements over the cross-section of a spray-formed billet of ESP 23 (Ø500 mm), measured by means of Spark Optical Emission Spectrometry (SOES) [17]

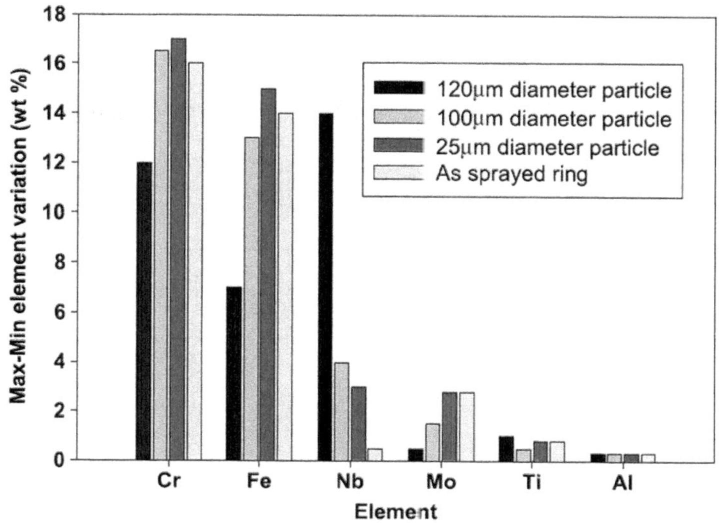

Fig. 9.10 Maximum to minimum concentration variations for alloying elements Cr, Fe, Nb, Mo, Ti, and Al in three Ni superalloy IN718 overspray particles of diameter 25, 100, and 120 μm and the corresponding spray-formed Ni superalloy ring manufactured at the same time, reprinted with permission from [18]

equilibration zone at the deposit surface during spray forming [18]. Figure 9.10 shows a histogram of the maximum to minimum variation in the concentration of alloying elements Cr, Fe, Nb, Mo, Ti, and Al in three Ni superalloy IN718 overspray particles of diameter 25, 100, and 120 μm, as well as the corresponding IN718 ring spray formed at the same time. Concentration variations were measured under identical electron probe microanalysis conditions where the incident electron beam was moved in 2 μm steps across a number of grain/dendrite boundary regions. This procedure was repeated many times on randomly selected areas of the microstructures [18]. It showed that as the particle diameter decreased and the cooling rate increased, the extent of alloying element microsegregation decreased. For all particles, Nb showed the greatest tendency for microsegregation. However, all the overspray particles showed greater Nb microsegregation than the as-spray-formed microstructure, despite the much slower cooling rate of the spray-formed ring.

9.3.3 *Cleanliness*

Rolling contact bearings belong to the most highly stressed components in mechanical engineering. This demands an extremely high quality of the bearing steels, especially high standard of cleanliness. The conventional manufacturing routes of bearing steels are ingot casting and continuous casting. The premium quality of the steels is achieved by refining technologies like electro slag remelting or vacuum arc remelting.

Table 9.1 Comparison of inclusions in conventional bearing steel and spray-formed bearing steel (*SS* sulfide stringers, *OA* oxide clusters, *OS* oxide stringer, *OG* globular oxides) [19]

Ingot casting, heat no. 645800, hot worked to dm. 180 mm | **SK2 V728 forged to a cross-section of 3573 mm²**

Number of inclusions on the test area (4815 mm²)

	Ingot casting, heat no. 645800, hot worked to dm. 180 mm									SK2 V728 forged to a cross-section of 3573 mm²								
	class 8	class 7	class 6	class 5	class 4	class 3	class 2	class 1	class 0	class 0	class 1	class 2	class 3	class 4	class 5	class 6	class 7	class 8
OG	0	0	0	0	1	1	2	17	53	17	8	1	1	0	0	0	0	0
OS	0	0	0	0	0	0	0	1	0	14	3	2	0	0	0	0	0	0
OA	0	0	0	0	2	19	59	131	181	4	0	1	0	0	0	0	0	0
SS	0	0	0	0	0	8	57	387	1186	69	2	0	0	0	0	0	0	0

Spray forming is another promising technology to achieve high standard of cleanliness since the process is normally conducted in a protective atmosphere and inclusions can be significantly refined and homogeneously distributed in the spray-formed deposits. Schulz et al. [19] investigated the number of inclusions in a spray-formed and hot worked bearing steel W7 (100CrMnMoSi8-4-6) according to German standard DIN 50602:1985-09 (microscopic examination of special steels using standard diagrams to assess the content of non-metallic inclusions). Table 9.1 classifies the inclusions according to their type and size by optical microscopic inspection at 100× magnification. The spray-formed W7 showed very few oxides and hardly any oxide agglomerations. The MnS stringers were hardly visible in the spray-formed material, too. The cleanliness of the spray-formed steel was much better than that of conventional steels. The values for oxide cleanliness of the spray-formed and forged steel are $K_0 = 0.9$ and $K_1 = 0.4$, respectively. In comparison, the values of an ingot casted and hot rolled W7 are $K_0 = 10.7$ and $K_1 = 8.3$, which are typical for high quality conventional W7.

9.4 Microstructure

9.4.1 Grain Structure and Morphology

The as-spray-formed materials are characterized by equiaxed grains and the complete absence of dendritic morphologies due to fragmentation of impinging droplets and remelting, mixing, and equilibration that occur in the mushy deposit top surface [18]. For example, a 26 kg Al-5.3Mg-1.2Li-0.28Zr alloy billet spray formed at Oxford University is shown in Fig. 9.11a, and the corresponding as-spray-formed microstructure is shown in the electron backscattered diffraction (EBSD) orientation map in Fig. 9.11b [18]. The as-spray-formed grain size was

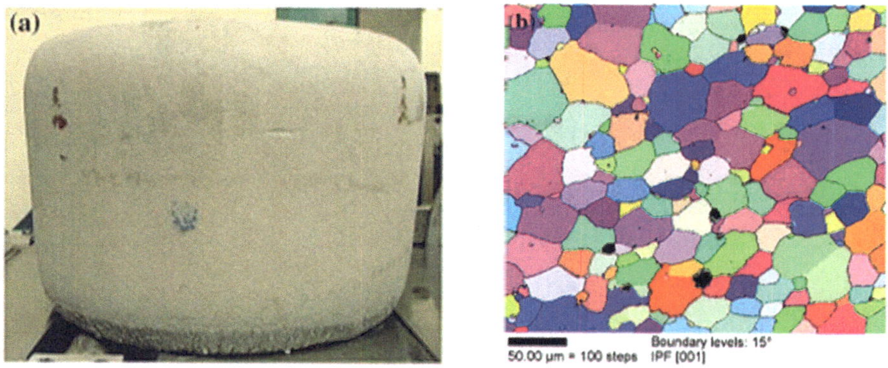

Fig. 9.11 (**a**) A typical 26 kg Al-5Mg-1.2Li-0.28Zr billet spray formed at Oxford University and (**b**) an Al-5Mg-1.2Li-0.28Zr as-sprayed microstructural orientation map obtained by EBSD showing noncolumnar/dendritic equiaxed polygonal grains characteristic of the spray forming process, reprinted with permission from [18]

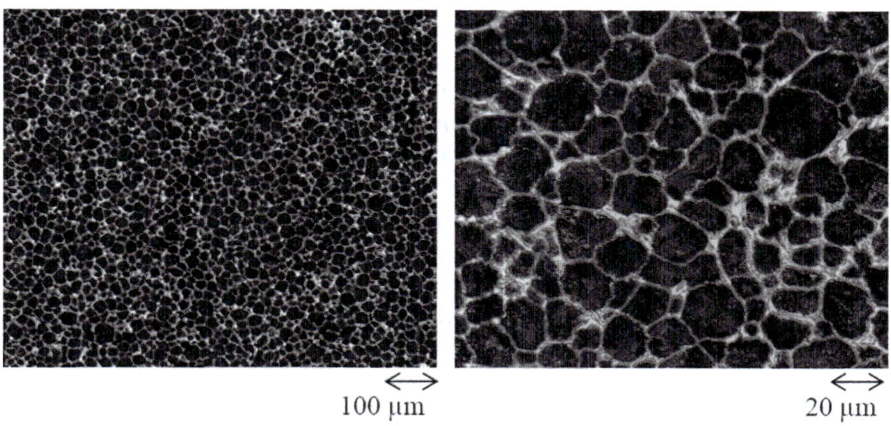

Fig. 9.12 Microstructure of the ledeburitic cold-work tool steel 1.2379 in as-sprayed condition [17]

around 15 µm. In another example, an optical micrograph of a 4 ton spray-formed billet of ledeburitic cold-work tool steel 1.2379 is shown in Fig. 9.12, which reveals a fine and equiaxied structure with a very fine ledeburitic carbide network of 5–40 µm [17].

9.4.2 Grain Size and Distribution

It has been reported that the average grain sizes of as-spray-formed materials fall mostly within the range of 10–50 µm and seldom reach 100 µm, although the spray forming experiments were conducted using different materials under various

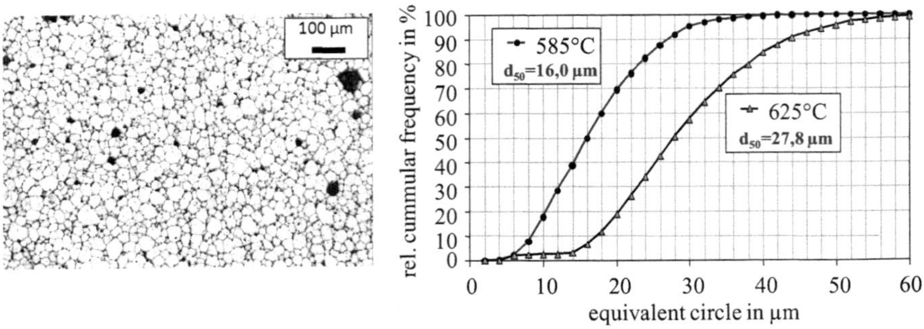

Fig. 9.13 Grain size distribution in the core of spray-formed billets of AlCu4SiMg (AA2014) [14]

processing conditions [6]. For example, Matthaei-Schulz et al. [14] used image analysis technique to measure the grain size and distribution of spray-formed AlCu4SiMg billets of about 10 kg in weight. The surface temperature of the billets was measured by a two color pyrometer. The thermal conditions of the spray forming process were adjusted via the gas to metal mass ratio (GMR). The GMR value was about 3.62 and 4.40 for the surface temperature of 625 and 585 °C, respectively. The grain size distribution in the central region of the spray-formed billets was measured by means of image processing technique. It is clearly seen that the grain size of the spray-formed materials decreased as the surface temperature decreased (see Fig. 9.13) [14].

The variation of microstructures in spray-formed deposits has also been the subject of a number of investigations [3, 20–22]. Generally, there are two distinct microstructural regions in the spray-formed deposits. In the vicinity of the cold collecting substrate there is always a chill zone (transient zone) in which the microstructure varies with the distance from the substrate. In the regions away from the substrate, however, the microstructural features remain relatively constant. Grant et al. [3], for example, studied the microstructure of a spray-formed Al-4 wt.% Cu alloy. It showed that through most of the thickness of the deposit, the microstructure was characterized by equiaxed grains. In the area close to the substrate there existed a chill zone, in which the grain boundaries were well defined. The grain size appeared to be larger in the central and upper regions than in the chill zone. Grant et al. also measured the size of equiaxied grains as a function of the distance from the substrate, as summarized in Fig. 9.14. In the area in the vicinity of the substrate, grain size increased with increasing distance. In the area away from the substrate, however, grain size became relatively constant. The transition between these two regimes occurred at a distance of approximately 20 mm. The average grain size found adjacent to the collecting substrate was approximately 30 µm, whereas that in the central region was around 75 µm [3, 6].

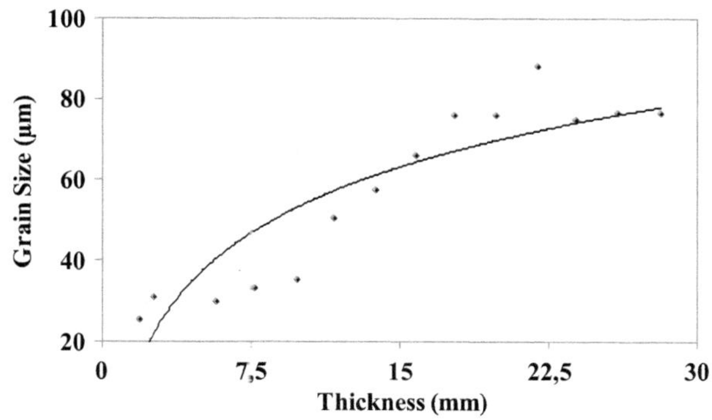

Fig. 9.14 Grain size in spray-deposited Al-4 wt.% Cu as a function of distance from the collecting substrate, adapted from [6]

9.4.3 Phase Structure and Morphology

The phases in spray-formed materials can be significantly refined, and in some cases completely suppressed. It is evident that the size of the intermetallic phases presented in spray-formed aluminum alloys is in an order of magnitude smaller when compare to those present in conventionally processed materials. Modification to the morphology of primary or secondary phases has been observed in a variety of alloys including Al-Si alloys, Mg Alloys, Ni-based superalloys, and Fe-based alloys [6].

Figure 9.15a shows an optical micrograph of a chill-cast Si-30 wt.% Al alloy comprising coarse and highly defective primary Si and α-Al/Si eutectic with significant shrinkage porosity associated with the wide freezing range of ~650 °C [18]. In contrast, Fig. 9.15b shows the microstructure of a spray-formed billet of the same alloy. The microstructure comprised a more globular, refined network of primary Si continuously penetrated by α-Al arising from a fully divorced eutectic [18, 23]. Figure 9.15c shows an EBSD orientation map for the same microstructure, indicating that the primary Si network consisted of randomly oriented Si grains with an average grain size of ~1 μm.

In another Study, Lotta and Hannula [24] investigated the hypoeutectic high chromium white iron (19% Cr, 2.5% C) processed by conventional casting and spray forming by means of scanning electron microscope (Zeiss LEO 1450) and X-ray diffraction. The microstructures of the spray-formed materials, as illustrated in Fig. 9.16a–c, were characterized by fine and evenly distributed carbides of M_7C_3-type in the matrix made of pearlite and ferrite. The carbides in the spray-formed material could be classified into two distinct groups based on morphology: (1) coarse carbides (>2 μm, occasionally up to ~30 μm) with high aspect ratio; (2) fine carbides (<2 μm) with low aspect ratio. The microstructures of the conventionally cast material, as illustrated in Fig. 9.16d, were represented by a coarse carbide network in a steel matrix. The length of the carbides ranged from micron level to up to about 100–200 μm, with the shape becoming increasingly elongated and irregular with size. The formation of carbide network structures led to large areas of matrix devoid

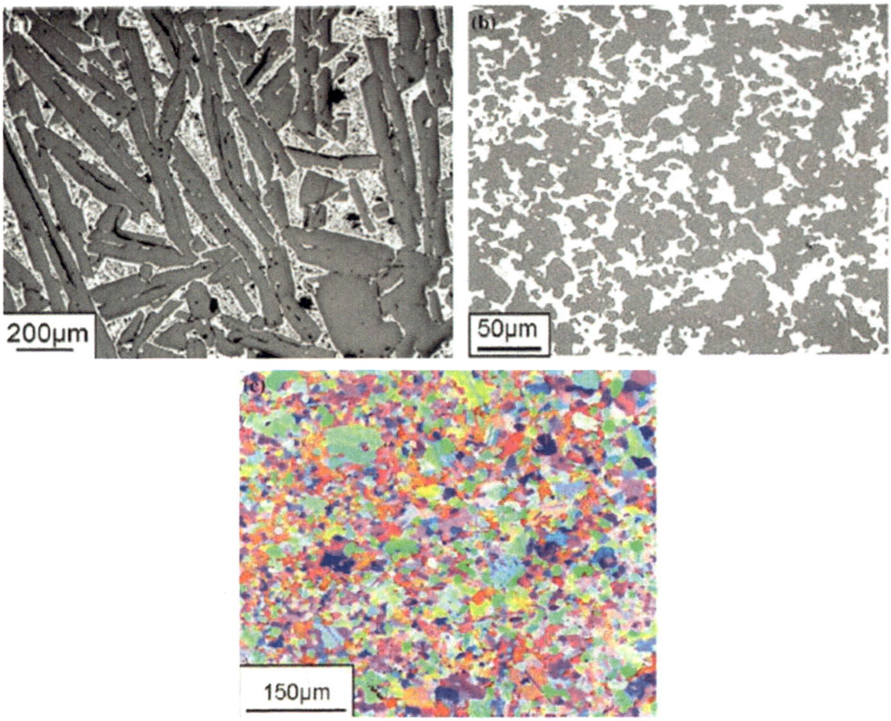

Fig. 9.15 (**a**) Optical micrograph of chill-cast Si-30 wt.% Al, (**b**) optical micrograph of as-spray-formed Si-30 wt.% Al, and (**c**) EBSD orientation map of (**b**), reprinted with permission from [18]

of carbides. X-ray diffraction confirmed that the carbides were also of M_7C_3-type (energy dispersive X-ray spectroscopy: Cr/Fe $\approx 3/2$) and indicated that the matrix consisted of retained austenite and some martensite.

The carbides in spray-formed steels can also be investigated by means of microprobe analysis. The distributions of N, C and V in spray-formed M2 high-speed steel (after soft annealing) were presented in Fig. 9.17 (bright colors represent high element concentrations). The corresponding X-ray images show that nitrogen is associated with MC type carbides but not found in carbides of M_6C type. The nitrogen concentration in this material was 810 ppm, mainly originating from a nitrogen atmosphere during melting.

Grohmann et al. [26] recently used a co-spray forming process to produce tube-shaped composite deposits to combine a cobalt-based alloy Stellite 21 with a hot working steel AISI H13. The interface between the two different materials was analyzed in terms of porosity, hardness and adhesive strength. Electron microprobe analysis of the composite deposits showed that there was a metallurgical bonding between the AISI H13 and the Stellite 21 (Fig. 9.18). This is evidently presented by the distributions of the elements Co and Fe. The diffusion zone of the elements at the interface was more than 20 μm thick. This explains the good bonding strength up to 700 MPa of the composite material.

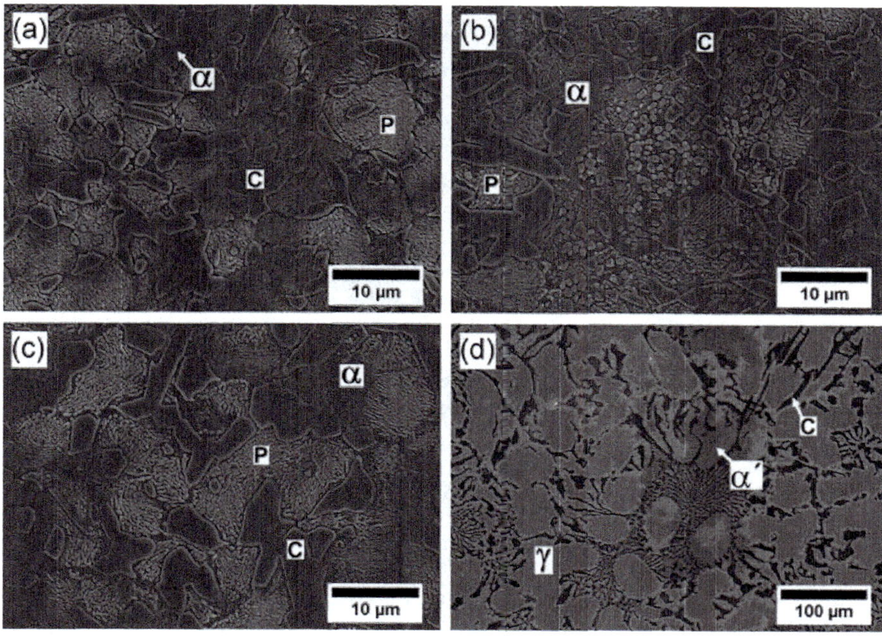

Fig. 9.16 Secondary electron micrographs showing the as-sprayed microstructure (Nital etching) of the materials sprayed with the GMRs of (**a**) 0.9, (**b**) 1.0 and (**c**) 1.1 (C = M_7C_3-type carbide, P = pearlite, α = ferrite). (**d**) A back scattered electron micrograph showing the as-cast microstructure (colloidal silica) of the conventionally cast material (C = M_7C_3-type carbide, γ = retained austenite, α′ = martensite), reprinted with permission from [24]

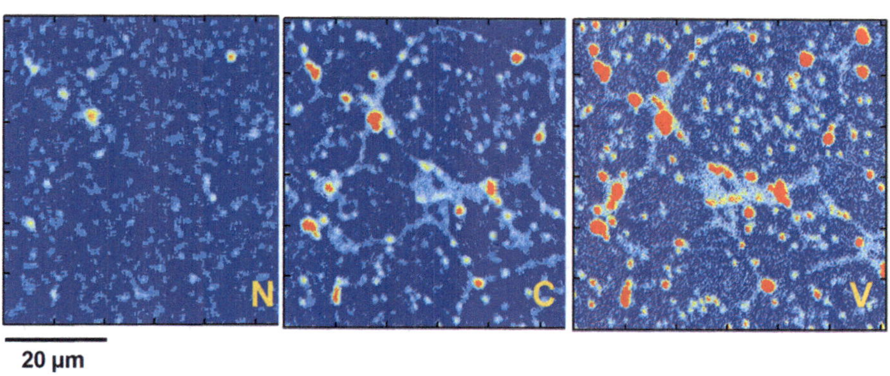

Fig. 9.17 Electron microprobe images of spray-formed AISI M2 (HS6-5-2C) melted and atomized with nitrogen as process gas [25]

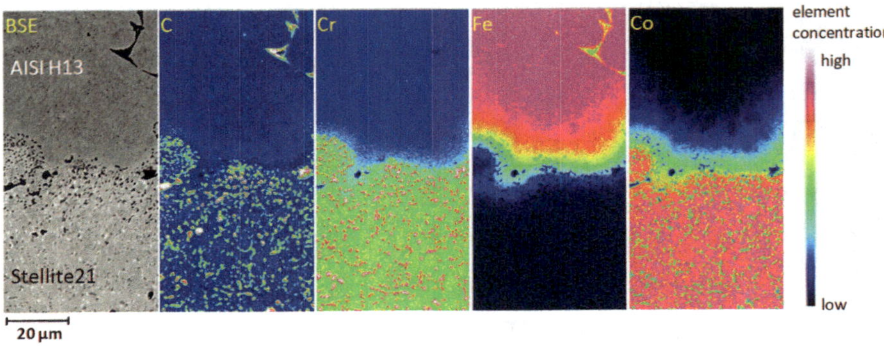

Fig. 9.18 Element distribution at the interface between AISI H13 (*top*) and Stellite 21 (*bottom*) in a spray-formed bimetallic tube, reprinted with permission from [26]

9.4.4 Phase Size, Shape and Distribution

The precipitates in spray-formed materials can be quantitatively analyzed by means of image processing technology in terms of size, shape and distribution. One example is the characterization of silicon phases in spray-formed and extruded hypereutectic Al–Si alloys by Cui et al. [27]. The image analysis of the silicon phases were made with the program Leica QWin V3.2.1 on a light microscope. Pores and silicon particles can be distinguished by grey value. The particle edges' determination is enhanced using delineation procedures. For the analysis of particle geometry, a minimum particle size of 8 pixels was applied.

In general, the silicon phases are greatly refined and uniformly distributed in the spray-formed Al-Si alloys. This improvement in the silicon phases is further facilitated by low thermal input as well as fast cooling conditions during spray forming. The silicon particles in the as-extruded Al-Si alloys appear more homogeneous and regular than those in the as-spray-formed Al-Si alloys but exhibit a certain amount of anisotropy and a tendency to preferred orientation. The silicon particles, depending on the particle size and shape, may fracture or coarsen during extrusion.

The quantitative analysis results of the size and shape of the silicon particles in the as-spray-formed Al-Si alloys and the as-extruded alloys are shown in Fig. 9.19 [27]. The median values d_{50} of the particle size and the particle shape factors, as well as the d_{16} and d_{84} values of the measurements given by the scatter bars, are plotted in the diagrams. Due to the presence of a large amount of extremely fine particles, the median length (the longest Feret) and the median equivalent diameter of the silicon particles are very small and only increase significantly for the AlSi35. The difference in particle size between the AlSi18 and AlSi25 alloys is not obvious. The influence of hot extrusion on the particle size is not clearly seen. Only for the largest aspect ratio (hot sprayed AlSi35) the value decreases after hot extrusion, which might be caused by fracture of parts of the Si particles [27].

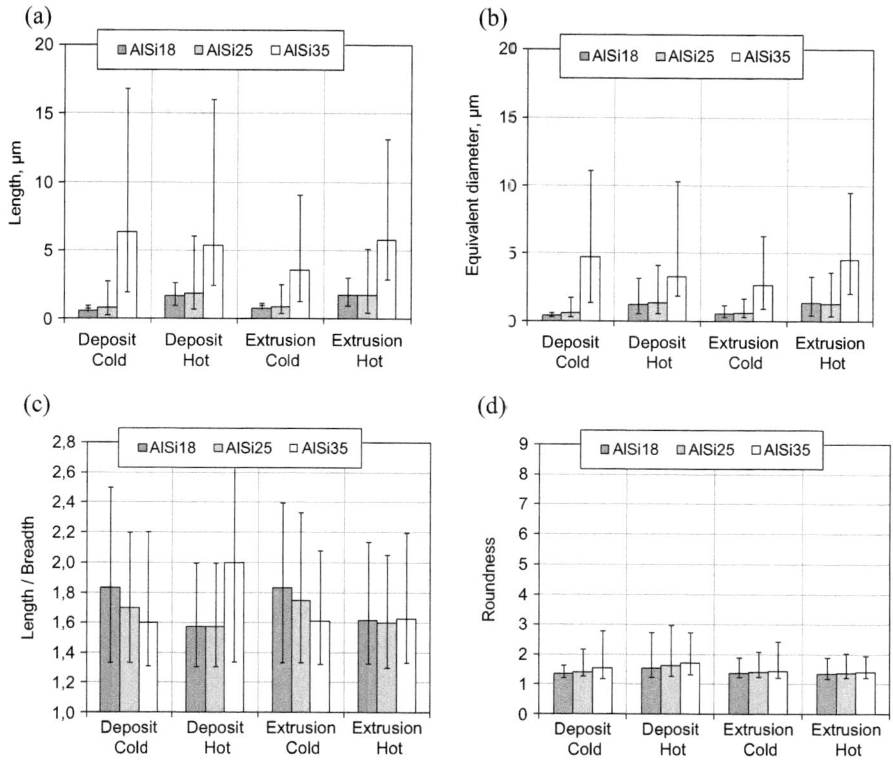

Fig. 9.19 Comparison of size and shape of all silicon particles in the as-deposited and as-extruded hypereutectic Al-Si alloys spray formed under different thermal conditions: (**a**) length, (**b**) equivalent diameter, (**c**) aspect ratio, (**d**) roundness, adapted from [27]

9.4.5 Phase Transformation

Phase transformations of spray-formed materials during subsequent hot processing have been frequently analyzed by means of differential scanning calorimetric (DSC) analysis and X-ray diffraction (XRD) [28, 29]. For example, Fig. 9.20a shows the DSC heat flow curves of spray-formed deposits and melt-spun ribbons of Al83Y5La5Ni5Co2 and Al85Y8Ni5Co2. The differential scanning calorimetric (DSC) analysis was carried out using a Perkin-Elmer diamond calorimeter at a heating rate of 40 °C/min under a continuous flow of purified argon [29]. The spray-formed Al85Y8Ni5Co2 alloy shows three exothermic peaks similar to the fully amorphous melt spun ribbons. Similarly, the spray-formed Al83Y5La5Ni5Co2 alloy shows two peaks. However, the crystallization energy for the first peak, for both deposits, is less compared to the melt-spun ribbons, which indicates partially amorphous condition of the deposits. In comparison with the first peak crystallization energy of melt-spun ribbons, the spray-formed Al83Y5La5Ni5Co2 and Al85Y8Ni5Co2 alloys exhibited an amorphous phase fraction of about 37%. Figure 9.20b shows the X-ray diffraction patterns of the melt spun ribbons, the

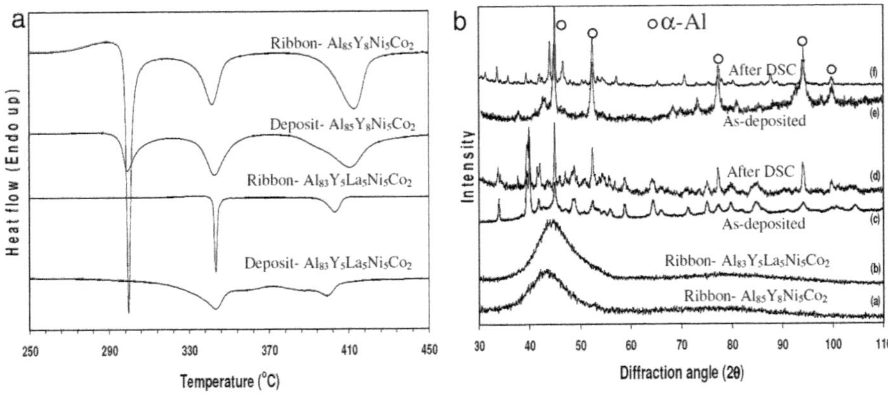

Fig. 9.20 (**a**) DSC scan curves for ribbons and deposits of Al83Y5La5Ni5Co2 and Al85Y8Ni5Co2, and (**b**) XRD diffraction patterns for Ribbon (**a**, **b**), Al83Y5La5Ni5Co2 (**c**, **d**) and Al85Y8Ni5Co2 (**e**, **f**) [29]

as-spray-formed deposits and the deposits after DSC scan up to 500 °C. The X-ray diffraction (XRD) analysis was done using Co-Kα-radiation for the as-deposited materials both before and after DSC analysis up to 500 °C [29]. The melt spun ribbons clearly shows the amorphous nature. The as-spray-formed Al85Y8Ni5Co2 alloy mainly shows the presence of α-Al and supersaturated Al_2Y and some unknown phases. The spray-formed Al83Y5La5Ni5Co2 shows α-Al, $Al_{11}La_3$, Al_3Y and some unknown phases. However, after the DSC test, the Al85Y8Ni5Co2 alloy indicated precipitation of Al_3Y and Al_9Co_2.

Differential scanning calorimetry (DSC) can also be used to measure the melting temperatures of spray-formed alloys, as reported by Cui et al. [30]. The specimens of various Al-Si based alloys were heated from room temperature to 650 °C at a heating rate of 10 K/min under argon atmosphere. The DSC curves of the spray-formed materials are shown in Fig. 9.21. The Al-Si-Zn based alloys showed single endothermic peak, and the Al-Si-Cu based alloys showed double endothermic peaks due to the eutectic precipitation of Al_2Cu. In this study, the highest endothermic peak temperature of the heating curve was defined as the liquidus temperature of the tested sample. It showed that the liquidus temperature of Al-Si-Zn based alloys decreased from 575 to 572 °C as the zinc content increased from 10 to 13 wt.%. As the copper content increased, the liquidus temperature of the Al-Si-Cu based alloys decreased too. The liquidus temperatures of the three alloys AlSi10Cu8Sn2Mg1, AlSi10Cu10Mg3, and AlSi7Cu20Sn2Mg1 were determined as 567, 549 and 528 °C, respectively. Since the heating process in the DSC analysis was in a non-equilibrium condition, the above determined liquidus temperatures of the spray-formed alloys were higher than their theoretical values. To prove this, the commercial AlSi12 filler material was also tested under the same testing condition. The eutectic AlSi12 alloy showed a melting temperature range from 583 to 597 °C, rather than the equilibrium eutectic point at 577 °C. Compared with the commercial AlSi12 alloy, the liquidus temperatures of the spray-formed alloys were at least 20 °C lower.

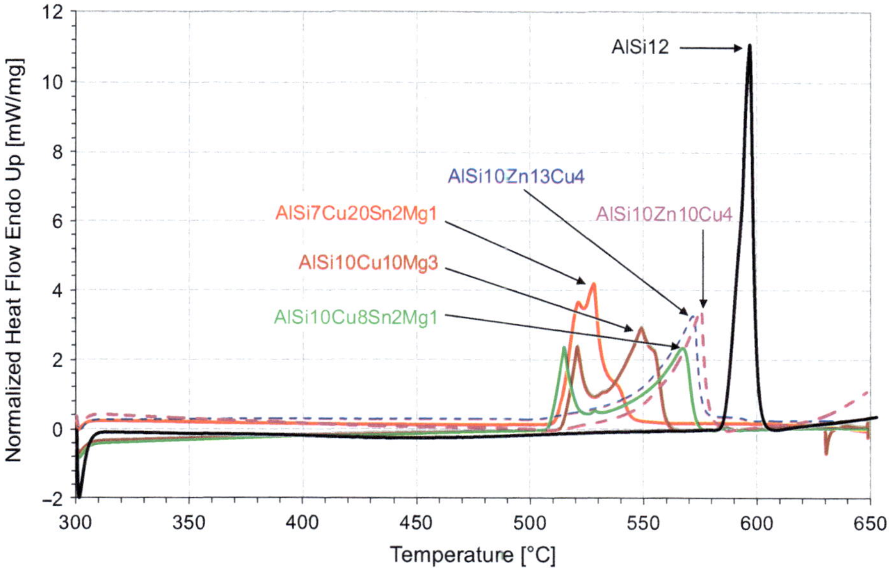

Fig. 9.21 DSC curves of spray-formed filler materials and commercial AlSi12 filler wire (heating curve, 10 K/min), reprinted with permission from [30]

9.4.6 Extension of Solid Solubility

Extension of solid solubility in a variety of spray-formed deposits has been reported by a number of investigators [6]. The results of these studies are summarized in Table 9.2. Ruhr et al. [31], for example, studied the solid solubility of Mn in α-Al in a spray-deposited Al-6.5Mn-2.35Li-0.8Zr alloy on the basis of X-ray lattice parameter measurements. The content of Mn was calculated from the dependence of the (331) Al lattice spacing on Mn content. The measured Mn content in the alloy was in the range of 2.2–2.5 wt.%, which was higher than the equilibrium solubility of Mn in α-Al of 0.9 wt.% at the eutectic temperature. Baram [32] also used X-ray crystal lattice measurements to determine the solid solubility in a spray-deposited Al-2.3Li-6.5Mn-0.65Zr alloy. He reported that in the as-spray-deposited condition, the α-Al matrix contained 2.1 wt.% of Mn which is higher than the equilibrium solid solubility of Mn in Al (0.9 wt.%). He further examined the volume fractions of the various phases in the as-spray-deposited and in the spray-deposited and isothermally heat-treated (for one hour) conditions. The volume fraction of the Al_3Zr phase increased from 1.3 to 2.6% and 3.1% following isothermal treatment at 430 °C and 530 °C, respectively. The Al_3Zr precipitation during annealing suggests the supersaturation of Zr in the as-spray-deposited condition.

Table 9.2 Solid solubility of alloying elements in primary phases as reported for spray-deposited materials, adapted from [6]

Alloy composition (wt.%)	Measured solid solubility (wt.%)	Maximum equilibrium solubility (wt.%)
Al-1.34Cu	0.7 (Cu)	5.56
Al-5.64Cu	3.05 (Cu)	5.56
Al-18.29Cu	3.42 (Cu)	5.56
Al-6.5Mn-2.35Li-0.8Zr	2.2-2.5 (Mn)	0.9
Al2.3Li-6.5Mn-0.65Zr	2.10 (Mn)	0.9
Al-Ti	2.3 (Ti)	0.74
Mg-Zn-Zr	5.2-5.4	6.2 (Zn)
Mg-Al-Zr	5.4-5.5	12.7 (Al)

9.4.7 Nitrogen Pick-Up

During spray forming process nitrogen is often used as protective gas for the spray forming chamber and as atomization gas due to low cost and easy availability. Nitrogen pick-up in steel melts in nitrogen atmosphere was reported by Schulz et al. [33]. The interaction of alloying elements with nitrogen in the melt and the changes in solubility due to alloy composition were investigated systematically in the study. The nitrogen content was measured by means of combustion technique (N/O determination). The plot of nitrogen solubility vs. the chromium equivalent is given in Fig. 9.22, showing the strong increase of solubility with increasing Cr_{eq} content. For high Cr_{eq} the solubility increases with decreasing melt temperature. The nitrogen content in the spray-formed steels is close to that limit in the case of melting in a nitrogen atmosphere, indicating that the melts were completely saturated during the melting process. In general, high nitrogen pick-up leads to reduction of toughness of the spray-formed steels due to additional precipitation of nitrides or carbonitrides [33]. Melting in an argon atmosphere can avoid nitrogen pick-up. The very short time of flight of the droplets (about 7 ms [34]) in a nitrogen atomization environment also allows very limited absorption of gas. Therefore, the nitrogen content in the spray-formed deposits from the steels melted in an argon atmosphere is drastically reduced. In this case the initial nitrogen content of the steels becomes important and it is partially conserved during the process (open dots in Fig. 9.22).

9.5 Deformation Behavior

The spray-formed deposits usually need to be deformed to eliminate the porosity and refine the microstructure. Therefore, the deformation behavior of the as-spray-formed deposits is of great interest. Compression test is one of the common methods applied to examine the deformation behavior of the spray-formed deposits. One example is the compression test of spray-formed hypereutectic Al-Si billets by

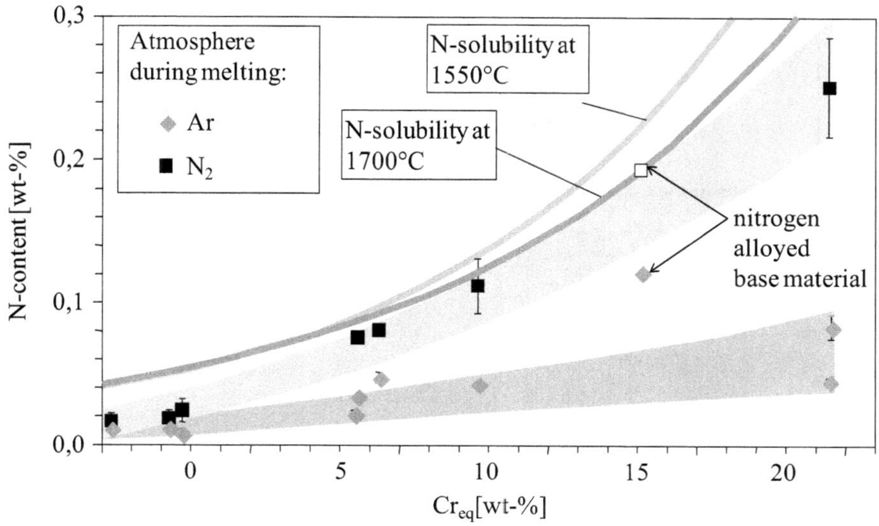

Fig. 9.22 Nitrogen concentration of the spray formed alloys vs. their Cr equivalent (for calculation of nitrogen solubility at 1550 and 1700 °C and 0.1 MPa N_2-pressure, reprinted with permission from [33]

the use of deformation dilatometer [35]. For all the spray-formed alloys their flow curves were determined in the temperature range from 300 °C to 500 °C with strain rates between 0.01 s^{-1} and 7.5 s^{-1}, i.e., the parameter range of hot extrusion (see Fig. 9.23). The flow stress increases with decreasing compression temperature and increasing strain rate. Higher silicon content in the alloys also leads to higher flow stress during deformation. The flow curves determined from the compression tests indicate that the deformation of the materials is controlled by two competing mechanisms: strain hardening, and flow softening. Particle damage during the deformation may have an influence on the flow curves of the alloys with large silicon particles. Based on the flow curves obtained from the compression tests and knowledge of aluminum extrusion, the spray-formed hypereutectic Al-Si alloy billets have been hot extruded into wires with a high area reduction ratio around 189. Since primary silicon particles were greatly refined and uniformly distributed in the spray-formed materials, the heavy deformations of the spray-formed Al–Si alloys containing high amount of silicon were successfully performed [35].

9.6 Residual Stresses and Strains

Residual stresses are also commonly present in spray-formed deposits. It is suggested that residual stresses in spray-formed deposits arise from two primary sources: (1) a thermal gradient in a spray-formed deposit during cooling from the

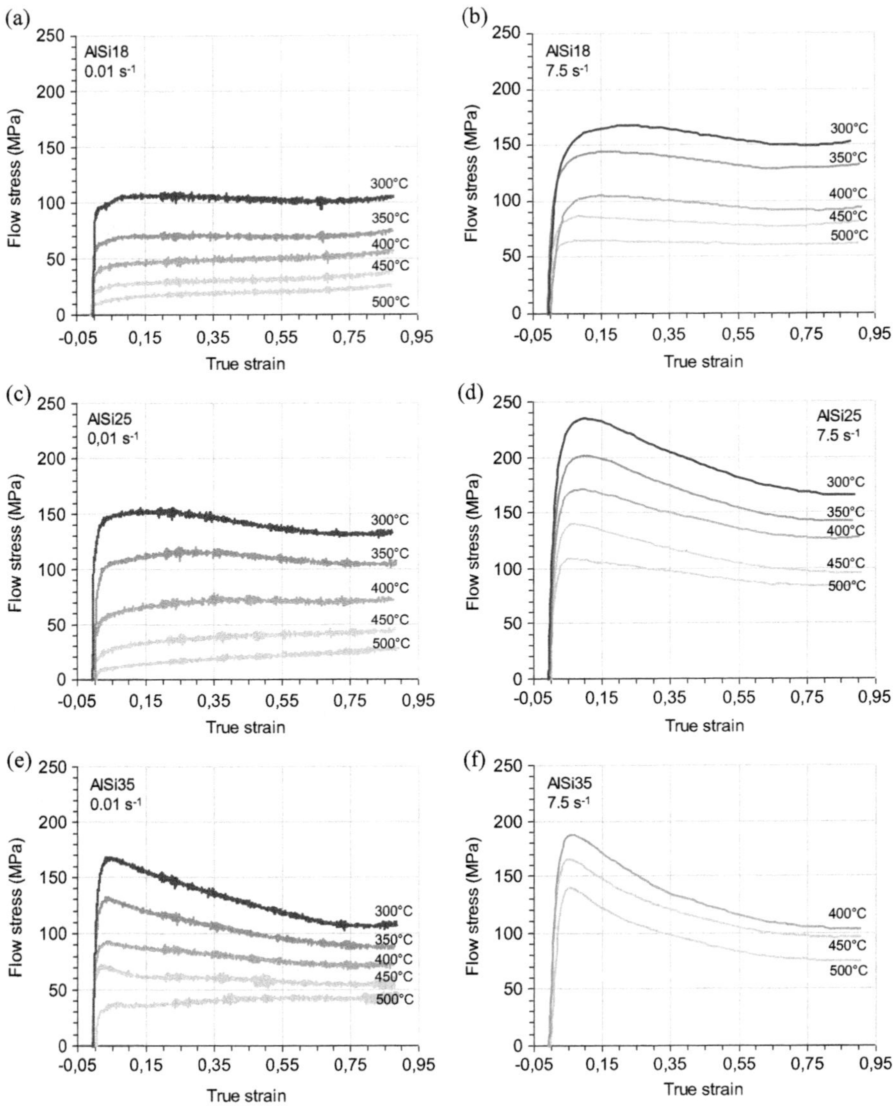

Fig. 9.23 Flow curves of spray-formed hypereutectic Al-Si alloys determined by compression tests at various temperatures and strain rates: (**a**) AlSi18, strain rate 0.01 s^{-1}; (**b**) AlSi18, strain rate 7.5 s^{-1}; (**c**) AlSi25, strain rate 0.01 s^{-1}; (**d**) AlSi25, strain rate 7.5 s^{-1}; (**e**) AlSi35, strain rate 0.01 s^{-1}; and (**f**) AlSi35, strain rate 7.5 s^{-1}, reprinted with permission from [35]

deposition temperature to ambient temperature may produce inhomogeneous plastic strain and thermal residual stresses in the deposit; (2) liquid–solid and solid–solid phase transformations during cooling process can also play an important role in the development of residual stresses due to non-uniform volumetric changes which vary with phase transformation time and fraction [36, 37]. Residual stresses

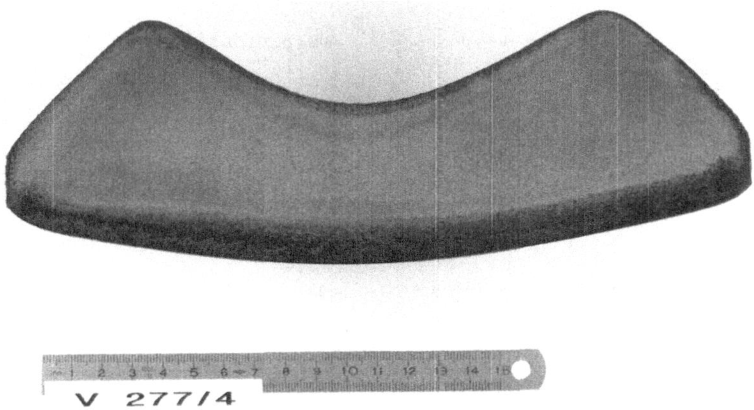

Fig. 9.24 Transverse section of a spray-formed plate of HS6-5-2C (GMR = 0.93) after macro-etching [38]

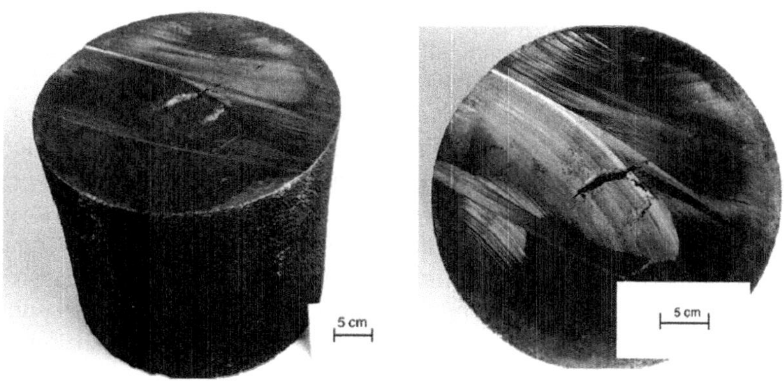

Fig. 9.25 Hot cracks in the top of a spray-formed 100Cr6 billet prepared under hot spray condition, reprinted with permission from [11]

in spray-formed deposits may cause undesirable effects such as shape distortion and cracking (see Figs. 9.24 and 9.25). It is therefore necessary to determine the residual stresses in spray-formed deposits and find appropriate approaches to control the residual stresses.

Residual stresses in spray-formed deposits can be investigated with both numerical calculation and experimental methods [36, 37, 39, 40]. For example, Hu et al. used finite-element method to calculate the thermal residual stresses in a spray-formed A2 tool steel deposit of Gaussian shape [36]. The finite-element analysis was conducted by using a commercial code ABAQUS to solve the coupled temperature and displacement problems. The von Mises stress, axial stress, and radial stress were calculated using a thermo-elasto-plastic finite-element method. Meanwhile, they also used X-ray diffraction method to determine the residual stresses in

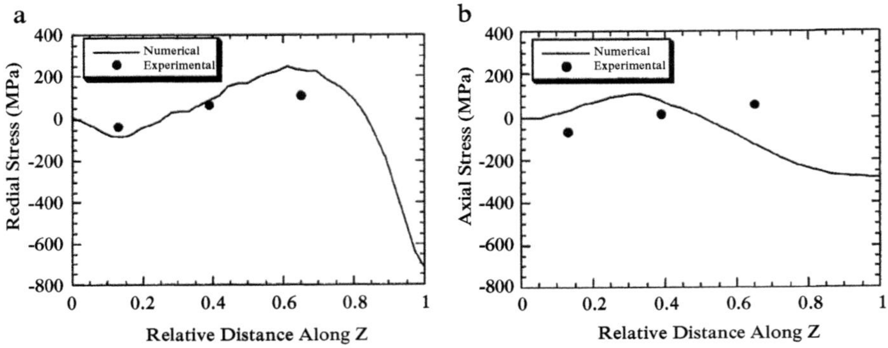

Fig. 9.26 Numerical and experimental residual stresses along the centerline as a function of relative distance from the top to bottom of the spray-formed material using a copper substrate: (**a**) radial stresses, (**b**) axial stresses, reprinted with permission from [36]

the deposit, based on the measurement of lattice spacing of particularly oriented sets of grains using Bragg's law.

Figure 9.26 shows the numerically and experimentally determined residual stresses along the centerline of the spray-formed deposit using a copper substrate [36]. The calculated results show that compressive radial stresses are present in the top and bottom regions of the deposit and tensile stresses in the middle region. The maximum compressive stress is present at the bottom of the deposit where the material experiences the highest cooling rate. The calculated radial stresses are tensile in the upper half region of the deposit and they change to compressive stresses in the lower half region. Similarly, the maximum compressive residual stresses occurs at the bottom due to rapid heat transfer to a cold copper substrate, and tensile stresses form in the middle region where the material experienced the slowest cooling rate. The calculation results are in general agreement with the experimental results.

Although the residual stresses in the spray-formed deposits can be examined by the use of X-ray diffraction techniques, the measurements were confined to the near-surface region due to the micrometer penetration depth of the X-rays [41]. Neutron diffraction (ND) offers more capability since neutrons can penetrate most metallic materials up to a few centimeters [42]. The recent development of third generation neutron strain scanners leads to at least an order of magnitude improvement in counting times, and such scanners have been successfully used to characterize the residual stress in "thick" engineering materials [43].

Most recently, Lee et al. [44] reported the use of neutron diffraction method to characterize the residual stress distribution through the deposit thickness. The residual stress distributions in an as-spray-formed tube-shaped steel deposit (20 mm thick) showed tensile residual stresses up to 220 MPa in dense regions and compressive stresses of up to 200 MPa in more porous regions, as the deposition conditions changed. There was a steep stress gradient in the dense-to-porous transition region. The residual stress distributions predicted by the FE model agreed

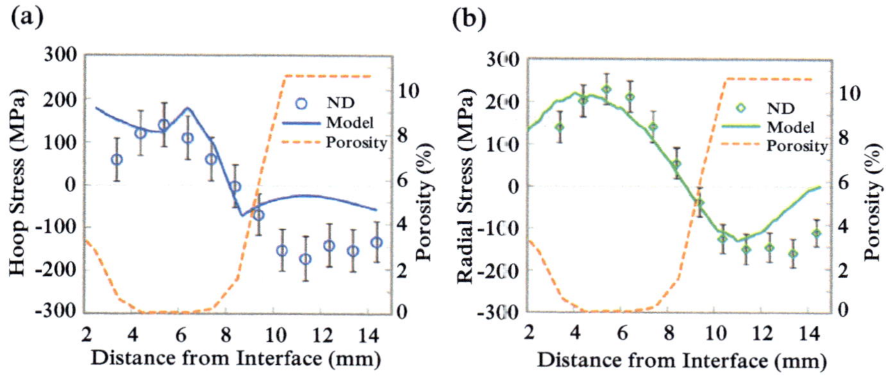

Fig. 9.27 (**a**) Hoop and (**b**) radial residual stress distributions in the as-sprayed preform from ND measurements and FE modeling with CT porosity measurements, reprinted with permission from [44] (http://dx.doi.org/10.1016/j.scriptamat.2014.12.019)

relatively well with the neutron diffraction measurement results, and highlighted the interdependence of local thermal history, phase fraction and porosity on residual stress profile development in as-spray-formed deposits, as shown in Fig. 9.27 [44].

9.7 List of Symbols

Greek

Symbol	Description
ρ_e	Experimentally measured density
ρ_t	Theoretical density
$\Delta\rho$	Relative density

References

1. Annavarapu, S., & Doherty, R. D. (1993). Evolution of microstructure in spray casting. *International Journal of Powder Metallurgy, 29*(4), 331–343.
2. Ellent, N., Stelling, O., Uhlenwinkel, V., von Hehl, A., & Krug, P. (2010). Influence of spray forming process parameters on the microstructure and porosity of Mg2Si rich aluminum alloys. *Materialwissenschaft und Werkstofftechnik, 41*(7), 532–540.
3. Grant, P. S., Kim, W. T., & Cantor, B. (1991). Spray forming of aluminum-copper alloys. *Materials Science and Engineering A, 134*(1), 1111–1114.
4. Gupta, M., Mohamed, F. A., & Lavernia, E. J. (1992). The Effect of ceramic reinforcements during spray atomization and codeposition of metal matrix composites: Part I. Heat Transfer. *Metallurgical Transactions A, 23*(3), 831–843.
5. Hu, H., Lee, Z. H., White, D. R., & Lavernia, E. J. (2000). On the evolution of porosity in spray-deposited tool steels. *Metallurgical and Materials Transactions A, 31*(3), 725–735.

6. Lavernia, E. J., & Wu, Y. (1996). *Spray atomization and deposition*. New York: John Wiley & Sons, Inc.
7. Mathur, P., Apelian, D., & Lawley, A. (1989). Analysis of the spray deposition process. *Acta Metallurgica, 37*(2), 429–443.
8. Uhlenwinkel, V., & Ellendt, N. (2007). Porosity in spray-formed materials. *Materials Science Forum, 534–536*, 429–432.
9. Liu, H., Lavernia, E. J., & Rangel, R. H. (1995). Modeling of molten droplet impingement on a non-flat surface. *Acta Metallurgica et Materialia, 43*, 2053–2072.
10. Cai, W. D., & Lavernia, E. J. (1997). Modeling of porosity during spray forming. *Materials Science and Engineering A, 226–228*, 8–12.
11. Cui, C., Fritsching, U., Schulz, A., Tinscher, R., Bauckhage, K., & Mayr, P. (2005). Spray forming of homogeneous 100Cr6 bearing steel billets. *Journal of Materials Processing Technology, 168*(23), 496–504.
12. Cui, C., Schulz, A., & Fritsching, U. (2007). Final Report of SFB570 "Distortion engineering", Project A2 Report, University Bremen (p. 154).
13. Achelis, L., Uhlenwinkel, V., Ristau, R., & Krug, P. (2009). Transient temperatures and microstructure of spray formed aluminium alloy AlSi sheets. In: *Proceedings of 4th International Conference on Spray Deposition and Melt Atomization and 7th International Conference on Spray Forming–SDMA 2009-ICSF VII*, 7–9 September 2009, Bremen, Germany (CD Version).
14. Matthaei-Schulz, E., Schulz, A., & Mayr, P. (2001). Gefügeauswertung an sprühkompaktierten Werkstoffen mit bildanalytischen Methoden. Kolloquium "Sprühkompaktieren" Band 5, SFB 372 an der Universität Bremen (pp. 179–191).
15. Cui, C., Schulz, A., & Zobel, F. (2010). Einbringung von Hartstoffen in Stahlmatrizes mittels Sprühkompaktieren zur Verwendung als Werkstoff für korrosionsbeständige Maschinenmesser. Final Report Project No. AiF 238ZN.
16. Grohmann, O., Meyer, C., Schulz, A., & Heinzel, C. (2014). Abscherwerkzeuge aus sprühkompaktierten Stelliten. Final report AiF-Project No. 17464 N.
17. Schruff, I., Schüler, V., & Spiegelhauer, C. (2003). Spray forming–the new technology or the production of high-grade tool steels. In: K. Bauckhage, U. Fritsching, V. Uhlenwinkel, J. Ziesenis, & A. Leatham (Eds.), *SDMA 2003 and ICSF V Vol. 2* (pp. 5–11). Bremen: Universität Bremen.
18. Grant, P. S. (2007). Solidification in Spray Forming. *Metallurgical and Materials Transactions A, 38*(7), 1520–1529.
19. Schulz, A., Trojahn, W., Meyer, C., & Uhlenwinkel, V. (2014). Bearing steel with high cleanliness produced by spray forming. *HTM Journal of Heat Treatment and Materials, 69* (6), 368–376.
20. Annavarapu, S., Apelian, D., & Lawley, A. (1988). Processing effects in spray casting of steel strip. *Metallurgical Transactions A, 19*(12), 3077–3086.
21. Bewlay, B. P., & Cantor, B. (1991). The relationship between thermal history and microstructure in spray-deposited tin-lead alloys. *Journal of Materials Research, 6*(7), 1433–1454.
22. Grant, P. S., & Cantor, B. (1991). Modelling of spray forming. *Cast Metals, 4*(3), 140–151.
23. Hogg, S. C., Lambourne, A., Ogilvy, A., & Grant, P. S. (2006). Microstructural characterisation of spray formed Si-30Al for thermal management applications. *Scripta Materialia, 55*(1), 111–114.
24. Lotta, J., & Hannula, S.-P. (2014). Microstructure and wear resistance of spray formed and conventionally cast high-chromium white iron. *Materialwissenschaft und Werkstofftechnik, 45* (8), 727–735.
25. Schulz, A., Spangel, S., Schneider, R., Viale, D., & Bertrand, C. (2004). The investigation and evaluation of the spray-formed states of tool steels to make a catalogue of materials properties available. *Final Report of ECSC Project No. 7210-PR-173, European Commission Technical Steel Research Series EUR 20948—Special and Alloy Steels*. Luxembourg: Office for Official Publications of the EC (ISSN: 1018-5593, ISBN: 92-894-7746-6).

26. Grohmann, O., Meyer, C., Schulz, A., Uhlenwinkel, V., & Heinzel, C. (2014). Analysis of hot forming tool generated via co-spray forming. *HTM Journal of Heat Treatment and Materials, 69*(4), 235–240.
27. Cui, C., Schulz, A., Matthaei-Schulz, E., & Zoch, H.-W. (2009). Characterization of silicon phases in spray-formed and extruded hypereutectic Al–Si alloys by image analysis. *Journal of Materials Science, 44*(18), 4814–4826.
28. Guo, M. L. T., Tsao, C. Y. A., Huang, J. C., & Jang, J. S. C. (2006). Microstructural evolution in spray-formed and melt-spun Al85Nd5Ni10 bulk hybrid composites. *Intermetallics, 14*(8–9), 1069–1074.
29. Srivastava, V. C., Surreddi, K. B., Scudino, S., Schowalter, M., Uhlenwinkel, V., Schulz, A., Eckert, J., Rosenauer, A., & Zoch, H.-W. (2009). Novel microstructural characteristics and properties of spray formed Al-RE-TM based alloys. In: *Proceeding of 4th International Conference on Spray Deposition and Melt Atomization and 7th International Conference on Spray Forming–SDMA 2009-ICSF VII*, 7–9 September 2009, Bremen, Germany (CD Version).
30. Cui, C., Schulz, A., Achelis, L., Uhlenwinkel, V., Leopold, H., Piwek, V., Tang, Z., & Seefeld, T. (2014). Development of low-melting-point filler materials for laser beam brazing of aluminum alloys. *Materialwissenschaft und Werkstofftechnik, 45*(8), 717–726.
31. Ruhr, M., Lavernia, E. J., & Baram, J. C. (1990). Extended Al(Mn) solution in a rapidly solidified Al-Li-Mn-Zr alloy. *Metallurgical Transactions A, 21*(6), 1785–1789.
32. Baram, J. (1991). Structure and properties of a rapidly solidified Al-Li-Mn-Zr alloy for high-temperature applications: Part II. Spray atomization and deposition processing. *Metallurgical Transactions A, 22*(10), 2515–2522.
33. Schulz, A., Uhlenwinkel, V., Bertrand, C., Kohlmann, R., Kulmburg, A., Oldewurtel, A., Schneider, R., & Viale, D. (2004). Nitrogen pick-up during spray forming of high-alloyed steels and its influence on microstructure and properties of the final products. *Materials Science and Engineering A, 383*(1), 58–68.
34. Ziesenis, J., & Bauckhage, K. (2001). *Partikelanalyse Mittels Streulichtmessungen im unbewegten und bewegten Sprühkegel. Kolloquium "Sprühkompaktieren", Vol. 5, SFB 372* (p. 97). Bremen: Universität Bremen.
35. Cui, C., Schulz, A., Epp, J., & Zoch, H. W. (2010). Deformation behavior of spray-formed hypereutectic Al–Si alloys. *Journal of Materials Science, 45*(10), 2798–2807.
36. Hu, H. M., Lavernia, E. J., Lee, Z. H., & White, D. R. (1999). Residual stresses in spray-formed A2 tool steel. *Journal of Materials Research, 14*(12), 4521–4530.
37. Ristau, R., Becker, A., Uhlenwinkel, V., & Kienzler, R. (2010). Simulation of temperatures, residual stresses, and porosity in spray formed super alloys tubes. In: *7th International Symposium on Superalloy 718 and Derivatives* (pp. 470–485). John Wiley & Sons, Inc., Hoboken, NJ
38. Schulz, A., Rabitsch, R., Stocchi, D., Viale, D., & Montero, M.C. (2007). Evaluation of the capability of spray forming for the production of very high alloyed steels without necessity of heavy hot working. *European Commission Technical Steel Research Series Report EUR 22438—Physical Metallurgy and Design of New Generic Steel Grades*. Luxembourg: Office for Official Publications of the European Communities. (ISBN 978-92-79-04955-2).
39. Ho, S., & Lavernia, E. J. (1997). The effect of heat transfer coefficient on thermal residual stresses in spray deposited materials. *Scripta Materialia, 36*(3), 283–290.
40. Ho, S., & Lavernia, E. J. (1997). Investigation of thermal residual stresses in layered composite using the finite element method and x-ray diffraction. *Metallurgical and Materials Transactions B, 28*(6), 969–978.
41. Withers, P. J., & Bhadeshia, H. K. D. H. (2001). Residual stress. Part 1—Measurement techniques. *Materials Science and Technology, 17*(4), 355–365.
42. Santisteban, J. R., Daymond, M. R., James, J. A., & Edwards, L. (2006). ENGIN-X: a third-generation neutron strain scanner. *Journal of Applied Crystallography, 39*(6), 812–825.

43. Zhang, S. Y., Evans, A., Eren, E., Chen, B., Pavier, M., Wang, Y., Pierret, S., Moat, R., & Mori, B. (2013). ENGIN-X—instrument for materials science and engineering research. *Neutron News, 24*(3), 22–26.
44. Lee, T. L., Mi, J., Zhao, S. L., Fan, J. F., Zhang, S. Y., Kabra, S., & Grant, P. S. (2015). Characterization of the residual stresses in spray-formed steels using neutron diffraction. *Scripta Materialia, 100*, 82–85.

Chapter 10
Spray Forming of Aluminium Alloys

Peter Krug

10.1 Introduction

As mentioned in Chap. 1, Prof. Singer claimed in 1990 four mean targets to make spray forming a success story, which are valid for aluminium alloys, too [1]. These four targets are:

- Spray forming must first concentrate on premium products.
- Spray forming may in some cases defeat the conventional product on cost because of *near-to-net shape manufacture*.
- Make products that have *better properties* than competitive ones.
- Make products that cannot be made by any alternative means. This target will be summarized under the headline "*Impossible alloys*".

In the following subchapters these four major points will be examined more closely. Another point is the intrinsic porosity due to the atomizing process which makes a decided *post processing* strategy mandatory. This issue will be treated first.

10.2 Post Processing

Since spray forming delivers a non-equilibrium material which contains large amounts of metastable phases one have to keep in mind, that reheating—depending on time and temperature—enables the re-establishment of equilibrium states, e.g. undesired morphologies, coarsening, precipitations etc., thus limiting post processing like hot extrusions or hot pressing. Since consolidation processes have

P. Krug (✉)
Faculty of Automotive Systems and Production, Cologne University of Applied Sciences, Köln, Germany
e-mail: peter.krug@fh-koeln.de

H. Henein et al. (eds.), *Metal Sprays and Spray Deposition*,
DOI 10.1007/978-3-319-52689-8_10

to be applied to eliminate porosity, a decrease of properties will be accompanied by these methods. Subsequent heat treating and fusing processes should be planned with a minimum heat input and limited duration. In case of welding, friction stir welding (FSW), electron beam welding or laser beam welding will be preferred, since melting of the material is completely suppressed or reduced to a minimum [2]. In contrast, FSW will refine microstructure additionally. Figures 10.1 and 10.2 show friction stir welded spray formed alloy and the resulting microstructure [3].

Fig. 10.1 Friction stir welded profiles (Al-20Si-5Fe-2Ni)

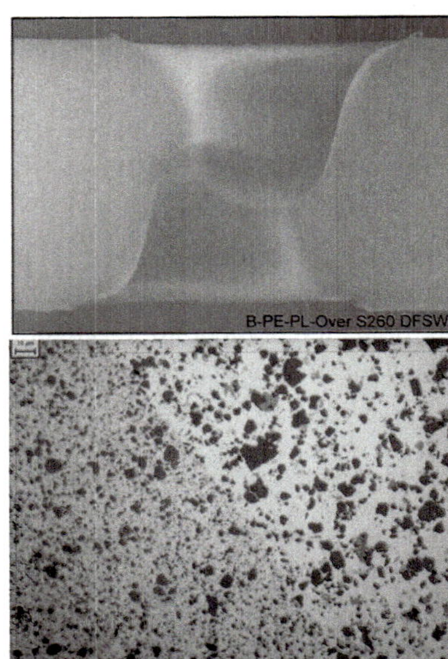

Fig. 10.2 Top overlapping weld seams; bottom microstructure of weld seam (*lower left*)

Fig. 10.3 Spray forming unit at PEAK France at St. Avold, France showing spray forms Al-25Si-4Cu-1Mg billet 2.5 m height and 300 mm diameter [5]

High consumption of atomizer gas (usually nitrogen) and deposition rate are two main factors contributing to high manufacturing cost. Both, gas consumption and the relationship of deposited material to overspray powder can be improved by reinjection of overspray powder into the atomizing process. Since the reinjected powder is not preheated it will serve as an additional heat sink, thus atomizer gas can be saved without decreasing the mechanical properties. Investigations at PEAK Werkstoff GmbH revealed no difference between spray formed material with and without overspray powder reinjection [4]. Figure 10.3 shows a spray formed aluminum billet, which is produced with such overspray powder reinjection.

Since porosity usually is measured by archimedian principle no distinction is made between atomizing gas entrapment—mainly nitrogen in aluminium spray forming, voids or micro shrinkage due to a lack of melt feeding, and hydrogen due to a dramatically reduced solubility in the solid state in comparison to the aluminium melt. These defects—either hot or cold porosity—is responsible for the need of a subsequent processing of the spray formed products since it will be detrimental to mechanical properties, especially to fatigue behavior based on the fine microstructure achieved by the high cooling rate. Usually porosity ranges between 1 to 12% [6–19] depending on spray forming parameters and/or alloy type. In Al-Si alloys lower porosity values were found for lower Si-content [20]. In industrial production the top and bottom section of spray formed billets will be cut off and scrapped. In case of flat deposits [21] has shown, that preheating of the deposition substrate and the substrate thickness itself will have an influence on porosity formation. Also the GMR (gas to metal flow rate) ratio is the most important parameter [22], but the variation of this ratio is limited to the cooling rate needed for certain alloys, which transfers to a high GMR. In addition, increasing melt superheat will lead to higher porosity, too [23]. The authors show also the influence of scanning frequency of atomizer nozzle. Optimum scanning frequency and rotational speed of the substrate plate will lead to a uniform deposition of the spray cone onto the substrate, thus avoiding periodic change in heat transfer. In addition, in [23] the enthalpy-flow-to-gas-flow ratio (EGR) is introduced which combines the GMR with the enthalpy difference of melt and solid state. They showed for a Mg_2Si rich aluminium alloy that lowering the EGR will reduce porosity from about 8 to 12 vol.% down to 2 vol.%.

Hydrogen induced porosity does not play such an important role like in casting since the atomizing of the melt show a kind of degassing effect [24]. Due to absence of hydrogen partial pressure in the atomizing gas, and although solidification is quick, the hydrogen can escape, a fact which is important for welding procedures, too. In fusion welding normally some amount of material is remelted which usually leads to pronounced porosity in the weld seam. Spray formed material shows only limited occurrence of gas porosity in the weldment [2].

Consolidation, to reduce porosity can be done by a variety of deformation processing, i.e. extrusion [25–35], hot pressing [36–40], rolling [9, 28] or hot isostatic pressing (HIP) [39, 41]. Hot pressing and extrusion are the main processes to close pores, thus reducing the porosity content below 1% [31, 42, 43], which is due to the high shear forces applied. HIP and cold rolling will affect the microstructure partially only or at least not uniformly, therefore a certain residual porosity will remain and has to be taken into account.

Further deformation processes are applied to achieve final product shape and size like e.g.; forging [3], rotary swaging (Fig. 10.4) [3, 29, 44, 45], backward extrusion [3] or drawing [29].

Batchwise production is also another disadvantage. There are design ideas giving the opportunity to have a continuous billet production but invest will be justified by high volume production like cylinder liners, only. Continuous spraying on flat substrates or on tubes will be much more easier to achieve, but industrial application is still pending.

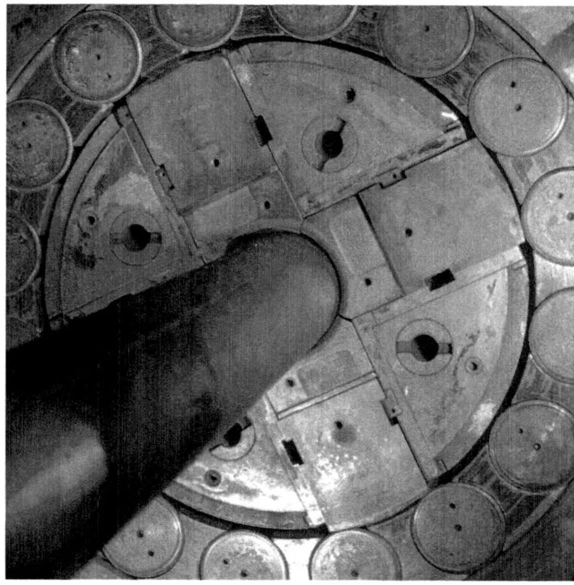

Fig. 10.4 Rotary swaging of extruded Al-25Si-4Cu-1Mg tube for production of cylinder liners [3]

All of the above disadvantages will contribute to cost and therefore one have to keep in mind the four hints, given by Singer in 1990 and focus efforts to these targets. So it is worth to take a closer look to the state of the art of each one of the four targets, which will be done in the following chapters.

10.3 Premium Products

"Spray forming must first concentrate on premium products where the throughput is moderate and where the benefits of scale of conventional products are not crippling" [1].

What products are meant, what products are premium? In most cases this will lead to better properties provided by spray formed products or components, but it is a combination of the performance of such parts in comparison to invest or cost. This will be true for e.g. aerospace, microelectronics or racing applications where performance is responsible for competitive advantage or safety, or both, and where invest plays a more or less subordinate role and uniform, homogeneous refined microstructure is paramount as well as the reproducibility of resulting properties. A number of examples will be shown in the following figures (refer to Figs. 10.5, 10.6, 10.7, 10.8).

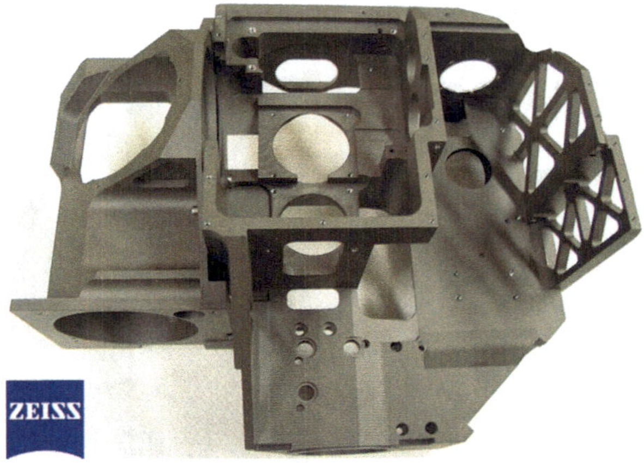

Fig. 10.5 Spectrometer housing for satellite. Alloy Al-35Si. Low thermal expansion and high stiffness is recommended because of steep changes in temperature in space [3]

Fig. 10.6 Al-Nd Sputter Target for PVD process for flat screens. Distribution of Nd has to be extremely homogenous to guarantee defect free sputtering process [46]

Fig. 10.7 Oil pump gears for racing application. Alloy: Al-17Si-4Fe-3Cu-1Mg-Zr, heat treated, providing high strength, wear resistance and reduced weight [picture by kind permission of Erbslöh Aluminium GmbH]

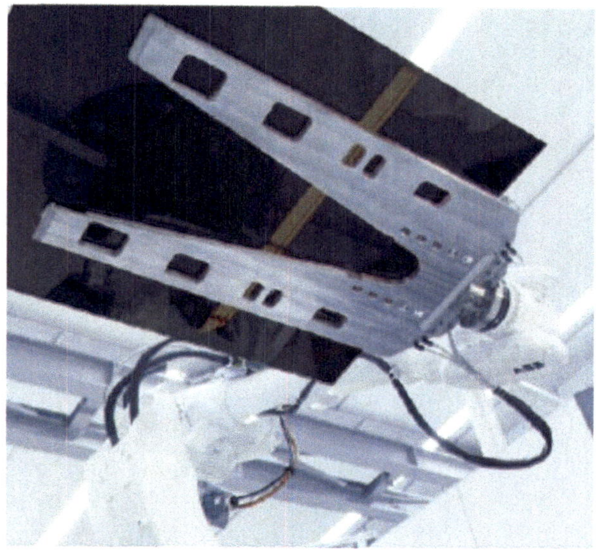

Fig. 10.8 Fork lifter (Al-20Si-5Fe-2Ni) for handling solar modules in/out of oven at 250°C. Exact position is required therefore high stiffness and yield strength at elevated temperature is mandatory [picture by kind permission of Erbslöh Aluminium GmbH]

10.4 Near-to-Net Shape Manufacture

"Spray forming may in some cases defeat the conventional product on cost because of *near-to-net shape manufacture*" [1].

Since the spray forming process increases overall cost due to the nitrogen consumption of several m^3 per kg melt atomized, cost reduction can be achieved by shortening the complete process chain. In the area of spray forming of aluminum alloys this target is not well developed and is still not in commercial production. There are some examples that will be described in the following. Focus should be on processes which allow continuous production in principle. Spray rolling is the best example for combining deposition and consolidation in one step. The atomized melt is directed to a pair of rolls where the final strip geometry is installed. A schematic drawing is given in Fig. 10.9.

A fine microstructure, shown in Fig. 10.10, with good properties at very low porosity levels could be achieved [47, 48]. With a GMR of 0.3 the authors achieved mechanical properties slightly higher than conventional produced aluminium alloy AA 2124 in the T851 state.

Another approach to reduce post processing steps is tube spraying either with one or two different materials. In production of spray formed cylinder liners made of Al-25Si-4Cu-1Mg the spray formed billet is extruded to a seamless tube and subsequently hot rotary swaged to the final inner diameter. To spray the alloy directly onto a carrier tube allows for a directly following swaging operation without cooling and reheating of billet or tubes. A schematic drawing is shown in Fig. 10.11. Furthermore a second alloy, e.g. Al-12Si can be sprayed on top of the

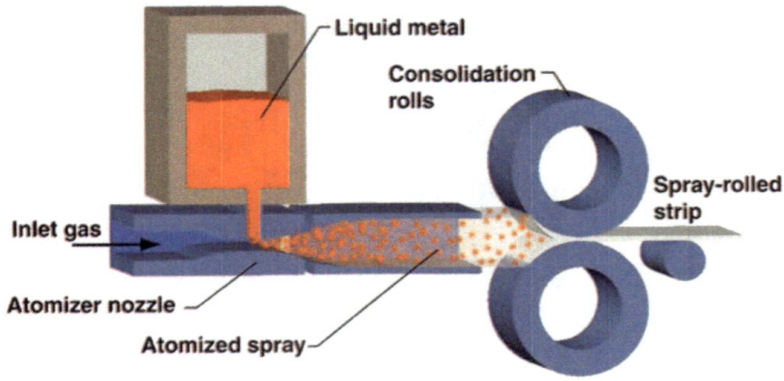

Fig. 10.9 Schematic of spray rolling approach [47]

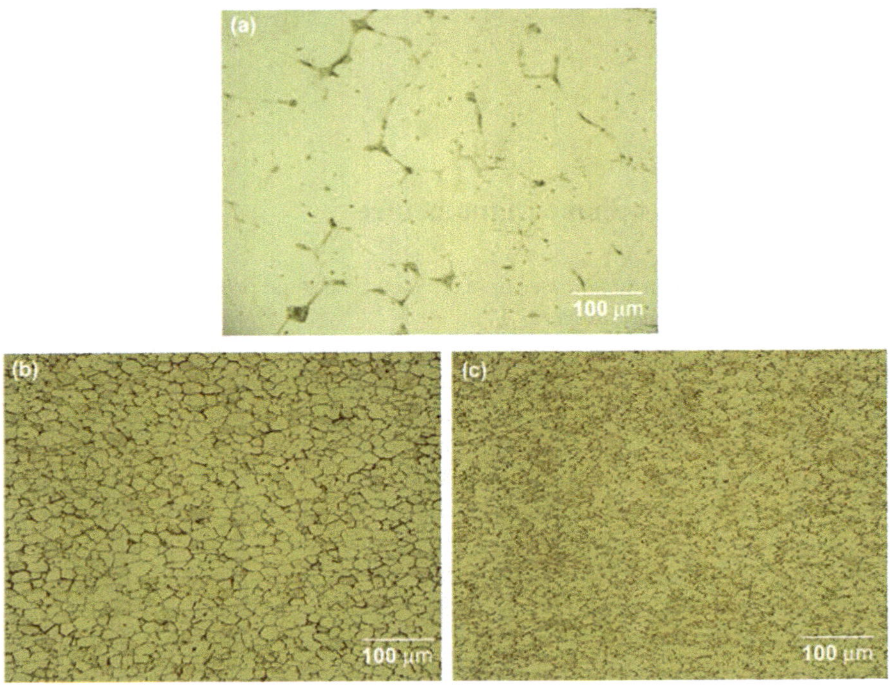

Fig. 10.10 Transverse photomicrographs of 2124 aluminum. (*top*) Ingot cast with slow solidification. (*lower left*) As-spray-rolled using G/M = 0.15. (*lower right*) As-spray-rolled using G/M = 0.30 [48]

Al-25Si-4Cu-1Mg to provide an improved bonding behavior during casting the liners in place.

For high performance engines, cylinder liners are coated by a thermal spray process using Al-12Si-wire and electric arc. The carrier tube could be manufactured from a low cost 6xxx series alloy with high extrudability. After casting the cylinder

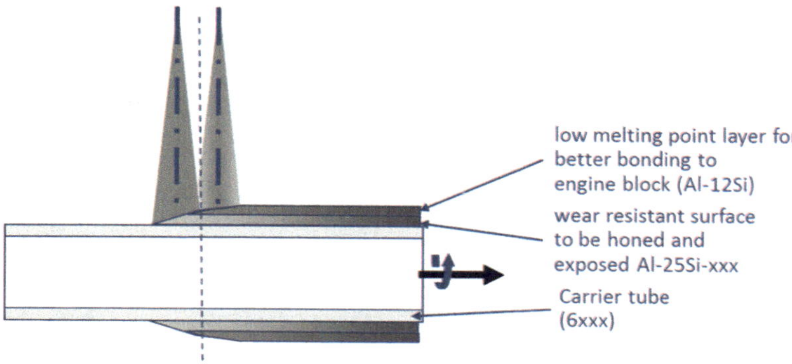

Fig. 10.11 Principle of spray forming of tubes onto a carrier tube with addition co-deposition of bonding layer [31]

Fig. 10.12 States of co-extrusion (from left to right) segment of spray formed billet with LPDC cast layer; co-extruded tube, co-swaged tube [45]

bore will be drilled to final shape. Since most of the removed material will be from the carrier tube, this process can help to save high cost spray formed material.

Tube spraying with co-deposition of bonding enhancement layer competes with another process which is already in production. In this case, the thermal spray process is replaced by a low pressure die casting operation just before extrusion. A approximately 10 mm thick layer of Al-12Si is cast around the spray formed billet, which is very challenging, since an aluminium billet is a perfect heat sink and achieving a sound and defect free hull is difficult. The advantage in comparison to the thermal spray coating is that approximately 500 cylinder liners (still in shape of a billet) will be coated at once. The consecutive production steps are given in Fig. 10.12 and the resulting microstructure of the cast in place liner in Fig. 10.13. The following extrusion process welds the cast and spray formed material together. No defects at the interface could be found. This process is also possible for all castable alloys, maybe to increase corrosion resistance or to save valuable spray formed material, where it is not recommended [3].

Last but not least spray forming of flat deposits with low porosities enables a subsequent rolling process to get geometries which cannot be achieved by extrusion due to restrictions of billet diameter or extrusion container providing only insufficient deformation ratio and therefore no satisfactory densification [24, 49–51].

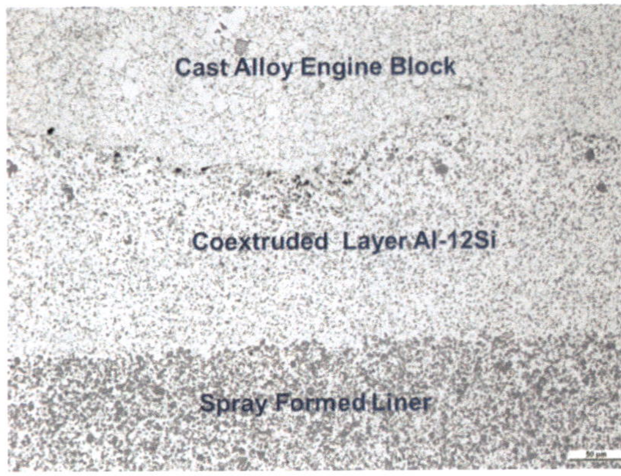

Fig. 10.13 Cast in place liner produced be co-extrusion process. Perfect fusion zones between Al-12Si layer, liner alloy and cast alloy, respectively [45]

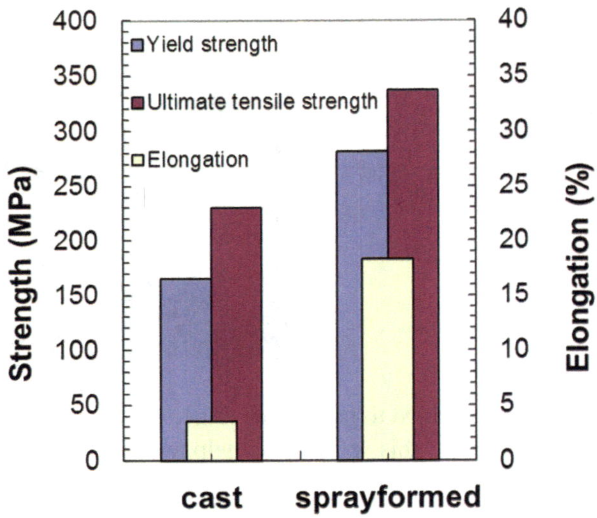

Fig. 10.14 Comparison of cast and spray formed alloy A356, data from [25, 35]

10.5 Better Properties

"Make products that have *better properties*, either physical, chemical or mechanical than competitive ones." [1]

High cooling rates, resulting in fine, homogeneous microstructures and as well as in desired morphologies of intermetallic precipitations and dispersoids, give a reasonable chance to transfer compositions based on casting alloys into wroughtable alloys. Furthermore, since restrictions of maximum solubilties are extended, completely new and unconventional alloy designs could be applied. These compositions lead to superior properties in one or another direction or combining a set of properties paramount to all other alloys, produced by conventional means. In Fig. 10.14 the mechanical properties of cast and spray formed A356 (Al-7Si-Mg) are compared for example.

These properties are usually mechanical values, like Young's modulus, yield or ultimate tensile strength, hot strength, elongation and fatigue behavior, but also wear resistance and thermal expansion has to be taken into account.

10.5.1 Young's Modulus

Young's modulus will be mainly affected by Silicon content, but Fe and Ni will also contribute. All of them are forming coarse pure or intermetallic precipitations when cast, resulting in poor mechanical properties in the final product. Due to a much finer microstructure one can get alloys with high amounts of these alloying elements without facing detrimental effects which are caused by the formation of either coarse primary silicon or by Al-Si-Fe phases in shape of platelets or needles [3, 32, 37, 52–56].

10.5.2 Thermal Expansion

High Si contents are leading to low coefficients of thermal expansion (CTE) down to 14×10^{-6} 1/K for an Al-40Si alloy. This can be achieved by Ni and Fe additions, too, therefore Silicon content may be reduced by 50% down to 20 wt.-% but CTE kept at nearly the same value in a spray formed Al-20Si-5Fe-2Ni alloy [5]. Machining is not too difficult, due to the small size of primary silicon particles thus avoiding excessive toll wear.

10.5.3 High Strength

High strength can be achieved by different base compositions. In conventional aluminium metallurgy alloys of 7xxx are high strength materials. Due to the enormous solubility of Zn, these alloys hardly keep their strength at elevated temperatures, so their application is limited to temperatures below well 100°C. As [57] has shown, yield strength has a close relationship with the total content of Zn, Mg and Cu. By increasing the alloying content and taking advantage of the spray forming process, strength levels up to 800 MPa can be achieved. Attempts to increase the alloying content failed so far due to dramatically reduced ductility, but innovative concepts [58, 59] maybe show a path to increase strength without increasing Zn, Mg and Cu content alloys (refer to Fig. 10.15).

[60, 69, 70] remelted chips from conventional alloy (AA7050) for spray forming. After extrusion higher properties than in AA7050 has been found. Additional ECAP-Processing did not improve the properties further. [58, 59] varied the chromium and the manganese, zirconium, chromium content, respectively.

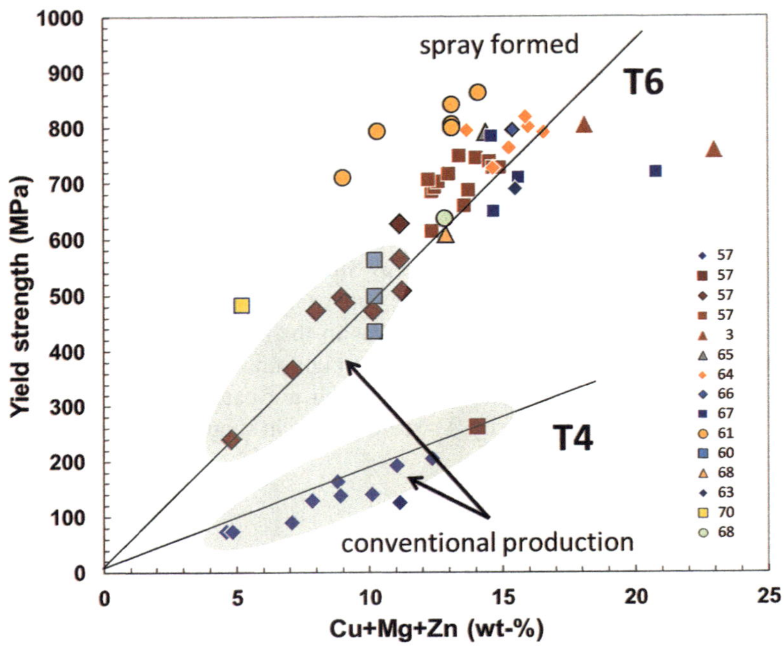

Fig. 10.15 High strength alloy on basis Al-Zn-Cu-Mg. Values taken from references as indicated. Data taken from [3, 57, 60, 61, 63–68, 70]

Scandium additions [61] improved strength further. Unfortunately no values for elongation are given in the article but due to the comparatively low total alloying content one can expect ductile behavior.

10.5.4 Low Density

Spray forming of aluminium-magnesium alloys has been done to improve strength on the one hand and to reduce density on the other. [27, 39, 41, 71] spray formed alloys with magnesium contents between 4 to 6% magnesium, with lithium additions in the range of 1–1.6% and some zirconium ranging from 0.2 to 0.4%. In the spray formed condition these alloys showed properties exceeding those of AA5081 and AA5083. Zirconium additions lead to Al_3Zr dispersoids, therefore no or only partial recrystallization occurred during extrusion. Magnesium additions as high as 12–20% lead together with small amounts of chromium and zirconium as well as with about 1% manganese to high strength alloys with UTS of about 500 MPa. Increasing magnesium content leads to a dramatic loss of ductility from 34 down to 7% [34].

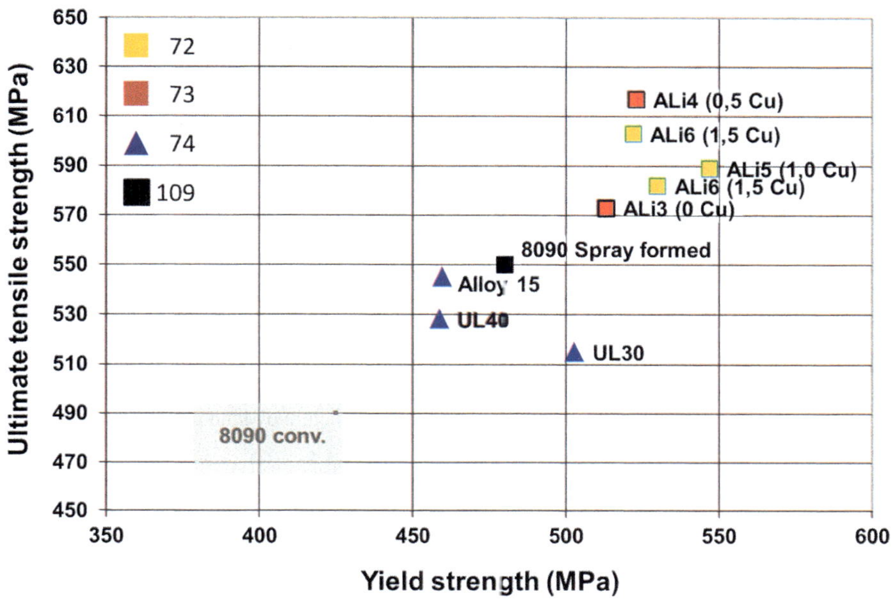

Fig. 10.16 Ultimate tensile and yield strength of Al-Li-based spray formed alloys. Data taken from [72–74, 109]

Lithium is another alloying element which can be added to improve strength and additionally decrease density. [72] tested several compositions with copper, germanium, zirconium and magnesium contents and achieved high strength material with strength values up to 600 MPa. Keeping at least a few percent of elongation makes these alloys advantageous for aerospace application. Figure 10.16 gives a overview over spray formed aluminium-lithium alloys.

Adding high amounts of magnesium and silicon with a near stoichiometric relation lead to low density alloys (below 2.5 g/cm^3) with increased stiffness, due to the formation of up to 35% Mg_2Si [12, 31]. This type of silicide has a very low density of approx. 1.9–2.0 cm/m^3, a high hardness up to 450 HV and a high Young's modulus of about 120 GPa [75]. Size of the silicides has been refined by additions of vanadium and other 3d metals [76]. High magnesium contents are the source of high hydrogen present in the melt and of nonmetallic inclusions. The high vapor pressure and oxidation rate of magnesium in aluminium melts makes it extremely difficult to reproduce alloy composition in useful ranges. As shown in Fig. 10.17 the high solidification rates change morphology of the Mg_2Si from faceted to non-faceted type.

10.5.5 Hot Strength

Hot strength is usually addressed by Al-Cu-Mg alloys. Spray forming of these alloys need high GMR's resulting in a high portion of small particles. Doping such

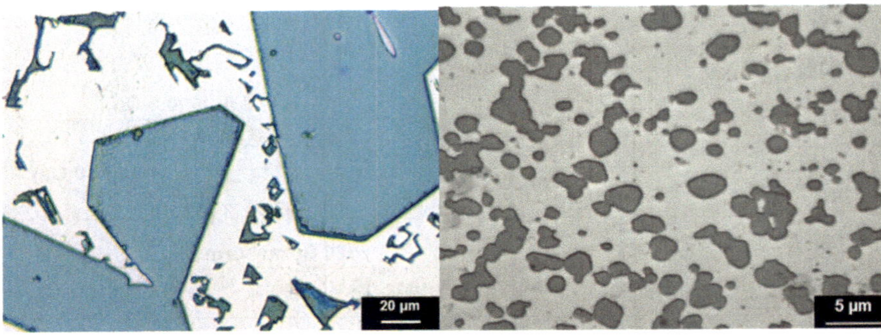

Fig. 10.17 Morphology of Mg_2Si rich alloy (Al-15Mg-7Si-2Cu). Low solidification rate leads to coarse, faceted primary crystals (*left*) compared to spray formed billet, with globular, not faceted occurrence (*right*), please note the different magnification [picture by kind permission of Erbslöh Aluminium GmbH]

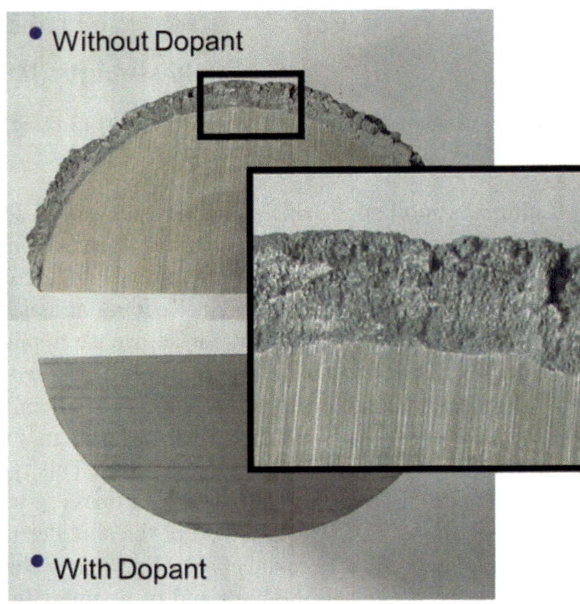

Fig. 10.18 Slices of spray formed Al–4.3Cu–0.36Mg–0.19Mn–0.44Ti–0.3Ag alloy. Without Sn dopant there is a highly porous and brittle layer (see magnification in the middle on the exterior which has to be machined away prior to extrusion. With dopant no layer will form [5]

alloys with surface active elements like Sn or In changes droplet size with beneficial effects, like reducing the highly porous surface layer on billets and the amount of overspray powder. The effect of 0.1 wt.-% of Sn as dopant is shown in Fig. 10.18.

In [77] the creep behavior was found to correspond to that of oxide or carbide dispersion strengthened aluminium. Spray Forming and hot rolling of an Al-10Cu alloy exhibits mechanical properties like PM-alloys [78]. [79] investigated the influence of Mn and Ag containing AlCuMg alloys. From cast alloys it is known, that Ag additions will change precipitation sequence and phase occurrence. Alloys with zirconium and manganese additons showed properties at elevated temperatures like the wrought alloy AA 2014. The Ag containing alloy kept the slightly higher room temperature strength up to 180°C which is comparable to the AlCuMgTi alloy investigated in [77]. In contrast AA 2014 show a marked decrease in strength beginning at 150°C.

To keep the good strength at elevated temperatures it is necessary to combine precipitation hardening providing strength approximately up to 250°C with dispersion hardening or use alloys with dispersion hardening only. This dispersoids will form during solidification or during post processing out of a supersaturated matrix when higher temperatures are applied. These alloys have to be sprayed with high GMR's up to 10. Postprocessing should not contain process steps with long heating periods or long soak times to avoid coarsening of dispersoids. By choosing elements with low diffusions rates in aluminium matrix will lead to long term stable alloys with good high temperature performance. Usually transition metals like Fe, Ni, Zr, Cr, Ti, Mn are used as alloying elements. Hata et al. showed superior hot tensile strength of an Al-8.1Cr-9.9Fe-3Ti alloy. Tensile strength of 340 MPa at 400°C and 1.3% elongation has been achieved [80]. A similar alloy, but with less alloying content (Al-2.7Fe-1.9Cr-1.8Ti) was spray formed by [81]. Due to melting problems a homogeneous microstructure could not be installed and mechanical properties are behind properties of specimens produced by conventional PM-route.

A group of metals within the transition metals, called refractory metals (Nb, W, Ta, Hf, Re), are promising alloying elements for hot strength via dispersion strengthening due to their high melting points. their low solubilities in aluminium and their ability to form a variety of stable intermetallic compounds. [82] showed the presence of ordered $L1_2$-Al_3Hf in Al-1.5Hf and Al-3.2Li-1.6Hf. Although promising, alloying of refractory metals is difficult because of reduced solution velocity when used as pure elements. Application of special master alloys will be helpful. Yield strength of 140 MPa and ultimate tensile strength of 200 MPa of an Al-20Si-5Fe-2Ni alloy alloyed with refractory metals are shown in [3] in comparison to base metal and derviates with rare earth metals. Additions of rare earth metals increase hot strength but not to such extend than refractory metals. Density of the later was increased up to 3.2 g/cm^3, due to the high specific weight of the refractory metals.

For intermediate temperatures between 150 and 250°C and moderate wear loading like e.g. cylinder liners, hypereutectic alloys with 20 to 25% Si are appropriate. A variety of experimental alloys were spray formed, extruded and round swaged to final cylinder liner geometry [83] Increasing amount of transition metals (Cr, Zr, Fe, Cu, Ti, Co, Mn) lead to superior mechanical properties and hot strength. [62, 84–87] show in their investigations a beneficial effect of Cr and Mn additions to Al–20Si–5Fe–3Cu and Al–25Si–5Fe–3Cu base alloy. They point out, that these additions will alter the morphology of Fe bearing phases like $Al_{15}Fe_3Si_2$ and that combined Cr + Mn addition is superior to single element additions. [35] uses Fe and Mn in an Al–20Si–3Cu–1 Mg alloy. Manganese changes shape of $Al_{15}Fe_3Si_2$ while combination of Fe and Mn is inefficient. The base alloy contains already a small amount of Fe from Silicon and Aluminum. Extraordinary Fe will diminish the effect of Manganese as a iron correction factor.

10.5.6 Fatigue

Due to the fine microstructure it is obvious that high fatigue strength can be achieved with spray formed materials but porosity has to be eliminated by a subsequent consolidation process. Small grain size and submicron precipitates/ dispersoids are responsible for an excellent fatigue behavior. To keep this advantageous structure is mandatory and will affect post processing significantly. The influence of extrusion parameters of Al-17Si-4Fe-3Cu-1Mg-Zr alloy was investigated by [88] an validated by ultrasonic fatigue testing. Crack initiation in the high cycle fatigue regime started at surface defects while in the very high cycle fatigue regime the cracks started at pores or at Si accumulations. It turned out, too, that a high deformation ratio during extrusion is important for good fatigue performance but extrusion temperature should kept as low as possible since a high deformation ratio will be compensated completely by a too high billet temperatures [3]. High supersaturation after spray forming and moderate extrusion temperatures will lead to superior fatigue strength. Usually aluminium alloys exhibit endurance limits of approximately 25% of the ultimate tensile strength. With spray formed material this can be increased to nearly 75% with an Al-Mg-Mn-Zr alloy [3]. Due to dispersoids with very low growth rates, fatigue resistance is also high at elevated temperatures. At 300°C the endurance limit ($P_{fr} = 50\%$, 5×10^7 cycles, $R = -1$) for a refractory metal containing Al-20Si-5Fe-2Ni is nearly twice as high as for AA 2618 T6 and still reaching 60% of ultimate tensile strength at this temperature as shown in Fig. 10.19 [3].

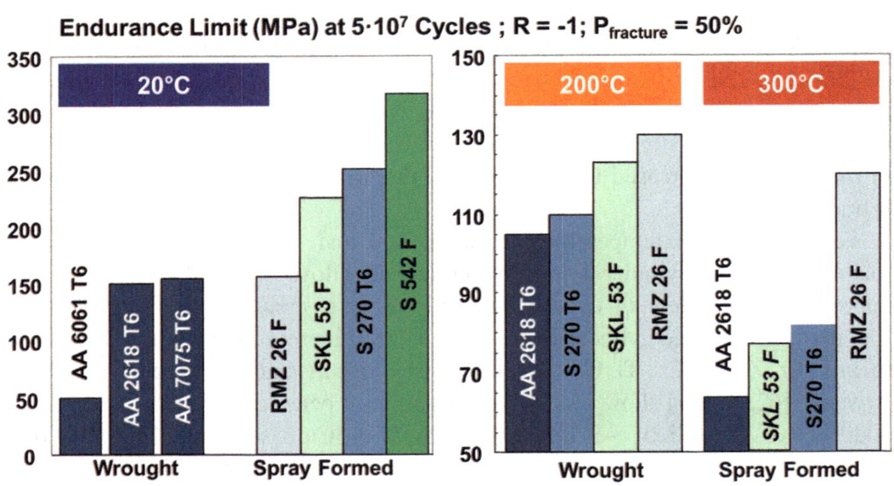

Fig. 10.19 Endurance limit for different conventional and spray formed alloys at room temperature (*left*) and at elevated temperatures (*right*) [3]

10.5.7 Wear Resistance

Wear resistance can be greatly improved by increasing silicon content. Spray forming is offering the possibility of high Si concentrations in combination with hot deformability. The higher the Si content the lower the wear rate and friction coefficient [36]. Due to smaller Si particles in comparison to cast alloys wear resistance will be increased in Al-22Si and Al–22Si–4Cu–1.7Mg [42, 43]. [89] revealed good wear resistance in Al–6.91Si and Al–10.1Si alloys with uniform Si particles of 2–5 µm size. Low wear rate and low friction coefficient can be achieved by smaller Si particles. [90] claimed that the smaller contact area between Si and friction partner is responsible for this, because coarser particles will be peeled out of the surface. The addition of 0.8 Sc in hypereutectic Al-Si alloys leads to better wear behavior due to the formation of $AlSi_2Sc_2$-phase and an average Si-size below 4 µm. Also hardness and mechanical properties are improved by 25% [91]. Similar findings could be found in [92]. Hypereutectic Al-25Si alloys with additional alloying elements show better wear resistance with smaller Si-particles. It is worth noting that lubrication medium will play a significant role. By changing the composition of the oil, by changing some friction modifying dopants, a dramatic decrease of wear rate can be achieved. In [38] an increase in wear resistance is shown by adding 5% Cu and 4% Fe to a spray formed Al-28Si alloy in comparison to cast material. Change in wear mechanism from delamination to abrasion at a comparison between cast and spray formed alloy with 30 wt.-% Mg_2Si [37]. Pb additions of 10% were added by [93, 94] via vortex melting technique. The spray formed Al-10Si-3.5Cu-0.15Fe-0.20Mg-0.01Mn alloy were tested at a pin-on-disk unit. The formation of a Pb film lowers friction and wear. An addition of 10–25% of Pb to Al-Si base alloy increases the amount of lead particles in the deposit [95]. In contrast to cast alloys with particle size of maximum 80 µm the spray formed deposit contained lead phases in the range of 5–12 µm. No data for wear resistance is given. [36] uses 25% Sn, which leads to a further reduction of wear rate and friction in comparison to a tin free Al–12.5Si. Although small Si particles show better wear behavior in special cases a minimum particle size is desired. The cylinder bore will be drilled and honed to provide the appropriate surface condition for the piston. The final preparation step will be a mechanical or chemical exposure operation, which takes away the matrix and leaves the Silicon particles standing slightly upright over the exposed surface. If exposure depth and particle size is in the same range it is very likely that the smaller Silicon particles will fall off the matrix. To avoid this cylinder liner material can be heat treated to coarsen the Si particles. This heat treatment should be applied after the last deformation process, because bigger Si particles will be cracked during extrusion while smaller ones will flow with the extrusion material [33]. A rapid coarsening treatment will be a high temperature anneal at 600°C [96]. Besides pore formation this treatment will lead to eutectic silicon because liquid phase formation occurs.

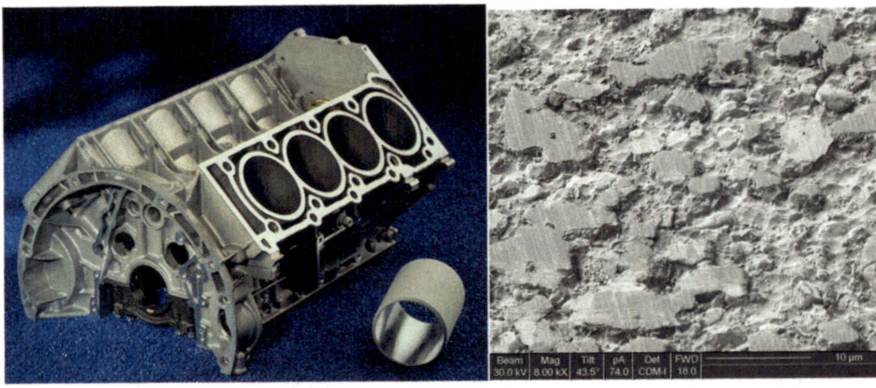

Fig. 10.20 HPDC V8 engine block with hyereutectic Al-Si liner (*left*) and honed and chemically exposed liner surface [37]

One of the most successful products in terms of volume and weight made of spray formed material is a cylinder liner for internal combustion engines. Development of this product was done collaboratively by the former company PEAK Werkstoff GmbH (now Erbslöh Aluminium GmbH) and Daimler Benz AG (now Daimler AG) at the end of last century. Changing from a hypereutectic Al-Si monolithic engine block to a heterogeneous solution with liners but keeping the good properties of the high Si content lead to the decision to use aluminum liners in an aluminum block. The latter has been made of a cost efficient high pressure die cast (HPDC) hypoeutectic alloy, while the hypereutectic Al-25Si-4Cu-1Mg liner assures a good thermal conductivity and low contribution to the total weight, compared to gray cast iron liners. High mileage with negligible wear and low friction losses paired with the excellent strength and thermal behavior enabled superior performance.

In Fig. 10.20 a V8 engine block is shown with a SEM picture detailing the running surface with its typical occurrence [3]. Due to an exposure treatment after the last honing step (either chemical or mechanical) one can clearly see the protruding silicon crystals with a flat surface and the aluminium matrix approximately 1.5 µm deeper. For more severe service conditions (higher output power or diesel engines) a variety of high strength high wear resistant alloys were developed [75, 83]. An overview over the mechanical properties is outlined in Fig. 10.21. Additional processing with electron beam modification of the surface is another sophisticated method to improve wear resistance. Spray formed, high strength alloys will help to support the integrity of the modified surface by avoiding cracking due to collapse of the underlying material [97, 98].

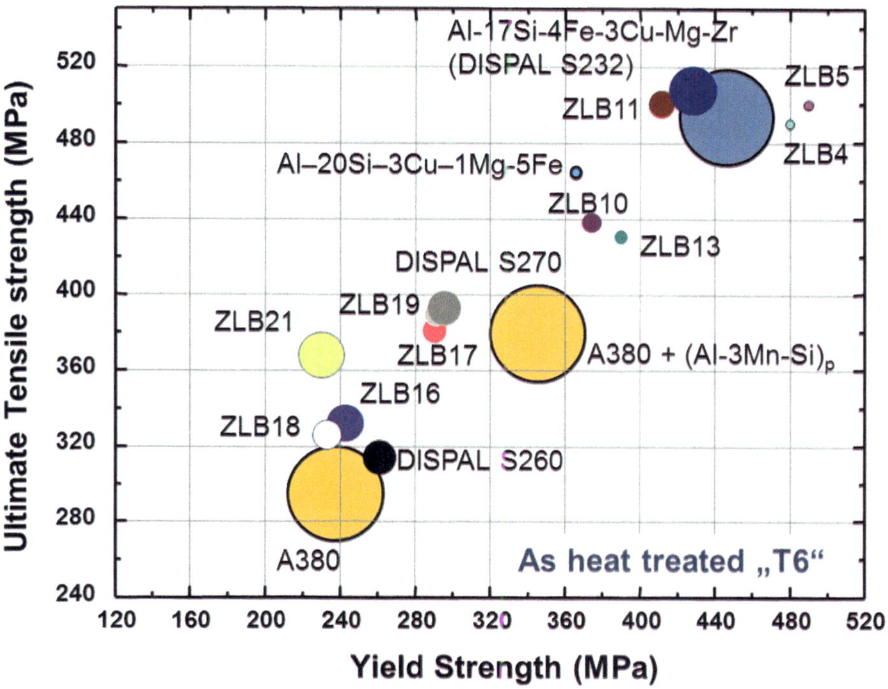

Fig. 10.21 Mechanical properties of spray formed hypereutectic Al-Si-alloys. The diameter of the circles gives the elongation. Data taken from [75, 83]

10.6 Impossible Alloys

"Make products that cannot be made by any alternative means at an affordable cost". [1]

Impossible alloys should be thought in two directions. First, materials combinations which which cannot be made by conventional casting processes like metal-matrix-composites, and second, alloy compositions leading to extraordinary microstructures.

Spray forming gives the advantage to co-inject particles of all kinds, thus enabling a broad range of alloy design to make metal-matrix-composites. Therefore unique properties will be available, depending only on the combination of the right combination of the composite. This is not only restricted to ceramic particles but can be extended to metallic materials as well. [25, 26] co-injected particles of Al–3Mn–4Si into an Al–8.9Si–3.2Cu–0.9Fe alloy which led to an increase of strength without significant reduction of elongation. Titanium particles were also introduced during spray forming into a Al-12Si alloy. Due to high cooling rates no interface reaction took place between the titanium particles and Al or Si. The material was post processed by accumulative roll bonding and ended up with a fibrous structure of the titanium [28]. Since bigger particles contribute more to a reduction of CTE,

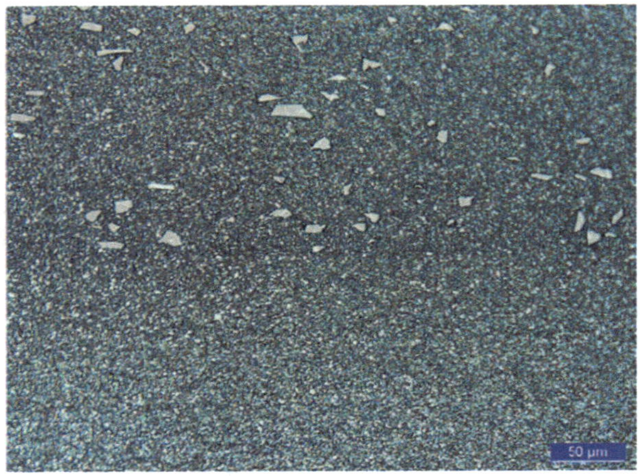

Fig. 10.22 Microstructure of Al-20Si-5Fe-2Ni + $(SiC)_p$ Silicon carbide reinforced material at the top of picture (big grey particles = SiC); part at the bottom is not reinforced material. Both parts were combined by a friction welding process [3]

silicon particles (40 µm) were introduced in an Al-35Si alloy, increasing the total content to 45% Si but with a bimodal distribution [75].

Fatigue resistance has been improved by injection of 15 vol.% SiC particles (4.5 µm) into a Al-7Si alloy, thus lowering the fatigue crack growth rate. Higher Young's modulus could be achieved with 20 µm particles, but ultimate tensile stress and elongation is slightly reduced with bigger particles [99]. The influence of particle content and temperature on the compressive deformation behavior was studied in [100]. An increase in strength and stiffness was found by [59, 101] when SiC (approx. 4 µm) was added between 12.7 and 14.5 vol.%. Although hard ceramic particles also contribute to strength and stiffness, but impact on wear resistance is dominant. Figure 10.22 shows a friction welded component between a SiC-reinforced aluminium alloy (Al-20Si-5Fe-2Ni) with the non-reinforced base material.

Aluminum highly alloyed with high transition metals together with high cooling rates of the spray forming process transfers to partially quasicrystalline and/or nanocrystalline or even amorphous microstructures. High contents of rare erath metals lead to the formation of amorphous or nanocrystalline microstructures. High Y (Al-21Y-9Ni-4Co) and high Y-La (Al-18La-12Y-8Ni-3Co) containing alloys show amorphous fraction up to 83%. With such compositions bulk amorphous alloys can be synthesized. Hardness of these alloys increase during annealing at 480 and 500°C respectively [50, 102]. An Al-21Y-9Ni-4Co alloy exhibit 925 MPa compressive strength [49]. Higher La contents in Al-24La-8Ni transfers in very high hardness but obvious no ductility [51, 103]. The influence of distance to substrate was studied by [104]. They achieved a high amorphous content of 92 vol.% at the bottom of a spray formed billet decreasing to 55 vol.% at top when spray forming an $Al_{86}Si_{0.5}Ni_{4.06}Co_{2.94}Y6La_{0.5}$ alloy. Ce in Al-8Fe-4Ce change morphology of Al_3Fe. Together with high cooling rate an elongation of 4.4% at 209 MPa ultimate tensile strength at 300°C is possible [105]. Substituting cerium by misch metal will reduce strength to 115 MPa at 350°C [40].

Spray formed billets with quasicrystalline (QC) phases has been investigated by [106]. He found in an $Al_{93}Fe_3Cr_2Ti_2$ alloy QC $Al_{13}Fe_4$ in the middle of the billet and nano-quasicrystalline precipitates in the external regions of the billet. No QC phases are present at the bottom, due to heat loading during spray forming. [107] produced a QC-$Al_{62.5}Cu_{25}Fe_{12.5}$ (Al-40Cu-17Fe) alloy. With tin additions the amount of QC phases may be increased, but the received alloy is very brittle.

10.7 Conclusion and Outlook

So let's have a final look to the targets mentioned in the beginning. There (still) is a market for and a lot of *premium products* already developed. Key markets are small but constant. One should keep in mind and as included in this target, too, with moderate throughput! In this area upscaling of production numbers are counterproductive. One has to keep to quality and not to quantity. In case of *near-to-net shape manufacture* promising development has been conducted, although transfer to production is scarce. This is mainly due to continuous improvement of conventional production methods, which gives spray forming process only limited chance to be competitive. Achieving *better properties* is not a big issue for spray forming, but improved properties cannot account completely for increased cost. For spray formed aluminium alloys situation is not too bad, but if price is too high, one may turn to conventional titanium alloys for elevated or high strength applications. On the other hand, having weight in mind, spray formed aluminium alloys have to compete with high sophisticated magnesium alloys or carbon fiber reinforced composites. Nevertheless, there are a number of alloys with outstanding properties but also a limited number of applications hard to be discovered and difficult to be accessed. Due to the high production cost—in comparison to normal wrought or cast alloys—other researchers are looking for substitution either by cheaper material or cheaper processing or at least both. To be economically successful with such substitutional efforts one will tunnel his visions to the cost driver and the high cost products. Since much more development capacities are present for conventional alloys and processes, sprayfoming will not be successful in the long term. So the only option left is to make *impossible alloys* which cannot be made by any alternative means at an affordable cost. This is the main target for spray forming. It is a unique process which allows one to produce material with absolutely outstanding properties. Many surprising effects will arise, when research and development will be more adventurous. Much more unusual alloy compositions should be tried out, limits are less tight as in conventional metallurgy. One should use co-injection of powders and particles more extended and not only for silicon carbide. One should also think about other metallic or alloy powders even with a completely other base metal or think about injection of mixtures of metallic and non-metallic particles and powders. Materials with complicated duplex, triplex or even multiplex microstructures can be generated with properties no one can predict yet. Therefore much more efforts should be laid on impossible alloys.

References

1. Singer, A. R. E. (1990) A future of spray forming. *Proceedings of 7th International Conference on Spray Forming (ICSF I), Swansea, United Kingdom, 17–19 Sept 1990.*
2. Gietzelt, T., Eichhorn, L., Wunsch, T., Gerhards, U., Przeorski, T., Weiss, H., & Dittmeyer, R. (2014). Contribution to the laser welding of wrought and spray formed aluminum alloys and the impact of the alloy composition on the welding microstructure. *Advanced Engineering Materials, 16*, 1052–1065. doi:10.1002/adem.201300497.
3. Krug, P. (2006). Neue pulvermetallurgische Al-werkstoffe und deren Anwendung. In H. Kolaska (Ed.), *Berichtsband 25. Hagener Symposium Pulvermetallurgie* (pp. 135–151). Witten: Heimdall Verlag.
4. PEAK Werkstoff GmbH 2004. Internal report, Velbert.
5. Commandeur, B., Krug, P., Loos, R., & Sinha, G. (2003). Spray compaction of aluminium alloys on an industrial scale. *Proceedings of 2nd International Conference on Spray Deposition and Melt Atomization (SDMA 2003) and 5th International Conference on Spray Forming (ICSF V), Bremen, Germany, 22–25 June 2003.*
6. Baram, J. (1991). Structure and properties of a rapidly solidified AI-Li-Mn-Zr alloy for high-temperature applications: part II. Spray atomization and deposition processing. *Metallurgical Transactions A, 22*(10), 2515–2522. doi:10.1007/BF02665017.
7. Chu, M. G., Denzer, D. K., Chakrabarti, A. K., & Billman, F. R. (1988). Evaluation of aluminum and nickel alloy materials produced by spray deposition. *Materials Science and Engineering, 98*, 227–232. doi:10.1016/0025-5416(88)90160-7.
8. Chaudhury, S. K., & Panigrahi, S. C. (2007). Role of processing parameters on microstructural evolution of spray formed Al-2Mg alloy and Al-2Mg-TiO_2 composite. *Journal of Materials Processing Technology, 182*(1–3), 343–351. doi:10.1016/j.jmatprotec.2006.08.013.
9. Chaudhury, S. K., & Panigrahi, S. C. (2007). Influence of TiO_2 particles on recrystallization kinetics of Al-2Mg-TiO_2 composites. *Journal of Materials Processing Technology, 182* (1–3), 540–548. doi:10.1016/j.jmatprotec.2006.09.014.
10. Duszczyk, J., Estrada, J. L., De Haan, T. L. J., Leatham, A. G., & Ogilvy, A. J. W. (1987). The Osprey process for the production of high performance aluminium alloys. *Proceedings of International Conference on PM Aerospace Materials, Lucerne, Switzerland, 2–4 Nov 1987.*
11. Duan, X., Hao, Y., Yoshida, M., Ando, T., & Grant, N. J. (1993). Liquid dynamic compaction of aluminium alloy 7150. *International Journal of Powder Metallurgy, 29*(2), 149–160.
12. Ellendt, N., Uhlenwinkel, V., Stelling, O., Irretier, A., & KEẞLER, O. (2007). Spray forming of Mg_2Si rich aluminum alloys. *Materials Science Forum, 534–536*, 437–440. doi:10.4028/www.scientific.net/MSF.534-536.437.
13. Grant, P. S., Kim, W. T., & Cantor, B. (1991). Spray forming of aluminium-copper alloys. *Materials Science and Engineering: A, 134*, 1111–1114. doi:10.1016/0921-5093(91)90935-G.
14. Lavernia, E. J., Ando, T., & Grant, N. J. (1986). Structure and properties of the lithium containing X2020 aluminum alloys produced by liquid dynamic compaction. In P. Lee & R. Carbonara (Eds.), *Rapidly solidified materials* (pp. 29–44). Materials Park: ASM.
15. Lengsfeld, P., Juarez-Islas, J. A., Cassada, W. A., & Lavernia, E. J. (1995). Microstructure and mechanical behavior of spray deposited Zn modified 7XXX series Al alloys. *International Journal of Rapid Solidification, 8*(4), 237–265.
16. Lavernia, E. J., Rai, G., & Grant, N. J. (1986). Liquid dynamic compaction of a rapidly solidified high strength aluminum alloy. *International Journal of Powder Metallurgy, 22*(1), 9–16.
17. Plies, J. B., & Grant, N. J. (1994). Structure and properties of spray formed 7150 containing Fe and Si. *International Journal of Powder Metallurgy, 30*(3), 335–343.
18. Underhill, R. P., Grant, P. S., & Cantor, B. (1993). Microstructure of spray-formed Al alloy 2618. *Materials and Design, 14*(1), 45–47. doi:10.1016/0261-3069(93)90045-W.

19. Zhou, J., Duszczyk, J., & Korevaar, B. M. (1991). As-spray-deposited structure of an Al-20Si-5Fe Osprey preform and its development during subsequent processing. *Journal of Materials Science, 26*(19), 5275–5291. doi:10.1007/BF01143222.
20. Raghukiran, N., & Kumar, R. (2013). Processing and dry sliding wear performance of spray deposited hyper-eutectic aluminum-silicon alloys. *Journal of Materials Processing Technology, 213*(3), 401–410. doi:10.1016/j.jmatprotec.2012.10.007.
21. Elis, L., Uhlenwinkel, V., Ristau, R., & Krug, P. (2010). Transient temperatures and microstructure of spray formed aluminium alloy AlSi sheets. *Materialwissenschaft und Werkstofftechnik, 41*(7), 498–503. doi:10.1002/mawe.201000635.
22. Cui, C., Schulz, A., Schimanski, K., & Zoch, H.-W. (2009). Spray forming of hypereutectic Al-Si alloys. *Journal of Materials Processing Technology, 209*(11), 5220–5228. doi:10.1016/j.jmatprotec.2009.03.009.
23. Ellendt, N., Stelling, O., Uhlenwinkel, V., von Hehl, A., & Krug, P. (2010). Influence of Spray Forming Process Parameters on the Microstructure and Porosity of Mg_2Si rich Aluminum Alloys. *Materialwissenschaft und Werkstofftechnik, 41*(7), 532–540. doi:10.1002/mawe.201000639.
24. Gurutze Pérez Artieda, M., Cortiella, A., Harlan, N. R., Zapirain, F., & Zubiri, F. (2010). Two step coating of a hypereutectic PM Al-Si alloy. *Materials Letters, 64*(13), 1458–1461. doi:10.1016/j.matlet.2010.03.056.
25. Bereta, L. A., Ferrarini, C. F., Botta F, W. J., Kiminami, C. S., & Bolfarini, C. (2007). Microstructure and mechanical properties of spray co-deposited Al-8.9 wt.% Si-3.2 wt.% Cu-0.9 wt.% Fe + (Al-3 wt.% Mn-4 wt.% Si)$_p$ composite. *Journal of Alloys and Compounds, 434–435*, 371–374. doi:10.1016/j.jallcom.2006.08.156.
26. Bereta, L. A., Ferrarini, C. F., Kiminami, C. S., Botta, W. J. F., & Bolfarini, C. (2007). Microstructure and mechanical properties of spray deposited and extruded/heat treated hypoeutectic Al-Si alloy. *Materials Science and Engineering: A, 449–451*, 850–853. doi:10.1016/j.msea.2006.02.241.
27. HOGG, S. C., Mi, J., Nilsen, K. E., Liotti, E., & Grant, P. S. (2010). Microstructure and property development in spray formed and extruded Al-Mg-Li-Zr alloys for aerospace and autosport applications. *Materialwissenschaft und Werkstofftechnik, 41*(7), 562–567. doi:10.1002/mawe.201000643.
28. Kelly, A. J., Mi, J., Sinha, G. V., Krug, P., Crosa, F., Audebert, F., & Grant, P. S. (2011). An Al-Si-Ti hierarchical metal-metal composite manufactured by co-spray forming. *Journal of Materials Processing Technology, 211*(12), 2045–2049. doi:10.1016/j.jmatprotec.2011.07.001.
29. Schulz, A., Cui, C., Uhlenwinkel, V., Buschenhenke, F., Seefeld, T. H., Vollertsen, F., et al. (2009). Spray forming of hypereutectic Al-Si filler materials for welding high strength aluminum alloys. *Proceedings of 4th International Conference on Spray Deposition and Melt Atomization (SDMA 2009) and 7th International Conference on Spray Forming (ICSF VII), Bremen, Germany, 7–9 Sept 2009.*
30. Stelling, O., Ellendt, N., Uhlenwinkel, V., von Hehl, A., Krug, P., & Zoch, H.-W. (2009). Coarsening of spray formed aluminium alloys with high Mg_2Si-content during subsequent processing. *Proceedings of 4th International Conference on Spray Deposition and Melt Atomization (SDMA 2009) and 7th International Conference on Spray Forming (ICSF VII), Bremen, Germany, 7–9 Sept 2009.*
31. Stelling, O., Keßler, O., Zoch, H.-W., Ellendt, N., & Uhlenwinkel, V (2006). New spray formed aluminum-copper alloys with high Mg_2Si-content. *Proceedings of 3rd International Conference on Spray Deposition and Melt Atomization (SDMA 2006) and 6th International Conference on Spray Forming (ICSF VI), Bremen, Germany, 4–6 Sept 2006.*
32. Srivastava, V. C., Mandal, R. K., Ojha, S. N., & Venkateswarlu, K. (2007). Microstructural modifications induced during spray deposition of Al-Si-Fe alloys and their mechanical properties. *Materials Science and Engineering: A, 471*(1–2), 38–49. doi:10.1016/j.msea.2007.04.109.

33. Srivastava, A. K., Srivastava, V. C., Gloter, A., & Ojha, S. N. (2006). Microstructural features induced by spray processing and hot extrusion of an Al-18% Si-5% Fe-1.5% Cu alloy. *Acta Materialia, 54*(7), 1741–1748. doi:10.1016/j.actamat.2005.11.039.
34. Sun, T., Sun, Y., Wang, F., Xin, H., Guo, M., & Zhu, X. (2006) The research of microstructures and mechanical properties of the spray-deposited high magnesium aluminum alloy. *Proceedings 3rd International Conference on Spray Deposition and Melt Atomization (SDMA 2006) and 6th International Conference on Spray Forming (ICSF VI), Bremen, Germany, 4–6 Sept 2006.*
35. Wang, F., Zhang, J., Xiong, B., & Zhang, Y. (2009). Effect of Fe and Mn additions on microstructure and mechanical properties of spray-deposited Al-20Si-3Cu-1 Mg alloy. *Materials Characterization, 60*(5), 384–388. doi:10.1016/j.matchar.2008.10.011.
36. Anil, M., Srivastava, V. C., Ghosh, M. K., & Ojha, S. N. (2010). Influence of tin content on tribological characteristics of spray formed Al-Si alloys. *Wear, 268*(11–12), 1250–1256. doi:10.1016/j.wear.2010.01.018.
37. Goudar, D. M., Raju, K., Srivastava, V. C., & Rudrakshi, G. B. (2013). Effect of secondary processing on the microstructure and wear behavior of spray formed Al–30Mg_2Si–2Cu alloy. *Materials and Design, 47*, 489–496. doi:10.1016/j.matdes.2012.11.010.
38. Goudar, D. M., Raju, K., Srivastava, V. C., & Rudrakshi, G. B. (2013). Effect of copper and iron on the wear properties of spray formed Al-28Si alloy. *Materials and Design, 51*, 383–390. doi:10.1016/j.matdes.2013.04.018.
39. Hogg, S. C., Palmer, I. G., Thomas, L. G., & Grant, P. S. (2007). Processing, microstructure and property aspects of a spraycast Al-Mg-Li-Zr alloy. *Acta Materialia, 55*(6), 1885–1894. doi:10.1016/j.actamat.2006.10.057.
40. Lei, H., Zhang, Y., Zhao, G., Yan, P., & Yan, B. (2014). Microstructure and elevated temperature mechanical properties of Al-8Fe-4Re alloy fabricated by spray forming. *Materialwissenschaft und Werkstofftechnik, 45*(8), 683–688. doi:10.1002/mawe.201400304.
41. Jagan Reddy, G., Srinivasan, N., Gokhale, A. A., & Kashyap, B. P. (2009). Processing map for hot working of spray formed and hot isostatically pressed Al-Li alloy (UL40). *Journal of Materials Processing Technology, 209*(18–19), 5964–5972. doi:10.1016/j.jmatprotec.2009.07.016.
42. Goudar, D. M., Raju, K., Srivastava, V. C., & Rudrakshi, G. B (2013). Effect of Cu and Mg on the wear properties of spray formed Al-22Si alloy. *Proceedings of 5th International Conference on Spray Deposition and Melt Atomization (SDMA 2013), Bremen, Germany, 23–25 Sept 2013.*
43. Raju, K., Goudar, D.M, Srivastava, V.C., & Rudrakshi, G.B (2013). Wear behavior of secondary processed spray formed Al-28Si-5Cu-4Mg alloy. *Proceedings of 5th International Conference on Spray Deposition and Melt Atomization (SDMA 2013), Bremen, Germany, 23–25 Sept 2013.*
44. Cui, C., Schulz, A., Achelis, L., Uhlenwinkel, V., Leopold, H., Piwek, V., Tang, Z., & Seefeld, T. (2014). Development of low-melting-point filler materials for laser beam brazing of aluminum alloys. *Materialwissenschaft und Werkstofftechnik, 45*(8), 717–726. doi:10.1002/mawe.201400308.
45. Krug, P., Commandeur, B., & Dan, T. (2009). What shall we do with spray formed aluminum alloys? - Examples of conventional and sophisticated post processing. *Proceedings of 4th International Conference on Spray Deposition and Melt Atomization (SDMA 2009) and 7th International Conference on Spray Forming (ICSF VII)*, Bremen, Germany, 7–9 September 2009.
46. Kobelco Research Institute, Inc 2016. Sputtering Target for Flat Panel Display. Kobelco Research Institute, Inc. [online]. Kobe, Japan, from http://www.kobelcokaken.co.jp/target/english/applicatioin/index.html. Accessed 28 June 2016.
47. Mchugh, K. M., Delplanque, J.-P., Johnson, S. B., Lavernia, E. J., Zhou, Y., & Lin, Y. (2004). Spray rolling aluminum alloy strip. *Materials Science and Engineering: A, 383*(1), 96–106. doi:10.1016/j.msea.2004.02.041.

48. Mchugh, K. M., Lin, Y., Zhou, Y., Johnson, S. B., Delplanque, J. P., & Lavernia, E. J. (2008). Microstructure evolution during spray rolling and heat treatment of 2124 Al. *Materials Science and Engineering: A, 477*(1–2), 26–34. doi:10.1016/j.msea.2007.04.130.
49. Srivastava, V. C., Surreddi, K. B., Scudino, S., Schowalter, M., Uhlenwinkel, V., Schulz, A., Eckert, J., Rosenauer, A., & Zoch, H.-W. (2010). Microstructure and mechanical properties of partially amorphous $Al_{85}Y_8Ni_5Co_2$ plate produced by spray forming. *Materials Science and Engineering: A, 527*(10–11), 2747–2758. doi:10.1016/j.msea.2010.01.057.
50. Srivastava, V. C., Surreddi, K. B., Scudino, S., Schowalter, M., Uhlenwinkel, V., Schulz, A., Eckert, J., Rosenauer, A., & Zoch, H.-W. (2013). Microstructural characteristics of spray formed and heat treated Al-(Y, La)-Ni-Co system. *Journal of Alloys and Compounds, 578*, 471–480. doi:10.1016/j.jallcom.2013.06.159.
51. Ted Guo, M.-L., Tsao, C. Y. A., Huang, J. C., & Jang, J. S. C. (2006). Microstructural evolution in spray-formed and melt-spun $Al_{85}Nd_5Ni_{10}$ bulk hybrid composites. *Intermetallics, 14*(8–9), 1069–1074. doi:10.1016/j.intermet.2006.01.041.
52. Cochrane, R. F., Newcomb, S. B., Evans, P. V., & Greer, A. L. (1989). Microstructural development in drop-tube processed Al-8wt.%Fe. *Key Engineering Materials, 38–39*, 21–42. doi:10.4028/www.scientific.net/KEM.38-39.21.
53. Ferrarini, C. F., Bolfarini, C., Kiminami, C. S., & Botta F, W. J. (2004). Microstructure and mechanical properties of spray deposited hypoeutectic Al-Si alloy. *Materials Science and Engineering: A, 375–377*, 577–580. doi:10.1016/j.msea.2003.10.062.
54. Juarez-Islas, J., Zhou, Y., & Lavernia, E. J. (1999). Spray atomization of two Al–Fe binary alloys: solidification and microstructure characterization. *Journal of Materials Science, 34* (6), 1211–1218. doi:10.1023/A:1004509022619.
55. Srivastava, V. C., Mandal, R. K., Ramachandra, C., Chatterjee, B., & Ojha, S. N. (1999). Microstructural and wear characteristics of spray deposited hypereutectic Al–Si alloy. *Transactions of the Indian Institute of Metals, 51*(1), 29–40.
56. Yang, B., Wang, F., Zhang, J. S., Xiong, B. Q., & Duan, X. J. (2001). The effect of Mn on the microstructure of spray-deposited Al–20Si–5Fe–3Cu–1Mg alloy. *Scripta Materialia, 45*(5), 509–515. doi:10.1016/S1359-6462(01)01051-X.
57. Mächler, R (1993). Höchstfeste sprühkompaktierte Aluminium-Zink-Magnesium-Kupfer-Legierungen, Dissertation No. 10332 (p. 72). Zurich, Switzerland: Eidgenössische Technische Hochschule. doi: 10.3929/ethz-a-000916722.
58. Sharma, M. M. (2008). Microstructural and mechanical characterization of various modified 7XXX series spray formed alloys. *Materials Characterization, 59*(1), 91–99. doi:10.1016/j.matchar.2007.01.013.
59. Sharma, M. M., Amateau, M. F., & Eden, T. J. (2005). Aging response of Al-Zn-Mg-Cu spray formed alloys and their metal matrix composites. *Materials Science and Engineering: A, 424* (1–2), 87–96. doi:10.1016/j.msea.2006.02.047.
60. Hyodo, A., Bolfarini, C., & Ishikawa, T. T. (2012). Chemistry and tensile properties of a recycled AA7050 via spray forming and ECAP/E. *Materials Research, 15*(5), 739–748. doi:10.1590/S1516-14392012005000097.
61. Sharma, M. M., Amateau, M. F., & Eden, T. J. (2006). Hardening mechanisms of spray formed Al-Zn-Mg-Cu alloys with scandium and other elemental additions. *Journal of Alloys and Compounds, 416*(1–2), 135–142. doi:10.1016/j.jallcom.2005.08.045.
62. Cai, Y., Liang, R., Hou, L., & Zhang, J. (2011). Effect of Cr and Mn on the microstructure of spray-formed Al-25Si-5Fe-3Cu alloy. *Materials Science and Engineering: A, 528*(12), 4248–4254. doi:10.1016/j.msea.2011.02.029.
63. Bai, P., Hou, X., Zhang, X., Zhao, C., & Xing, Y. (2009). Microstructure and mechanical properties of a large billet of spray formed Al-Zn-Mg-Cu alloy with high Zn content. *Materials Science and Engineering: A, 508*(1–2), 23–27. doi:10.1016/j.msea.2008.12.010.
64. Eden, T. J. (2005). High strength aluminum alloys produced by spray metal forming. *iMAST Quarterly, 2*, 3–6.
65. Sun, M 2006. Private communication

66. Wang, H., Huang, J., Yang, B., Cui, H., Zhang, J., Zhu, B., et al. (2003) Homogenization and aging behavior of spray deposited high strength Al-Zn-Mg-Cu alloys. *Proceedings 2nd International Conference on Spray Deposition and Melt Atomization (SDMA 2003) and 5th International Conference on Spray Forming (ICSF V), Bremen, Germany, 22–25 June 2003*
67. Yu, H., Wang, M., Jia, Y., Xiao, Z., Chen, C., Lei, Q., Li, Z., Chen, W., Zhang, H., Wang, Y., & Cai, C. (2014). High strength and large ductility in spray-deposited Al-Zn-Mg-Cu alloys. *Journal of Alloys and Compounds, 601*, 120–125. doi:10.1016/j.jallcom.2014.02.126.
68. Yu, H., Wang, M., Sheng, X., Li, Z., Chen, L., Lei, Q., Chen, C., Jia, Y., Xiao, Z., Chen, W., Wei, H., Zhang, H., Fan, X., & Wang, Y. (2013). Microstructure and tensile properties of large-size 7055 aluminum billets fabricated by spray forming rapid solidification technology. *Journal of Alloys and Compounds, 578*, 208–214. doi:10.1016/j.jallcom.2013.05.117.
69. Godinho, H. A., Beletati, A. L. R., Giordano, E. J., & Bolfarini, C. (2014). Microstructure and mechanical properties of a spray formed and extruded AA7050 recycled alloy. *Journal of Alloys and Compounds, 586*(1), S139–S142. doi:10.1016/j.jallcom.2012.12.122.
70. Mazzer, E. M., Afonso, C. R. M., Galano, M., Kiminami, C. S., & Bolfarini, C. (2013). Microstructure evolution and mechanical properties of Al-Zn-Mg-Cu alloy reprocessed by spray-forming and heat treated at peak aged condition. *Journal of Alloys and Compounds, 579*, 169–173. doi:10.1016/j.jallcom.2013.06.055.
71. Moore, K. L., Sykes, J. M., Hogg, S. C., & Grant, P. S. (2008). Pitting corrosion of spray formed Al-Li-Mg alloys. *Corrosion Science, 50*(11), 3221–3226. doi:10.1016/j.corsci.2008.08.012.
72. Del Castillo, L., Wu, Y., Hu, H. M., & Lavernia, E. J. (1999). Microstructure and Mechanical Behavior of Spray-Deposited High-Li Al-Li Alloys. *Metallurgical and Materials Transactions A: Physical Metallurgy and Materials Science, 30*(5), 1381–1389. doi:10.1007/s11661-999-0286-3.
73. Dai, S. L., Wu, Y., Del Castillo, L., & Lavernia, E. J. (1997). Microstructure and properties of spray deposited high Li Al-Li-Mg-Ge-Zr alloys. *Scripta Materialia, 37*(3), 265–270. doi:10.1016/S1359-6462(97)00116-4.
74. Grant, P., Palmer, I., & Stone, I. (1999). Spray formed aerospace alloys are high flyers. *Materials World, 7*(6), 331–333.
75. Krug, P. & Commandeur, B. (2005) Spray forming of advanced high strength aluminum alloys. *Continuous Casting: Proceedings of the International Conference on Continuous Casting of Non-Ferrous Metals*. Neu-Ulm: Wiley-VCH.
76. Stelling, O. (2011). *Einfluss der Prozessparameter auf die Mikrostruktur und die Eigenschaften Sprühkompaktierter, Hochlegierter Al-Mg-Si-Legierungen, Dissertation*. University of Bremen, Germany.
77. Eddahbi, M., Carreño, F., & Ruano, O. A. (2006). Deformation behavior of an Al-Cu-Mg-Ti alloy obtained by spray forming and extrusion. *Materials Letters, 60*(27), 3232–3237. doi:10.1016/j.matlet.2006.02.085.
78. Schimanski, K., Schulz, A., Vetters, H., & Zoch, H.-W. (2007). Optimisation of Al-Cu alloys by spray forming. *HTM–Härterei-Technische Mitteilungen, 62*(2), 53–57.
79. Del Castillo, L., & Lavernia, E. J. (2000). Microstructure and mechanical behavior of spray-deposited Al-Cu-Mg(-Ag-Mn) alloys. *Metallurgical and Materials Transactions A: Physical Metallurgy and Materials Science, 31*(9), 2287–2298. doi:10.1007/s11661-000-0145-8.
80. Hata, H., Kajihara, K., Takagi, T., Takahara, T., & Ehira, M. (2006). Spray forming of Al alloy of high temperature strength. *Proceedings of 3rd International Conference on Spray Deposition and Melt Atomization (SDMA 2006) and 6th International Conference on Spray Forming (ICSF VI), Bremen, Germany, 4–6 Sept 2006.*
81. Banjongprasert, C., Hogg, S. C., Liotti, E., Kirk, C. A., Thompson, S. P., Mi, J., & Grant, P. S. (2010). Spray forming of bulk ultrafine-grained Al-Fe-Cr-Ti. *Metallurgical and Materials Transactions A: Physical Metallurgy and Materials Science, 41*(12), 3208–3215. doi:10.1007/s11661-010-0386-0.

82. Antipas, G. S. E., Lekakou, C., & Tsakiropoulos, P. (2011). Microstructural characterisation of Al-Hf and Al-Li-Hf spray deposits. *Materials Characterization, 62*(4), 402–408. doi:10.1016/j.matchar.2011.02.001.
83. Krug, P., Kennedy, M., & Foss, J. (2006). New aluminium alloys for cylinder liner applications. SAE Technical Paper 2006-01-0983. doi: 10.4271/2006-01-0983.
84. Hou, L. G., Cui, H., Cai, Y. H., & Zhang, J. S. (2009). Effect of (Mn + Cr) addition on the microstructure and thermal stability of spray-formed hypereutectic Al-Si alloys. *Materials Science and Engineering: A, 527*(1–2), 85–92. doi:10.1016/j.msea.2009.07.041.
85. Huang, H. J., Cai, Y. H., Cui, H., Huang, J. F., He, J. P., & Zhang, J. S. (2009). Influence of Mn addition on microstructure and phase formation of spray-deposited Al-25Si-*x*Fe-*y*Mn alloy. *Materials Science and Engineering: A, 502*(1–2), 118–125. doi:10.1016/j.msea.2008.10.005.
86. Hou, L. G., Cui, C., & Zhang, J. S. (2010). Optimizing microstructures of hypereutectic Al-Si alloys with high Fe content via spray forming technique. *Materials Science and Engineering: A, 527*(23), 6400–6412. doi:10.1016/j.msea.2010.06.066.
87. Jia, Y., Cao, F., Ning, Z., Sun, X., & Sun, J. (2011). Hot deformation behavior of spray formed Al-22Si-5Fe-3Cu-1Mg alloy. *Transactions of Nonferrous Metals Society of China, 21*(2), s299–s303. doi:10.1016/S1003-6326(11)61595-1.
88. Stanzl-Tschegg, S. E., Mayer, H., Schuller, R., Przeorski, T., & Krug, P. (2012). Fatigue properties of spray formed hypereutectic aluminium silicon alloy DISPAL® S232 at high and very high numbers of cycles. *Materials Science and Engineering: A, 538*, 327–334. doi:10.1016/j.msea.2012.01.052.
89. Ojha, K. V., Tomar, A., Singh, D., & Kaushal, G. C. (2008). Shape, microstructure and wear of spray formed hypoeutectic Al-Si alloys. *Materials Science and Engineering: A, 487*(1–2), 591–596. doi:10.1016/j.msea.2007.10.032.
90. Raju, K., & Ojha, S. N. (2014). Effect of spray forming on the microstructure and wear properties of Al-Si alloys. *Procedia Materials Science, 5*, 345–354. doi:10.1016/j.mspro.2014.07.276.
91. Raghukiran, N., & Kumar, R. (2015). Effect of scandium addition on the microstructure, mechanical and wear properties of the spray formed hypereutectic aluminum-silicon alloys. *Materials Science and Engineering: A, 641*, 138–147. doi:10.1016/j.msea.2015.06.027.
92. Krug, P (2009). Reibungsoptimierte Werkstoffe für motorische Anwendungen. Reibungsverringerung in Antrieben—ganzheitliche Lösungswege und Strategien. *Oral Presentation at Inhouse-Seminar, IWM Freiburg, Germany, 12 Feb 2009.*
93. Rudrakshi, G. B., Srivastava, V. C., & Ojha, S. N. (2006). A comparative study of wear behavior of spray formed Al-3.5Cu-10Si-20Pb alloy in air and vacuum. *Proceedings of 3rd International Conference on Spray Deposition and Melt Atomization (SDMA 2006) and 6th International Conference on Spray Forming (ICSF VI), Bremen, Germany, 4–6 Sept 2006.*
94. Rudrakshi, G. B., Srivastava, V. C., & Ojha, S. N. (2007). Microstructural development in spray formed Al–3.5Cu–10Si–20Pb alloy and its comparative wear behaviour in different environmental conditions. *Materials Science and Engineering: A, 457*(1–2), 100–108. doi:10.1016/j.msea.2006.12.024.
95. Mittal, R., Tomar, A., & Singh, D. (2012). Thickness uniformity and microstructure of disc shape spray formed Al-Si-Pb alloys. *Advanced Materials Letters, 3*(1), 55–63. doi:10.5185/amlett.2011.5261.
96. Zhu, X., Wang, R., Peng, J., & Peng, C. (2014). Microstructure evolution of spray-formed hypereutectic Al-Si alloys in semisolid reheating process. *Transactions of Nonferrous Metals Society of China, 24*(6), 1766–1772. doi:10.1016/S1003-6326(14)63251-9.
97. Dalke, A., Buchwalder, A., Spies, H.-J., Biermann, H., & Klemm, M. Erhöhung der tribologischen beanspruchbarkeit von aluminiumwerkstoffen durch die kombination von randschicht-umschmelzlegieren und nitrieren. *HTM Journal of Heat Treatment and Materials, 67*(1), 13–21. doi:10.3139/105.110131.

98. Klemm, M., Rose, A., von Hehl, A., Haase, I., Zenker, R., Franke, R., & Franke, A. Lokales Werkstoffengineering zur Modifizierung der Randschichteigenschaften von Aluminiumlegierungen mittels moderner Elektronenstrahl-Ablenktechniken *HTM Journal of Heat Treatment and Materials, 67*(1), 4–12. doi:10.3139/105.110132.
99. Li, W., Liang, H., Chen, J., Zhu, S. Q., & Chen, Y. L. (2014). Effect of SiC particles on fatigue crack growth behavior of SiC particulate-reinforced Al-Si alloy composites produced by spray forming. *Procedia Materials Science, 3*, 1694–1699. doi:10.1016/j.mspro.2014.06.273.
100. Srivastava, V. C., Jindal, V., Uhlenwinkel, V., & Bauckhage, K. (2008). Hot-deformation behaviour of spray-formed 2014 Al + SiC_P metal matrix composites. *Materials Science and Engineering: A, 477*(1–2), 86–95. doi:10.1016/j.msea.2007.06.086.
101. Sharma, M. M., Ziemian, C. W., & Eden, T. J. (2011). Fatigue behavior of SiC particulate reinforced spray-formed 7XXX series Al-alloys. *Materials and Design, 32*(8–9), 4304–4309. doi:10.1016/j.matdes.2011.04.009.
102. Srivastava, V. C., Surreddi, K. B., Uhlenwinkel, V., Schulz, A., Eckert, J., & Zoch, H.-W. (2009). Formation of nanocrystalline matrix composite during spray forming of $Al_{83}La_5Y_5Ni_5Co_2$. *Metallurgical and Materials Transactions A: Physical Metallurgy and Materials Science, 40*(2), 450–461. doi:10.1007/s11661-008-9737-5.
103. Ted Guo, M.-L., Tsao, C. Y. A., Chang, K. F., Huang, J. C., & Jang, J. S. C. (2007). Thermal stability and mechanical properties of spray-formed and melt-spun $Al_{89}La_6Ni_5$ metallic glass matrix composites. *Materials Transactions, 48*(7), 1717–1721. doi:10.2320/matertrans.MJ200767.
104. Zhuo, L., Yang, B., Wang, H., & Zhang, T. (2011). Spray formed Al-based amorphous matrix nanocomposite plate. *Journal of Alloys and Compounds, 509*(18), L169–L173. doi:10.1016/j.jallcom.2011.02.125.
105. Yaney, D. L., & Nix, W. D. (1987). Elevated Temperature Deformation Behavior of an AI-8.4 Wt Pct Fe-3.6 Wt Pct Ce Alloy. *Metallurgical Transactions A, 18*(6), 893–902. doi:10.1007/BF02646930.
106. Garcia-Escorial, A., Echevarria, M., Lieblich, M., & Stone, I. (2010). Characterisation of an $Al_{93}Fe_3Cr_2Ti_2$ alloy obtained by spray forming. *Journal of Alloys and Compounds, 504*(1), S519–S521. doi:10.1016/j.jallcom.2010.03.031.
107. Srivastava, V. C., Huttunen-Saarivirta, E., Cui, C., Uhlenwinkel, V., Schulz, A., & Mukhopadhyay, N. K. (2014). Bulk synthesis by spray forming of Al-Cu-Fe and Al-Cu-Fe-Sn alloys containing a quasicrystalline phase. *Journal of Alloys and Compounds, 597*, 258–268. doi:10.1016/j.jallcom.2014.01.241.
108. Cai, Y., Liang, R., Su, Z., & Zhang, J. (2011). Microstructure of spray formed Al-Zn-Mg-Cu alloy with Mn addition. *Transactions of Nonferrous Metals Society of China, 21*(1), 9–14. doi: 10.1016/S1003-6326(11)60671-7
109. Hort, N., & Kainer, K. U. (2004). Pulvermetallurgisch hergestellte Metall-Matrix-Verbundwerkstoffe, In K. U. Kainer (Ed.), *Metallische Verbundwerkstoffe*. Weinheim: Wiley-VCH Verlag GmbH & Co. KgaA. doi: 10.1002/3527602062

Chapter 11
Spray Forming of Copper Alloys

Hilmar R. Müller and Igor Altenberger

11.1 Introduction

Compared with other forming processes, such as continuous casting, spray forming is a fairly new method of manufacturing semi-finished products. In this process, a metal melt is converted into solid state by the intermediate step of atomisation. Already in the early fifties, some proposals were published for atomising metal, collecting it on a substrate and solidifying it [1]. The current method of spray forming is based on developments by Prof. Singer, Wales/UK [2]. At the beginning of the seventies, initial publications and patents came out. Efforts were focused on the production of strip directly from the melt [3]. As a result of these activities at the University of Swansea, the company Osprey Metals was founded dealing with the development and marketing of the spray forming process. They give licenses to manufacturers of plants and users in the steel, aluminium- and copper-alloy industry. The atomisation in droplets and the subsequent fast cooling with rapid solidification opens up new opportunities for the production of materials which, up to now, could not be produced or produced only with strong segregation.

H.R. Müller (✉) • I. Altenberger
Wieland-Werke AG, Graf-Arco-Str. 36, 89079 Ulm, Germany
e-mail: hilmar.mueller@wieland.de; igor.altenberger@wieland.de

H. Henein et al. (eds.), *Metal Sprays and Spray Deposition*,
DOI 10.1007/978-3-319-52689-8_11

11.2 Spray Forming Plant for Production of Copper Alloy Billets

11.2.1 Classification of the Process

The designation spray casting, which is also used, indicates that the process can be associated neither with powder metallurgy nor with classical casting methods. While the casting process deals with the production of semi-finished products made of castable alloys and powder metallurgy with the production of semi-finished or finished products made of powders of special alloys, spray forming is a combination of both.

Main applications are the production of semis made of

- Materials not castable or castable only with difficulty
- High-alloyed materials
- Strongly segregating materials
- Composite materials (metal matrix composites, MMC)

11.2.2 Process Description

The melt is first guided from a tundish via a ceramic casting nozzle to a concentric atomising gas nozzle. The melt jet is then dispersed to small droplets by an inert gas according to Fig. 11.1. The average droplet diameter is about 60 μm. Compared to the metal powder production, the mass flow rate is significantly higher, namely up to 35 kg/min with a single atomiser and up to 70 kg/min with a twin atomiser. The primary gas nozzle runs with slightly elevated pressure and is used for guiding the melt jet. Figure 11.2 shows a typical spray forming arrangement. The melt is prepared in a melting furnace not being shown and after that transported to the spray forming plant with a holding furnace. Then it flows through an inert gas shrouded tube into the tundish and from there through the casting nozzle to the atomiser.

A controlled stopper rod keeps the melt level in the tundish and therefore the mass flow rate constant. Bottom pouring and shrouding of the melt surface keep off oxygen coming from the surrounding air.

The spray chamber is also flooded with an inert gas, usually nitrogen. The atomisation is similar to the production of metal powder but the flight distance of the droplets during spray forming is significantly shorter. Just before complete solidification the droplets hit a suitable substrate, which rotates and can be moved in a vertical direction, and solidify. A metal layer is formed and grows according to the flow rate of the droplets. Proportional to the growing layer, the substrate is withdrawn in order to keep the distance between the atomiser and the tip of the billet constant. The angle between the axis of the atomiser and the billet allows the diameter to be controlled merely by adjusting the withdrawal speed. Figure 11.3

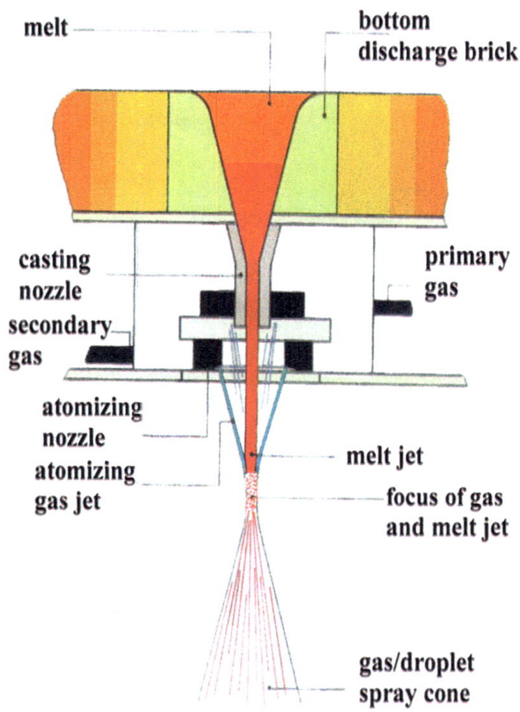

Fig. 11.1 Atomiser

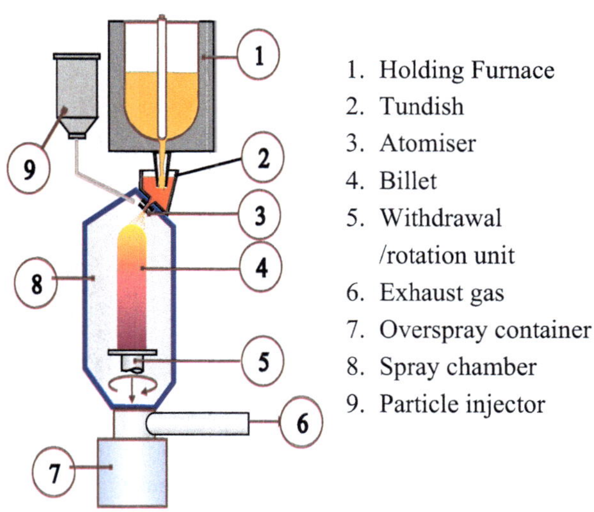

Fig. 11.2 Principle of the Wieland spray forming unit

shows the finished billet after the end of the spraying process with opened spray chamber, and Fig. 11.4 the billet being formed shortly after the beginning of the spraying process. The maximum diameter reached up to now is about 500 mm. Figure 11.5 shows an example.

Fig. 11.3 Finished billet in the spray chamber

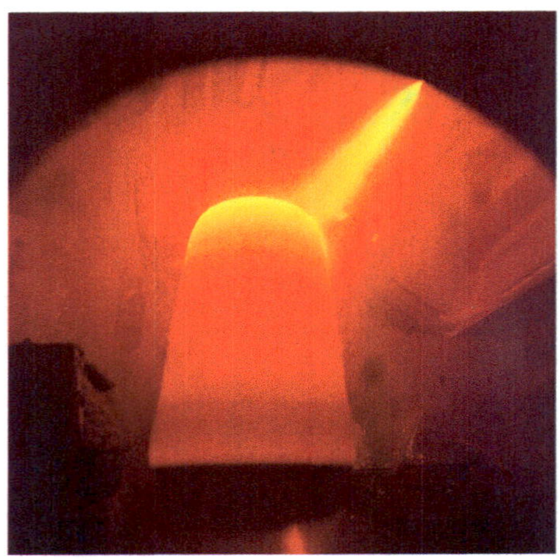

Fig. 11.4 View inside the spray chamber at the beginning of the process

Also other particles can be added to the primary gas flow. So it is possible to produce dispersion materials with big density differences between the particles and the matrix. The particles are homogeneously embedded in the solidifying metal.

Part of the solidifying droplets do not stick to the billet surface but follow the gas flow. This so-called overspray means a loss of metal in the order of 10–40%. Most of the droplets are separated from the gas already at the bottom of the chamber, the remainder in a cyclone and bag filter. In Table 11.1 some typical technical data of a spray forming plant are collected.

Fig. 11.5 Billet CuCrZr, dia. 500 mm

Table 11.1 Technical data of the Wieland spray forming unit

Semi-finished product	Billet
d_{min} (mm)	160
d_{max} (mm)	500
l_{max} (mm)	2200
m_{max} (kg)	2400
$Input_{max}$ (kg)	3200
Production rate, single-atomiser (kg/min)	35
Production rate, twin-atomiser (kg/min)	70
Overspray (%)	10–40
Additional equipment	Particle injector

11.2.3 *Differences from Conventional Casting Processes*

The mere comparison of the macrostructure of continuously cast and spray-formed billets shows big differences. In Fig. 11.6, the cross section of a continuously cast billet (on the left) is compared with that of a spray-formed counterpart (on the right). Both are made of the alloy Cu-Cr0.8%-Zr0.1%. The diameter is just the same. While on the left the typical cast structure is visible, it totally disappears on the right. The average grain size is about 60 μm [4].

Microstructures and segregation are also clearly influenced by the process. Figure 11.7 compares the microstructures of permanent mould cast and spray-formed Cu-Sn16%. The permanent mould cast is of dendritic structure. Between the dendrites the tin-rich δ-phase is enriched. The high fraction of this brittle, low melting phase prevents cold and hot forming. In the spray-formed structure (on the right) the fraction of δ-phase is much lower and not linked. This structure allows cold and hot forming. While prematerial made of conventional—permanent mould or continuous—cast is formable only up to 8% Sn, spray-formed material can be hot

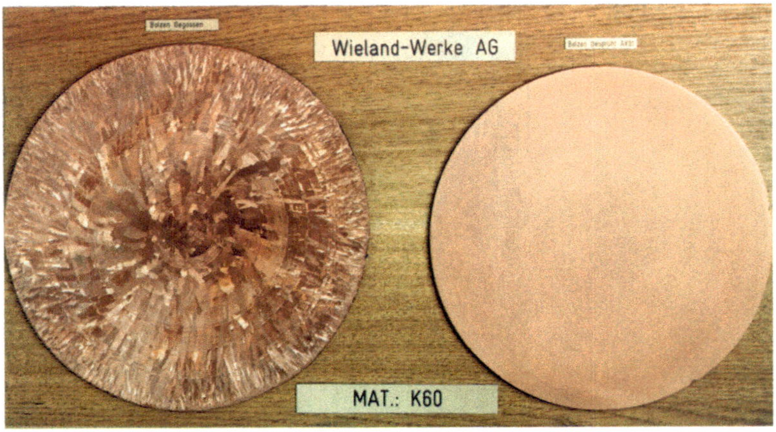

Fig. 11.6 Macrostructure of continuously cast (*left*) and spray-formed (*right*) billet, dia. 250 mm, CuCrZr

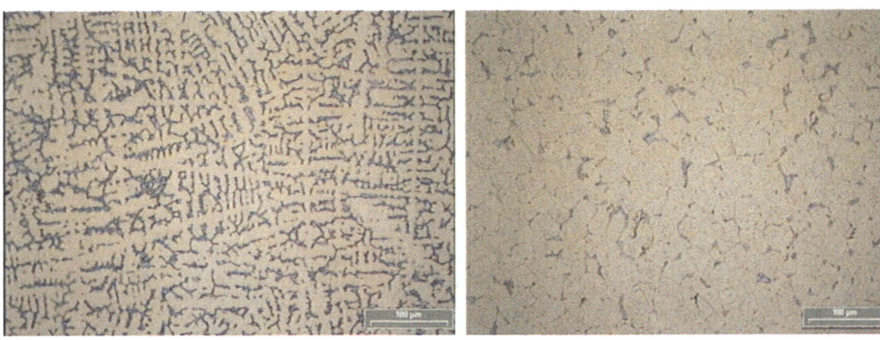

Fig. 11.7 Microstructure of permanent mould cast (*left*) and spray-formed (*right*) CuSn16

and cold formed by extrusion and drawing up to 16% Sn without any prior homogenisation. The classical border between wrought and cast alloys is shifted to considerably higher tin contents [4].

11.3 Materials Groups

11.3.1 High Alloyed Copper Materials

11.3.1.1 Tin Bronze

Homogeneous Element Distribution in Spray-Formed Bronze

One of the main challenges in high tin bronzes is to achieve a homogenous distribution of the tin content through the complete volume of a large billet [5].

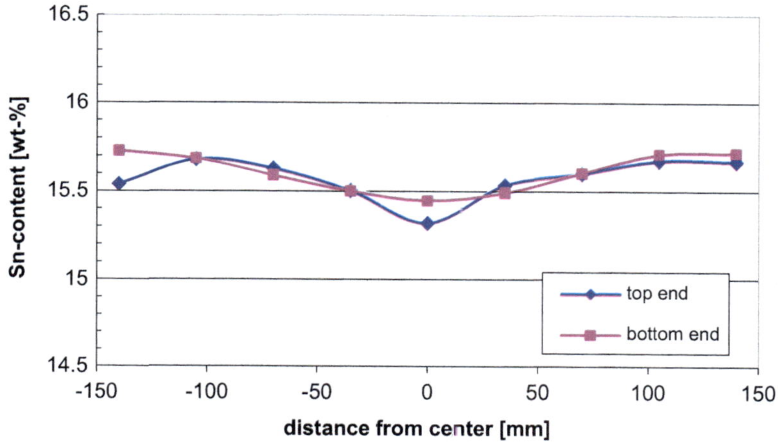

Fig. 11.8 Tin concentration across the cross section of a spray-formed CuSn15.5Ti billet diameter 298 mm and over the billet length. Maximum measured fluctuation is 0.4 wt.%

Here the spray forming process shows its advantage being able to keep macro segregation of tin at a minimum below 1 wt.-% [4]. Figure 11.8 shows the tin concentration measured across the cross section of a CuSn15.5Ti billet with a diameter of 298 mm. The fluctuation in tin content is 0.4 wt.-% only. The tin fluctuation over the length of a billet is in the same order, see data of top end and bottom end in Fig. 11.8. In comparison to this, cast bronzes show a considerably higher tin concentration fluctuation [4, 6].

Homogeneous Microstructure of Spray-Formed Bronze

Another big advantage of spray forming is its ability to generate material with a homogeneous microstructure [4–7]. This includes both, a fine and homogeneous grain structure and a fine dispersed and homogeneous intermetallics structure.

Grain Structure

The grain structure in spray-formed bronze is very fine and homogeneous. The grains have a spherical shape with an average diameter of around 60 μm, see Fig. 11.9. This is a big advantage for further workability. In comparison, cast bronzes have course and inhomogeneous grains, published in detail in [6].

Intermetallics Structure

Figure 11.9 presents the microstructure in the condition after spray forming prior to further hot and cold working of two commercial superconductor bronzes. Their tin content exceeds 14% and thus these bronzes exhibit precipitations in their microstructure. Delta phase particles are segregated at grain boundary triple points in the bronze matrix. Additionally, small Ti-rich precipitates at grain boundaries and inside the grains are present. X-ray measurement results show that the particles probably are Ti_6Sn_5.

Figure 11.9a, b show an overview over the delta phase distribution at a low magnification in a cross section of spray-formed billets of CuSn14.5Ti and CuSn16Ti0.3, respectively. The higher the tin content is the higher is the volume fraction of delta phase. These delta phase configurations are considered to be the

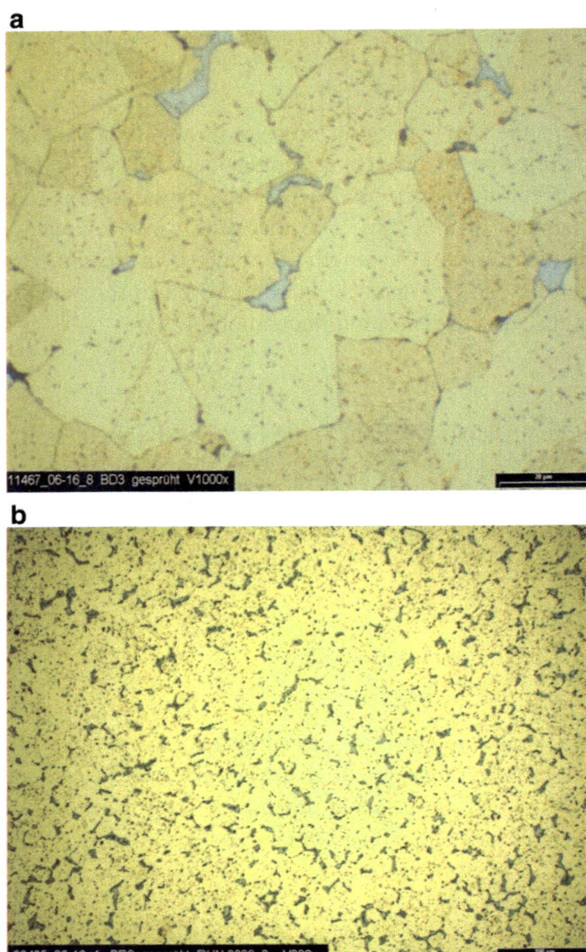

Fig. 11.9 Microstructure of high-tin bronze after spray forming, cross sections. Bronzes are commercially applied in superconductor wire and contain (**a**) 14.5% Sn and (**b**) 16.0% Sn

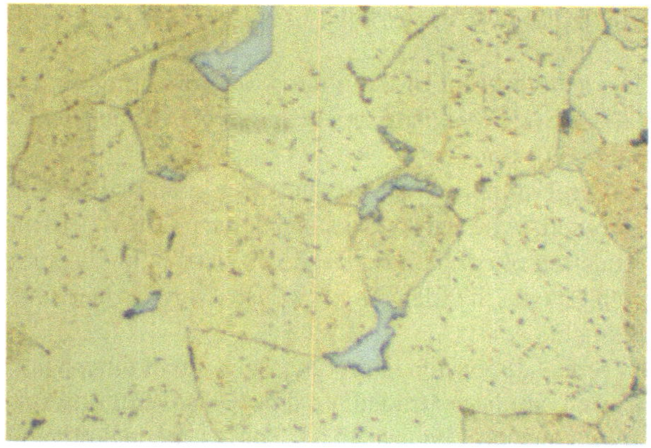

Fig. 11.10 Delta phase particles (grey) and small Ti_6Sn_5 particles (black dots) in CuSn 14.5Ti, condition after spray forming

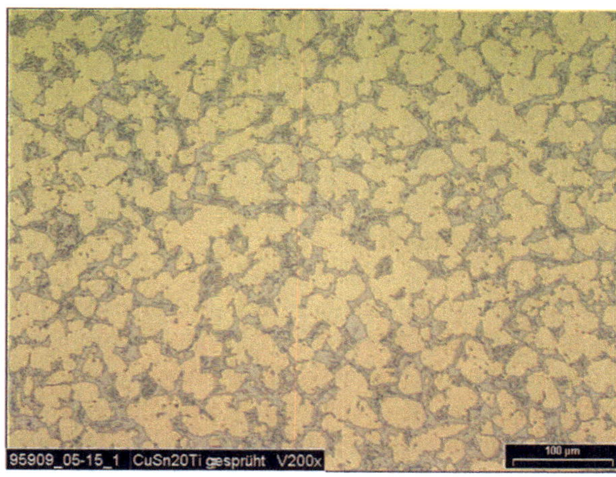

Fig. 11.11 Microstructure of 20 % tin containing bronze after spray forming. Cross section

optimum microstructure for further working processes. In comparison, cast bronzes exhibit dendrite structures which have to be removed by an expensive annealing process prior to further working [6]. Figure 11.10 shows the intermetallics structure in CuSn14.5Ti at a high magnification.

Figure 11.11 presents the microstructure of 20% Sn bronze in the condition after spray forming. Bronzes with 17 and 20% tin now are subject of investigations for the production of future Nb_3Sn-based superconducting magnets with increased magnetic field strength. Their content of precipitations inside the microstructure is even higher than in the commercial alloys. Hot and cold working of such a material is a challenge. It is possible to heat treat the material and transfer it into a alpha/beta microstructure in order to achieve better working properties [8].

Microstructure Development During Production Process

The development of the microstructure during the production of the semi-finished product is dependent on the product's dimension and whether cold forming is applied or not.

Large-dimensioned products, e.g. gun-drilled billets and sheath tubes, are hot extruded. After hot extrusion, delta phase is still present (see also [6]). Further reduction of delta phase fraction by annealing is possible. In gun-drilled billets this counteracts the danger of constricting (squeezing) or rupturing the niobium filaments.

Small-dimensioned products, e.g. small tubes with outer diameter in the order of 25 mm, are hot extruded, cold drawn several times and subsequently recrystallization annealed. Figure 11.12 shows the development of the microstructure in CuSn16Ti0.3 during the production process. After spray forming, the delta-phase particles show the typical configuration at grain boundary triple points and a volume fraction of about 10–15% (Fig. 11.12a), an optimum precondition for further working processes. After hot extrusion delta phase is arranged in longitudinal rows with different density having a fraction of about 15–20% (Fig. 11.12b). After several cold drawing steps and heat treatments, delta phase volume fraction drops to a value below 5%. Besides delta phase, the small-sized Ti_6Sn_5 particles are present (Fig. 11.12c).

11.3.1.2 Cu-Mn-Ni

For a number of applications, such as components for deep-well drilling, antimagnetic copper base alloys with very high strength and low scuffing tendency are required. The system CuMnNi basically has the potential to meet these requirements. However, also in this case, production with conventional casting methods is very difficult due to the high alloy contents. The phase diagram Fig. 11.13 shows the range of an age-hardened alloy. The position of the alloy Cu-20Mn-20Ni is marked with a blue dot. Similarly, the spinodal alloy Cu-15Ni-8Sn, normally produced by powder metallurgy, can be produced this way. In subsequent production, processing of CuMnNi is, however, somewhat easier, so that this alloy with similar mechanical properties is favoured. Figure 11.14 compares both alloys in the A5–R_m-diagram to beryllium copper and different steel grades. The entire range can be covered by Cu-20Mn-20Ni. In mechanical properties of the high-strength material Cu-20Mn-20Ni the standard values of the mechanical properties for three different conditions are listed in Table 11.2.

The alloy CuMn20Ni20 is characterized by

- High strength through final ageing up to more than 1100 MPa
- Non-magnetizability
- Good ductility
- No tendency to overageing and, therefore,
- Maintenance of the mechanical properties during the entire operation time.

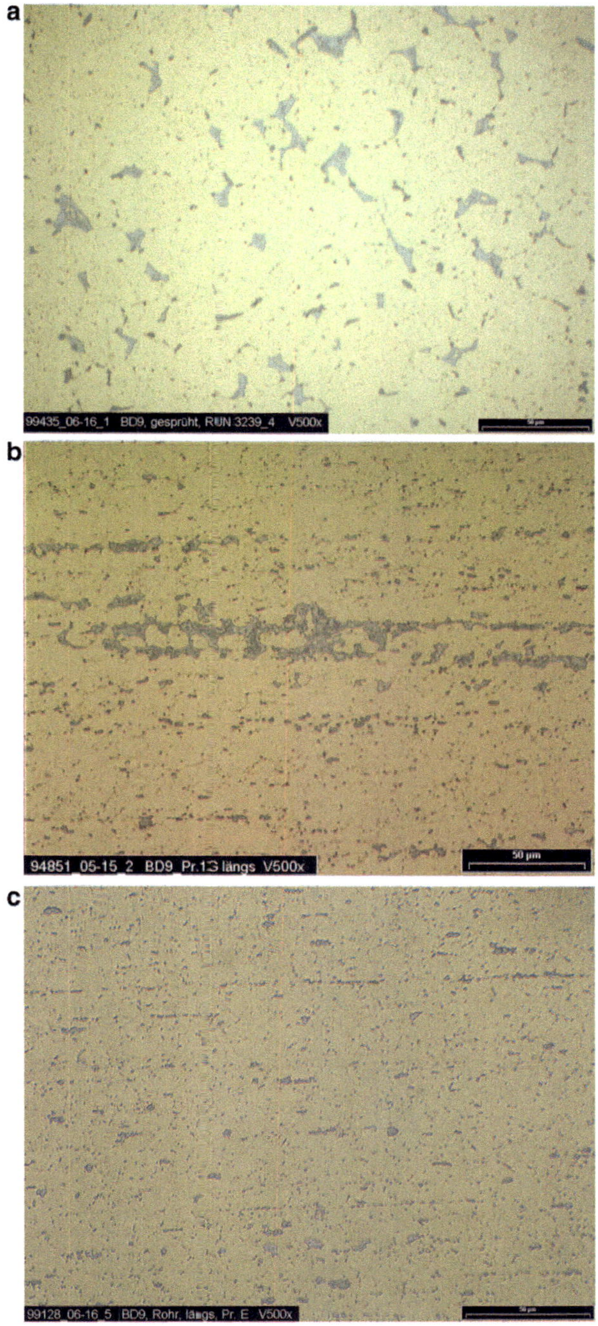

Fig. 11.12 Development of microstructure in CuSn16Ti0.3 superconductor bronze during the production of small-dimensioned pre-material (outer tube diameter 23 mm) in the condition (**a**) after spray-forming, (**b**) after hot extrusion, and (**c**) after several drawing and annealing processes

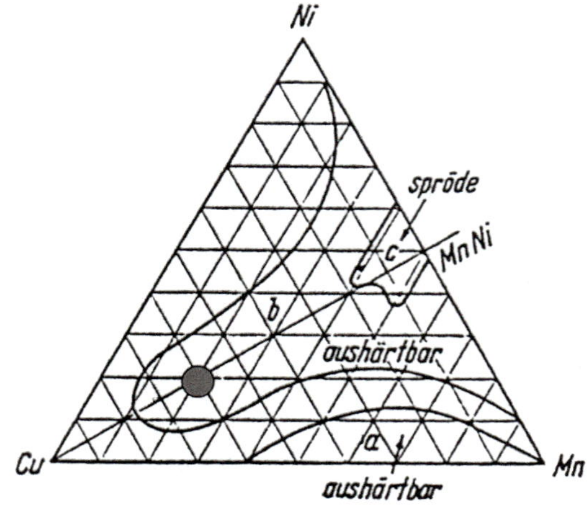

Fig. 11.13 Ternary phase diagram CuMnNi, with permission from [9]

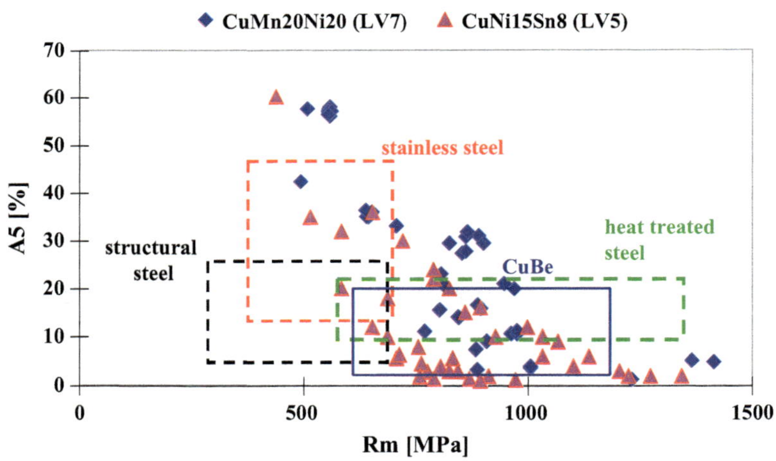

Fig. 11.14 Mechanical properties of Cu-20Mn-20Ni and Cu-15Ni-8Sn compared with other high strength materials

Table 11.2 Mechanical properties of the high-strength material Cu-20Mn-20Ni

Material	R_m (MPa)	$R_{p0.2}$ (MPa)	A5 (%)	HV
Cu-20Mn-20Ni, soft	450	180	35	120
Cu-20Mn-20Ni, 30% cold worked	750	650	5	240
Cu-20Mn-20Ni, 30% cold worked + aged	1100	950	0.5	350

11.3.1.3 Aluminium Bronze

Standard multi-phase aluminium bronzes contain up to 12 wt.% aluminium and an addition of iron, nickel and manganese [10, 11]. Multi-phase Cu–Al–Fe alloys with Al contents between 8 and 10 wt.% are well known as cast materials. The standard Cu–Al–Fe cast alloy Cu–10Al–3Fe is produced mainly by die casting. Although it may contain as much as 1 wt.% each of nickel and manganese, it is actually a ternary alloy with a duplex $\alpha + \beta$ structure. The influence of alloying elements and heat treatment on the microstructure and mechanical properties of these Al–bronzes has been investigated by a large number of authors [12–16]. Alloys containing 9–12 wt.% aluminium with additions of up to 6 wt.% each of iron and nickel represent a major group of commercial wrought aluminium bronzes. Standard alloys normally containing 3–6 wt.% each of iron and nickel were analysed in terms of microstructure and mechanical properties [11, 12, 14, 17, 18]. In this type of bronze a hardness of approximately 280 HB can be achieved. For some applications such as tube-bending tools, deep-drawing tools, slide bars and moulds a higher hardness level is required. Copper–aluminium–iron alloys containing 12–15 wt.% aluminium are characterised by exceptionally high hardness in the range of 300–400 HB and low ductility [19]. Unlike information on standard aluminium bronze, information on aluminium bronze with an aluminium content of more than 12 wt.% is limited [4, 20, 21]. In [4] it has been reported that hardness values of more than 400 HB and compressive strength up to 1350 MPa can be achieved in spray-formed high-aluminium bronzes. Spray-formed aluminium bronzes have a considerably improved machinability compared to cast high-aluminium bronzes. Here, the microstructure and properties of spray-formed wrought copper–aluminium–iron alloys with high aluminium content is characterised. The microstructure and properties of three spray-formed aluminium bronze grades are analysed.

Experimental Procedure and Investigated Materials

A vacuum induction furnace with a melt capacity of 3 t is used [22]. In the first step the pure copper is melted and degassed through vacuum treatment. In the second step the additional elements in pure condition (Al) or as master alloys (Cu–Fe, Cu–Co, and Cu–Mn) are charged. After homogenisation through inductive stirring and after the pouring temperature has been reached, the melt is transferred to the holding furnace in the spray forming unit (see Fig. 11.2). The spray forming unit and process are described in detail in [4]. The outlet of the holding furnace is at the bottom and prevents the melt from picking up oxygen from the ambience. The melt stream is dispersed into droplets by nitrogen in a single-atomiser unit. During the flight time the droplets partly solidify and subsequently hit a rotating target where they are compacted into the shape of a billet. The analysed billets had a diameter of 320 mm and a length of 2200 mm. Before extrusion the billets are machined into a

Table 11.3 Nominal chemical composition of spray-formed aluminium bronze

Alloy	Cu	Al	Fe	Co	Mn
Cu–12.5Al–4Fe–Mn–Co	Balance	12.5	4.5	1.0	1.0
Cu–13.5Al–4Fe–Mn–Co	Balance	13.5	4.5	1.0	1.0
Cu–14.5Al–4Fe–Mn–Co	Balance	14.5	4.5	1.0	1.0

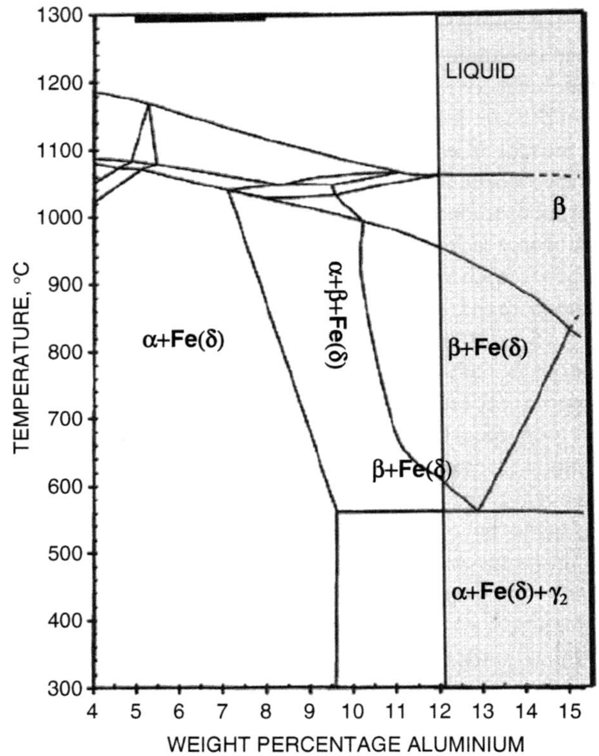

Fig. 11.15 Vertical section of the Cu–Al–Fe system at 5 wt.% Fe, by Yutaka [23]. Area of spray-formed Al–bronze is marked in grey colour

diameter of 298 mm and cut into a length of 450 mm. These billets are reheated and hot extruded at temperatures above the eutectoid transformation temperature. The rods had a diameter of 26 and 50 mm. The microstructure and properties of three spray-formed aluminium bronze grades were analysed. The nominal chemical composition of these three grades is presented in Table 11.3.

The vertical section of the copper–aluminium–iron system with 5 wt.% Fe of [23] shows that aluminium bronzes containing 12–15 wt.% Al have little or no α-phase and are therefore very susceptible to grain growth at high temperatures (see Fig. 11.15). Therefore, iron is added, and occasionally also certain amounts of Co and Mn in order to improve grain refinement. The addition of cobalt and manganese eliminates the tendency towards eutectoid formation in these alloys, which otherwise may occur during heat treatment or after thermal stress [21].

The eutectoid structure consists of α-phase and γ_2-phase formed as a result of the decomposition of β-phase. The eutectoid structure is brittle and exhibits low ductility and poor machinability.

Results

Microstructure Investigation

Spray forming is known to generate material with a homogeneous distribution of elements, a fine and homogeneous grain structure as well as a fine and homogeneous distribution of intermetallic particles [4, 6, 7]. The structure of spray-formed Al–bronze with an aluminium content of 14 wt.% and the structure of cast Al–bronze containing 14 wt.% aluminium are compared to each other in Figs. 11.16 and 11.17. Compared to mould-cast bronze, spray-formed bronze shows a fine-grained matrix structure (see Figs. 11.16a and 11.17a). In both cases, the β-phase is assumed as matrix, since eutectoid transformation within

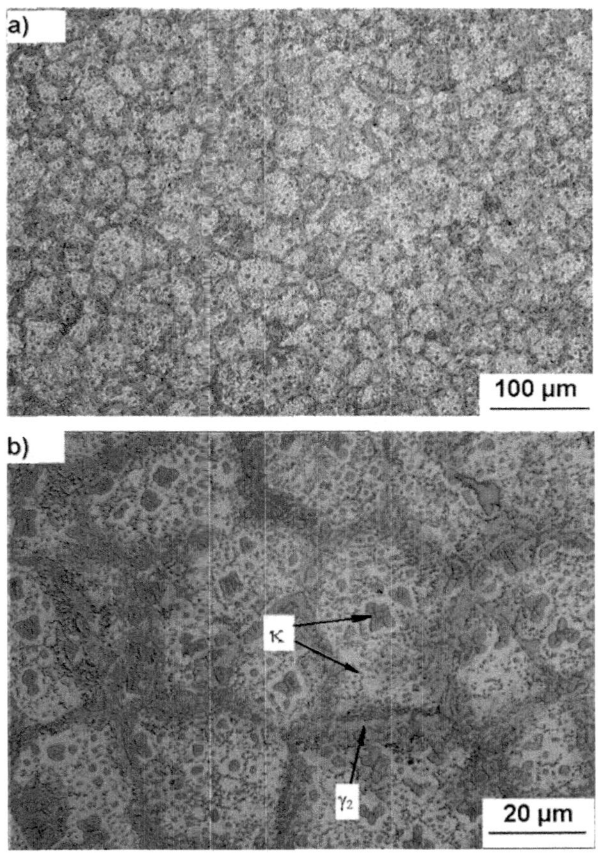

Fig. 11.16 Structure of spray-formed Al-bronze Cu–14.5Al–4Fe–Mn–Co

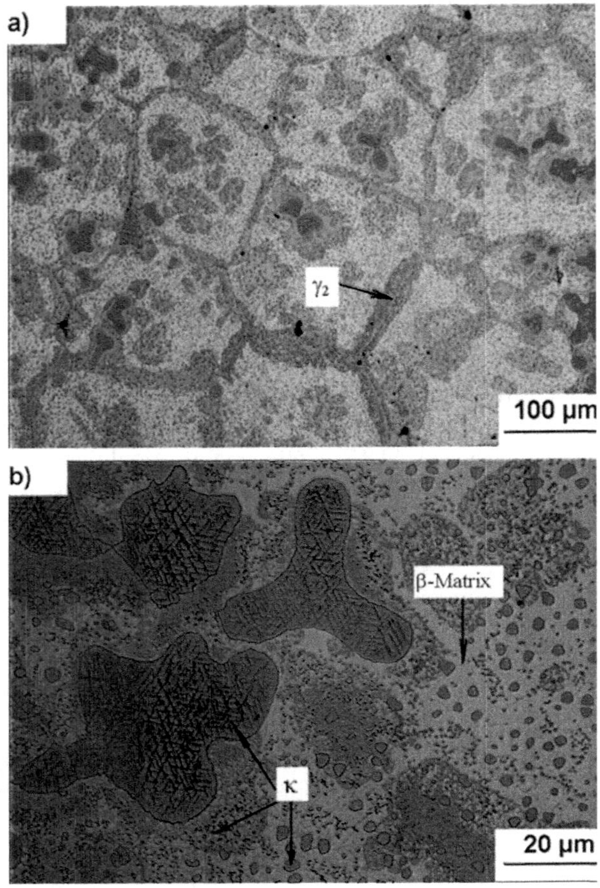

Fig. 11.17 Structure of permanent-mould cast Al-bronze Cu–14.5Al–4Fe–Mn–Co

grains is prevented by adding Fe, Co and especially Mn [10, 24]]. In the case of spray-formed material, the κ-particles (Cu–Al–Fe–Co) are finer (see Figs. 11.16b and 11.17b). In both bronzes the brittle γ_2-phase (Cu_9Al_4) and the brittle eutectoid $\alpha + \gamma_2$ are only located at the grain boundaries (see Figs. 11.16a and 11.17a). Meigh [11] reported that in high-aluminium bronzes γ_2-phase and eutectoid $\alpha + \gamma_2$ also occur at the primary grain boundaries where they create an area of weakness along which fracture would tend to occur. Unlike mould-cast bronze which can only be used in the cast condition, spray-formed high-aluminium bronzes are able to be hot formed. The distribution of **κ**- and $\boldsymbol{\gamma}_2$-phases in spray-formed aluminium–bronzes is modified by the hot forming process. Figure 11.18a, b shows the structures of hot-formed aluminium bronze with an aluminium content of 12.5 and 14.5 wt.%. After hot working, the γ_2-phase is located in the β-matrix and no longer at the grain boundaries. Moreover, the particles of γ_2-phase are no longer connected to each other (compare Figs. 11.16a–11.18a with Figs. 11.16b–11.18b). A fine and homogeneous distribution of κ- and γ_2-phases according to the literature [11] has a favorable effect on the mechanical properties and corrosion resistance of multiphase high-aluminium bronze.

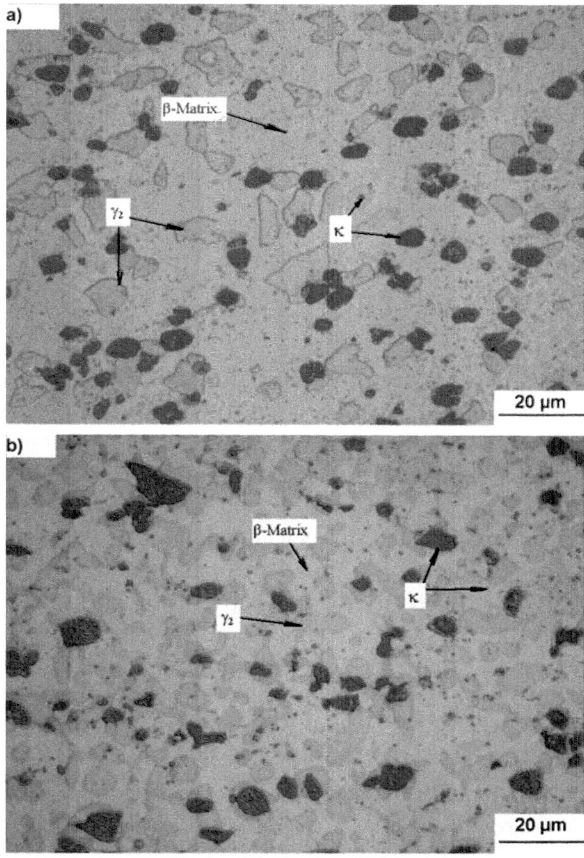

Fig. 11.18 Structure of spray-formed and extruded Al-bronzes, **a**) Cu–12.5Al–4Fe–Mn–Co, **b**) Cu–14.5Al–4Fe–Mn–Co

Table 11.4 Properties of spray-formed extruded aluminium bronze

Alloy	Hardness HB2.5/167.5	Young's modulus (GPa)	Density (g/cm^3)	Thermal conductivity (W/(m·K)) 25°C	300°C
Cu–12.5Al–4Fe–Mn–Co	260–320	96	7.2	43	8
Cu–13.5Al–4Fe–Mn–Co	320–360	109	7.1	34	60
Cu–14.5Al–4Fe–Mn–Co	360–400	122	7.1	30	50

Mechanical and Physical Properties

Table 11.4 shows a strong dependency of physical properties and hardness on the aluminium content. Hardness rises by approximately 100 HB with an increase in aluminium content by approximately 2 wt.%. The Young's modulus of the aluminium bronze presented in Table 11.4 increases substantially with the increase in aluminium content. In commercial copper–aluminium alloys with α-structure the Young's modulus decreases with the increase in aluminium content and increases

Table 11.5 Comparison of mechanical properties of spray-formed and mould-cast Al-bronzes

Alloy	Compressive strength (MPa)	0.2% yield strength (MPa)	Elongation at fracture (%)	Hardness (HB)
CuAl12.5Fe4.5CoMn	1200	650	5	300
CuAl14.5Fe4.5CoMn	1600	970	5	380
CuAl10Fe3 [20]	980	235	14	200
CuAl10Ni5Fe4 [20]	1200	700	8	280
CuAl13 [20]	1100	343	1.5	280
CuAl14 [20]	1200	540	0.5	300
CuAl15 [20]	1450	720	0.2	350

with the occurrence of γ_2-phase [10, 25]. According to [26], the Young's modulus reaches a maximum value of approximately 200 GPa if the composition of aluminium bronze equals γ_2. Bronze with 14.5 wt.% Al exhibits a clearly higher fraction of γ_2-phase compared to bronze with 12.5 wt.% Al (see Fig. 11.18a, b). Thermal conductivity decreases with the increase in Al content (see Table 11.4). Thermal conductivity of spray-formed aluminium bronze rises substantially with an increase in temperature. For example, thermal conductivity at 300°C is already more than 50% higher than thermal conductivity at room temperature (RT) (see Table 11.4). Table 11.5 shows the compression test data for two grades of spray-formed and several conventional aluminium bronzes [20]. The average values of the conventional bronzes presented in Table 11.5 are given in [20]. The mechanical properties of the spray-formed alloy Cu–12.5Al–4.5Fe–Co–Mn are similar to those of the standard alloy Cu–10Al–5Ni–4Fe. However, the spray-formed alloy offers a better combination of characteristics at elevated temperatures (see Sect. 11.3.1.3.2.3). Compared to the alloy Cu–13Al the spray-formed bronze with 12.5 wt.% aluminium exhibits a clearly higher 0.2% compressive strength and a higher elongation at fracture. Compared to other conventional Al-bronzes from Table 11.5 the spray-formed material Cu–14.5Al–4.5Fe–Co–Mn shows the highest compressive strength and an elongation at fracture of approximately 5%. The better elongation of spray-formed materials can be explained by homogeneous distribution of brittle γ_2-phase in the β-matrix. Due to the fine and homogeneous distribution of κ- and γ_2-phases the spray-formed bronzes exhibit a higher 0.2% compressive strength than conventional aluminium bronzes with a comparable Al content.

Properties at Elevated Temperatures

Figure 11.19 shows the results of the tensile tests at elevated temperatures between room temperature and 700°C. For comparison, Fig. 11.19 also includes published data from [24] for a standard high-strength bronze Cu–10Al–5Ni–4Fe. Spray-formed bronze shows a considerable decrease in strength only from 300°C, whereas the standard bronze Cu–10Al–5Ni–4Fe shows a substantial loss of strength already at 100°C. Elongation at fracture of spray-formed bronze at low temperatures increases with an increase in temperature. Then it drops in the temperature range

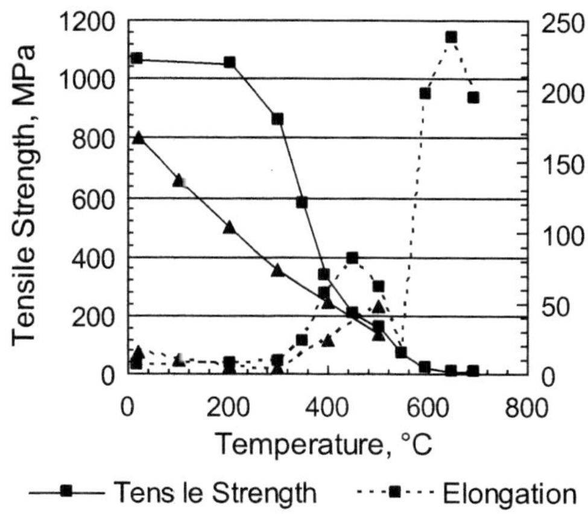

Fig. 11.19 Tensile properties vs. test temperature of spray-formed bronze CuAl12.5Fe4.5CoMn (square) and of conventional bronze CuAl10Ni5Fe4 drawn and annealed at 850°C/0.5 h [18](triangle)

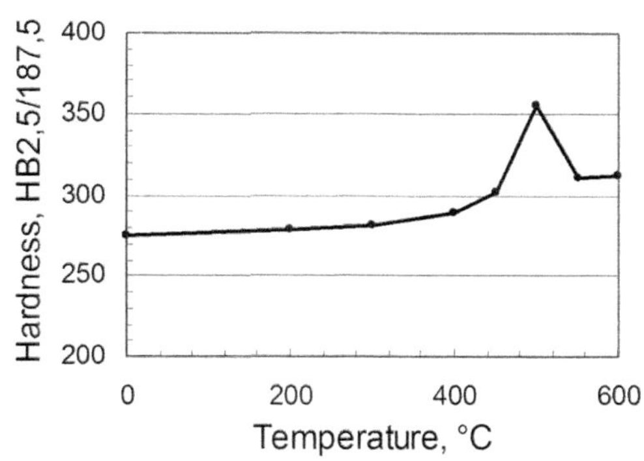

Fig. 11.20 Hardness of CuAl12.5Fe4.5CoMn at room temperature after heat treatment at different temperatures for 1 h

between 450 and 550°C. In this temperature range the eutectoid decomposition $\beta \rightarrow \alpha + \gamma_2$ occurs. The increase in γ_2–fraction leads to a decrease in elongation at fracture. At temperatures above approximately 560°C (see Fig. 11.15) the eutectoid $\alpha + \gamma_2$ transforms into β-phase and elongation at fracture increases substantially. At temperatures above 700°C grain coarsening is probably responsible for the decrease in elongation. For assessing the thermal stability of the structure and consequently the properties after thermal stress, the Al–bronze Cu–12.5Al–4.5Fe–Mn–Co was heat treated for one hour in the temperature range between 200 and 600°C. The hardness at room temperature after heat treatment is shown in Fig. 11.20. The analysed specimens did not show a decrease in hardness throughout the entire temperature range. Up to 450°C hardness increases slightly, and at 500°C the maximum hardness is achieved. This maximum hardness is related to the increase in volume fraction of γ_2-phase in the microstructure due to the decomposition of β-phase.

Overlay Welding

Figure 11.21a and b show overlay welds and a cross-section of a longitudinal weld, respectively. The analysed welds are free from pores and cracks. The hardness curves across weld, heat transfer area and base material in a specimen are presented in Fig. 11.22. Compared to the base material (measuring points 4–6), the weld and heat-affected zone (measuring points 1–3) exhibit a slightly lower hardness. In the base material the hardness HV5 is approximately 280–310. The hardness in the welds (230–240) is considerably higher than the hardness of the electrode material Cu–8Al. According to the data sheet the hardness is approximately 100 HB. The filler material is well connected with the base material. The welded areas are sufficiently hard for further processing.

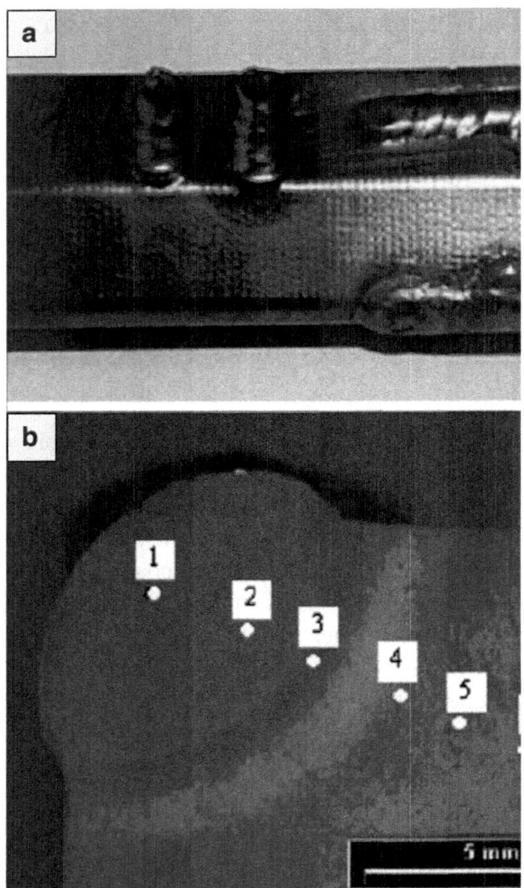

Fig. 11.21 (**a**) Specimen with overlay welds and (**b**) cross-section of a longitudinal weld

Fig. 11.22 Hardness curve in the weld from Fig. 11.21

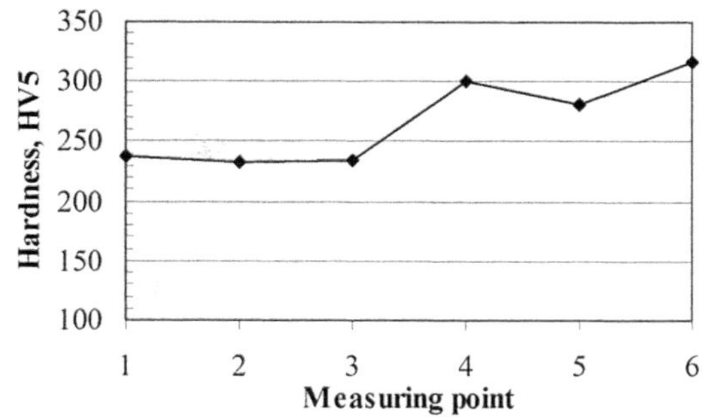

Machinability

Unlike standard cast grades, spray-formed aluminium bronzes exhibit good machinability. Good machinability has a direct influence on the surface quality of the work pieces. The cutting edges of machined conventional aluminium bronze containing 14.5 wt.% aluminium show substantial material breakage (see Fig. 11.23a). In contrast, spray-formed aluminium bronze with the same aluminium content exhibits considerably reduced material breakage at the cutting edges (see Fig. 11.23b). Figure 11.23c, d shows the reversible plate tool after these tests. After a test run with conventional bronze the tool edge is damaged and the tool can no longer be used. After the test run with the same cutting parameters, but spray-formed bronze with identical Al content, the tool edge shows no damage (Fig. 11.23d) and can be further used. If the spray-formed bronze with 14.5 wt.% Al is machined with optimum parameters, the machined surface is smooth and the cutting edge does not show any material breakages (see Fig. 11.24).

New developed types have considerably improved machinability, which is realised in better surface quality, lower tool costs due to reduced wear and in less roughness of the chips, because they break more easy during the machining process (see Fig. 11.25).

11.3.1.4 Copper-Nickel-Silicon Alloys

For low-alloyed Cu-Ni-Si (such as CuNi2Si or CuNi3Si) alloys there is no usually need for spray-forming since conventional continuous casting can provide a sufficiently homogeneous pre-material. However, for high-alloyed Cu-Ni-Si alloys such as CuNi7Si2Cr spray forming has found its way into production. In the past decade, the European automotive industry's interest in replacing high-strength beryllium-copper by beryllium-free copper alloys has increased significantly. Here, spray-formed CuNi7Si2Cr is an excellent alternative to Cu-Be, Fig. 11.26. Spray-formed CuNi7Si2Cr is an age-hardenable copper alloy which can be hot-extruded or forged into rods, tubes or more complex shapes.

In this alloy, precipitation hardening gives rise to Ni- and Cr-silicides, leading to excellent strength at ambient as well as at elevated temperatures. Moreover, the

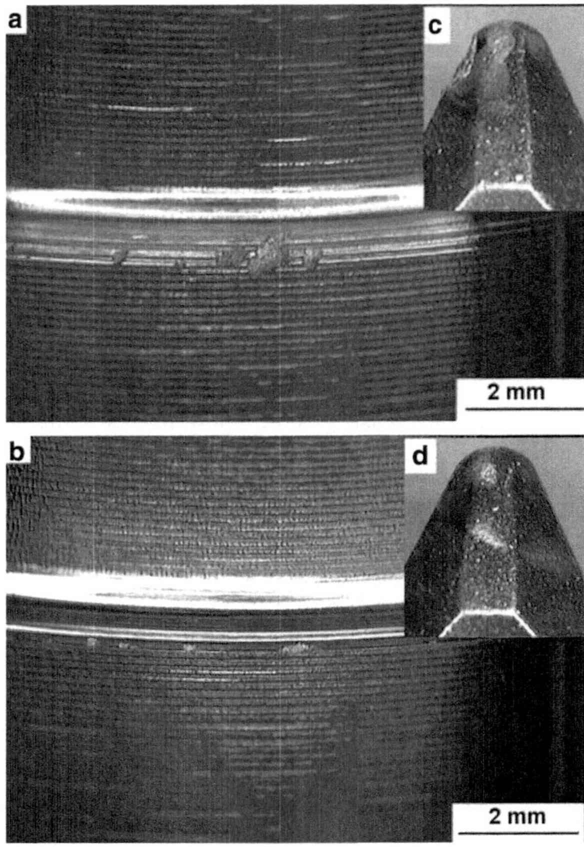

Fig. 11.23 Cutting edge of (**a**) conventional and (**b**) spray-formed 14.5% Al-containing aluminium bronze, machined with same setting of the turning machine (1250 rpm, feed 0.15 mm, depth of cut 1 mm) and the reversible plate (**c**) conventional and (**d**) spray-formed 14.5% Al-containing aluminium bronze after this test

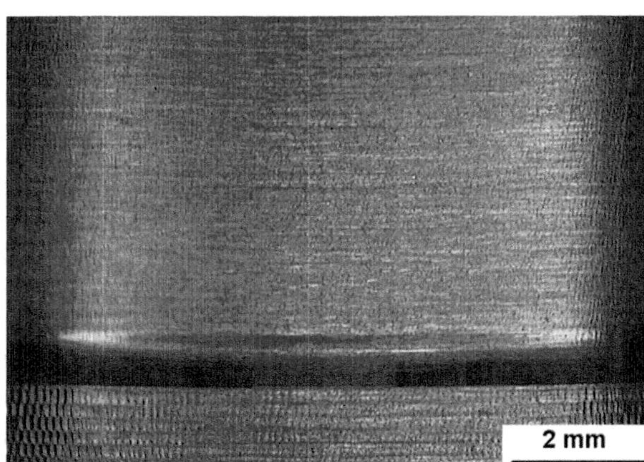

Fig. 11.24 Smooth surface without material breakages and cutting edge with optimized machining parameters

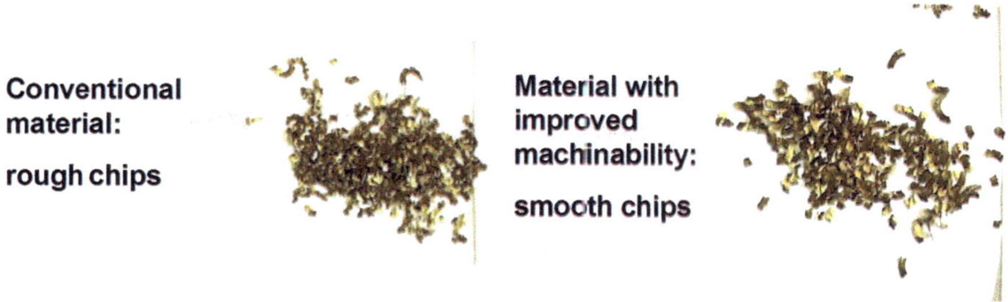

Fig. 11.25 Al-Bronze chips of conventional and improved material at same machining conditions

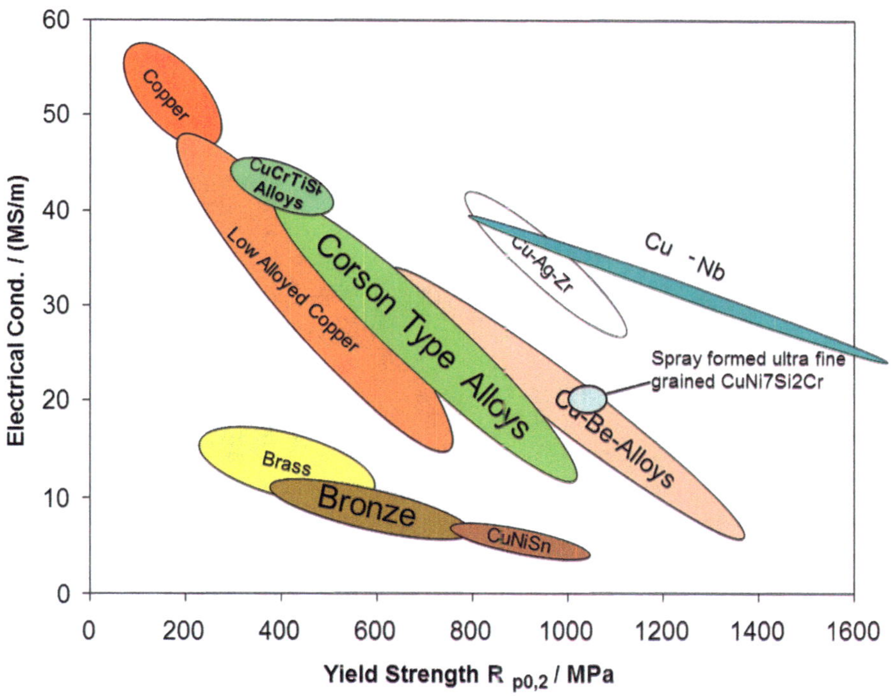

Fig. 11.26 Ashby-map (here: conductivity vs. yield strength) of different copper alloys. Spray-formed CuNi7Si2Cr shows properties comparable to Cu-Be-alloys

yield strength at elevated temperatures is superior to austenitic stainless steel and conventional high performance copper alloys up to 700°C. Hence, spray-formed and hot-extruded CuNi7Si2Cr is an excellent material for high-temperature applications, e.g. for die-casting plungers.

Due to the presence of Cr- and Ni-silicides (predominantly $CrSi_2$ and Ni_2Si) CuNi7Si2Cr exhibits excellent thermal stability as well as stress relaxation resistance, Fig. 11.27. Additional cold working (such as swaging) prior to optimized

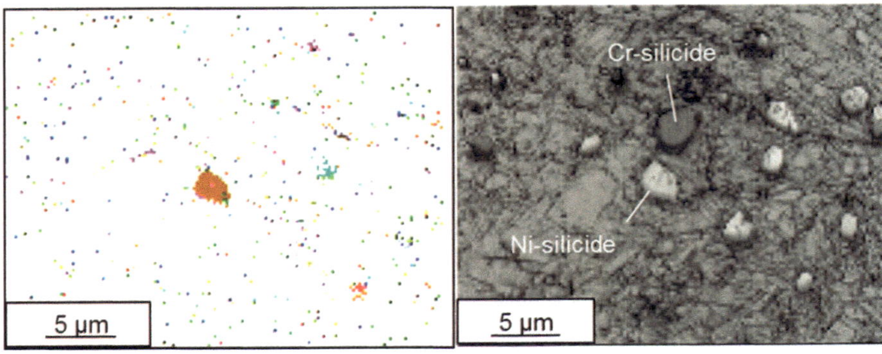

Fig. 11.27 Microstructure of spray-formed, hot extruded and subsequently swaged and precipitation hardened CuNi7Si2Cr as characterized by electron backscatter diffraction (EBSD)-mapping (*Left*: location of coarse and nanoscopic orthorhombic Ni-silicides, *Right*: band contrast showing the grain- and deformation structure)

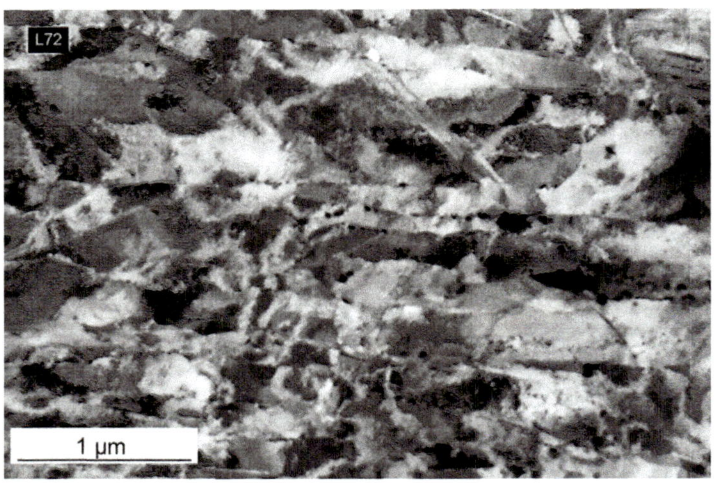

Fig. 11.28 Electron channeling contrast imaging (ECCI) picture of spray-formed, hot extruded and subsequently swaged, precipitation hardened and cold worked CuNi7Si2Cr. Precipitates (*arrow*) are able to stabilize the ultra fine grained structure during thermal exposure

aging results in ultimate tensile strengths of 1050–1100 MPa and yield strengths above 1000 MPa at elongations of about 6%. In severely plastically deformed and subsequently aged CuNi7Si2Cr a localized precipitation of silicides at grain- and sub-grain boundaries is able to very effectively pin the grain and sub-grain boundaries during thermal exposure and thus enhance the thermal stability of the microstructure (Fig. 11.28, see also [27, 28]). In respect to fatigue strength (10^7 cycles endurance limit) of CuNi7Si2Cr fatigue strength values of 380 without surface treatment and 450 MPa with surface treatment (optimized shot peening or deep rolling) are routinely possible [29], the latter being associated with achievable near-surface hardness of up to 450 HV10.

11.3.2 MMC

11.3.2.1 Injection of a Second Component

To enhance the friction characteristics as well high temperature stability sliding materials are designed preferably with a multiphase structure. Beside the possibility to precipitate further phases, spray forming offers the embedding of particles [7]. Figure 11.29 shows for example graphite particles embedded in the known sliding alloy Cu-6Ni-6Sn, originally designed for high loaded, oscillating bearings.

The fraction of graphite particles can be increased up to 1%. In production processes with a preferred direction, as extrusion or drawing, an in line order of the particles can not be avoided. This impairs the original excellent isotropy of spray-formed materials. So it is necessary to take care of the load direction in the application. Therefore friction properties are no materials but system properties, depending on the sliding counterpart, lubricant, test method, etc., no friction coefficient will be specified here. In principle it is possible to spray graphite in all alloys but other particles like boron nitride or alumina can be embedded also. The particle size should not be less than 10 μm, because the efficiency will be strongly diminished. Smaller particles follow the gas flow and are not embedded.

Instead of lead graphite can be used as chip breaking mean. Figure 11.30 shows the chip length dependent on the graphite content for Cu-2Ni-0.5Si as an example, [30, 31]. Only little additions up to a volume fraction of about 2.5% improve the machinability significantly.

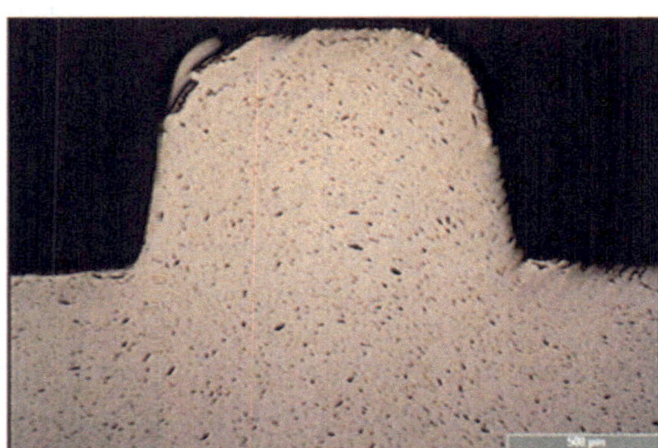

Fig. 11.29 Cu-6Ni-6Sn with embedded graphite particles, about 0.9% graphite

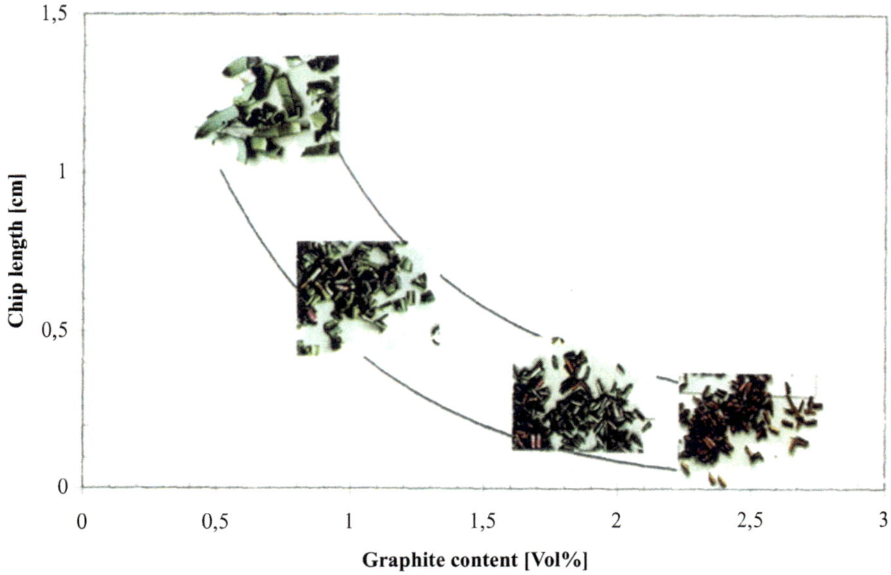

Fig. 11.30 Chip length vs. volume fraction of graphite in Cu-2Ni-0.5Si

11.3.2.2 Reactive Spray Forming

During spray forming, a melt is atomized into small droplets by means of an inert gas. Depending on the spraying conditions, droplets with average diameters of 60 μm are yielded. The maximum diffusion path is thus roughly 30 μm. The liquid medium ensures high diffusivities of gaseous compounds and each alloy component. Spray forming is therefore considered to have a high potential for the production of ODS-alloys [32]. Several experiments have been conducted using different reactive systems (e.g. (CuAl) melt in (N_2–O_2) gas mixture [33]). The idea is, to utilize the availability of large surface areas, small diffusion paths, high diffusivities and high negative enthalpies for the formation of the oxide according to Eq. (11.1).

$$2\,Al_{liq.in\ Cu} + \frac{3}{2}O_{2\ gas} \rightarrow Al_2O_{3\ sol} \tag{11.1}$$

Although in principle this chemical reaction has been proven, a distribution of fine dispersoids with particle sizes in the range of 20–40 nm was not found. Possible explanations are short reaction times, reduced diffusivities through growing oxide scales, non-isothermal conditions (i.e. steadily decreasing temperatures), high differences in density and the formation of copper oxides due to the high oxygen partial pressure used by the authors.

The fundamental difference between Eq. (11.1) and the production of dispersion strengthened copper alloys using the GlidCop®-process is the physical state of the alloy during the oxidation process. In contrast to Eq. (11.1) the aluminium atoms are more or less “fixed” in the fcc-lattice of the dilute CuAl-powder. Buoyancy forces, particle instability and insufficient resistance against coarsening can therefore be neglected. The presence of aluminium in a liquid state changes most of the conditions in a negative way. When N_2/O_2-mixtures are used, oxygen adsorbs on the surface of the droplet during the flight and dissociates into highly reactive atoms. If thermodynamic conditions favour the formation of Al_2O_3 only, the oxygen atoms have the following principal possibilities:

A) Dissolve into the droplet and react with “floating“ aluminium
B) React with aluminium on the surface of the droplet.
A) Dissolution and Subsequent Reaction
The oxygen atoms are thought to be highly mobile within the liquid alloy. Calculations have shown, that the total diffusion path can be in the order of a few microns [34]. The oxidation of all aluminium within small droplets can therefore be expected. Upon formation of the oxide, the large differences in density ($Al_2O_3 \rightarrow 3.4$ g/cm^3, CuAl-alloy $\rightarrow 7.5$ g/cm^3) result in a buoyancy force that sets the Al_2O_3-particle in motion. Individual particles may become trapped by the simultaneous occurring solidification process. However, a significant statistical distribution should not be expected since the particles are located either interdendritically or on grain boundaries. Moreover, the high mobility supports the coarsening rate of each particle.
B) Surface Reaction
Aluminium oxide, formed on the surface of the droplets, has no net force to penetrate or dissolve into the liquid droplet. However, the oxide film might be easily torn apart as long as the underlying alloy is liquid. In this case areas of melt are exposed for further reactions. Upon solidification, the floating oxide film is fixed and further growth is limited by solid state diffusion of either oxygen or aluminium. In any case, the diffusion rates are decreased by orders of magnitude. The growth depends on the defect structure of the oxide. The number and type of defects are responsible for the growth rate of the oxide. SiO_2 and Al_2O_3 have shown to form protective oxide scales. This is explained by the low concentration of defects. Continuous growth through preformed oxide scales can thus be neglected.

Several experimental approaches indicate, that the formation of nano-scale particles require a solid state reaction (like the long established GlidCop®-process). The “advantages” of the spray forming process (high surface area, short diffusion paths, high diffusivities) turn out to be major obstacles. However, in order to avoid the sequence of expensive manufacturing steps as described above, the distribution of dispersoids (or their preforms) and subsequent compaction into a billet should stimulate scientists or metallurgical engineers to search for promising systems.

The atomization into small droplets and simultaneous creation of high surface areas offers the opportunity to distribute reactants throughout the billet. The following chemical reaction into a dispersoid should be postponed until the solidification has been completed. Temperature induced diffusion should yield the desired formation of fine particles.

11.4 Applications

11.4.1 Low Temperature Superconductor (CuSn)

The application of spray-formed high tin bronzes for superconducting wires is described in detail elsewhere [4]. One example is the ITER-Project for using the nuclear fusion process for power generation. Figure 11.31 shows, where this material is used. The tin bronze is used as a carrier of tin for the final reaction with Nb to the superconducting Nb_3Sn-phase. The production process can be splitted in 6 steps (Fig. 11.32):

1. Spray forming of bronze billet.
2. Extrusion and cold drawing of bronze sheath tubes and hexagon tubes.
3. Assembling bronze tubes, Nb-rods.
4. Co-working of Niobium and bronze (sequences of extrusion, cold drawing, heat treatment).
5. Coiling of wire to final shape on the magnet.
6. Formation of Nb_3Sn by final annealing process.

During this process a multi-filamentary composite material is generated. Figure 11.33 shows a cross section through a non-stabilised type of superconductor

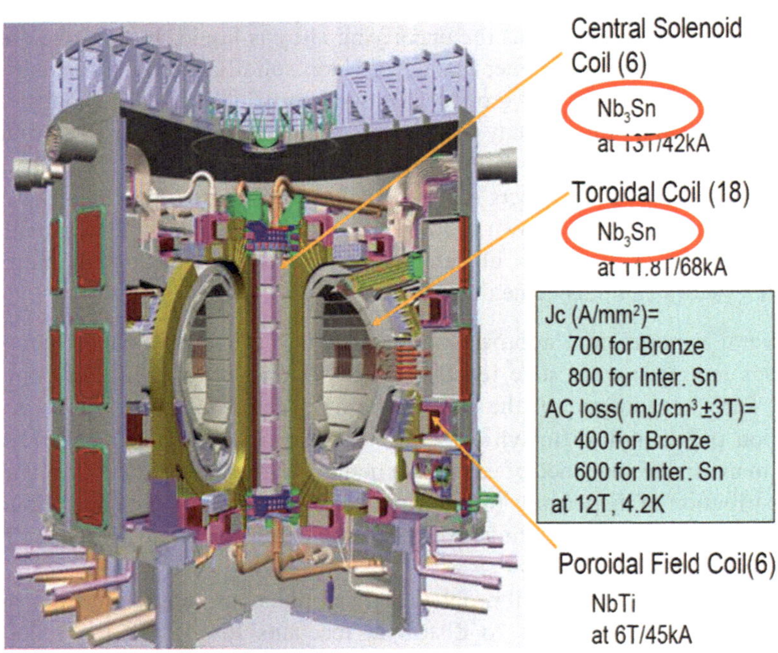

Fig. 11.31 Application of high-tin bronze in International Thermo-Nuclear Experimental Reactor (ITER), adapted from [35]

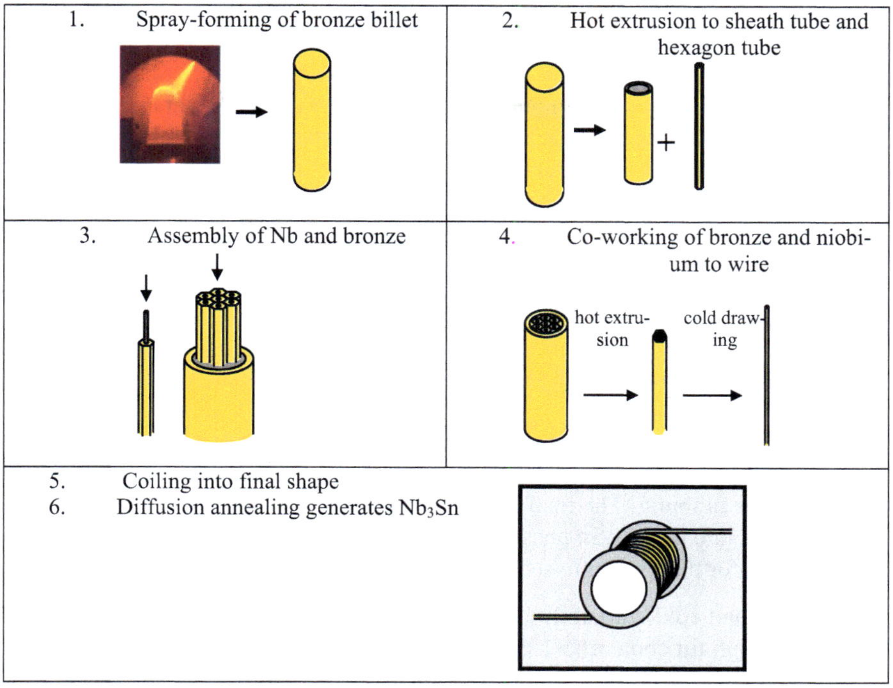

Fig. 11.32 Production process of bronze for super-conducting Nb_3Sn wire (scheme) [4]

Fig. 11.33 Cross-section of superconductor wire in production stage after hot extrusion. Bronze matrix contains 120 x 84 = 10080 Nb-filaments, with courtesy to Bruker EAS, Hanau, Germany

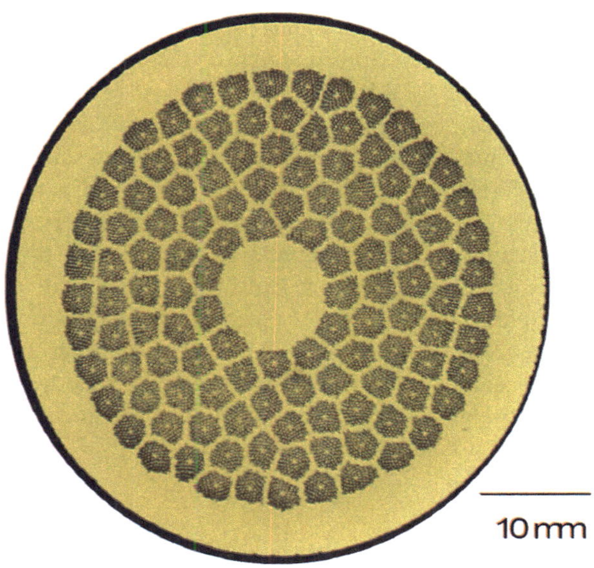

in a production stage after extrusion and before cold-drawing of the niobium-bronze-composite (production step 4). At a first glance 120 hexagonal grey dots surrounded by bronze are visible. Each dot consists of 84 niobium filaments which also are surrounded by a bronze matrix. This matrix has been generated during the extrusion of 84 hexagon tubes of bronze (e.g. hexagonal dimension of 19 mm) filled with niobium rods (e.g. diameter of 9 mm).

The following production steps are numerous cold-drawing steps and heat treatments. The dimensions of niobium filaments and bronze channels finally reach the size of 5–20 μm.

11.4.1.1 Requirements to High-Tin-Bronze Semi-Finished Material

The manufacturers of super-conducting wire need a good workability of the tin-carrying bronze. By means of forming processes the tin is brought as near as possible to the niobium. The final heat treatment lets tin diffuse to the niobium and transfers both into the Nb_3Sn phase. Accordingly the manufacturers state two main requirements for the semi-finished bronze material:

- Good hot and cold workability.
- Suitable high tin content (>13%) and homogeneous tin distribution.

11.4.1.2 Advantages of Spray-Formed High-Tin-Bronze

The requirements to the material are a challenge to the material manufacturer. Bronzes produced by casting processes have a high tendency to grain growth, tin-segregation and precipitation of the inter-metallic delta-phase, a copper-tin-phase with 32–33% tin. These two effects are detrimental to the further production processes of the super-conducting material. Cast bronzes therefore have to be heat treated time- and cost-intensively before being suitable for the production process. Spray-formed high-tin bronzes have various advantages compared to cast high-tin bronzes. The grain structure of spray-formed material is fine and homogeneous. The contents up to 13.5% tin are single-phased. Those with tin content of more than 13.5% have a homogeneously distributed and easily deformable structure of small delta-phase particles. Inverse segregation is minimised to 1% tin concentration difference across the billets. All these advantages are beneficial to a good workability and to process stability as well as they help to simplify the production process, e.g. help to make initial heat treatment unnecessary. Therefore, they contribute to reduce production cost.

11.4.2 Oil Drilling Equipment (Cu-20Mn-20Ni)

Spray-formed Cu-20Mn-20Ni is a high strength, non-magnetic alloy and offers an interesting combination of properties for materials used in oil drilling equipment [36]. It is used as a construction material for rods in offshore/onshore oil drilling exploration and has comparable mechanical properties to high strength steel, titanium alloys or copper-beryllium alloys [37]. Since it is non-magnetic, it can be used for rods containing sensors and electronics for orientation purposes. In Fig. 11.34 components are marked with yellow circles. Cu-20Mn-20Ni is an age-hardenable alloy, therefore appropriate heat treatment can tailor the desired mechanical properties. Figure 11.14 illustrates the mechanical properties of spray-formed and extruded Cu-20Mn-20Ni in different aging conditions. Although stainless steels and copper-beryllium alloys exhibit comparable or higher ductility and comparable tensile strength, respectively, Cu-20Mn-20Ni offers a better combination of strength and ductility. The basic strengthening mechanism in this alloy is precipitation hardening by ordered tetragonal MnNi-precipitates [38] serving as dislocation obstacles and thus improving the tensile- and yield strength. The heat treatment comprises a solution treatment at 700°C followed by artificial aging at 400°C. Spray-formed and extruded Cu-20Mn-20Ni offers a good corrosion resistance and resistance against stress corrosion cracking, thermal long term stability up to 300°C, good impact energy of >20 J and a low tendency to over-aging. Moreover, it can be cold worked easily and machined in the soft (solution heat treated) and medium hard (partially aged) condition. Tensile strength varies from 500 to 1300 MPa, depending on the aging condition of the material. Fine wire (dia. 50–60 μm) can be drawn and hardened to a tensile strength up to 1600 MPa.

11.4.3 Cold Working Tools and Injection Moulds (Al-Bronze)

The higher procurement costs of spray-formed aluminium bronzes are often not only compensated but even turned into an advantage by the cost savings resulting from good machinability. Therefore, spray-formed aluminium bronzes are widely used for a number of industrial applications such as tube-bending tools, see Fig. 11.35, and deep-drawing tools. Another field of application of spray-formed aluminium bronzes is plastic injection moulding, for example, for the production of bottle cases in the beverage industry. The moulds consist of several bars (see Fig. 11.36). Today, the bars are increasingly being made of spray-formed aluminium bronzes. This application requires a certain level of ductility because bending stresses in the bars may occur during mounting and field operation. It also requires the ability to be weld repaired. Spray-formed aluminium bronze grades containing up to 12.5 wt.% aluminium fulfil both requirements.

Fig. 11.34 CoilTrak drilling equipment (BHI Celle)

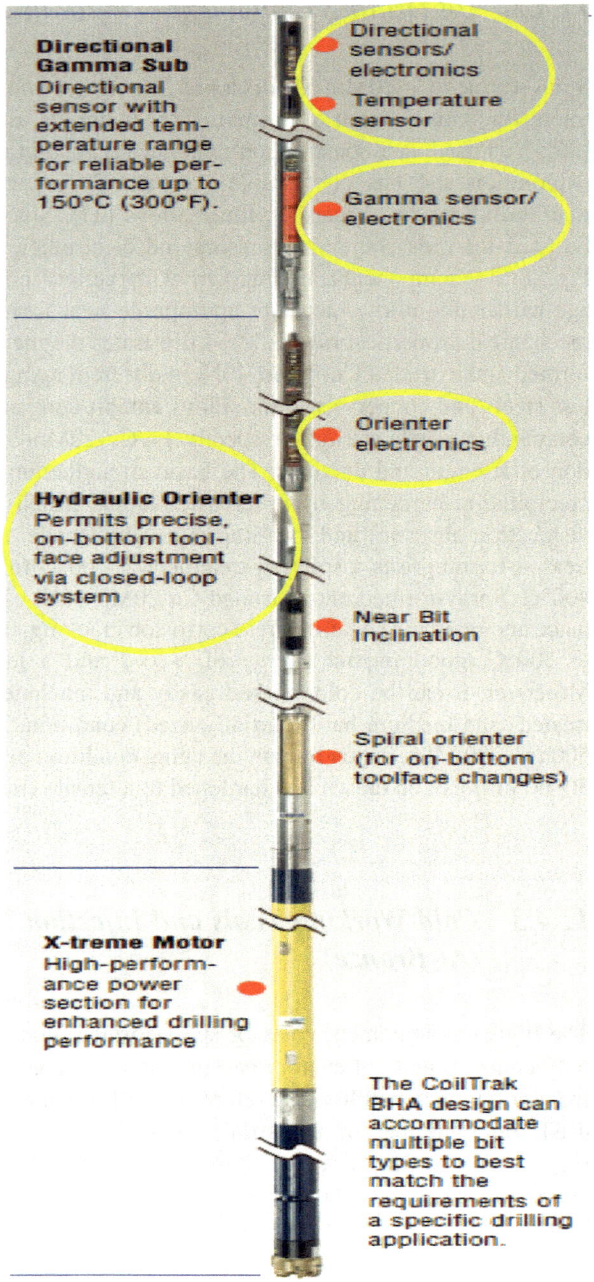

Fig. 11.35 Tube bending tool made of spray-formed aluminium bronze

Fig. 11.36 Assembled bars of plastic injection mould for a bottle case. Bars made of spray-formed aluminium bronze Cu-12.5Al4.5Fe-Co-Mn which exhibits moderate ductility and is weldable

11.5 Quality

11.5.1 Segregation

In continuous casting of high tin containing copper alloys the so called inverse segregation is well known. In the early days of the continuous casting development various publications dealt with this phenomenon [39–43]. Voßkühler [39] summarized and commented possible explanations:

- Pressure caused by growing crystals
- Vapour pressure in the enriched liquid phase
- Pressure of dissolved gas
- Volume change by shrinkage during solidification creating a pressure in the sump
- Metallo-static pressure
- Volume contraction during solidification starts liquid flow by sucking
- Capillary effect starts liquid flow

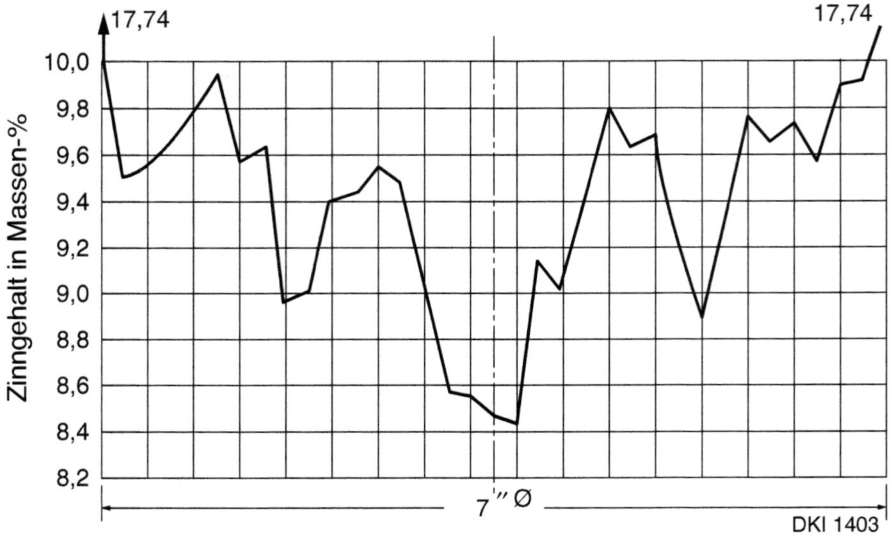

Fig. 11.37 Sn-concentration in a cross section of a continuously cast CuSn8 billet, with permission from [58]

Sucking by volume contraction and metallo-static pressure are the most likely explanations. The strong influence of metallo-static pressure was impressively demonstrated by Ohm [43]. Figure 11.37 gives an example for the Sn-concentration in the cross section of a comparable small billet (diameter 7″). Near the surface the tin concentration reaches more then 17% at a nominal composition of 8% (all compositions in wt.%). Due to the phase diagram Fig. 11.39 the tin concentration should be higher in the centre than at the surface, because the solidification starts at the surface and runs to the centre. Figure 11.38 shows what happens in a continuous cast mould. If there is no air gap formed by shrinkage between billet surface and mould wall as shown in the right half, segregation will follow the phase diagram. But in reality the billet surface loses contact to the cooled mould wall by shrinkage and air gap formation. So the solidified shell is re-heated and the mushy zone extends to the surface. Now it is possible, that tin-rich phase, driven by the metallostatic pressure, flows through inter-dendritic channels to the surface and enriches the outer parts of the billet with tin. If the air gap is very large, tin-rich tears can be formed on the surface.

In spray forming there is no sump of liquid metal and therefore no high metallo-static pressure. But in principle the same Sn-concentration pattern is found as demonstrated in Fig. 11.40. The lower curve shows the segregation across the diameter at a low, and the upper at a high gas flow rate at the same melt flow rate. Obviously the segregation is influenced by the gas flow rate or the ratio between gas- and melt flow rate (G/M in m^3/kg, volume rates at standard conditions).

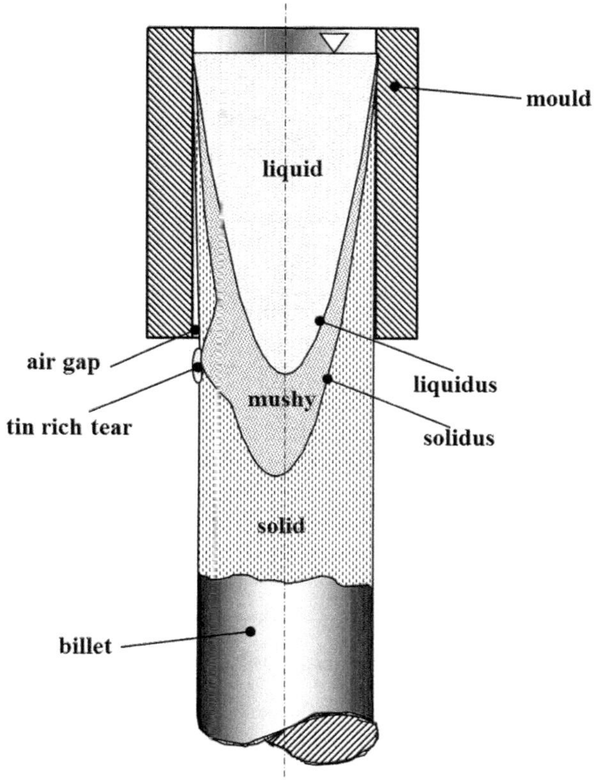

Fig. 11.38 Inverse segregation in continuous casting

What is the reason for the inverse segregation in spray-formed CuSn-billets? The temperature profile of a spray-formed billet is very similar to the pattern in a continuous cast billet as the calculated temperature field in Fig. 11.41 demonstrates [44]. The only differences are the absence of a liquid sump and an insulating air gap. Only a very thin liquid metal layer is permanently fed by the spray cone. Figure 11.42 describes the process schematically. The partly solidified droplets hit the mushy surface of the billet. Now follows the first separation into copper rich solid and into tin-rich liquid parts. The solid sticks at the previously deposited particles and the liquid is driven by centrifugal force to the outer diameter. The increase of the tin content is limited by mixing with incoming droplets, which are in average of the nominal composition. The same force forms a pressure difference Δp inside the interdendritically channels [45] and moves the liquid fraction to the outside (ρ: mass density, R: radius of the billet, f: rotation frequency).

$$\Delta p = 0.5\rho R^2(2\pi f)^2 \tag{11.2}$$

On its path the liquid cools down (see Fig. 11.41) and loses copper to the solid particles. The remaining liquid accumulates tin. The increasing tin content lowers solidus temperature and keeps the liquid moving. When this tin-rich phase reaches

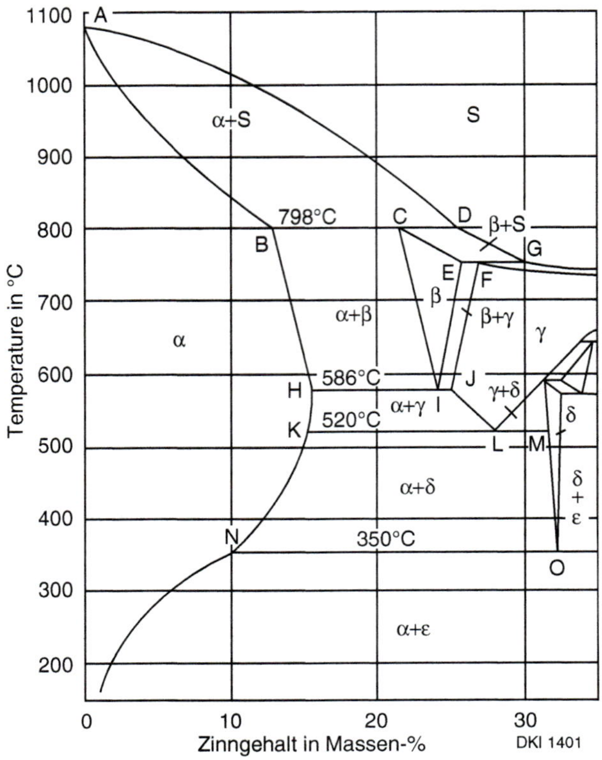

Punkt	Temp °C	Zinngeh. Mass.-%
A	1083	0
B	798	13,5
C	798	22,0
D	798	25,5
E	755	25,9
F	755	27,0
G	755	30,6
H	586	15,8
I	586	24,6
J	586	25,4
K	520	15,8
L	520	27,0
M	520	32,4
N	350	11,0
O	350	32,55

Fig. 11.39 Phase diagram of Copper—Tin [58]

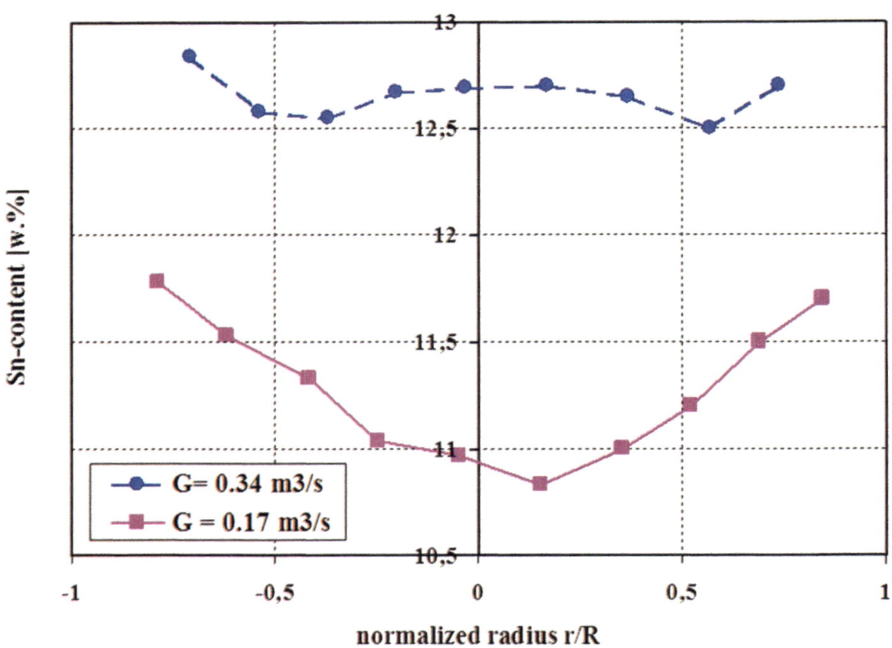

Fig. 11.40 Sn-concentration in the cross section of CuSn12 billets spray-formed with different gas flow rates G in m^3/s and constant melt flow rate

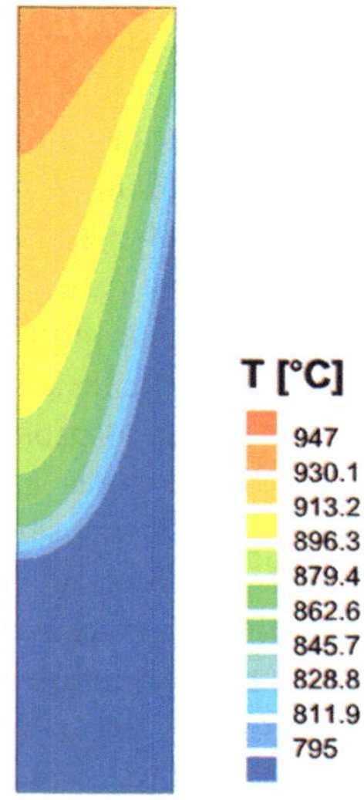

Fig. 11.41 Calculated temperature field in a spray-formed CuSn15.5Ti0.2 billet

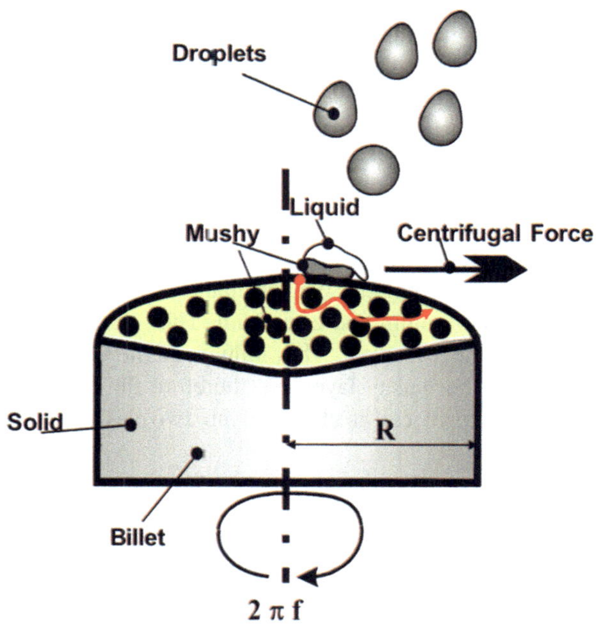

Fig. 11.42 Schematic diagram showing the driving forces for segregation

the surface, it is mixed with the depositing droplets as described above. This is the reason for the limitation of segregation in spray forming (compare Fig. 11.37 with Fig. 11.40). If G/M is higher, the droplets are cooled down more intensively. In this case the segregation decreased further on. So it seems to be very easy to avoid segregation by increasing gas flow rate—if there would not be the phenomenon of cold porosity.

11.5.2 Porosity

Quality issues require limitation of as-sprayed porosity for subsequent processing of spray-formed copper-alloy billets to a certain level. This porosity can only be minimised when the parameters influencing porosity are identified. These are described previously by various authors. Pores can be formed inside the droplets by gas entrapment during droplet formation, dissolution of gas from molten metal, solidification shrinkage or collision between larger liquid and smaller solidified droplets.

During deposition the fraction of solid f_s is most important. At a high fraction of solid f_s the pores arise from poor spreading of droplets and insufficient liquid feeding. Interstices between adjacent droplets arriving at the surface leave irregular cavities. This effect can be intensified by a high heat extracting substrate. The layers near the substrate are more porous than the bulk material.

At a low fraction of solid f_s the interstices between solid particles can be fed by liquid metal and porosity decreases. There are some contrary results reported by various authors, which is discussed in [46] more detailed At a high fraction of liquid f_l the mushy layer is thick and hot. Therefore, viscosity is low. The mushy layer is continuously disturbed by the high velocity gas jet and atomised droplets. When the droplets hit the mushy surface, they entrain the surrounding gas. It is therefore expected that f_l has an upper and a lower limit for preventing either gas porosity or cold porosity.

Another type of porosity is caused by the so-called "cauliflower-effect". It occurs during spray deposition in the centre of the billet when many solidified droplets are collected.

The minimising effect on porosity by reactive elements is reported by Watson [47, 48] and Cookey [49]. The latter recommends strong nitride formers such as silicon and chromium. Watson named as preferred elements aluminium, silicon, titanium, chromium and zirconium. The formation of nitrides changes the surface tension of the droplets. This influences the behaviour of the droplets during the impact of the mushy layer and therefore the entrainment of gas. All the reported types of porosity can be classified into two classes: cold and hot porosity (Fig. 11.43).

11.5.2.1 Definition and Measurement of Porosity

Porosity is defined by the following equation:

$$Pt = \left(1 - \frac{\overline{\rho}}{\rho_0}\right) * 100 \tag{11.3}$$

Pt: Porosity in %

ρ_0: Mass density of material without pores. Not the theoretical but the measured density of hot and cold worked material is used.

$\overline{\rho}$: Mass density as measured in the deposit

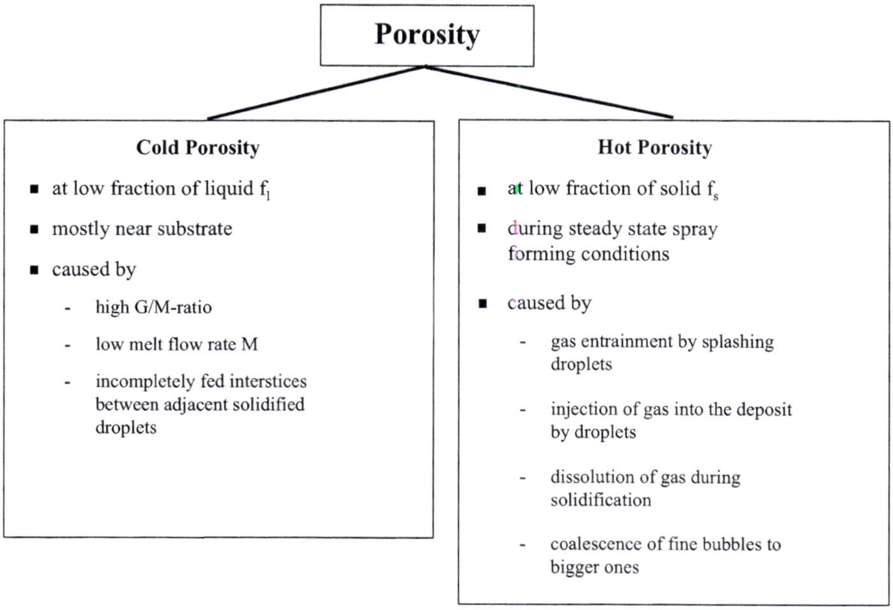

Fig. 11.43 Classification of porosity in spray-formed deposits

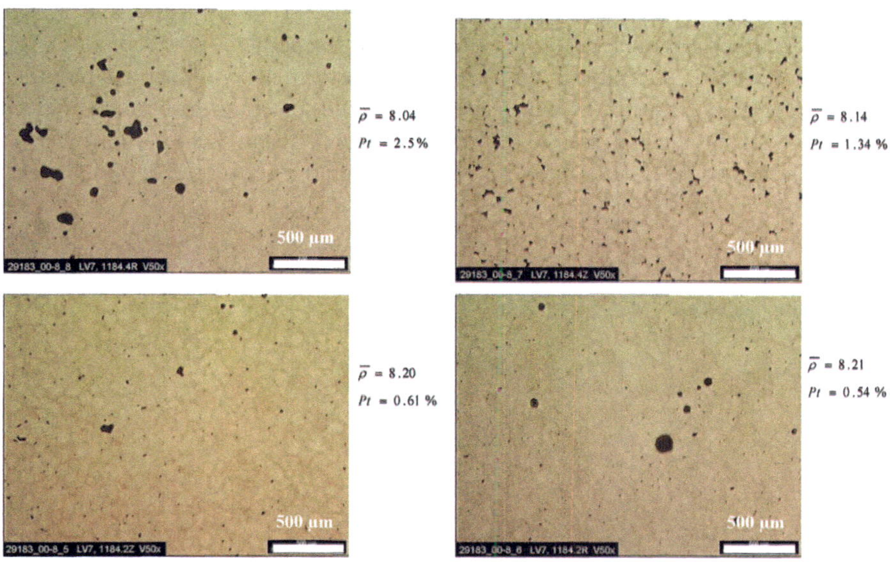

Fig. 11.44 Microstructure and porosity of Cu-20Mn-20Ni, $\rho_0 = 8.25$ g/cm^3

The density is measured with a buoyancy weighing-machine according to DIN EN 6018. The specimens are 10 × 10 × 10 mm^3 cubes [50]. The results are compared with pictures of the microstructure at 50× magnification. Figure 11.44 gives an impression of the measured data for Cu-20Mn-20Ni. Especially samples with low porosity could show poorer consistency between photograph and measured figures. Measurement by buoyancy takes the whole sample volume into account. In the case of the photograph, only the surface is evaluated. Some isolated pores inside the sample can influence the measured result but they are not visible on the surface.

The most important parameter is the Gas-to-Metal flow-rate ratio G/M as reported elsewhere [51]. It is not only important for porosity but for segregation too. In production plants the minimum G/M ratio is not limited by an increasing porosity caused by the lower viscosity of the mushy layer and therefore an increased gas entrainment. The limiting factor is damage to the billet by centrifugal forces. Large pieces of partly solidified material are expelled.

Further parameters are metal flow rate, melt temperature, flying distance of the droplets, gas and droplet velocity, spray cone angle and scanning angle. Beside these parameters the effect of reactive elements is not negligible.

11.5.2.2 Reactive Elements Influencing Porosity

The entrainment of gas by the droplets hitting the liquid surface of the deposit is the main reason for hot porosity. This effect is boosted by cavities in the solidified droplets. Partly solidified droplets collide with rigid particles. Some of them are embedded, others break out and leave cavities. Figure 11.45 at position a, a droplet is embedded and completely welded. At position b, a particle broke out and left a crater. It cannot be identified if this happened during flight or after preparation of the sample. At position c, the particle which formed the crater lost contact before collision with particle a. Crater c is deformed by this collision. Such craters increase gas entrainment when the droplets dive into the liquid layer on the deposit.

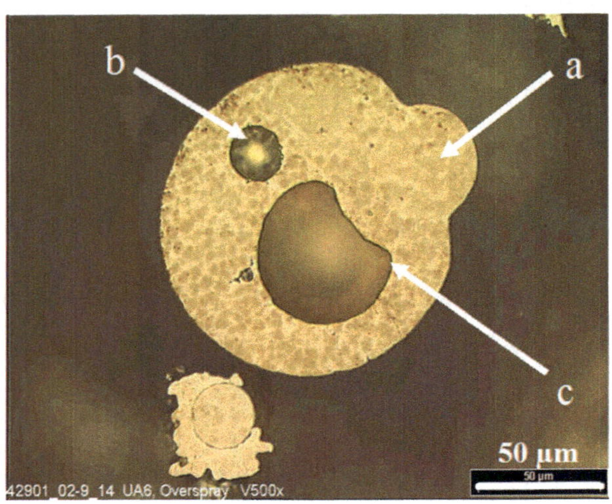

Fig. 11.45 Overspray particle with embedded particle and two craters

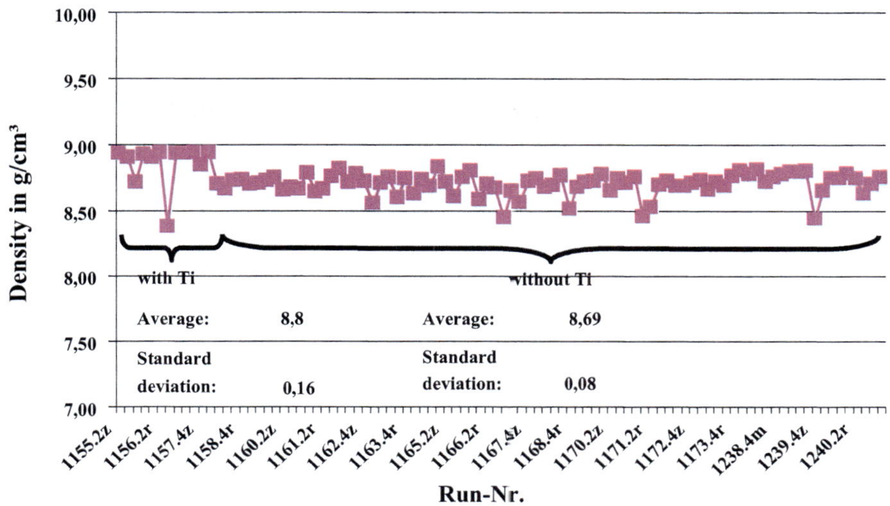

Fig. 11.46 Density of Cu-13.5Sn with and without Ti

To reduce porosity, the reactive elements zirconium and titanium were tested in preliminary trials. Their effect on porosity is evaluated as equivalent. So for the following tests only titanium was added to the melt during normal production of numerous billets. One slice was cut from each end and density was measured at three radial positions (edge, median, centre). This was carried out on three different alloy groups represented by Cu-13.5Al-4.5Fe-Co-Mn, Cu-13.5Sn and Cu-20Mn-20Ni. All results are collected in data files, the average and standard deviation is computed and plotted in diagrams. The addition of 0.15 wt.% titanium in the melting furnace reduces porosity as shown for example in Fig. 11.46 for the alloy Cu-13.5Sn.

11.5.2.3 The Effect of Titanium on Porosity

The method for analysing nitrides is described in [46]. The following assumptions are made for a calculation of the entrained nitrogen and reaction with titanium:

1. The entrained gas bubbles adapt their temperature to the surrounding metal immediately.
2. Solidification time of the mushy layer allows the reaction of titanium and nitrogen. Watson [47] indicating 5–22 s for a strip and Doherty [52] 10–200 s for a billet. Compared with this, the reaction time during flight is only a split-second (droplet speed 50–100 m/s, flight distance 600 mm).
3. During solidification, the bubble pressure is constant. Because of the relative thin liquid layer the metallostatic pressure is neglected. The initial pressure in the bubble is equal to the spray chamber pressure.

Fig. 11.47 Scheme of the mushy layer on top of the deposit with embedded gas bubbles

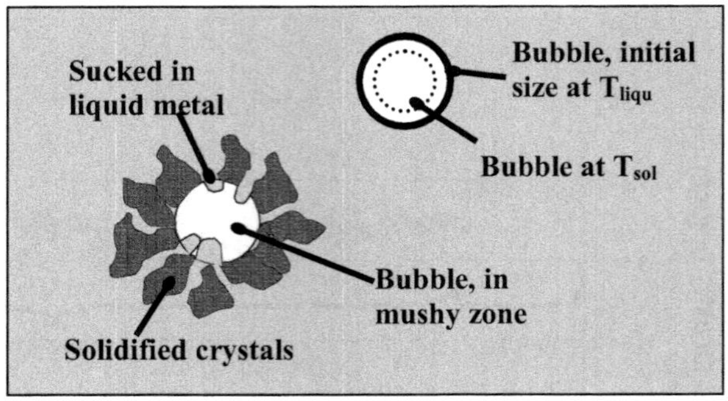

4. The bubble volume changes isobarically until solidus temperature is reached. When they are enclosed in regions of higher solid fraction fs, they are fed with liquid metal through channels with low melting phases (Fig. 11.47). With further cooling, the bubble (pore) volume changes only by shrinkage of the surrounding metal.
5. The pores are regularly (like a ball) or irregularly shaped.
6. The gas in the bubble behaves ideally.
7. The overall volume V_P of the pores in a deposit of the mass m_D can be calculated from the measured density, using the definition of porosity in Eq. (11.3).

$$V_D = m_D/\overline{\rho} \tag{11.4}$$

$$V_P = Pt{*}V_D \tag{11.5}$$

V_D: Volume of the deposit
m_D: mass of the deposit
V_P: volume of all pores in the deposit
Therefore, density of the gas is kept constant during cooling to solidus temperature, density at ambient temperature can be calculated taking into consideration the shrinkage of the surrounding metal during cooling from solidus to ambient temperature:

$$\rho_a = \rho_{sol}(1 + 3\alpha(T_{sol} - T_E)) \tag{11.6}$$

ρ_a: density of gas at ambient temperature
ρ_{sol}: density at solidus temperature
T_{sol}: absolute solidus temperature
α: linear thermal expansion coefficient
The cooling from solidus to ambient temperature decreases the pressure in the bubbles. This effect is counteracted by the shrinkage of the surrounding metal. According to the law of ideal gas the pressure is:

$$P_a = P_{sol}\frac{\rho_a T_a}{\rho_{sol} T_{sol}} \tag{11.7}$$

P_a: pressure in the pores at ambient temperature
T_a: absolute ambient temperature
The number of moles N_2 in the deposit is now defined:

$$n = \frac{P_a V_P}{RT_a} \tag{11.8}$$

n: number of moles N_2
R: molar gas constant
The mass and concentration of N_2 in the deposit can now be calculated using the molar mass of N:

$$m_{N_2} = 2nM_N \tag{11.9}$$

$$c_{N_2} = \frac{m_{N_2}}{m_D} \tag{11.10}$$

m_{N_2}: mass of nitrogen in the deposit
M_N: molar mass of nitrogen
c_{N_2}: concentration of nitrogen in the deposit
The concentration was measured by hot extraction with the LECO ON-Analysator TC 600 [53]. In Table 11.6 the measured concentrations $c_{N_2,m}$ (grey shaded column) are compared with the calculated concentration $c_{N_2,\min}$ and $c_{N_2,\max}$.

As the samples for density measurement and nitrogen analysis are from the same billet but not identical, the nitrogen concentration is calculated with the minimum and maximum density measured in the billet. The measured nitrogen concentration $c_{N_2,m}$ is nearly as high as the calculated maximum $c_{N_2,\max}$. This confirms the theory that in copper-base alloys without reactive elements the nitrogen is entrapped in the pores.

The reaction of titanium with nitrogen can be proved by a similar calculation. Using Eq. (11.5) the pore volume is calculated by inserting in the measured density with and without titanium. The volume difference is the amount of nitrogen, which reacted with titanium to TiN.

$$\Delta V_P = V_P - V_{P,Ti} \tag{11.11}$$

$$m_{Ti} = 2nM_{Ti} \tag{11.12}$$

$$c_{Ti} = \frac{m_{Ti}}{m_D} \tag{11.13}$$

Table 11.6 Comparison of calculated with measured nitrogen concentration in the deposit

Alloy	Run	ρ_0 (g/cm^3)	$\overline{\rho}_{\min}$ (g/cm^3)	$\overline{\rho}_{\max}$ (g/cm^3)	$Pt_{\min}$ (%)	$Pt_{\max}$ (%)	$c_{N_2,\min}$ (%)	$c_{N_2,\max}$ (%)	$c_{N_2,m}$ (%)
Cu-13.5Sn-Pb	1528	8.95	8.60	8.71	2.68	3.91	0.00010	0.00015	0.00015
Cu-13.5Sn	1736	8.93	8.65	8.81	1.34	3.14	0.00005	0.00012	0.00020

ΔV_P: pore volume difference between deposits with and without titanium
V_P: pore volume without titanium
$V_{P,Ti}$: pore volume with titanium
m_{Ti}: mass of titanium in the deposit
c_{Ti}: concentration of titanium, which is necessary for reducing the porosity from Pt to Pt_{Ti}

The mass and the concentration of TiN are:

$$m_{TiN} = 2n(M_{Ti} + M_N) \tag{11.14}$$

$$c_{TiN} = \frac{m_{TiN}}{m_D} \tag{11.15}$$

m_{TiN}: mass of TiN in the deposit
c_{TiN}: concentration of TiN in the deposit

Table 11.7 shows an example calculation for Cu-20Mn-20Ni. The grey coloured fields are input data. The difference in density with and without Ti seems to be very small but in terms of porosity there is a factor of 3. The minimum concentration of Ti in the melt for this reduction of porosity is very low. The presence of TiN is proved (Fig. 11.48).

The quantitative analysis of the TiN concentration in Cu-20Mn-20Ni-Ti was performed with the experimental set described elsewhere [46]. The concentration of TiN in the sample determined by this method was 0.0008%. Compared with the calculated value in Table 11.7 the measured figure is 4 times higher than the calculated one but of the same order of magnitude. The calculated figure is based on statistical average values. The measured figure is based on a single sample. Therefore a perfect match cannot be expected.

The probability of formation of TiN is estimated by an equilibrium calculation with the Outokumpu-Software HSC [54]. The main input parameters and results are given in Fig. 11.49. The atomized alloy is Cu-20Mn-20Ni with 0.15 wt.% Ti addition. The atomising gas is nitrogen with a little amount of oxygen impurities. Unfortunately, the kinetics cannot be calculated. But the equilibrium calculation shows, that, at melt temperature of the atomisation process of about 1100°C, the TiN dominates. At lower temperatures, the equilibrium changes to Mn-nitrides and Mn-oxide. At these temperatures the solidification process is finished and the reaction velocity is reduced, so that the Mn-nitride formation is negligible.

11.5.3 Detection of Pores and Cracks

11.5.3.1 Porosity Measurement

State of the art density measurements are used for the evaluation of powder metallurgical products. It was therefore obvious to adopt this method for spray forming. The influence of reactive elements and the gas to metal flow rate ratio

Table 11.7 Example calculation for Cu-20Mn-20Ni with and without Ti

ρ_0	8.20 (g/cm^3)	Density of pore free deposit
$\overline{\rho}$	8.05 (g/cm^3)	Measured mean value of deposit without Ti
$\overline{\rho}_{Ti}$	8.15 (g/cm^3)	Measured mean value of deposit with Ti
m_D	1500 (kg)	Mass of deposit
ρ_N	1.17 (kg/m$^{3)}$	Density of gas at ϑ_N, 1 bar
ϑ_N	15 (°C)	Standard temperature
ϑ_{sol}	1005 (°C)	Solidus temperature
α	1.86 × 10^{-5} (1/K)	Thermal linear expansion coefficient
ϑ_a	25 (°C)	Ambient temperature
P_{liqu}	0.1 (MPa)	Pressure at entrainment of bubble (chamber pressure)
R	8.31 (J/mol K)	Molar gas constant
M_{Ti}	47.90 (g/mol)	Molar mass of Ti
M_N	14.01 (g/mol)	Molar mass of N
Pt	1.83 (%)	Porosity without Ti
Pt_{Ti}	0.61 (%)	Porosity with Ti
ρ_{sol}	0.000264 (g/cm^3)	Density of gas in pore at solidus temperature
ρ_a	0.000278 (g/cm^3)	Density of gas in pore at ambient temperature
V_D	186,335.40 (cm^3)	Volume of deposit without Ti
$V_{D,Ti}$	184,049.08 (cm^3)	Volume of the deposit with Ti
V_P	3409 (cm^3)	Volume of the pores in the deposit without Ti
$V_{P,Ti}$	1122 (cm^3)	Volume of the pores in the deposit with Ti
ΔV	2286 (cm^3)	Volume of N_2 reacted with Ti
P_a	0.0246 (MPa)	Pressure in pore at ambient temperature
n	0.023 (mol)	Number of moles N_2 reacted with Ti
n_{Ti}	0.045 (mol)	Number of moles Ti reacted with N
m_{Ti}	2.17 (g)	Mass of Ti reacted with N_2
c_{Ti}	0.00014 (%)	Concentration of Ti needed for reaction
n_N	0.068 (mol)	Total number of moles N in pores of the deposit
m_{N_2}	0.948 (g)	Total mass N_2 in pores of the deposit
c_{N_2}	0.000063 (%)	Total concentration N_2 in the deposit
m_{TiN}	2.810 (g)	Mass of TiN in the deposit
c_{TiN}	0.00019 (%)	Concentration of TiN in the deposit

(GMR) on density and porosity were reported in Sect. 11.5.2. Monitoring the density of a number of spray-formed billets for example makes the influence of reactive elements visible Fig. 11.46. Such evaluations are not only useful for quality assurance but also for process improvement. In practice the measurement of density with a buoyancy weighing machine according to DIN EN 6018 has some disadvantages. It is not possible to check the whole volume of the billet. The method is time-consuming and therefore expensive.

To overcome these disadvantages the possibilities of ultrasonic testing were examined. Preliminary tests with continuously cast billets were not successful as a result of the coarse structure of conventional castings (Fig. 11.6, left). The

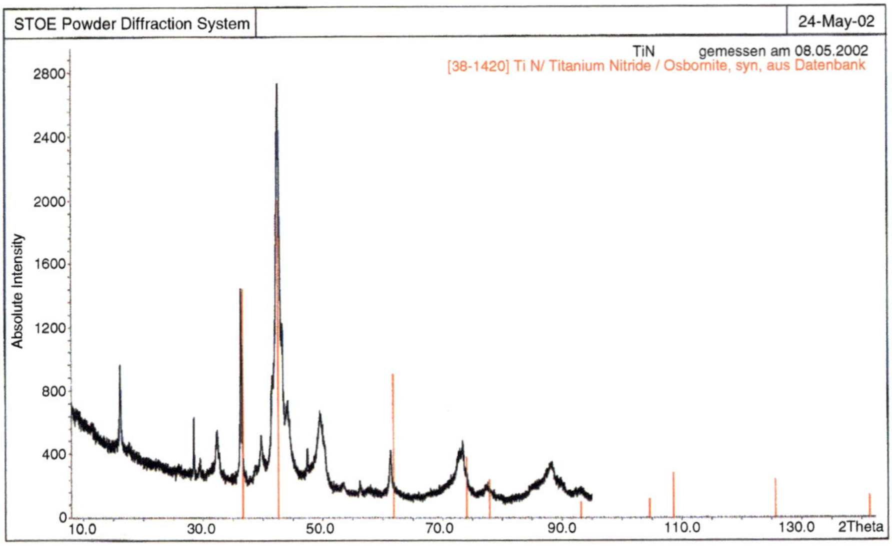

Fig. 11.48 X-ray diffraction and theoretical diffraction diagram for TiN (Osbornite)

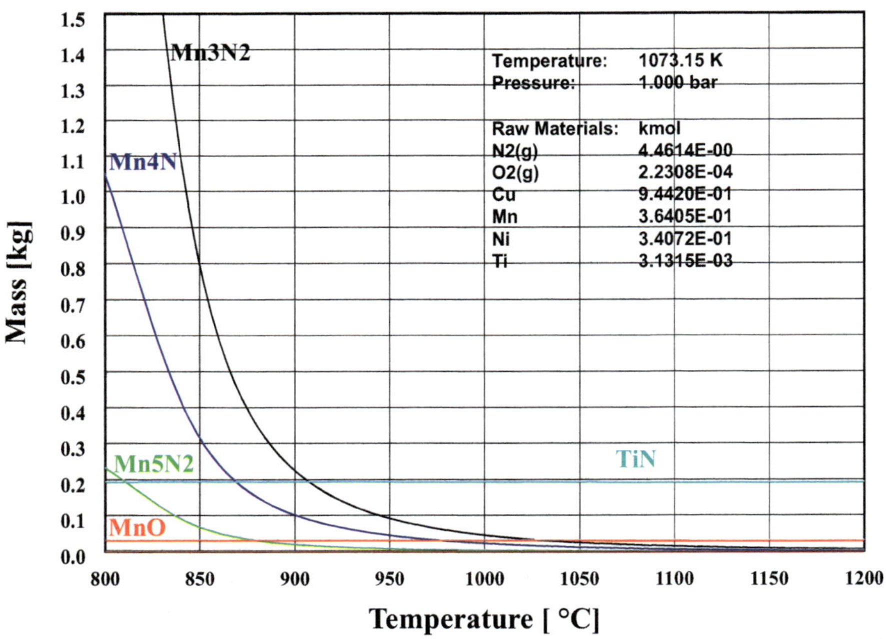

Fig. 11.49 Mass of reaction products vs. temperature

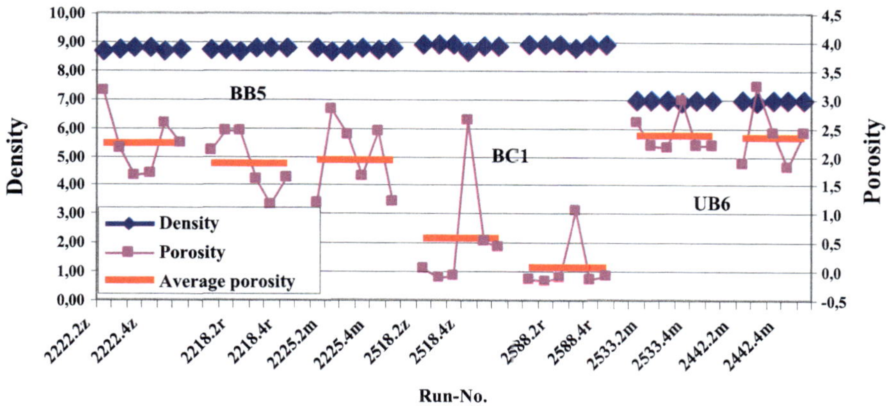

Fig. 11.50 Density and porosity of different billets at centre (z), median (m) and edge position (r)

ultrasonic signal is scattered at each crystal which makes the differentiation of noise and signal difficult or impossible. But the spray-formed structure (Fig. 11.6, right) is so fine that ultrasonic testing could be successful.

The typical sample size is a cube of 1 cm^3. The samples are cut from discs at the top and the bottom of a 2 m long billet. The main volume in between remains uninspected. The disc samples are cut from the edge, the median and the centre position. The median radius divides the circular area of the disc into two equally sized areas. Figure 11.50 shows the measured density for 3 different alloys BB5 (CuSn13.5), BC1 (CuSn15.5Ti0.25) and UB6 (CuAl13Fe4.5CoMn). Due to the composition the density of BB5 and BC1 is similar and very different from the aluminium-bronze UB6.

The measured density is converted into the calculated porosity according Eq. (11.3). The density of different copper alloys varies between 7.0 and 8.9 g/cm^3 (Fig. 11.50). Standardization according to Eq. (11.3) based on the density of material formed by extrusion or forging makes the results of different alloys comparable.

Although the individual points vary strongly when measured in different places as described, the average density of BB5 and BC1, whose composition is nearly the same except for 0.25% Ti in BC1, shows the strong influence of Ti in density or porosity. These examples demonstrate that density or porosity measurement makes the evaluation of process parameters possible and can be used to improve the process.

11.5.3.2 Ultrasonic Test

Preliminary Tests

For initial tests a piece of billet with rather high porosity was picked out by manual ultrasonic testing. Therefore, the size of the test setup was limited and the original diameter of the sample was reduced to 100 mm by machining. After the ultrasonic

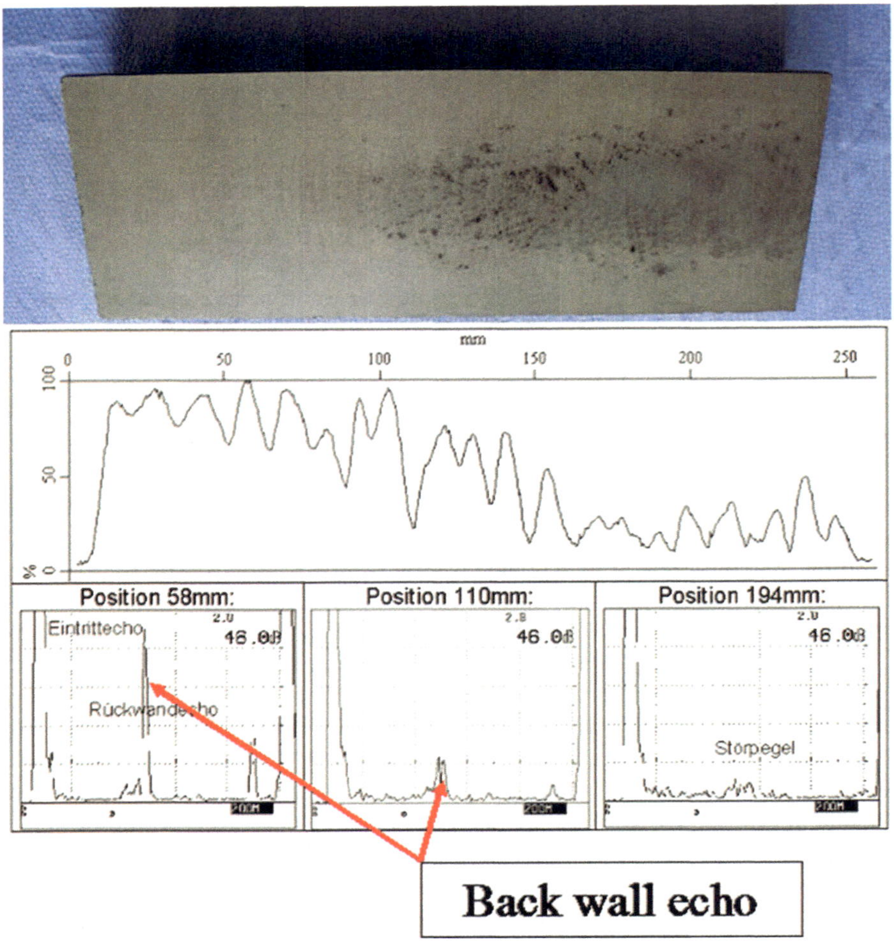

Fig. 11.51 Longitudinal section of partly porous CuMn20Ni20- billet, dia. 100 mm, line-scan, with permission from [55]

test the sample was cut longitudinally. Figure 11.51 shows the macro-etching and the back wall echoes at different positions. At the porous parts on the right side the back wall echo is damped from nearly 100% cathode ray tube height (CRTH) to less than 20%. The test performed at a few specific points provides only a limited view into the material. The line-scan represents the porosity distribution inside the billet much better. The mapping can be improved, if in addition to the line-scan the winding off and a colour coding of different thresholds are used (Fig. 11.52). Porous areas are red and dense areas are green coloured. This graph makes the typical layer structure of cold sprayed deposits visible.

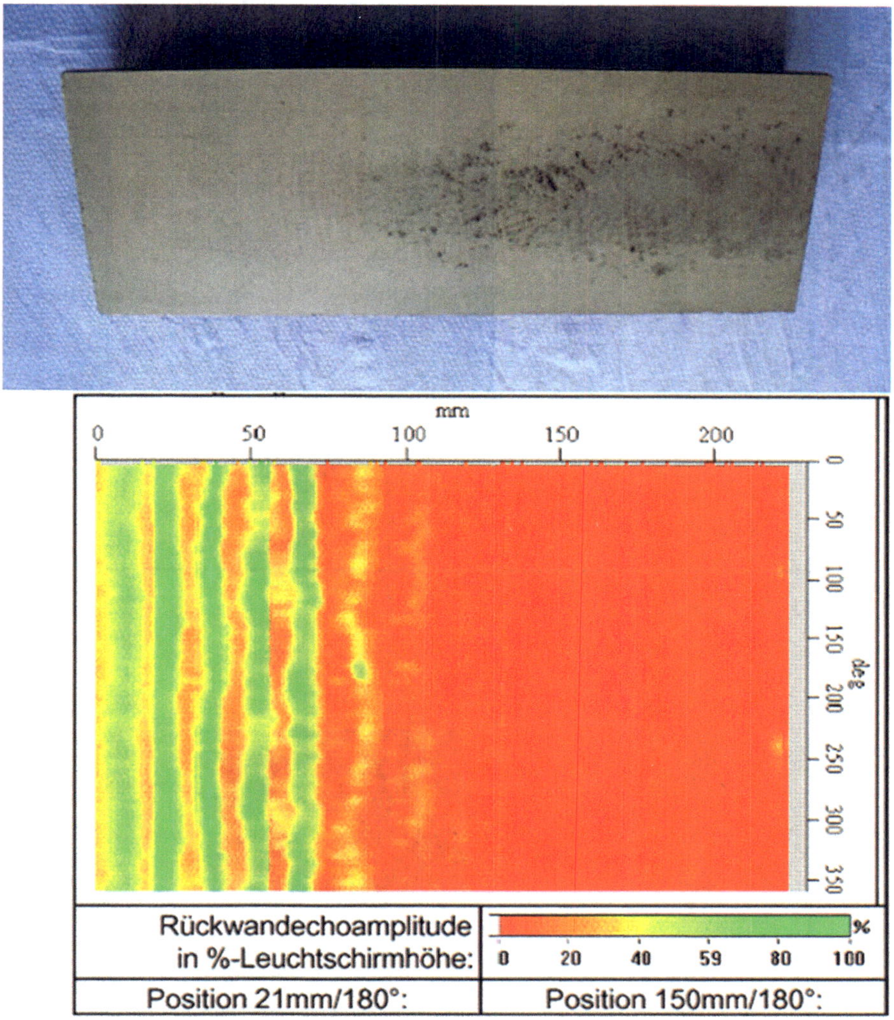

Fig. 11.52 Longitudinal section of partly porous CuMn20Ni20 billet, dia. 100 mm, C-picture, with permission from [56]

Ultrasonic Test Setup

After these promising results the tests were continued on the turning lathe. For this purpose contactless probes with running water coupling were used [57]. The basic test setup and the different test samples are shown Fig. 11.53. The resulting C-picture in Fig. 11.54 combines the back wall echo and for additional information the flaw echo. The drilled holes are visible in both pictures. The minimum size of defects was not identified, because the flaw echo was not in the focus of these experiments. During standard test routine it will be used for more detailed information on large defects such as cracks.

Alloy	Diameter [mm]	Sound velocity [m/s]
LV7 (CuNi20Mn20)	243	5900
UB6 (CuAl13Fe4.5CoMn)	197	5250
BB5 (CuSn13.5)	138	4680

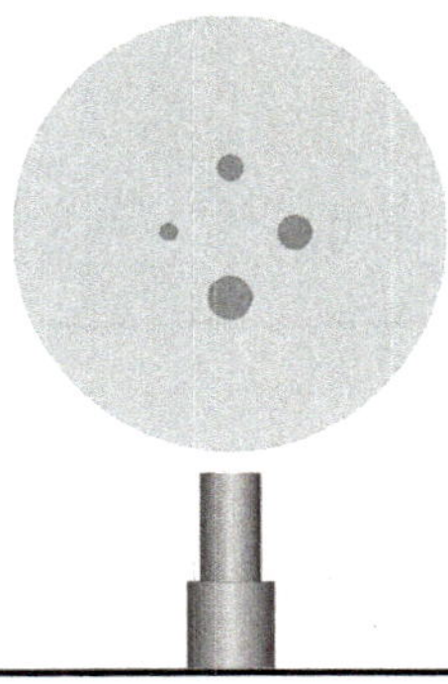

On the front end 4 drilled holes with different diameters (2, 2.5, 3 und 4 mm; length about 40 mm).

Sampling with running water probes H2KF resp. H1NF and the USD15-test device.

Fig. 11.53 Test samples

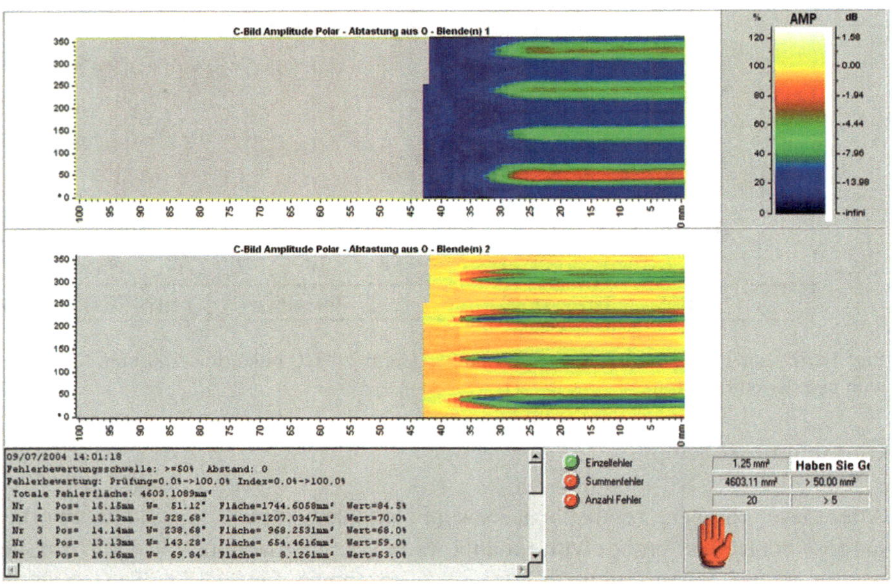

Fig. 11.54 Flaw and back wall echo of the test sample

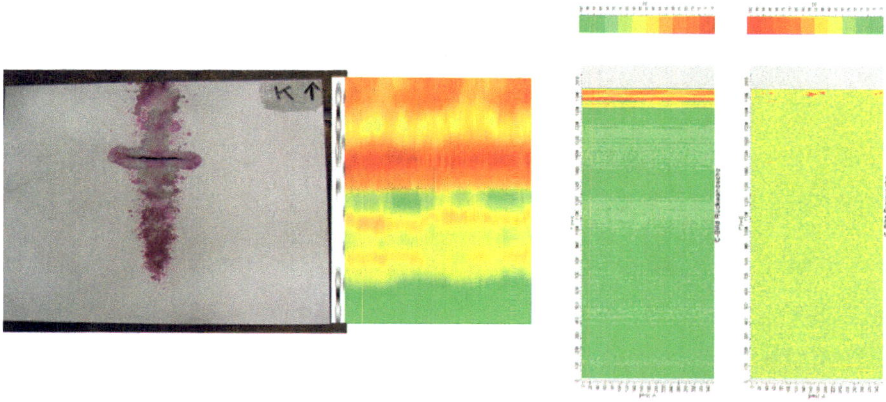

Fig. 11.55 *Right*: US-picture of the whole billet, *left*: section of the billet's head compared with Met-L-Chek (longitudinal section)

Standard Test Routine

Every spray-formed billet is tested by this method. As described before not only porosity but also cracks and changes in macro structure can be detected. Figure 11.55 shows on the right the US-picture of the whole length of the billet (backwall and failure echo). Near the top are failures indicated by red colour. This part was longitudinal cut and tested with a contrast penetrant system. On the left side the result is compared with the zoomed area of the US-picture. The dark red area indicates the crack, green areas are sound an porous areas are between. Figure 11.56 shows a failure at 600 mm from the left side. This area was longitudinally cut and checked with a contrast penetrant system, but no failure was visible (Fig. 11.57a). The macro-etching (Fig. 11.57b) shows a change in the macro structure, which follows the shape of the growing billet top during spray forming. This example demonstrates that US-testing is sufficient sensitive for quality assurance.

11.6 Summary and Outlook

Spray forming is a fairly new process. Introduction into industrial production has reached a considerable level. The process offers interesting production possibilities on the one hand for materials already known, which cannot be cast or are difficult to cast, and on the other hand for innovative materials with special properties. So tin-bronze up to 16% Sn can be produced by spray forming exhibiting excellent ductility and high mechanical strength. Workability is excellent without the need for preceding homogenisation.

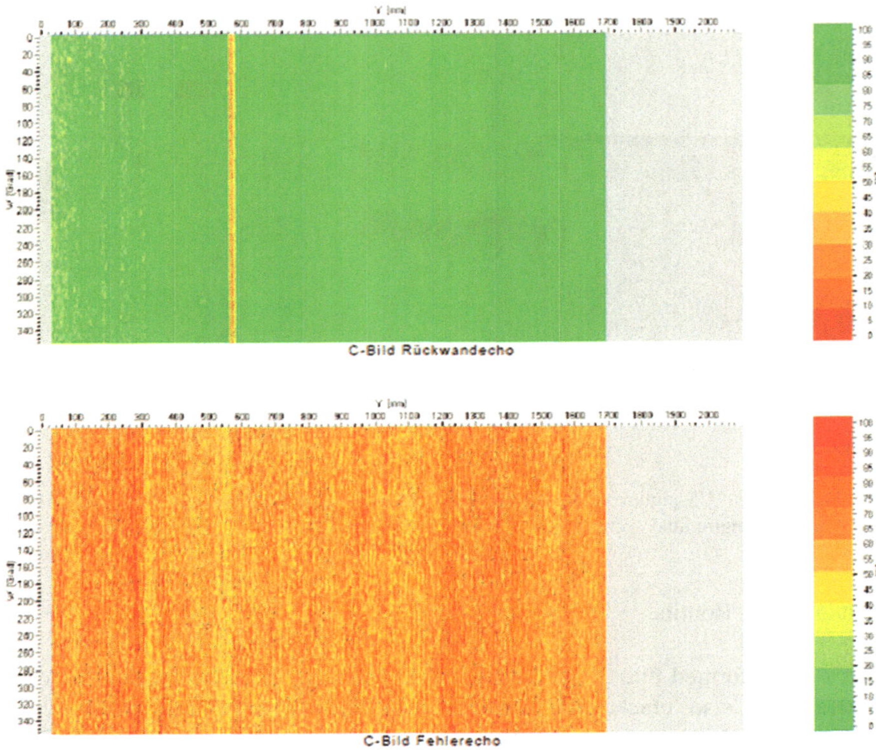

Fig. 11.56 Backwall- and flaw-echo Run 5468: failure detection at 600 mm

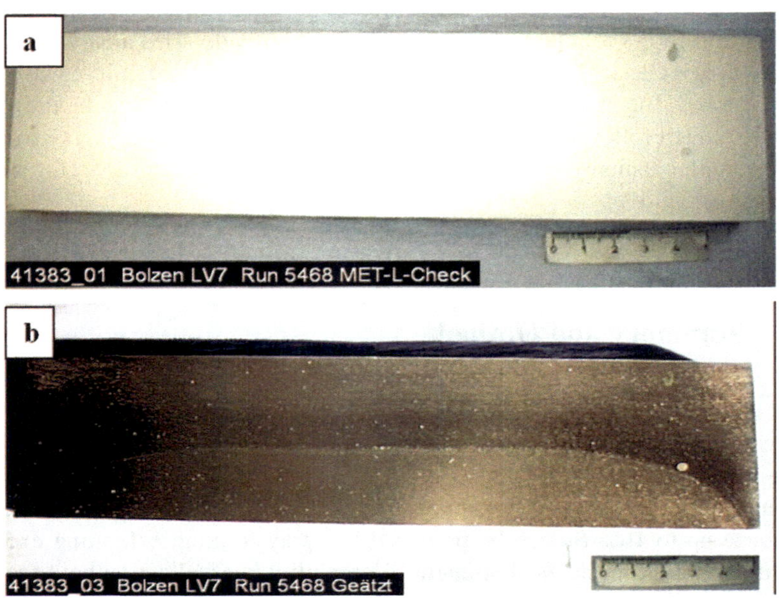

Fig. 11.57 Met-L-Chek (**a**) and macro-etching (**b**) of a longitudinal section, Cu-20Mn-20Ni (LV7), Run 5468

Above about 14% Sn, a brittle phase appears, which is much lower and finer distributed compared to permanent mould or continuously cast material and, therefore, allows good hot and cold working. The spray-formed tin bronze covers a field of applications in sliding elements and superconductive wire.

Cu-20Mn-20Ni is a high-strength construction material which mechanical properties can compete with those of beryllium copper and high strength steels. This and the non-magnetic property introduced it in on- and off-shore industry for oil-drilling equipment.

Spray-formed complex aluminium bronzes are characterised by high hardness, high compressive strength and uniform distribution of the properties in the work piece. Additionally the machining properties are improved considerably. So this type of material now is widely applied in forming and deep drawing tools for steel and stainless steel.

11.7 List of Symbols

Latin

Symbole	Description
c_{N_2}	Concentration of nitrogen in the deposit
c_{Ti}	Concentration of titanium, which is necessary for reducing the porosity from Pt to Pt_{Ti}
c_{TiN}	Concentration of TiN in the deposit
f	Rotation frequency
m_D	Mass of the deposit
m_{N_2}	Mass of nitrogen in the deposit
m_{Ti}	Mass of titanium in the deposit
m_{TiN}	Mass of TiN in the deposit
M_N	Molar mass of nitrogen
n	Number of moles N_2
P_a	Pressure in the pores at ambient temperature
Pt	Poriosity in %
R	Radius of the billet
R	Molar gas constant
T_a	Absolute ambient temperature
T_{sol}	Absolute solidus temperature
V_D	Volume of the deposit
V_P	Volume of all pores in the deposit [for 11.5 and 11.8]
V_P	Pore volume without titanium
$V_{P,Ti}$	Pore volume with titanium

Greek

Symbole	Description
α	Linear thermal expansion coefficient
Δp	Pressure difference
ΔV_P	Pore volume difference between deposits with and without titanium
$\overline{\rho}$	Mass density as measured in the deposit
ρ_0	Mass density of material without pores. Not the theoretical but the measured density of hot and cold worked material is used
ρ_a	Density of gas at ambient temperature
ρ	Mass density
ρ_{sol}	Density at solidus temperature

References

1. Herrmann, E. (1958). *Handbuch des Stranggießens*. Düsseldorf: Aluminium-Verlag, Fig. 361.
2. Singer, A. R. E. (1970). The principles of spray rolling of metals. *Metals and Materials, 4*, 246–257.
3. Cramb, A. W. (1988). New steel casting processes for thin slabs and strip–a historical perspective. *Iron and Steelmaker, 15*, 45–60.
4. Müller, H. R., & Zauter, R. (2003). Spray-formed copper alloys–process and industrial applications. *Erzmetall, 56*(11), 643–650.
5. Zauter, R., Mueller, H. R., & Kudashov, D. (2006). Spray-formed high-tin bronze—a homogeneous prematerial for Nb_3Sn-based superconductor wire. *Vortrag ASC 2006 Conference in Seattle, Washington, August 27–September 1, 2006.*
6. Zauter, R., Ohla, K., Müller, H.R., & Maier, J.. (2003). Spray-formed materials for low temperature superconductors. *Proceedings of the 5th International Conference on Spray Forming (SDMA 2003 /ICSF V)* (pp. 5–122). Bremen: Universität Bremen. ISBN 3-8330-0571-8.
7. Hummert, K., Müller, H. R., & Spiegelhauer, C. (2003). Spray forming. In *Landolt-Börnstein, group VIII: advanced materials and technologies* (Vol. 2, pp. 4-43–4-61). Berlin: Springer ISBN 3-540-42942-5.
8. Abächerli, V. (2005). Improvement of workability and superconducting properties of high tin content (Nb, Ta, $Ti)_3Sn$ bronze route wires. *Doctorate Thesis at the Université de Genève, Switzerland.*
9. Dies, K. (1961). Mangan-Bronze. *Metall, 15*(12), 1161–1172.
10. Rabald, E. (1958). *Die Aluminiumbronzen*. Fachbuch. Berlin: Deutsches Kupfer-Institut.
11. Meigh, H. (2000). Cast and wrought aluminium bronzes. University Press, Cambridge.
12. Brezina, P. (1982). *International Metals Reviews, 27*, 77–120.
13. Hasan, F., Igbal, J., & Ridley, N. (1985). *Materials Science and Technology, 1*, 312–314.
14. Dies, K., Heubner, U., König, W. J., & Wincierz, P. (1965). *Zeitschrift fuer Metallkunde, 56.*
15. Benkißer, G., & Horn-Samodelkin, G. (1993). *Metall, 47*, 1033–1037.
16. Stenger, H. (1969). *Metall, 23*, 431–443.
17. Cook, M., Fentiman, W. P., & Davis, E. (1951/1952). *Journal of the Institute of Metals, 80*, 419–429.
18. Macken, P. J., & Smith, A. A. (1966). *The aluminium bronzes*. UK: CDA.
19. Roucka, J., Macasek, I., Rusin, K., & Svejcar, J. (1983). Possibilities of applying aluminium bronze in the production of cast tools for sheet drawing. Solidification technology in the foundry and casthouse. *The Metals Society*, 392–397.
20. Glas, F. (2005). *Tribologie und Schmierungstechnik, 52*, 55–63.
21. Klement, J. F. (1961). US Patent 2,979,397.

22. Kudashov, D. V., Zauter, R., & Müller, H. R. (2008) Spray-formed high-aluminium bronzes. *Materials Science and Engineering A, 477*, 43–49.
23. Yutaka, A. (1941). *Nippon Kinzoky. Gakkai-Si, 5*, 136–155.
24. Dies, K., & König, W. J. (1960). *Metall, 14*, 1085–1093.
25. Köstner, W., & Rauscher, W. (1948). *Zeitschrift fuer Metallkunde, 39*, 11–120.
26. Villasenor, C.T., & Radcliffe, S.V. (1974). INCRA Report No. 167, pp. 18–32.
27. Kuhn, H.-A., Altenberger, I., Riedle, J., & Hölzl, H. (2013). Microstructure and mechanical properties of ultra fine grained high performance copper alloys. *Proceedings Copper 2013* (pp. 129–138). Santiago: Chilean Institute of Mining Engineers.
28. Altenberger, I., Kuhn, H.-A., Gholami, M., Mhaede, M., & Wagner, L. (2014). Characterization of ultra-fine-grained Cu-Ni-Si alloys by electron backscatter diffraction (EBSD). *IOP Conference Series: Materials Science and Engineering, 63*, 012135.
29. Altenberger, I., Kuhn, H.-A., Müller, H. R., Mhaede, M., Gholami-Kermanshahi, M., & Wagner, L. (2015). Material properties of high-strength-beryllium-free copper alloys. *International Journal of Materials and Product Technology, 50*, 124–146.
30. Müller, H.R. (1996). *SFB Kolloquium, Bremen, Band 1* (pp. 33–41). Bremen: Universität Bremen. ISBN 3-88722-363-2.
31. Füller, K.-H., & Stock, D. (1995). *Metall, 49*, 274–277.
32. Ohla, K., Müller, H. R., & Riedle, J. (2000). Principle considerations for the production of dispersion strengthened copper. *SDMA (Spray Deposition and Melt Atomization) Proceedings of the International Conference on Spray Deposition and Melt Atomization* Vol. 1 (pp. 181–190). Bremen: Deutsche Forschungsgemeinschaft.
33. Perez, J. F., & Morris, D. G. (1994). *Scripta Metallurgica et Materialia, 31*(3), 231–235.
34. Liang, X., Earthman, J. C., & Lavernia, E. J. (1992). *Acta Metallurgical and Materials Transactions, 40*, 3003.
35. Osamura, K. (2005). Role of copper and copper alloys in advanced composite superconductors. *Department of Materials Science and Engineering, Kyoto University, IWCC Technical Seminar, Tokyo, Nov 2005.*
36. Altenberger, I., Müller, H. R., & Zauter, R. (2010). Spray-formed copper alloys have become mature. *Proceedings Copper 2010, Vol. 1* (pp. S3–S12). Hamburg: GDMB Informationsgesellschaft mbH. ISBN 978-3-940276-25-4.
37. Altenberger, I. , Müller, H.R., Zauter, R., & Kudashov, D.V. (2009). Microstructures of spray-formed copper alloys. *Proceedings of the 7th International Conference on Spray Forming, Bremen, Germany.*
38. Shapiro, S., Tyler, D.E., & Lanam, R. (1972). Phenomenology of Precipitation in Copper-20%-Nickel-20%-Manganese, In: Proc. CDA-ASM Conference on Copper, Oct. 16-19, Cleveland, Ohio (USA).
39. Voßkühler, H. (1949). Beitrag zur Frage der umgekehrten Blockseigerung bei Aluminium-Kupfer-Magnesium-Legierungen. *Zeitschrift fuer Metallkunde, 40*(8), 305–311.
40. Roth, W. (1949). Stranggießen von Leichtmetall nach dem Wassergießverfahren. *Zeitschrift fuer Metallkunde, 40*(12), 445–460.
41. Kästner, H. (1950). Die umgekehrte Blockseigerung bei Strangguß I. *Zeitschrift fuer Metallkunde, 41*(8), 193–205.
42. Kästner, H. (1950). Die umgekehrte Blockseigerung bei Strangguß II. *Zeitschrift fuer Metallkunde, 41*(8), 247–254.
43. Ohm, L., & Engler, S. (1989). Treibende Kräfte der Oberflächenseigerungen beim NE-Strangguß. *Metall, 43*(4), 520–524.
44. Schröder, R., & Uhlenwinkel, V. (2000). *Persönliche Information.* Bremen: IWT Stiftung Institut Werkstofftechnik.
45. Hansmann S., & Müller, H.R. (1999). Hochzinnhaltige Bronzen mittels Sprühkompaktieren seigerungsarm hergestellt. *SFB 372 Kolloquium Band 4* (pp. 1–6). Bremen: Universität Bremen. ISBN 3–88722-440-X.

46. Müller, H. R., Ohla, K., Zauter, R., & Ebner, M. (2004). Effect of reactive elements on porosity in spray-formed copper-alloy Billets. *Materials Science and Engineering A, 383*, 78–86.
47. Watson, G. (1990). Thermal and microstructural characterization of spray cast copper alloy strip. *Proceedings of the First International Conference on Spray Forming, 17–19 Sept 1990, Swansea, UK.*
48. Watson, W.G., Ashok, S., & Cheskis, H.P. (1990). Method to reduce porosity in a spray cast deposit. U.S. Patent No. 4,961,457, 9 Oct 1990.
49. Cookey, R.H., & Wood, J.V. (1990). Production and development of copper-base alloys by the osprey process. *Proceedings of the First International Conference on Spray Forming, 17–19 Sept 1990, Swansea, UK.*
50. Mathei-Schulz, E., Schulz, A., & Mayer, P. (2001). Gefügeauswertung an sprühkompaktierten Werkstoffen mit bildanalytischen Methoden. *Kolloquiumsband Sprühkompaktieren, Band 5* (pp. 179–191). Bremen: Universität Bremen.
51. Müller, H.R., Hansmann, S., & Ohla, K. (2000). Influence of process parameters on segregation and porosity in spray-formed Cu-Sn-billets. *Spray Deposition and Melt Atomization Conference* (pp. 205–218). Bremen: Universität Bremen.
52. Doherty, R., Annavarapu, S., Cai, C., & Kohler, K. (1997). Modelling based studies for control and microstructure development in spray forming. *Kolloquiumsband Sprühkompaktieren, Band 2* (pp. 45–78). Bremen: Universität Bremen.
53. Ebner, M. (2002). Nachweis von TiN-phasen im sprühkompaktierten Material BC1. *Interner Laborbericht Nr. 1859 der Wieland-Werke AG, Ulm, 22 Aug 2002.*
54. Roine, A. (1999). Outokumpu HSC chemistry for windows, chemical reaction and equilibrium software with extensive thermochemical database. *User's guide version 4.0 30 Jun. 1999.* Helsinki: Outokumpu. ISBN 952-9507-05-4.
55. Schulz, S. (2001). *Aufnahme eines Line-Plots an Ms-Halbzylinder und eines zusätzlichen Musters*. Hürth, Germany: Interne Mitteilung Krautkraemer GmbH & Co. oHG.
56. Heinrich, M. (2002). *US-Untersuchung von sprühkompaktiertem Material auf Porosität*. Ulm, Germany: Interne Mitteilung Wieland-Werke AG.
57. Müller, H. R., Heinrich, M., Zauter, R., & Kudashov, D. (2006). Non-destructive testing of spray-formed copper-alloy billets. *Vortrag International Conference on Spray Deposition and Melt Atomization (SDMA 2006), 4–6, Sept 2006 Bremen.*
58. DKI (Deutsches Kupferinstitut). (1965). Legierungen des Kupfers mit Zinn, Nickel, Blei und anderen Metallen. Berlin Düsseldorf, 14.

Chapter 12
Spray Forming of Steels

Juho Lotta, Claus Spiegelhauer, and Simo-Pekka Hannula

12.1 Introduction

Spray forming of a wide range of steels and iron based alloys have been investigated since the 1970s. These range from low-alloy carbon steels to high-carbon, high-alloy tool steels. The preform types include round billets, flat deposits, tubular preforms, clad structures, gradient deposits, and molds/dies. While the size of the deposits produced in pilot-scale plants is typically less than 100 kg, the industrial plants are in some cases capable of producing preforms with weight up to several tons. Microstructure and properties of the spray formed steels are usually far superior to those of cast material, typically resembling those of the equivalent powder metallurgy steels. The main advantage of spray forming over powder metallurgy route is the possibility to eliminate powder handling steps. This not only minimizes the risk of contamination but also results in cost savings.

In this chapter we review the status of the spray forming of steels. The chapter is divided into two main parts. The first part focuses on the processing aspects of spray forming of steels. The latter part provides an overview of the spray formed steel grades and other iron based alloys.

J. Lotta • S.-P. Hannula (✉)
Department of Chemistry and Materials Science, Aalto University, Espoo, Finland
e-mail: juho.lotta@aalto.fi; simo-pekka.hannula@aalto.fi

C. Spiegelhauer
DanSpray A/S, Frederiksvaerk, Denmark
e-mail: csp@danspray.dk

H. Henein et al. (eds.), *Metal Sprays and Spray Deposition*,
DOI 10.1007/978-3-319-52689-8_12

12.2 Processing of Steels by Spray Forming

This section provides a short comparison between spray forming and other primary processing routes for steels and an overview of spray forming of steel preforms, with emphasis on preform types that have achieved most industrial success.

12.2.1 Comparison to Other Manufacturing Routes

Processing of steels is usually based on either ingot metallurgy or continuous casting. Remelting and PM processes are preferred with high-carbon, high-alloy grades. Table 12.1 identifies some of the key advantages of spray forming over casting and remelting processes. Most of the advantages of spray forming are due to rapid cooling during atomization and partial remelting at the deposition zone [1]. Rapid cooling suppresses diffusion, which in turn eliminates macrosegregation and promotes microstructural refinement. This, together with the partial remelting of solidified material, gives rise to a fine and uniform microstructure, which is characterized by equiaxed grains and even distribution of second phase particles. This not only improves the materials properties, but also minimizes the need of secondary processing (e.g., forging operations).

Melting routes for high-carbon, high-alloy steel grades may consist of multiple melting operations to meet the higher demands set on properties such as ductility, homogeneity, and cleanliness [2]. The primary melting operation such as Vacuum-Arc-Melting (VIM) produces desired composition and residual element control. The secondary melting operation, which is based on either Electro-Slag-Remelting (ESR) or Vacuum-Arc-Remelting (VAR) process, is carried out to refine composition and reduce segregation of alloying elements. Triple melting (1 × primary melting, 2 × remelting) can be applied to further improve materials properties. The upper limit of producible steel composition is set by segregation of alloying elements during solidification in all of these melting routes.

Table 12.1 Advantages of spray forming over casting and remelting processes

Category	Advantage
Processing	Higher degree of compositional freedom
	Possibility to produce better quality composites
	Fewer processing steps
	Possibility to manufacture near-net-shapes
	Reduced need for thermomechanical treatments
	Enhanced hot workability
Microstructure	Macro-segregation free microstructure
	Greater microstructural uniformity
	Finer grain and/or carbide size
Properties	Higher degree of isotropy
	Improved mechanical properties

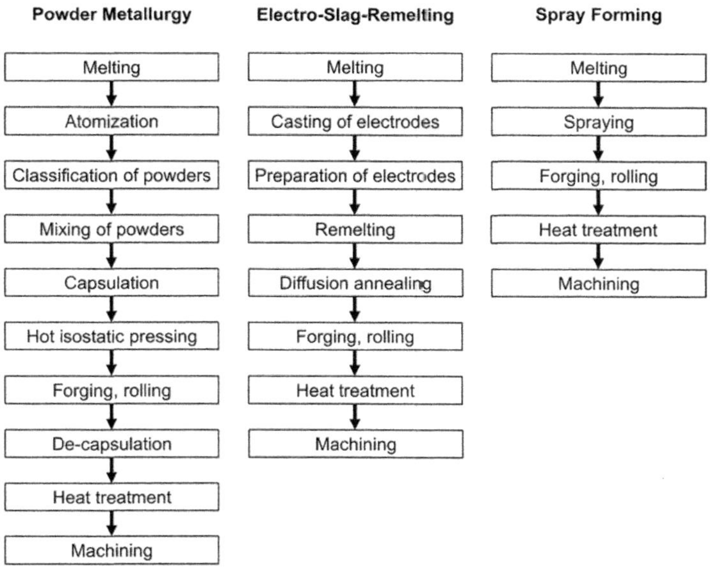

Fig. 12.1 Comparison of production routes for high-alloy tool steels (adapted from [3])

Development of powder metallurgical (PM) processes (e.g., hot-isostatic-pressing) has provided a mean to overcome the compositional limitations. This is mainly because PM route is based on consolidation of powder particles, which due to rapid solidification are characterized by very low level of segregation. In addition to widening the alloy selection, PM processes are very efficient at refining microstructure and improving mechanical properties. Spray forming more or less shares the advantages of PM route despite including fewer process steps, as illustrated in Fig. 12.1. The possibility to eliminate the powder handling steps with spray forming not only minimizes the related costs but also decreases the risk of oxidation and contamination. Microstructure and mechanical properties of spray formed high-alloy steels are typically reported to be intermediate between the equivalent steel produced by ingot metallurgy and PM route (see Sect. 12.3.1).

There is little concrete information available on how the cost of spray forming of steels compares to that of conventional processing of steels. Chesney et al. [4] estimated the cost of spray formed bimetal mill roll to be comparable to the cost of cast bimetal roll and half the cost of PM bimetal roll. The cost of stainless steel clad carbon steel bar produced by spray forming was estimated to be 50% of the solid stainless steel equivalent.

The disadvantages of spray forming are primarily due to material losses. These arise from the following factors:

- Some of the particles (gas atomized droplets) do not impact the preform surface;
- Some of the impacting particles bounce off from the preform surface;
- Machining losses (e.g., removal of porous surface layer) [5].

12.2.2 Spray Forming of Steel Preforms

A considerable part of the literature on spray forming of steels concerns deposition of round billets. Other deposit types include flat preforms [6–17], tubular preforms [4, 18–37], clad structures [4, 21, 23, 25, 26], and gradient structures [38–41]. Spray forming has been employed in making of molds and dies as well [42–55]. This section provides short overviews on spray forming of round billets, mill rolls, tubes, and molds/dies. These are the application areas which have demonstrated the most industrial interest and promise.

12.2.2.1 Round Billets

Spray forming of round steel billets involves deposition of gas atomized steel droplets onto a rotating substrate, which moves either in vertical or horizontal direction to accommodate the growth of the deposit and keep the spray distance constant. The pilot plants are typically based on single atomizer technology, whereas the industrial plants tend to prefer twin-atomizer technology to improve deposition rates and yield. The billets produced for research purposes weigh typically less than 100 kg, while the industrial scale billets weigh up to several tons. The selection of steels spray formed into round billets covers most of the alloys discussed in Sect. 12.3, although the commercial activity has been concentrating primarily on high-alloy tool steels.

Regardless of billet diameter, the material in the surface region of a spray formed billet features substantially finer microstructure than the material in the interior. This is because the surface layer is subject to much faster cooling rate than the interior. Besides promoting microstructural refinement, the fast cooling rate results in incomplete filling of interstices between pre-solidified droplets, giving rise to a highly porous surface layer. The amount of surface porosity is not only dependent on the thermal conditions, but also on the composition of the deposited steel. Carbon content especially has a significant impact on surface porosity, because it has considerable control over the temperature range at which steel remains in semisolid state. The longer time steel remains in semisolid state at deposition zone, the more time there is for melt to fill interstices. Consequently, steel grades with higher carbon content tend to exhibit less surface porosity than steel grades with lower carbon content.

Schulz et al. [56] estimated that more than 80% of a medium sized (billet: ø200 mm, 300–400 mm in length) high-speed steel billet is dense enough to be processed by forging. The percentage is even higher for larger diameter billets since the ratio between surface area and volume decreases as the size of the billet increases. Interior of a billet is not susceptible to cold-porosity, but may instead suffer from gas-porosity and/or hot-cracks. Formation of hot-cracks occur when excessive build-up of residual melt inside the top part of billet contracts on solidification. Gas-porosity is due to entrapment of atomization gas during deposition.

Dan Spray Ltd., established by Danish steel works Ltd. in 1998, was the first company to start commercial spray forming of specialty steel billets in 1999 [57]. Prior to this there had been two projects exploring the industrial potential of spray forming in production of specialty steels, both of which were carried out with the financial aid of the European coal and steel community. The first project (1989–1991) used a horizontal pilot plant at Osprey Metals Ltd. to spray form billets of stainless steels, hot-work tool steels, and high-speed steels [58, 59]. Although materials investigation produced promising results, the project identified several challenges in process control. The second project (1993–1997) aimed to demonstrate the applicability of spray forming as a viable process for industrial scale production of specialty steels [2, 59–61]. As part of the project the melting capacity of an existing horizontal billet plant at Osprey Metals Ltd. was increased from 300 to 1200 kg. Majority of the process optimization was carried out with the cold-work tool steel grade AISI D2, although spray forming of several other tool steel and stainless steel grades was investigated as well. The project was successful in increasing the diameter of billets from 250 to 400 mm (length up to 1300 mm), reducing consumption of nitrogen, reducing surface porosity, improving microstructural homogeneity, and increasing both deposition rate and yield [2]. The improvements were attributed to the use of twin atomizer technology.

The plant operated by Dan Spray Ltd. in Denmark was based on a vertical system due to technical difficulties experienced with the horizontal system at the Osprey Metals Ltd. [62]. Twin atomizer was employed to produce tool steel billets with a maximum diameter of 500 mm, length of 2500 mm, and weight of approximately 4 tons [62], typical yield being around 85% [63]. The annual production of the plant was estimated at 2000 tons in one shift [62]. The selection of spray formed steel grades included standard tool steels grades (e.g., AISI D2, AISI T15) as well as high-alloy tool steel grades developed in cooperation with Uddeholm Tooling AB (Touchtec [64, 65], Weartec [63, 64], and Roltec [63, 64]) and Edelstahl Witten-Krefeld GmbH (ESP23, ESP32 [3]).

Until recently, Peak Werkstoff GmbH was operating a plant, which used single atomizer to produce round steel billets with a maximum diameter of 400 mm and a length of 1100 mm. This, as well as the earlier plant operated by Dan Spray Ltd. and later by Uddeholms AB, Sweden, are no longer in operation. There is currently only one plant in Europe concentrating on spray forming of steel billets. This recently established plant is operated by DanSpray Ltd. in Denmark. The plant uses twin atomizer technology (100 kg/min) to produce billets up to 5 tons in weight, 500 mm in diameter, and 3000 mm in length. It includes a future option to add a second induction melting unit so that four atomizers can be used to produce billets up to 8–9 tons in weight and 1000 mm in diameter. Adding a second melting unit would also enable production of bimetallic products. In China, there are several more commercial spray forming plants operating in large scale (volume) and variety in materials [66]. The details, however, are not public.

12.2.2.2 Mill Rolls and Tubes

Spray forming of mill rolls and tubes involves deposition of gas atomized metal onto a rotating cylindrical collector. Hollow cylindrical collectors are used in spray forming of tubes, while spray forming of mill rolls uses either solid or hollow collectors. The collector is either withdrawn (single-pass technique) or reciprocated (multi-pass technique) under the spray to build up the deposit layer [26]. The former technique is better suited for production of long preforms (i.e., tubes), while the latter one is better suited for production of thick deposits (e.g., ring-like preforms for mill rolls). The length of the spray formed mill rolls is measured in tens of centimeters, whereas the length of the spray formed tubes has been reported to be up to 8 m [21].

Conventional manufacturing of mill rolls is based on ingot casting, which, depending on the alloy type and application area, may be followed by forging [67]. Spin casting and continuous pouring processes have been used for manufacturing of composite rolls. Sleeve rolls, which are fitted over a roll mandrel, can be also manufactured by spin casting. High-alloy mill rolls have been produced by PM route as well. The spray formed preforms for mill rolls are typically produced in the form of rings, which are subsequently fitted onto a roll mandrel. In another approach the atomized roll alloy is directly bonded with a preheated steel mandrel that acts as a collector.

Sumitomo heavy industries Ltd. started investigating spray forming of mill rolls in 1986 in a pilot plant (Fig. 12.2) built in Japan [22]. Production of mill rolls was started in 1987. A commercial plant capable of producing rolls with maximum diameter of 800 mm and length of 500 mm was built in 1991. Sumitomo's process used multi-pass technique to deposit rings on a thin collector tube. After downstream processing (e.g., machining operations, HIPping) the rings were fitted onto a roll mandrel for use in bar, flat bar, wire rod, and section mills. The spray formed alloys included high-chromium white irons, high-carbon/high-vanadium tool steels, and high-carbon/high-vanadium tools steels [19, 22], of which the latter two represents alloys developed by Sumitomo.

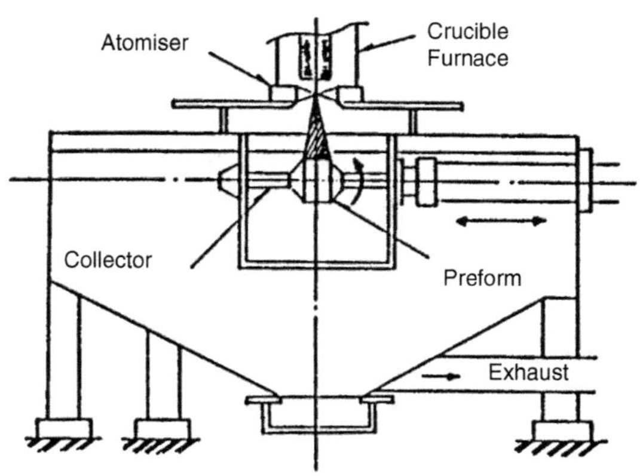

Fig. 12.2 A schematic of a pilot plant used for spray forming of mill rolls by Sumitomo heavy industries Ltd. [19]

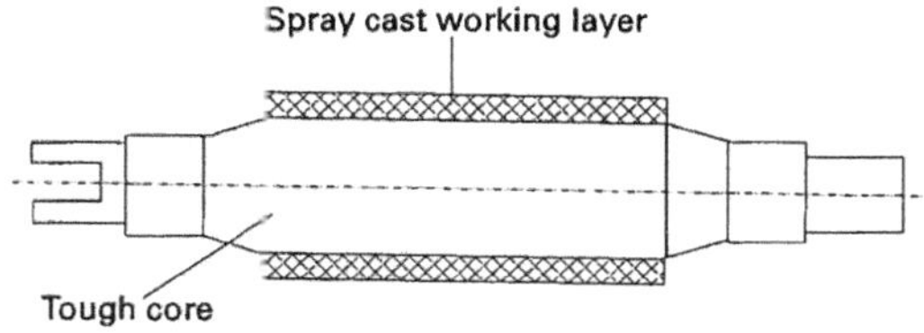

Fig. 12.3 A schematic illustrating the structure of a clad mill roll produced by direct spray forming of roll alloy onto a solid steel mandrel [28]

According to Itami and Kawashima [22] the high costs and technical challenges of PM route give spray forming a clear advantage over PM route in manufacturing of large size mill rolls. Ikawa et al. [18] noted spray forming to have the following advantages over conventional casting in roll production:

- Refinement of grain and carbide size,
- elimination of macro-segregation and shrinkage defects,
- improved hot workability and machinability, and
- possibility to produce rolls with higher alloy content.

Ikawa et al. [18] reported a comparison between the microstructure and mechanical properties of cast mill rolls and spray formed high-carbon, high speed steel (2.5%C, 6%V) mill rolls manufactured by Sumitomo. The spray formed material was observed to be superior to cast material with respect to both bending strength and wear resistance. The improved performance was attributed to the more uniform distribution of carbides, which were significantly smaller (<10 vs. 10–30 μm) than in the cast rolls. The lifetime of the spray formed rolls was 2–3 times longer than that of the cast rolls. According to Itami and Kawashima [22] the lifetimes are 3.6–10 times longer in comparison to the conventional rolls when the spray formed rolls are made of the alloys developed by Sumitomo.

A project to assess the feasibility of direct spray forming of roll alloys onto a solid steel mandrel (Fig. 12.3) was carried out in UK in the 1990s [26]. The existing plant at Ospreys Metals Ltd. was modified to include induction heater for preheating of the roll mandrel. Twin atomizer technology was utilized to permit deposition of thick layers in a single pass. The spray formed materials included standard industry cold (e.g., 0.8%C–3%Cr, 0.8%C–5%Cr) and hot (e.g., 2.5%C–17%Cr, high speed steels) rolling mill work roll qualities. The produced clad rolls were up to 400 mm in diameter and 1000 mm in length, while clad thickness was in the range 25–100 mm [68].

Hanlon et al. [27, 28] reported direct bonding of cold rolling mill work roll steels, 0.8%C/3%Cr and 0.8%C/5%Cr, onto a mild steel mandrel. The mean carbide size was finer in the spray formed material than either in heavily forged conventional material or lightly forged commercial material. The wear performance (rolling/sliding configuration) was similar to that of the conventionally processed material. With substrate preheating a high integrity metallurgical bond between the deposit and substrate [28] was achieved, the cross-bond tensile strength being ~85% of the ultimate strength recorded for the material taken from the bulk of the deposit. Other studies by Hanlon et al. report detailed investigations of

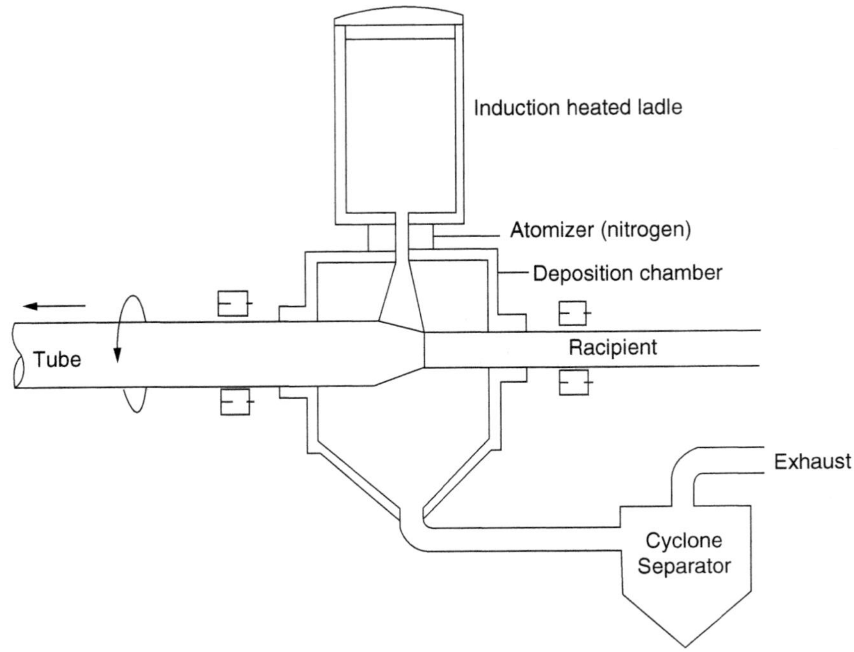

Fig. 12.4 A schematic illustrating the principle of spray forming of tubes [69]

high-chromium white iron [32, 33] (see Sect. 12.3.4) and high-speed steel [36] (see Sect. 12.3.1.1) deposited on a cold tubular collector. Deposition of high-chromium white iron on a tubular collector is also reported by Guo et al. (see Sect. 12.3.4) [34]. Cui et al. [31] investigated the microstructure of spray formed AISI 52000 layer (thickness: ~30 mm) deposited onto a solid collector bar (ø40 mm) made of austenitic steel. The grain size of the deposit was observed to be at least one ASTM grain size class finer than in the spray formed billet of the same material. Excessive porosity at the substrate-deposit interface, which occurred despite of the preheated (1100 °C) collector, was attributed to the cooling effect of the atomizing gas.

The most notable example of industrial scale plant for production of tubes is the facility (1-ton capacity) which was operated by Sandvik steel in Sweden [20, 21, 24]. The plant, which started its operation in 1986, was capable of producing tubular preforms with length up to 8 m, outer diameter up to 0.4 m (±1%), and wall thickness in the range 25–50 mm. Thin walled mild steel tube collector was rotated and withdrawn (single pass technique) under the spray by a chuck located on a carriage, as illustrated in Fig. 12.4. Fully dense tube was achieved after removing less than 5 mm of the inner porous surface and 4 mm of the outer surface. Examples of stainless steel grades spray formed by Sandvik steel include AISI 304, AISI 304L, AISI 310S, SAF 2205 (AISI 318LN), and Sanicro 28 [21]. Yaman [21] reported the tensile and impact properties of spray formed

SAF 2205 (AISI 318LN), 3R12 (AISI 304L), and Sanicro 28 to exceed the required minimum levels, while the corrosion resistance is similar or better than that of conventionally processed material.

Another industrial scale spray forming plant for production of tubes was operated by Babcox and Wilcox Nuclear Equipment Division [4, 25, 29]. The plant, equipped with 5 ton melting capacity, was capable of manufacturing tubes up to ~1 m in diameter and ~6 m in length [25]. According to Chesney et al. [29] the mechanical properties of the spray formed AISI 304L tubes produced by the plant met the requirements set by the ASTM standards for wrought material.

Excessive porosity near the substrate is one of the major problems in spray forming of tubes, because of the material losses and manufacturing costs associated with the removal of the porous layer. The porosity is typically caused by the chilling effect of substrate. Solutions to the problem include substrate preheating and use of multi-atomizers [70]. Chesney et al. [29], for instance, observed a 50% reduction in porosity when twin atomizer was used instead of single atomizer. Wahlroos and Liimatainen [23] investigated the effect of substrate preheating (600–1200 °C) on interface quality between spray formed AISI 304 (thickness: 8.5–11 mm) and carbon steel base tube (ø95 mm, wall thickness 5–6 mm, length: 520 mm). The lowest preheating temperature resulted in poor interface adherence, whereas a very high and constant bond strength was achieved with the preheating temperatures of 1000, 1100, and 1200 °C; use of preheating temperature of 1200 °C reduced porosity at the substrate-deposit interface to 0.65 vol.%.

12.2.2.3 Molds and Dies

Conventional making of molds and dies is characterized by considerable time lag between design and finalized tool [45]. This is due to the need to qualify the part design by making prototypes and production tooling lead time after approval of the prototype. Making of production tooling uses such unit operations as CNC machining, sink EDM, benching, polishing, engraving, and heat treatments to transform a forged block of tool steel into a finished tool. The process is both expensive and time consuming. This is problematic particularly with low-volume products and rapidly changing high-volume products. Rapid tooling technologies have gained interest because they offer the possibility to shorten the design-to-part cycle and thus reduce the cost of molds and dies.

Rapid Solidification Process (RSP) and Precision Spray Forming (PSF) are spray forming based rapid tooling methods. The RSP tooling was developed at the Idaho National Engineering and Environmental Laboratory (INEEL) in the 1990s for making of molds and dies [43, 44]. The process was commercialized in 2002 with the formation of RSP Tooling, LLC [51]. In RSP tooling molten metal is gas atomized and subsequently deposited (horizontal system) onto a ceramic pattern, as illustrated in Fig. 12.5. The PSF process, which was developed at the VTT Industrial Technologies (Finland) in the early 2000s, uses a modified Osprey™ spray-forming machine (vertical system) to make molds and dies by depositing

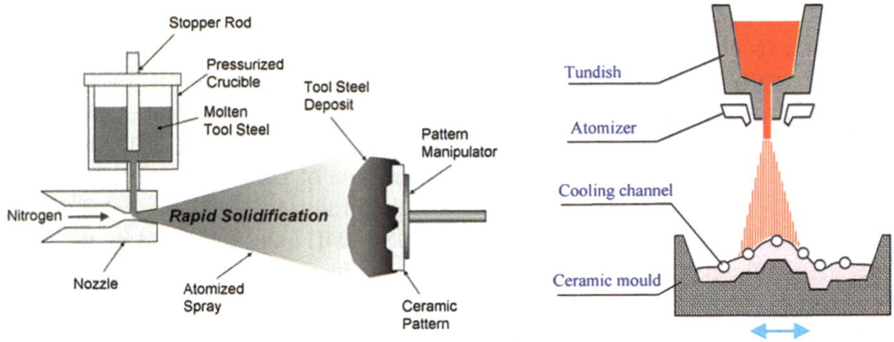

Fig. 12.5 Schematics illustrating the principle of rapid solidification process (*left*, [51]) and precision spray forming process (*right*, [54])

steel onto a ceramic mold, as illustrated in Fig. 12.5 [47]. The PSF process has been commercially used by Dongfeng Motor in China [71].

The primary process steps of RSP tooling include pattern making, ceramic mold making, spray deposition, and finishing operations [51]. The tooling master is typically converted from a CAD file to a physical tooling master using a suitable rapid prototyping technology such as stereolithography. The pattern is then transferred (e.g., slip casting, freeze casting) into a ceramic mold made of alumina or fused silica. This is followed by deposition of thick layer of desired alloy on to the ceramic pattern to capture the shape, pattern, and surface texture. The deposit is separated from the pattern once it has cooled to room temperature. The finishing operations include adding bolt holes and water lines as well as EDM machining the exterior walls square. PSF Tooling includes the same primary process steps as RSP tooling [47], although in PSF process the ceramic pattern is both rotated and moved sideways along a predetermined path during the deposition. There is also a significant difference in the deposition rates of the two methods. In PSF process the deposition rate is approximately 30 kg/min [47], while in RSP tooling the deposition rate is at most 250 kg/h (~4 kg/min) [45]. Examples of die inserts produced by RSP tooling and PSF process are shown in Fig. 12.6.

The literature on RSP tooling and PSF process focuses primarily on deposition of hot-work tool steel AISI H13 (see Sect. 12.3.1.3). In addition to AISI H13, RSP tooling has been used for deposition of several tool steel grades, cast irons, copper alloys, and aluminum alloys [51]; RSP tooling of AISI A2, modified AISI H13, and AISI M2 is reported in [50]. The literature reports on alloys processed by PSF process include AISI H13 [46, 48], novel compositions resembling AISI H13 [46, 48], and white cast iron [54].

Possibility to minimize cost and turnaround times without sacrificing quality or accuracy is the main advantage of spray forming over conventional manufacturing of molds and dies. The advantage is based on the fact that many of the steps in normal mold making operations (e.g., milling, benching, polishing, and engraving) can be eliminated, because the deposited metal replicates the features on the

Fig. 12.6 A die cast insert produced by RSP tooling (*left*, [51]) and a forging die insert produced by PSF process (*right*, [48])

ceramic pattern accurately. For instance, RSP tooling is reported to replicate as small details as 75 μm [51]. The elimination of the process steps has the potential to reduce the design-to-part cycle from months to days [47, 51].

The literature reports on RSP tooling of AISI H13 show spray forming to promote enhanced materials properties (see Sect. 12.3.1.3). This leads to cost savings by extending the lifetime of spray formed tooling. McHugh et al. [55] reported AISI H13 steel die produced by RSP tooling to outlast its conventional counterpart by more than 250% in open-die forging of aluminum. Considerable increase in tool-life (7–67%) was reported in extrusion and die casting trials of aluminum as well. Prolonged tool lifetime has been reported in case of PSF process as well [48]. The possibility to encapsulate conformal cooling lines (see Fig. 12.5) inside the deposit enable further cost savings by reducing part cycle time [51, 54]; despite of reducing part cycle time by 15–50% in plastic injection molding, conformal cooling is rarely implemented in conventional tools because of the challenges in machining the cooling channels [45].

Size of the inserts produced by RSP tooling and PSF process (twin atomizer) has been limited to 200 mm [51] and 400 mm [54], respectively. However, these limitations are not inherent to the processes, as it is possible to further scale up the insert size by increasing the number of spray heads. The main limitation of both methods relates to the aspect ratio of standing features on die surface: RSP tooling and PSF process cannot replicate small features with aspect ratio of 4:1 or higher [47, 51]. The limit is set by the tendency of molten metal to bridge across the features with high aspect ratio before they are entirely filled.

12.3 Spray Formed Steels and Iron Based Alloys

This section provides an overview of spray formed steels and iron based alloys. Discussion is made on tool steels, low-alloy steels, stainless steels, high-chromium white irons, and steel matrix composites. The sub-section on tools steels is furthermore divided between high-speed steels, cold-work tool steels, and hot-work tool steels, while the sub-section on low-alloy steels is divided between low- and medium-carbon, high-carbon, and ultrahigh carbon steels. Selected research results from literature are presented in each sub-section.

12.3.1 Tool Steels

12.3.1.1 High-Speed Steels

High-speed steels are mostly used for cutting tools that generate considerable amount of heat during high-speed machining of steels and other metals [72]. The high wear resistance and red-hardness required in these applications is achieved with a microstructure containing a large volume fraction of carbides in a high alloyed martensitic matrix with secondary carbides. Typical metallic alloying elements include W, Mo, V, Cr, and Co. The spray formed high-speed steel grades include AISI M2 [14, 56, 73–81], AISI M3:2 [76, 82–87], AISI M4 [88], AISI M15 [74], HS 6-5-3-8 [89–93], AISI M50 [61], AISI T15 [60, 76, 94–96], HS 3-3-4 [36], and ESP32 [3]; ESP32 (1.8%C–4.0%Cr–4%Mo–4.8%V–9.5%W–10%Co–1.3% Nb) is a high-speed steel grade developed for commercial spray forming (see Sect. 12.2.2.1). The research activities have been primarily focused on spray forming of round billets. There are also reports on spray forming of flat [14] and tubular preforms [18, 19, 22, 36]. Current industrial spray forming of high-speed steels includes at least production of round billets (see Sect. 12.2.2.1). Past activities have also included industrial spray forming of mill rolls (see Sect. 12.2.2.2).

Igharo and Wood [75] investigated microstructure and mechanical properties of spray formed AISI M2 (billet: ø130 mm) in both as-sprayed and forged condition. The as-sprayed microstructure featured fine grain size (~12 μm) and cellular carbide (MC, M_6C) network, which was transformed into a uniform dispersion of fine carbides (0.5–2 μm) in subsequent hot forging. The spray formed material was similar to extensively hot worked cast material with respect to hardening response. Bending strength was concluded to be superior to that of the equivalent cast material, being comparable to the bending strength of PM material. In another study Igharo et al. [76] observed the eutectic carbide cell size of spray formed AISI M2 to be one tenth of that in cast equivalent grades. Tempering peak hardness was observed to be higher than that of the equivalent cast and wrought alloys.

Schulz et al. [56] compared the microstructure and properties of spray formed (billet: ø200 mm) AISI M2, cast AISI M2, and PM AISI M3. The PM material featured the finest and most uniform distribution of carbides, while the spray formed material was set apart from the cast material by its more uniform distribution of carbides rather than carbide size. In a pin-on-disk test (SiO_2 paper) the wear performance of spray formed material was comparable to that of the cast and PM materials, while in a dry sand rubber/wheel test the spray formed material performed slightly better than the cast material, but considerably worse than the

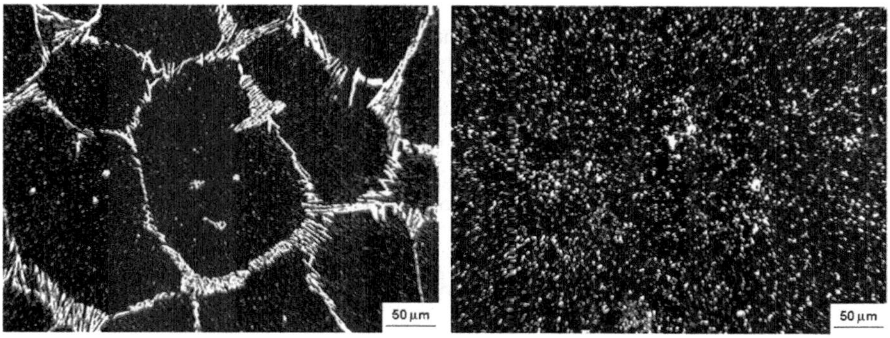

Fig. 12.7 Optical micrographs illustrating the difference between the carbide morphologies of conventionally cast (*left*) and spray formed (*right*) high speed steel AISI M3:2 [83]

PM material. Impact toughness of the spray formed material was intermediate between that of the cast and spray formed material. Schulz et al. furthermore observed slightly lower impact toughness and wear resistance in material melted under nitrogen atmosphere instead of argon atmosphere; in another study Schulz et al. [78] found melting under nitrogen atmosphere to increase the nitrogen content of spray formed AISI M2 by a factor of 1.7.

Mesquita and Barbosa [83] compared the microstructure and mechanical properties of spray formed (billet: ø400 mm) and conventionally cast AISI M3:2. The microstructure (Fig. 12.7) of the spray formed material was characterized by a uniform distribution of fine carbides (MC, M_6C), whereas the microstructure of the cast material featured a coarse carbide network. Bending strength of the spray formed material was superior to that of the cast material in transverse direction and comparable in longitudinal direction. Higher degree of isotropy in bending strength was attributed to the more uniform microstructure of the spray formed material. In another study Mesquita and Barbosa [82] observed the microstructure and bending strength (Table 12.2) of spray formed AISI M3:2 (billet: ø400 mm) to be intermediate between cast AISI M3:2 and PM AISI M3:2.

Ernst and Duh [88] compared microstructure and mechanical properties of HIPped TSP4 (AISI M4) to those of spray formed equivalent steel grade (billet: ø500 mm). The typical carbide size in the spray formed and PM material was in the range 1–6 μm and 1–3 μm, respectively. Although spray forming produced a fairly uniform and fine microstructure, the PM material was superior to the spray formed material with respect to both bending strength (Table 12.2) and impact toughness.

Spiegelhauer and Davin [94] compared microstructure and mechanical properties of AISI T15 (forged and heat treated) produced via spray forming (billets: ø250 mm, ø375 mm), HIPping, and vacuum sintering. Carbide sizes for the spray formed, HIPped, and vacuum sintered material were 3–5, 1–3, and 1–4 μm, respectively. Bending strength (Table 12.2) of the spray formed material was intermediate between the vacuum sintered and HIPped material. No major difference was observed in hardness between the materials. Zhang et al. [96] investigated the

Table 12.2 Bending strength of spray formed high-speed steels in comparison to high-speed steels processed via other routes (L = longitudinal, T = transverse)

Ref.	Material	Processing route[a]	Bending strength (MPa)
[82]	AISI M3:2[b]	Sintered (as-sintered)	L: 3490, T: 3600
		Spray formed (forged, rolled)	L: 2420, T: 2100
		Cast (wrought)	L: 2550, T: 1370
[88]	AISI M4[b]	Spray formed (forged)	L (edge): 4460 L (center): 4400 T (center): 3240
		HIPped (as-sintered)	L (edge): 4830 L (center): 4830 T (center): 4770
[94]	AISI T15	Spray formed (ø375 mm, forged)	L: 3850, T: 2640
		Spray formed (ø250 mm, forged)	L: 3830, T: 2722
		HIPped (forged)	L: 4537, T: 3527
		Vacuum sintered (forged)	L: 3623, T: 2104

[a]All the materials were tested in hardened and tempered condition
[b]The values for bending strength and impact toughness were read from graphs

influence of heat treatments, hot forging, and HIPping on the microstructure and mechanical properties of spray formed AISI T15 (billet: ~ø280 mm). HIPping of the spray formed materials resulted in a very slight improvement of impact toughness (~10 J/cm^2) and bending strength (~2500 MPa). A much more substantial increase in impact toughness (~25 J/cm^2) and bending strength (~4600 MPa) was achieved by subjecting the spray formed material to hot forging. According to Zhang et al. the achieved bending strength was considerably higher than that of conventionally processed AISI T15 (3800 MPa), being comparable to the bending strength of PM AISI T15 (4400 MPa).

Itami et al. [89] compared cutting performances of spray formed HS 6-5-3-8, PM HS 6-5-3-8, and cast high-speed steel with composition close to that of HS 6-5-3-8. The spray formed material was intermediate to the cast and PM material with respect to microstructure, while it possessed a combination of hardness and transverse rupture strength equivalent to that of the PM material. Cutting performance of the spray formed material was comparable to that of the PM material. Yabuuchi et al. [90], in their respective study of HS 6-5-3-8, found the tool life of spray formed material to be superior to both the PM and cast materials, although its Charpy V impact toughness was substantially lower than that of the PM material.

Hanlon et al. [36] compared microstructure and properties of spray formed (tubular preform: ø80 mm) and conventionally processed high-speed steel HS 3-3-4 (spray formed: 1.2%C–3.4%W–8.9%Cr–4.3%V–2.7%Mo; conventionally processed: 1.5%C–3.0%W–8.0%Cr–4.0%V–2.7%Mo). Spray forming was observed to eliminate continuous carbide (VC, M_2C, M_6C) network and result in substantial refinement of grain size (11 vs. 60 μm) in comparison to cast material. Wear performance (rolling/sliding configuration) of the spray formed material against AISI M2 in the temperature range 20–650 °C was superior to that of the cast material. The improvement in wear resistance was attributed to refinement of carbide size: Fractured carbides were almost absent in the spray formed materials, whereas in the conventionally processed material there was widespread fracturing of carbides up to a depth of ~40 μm.

12.3.1.2 Cold-Work Tool Steels

The literature on spray forming of cold-work tool steels deals mainly with high-carbon grades alloyed with considerable amount of Cr and/or V. The high carbon and alloy content produces large volume fraction of carbides, which gives these alloys an excellent resistance against wear. Consequently, these type of steels are widely used for blanking and cold-forming punches and dies, where good wear resistance is needed [72]. The spray formed high-carbon, high-chromium cold-work tool steels include X110CrMoVAl8-2 [14, 78, 97], AISI D2 [3, 56, 60, 61, 79, 97–99], AISI D3 [14], AISI D3 resembling alloys with additional carbon [14], and AISI D7 [14]. Studies have been carried out on spray forming of vanadium alloyed high-carbon, cold-work tool steels as well [30, 100–102]. Roltec [64], Weartec [64], and ESP23 [3] represent high-carbon, high-alloy cold-work tool steels, which were developed for commercial spray forming (see Sect. 12.2.2.1).

Spray forming of AISI D2 has been investigated more extensively than any other high-carbon, high-alloy cold-work tool steel. Spiegelhauer [60] observed the microstructure of spray formed AISI D2 to be free of macro-segregation and feature carbide structure that is both finer and more uniform than that in cast and remelted AISI D2. While the wear resistance (pin-on-disk) of the spray formed material was observed to be slightly lower than that of cast material, the spray formed material was superior to the cast material with respect to both hardness (65 vs. 63 HRC) and impact toughness (Table 12.3).

Ernst and Duh [99] compared microstructure and mechanical properties of spray formed (billet: ø500 mm) and cast X155CrVMo12-1 (AISI D2). Although the microstructure of the as-sprayed material featured a carbide network, it was considerably finer than that in the as-cast material. After hot rolling, the spray formed material contained carbides with length ranging from 25–30 to 90 μm. In the cast material carbide length ranged from 35–50 to 190 μm even though the deformation ratio was twice as high as that applied to the spray formed material. Toughness determined by impact bending test (Table 12.3) was twice as high for the spray formed material as for the cast material. The spray formed material outperformed the cast material in static bending test as well. The superiority was attributed to lack of macrosegregation, microstructural refinement, and more uniform distribution of carbides.

Schulz et al. [56] observed spray formed (billet: ø200 mm) AISI D2 to be set apart from cast AISI D2 by more uniform distribution of carbides rather than finer carbide size. Finest and most uniform carbide structures were observed in PM processed material. Impact toughness (Table 12.3) of the spray formed material was intermediate between the impact toughness of cast and PM material. Melting

Table 12.3 Impact toughness of spray formed cold-work tool steels in comparison to cold-work tool steels processed via other routes (L = longitudinal, T = transverse)

Ref.	Material	Processing route[a]	Impact toughness (J)
[56]	AISI D2[b]	Spray formed (forged/rolled, low N_2)	L: 20.8, T: 10.7
		Spray formed (forged/rolled, high N_2)	L: 17.6, T: 7.9
		Cast (wrought)	L: 14.5, T: 4.4
		PM (wrought)	L: 26.9, T: 14
[60]	AISI D2	Spray formed (forged)	L: 30, T: 10
		Cast (forged)	L: 20, T: 8
[99]	AISI D2[b]	Spray formed (forged)	Edge: 45–57, Core: 32–52
		Cast (hot rolled)	Edge: 18–27, Core: 14–20
[97]	X110CrMoVAl8-2[b]	Spray formed (forged, low N_2)	L: 116, T: 39
		Spray formed (forged, high N_2)	L: 61, T: 13
		ESR (wrought)	L: 24, T: 7

[a]All the materials were tested in hardened and tempered condition
[b]The values for impact toughness were read from a graph

under nitrogen instead of argon resulted in a noticeable decrease in impact toughness (Table 12.3), whereas wear performance was not significantly influenced by melting under nitrogen. The spray formed material outperformed the cast and PM materials in both dry sand/rubber wheel test and pin-on-disk test (SiO_2 paper).

Schulz et al. [97] also investigated the effect of nitrogen pick-up on microstructure and mechanical properties of spray formed (billet: ø200 mm) X110CrMoVAl8-2. With this alloy melting under nitrogen instead of argon resulted in two- and threefold decrease in longitudinal and transverse impact toughness (Table 12.3), respectively. This was attributed to formation of aluminum nitride precipitates in the microstructure. Nevertheless, even the spray formed material with lower toughness had an impact toughness 2–3 times higher than the ESR processed X110CrMoVAl8-2. The superior toughness of the spray formed material was attributed to lower level of macrosegregation and finer carbide (M_7C_3) size.

Yan et al. [100, 101] investigated the microstructure of spray formed (billet: ø130 mm, 30 mm in length) high-carbon, high-vanadium, cold-work tool steel Vanadis 4 (1.5%C–8.0%Cr–4%V–1.5%Mo) in as-sprayed and hot rolled/annealed condition. Comparison between the as-sprayed and as-cast microstructure showed spray forming to result in substantial refinement in both grain (8–10 μm) and carbide size (carbides at the grain boundaries: 0.5–2 μm; carbides inside grains: ~0.18 μm [101]). The carbide (V-rich MC, Cr-rich M_7C_3) size of the hot rolled/annealed material was slightly finer than that in the commercial Vanadis 4 - (PM-route) [101]. Ni et al. [102] investigated the as-sprayed (billet: ø230, 330 mm in length) microstructure of high-carbon, high-vanadium cold-work tool steel (2.4%C–10.7%V–5.4%Cr–1.4%Mo), which was compositionally very close to AISI A11. Spray forming produced a microstructure consisting of equiaxed grains with a diameter ~20 μm. Size of the uniformly distributed V-rich MC-type carbides was 3–5 and 1 μm at the grain boundaries and inside the grains, respectively.

Sandberg and Jönson [64] compared the microstructure and properties of spray formed (billets: ø500 mm) Roltec (1.4%C–4.6%Cr–3.2%Mo–3.7%V) and Weartec (2.8%C–7.0%Cr–2.3%Mo–8.9%V) to conventionally processed AISI D2 and AISI D6. The carbide size in Roltec (9 vol.% MC) and Weartec (15 vol.% MC, 7 vol.% M_7C_3) ranged from 1 to 15 μm, whereas in AISI D2 and AISI D6 carbide size was 100 and 20 μm in longitudinal and transverse directions, respectively. Despite their higher hardness (61–63 vs. 58–61 HRC), the spray formed materials showed better impact toughness than the conventionally processed materials. The abrasion resistance (pin-on-disc test with SiO_2 paper) of the Weartec was superior to those of both AISI D2 (15 vol.% M_7C_3) and AISI D6 (21 vol.% M_7C_3), while Roltec was equal to AISI D2 in abrasion resistance. According to Sandberg and Jönson the Roltec (punching, blanking) and Weartec (form rolls for tube manufacturing) showed considerable performance improvement over AISI D2 in application tests.

Schruff et al. [3] compared the microstructure and properties of spray formed (billet: ø500 mm) ESP23 (1.55%C–9.0%Cr–2.0% Mo–2.0% V–1.0%Nb) to conventionally processed AISI D2. The ESP23 contained a uniform distribution of carbides with size in the range 4–8 μm. Despite substantially higher carbide content and similar hardness (14 vs. 26 vol.%, 59–61 HRC), the ESP23 outperformed the AISI D2 in both impact toughness and bending strength. Abrasive wear (pin-on-disc test) performance of the ESP23 was similar (abrasive particle size: 80 mesh) or superior (abrasive particle size: 220 mesh) to that of the AISI D2 depending on the size of the abrasive particles. According to Schruff et al. the spray formed ESP23 outperformed conventional materials solutions in several industrial applications (e.g., thread rolling tools, punches for cutting steels).

12.3.1.3 Hot-Work Tool Steels

The literature on spray forming of chromium hot-work tool steels focuses almost exclusively on AISI H13. This steel grade is widely used in die-casting dies, hot extrusion tooling, and forging dies because it has a good combination of hot strength, toughness, oxidation resistance, and resistance to thermal fatigue [72]. The studies on spray forming of AISI H13 can be divided between those focusing on spray forming of billets [14, 56, 78, 79, 97, 103, 104] and those focusing on rapid tooling by either RSP or PSF method (see Sect. 12.2.2.3); RSP [50] and PSF [46, 48] methods have been also used for spray forming of hot-work tool steels, whose composition deviates from that of AISI H13. Touchtec (1.6%C–5.0%Cr–2.3%Mo–7.3%V) represents a wear resistant hot-work tool steel which was developed for commercial spray forming (see Sect. 12.2.2.1) [64, 65].

All of the studies on spray forming of AISI H13 show spray formed AISI H13 to feature a fine and uniform microstructure. Several of the studies furthermore report spray formed AISI H13 to achieve higher tempering hardness than conventionally processed AISI H13 [42, 49, 55, 103]. For instance, McHugh and Wickham [42]

Table 12.4 Tensile properties of spray formed and conventionally processed AISI H13

Ref.	Processing route[a]	Testing temperature (°C)	$R_{p0.2}$ (MPa)	R_m (MPa)	A (%)
[45]	RSP tooling	22	1158	1358	–
	Cast		–	882	–
	Commercial (forged)		1681	1799	–
	RSP tooling: tempered at 540 °C	550	1475	1647	–
	Commercial (forged)		1247	1323	–
[104]	Spray formed (forged)	Room temperature	1421	1621	12.8
	Cast (forged)		1379	1593	11.2
	Spray formed (forged)	650	594	745	27.8
	Cast (forged)		547	610	24.2

[a]All the materials were tested in hardened and tempered condition unless otherwise mentioned

reported the spray formed (RSP tooling) AISI H13 to achieve a maximum hardness of 59 HRC, while the maximum hardness of commercial forged AISI H13 was limited to 53 HRC. Lin et al. [49] found the Rockwell C hardness of spray formed AISI H13 (RSP tooling) to be up to 3 units higher (e.g., 56.6 vs. 53.3 HRC) than that of the commercially forged AISI H13. The spray formed AISI H13 is explained to attain higher hardness, because its matrix contains more carbon than the matrix of conventionally processed material due to high austenization temperature it experiences during spray forming [55]; high austenization temperature promotes greater dissolution of proeutectoid carbides, which in turn increases hardness by maximizing the precipitation of fine secondary carbides during tempering.

The superior mechanical properties of spray formed AISI H13 are not limited to hardness. McHugh and Folkestad [45] found spray formed AISI H13 to compare favorably to commercial AISI H13 with respect to tensile properties as well (Table 12.4). The spray formed material was also noted to better retain its strength at elevated temperature than the commercial AISI H13. Zhang et al. [104] confirmed these findings in their respective study of spray formed (billet: ø120 mm, 150 mm in length) and cast AISI H13 (Table 12.4). Zhang et al. furthermore observed spray formed material to be superior to cast material with respect to impact toughness (45.6 vs. 15.2 J). Similar observation was reported by Schulz et al. [97], who attributed this to the lower level of macrosegregation in the spray formed (billet: ø200 mm) AISI H13. Schulz et al. additionally observed ESR processed material to have higher impact toughness than the spray formed material. Nitrogen pick-up during spray forming was considered as a possible explanation for the lower impact toughness as the spray formed material had 2.3–2.7 times higher nitrogen content than the ESR processed material. Another study by Schulz et al. [78] did not find clear correlation between the impact toughness and nitrogen content in spray formed (billet: ø200 mm) AISI H13, but, on the other hand, observed deterioration of the high temperature (550 °C) tensile properties with increasing nitrogen content.

Yang and Hannula [46, 48] investigated PSF processing of alloys, which contain higher amount of carbon (0.35–0.90%) and/or vanadium (up to 3.5%) than AISI H13. The alloying additions were observed to improve hardenability of the material, thus making it possible to temper the material in as-sprayed condition [46]. This is beneficial because it shortens lead times and eliminates the risk of dimensional distortion involved with quenching. Increase in the vanadium content was furthermore observed to be an effective way to minimize porosity [48].

12.3.2 *Low-Alloy Steels*

Spray forming of a wide range of low-alloy steels has been investigated. These can be divided into three groups based on carbon content: low- and medium carbon steels, high-carbon steels, and ultrahigh carbon steels.

12.3.2.1 Low- and Medium-Carbon Steels

Spray forming of the following low- and medium carbon steel grades has been reported in the literature: AISI 1015 [9, 13], AISI 1020 [8, 10, 73], AISI 1035 [105], AISI 1040 [73], AISI 1045 [9, 11, 12], EN16 [73], EN33 [73], AISI 4340 [73], AISI 5120 [106, 107], AISI 4140 [73], and AISI 4140H [106]. The spray formed preform shapes include round billets and flat preforms. The mechanical properties of the spray formed low- and medium alloys are reported to be equal or superior to those of the conventionally processed alloys. Examples of tensile properties of spray formed and conventionally processed low- and medium carbon steels are listed in Table 12.5.

Ibrahim et al. [8] investigated the process-microstructure relationship in spray forming of plate-like deposits of AISI 1020 (0.2%C). Optimization of deposition parameters resulted in equiaxed ferrite-pearlite microstructure with grain size in the range 4–20 μm. Full density material was achieved either by warm rolling the material at 700 °C (70% reduction) or hot rolling it at 1050 °C (55% reduction). Matsuo et al. [10] investigated the microstructure and tensile properties of plain and micro-alloyed (0.05% N, 0.06% Al) AISI 1020 spray formed into plate-like deposits. Hot rolling at 850 °C for a thickness reduction of 70% and subsequent normalizing treatment resulted in a fully dense material with uniform ferrite-pearlite structure. The grain sizes were 3–4 μm and 5–15 μm for the micro-alloyed and plain grades, respectively. Both of the spray formed materials were superior to the conventional micro-alloyed material with respect to tensile strength (Table 12.5).

Cui et al. [107] investigated microstructure and distortion behavior of steel grade AISI 5120 (0.2%C) spray formed (billet: ~ø110 mm, ~200 mm in length) under different spray conditions. The spray-formed material, subjected to machining operations and heat treatments, showed lower distortion potential than the

Table 12.5 Tensile properties of spray formed and conventionally processed low- and medium carbon steels (L = longitudinal, T = transverse)

Ref.	Material	Processing route/condition	$R_{p0.2}$ (MPa)	R_m (MPa)	A (%)
[10]	AISI 1020	Spray formed (hot rolled)/Normalized (Al-micro-alloyed)	550	630	23
		Spray formed (hot rolled)/normalized (plain)	460	590	26
		Commercial grade/normalized	340	440	36
[73]	EN8 (AISI 1040)	Spray formed (forged)/normalized	357	567	25
		Conventional (forged)/normalized[a]	360	543	30
[73]	EN24 (AISI 4340)	Spray formed (forged)/hardened and tempered	772	869	19
		Conventional (forged, L)/hardened and tempered	772	896	25
		Conventional (forged, T)/hardened and tempered	717	855	16

[a]Typical published results according to Brooks et al.

conventional material. This was attributed to improved metallurgical homogeneity. Schumacher et al. [106] investigated microstructure and properties of spray formed (billets: ø140–210 mm) sulfur alloyed (0.2%) variants of AISI 5120 and AISI 4140H (0.4%C). The spray formed material contained uniform distribution of finer manganese sulfides than conventionally produced steels. Despite the significantly higher sulfur content, the spray formed steels reached a similar endurance limit in fatigue testing (rolling direction) as the conventionally produced steels.

12.3.2.2 High-Carbon Steels

The literature on spray forming of high-carbon steels focuses almost exclusively on steel grade AISI 52100 [13, 26, 31, 73, 106, 108–114], which is a widely used steel grade in bearing applications. There are also reports on spray forming of AISI W110 [9, 13] and AISI W7 [37]. Although most of the studies concentrate on spray forming of round billets, there are reports on spray forming of flat [13], tubular [31], and ring-like [37] preforms as well. Spray forming is considered as a promising alternative to the conventional processing of high-carbon steels, because it produces microstructure that is both uniform and free of macro-segregation. This is of significance because metallurgical inhomogeneity is one of the main causes for dimensional instability in bearing applications.

Tinscher et al. investigated microstructure, tensile properties, and fatigue resistance of spray formed (billets: ø160–180 mm, 350–450 mm in length) AISI 52100 [108]. The spray formed microstructure was characterized by a high degree of isotropy and lack of macrosegregation. The tensile properties of the material were very close to those reported for conventionally cast material while its endurance limit in fatigue testing was slightly higher than that reported for conventionally cast material. Cui et al. [111, 112] subjected spray formed AISI 52100 billets (ø170 mm, 300–350 mm in length) to hot rolling and ring rolling to produce ring preforms. Microstructure and distortion behavior following machining and heat treatment operations were studied in comparison to ring preforms made of continuous cast AISI 52100. Microstructure of the rings made of the spray formed material was homogeneous across the cross-section of the ring preform, while the microstructure of the ring preforms made of cast material showed high degree of macrosegregation. The lower distortion potential of the spray formed material in quenching was attributed to its better metallurgical homogeneity.

Schulz et al. [37] investigated microstructure, hot workability, and rolling-contact fatigue of spray formed (billet: ~80 kg) 100CrMnMoSi8-4-6 (AISI W7). Both homogeneity and cleanliness of the spray formed material were superior to those of the conventional material. No carbide network was present in the spray formed material despite low degree of deformation. Its performance in rolling-fatigue test was similar to that of the conventional material.

12.3.2.3 Ultrahigh Carbon Steels

Ultrahigh carbon steels (UHCS) are hypereutectoid steels that contain 1.0–2.1% carbon and small amounts of other alloying elements [115]. UHCSs with fine grain and carbide size exhibit high strength and super plasticity at temperature range 650–800 °C. Conventional processing of UHCS includes extensive thermo-mechanical treatments to break down the coarse carbide structures which form during casting. Spray forming provides a promising alternative to the conventional processing route, because the high cooling rate prevents formation of coarse carbide structures, thus minimizing the need for thermomechanical treatments [116–118].

Zhang et al. [116] investigated the microstructure and superplasticity of spray formed (charge weight 5 kg) ultrahigh carbon steel (1.25%C–3.0%Si–1.5%Cr) in as-sprayed condition. The microstructure was observed to consist of fine pearlite with interlamellar spacing of 0.2 μm. Testing at 820 °C with initial strain rates of 2.5×10^{-4} and 8×10^{-4} s^{-1} resulted in total elongations of ~375 and ~275%, respectively. Luo et al. [117] compared the microstructure and tensile properties of the spray formed and cast ultrahigh-carbon steel (1.28%C–3.0%Si–1.50%Cr) in as-sprayed/as-cast and hot rolled conditions, respectively. The spray formed material was set apart from the cast material by the lack of continuous proeutectoid ferrite and carbide networks. Furthermore, the as-sprayed material had much finer grain size (20 vs. 200 μm) and smaller interlamellar spacing of pearlite than the cast material (0.15 vs. 0.36 μm). The tensile properties of the spray formed material were superior to those of the cast material regardless whether the material had been hot rolled or not.

Table 12.6 Classification of spray formed stainless steel grades

Austenitic	AISI 304 [21, 29], AISI 304L [21], AISI 316 [73, 74], AISI 316L [6, 7, 119], AISI 310S [21], Sanicro 28 [21], Rex734 [61]
Ferritic	AISI 430 [14, 97], G–X70CrMo29 2 [120]
Martensitic	AISI 420 [105], AISI 420HC [119, 121], martensitic 12%Cr [73], X5CrNiMo12-5-1 [15], Jethete M152 [61], FV 535 [61]
Duplex	AISI 318LN [14, 21, 97], UNS J93380 [120]

12.3.3 Stainless Steels

Spray forming of a wide range on stainless steel grades has been investigated (Table 12.6). The type of spray formed preforms include round billets, flat preforms, and tubes (see Sect. 12.2.2.2). The studies show spray formed stainless steels to exhibit mechanical properties similar or superior to those of conventionally processed stainless steel.

Rickinson et al. [74] reported the mechanical properties and corrosion resistance of spray formed stainless steels (e.g., AISI 316) to be comparable to those of equivalent conventionally processed stainless steel grades. Üçok et al. [6, 7] produced plate-like deposits of AISI 316L by spray forming. The as-sprayed grain size was in the range 10–40 μm, while density varied from 93 to 99%. Hot rolling for a 70% reduction resulted in full density [6]. Cryogenic rolling at -196 °C in excess of 80% followed by annealing at 600 °C and 700 °C resulted in average grain sizes of 0.1 and 0.3 μm, respectively. Yield strength and ultimate strength of spray formed material were 4.8–5.1 and 2.4–2.8 times higher, respectively, than those of the conventionally processed material (Table 12.7) [7].

Schulz et al. [97] investigated the mechanical properties of spray formed (billet: ø250 mm) ferritic and duplex stainless steels. The tensile properties and impact energies of the spray formed duplex stainless steel X2CrNiMoN22-5-3 (AISI 318 LN) were comparable to the literature values (Table 12.7), while the elongation to fracture was considerably higher than that reported in the literature (>20 vs. 40%). The tensile properties (Table 12.7) of spray formed ferritic stainless steel X6Cr17 (AISI 430) were slightly lower than those of conventionally processed material (continuous casting), but nevertheless above the minimum values requested by the standard EN 10088-3.

Brooks et al. [73] reported the tensile properties (Table 12.7) of spray formed martensitic stainless steel (12%Cr) to be comparable to those of conventional forged material in longitudinal direction. In transverse direction the spray formed material was superior to the conventional material with respect to elongation. Impact toughness, although somewhat lower than that of the conventional material, was clearly above the specified minimum value.

Spray forming (billet: ø250 mm) of creep resistant martensitic stainless steel Jethete M152 (max. 0.15%C, 12%Cr) was investigated as part of a project which aimed to demonstrate the industrial applicability of spray forming of specialty steel billets (see Sect. 12.2.2.1). The mechanical properties (Table 12.7) of the material

Table 12.7 Tensile properties of spray formed and conventionally processed stainless steels (L = longitudinal, T = transverse)

Ref.	Material	Processing route/condition	$R_{p0.2}$ (MPa)	R_m (MPa)	A (%)
[29]	AISI 304	Spray formed: annealed	311	630	57
		Standard ASTM E8-96A	>240	400–630	>20
[73]	AISI 316	Spray formed (forged)	221	552	65
		Conventional (forged)[a]	215	524	55
[74]	AISI 316	Spray formed (wrought alloy)	545	661	39.5
		Conventional (wrought alloy)	519	664	40
[7]	AISI 316L	Spray formed (hot rolled, 50%, 1000 °C)	580	770	46
		Spray formed (cold rolled, 80%, −196 °C)	1670	1920	2.9
		Spray formed (cold rolled, 80%, −196 °C)/annealed at 600 °C	1280	1610	3.2
		Spray formed (cold rolled, 80%, −196 °C)/annealed at 700 °C	1190	1400	5
		Conventional/Annealed at 1100 °C	250	580	50
[97]	AISI 430	Spray forming (forged)/annealed at 800 °C, rapid cooling[b]	L: 270, T: 280	L: 415, T: 439	L: 38, T: 34
		Continuous casting (hot rolled)[b]	293	455	41
		Standard EN 10088-3	>240	400–630	>20
[73]	Martensitic 12%Cr	Spray formed (forged)/hardened and tempered	608	745	19
		Conventional (forged, L)/hardened and tempered	590	739	23
		Conventional (forged, T)/hardened and tempered	614	757	7
[60]	Jethete M152	Spray formed (as-sprayed, T)	1031	1194	16
		VIM/VAR	990	1141	17
[97]	AISI 318LN	Spray formed (forged)/annealed at 1080 °C, rapid cooling[b]	L: 495, T: 500	L: 740, T: 740	L: 39, T: 34
		Literature values reported by Schulz et al.	450	650–880	>20

[a]Typical published results according to Brooks et al. [73]
[b]The values for $R_{p0.2}$, R_m, and A were read from a graph

were comparable to those of typical remelted and forged products, while the hot formability was superior in comparison to both cast and wrought material [60].

Zepon et al. [15] investigated the effect of boron additions on spray formed supermartensitic stainless steel (0.05%C–11%Cr–5%Ni–1%Mo) with and without minor boron additions (0.3, 0.7%). Boron was observed to form M_2B borides at the grain boundaries. This inhibited grain growth which in turn resulted in higher hardness and improved abrasive wear resistance as compared to material without boron addition. Abrasive wear resistance of the alloy with 0.7% of boron addition was higher than that of AISI D2. These observations were confirmed in another study by Zepon et al. [16]. In this study it was furthermore showed that deterioration of corrosion properties, which is related to the formation of borides, can be minimized by increasing chromium content of base alloy, while maintaining the improved wear properties. Nascimiento et al. [120] investigated the spray forming of duplex stainless steel UNS J93380 with 3.5% boron addition. The microstructure of the material contained 35 vol.% M_2C borides. This doubled the hardness and resulted in substantial improvement in abrasive wear resistance in comparison to wrought UNS J93380 without boron addition.

12.3.4 High-Chromium White Cast Irons

The literature on spray forming of cast irons focuses on high-chromium white cast irons, although there are reports on spray forming of white [35, 54] and gray [122] cast iron as well. High-chromium white irons are ferrous alloys that typically contain 11–30% chromium, 1.8–3.6% carbon and minor additions of other alloying elements [123]. Their excellent wear resistance is due to high volume fraction of Cr rich eutectic and/or primary carbides (M_7C_3, M_3C). Conventional casting of these materials produces coarse carbide structures, which are detrimental to their mechanical properties. Modification of casting conditions and alloying additions are not efficient at preventing the formation of coarse carbides structures, while heat treatments are inefficient at refining the structures. Spray forming, on the other hand, is very efficient at promoting refinement of carbide morphology [17, 32–34, 124–126], as illustrated in Fig. 12.8. Billets [125], tubular preforms [32–34], and flat preforms [17, 124] of high-chromium white cast iron have been produced for research purposes. Industrial activity is represented, for instance, by Sumitomo's spray formed mill rolls (see Sect. 12.2.2.2).

Hanlon et al. [33] compared the rolling/sliding wear resistance of conventionally cast and spray formed high-chromium white cast irons (2.5%C–17%Cr; tubular

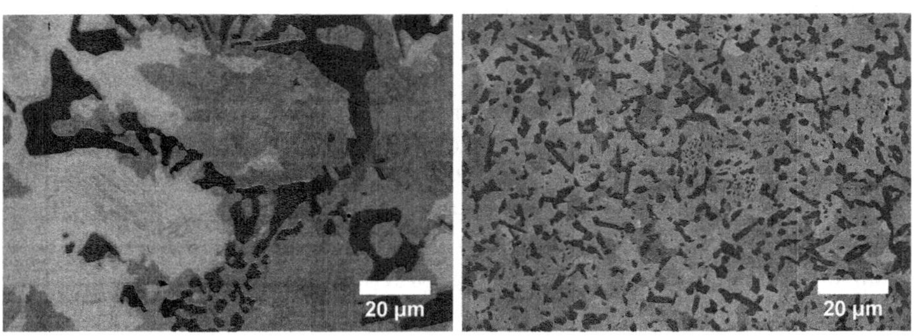

Fig. 12.8 SEM micrographs illustrating the difference between the carbide morphology of cast (*left*) and spray formed (*right*) high-chromium white cast iron (2.6%C–19%Cr)

collector: ø80 mm) against a tool steel counterpart. The wear rate of the cast material was significantly higher than that of the spray formed material when testing was carried out in the temperature range 20–500 °C. Testing in the temperature range 600–700 °C resulted in similar wear rates between the materials due to softening of the steel matrix. Hanlon et al. attributed the poor performance of cast material (<500 °C) to extensive cracking of the coarse carbides. In another study Hanlon et al. [32] showed fine carbides in spray formed material (2.5%C–17%Cr; tubular collector: ø80 mm) to improve the flow of matrix around the carbides during forging. Unlike the conventionally cast material, the spray formed material could be forged without carbide fracture or void formation. The transverse rupture stress and work of fracture were 50% higher for the spray formed than for the conventionally cast material.

Guo et al. [34] compared sliding wear performance of spray formed (2.1%C–23%Cr; tubular preform: ø260 mm) and cast (2.6%C–21%Cr) high-chromium white cast iron against steel counterpart (AISI 4145, AISI M2) in a pin-on-disk test. Although no significant difference was observed in the wear performances, the spray formed material was observed to have much better fracture resistance than the cast material. The improvement in the fracture resistance was attributed to the finer carbide morphology of the spray formed material. Matsuo et al. [17] reported similar findings in their respective study of spray formed (several compositions, 4 kg charge) and cast high-chromium white cast iron.

Kasama et al. [124] investigated abrasive wear resistance of spray formed (2.8% C–22.5%Cr, 4 kg charge) and cast materials. The abrasive wear performance (dry sand/rubber wheel test, abrasive size 160–250 μm) of spray formed material was not significantly different from that of the cast material, although there was some variation depending on the spraying conditions and subsequent heat treatments. The respective study by Lotta and Hannula [125] found cast material to clearly outperform spray formed material (2.6%C–19.5%Cr; billet: ø200 mm, 460 mm in length) in abrasive wear resistance when coarser (100–600 μm) abrasive is used.

12.3.5 Steel Matrix Composites

Spray forming of steel matrix composites has attracted some interest from time to time as it is possible to feed particles into the melt or to the droplet stream during atomization. Spiegelhauer et al. [127] investigated the feasibility of spray forming steel-alumina (Al_2O_3) composite billets by injecting alumina particles to the droplet spray of medium carbon steel (0.5%C; 12 vol.% alumina, mean size 230 μm) and high carbon steels (0.7%C; 6 vol.% alumina, mean size 53 μm). Petersen et al. [128] studied the microstructure and properties of low alloy (0.2%C) boron steel billet injected either with fine (5 vol.%, mean size 46 μm) or coarse (6.5 vol.%, mean size 134 μm) alumina particles. Uniform distribution of alumina particles in steel matrix was observed in both of these studies.

Wei et al. [77] investigated injection of WC particles (2–10 μm, 0–70 wt.%) into AISI M2 during spray forming (5 kg charge). The increase in WC content from 0 to 70 wt.% coincided with a decrease in bending strength from ~2000 to ~1200 MPa and an increase in hardness from ~65 to ~75 HRC. Banhart et al. [105] investigated injection of small amounts of alumina, SiC, and WC particles into low-alloy steel Ck35 (AISI 1035) and martensitic stainless steel X20Cr13 (AISI 420) during spray forming. The microstructures were characterized by a uniform distribution of injected particles in all of the cases. The most promising results were achieved by injecting silicon carbide particles into the martensitic stainless steel as it resulted in dramatic increase of wear resistance in a pin-on-disk test.

Cui et al. [119] investigated injection of TiC and VC particles (1–10 μm) into austenitic stainless steel X2CrNiMo17-12-2 (AISI 316L) and martensitic stainless steel X46Cr13 (AISI 420HC) during atomization (billets: ø160–180 mm, 400–500 mm in length) billets. The embedded TiC particles were largely uninfluenced by the surrounding steel matrix, while the VC particles went through partial dissolution. Hot forging and hot rolling was performed successfully in both cases. Porosity level, which tended to increase with feeding rate of the particles, was brought down to less than 1 vol.% by optimizing the feeding and spray forming conditions. Although introduction of TiC particles into AISI 316L resulted in massive reduction of toughness, it also coincided with increase in hardness and wear resistance. A more fundamental investigation into the microstructural evolution of spray formed X46Cr13–TiC/VC composites is reported in another study by Cui et al. [121].

12.4 Summary

Spray forming of a wide range of steels has been studied intensively since the 1970s. In most of the cases, the obtained properties are similar or superior to those of conventionally processed materials, often resembling the properties of materials processed by powder metallurgical route. In addition to spray forming of existing steel grades, the flexibility of the process has been utilized in development of new materials solutions (e.g., steel composites, gradient materials).

The spray formed steel preforms include round billets, flat deposits, tubular preforms, clad structures, gradient deposits, and molds/dies. Most of these types either are or have been in commercial production. In China, there are currently several commercial spray forming plants operating in large scale and variety of materials. The details of these operations, however, are not public. In Europe, there is currently one spray forming plant in commercial operation. The plant, which is operated by DanSpray Ltd. in Denmark, uses twin atomizer (100 kg/min) to produce billets up to 5 tons in weight, 500 mm diameter, and 3000 mm in length. The production plant has been designed with the future capability of adding a second induction melting system to the spray forming chamber. This, in addition to enabling production of larger billets, would provide the possibility to spray form bimetallic products and even composites.

References

1. Zepon, G., Ellendt, N., Uhlenwinkel, V., & Bolfarini, C. (2016). Solidification Sequence of Spray-Formed Steels. *Metallurgical and Materials Transactions A, 47*(2), 842–851.
2. Shaw, L. H. (1997). Spray forming of special steels and nickel alloys gears up for commercialization. *Powder Metallurgy, 40*(1), 28–31.
3. Schruff, I., Schüler, V., & Spiegelhauer, C. (2003). Spray forming—The new technology for the production of high-grade tool steels. In: *Proceedings of the 2nd International Conference on Spray Deposition and Melt Atomization (SDMA 2003) and 5th International Conference on Spray Forming (ICSF V), Vol. 1* (pp. 5.11–5.25). Bremen: Universität Bremen.
4. Chesney, P. F., Madden, C., & Mascolino, J. (1995). Economics of spray forming pipe, rolls and bi-metallic billets in advances in powder metallurgy & particulate materials. *Proceedings of the 1995 International Conference and Exhibition on Powder Metallurgy & Particulate Materials* (pp. 7.47–7.56). Princeton: Metal Powder Industries Federation.
5. Grant, P. (1995). Spray forming. *Progress in Materials Science, 39*(4), 497–545.
6. Üçok, I., Ando, T., & Grant, N. J. (1991). Structure and Properties of Spray Formed Stainless Steel. *International Journal of Powder Metallurgy, 27*(3), 237–247.
7. Üçok, I., Ando, T., & Grant, N. J. (1991). Property enhancement in Type 316L stainless steel by spray forming. *Materials Science and Engineering A, 133*, 284–287.
8. Ibrahim, I. A., Ando, T., & Grant, N. J. (1992). Processing-Structure Relationships in Spray Formed SAE 1020 Steel. *International Journal of Rapid Solidification, 7*(1), 35–50.
9. Brinksmeier, E., Brockhoff, T., & Schünemann, M. (1999). Spray forming and rolling of low carbon steel. *Annals German Academic Society for Production Engineering, 6*(1), 7–10.
10. Matsuo, S., Ando, T., & Grant, N. J. (2000). Grain refinement and stabilization in spray-formed AISI 1020 steel. *Materials Science and Engineering A, 288*(1), 34–41.
11. Schünemann, M., & Brinksmeier, E. (2000). Umformen und sprühkompaktieren von flachprodukten. *HTM–Härterei-Technische Mitteilungen, 55*(4), 235–240.
12. Brinksmeier, E., & Schünemann, M. (2001). Generation and forming of spray-formed flat products. *Journal of Materials Processing Technology, 115*(1), 55–60.
13. Spangel, S., Matthaei-Schulz, E., Schulz, A., Vetters, H., & Mayr, P. (2002). Influence of carbon and chromium content and preform shape on the microstructure of spray formed steel deposits. *Materials Science and Engineering A, 325*(1), 26–39.
14. Schulz, A., Rabitsch, R., Stocchi, D., Viale, D., & Montero, M. C (2007). Evaluation of the capability of spray forming for the production of very high alloyed steels without necessity of heavy hot working. *European Commission Technical Steel Research Series: Physical Metallurgy and Design of New Generic Steel Grades (EUR 22438).*
15. Zepon, G., Kiminami, C. S., Botta Filho, W. J., & Bolfarini, C. (2013). Microstructure and wear resistance of spray-formed supermartensitic stainless steel. *Materials Research, 16*(3), 642–646.
16. Zepon, G., Nascimento, A. R. C., Kasama, A. H., Nogueira, R. P., Kiminami, C. S., Botta, W. J., et al. (2015). Design of wear resistant boron-modified supermartensitic stainless steel by spray forming process. *Materials and Design, 83*, 214–223.
17. Matsuo, T. T., Kiminami, C. S., Botta Fo, W. J., & Bolfarini, C. (2005). Sliding wear of spray-formed high-chromium white cast iron alloys. *Wear, 259*(1–6), 445–452.
18. Ikawa, Y., Itami, T., Kumagai, K., Kawashima, Y., Leatham, A. G., Coombs, J. S., et al. (1990). Spray deposition method and its application to the production of mill rolls. *ISIJ International, 30*(9), 756–763.
19. "Sumitomo 'Ospreys' Rolling Mill Rolls," Article in Magazine Metal Powder Report, 45(3), pp. 813–814 (1990).
20. Yaman, M., & Widmark, M. (1990). Modelling and applications on spray deposition of tubes. In: *Proceedings of the 1st International Conference on Spray Forming (ICSF I), Swansea, Wales, UK, September 17–19, 1990.*

21. Leatham, A. G., Elias, L. G., Yaman, M., Itami, T., Kawashima, Y., Brooks, P. J. S., et al. (1992). Spray forming-commercialisation and applications. In: *Proceedings of the Powder Metallurgy World Congress* (pp. 59–76). San Francisco: EPMA.
22. Itami, T., & Kawashima, Y. (1992). The production of rolling mill rolls and high speed cutting tools at sumitomo heavy industries. In: *Spray Forming: Science, Technology and Applications* (pp. 77–91). Princeton: Metal Powder Industries Association.
23. Wahlroos, J., & Liimatainen, T. (1993). Interface adherence of spray formed compound tube. In: *Proceedings of the 2nd International Conference on Spray Forming (ICSF II), Swansea, UK, September 13–15, 1993.*
24. Forsberg, U., Wilson, A., Nylof, L., & Yaman, M. (1996). Sandvik Sanicro 65 composite tube for municipal waste incinerators. In: *Proceedings of the 3rd International Conference on Spray Forming (ICSF III), Cardiff, UK, September 9-11, 1996.*
25. Chesney, P. F., & Mascolino, J. J. (1997). Markets and opportunities for tubular and bimetallic spray formed products. *Powder Metallurgy, 40*(1), 31–33.
26. Forrest, J., Price, R., & Hanlon, D. (1997). Manufacturing Clad Products by Spray Forming. *International Journal of Powder Metallurgy, 33*(3), 21–29.
27. Hanlon, D. N., Rainforth, W. M., & Sellars, C. M. (1997). The effect of processing route, composition and hardness on the wear response of chromium bearing steels in a rolling-sliding configuration. *Wear, 203–204*, 220–229.
28. Hanlon, D. N., Rainforth, W. M., & Sellars, C. M. (1998). The structure and properties of spray formed cold rolling mill work roll steels. *Journal of Materials Science, 33*(13), 3233–3244.
29. Chesney, P., Mascolino, J., & Madden, C. (1999). Commissioning trials on a US Navy-Mantech five ton spray forming plant for large diameter thick wall pipe and rolls. In: *Proceedings of the 4th International Conference on Spray Forming (ICSF IV), Baltimore, Maryland, USA, September 13–15, 1999.*
30. Forbes Jones, R. M., & Kennedy, R. L. (1999). Processing of spray formed tool steel rolling mill sleeves. In: *Proceedings of the 4th International Conference on Spray Forming (ICSF IV), Baltimore, Maryland, USA, September 13–15, 1999.*
31. Cui, C., Schulz, A., Fritsching, U., Bauckhage, K., & Mayr, P. (2004). Spray forming of tubular bearing steel preforms. *International Journal of Powder Metallurgy, 40*(5), 49–53.
32. Hanlon, D. N., Rainforth, W. M., & Sellars, C. M. (1999). The effect of spray forming on the microstructure and properties of a high chromium white cast iron. *Journal of Materials Science, 34*(10), 2291–2301.
33. Hanlon, D. N., Rainforth, W. M., & Sellars, C. M. (1999). The rolling/sliding wear response of conventionally processed and spray formed high chromium content cast iron at ambient and elevated temperature. *Wear, 225*, 587–599.
34. Ted Guo, M., Chiang, C.-H., & Tsao, C. Y. A. (2002). Microstructure and wear behavior of spray-formed and conventionally cast rolls of 18Cr–2.5 Mo–Fe alloy. *Materials Science and Engineering A, 326*(1), 1–10.
35. Yang, Y., Hirvonen, A., Virta, J., & Hannula, S.-P. (2003). Microstructures and mechanical properties of spray formed white irons. *International Journal of Cast Metals Research, 16* (1–3), 333–337.
36. Hanlon, D. N., & Rainforth, W. M. (2003). The rolling sliding wear response of conventionally processed and spray formed high speed steel at ambient and elevated temperature. *Wear, 255*(7), 956–966.
37. Schulz, A., Trojahn, W., Meyer, C., & Uhlenwinkel, V. (2014). Spray-formed bearing steel with high oxide cleanliness and small and fine dispersed inclusions. *HTM Journal of Heat Treatment and Materials, 69*(6), 368–376.
38. Cui, C., Schulz, A., & Uhlenwinkel, V. (2013). Co-spray forming of gradient deposits from two sprays of different tool steels using scanning gas atomizers. *Steel Research International, 84*(11), 1075–1084.

39. Cui, C., Schulz, A., & Uhlenwinkel, V. (2014). Materials characterization and mechanical properties of graded tool steels processed by a new co-spray forming technique. *Materialwissenschaft und Werkstofftechnik, 45*(8), 652–665.
40. Grohmann, O., Meyer, C., Schulz, A., Uhlenwinkel, V., Heinzel, C., & Brinksmeier, E. (2014). Analysis of hot forming tool generated via co-spray forming. *HTM Journal of Heat Treatment and Materials, 69*(4), 235–240.
41. Cui, C., Schulz, A., Steinbacher, M., Moumi, E., Kuhfuss, B., Böhmermann, F., et al. (2015). Development of micro rotary swaging tools of graded tool steel via co-spray forming. *Manufacturing Review, 2*(22), 1–8.
42. McHugh, K. M., & Wickham, B. R. (2000). Spray-formed tooling for injection molding and die casting applications. In: *Proceedings of the International Conference on Spray Deposition and Melt Atomization (SDMA 2000)* (pp. 121–134). Bremen: Universitat Bremen.
43. Knirsch, J. R. (2001). Spray-formed tooling—an update. In: *Proceedings of the 21st International Die Casting Congress & Exposition*, Cincinnati (pp. 423–426). Rosemont, IL: North American Die Casting Association.
44. Knirsch, J., Folkestad, J., & McHugh, K. (2002). RSP tooling—a revolutionary new process to manufacture die cast production tooling in prototype timing. *Die Casting Engineer, 46*(3), 56–60.
45. McHugh, K. M., & Folkestad, J. E. (2003). Production of Molds and Dies Using the RSP Tooling Approach. In: *Proceedings of the 2nd International Conference on Spray Deposition and Melt Atomization (SDMA 2003) and 5th International Conference on Spray Forming (ICSF V) Vol. 1* (pp. 5.123–5.134). Bremen: Universität Bremen.
46. Yang, Y., & Hannula, S.-P. (2003). Sprayformed hot work steels for rapid tooling. *Journal of Materials Science and Technology, 19*(1), 169–172.
47. Hannula, S.-P., & Yang, Y. (2004). Potential and perspectives of spray forming of near-net shape tools and components. *Kolloquium des SFB 372*, 7, 187–194.
48. Yang, Y., & Hannula, S.-P. (2004). Soundness of spray formed disc shape tools of hot-work steels. *Materials Science and Engineering A, 383*(1), 39–44.
49. Lin, Y., McHugh, K. M., Zhou, Y., & Lavernia, E. J. (2006). Microstructure and hardness of spray-formed chromium-containing steel tooling. *Scripta Materialia, 55*(7), 581–584.
50. McHugh, K. M., & Lavernia, E. J. (2006). Development and demonstration of advanced tooling alloys for molds and dies. *Idaho National Laboratory, Final Technical Report (INL/EXT-06-01079).*
51. Knirsch, J. R. (2007). Faster, less expensive dies using RSP tooling. *Journal of Materials Engineering and Performance, 16*(4), 432–439.
52. Lin, Y., McHugh, K. M., Zhou, Y., & Lavernia, E. J. (2007). Modeling the spray forming of H13 steel tooling. *Metallurgical and Materials Transactions A, 38*(7), 1632–1637.
53. Lin, Y., McHugh, K. M., Zhou, Y., & Lavernia, E. J. (2008). Evolution of Carbides during Aging of a spray-formed chromium-Containing tool steel. *Metallurgical and Materials Transactions A, 39*(2), 473–476.
54. Yang, Y., & Hannula, S.-P. (2008). Development of precision spray forming for rapid tooling. *Materials Science and Engineering A, 477*(1), 63–68.
55. McHugh, K. M., Lin, Y., Zhou, Y., & Lavernia, E. J. (2008). Influence of cooling rate on phase formation in spray-formed H13 tool steel. *Materials Science and Engineering A, 477* (1–2), 50–57.
56. Schulz, A., Uhlenwinkel, V., Bertrand, C., Kohlmann, R., Kulmburg, A., Oldewurtel, A., et al. (2002). Highalloyed tools steels spray-formed to medium size billets and their performance in comparison to conventional route material. *Kolloquium des SFB 372, 6*, 149–172.
57. Kjeldsteen, P. (2003). Development of high alloyed tool steels using spray forming on an industrial scale. *Powder Metallurgy, 46*(4), 297–298.
58. Overgaard, J., Andersen, T., & Schwarz, K. (1996). Feasibility study of a continuously operating horizontal billet spray deposition plant. *Technical Steel Research: Pilot and Demonstration Projects (EUR 15593).*

59. Spiegelhauer, C. (2000). Spray forming of tool steels–advanced manufacturing technology for the new millennium. *Steel World (UK), 5*(1), 30–32.
60. Spiegelhauer, C. (1998). Properties of spray formed tool and high speed steels. In: *Proceedings of the 3rd Pacific Rim International Conference on Advanced Materials and Processing (PRICM-3)* (pp. 1653–1659). Honolulu: The Minerals, Metals and Materials Society.
61. Spiegelhauer, C., Andersen, T., Shaw, L. H., & Oakes, G. (1999). Pilot and demonstration project for spray forming round steel products. *Technical Steel Research: Pilot and Demonstration Projects (EUR 18798).*
62. Spiegelhauer, C. (2001). State of art for making tool steel billets by spray forming. *Kolloquium des SFB 372, 5*, 63–68.
63. Spiegelhauer, C. (2002). Industrial production of tool steels using the spray forming technology. In: *Proceedings of The 6th International Tooling Conference. The Use of Tool Steels: Experience and Research, Vol. 2* (pp. 923–930). Karlstad: Karlstad University.
64. Sandberg, O., & Jönson, L. (2002). New generation of tool steels made by spray forming. In: *Proceedings of the 6th International Tooling Conference. The Use of Tool Steels: Experience and Research, Vol. 2* (pp. 961–972). Karlstad: Karlstad University.
65. Kjeldsteen, P., Spiegelhauer, C., & Sandberg, O. (2003). Investigation of the homogeneity of a spray formed tool steel billet. In: *Proceedings of the 2nd International Conference on Spray Deposition and Melt Atomization (SDMA 2003) and 5th International Conference on Spray Forming (ICSF V), Vol. 1* (pp. 5.27–5.35). Bremen: Universität Bremen.
66. Yang, Y (2015). Personal communication.
67. Davis, J. R. (1995) *ASM specialty handbook: tool materials*. Materials Park: ASM International.
68. Leatham, A. (1999). Spray forming: alloys, products and markets. *Metal Powder Report, 54* (5), 28–37.
69. Dowson, G. (1996). Commercial products aid spray forming's growing maturity. *Metal Powder Report, 51*(2), 10–11.
70. Leatham, A. G., & Ogilvy, J. W. (1997). Commercial-scale application of spray-formed materials. *Journal of Materials Synthesis and Processing, 5*(1), 5–10.
71. First Prize of China Automotive Industry Science and Technology Award, Chinese Society of Automotive Engineers, November 26, 2013.
72. Robertson, G. A., Kennedy, R., & Krauss, G. (1998). *Tool steels*, 5th ed. Materials Park: ASM International.
73. Brooks, R. G., Leatham, A. G., Coombs, J. S., & Moore, C. (1977). The Osprey process: a novel method for the production of forgings. *Metallurgia and Metal Forming, 44*(4), 157–163.
74. Rickinson, B. A., Kirk, F. A., & Davies, D. R. G. (1981). CSD: a novel process for particle metallurgy products. *Powder Metallurgy, 24*(1), 1–6.
75. Igharo, M., & Wood, J. V. (1989). Investigation of M2 high speed steel produced by Osprey process. *Powder Metallurgy, 32*(2), 124–131.
76. Igharo, M., Kirby, T., & Wood, J. V. (1989). Development of High-Speed Steels Produced by the Osprey Process. In: *Advanced Materials and Processes: Proceedings of the First European Conference on Advanced Materials and Processes (EUROMAT'89), Vol. 1* (pp. 255–260). Aachen, Netherlands: DGM Informationsgesellschaft.
77. Wei, Y., Mu, D., Zhang, L., & Wu, C. (1999). Microstructures and properties of tungsten carbide particle-reinforced high-speed composites fabricated by spray forming. *Powder Technology, 104*(1), 100–104.
78. Schulz, A., Uhlenwinkel, V., Bertrand, C., Kohlmann, R., Kulmburg, A., Oldewurtel, A., et al. (2004). Nitrogen pick-up during spray forming of high-alloyed steels and its influence on microstructure and properties of the final products. *Materials Science and Engineering A, 383*(1), 58–68.

79. Schulz, A., Spangel, S., Schneider, R., Viale, D., & Bertrand, C. (2004). The investigation and evaluation of the spray-formed states of tool steels to make a catalogue of materials properties available. *European Commission Technical Steel Research Series: SPECIAL and Alloy Steels (EUR 20948).*
80. Jesus, E. R. B., Jesus Filho, E. S., & Rossi, J. L. (2006). Evaluation of a spray formed high-speed steel AISI M2 as machining tools. In: *Proceedings of the 3rd International Conference on Spray Deposition and Melt Atomization (SDMA 2006) and 6th International Conference on Spray Forming (ICSF VI), Bremen, Germany, September 4–6, 2006.*
81. Serna, M. M., & Rossi, J. L. (2009). MC complex carbide in AISI M2 high-speed steel. *Materials Letters, 63*(8), 691–693.
82. Mesquita, R. A., & Barbosa, C. A. (2002). High speed steel produced through conventional casting, spray forming and powder metallurgy. In: *Proceedings of the 6th International Tooling Conference. The Use of Tool Steels: Experience and Research, Vol. 1* (pp. 325–338). Karlstad: University of Karlstad.
83. Mesquita, R. A., & Barbosa, C. A. (2004). Spray forming high speed steel—properties and processing. *Materials Science and Engineering A, 383*(1), 87–95.
84. Rodenburg, C., Krzyzanowski, M., Beynon, J. H. & Rainforth, W. M. (2004). Hot workability of spray-formed AISI M3:2 high-speed steel. *Materials Science and Engineering A, 386*(1), 420–427.
85. Yu, Y., Huang, J., Cui, H., Cai, Y., & Zhang, J. (2012). Effect of Nb on the microsructure and properties of spray formed M3 high speed steel. *Acta Metallurgica Sinica, 48*(8), 935–940.
86. Schulz, A., Uhlenwinkel, V., Bertrand, C., Kohlmann, R., Kulmburg, A., Oldewurtel, A., et al. (2004). Nitrogen pick-up during spray forming of high-alloyed steels and its influence on microstructure and properties of the final products. *Materials Science and Engineering A, 383*(1), 58–68.
87. Lu, L., Hou, L. G., Zhang, J. X., Wang, H. B., Cui H., Huang, J. F., et al. (2016). Improved the microstructures and properties of M3: 2 high-speed steel by spray forming and niobium alloying. *Materials Characterization, 117*, 1–8.
88. Ernst, I. C., & Duh, D. (2004). ESP4 and TSP4, a comparison of spray formed with powder metallurgically produced cobalt free high-speed steel of type 6W-5Mo-4V-4Cr. *Journal of Materials Science, 39*(22), 6831–6834.
89. Itami, T., Ikawa, Y., & Kumagai, K. (1991). Production of high speed steel endmills by spray forming technique (Osprey process). In: *Advanced materials for future industries: Needs and Seeds. Proceedings of the 2nd Japan International SAMPE Symposium and Exhibition*, Chiba (pp. 395–401). Tokyo, Japan: International Convention Management, Inc.
90. Yabuuchi, E., Kuroshima, Y., Masuda, M., Kubokuti, M., & Ikawa, Y. (1993). The cutting performance of high-speed steel end mill produced by spray forming process. *International Journal of the Japan Society for Precision Engineering, 27*(4), 327–332.
91. Lee, E.-S., Park, W.-J., Baik, K.-H., & Ahn, S. (1998). Different carbide types and their effect on bend properties of a spray-formed high speed steel. *Scripta Materialia, 39*(8), 1133–1138.
92. Lee, E.-S., Park, W.-J., Jung, J. Y., & Ahn, S. (1998). Solidification microstructure and M2C carbide decomposition in a spray-formed high-speed steel. *Metallurgical and Materials Transactions A, 29*(5), 1395–1404.
93. Zhao, S., Fan, J., Zhang, J., Chou, K., & Le, H. (2016). High speed steel produced by spray forming. *Advanced Manufacturing, 4*(2), 115–122.
94. Spiegelhauer, C., & Davin, H. (1998). Properties of spray formed high speed steels. In: *Powder Metallurgy World Congress and Exhibition, Granada, Spain, October 18–22, 1998.*
95. Zhang, Y., Zhang, G. Q., Zhou, L., Li, Z. D., Hua, Y. U. A. N., Xu, W. Y., et al. (2007). Analysis of twin-nozzle-scanning spray forming process and spray formed high speed steel (HSS). *Journal of Iron and Steel Research International, 14*(5), 7–10.
96. Zhang, G., Yuan, H., Jiao, D., Li, Z., Zhang, Y., & Liu, Z. (2012). Microstructure evolution and mechanical properties of T15 high speed steel prepared by twin-atomiser spray forming and thermo-mechanical processing. *Materials Science and Engineering A, 558*, 566–571.

97. Schulz, A., Uhlenwinkel, V., Escher, C., Kohlmann, R., Kulmburg, A., Montero, M. C., et al. (2008). Opportunities and challenges of spray forming high-alloyed steels. *Materials Science and Engineering A, 477*(1–2), 69–79.
98. Zhang, J. G., Xu, H. B., Shi, H. S., Wu, J. S., & Sun, D. S. (2001). Microstructure and properties of spray formed Cr12MoV steel for rolls. *Journal of Materials Processing Technology, 111*(1–3), 79–84.
99. Ernst, I. C., & Duh, D. (2004). Properties of cold-work tool steel X155CrVMo12-1 produced via spray forming and conventional ingot casting. *Journal of Materials Science, 39*(22), 6835–6838.
100. Yan, F., Xu, Z., Shi, H., & Fan, J. (2008). Microstructure of the spray formed Vanadis 4 steel and its ultrafine structure. *Materials Characterization, 59*(5), 592–597.
101. Yan, F., Shi, H., Jin, B., Fan, J., & Xu, Z. (2008). Microstructure evolution during hot rolling and heat treatment of the spray formed Vanadis 4 cold work steel. *Materials Characterization, 59*(8), 1007–1014.
102. Ni, X. L., Li, Z., Yuan, H., Xu, W. Y., Zhang, Y., & Zhang, G. Q. (2014). As deposited microstructure of spray formed 10V high speed steel. *Materials Research Innovations, 18* (S4), S4-295–S4-300.
103. Huang, J. F., Yu, Y. P., Cui, H., Li, G. N., & Zhang, J. S., Microstructural characterization and mechanical properties of spray-formed H13 tool steel. In: *Proceedings of the 4th International Conference on Spray Deposition and Melt Atomization (SDMA 2009) and 7th International Conference on Spray Forming (ICSF VII), Bremen, Germany, September 7–9, 2009.*
104. Zhang, J., Huang, J., Wang, H., Lu, L., Cui, H., & Zhang, J. (2014). Microstructures and mechanical properties of spray formed H13 tool steel. *Acta Metallurgica Sinica, 50*(7), 787–794.
105. Banhart, J., Grutzner, H., & Knuwer, M. (1999). Manufacture of particle reinforced steels by spray forming. *Advances in Powder Metallurgy and Particulate Materials, 2*, 4-217–4-224.
106. Schumacher, J., Bomas, H., & Zoch, H.-W. (2010). Microstructure and mechanical properties of sulphur-alloyed spray-formed steels. *Materialwissenschaft und Werkstofftechnik, 41*(7), 585–596.
107. Cui, C., Schulz, A., Fritsching, U., & Kohlmann, R. (2006). Spray forming of homogeneous 20MnCr5 steel of low distortion potential. *Materialwissenschaft und Werkstofftechnik, 37*(1), 34–39.
108. Tinscher, R., Bomas, H., & Mayr, P. (2002). Mechanical properties of a spray deposited bearing steel. *Materials Science and Engineering A, 326*(1), 11–19.
109. Zhang, J. G., Sun, D. S., Shi, H. S., Xu, H. B., Wu, J. S., & Wu, X. F. (2002). Microstructure and continuous cooling transformation thermograms of spray formed GCr15 steel. *Materials Science and Engineering A, 326*(1), 20–25.
110. Cui, C., Fritsching, U., Schulz, A., Bauckhage, K., & Mayr, P. (2004). Control of cooling during spray forming of bearing steel billets. *Materials Science and Engineering A, 383*(1), 158–165.
111. Cui, C., Fritsching, U., Schulz, A., Bauckhage, K., & Mayr, P. (2004). Spray formed bearing steel insensitive to distortion. Part I. Material characterization. *Journal of Materials Science, 39*(18), 5639–5645.
112. Cui, C., Schulz, A., Fritsching, U., Bauckhage, K., & Mayr, P. (2005). Spray formed bearing steel insensitive to distortion. Part II. Distortion behavior. *Journal of Materials Science, 40* (7), 1673–1680.
113. Cui, C., Fritsching, U., Schulz, A., Tinscher, R., Bauckhage, K., & Mayr, P. (2005). Spray forming of homogeneous 100Cr6 bearing steel billets. *Journal of Materials Processing Technology, 168*(3), 496–504.
114. Schumacher, J., Bomas, H., & Zoch, H.-W. (2012). Endurance limit prediction for sulphur-alloyed spray-formed steels. *International Journal of Fatigue, 41*, 119–129.
115. Sherby, O. D., Oyama, T., Kum, D. W., Walser, B., & Wadsworth, J. (1985). Ultrahigh carbon steels. *JOM, 37*(6), 50–56.

116. Zhang, J. G., Lin, Y. J., Hillert, M., Selleby, M., Shi, H. S., Yan, B., et al. (2004). Microstructure and mechanical properties of spray formed ultrahigh-carbon steels. *Materials Science and Engineering A, 383*(1), 45–49.
117. Luo, G. M., Wu, J. S., Fan, J. F., Shi, H. S., Lin, Y. J., & Zhang, J. G. (2004). Microstructure and mechanical properties of spray-deposited ultra-high carbon steel after hot rolling. *Materials Characterization, 52*(4–5), 263–268.
118. Luo, G. M., Wu, J. S., Fan, J. F., Shi, H. S., Lin Y. J., & Zhang, J. G. (2004). Excellent mechanical properties of a spray deposited ultrahigh carbon steel after hot rolling. *Journal of Materials Science, 39*(14), 4679–4681.
119. Cui, C., Schulz, A., Uhlenwinkel, V., Zobel, F., & Zoch, H.-W. (2010). Spray forming of stainless steel matrix composites with injection of hard particulates. *Materialwissenschaft und Werkstofftechnik, 41*(7), 524–531.
120. Nascimento, A. R. C., Zepon, G., Kiminami, C. S., Botta, W. J., Kasama, A. H., & Bolfarini, C. (2013). Abrasive wear resistance of different spray-formed iron-based alloys. In: *Proceedings of the 5th International Conference on Spray Deposition and Melt Atomization (SDMA 2013), Bremen, Germany, September 23–25, 2013.*
121. Cui, C., Schulz, A., Uhlenwinkel, V., & Zoch, H.-W. (2011). Spray-formed stainless steel matrix composites with co-injected carbide particles. *Metallurgical and Materials Transactions A, 42*(8), 2442–2455.
122. Ebalard, S., & Cohen, M. (1991). Structural and mechanical properties of spray formed cast-iron. *Materials Science and Engineering A, 133*, 297–300.
123. Tabrett, C. P., Sare, I. R., & Ghomashchi, M. R. (1996). Microstructure-property relationships in high chromium white iron alloys. *International Materials Review, 41*(2), 59–82.
124. Kasama, A. H., Mourisco, A. J., Kiminami, C. S., Botta Fo, W. J., & Bolfarini, C. (2004). Microstructure and wear resistance of spray formed high chromium white cast iron. *Materials Science and Engineering A, 375*, 589–594.
125. Lotta, J., & Hannula, S.-P. (2014). Microstructure and wear resistance of spray formed and conventionally cast high-chromium white iron. *Materialwissenschaft und Werkstofftechnik, 45*(8), 727–735.
126. Lotta, J., & Hannula, S.-P. (2015). Microstructural comparison of spray-formed and conventionally cast 2.5 C–19Cr high-chromium white iron. *Metallography, Microstructure, and Analysis, 4*(4), 261–272.
127. Spiegelhauer, C., Andersen, T., Overgaard, J., & Brooks, R. G. (1993). Steel-alumina MMC billets manufactured by sprayforming. *Key Engineering Materials, 79–80*, 145–154.
128. Petersen, K., Pedersen, A. S., Pryds, N., Thorsen, K. A., & List, J. L. (2002). The effect of particles in different sizes on the mechanical properties of spray formed steel composites. *Materials Science and Engineering A, 326*(1), 40–50.

Chapter 13
Spray Forming of Nickel Superalloys

William T. Carter, Robin M. Forbes Jones, and Ramesh S. Minisandram

13.1 Introduction

Superalloys are prepared for service as castings or wrought products. When cast, they are usually precision-cast to final shape using the lost wax investment casting process. Control of solidification can yield directionally solidified microstructures or single crystals. Turbine blades are cast in this way to minimize creep rates in the hot, high-stress environments of modern turbine engines. Spray forming is inherently imprecise due to the wide spray pattern, and it cannot produce a directionally solidified structure or a single crystal. Thus, spray forming does not compete with the precision casting market segment.

Wrought superalloys start as cast ingots that are thermomechanically processed using upset and draw open-die forging operations to generate a microstructure that is more homogeneous and ductile than the as-cast microstructure. This process sequence is termed the cast & wrought (C&W) process. Resulting billets are forged to a near-net shape and machined to a final shape. Some wrought superalloys have very complex chemistries that cannot be cast into ingots because the alloying elements segregate during solidification to cause inhomogeneous structure and properties. To avoid segregation, these alloys are gas atomized into powder that is reconsolidated via solid state processes to obtain billets as input stock for forging.

W.T. Carter (✉)
GE Global Research, Additive Manufacturing Laboratory, One Research Circle, NY 12309, USA
e-mail: carter@ge.com

R.M. Forbes Jones
Retired

R.S. Minisandram
ATI Specialty Materials, Monroe, NC 28110, USA
e-mail: ramesh.minisandram@atimetals.com

H. Henein et al. (eds.), *Metal Sprays and Spray Deposition*,
DOI 10.1007/978-3-319-52689-8_13

These alloys are termed powder metallurgy (PM) alloys. Spray forming offers a high-production-rate alternative that bypasses many of the steps associated with both cast & wrought and powder metallurgy processes.

This chapter gives a historic perspective dating to the early 1980s on the development of spray forming to compete with conventional superalloy manufacturing processes, primarily for aircraft engine applications including turbine disks and casings. Ambitious research programs were undertaken by independent teams in the United States, Europe, and China over many years, and resulted in innovations in melting, pouring, atomization, and collection. Atomization was performed with argon, nitrogen, or in vacuum with process rates that spanned orders of magnitude. Though technical feasibility for spray forming of many superalloys was shown repeatedly, the process has not yet been commercialized due to the high capital equipment cost for a production-scale spray forming plant and also due to the high qualification cost associated with introducing new materials processes into the aerospace industry.

13.2 Microstructure of Ni-Based Superalloys

All nickel-base superalloys contain a face-centered cubic solid solution matrix of nickel and other elements that include cobalt, chromium, iron, molybdenum, and tungsten. This matrix, known as the gamma (γ) phase, is not particularly creep-resistant, but it can be strengthened to withstand high stress at high temperatures for extended periods of time. Alloy additions give corrosion and oxidation resistance. Alloys that are strengthened in this manner are used extensively in the hot sections of turbine engines and are termed *superalloys*. Superalloys cover a large spectrum of chemistries and are both solid solution strengthened and precipitation hardened. Hardening phases include gamma prime (γ'), gamma double prime (γ''), and a number of carbides. Unwanted phases include topologically close-packed phases such as Laves, σ, and, μ which embrittle the alloy and reduce creep strength [1, 2]. The intermetallic compound Ni_3Al forms the basis for γ', though other elements including Nb, Ta, and Cr are also found in γ'. Examples of γ' alloys include Alloy 720, Rene' 95™, Rene' 88™ and Rene'65™. Typical billet microstructures for γ'-strengthened superalloys processed via C&W and PM processes are shown in Fig. 13.1 for alloys of almost-identical composition. Ni_3Nb forms the basis for the phases δ and γ''. Heat treatments must be chosen to avoid precipitation of δ, which does not improve properties, in favor of γ'', which does, though δ can be an important phase for grain size control. Alloy 718 is the predominant γ''-strengthened superalloy.

Grain size depends on processing history and can vary greatly; care must be taken to achieve the required grain size in both billet and forged part because this strongly influences hot workability in the billet and strength, creep, and fatigue properties in the final part. Generally, a fine grain structure is desired in the billet. After forging of the billet, an even finer grain size may be required to give sufficient

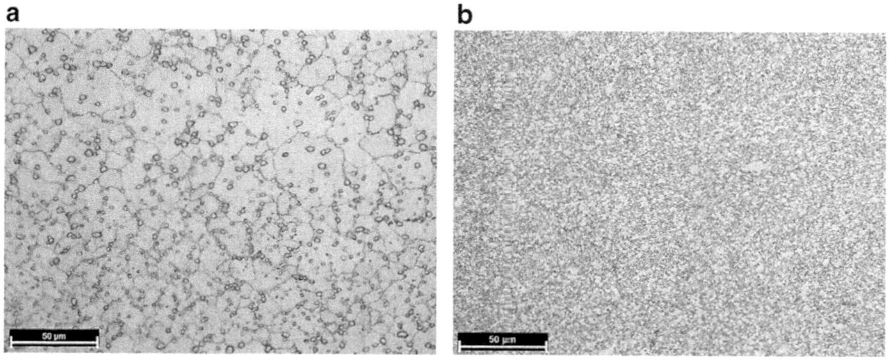

Fig. 13.1 Typical billet microstructures for conventionally processed superalloys. A cast & wrought γ' alloy after conversion via upset and draw (**a**), and a powder metallurgy γ' alloy of similar composition (**b**). The particles in both micrographs are primary γ'

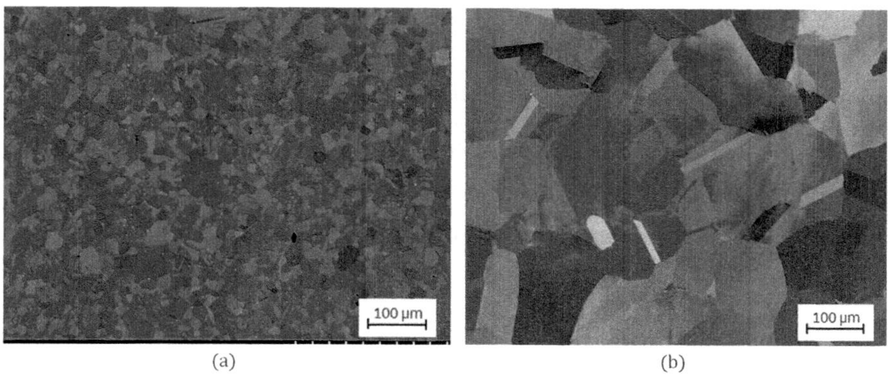

Fig. 13.2 Typical microstructures for conventionally processed superalloys. A PM γ' alloy after consolidation and forging (**a**), and a C&W γ'' alloy after forging (**b**)

resolution for ultrasonic inspection of forged part. Finally, a supersolvus heat treatment is applied to grow the grains to the desired size for service. Typical microstructures for γ'- and γ''-strengthened superalloys ready for service are shown in Fig. 13.2.

The product of spray forming compares well with billet product of either C&W or PM processes. The as-sprayed grain size can be finer than that obtained by C&W processing, and equivalent to that obtained by PM processing; consequently economy can be gained by reducing the number of required processing steps to obtain billet. Upset and draw operations after spray forming may be needed in cases where thermomechanical work is required to achieve ultrasonically inspectable structure, though forging alone may be sufficient. In all cases, C&W, PM, or spray forming, a final forging operation is required to give a near-net shape.

As clarification regarding current technical jargon: The product of a casting process is termed an *ingot*; the ingot is termed an *electrode* if it used as input material for ESR or VAR; the product of *conversion* is termed a *billet* or a *bar*; and the product of spray forming is termed a *preform*. Ingots, billets, and preforms all require thermomechanical processing or heat treatment to achieve required service properties. A bar is understood to require only machining before being placed into service. A summary of C&W, PM and spray form processing is given in Table 13.1 and more detail is provided in the subsections below. A detailed overview of the microstructural evolution of typical superalloys during hot working is given by Forbes Jones and Jackman [3].

Table 13.1 Processing of superalloys

Process		Product of process
Cast & wrought processing		
– *Double melt* includes VIM and VAR or VIM and ESR		
– *Triple melt* includes VIM, ESR and VAR		
Vacuum induction melting (VIM)	Raw metal is melted in vacuum to achieve alloy chemistry and poured into ingot molds.	Ingot or electrode
Electroslag remelting (ESR)	A metal electrode is remelted in slag to remove oxide inclusions and achieve homogeneity and solidified in a copper crucible.	Ingot or electrode
Vacuum ARC remelting (VAR)	A metal electrode is remelted in vacuum to remove dissolved gasses and oxides and achieve homogeneity and solidified in a copper crucible.	Ingot
Conversion	Metal ingots are mechanically worked to refine grain structure by imparting strain.	Billet
Powder metallurgy		
Atomization	Molten metal is gas atomized to produce metal powder.	Powder
Sieving	Metal powder is separated by size.	
Consolidation	Metal powder is consolidated by HIP or extrusion.	Billet
Spray forming	Molten metal is atomized and the spray is collected before solidification.	Preform
	Option 1: The molten metal is tilt-poured or bottom-poured from a VIM.	
	Option 2: The molten metal is bottom-poured from a cold crucible.	
	Option 3: The molten metal is bottom-poured from an ESR furnace.	
Hot isostatic pressing (HIP)	Metal is subject to high temperature and pressure to close porosity or consolidate powder.	
Forging of part	A billet or preform is formed to near-net shape by pressing rolling or ring rolling. Grain size may be decreased.	
Heat treatment	A time-temperature scheduled in a furnace dissolves then precipitates and grows hardening phases. Grain size may be increased.	

13.2.1 Cast & Wrought Processing of Superalloys

Both γ'- and γ''-strengthened alloys are manufactured via conventional C&W processing as shown in Fig. 13.3. The first step is Vacuum Induction Melting (VIM), where melting is done in an induction-heated ceramic crucible inside a vacuum chamber that includes an airlock to allow sampling and alloy additions to the liquid metal to achieve the required alloy chemistry. The alloy is stirred using induction coils to give a uniform chemistry before tilt-pouring from the crucible into ingot molds. This process is vital for the economy of superalloy processing because it can melt both raw stock and recycled metal with miscellaneous shapes. However, solidification in the ingot mold often results in inhomogeneous ingots, sometimes with large central voids referred to as *pipe*, and remelting is required to achieve fully dense ingots. Remelting processes include Electroslag Remelting (ESR) and Vacuum Arc Remelting (VAR) or a combination of both. These processes are termed *consumable electrode* processes, where the raw material serves as an electrical conductor for heating the furnace; it is melted from bottom to top in a controlled manner such that the melt rate is approximately equal to the solidification rate of the resulting ingot. The consumable electrode is the product of a VIM furnace or a prior remelting step, and is provided in the form of a cylindrical casting.

In the case of ESR, the electrode is fed into a pool of hot liquid slag (e.g., calcium fluoride, calcium oxide, and alumina in various proportions) where it melts. As it does so, any oxide impurities are exposed to the slag where they dissolve and remain, thus the process serves to remove unwanted ceramic inclusions. The slag is kept hot by passing an electric current from the electrode, through the slag, and into the liquid metal pool. Very high current ranging from 10,000 to 50,000 amps is often required to maintain temperature. Under the intense heat of the liquid slag (~1800 °C), a thin film of liquid metal forms on the bottom of the electrode, thickens, and accumulates into droplets that eventually fall through the slag to form a molten metal pool beneath the slag. A flat-bottomed electrode with multiple droplet sites is desired. The electrode is continuously weighed to provide feedback to a melt rate control system. Melt rate control may involve continuously advancing

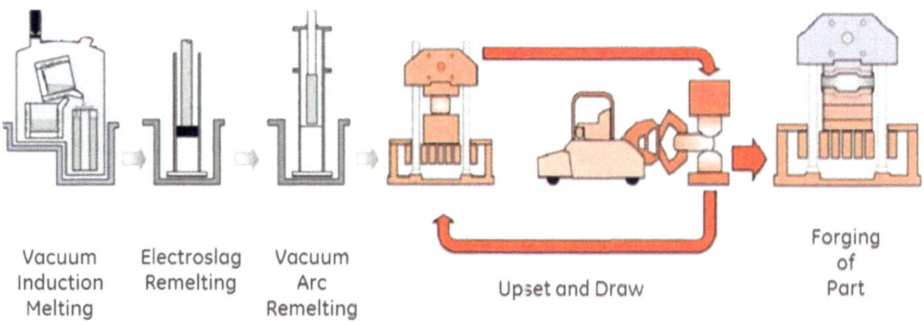

Fig. 13.3 Conventional processing of cast & wrought nickel superalloys

and retracting the electrode to track the voltage, while adjusting the electric current applied to the system. Below the melt pool, the solidifying metal is contained in a water-cooled copper crucible. Copper is chosen because of its high thermal and electrical conductivities. Because the copper is maintained at a low temperature, it freezes the liquid metal pool to some thickness. This solid metal is termed a *skull*; thus the liquid metal is retained in a skull of solid metal of its own composition.

VAR is a similar process where a consumable electrode is melted from bottom to top in a water-cooled copper crucible in which an ingot is solidified at approximately the same rate as the electrode is melted. However, this process is done under vacuum (0.001–0.1 mm Hg), and the heat source is a DC arc rather than electrically heated slag as in ESR. The arc is caused to span the gap between the electrode and the liquid metal pool, and this arc provides the heat to melt the electrode. A uniform and narrow arc gap is desired to provide a stable and diffuse arc to prevent segregation in the ingot. The gap is regulated by controlling the rate at which droplets form and fall from the electrode face to the molten melt pool. VAR is known to remove dissolved gases such as hydrogen and nitrogen from the metal, and it removes undesired trace elements of high vapor pressure. VAR is also known to remove oxide inclusions by various mechanisms including dissociation and floatation. The product of VAR is fully dense and homogeneous except for an outer skin where the liquid metal was quenched by the cold copper crucible. This outer skin is mechanically removed from the ingot after VAR.

Applications may require *double-melt,* which is a combination of VIM and ESR or VIM and VAR, or *triple melt*, which requires VIM, ESR and VAR. Triple-melt is usually required for critical applications such as rotating disks for turbomachinery.

After casting, the ingot is *converted* into a billet through a thermomechanical sequence that usually involves multiple upset and draw operations in an open-die forging press and/or a radial forge. Thermomechanical processing can result in significant yield loss because ends are cropped from the billet after each step and reverted to VIM, imposing the cost of remelting the alloy. Finally, after conversion, the product is forged or rolled to a rough shape before machining to final shape and heat treatment to achieve the required properties.

13.2.2 Powder Metallurgy Processing of Superalloys

The aircraft engine industry continuously drives for higher operating temperatures and stresses to increase fuel efficiency and performance while decreasing emissions. The demand for higher strength materials and oxidation resistance materials has led to the need for higher alloy content in superalloys. Increased strength is obtained by increasing the volume fraction of γ'. However, the high alloy content leads to a propensity for macrosegregation and a decrease in hot workability, making the C&W process route impossible. For such alloys, powder metallurgy offers a viable production route.

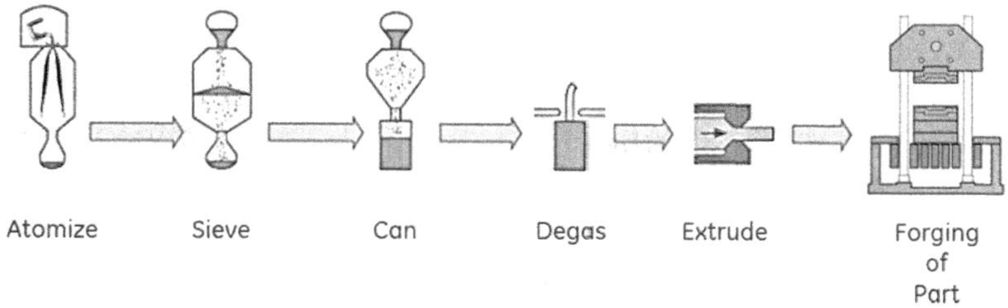

Fig. 13.4 Conventional powder metallurgy processing of nickel superalloys

PM processing is a multi-step process shown in Fig. 13.4. Molten metal is poured from a VIM into a close-coupled gas atomizer and the resulting liquid metal spray is allowed to fully solidify while in flight. Metal particles that are either too large or too small are removed using multiple sieve operations before the metal powder is transferred into a *container*. Gas is removed from the container before the sealed assembly is hot isostatically pressed (HIPed) or extruded to consolidate the powder into a fully dense billet for forging, machining, and heat treatment. PM has the advantage that it produces microstructures that are free of macrosegregation, but extreme care must be taken in powder handling to avoid the introduction of unwanted particles or inclusions that could limit the life of parts. All issues with oxygen pickup and oxide inclusions have been addressed, resulting in a robust PM industry.

13.2.3 Spray Forming of Superalloys

Spray forming of superalloys results in desirable properties because chemically homogeneous, fine-grained material is produced directly as-sprayed. Various melting schemes have been demonstrated to feed a continuous stream of liquid metal into an atomizer. When the metal is atomized into a fine spray, the high relative velocity between the cool gas and the hot liquid metal droplets results in rapid heat exchange, and the droplets cool quickly. Where in the competing technology of powder atomization the droplets are allowed to fully solidify in flight, in spray forming the droplets of the spray are *partially* solidified when they strike the preform. Large droplets are fully liquid, small droplets are fully solid, and medium-sized droplets are partially solidified. Manipulation of the spray nozzle in coordination with a rotating and withdrawing preform results in fully-dense, homogeneous product. After deposition onto the preform, nucleated grains will grow or shrink depending on local thermal conditions, but any grain will grow only until it consumes all local liquid metal and meets adjacent growing grains. The large number of nucleation sites initiated in the spray results in a fine-grained microstructure in the preform.

Several processing steps can be avoided by introducing spray forming into either the C&W or PM process sequence. Remelting and conversions steps of C&W processing may be avoided, or the sieving, canning, HIPing and extrusion steps of PM processing can be avoided. Thus, spray forming offers a low-cost alternative because of high production rates for billet preforms and near-net-shape components. However, when dropping process steps for economic reasons, it is extremely important that the technical benefits of the process steps are not also eliminated. This is particularly true from the oxide cleanliness perspective, as will be discussed in detail in sections below.

13.3 History of Superalloy Spray Forming

In published work as early as 1986, Bricknell [4] of General Electric explored spray forming of the superalloy René 80™ using a system purchased from Osprey Metals. He found that as-sprayed densities exceeding 99% could be achieved with either argon or nitrogen atomization. No evidence of layering (visible inhomogeneities between spray passes) was found except at the chill layer near the collector. Yield strengths and ultimate tensile strengths were excellent. He noted that ceramic inclusions were present in the sprayed product and of a sufficient size that they would limit use of spray formed superalloys in some specific applications such as aircraft engine turbine disks because these brittle ceramic inclusions would decrease fatigue life. He recommended coupling the process with a clean-melting system to mitigate the issue. Continuing his work at General Electric, Fiedler et al. [5, 6] investigated spray formed billets of René95™ and Alloy 718 in 1987. Preforms with densities approaching 100% could be sprayed with nitrogen, while argon-atomized preforms showed porous structure that had to be closed with forging or with HIPing. Low Cycle Fatigue (LCF) results indicated that fatigue cracks initiated from ceramic particles, as predicted by Bricknell. The particles were identified as originating from the VIM crucible, or from the cement used to bond the spray forming nozzle to the crucible. Specimens from argon-atomized samples that were forged but not HIPed showed fatigue crack initiation sites at voids in the samples, concluding that forging alone was not sufficient to close all porosity.

In the period 1988–1996, several research teams announced preliminary projects for commercialization of spray forming of superalloys for several applications. Moran [7] announced in 1988 that the U.S. Navy was developing spray forming of Inconel 625™ for large-diameter tubular applications that were later identified as torpedo tubes. This work was continued for several years [8–11]. The next year, Kennedy [12] reported an evaluation of spray formed Alloy 718 demonstrating promising properties, comparing favorably to C&W product. The question of nitrogen pickup resulting from nitrogen atomization was addressed. Nitrogen-atomized metal showed 230 ppm nitrogen, where argon-atomized metal showed 72 ppm. Nitrogen was shown to promote intergranular precipitation of Ti-rich carbides and carbonitrides during high temperature thermal treatments. These results were duplicated in later work on René 95™ by Benz [13].

Prichard and Dalal [14, 15] announced in 1992 that Howmet was developing the *Spraycast-X®* process for ring applications in Inconel 718™, René 41™, Waspaloy™ and several other alloys. A unique fine-grained (ASTM 6-8) nondendritic structure without macrosegregation was achieved. The process was developed aggressively and by 1994 a prototype high pressure turbine casing fabricated in Inconel 718™ by spray + HIP + ring rolling was successfully fabricated for aircraft engine testing at Pratt & Whitney [16, 17].

Shifan, et al. [18], announced in 1996 that research was under way in China at the Beijing Institute for Aeronautical Materials in several superalloys including Nimonic 115™, Inconel 718™ and René 95™.

Continuing the work at General Electric, Huron [19] published a paper in 1993 on the properties of spray formed René 41™ rings. He concluded that spray + HIP, spray + HIP + ring rolling, or direct ring rolling of sprayed rings were feasible manufacturing approaches. However, he noted that subsequent welding of the argon-atomized metal was not recommended because porosity which may have been previously closed by HIP, would reopen when the metal was re-heated, perhaps during repair welding. This was a key finding because it indicated that argon-atomized parts were essentially unrepairable by welding; any damage in service or manufacturing would result in scrapping of the part. This observation was not true in nitrogen-atomized metal and it was concluded that any trapped nitrogen in the sprayed product reacted with elements to form solid nitrides, as shown earlier by Kennedy and Benz, whereas trapped argon remained as small inert gas pores

In 1990, Benz et al. [20] published a paper entitled *Properties of Superalloys Spray Formed at Process Flow Rates of Less than 20 cm³/s* and reported that acceptable properties could be obtained at this low flow rate. The title of the article was curious, but the reason for exploring operating conditions at such low spray process rates became evident in a patent publication two years later, *Direct Processing of Electroslag Refined Metal* [21], in which the concept of bottom-pouring from an ESR furnace using an all-copper induction-heated funnel was introduced. The spray rate had to be matched to feasible ESR melt rates (10 kg/min) to efficiently remove oxide inclusions from incoming VIM electrodes. This invention would solve the oxide inclusion issue highlighted by prior investigators and several years of development of this concept ensued in a research partnership that included General Electric and Teledyne Allvac, now ATI Specialty Materials. These developments will be discussed in greater detail below.

Cantor and Grant at Oxford University in England have a long history of working on spray forming of a variety of material systems spanning almost the entire period of time mentioned above. Their work included sprayforming composites, carbon nanotubes [22–24], aluminum alloy systems [25–30], and superalloys. In 1995, Grant et al. [31–34] investigated the effects of time and temperature for the alloys UDIMET 720™ and MAR-M-002™ over a fraction solid range spanning 78–100%. This work was aimed at confirming qualitative modeling investigations of microstructural evolution. The resulting fine grain size and hence lower segregation in the sprayed preforms was compared to those formed during more conventional processing techniques such as ESR and VAR. During spray forming, the molten metal spray impacting the semi-solid preform surface (>80% solid) and qualitative models of microstructural evolution showed that the fully solid smaller droplets seed the semisolid top surface layer resulting in a high nucleation density and hence fine grain size. Numerical models [35–37] were developed to simulate dynamic shape evolution during the spray forming of nickel superalloy rings. These models were used to calculate the optimum thermal conditions in terms of alloy liquid fraction (40–70%) inside the preform for the spray forming of large diameter rings with low microporosity and freedom from microcracks. The models were also used to calculated overall process yield through a consideration of splashing and droplet redeposition during spray deposition. The results of the models compared favorably with those from experiments.

In the 10-year period starting 1996, many research activities showed that superalloys could be produced with reduced post-processing and improved process yield over conventional processing while meeting property requirements. As a result, all attempts to commercialize spray forming of superalloys were introduced as cost reduction projects, and research expenses were quite high. However, these attempts were thwarted by the high investment cost of industrial-scale spray forming facilities. A robust conventional superalloy industry was able to reduce costs through process optimization to remain competitive. By 2006, all attempts to commercialize spray forming of superalloys in the United States were abandoned.

International researchers remain optimistic. In more recent work, for example, a Chinese research team led by Changchun Ge [38, 39] duplicated the results of the GE-Teledyne Allvac team to show that the aircraft engine disk alloy FGH4095, a γ'-strengthened alloy currently manufactured from powder, can be spray formed using nitrogen gas. The resulting preform shows a nitrogen increase of about 180 – 200 ppm, with porosity at 0.6%. The nitrogen is present in the form of very fine TiN particles. The preform was HIPed to full density, isothermally forged, and ultrasonically inspected. During HIP and isothermal forging, the as-sprayed grain size of 40 μm was refined to approximately 12 μm. See Fig. 13.5. After final heat treatment,

(a) (b)

Fig. 13.5 Microstructure of FGH4095 superalloy as-sprayed (**a**) and after HIP and isothermal forging (**b**), adapted from [38] resulting properties compared very favorably to conventionally-processed powder metal, exceeding the Chinese Aviation Material Standard for PM FGH4095. Similar work on other superalloys including FGH102L [40, 41] also shows promise.

13.4 Spray Forming of Superalloy Rings

Aircraft engine turbine casings are among the most costly items to manufacture in the aircraft engine business. Conventional casings are machined from rings that are first cast into solid cylindrical shapes, upset-forged, pierced, and ring rolled to shape as shown in Fig. 13.6. This processing is expensive and results in yield loss that adds to cost. During the 1990s and 2000s, several investigators looked at spray forming ring shapes as preforms for ring rolling to reduce processing. The goal was to replace steps of casting, upsetting, punching, and piercing with a single spray forming process that would yield a ring preform ready for ring rolling.

Spraying of rings may take two forms as shown in Fig. 13.7: spraying onto the outer diameter of a mandrel, as was demonstrated by the Howmet team [16, 17], and a team organized by INASMET Tecnalia [42]; or spraying onto the inner diameter of a mandrel, as was demonstrated by a team at the University of Birmingham [43, 44].

Over several years of development, the Howmet team built three spray forming facilities for the SprayCast-X® process with increasing capability. The capacity of their largest system is given in Table 13.2 below. The results of a demonstration program at Pratt & Whitney in cooperation with Howmet are shown in Fig. 13.8, which shows an aircraft engine turbine casing as sprayed, as ring-rolled, and ready for flight testing. Ring rolling experiments showed that the sprayed + HIP preform exhibited improved forgeability and reduced edge cracking relative to standard C&W billets of the same composition. This was attributed to a uniform fine grain

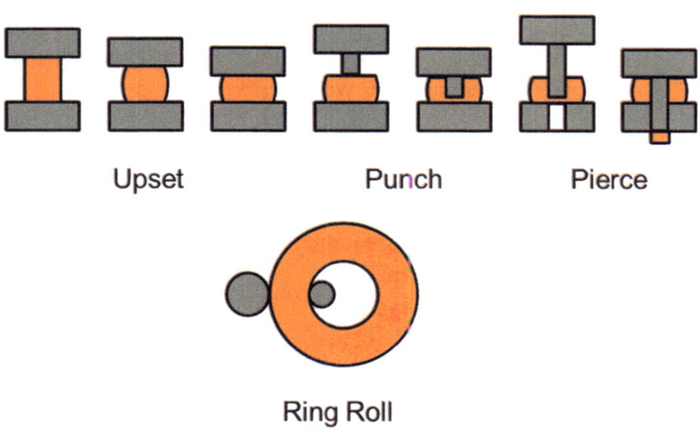

Fig. 13.6 Conventional processing of seamless rings

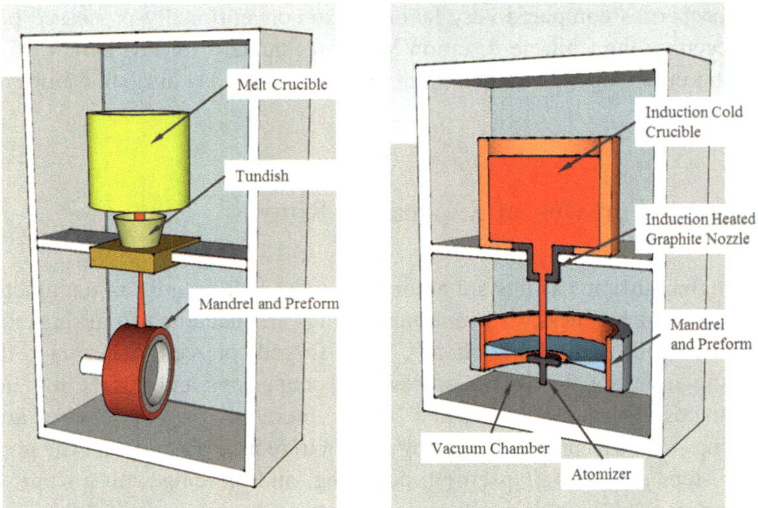

(a) (b)

Fig. 13.7 Spraycast-X® process of Howmet (**a**) and the centrispray system of the University of Birmingham (**b**)

Table 13.2 Details of the Howmet spray forming unit

Melt capacity	2721 kg (6000 lb.)
Deposit mass	2177 kg (4800 lb.)
Maximum ring size	1524 mm (60 in)
Maximum ring length	1524 mm (60 in)
Cycle time	4 h
Expected annual throughput	453,592 kg (1,000,000 lb.)
Spray chamber size	4.88 m × 5.49 m × 7.92 m (16 ft. × 18 ft. × 26 ft.)
Alloys demonstrated	Inconel 100™, Inconel 713C™, Alloy 718, Inconel 783™, Inconel 939™, Haynes 242™, MAR-M247™, MERL 76™, René 41™, René 77™, Thermospan™, Waspaloy™

structure and the absence of coarse MC carbides in the sprayed + HIP preform. The casing performed without evidence of distress in over 1000 hot endurance cycles.

Similar results in sprayed superalloy rings for engine casings were obtained by a European team that included INASMET-Tecnalia, ITP, Turbomeca, and MTU Aero Engines. This team termed their process *OPTISPRAY* [42], and trials in Inconel 718™ and Udimet 720™ were undertaken. The team performed a battery of room temperature and high temperature tests on spray formed material processed via spray + forge, spray + ring roll, spray + HIP + forge and spray + HIP + ring-rolling. After solution heat treatment and aging, the material met the specifications

Fig. 13.8 High pressure turbine case for an aircraft engine fabricated from spray formed Alloy 718. As-sprayed, ring rolled, and machined, adapted from [17]

for the materials in tensile tests, stress rupture, creep, fracture toughness, and LCF, often with large margin. The process was deemed a viable low-cost process for manufacturing compressor casings and low-pressure and intermediate-pressure turbines.

Researchers at the University of Birmingham published results with parts sprayed in Waspaloy™ and Inconel 718™ using a centrifugal atomizing system termed *Centrispray*, which was unique because the metal was atomized with a spinning water-cooled copper plate instead of gas. Melting was performed in a cold copper crucible with a 7000 cm^3 capacity, and the metal was bottom poured through a graphite nozzle, which was also induction heated. The metal fell from the nozzle by gravity onto a rotating copper plate which served as the atomizer. A horizontal spray was produced and directed onto the inner diameter of a 400-mm diameter mandrel. The mandrel was oscillated vertically to distribute the spray along its axis. With such a system the researchers could avoid contamination of the deposits with any oxide inclusions that might otherwise erode from crucibles or nozzles. Oxide-free deposits would be expected if double or triple melted feedstock were used for this system. It was noted that carbon pickup resulted from erosion of the graphite nozzle, suggesting that a cold induction guide such as that developed by the GE team discussed below, would be needed for industrialization. Most importantly, this approach made it was possible to atomize and deposit in vacuum. As a result, any as-sprayed porosity (reported at 3–8%) did not contain trapped gas. Thus porosity could be closed with HIP and would not reappear during subsequent heat treatment. This finding presented a feasible solution to the thermally induced porosity concerns of Huron [19], and the resulting casings would be repairable.

All three of these research activities aimed at reducing the cost of aircraft engine casings showed technical promise and feasibility, but none were commercialized. The reasons for abandoning the efforts are not clear from the technical literature, but it can be assumed that economic pressures from established manufacturing techniques made the required investment in spray forming facilities unattractive to investors.

13.5 Spray Forming of ESR Liquid Metal

Ceramic inclusions may be present in the raw materials used for melting superalloys, or they may be introduced to the metal from erosion of the crucible liner in VIM; these inclusions play a significant role in the LCF life of components made from superalloys [45, 46]. A ceramic inclusion is far more brittle than surrounding metal and will crack early in life, possibly as early as the first load cycle, and the cracked inclusion acts as a fatigue crack starter for surrounding metal. Since a large fraction of the fatigue life of any part occurs while cracks are short, it is important to eliminate or limit the size of the ceramic inclusions. This requirement poses a metals processing challenge that has been aggressively attacked in the C&W industry and in the PM industry. In the C&W industry, a high level of oxide cleanliness is achieved using the triple melt procedure discussed above. In the PM industry, cleanliness is achieved by sieving the powder. The sieve size is chosen to limit the size of oxide inclusions in the product. Since any metal powder that exceeds the sieve size is rejected with the oxides, reducing yield, high-yield atomizers have been developed to produce the desired small powder size. After sieving, careful handling during subsequent canning, vacuum degassing, and extrusion is required to avoid reintroducing oxide inclusions or other contaminants. The primary oxide removing process in C&W processing is the ESR step and the primary oxide removing step in PM is the sieve step.

As discussed above, spray forming represents an economical alternative to powder processing for billet making because of the reduced number of processing steps and increased yield. This possibility had not been exploited as a cost-effective alternative for superalloys to be used in critical fatigue-limited applications because of the lack of a primary oxide removing process. A solution for the problem was proposed and demonstrated by Benz and a team that included the co-authors of this chapter during a period 1992–2006. The process is shown in Fig. 13.9, in which spray forming is performed directly from an ESR melt pool.

The ESR portion of the system is conventional. An electrode is fed into a hot liquid slag where it is caused to melt in a controlled manner. Any inclusions in the electrode are exposed to the slag where they are dissolved, and refined liquid metal forms a pool beneath the slag.

A bottom-pouring system referred to as the *Cold-Walled-Induction Guide*, (CIG) was developed to transfer the liquid metal from the ESR crucible into the spray forming chamber as a steady stream. As described by Hohmann [47], the CIG system is a water-cooled copper funnel with induction heating to maintain superheat and avoid freezing of the liquid metal as it flows through the funnel. It was first applied to powder production using a plasma melting system for the production of

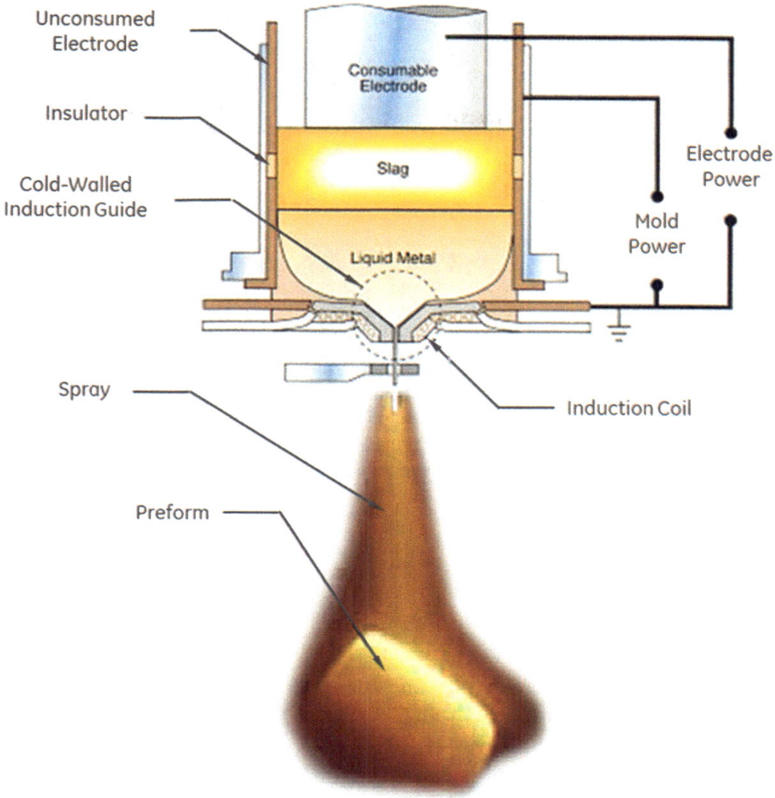

Fig. 13.9 Clean metal spray forming concept

ceramic-free titanium powder [48]. The CIG system is made from copper to avoid introduction of ceramic inclusions that would otherwise be introduced to the melt from conventional ceramic transfer nozzles such as those used to feed powder atomizers or conventionally spray forming atomizers. Similar systems were available using induction-heated graphite nozzles, but these were not selected because of limitations in carbon pickup in the alloys.

A photograph of the output stream of the ESR-CIG system is shown in Fig. 13.10. The copper funnel is slotted and surrounded by an induction coil. The slotting allows penetration of the induction field to the flowing metal stream in a manner similar to the design of induction skull melting systems. High frequencies (>100 kHz) are required to couple efficiently to the thin metal stream [49]. Also shown in Fig. 13.10 is the spray and preform. In this case the preform is approximately 25-cm (10-in) in diameter and is approximately 50-cm (20-in) below the atomizer.

The combination of ESR, CIG, and spray forming systems resulted in a new process, termed *Clean Metal Spray Forming* [50–54]. A pilot plant was constructed

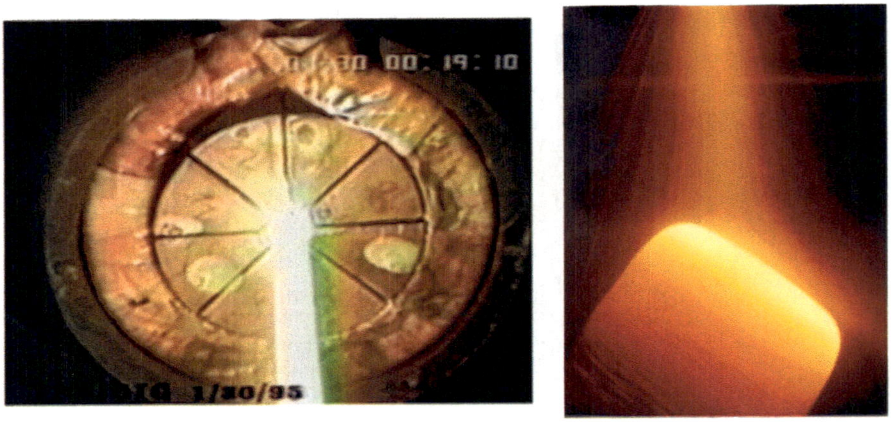

Fig. 13.10 The output superalloy stream from the ESR-CIG system. The stream is approximately 6 mm in diameter. The 8-segment CIG is surrounded by a 100-kHz, 100 kW induction coil

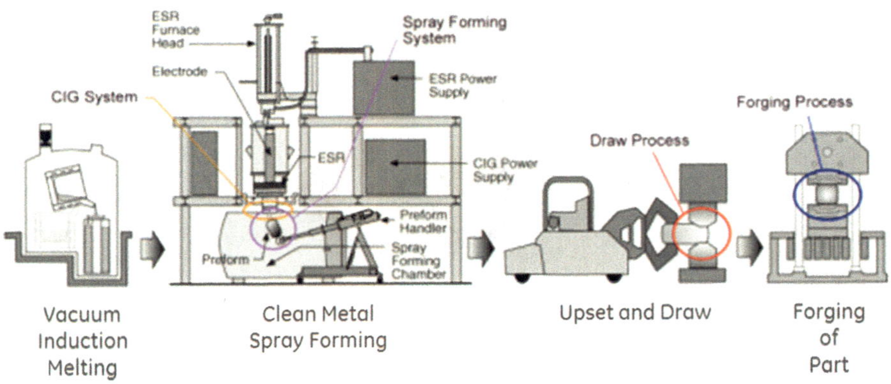

Fig. 13.11 The clean metal spray forming process as integrated in a production process for turbine disk forgings

to demonstrate the concept and to verify the premise that material generated was satisfactory for commercial purposes and to substantiate economic viability. These goals lead to the requirement that the pilot plant operate at production processing rates using production electrode and preform diameters. Production sized power supplies were required. A schematic of the pilot plant is shown in Fig. 13.11, and a photo of the top of the system while loading an electrode is given in Fig. 13.12. The pilot plant generated forging preforms up to 260-kg (580 lb.) for evaluation as rotating components in aircraft engines.

Controlling the pour rate required modifications to the system to allow pressurization of the gas above the slag to accommodate variations in metal height that occurred as a result of varying melt rate. As the metal height decreased, pressure was increased to maintain a constant hydrostatic head at the nozzle outlet. Further

Fig. 13.12 A view of the top of the GE spray forming facility. Here workers load a 1-ton electrode of superalloy into the ESR furnace

modifications were required to allow restarting of the ESR system without consumption of metal from the consumable electrode; this required the development of an unconsumed electrode, which was integrated into the ESR mold as a second electric circuit in parallel with the primary circuit as shown in Fig. 13.9.

In an effort to increase process yield, the process was further adapted to spray into a mold rather than onto a preform. The resulting process was termed *Clean Metal Nucleated Casting* [55–58].

Several benefits were demonstrated in the clean metal spray forming and clean metal nucleated casting programs:

- The liquid slag dissolved ceramic inclusions that may have been present in the VIM electrode, so that outgoing material was cleaner than incoming material.
- Re-contamination of the metal that would otherwise originate from a ceramic transfer tube was avoided through the use of the copper CIG system.
- The process was a "melt-as-needed" process in which large amounts of metal were processed while only a small amount of metal was liquid at any one time. Chemistry problems generally associated with large melts were eliminated.

- The economy of the process when compared to powder metal processing was apparent. The number of processing steps between atomization and the final preform were significantly reduced. Spray forming yield was typically higher than P/M sieve yield as well.

Despite the technical advantages, the process was not adopted commercially because of the large economic investment required to build a commercial plant. The technology remains available for possible future commercialization.

13.6 Electric Arc Spray

An interesting spray forming process, termed the electric arc spray process was used to manufacture dies out of tool steel [59–64]. In this process, two wires are fed through a tool, converging toward each other at an angle. A power supply is connected to the wires and an electric arc between the wires is generated as shown in Fig. 13.13. The arc provides heat that melts the tips of the wires as they are fed through the tool. A high pressure gas then atomizes the molten material and propels the resulting droplets toward a substrate at high velocities. This process was used to make thick, free-standing tool steel molds on three dimensional ceramic patterns by using four electric arc spray guns mounted on a six axis programmable robot. Demonstrator tools for use in high volume press tool applications exhibited acceptable to outstanding performance. The process is fit for fabrication of large components and difficult-to-work materials. Electric arc spray forming has some drawbacks, namely low production rates and the need to produce the feedstock in the form of a wire.

Inconel 718™ was arc sprayed onto 4-mm steel substrates under a range of processing conditions and the microstructure of the resulting material was characterized using a variety of microscopy techniques. Control of the preform thermal conditions was required to achieve desirable, equiaxed grain microstructures. Low spray temperatures resulted in more chemically homogeneous deposits, whereas

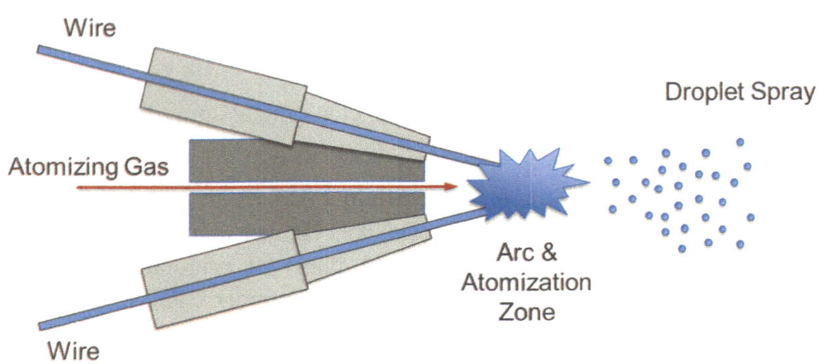

Fig. 13.13 The arc spray process

high spray temperatures gave more equiaxed microstructures, albeit with excessive microsegregation and the formation of deleterious Laves phase. Although the work showed that electric arc spraying was a convenient method for making small preforms, without the need for extensive and expensive liquid metal systems, the process operated at an order of magnitude lower flow rate than spray forming systems that were fed from liquid metal sources.

More recently, a hybrid arc spray forming technique [65] was introduced for Inconel 617™ in which a conventional arc spray gun was combined with a secondary atomizer to focus the spray cone. The hybrid system reduced porosity to about 2% and encouraging properties were measured after HIP. Most importantly, the distinct layered structure of standard arc spray, in which each metal droplet fully solidifies before arrival of the next, was successfully eliminated through control of the environment and process conditions.

13.7 Summary and Outlook

Spray forming of superalloys continues to be a topic of active research in Europe and China, but not in the United States. U.S. companies made preliminary attempts to industrialize the process in the late 1990s and early 2000s, but abandoned efforts before commercialization, primarily due to economic pressures. The process remains attractive from a technical perspective and the likelihood that the process can compete with established industry on an economic basis remains high.

A conclusion can be drawn from the experience of the PM superalloy industry. Originally developed as both a cost reduction and performance improvement over C&W alloys, powder processing proved to be significantly more expensive. Today, powder metal processing for turbine disk applications remains significantly more expensive than wrought processing, but the performance benefit over wrought alloys offsets the additional cost. Current cost reduction efforts in the superalloy industry are aimed at modifying alloy chemistries to render the PM alloys amenable to C&W processing. No program for modifying these chemistries to make the alloys more amenable to spray forming has been undertaken. If such a program were to be initiated, it is likely that spray forming will be quickly adopted only if the performance of the newly-developed alloys eclipses C&W properties at a process cost that is significantly lower than PM costs.

Acknowledgements The Clean Metal Spray Forming project was partially funded by the Defense Advanced Research Projects Agency (DARPA) under Cooperative Agreement F33615-96-2-2565. The Clean Metal Nucleated Casting Project was partially sponsored by the U.S. National Institute of Standards and Technology (NIST) under the Advanced Technology Program (No. 70NANB1H3042). The authors gratefully acknowledge their technical and financial support. Further technical contributions from several individuals, particularly Lawrence A. Jackman, are gratefully acknowledged.

Trademark Identification

CRS Holdings: A Subsidiary of Carpenter Technology Corporation

Thermospan

General Electric Corporation

René 41, René 77, René 80, René 88, René 95

Haynes Alloys International

Haynes 242

Howmet Corporation

Spraycast-X

Martin Metals Corporation

MAR-M-002, MAR-M247

Special Metals Corporation: A Wholly Owned Subsidiary of Precision Castparts Corp.

Inconel 100, Inconel 617, Inconel 625, Inconel 707, Inconel 713C, Inconel 718, Inconel 783, Inconel 939, Nimonic 115, Udimet 720

United Technologies Corporation

Waspaloy

References

1. Walter, J. L., Jackson, M. R., & Sims, C. T. (1988). *Alloying*. Metal Park, OH: ASM International.
2. Bowman, R. Superalloys: A primer and history, TMS, http://www.tms.org/meetings/specialty/superalloys2000/superalloyshistory.html
3. Forbes Jones, R. M., & Jackman, L. A. (1999). The structural evolution of superalloy ingots during hot working. *Journal of Metals, 51*(1), 27–31.
4. Bricknell, R. H. (1986). The structure and properties of a nickel-based superalloy produced by osprey atomization process. *Metallurgical Transactions, 17A*, 583–591.
5. Fiedler, H. C. Sawyer, T. F., Kopp, R. W., Leatham, A. G. (1987). The spray forming of superalloys. *Journal of The Minerals, Metals & Materials Society, 39*(8), 28–33.
6. Chang, K. M., & Fiedler, H. C. (1988). Spray-formed high-strength superalloys. In S. Reichman, et al. (Eds.), *Superalloys 1988, Proceedings of the Sixth International Symposium on Superalloys* (pp. 485–493). TMS-AIME.
7. Moran, A. L., & Palko, W. A. (1988). Spray forming alloy 625 marine piping. *Journal of Metals, 40*(12), 12–15.
8. Moran, A. L., Palko, W. A., Madden, C. J., Kelley, P. (1990). Thick section alloy 615 tubulars produced via spray forming. In *Proceedings of the 1990 Powder Metallurgy Conference & Exhibition Sponsored by the Metal Powder Industries Federation and the American Powder Metallurgy Institute*, Pittsburgh, PA, May 20–23.
9. Moran, A. L., Palko, W.A., Madden, C.J., & Kelly, P. (1991). Spray forming thick section alloy 625 tubulars. In *Proceedings of the P/M in Aerospace and Defence Technologies Symposium Sponsored by the Metal Powder Industries Federation*, March 4–6, Tampa FL.
10. Moran, A. L., Madden, C. J., Rebis, R., Payne, R., Matteson, M. A. (1994). Spray forming technology for military applications. *Journal of Thermal Spray Technology, 3*(2), 197–198.
11. Moran, A. L., Madden, C. & Rebis, R. (1993). Industrialization and certification of large diameter spray formed superalloy components for military applications. In *Proceedings of the 2nd International Conference on Spray Forming*, September 13–15.
12. Kennedy, R. L., Davis, R. M., & Vaccaro, F. P. (1989). An evaluation of spray formed alloy 718. In E.A. Loria (Ed.), *Superalloy 718—Metallurgy and applications* (pp. 97–108). TMS-AIME.
13. Benz, M. G., Sawyer, T. F., Clark, F. W., & Dupree, P. L. (1993). Nitrogen in spray formed superalloys. In *Proceedings of the Second International Conference on Spray Forming*, Swansea, Wales, UK, 13–15 September.
14. Prichard, P. D., & Delal, R. P. (1992). Spraycast-X superalloy for aerospace applications. In S.D. Antolovich, et al. (Eds.), *Superalloys 1992, Proceedings of the Seventh International Symposium on Superalloys* (pp. 205–214). TMS-AIME: Warrendale, PA.
15. Dalal, R. P., & Prichard, P. D. (1993). Thermomechanical processing of spraycast-X superalloys. In *Proceedings of the 2nd International Conference on Spray Forming,* September 13–15.
16. Paton, N., Cabral, T., Bowen, K., Thomas, T. (1997). Spraycast-X IN718 processing benefits. In E.A. Loria (Ed.), *Superalloys 718, 625, 706 and various derivatives*. The Minerals Metals & Materials Society.
17. Cabral, A. C., & Haynes, A. L. (1999). Evaluation of spray formed nickel base superalloys for gas turbine engine components. In *Proceedings of the Fourth International Conference on Spray Forming,* Baltimore, MD, September.
18. Shifan, T., Zhou, L., Xianguo, Z., Zhikai, L., LiPing, R., Guofa, M., & Minggao, Y. (1996). The spray atomized and deposited process. In *Proceedings of the 3rd International Conference on Spray Deposition and Melt Atomization*, September 23–25.
19. Huron, E. S. (1993). Properties of sprayformed superalloy rings. In *Proceedings of the Second International Conference on Spray Forming,* Swansea, September 13–15.

20. Benz, M. G., Sawyer, T. F., Clark, F. W., & Dupree, P. L. (1990). Properties of superalloys spray formed at process flow rates of less than 20 cm^3/s. In *Proceedings of the First International Conference on Spray Forming*, Swansea, Wales, UK, 17–19 September.
21. Benz, M. G., & Sawyer, T. F., Direct Processing of electroslag refined metal. US Patent 5,160,532, November 1002.
22. Zhao, X., Zhao, X., Chu, B. T. T., Ballesteros, B., Wang, W., Johnston, C., et al. (2009). Spray deposition of steam treated and functionalized single and multi-walled carbon nanotube films for supercapacitors. *Nanotechnology, 20*(6), 065605.
23. Zhao, X., Koos, A. A., Chu, B. T. T., Johnston, C., Gorbert, N., Grant, P. S. (2009). Spray deposited fluoropolymer/multi-walled carbon nanotube composite films with high dielectric permittivity at low percolation threshold. *Carbon, 47*, 561–569.
24. Zhao, X., Hinchliffe, C., Johnston, C., Dobson, P. J. Grant, P. S. (2008). Spray deposition of polymer nanocomposite films for dielectric applications. *Materials Science and Engineering B, 151*, 140–145.
25. Hogg, S. C., Palmer, I. G., Thomas, L. G., & Grant, P. S. (2010). Spray forming of bulk ultrafine grained Al–Fe–Cr–Ti. *Metallurgical and Materials Transactions A, 41*, 3208–3215.
26. Kelly, A. J., Mia, J., Sinha, G. V., Krug, P., Crosa, F., Audebert, F., et al. (2011). An Al–Si–Ti hierarchical metal-metal composite manufactured by co-spray forming. *Journal of Materials Processing Technology, 211*, 2045–2049.
27. Hogg, S. C., Palmer, I. G., Ghomas, L. G. & Grant, P. S. (2007). Processing, microstructure and property aspects of a spray cast Al–Mg–Li–Zr alloy. *Acta Materialia, 55*, 1885–1894.
28. Moore, K. L., Sykesa, J. M., Hobb, S. C., Grant, P. S. (2008). Pitting corrosion of spray formed Al–Li–Mg alloys. *Corrosion Science, 50*, 3221–3226.
29. Hogg, S., Lambourne, A., Ogilvy, A., & Grant, P. S. (2006). Microstructural characterisation of spray formed Si-30Al for thermal management applications. *Scripta Materialia, 55*, 111–114.
30. Hogg, S. C., Palmer, I. G., & Grant, P. S. (2006). An investigation of novel spraycast Al–Mg–Li–Zr–(Sc) alloys. *Materials Science Forum, 519–521*, 1629–1633.
31. Mi, J., & Grant, P. S. (2008). Modeling the shape and thermal dynamics during the spray forming of Ni superalloy rings. Part 1: droplet deposition, splashing and re-deposition. *Acta Materialia, 56*, 1588–1596.
32. Mi, J., & Grant, P. S. (2008). Modelling the shape and thermal dynamics during the spray forming of Ni superalloyrings. Part 2: heat flow and solidification. *Acta Materialia, 56*, 1597–1608.
33. Hedges, M. K., Newbery, A. P., & Grant, P. S. (2002). Characterization of electric arc spray formed Ni superalloy IN718. *Materials Science and Engineering, A326*, 79–91.
34. Underhill, R. P., Grant, P. S., Bryant, D. J., & Cantor, B. (1995). Grain growth in spray-formed Ni superalloys. *Journal of Materials Synthesis and Processing, 3*(3), 171–179.
35. Mi, J., Grant, P.S., Fritsching, U., Belassam, O., Garmendia, I., Landerberea, A. (2008). Multiphysics modelling of the spray forming process. *Materials Science and Engineering A, 477*, 2–8.
36. Grant, P. S. (2007). Solidification in spray forming. *Metallurgical and Materials Transactions A, 38A*, 1520–1529.
37. Jones, P. D. A., Duncan, S. R., Rayment, T., Grant, P. S. (2007). Optimal robot path for minimizing thermal variations in a spray deposition process. *IEEE Transactions on Control Systems Technology, 15*, 1–11.
38. Ge, C. C., Xu, Y., Jia, C. L., Guo, B., Wang, J. (2013). Microstructure and mechanical properties of spray formed FGM4095 superalloy by nitrogen atomization. In *Proceedings of the 5th International Conference on Spray Deposition and Melt Atomization*, Bremen, September 23–25.
39. Jia, C. L., Ge, C. C., Jie, W., Biao, G., Min, X., Tian-Tian, W., Hong-Kai, Z., Yu-Chun, Z., Hao, W. (2013). Research on metallurgical quality and mechanical properties of spray formed

superalloy FGH4095M. In *Proceedings of the 5th International Conference on Spray Deposition and Melt Atomization*. Bremen, September 23–25.

40. Ge, C. C., Jie, W., Jia, C.L., Biao, G., Min, X., Tian-Tian, W., Hong-Kai, Z., Yu-Chun, Z., Hao, W. (2013). Research on hot compressive deformation behavior of spray formed superalloy FGH102L. In *Proceedings of the 5th International Conference on Spray Deposition and Melt Atomization*. Bremen, September 23–25.
41. Kang, F., Cao, F., Zhang, X., Honyan, Y., Feng, Y. (2014). Microstructure and mechanical properties of a spray formed superalloy. *Acta Metallurgica Sinica, 27*(6), 1063–1069.
42. Atxaga, G. Caballero, O., Fournier, D., Smarly, W. (2006). Mechanical characterization of sprayformed components. In *Proceedings of the 3rd International Conference on Spray Deposition and Melt Atomization,* Bremen, September 4–6.
43. Barratt, M. D., Dowson, A. L., & Jacobs, M. H. (2004). The microstructure and properties of IN718 rings produced by centrifugal spray deposition. *Materials Science and Engineering A, 383*, 69–77.
44. Whyman, D, Young, J. M., & Jacobs, M. H. (1995). Processing and evaluation of waspaloy ring preforms prepared by centrifugal spray deposition. In *Proceedings of the 1995 4th International conference on Powder Metallurgy in Aerospace, Defense and Demanding Applications*, May 8–10.
45. Chang, D. R., Kreuger, D. D., & Sprague, R. A. (1984). Superalloy powder processing, properties and turbine applications. In *Superalloys 1984, Proceedings of Fifth International Symposium on Superalloys* (pp. 245–273). Warrendale, PA: TMS-AIME.
46. Fiedler, H.C., Sawyer, T.F., & Kopp, R.W. (1986). Spray forming-an evaluation using IN718. In *Proceeding of the 1986 Vacuum Metallurgy Conference on Specialty Metals Melting and Processing* (pp. 157–165).
47. Hohman, M., Ertl, M., Choudhury, A., & Ludwig, N.. Experience with ceramic-free powder production methods. In *1991- P/M in Aerospace and Defense Technologies, Metal Powder Industries Federation,* Princeton, NJ (pp. 261–272).
48. Gerling, R., Schimansky, F. R., & Wagner, R., Progress in atomizing high melting intermetallic titanium based alloys by means of a novel plasma melting induction guiding gas atomization facility. In (Vol. 1), *Powder Metallurgy World Congress, PM'94* (pp. 387–390).
49. Carter, W. T., Miller, R. S. (2013). The cold-induction guide: a ceramic-free device for melt stream delivery. In *Proceedings of the 5th International Conference on Spray Deposition and Melt Atomization,* Bremen, September 23–25.
50. Carter, W. T., Jr., Benz, M. G., Müller, F. G., Forbes Jones, R. M., & Leatham, A. G. (1995). Electroslag remelting as a liquid metal source for spray forming. In *1995 International Conference on Powder Metallurgy and Particulate Processing,* Seattle, May 14–17.
51. Carter, W. T., Jr., Benz, M. G., Müller, F. G., Forbes Jones, R. M. (1995). Electroslag refining as a liquid metal source for spray forming. In *European Conference on Advanced PM Materials*, Birmingham, October 23–25.
52. Benz, M. G., Carter, W. T., Zabala, R. J., Knudsen, B. A., Dupree, P. L., Muller, F. G., Forbes Jones, R. M. (1996). Electroslag refining as a liquid metal source for powder atomization, spray deposition, investment casting, melt-spinning, strip casting, and slab casting. In *Spring 1996 TMS Annual Meeting* Anaheim. Feb. 4–8.
53. Carter, W. T., Koca, J. M., Benz, M. G., Mourer D. P., Forbes Jones, R. M., Davis, R. M., Kennedy, R. L.. (1996). A clean metal spray forming system for superalloys. In *Third International Conference on Spray Forming,* Cardiff.
54. Forbes Jones, R. M., Davis, R. M., Kennedy, R. L., Carter, W. T., Benz, M. G. (1996). Characteristics of clean metal spray formed material. In *Third International Conference on Spray Forming*, Cardiff, September 9–11.
55. Carter, W. T., Forbes Jones, R. M., & Minisandram, R. S. (2003). Clean metal nucleated casting. In *Proceedings of the International Symposium on Liquid Metal Processing and Casting,* Nancy, September 21–24.

56. Carter, W. T., Forbes Jones, R. M., & Minisandram, R. S. (2003). Nucleated casting of superalloys. In *International Conference on Spray Deposition and Melt Atomization*, Bremen, June 22–25.
57. Carter, W. T., & Forbes Jones, R. M. (2004). Nucleated casting for the production of large diameter superalloy ingots. In *TMS Annual Meeting*, March 14–18, Charlotte.
58. Carter, W. T., et al. (2005). Clean metal nucleated casting of superalloys. In J. Groh (Ed.), *Superalloys 718, 625, 706 and derivatives*. TMS (The Minerals, Metals & Materials Society).
59. Roche, A., Duncan, S. R., & Grant, P. S. (2006). Scientific, technological and economic aspects of rapid tooling by electric arc spray forming. *Journal of Thermal Spray Technology, 15*, 796–801.
60. Rayment, T., & Grant, P. S. (2006). Modelling the heat flow in spray formed steel shells for tooling applications. *Metallurgical and Materials Transactions B, 37B*, 1037–1047.
61. Hoils, S., Rayment, S., Grant, P., Roche, A. (2003), Oxide formation in the sprayform tool process. In *Proceedings of SDMA* (pp. 4–11).
62. Newbery, P., Rayment, T., & Grant, P. (2003). A particle image velocimetry investigation of in-flight and deposition behavior of steel droplets during electric arc sprayforming. In *Proceedings of SDMA* (pp. 7–45).
63. Newbery, A. P., & Grant, P. S. (2006). Oxidation during the electric arc spray forming of steel. *Journal of Materials Processing Technology, 178*, 259–269.
64. Newbery, A. P., & Grant, P. S. (2009). Arc sprayed steel: microstructure in deep substrate features. *Journal of Thermal Spray Technology, 18*, 256–271.
65. Sato, A., Taneike, M., Okada, I., Grant, P. S. (2014). A hybrid arc spray forming technique for the manufacture of Nickel superalloy IN617. *Materialwissenschaft und Werkstofftechnik, 45* (8), 758–764.

Chapter 14
Spray Forming of Novel Materials

Bulk Processing of Glass-Forming Alloys by Spray Deposition

Claudemiro Bolfarini and Vikas Chandra Srivastava

14.1 Introduction

In the last few decades, the world inclination towards newer materials to cater to the presently stringent and specific requirements has led to unprecedented innovations and the development of a number of new classes of material and processes e.g. metal matrix composites (MMCs), nanomaterials, amorphous materials/metallic glasses, quasicrystals, superalloys, intermetallics, nanocomposites, grapheme aided coatings etc. Among them, the amorphous alloys and metallic glasses have attracted considerable interest and application possibilities as they show high strength as well as stiffness far above the conventional material classes of similar or closely related compositions. Despite such incremental developments, a paradigm shift has been observed in the design of new alloys with low cost alloying elements such as iron, aluminium and magnesium, instead of the costly Pd-, Zr- and La-alloy systems, and in the development of viable processing routes. However, the lower Glass Forming Ability (GFA) of many of these alloys poses challenge on the process selection and/or modification thereof. In the last few years, a few research endeavors have demonstrated the development of bulk amorphous, nanocrystalline or a combination of amorphous-nanocrystalline-crystalline materials by spray forming, which is one of the variants of rapid solidification process. However, a coordinated effort by researchers as well as the industrial establishment for the

C. Bolfarini (✉)
Materials Engineering Department, Federal University of Sao Carlos,
Rod. Washington luiz-km235-, 13565-905 São Carlos, SP, Brazil
e-mail: cbolfa@ufscar.br

V.C. Srivastava
National Metallurgical Laboratory, Near Tata Steel gate No-1,
Burma Mines, Jamshedpur, Jharkhand 831007, India
e-mail: vcsrivas@nmlindia.org

H. Henein et al. (eds.), *Metal Sprays and Spray Deposition*,
DOI 10.1007/978-3-319-52689-8_14

utilization of these initiatives for the further development of such materials is still lacking. Therefore, this chapter is an attempt to consolidate the results reported so far on spray forming of aluminum- and iron-based alloys, whose compositions are derived from rapid solidification studies aimed at obtaining amorphous structures. Due to the unique effect of the combinations of various process parameters, the processed aluminium-based glass-forming alloys show the formation of amorphous phase throughout the deposit of the Al-based alloy at high Gas to Metal ratio. This is generally not observed in the Fe-based alloys. However, some iron-based compositions displaying the highest glass-forming ability showed a high volume fraction of amorphous phase up to 4 mm thickness of the deposit. A similar value is obtained for this class of material when processed by copper mold casting.

The first solid amorphous structure was developed in 1960 by Paul Duwez from CalTech by imposing a very high cooling rate, of the order of 10^9 K/s, to the molten metal droplets of Au–Ge using two splatting copper plates [1]. With the advent of melt spinning process in 1973, further research led to the production of amorphous ribbons (thickness < 25–30 μm) for several different alloy combinations. Despite the successful processing, the product geometry was restricted to ribbons, as a high cooling rate in bulk was difficult to attain, and forming of the consolidated materials was difficult due to the thermal behavior of glass forming alloys. These conventional glasses suffered crystallization and loss of amorphous structure on heating, as the onset of the crystallization temperature was attained before the glass transition temperature. However, Professor Inoue in Japan and Professor Johnson in USA gave the breakthrough concept of bulk metallic glasses (BMG) in their pioneering discoveries in the late 1980s [2–4]. The new compositions, typically identified as the materials with high GFA, could be processed to fully amorphous structure even at cooling rates as low as 1–100 K/s allowing the use of conventional copper mold casting and the critical dimensions reached to even several millimeters [2–4].The main feature of these bulk metallic glasses is to present glass transition temperatures (T_g) below the onset of the crystallization temperature (T_x), presenting this way a supercooled liquid region between T_g and T_x, which allows forming the materials without crystallization, as above T_g the viscosity drops rapidly.

For the last 20–25 years, there is an exponential increase in the number of new BMG discoveries based on Pd-, Zr- and La-alloy systems, which have branched out into several alloy families and are presenting a pronounced glass forming ability. The bulk amorphous/glassy materials, possessing outstanding specific mechanical properties, resistance to wear and to corrosion hold the promise for future materials requirements [5–7].

The high elastic strain limit of metallic glasses combined with a high yield strength and fracture toughness make metallic glasses ideal for applications, where the storage of high densities of elastic energy is needed. The maximum elastic energy density is ca. four times higher than that of conventional crystalline materials. This property has led to the use of bulk metallic glasses in golf-club heads, high-performance springs etc. The combination of high strength, elasticity, hardness, and wear resistance has opened markets in a diverse spectrum of products. The applications include cases for electronic products, cell phone cases, sporting goods, surgical instruments, implants for bone replacement and other medical devices [8–12].

Furthermore, the ability to process amorphous metals throughout the undercooled liquid state over practical time scales opens opportunities for thermo-plastic processing not available for conventional metallic materials. The relatively low viscosity of the undercooled melt permits shaping and forming processes as well as molding into precision net-shaped components, similar to processes used for plastics. Thus, with similar elastic strain limits and comparable processability, but with considerable higher yield strengths, bulk metallic glasses will in some areas replace the application of polymers. In addition, due to the lack of crystallinity and solidification shrinkage, amorphous materials can replicate very fine microstructures down to the nanometer range. This is of great interest for micro-electro-mechanical systems (MEMS) and other areas where high precision parts are needed. Furthermore, many bulk metallic glasses show a relatively high corrosion resistance [13–15].

In the present review, therefore, we look into the alloy systems, their glass formability and the efficacy of spray forming to produce bulk metallic amorphous materials, in particular. The chapter also brings out the prevailing mechanisms during the development of amorphous phase in the bulk deposits and the critical assessment of the process characteristics that points towards new directions for future developments.

14.2 Essence of Amorphous Metallic Alloys

Amorphous materials are defined as having no long range atomic order as in crystalline solids. The amorphous metallic alloys can be distinguished from the conventional oxide glasses by a clearly defined glass transition temperature (T_g). A thermal analysis of a metallic glass shows three distinct characteristic temperature i.e. glass transition temperature (T_g), crystallization temperature (T_x) and the melting temperature (T_m) [16–19]. During a continuous cooling transformation, if the liquid alloy cools below T_g, without witnessing crystallization, the viscosity of liquid increases leading to very sluggish atomic diffusion [19]. This kinetic arrest of the liquid structure remains intact in the finite time scale below T_g and gives rise to an amorphous structure. At the same time, an appropriate alloy composition can shift the continuous cooling transformation curve to extreme right enabling amorphization at decreased cooling rates (Fig. 14.1). The absence of long range order in the solid imparts high strength and young's modulus to metallic glasses, but low ductility that limits the application of such materials. Since the birth of the amorphous alloys in 1960s to the development of the bulk metallic glasses in the late 1980s, due to the continuous development in the discipline, several different classes of alloys have been developed so as to reach to a glassy state in a given processing condition, i.e. copper mold casting, which offers a cooling rate up to the order of 10^1–10^2 K/s. Most of the studies in metallic glasses were concentrated on

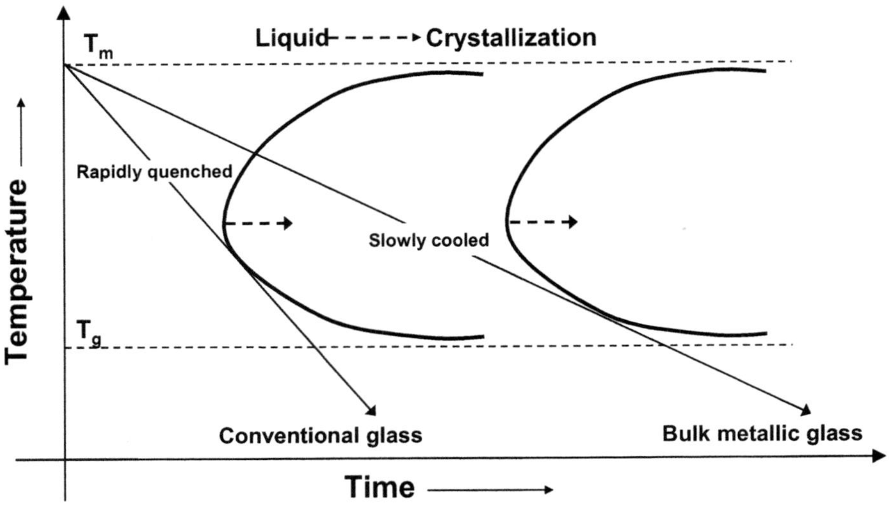

Fig. 14.1 Cooling transformation curve of a conventional glassy alloy and of a bulk metallic glass

the alloy development and employing high cooling rates to measure the glass forming ability, which could be expressed by either the critical cooling rate to produce the amorphous state or the maximal diameter obtained with a full amorphous structure. The requirement of high cooling rate, therefore, potentially depends upon kinetic suppression of nucleation and growth from an undercooled melt. However, the relative stability of liquid in undercooled state and the resistance to crystallization, which are also the prime factors for determining the ease of amorphization, depends upon the chemical composition of the alloy [20, 21]. Therefore, the current emphasis the world over is to address the challenges involved with the design of composition for various alloy classes to obtain low critical cooling rates and the modified processing routes that can engender amorphization in bulk materials.

The synthesis of amorphous metallic materials can be accomplished during liquid to solid transformation or by solid state amorphization. Commonly known processes of water quenching, splat quenching, melt spinning and mechanical alloying have been initially used for getting thin sectioned amorphous materials. However, with the advent of new compositions, processes like copper mold casting, electromagnetic levitation followed by casting, zone melting, arc melting followed by injection molding or suction casting, high pressure gas atomization [22], thermal spraying, spray forming [23–32] have also been considered to be of promise. The conventional bulk amorphous materials' synthesis processes are shown in Fig. 14.2. Fluxing the liquid alloy to trap heterogeneous impurities, which may act as potential nucleation sites during solidification, is also one of the well tested process routes for making bulk metallic glasses. The major limitation of the processes dependent on high cooling rate for amorphization is the restricted size of the amorphous materials, particularly for the alloys with low glass forming ability.

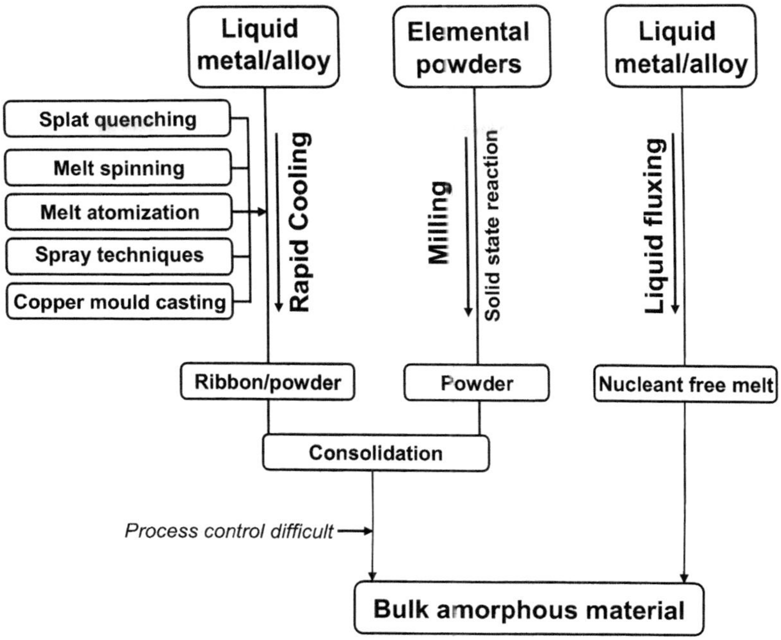

Fig. 14.2 Conventional bulk amorphous materials synthesis processes

In general, these processes require further consolidation of amorphous ribbons or powders to bulk, which suffers from several difficulties due to increased number of process steps and possibility of crystallization of the amorphous materials at consolidation temperatures, thus, losing the characteristic attributes of metallic glasses.

One of the other major technological challenges of the bulk metallic glasses is the ductility. Commonly observed tensile elastic strain limit of metallic glasses in around 2%, much higher than that of the crystalline alloys. Therefore, the yield strength of metallic glasses is generally high both in compression and tension. However, they show a work softening during plastic flow, instead of work hardening like the crystalline materials, leading to shear localization after yielding. This shear localization limits the ductility and mostly leads to catastrophic failures. This limiting ductility was addressed by several investigators [33–35] suggesting that a uniform dispersion of ductile nanocrystals in an amorphous matrix could lead to higher strength as well as remarkable plasticity. This has been shown by Kawamura et al. [33] and Hofmann et al. [35] for the Al- and Zr- alloys, respectively. The ideas behind such composites are twofold (1) amorphization of material followed by controlled crystallization so as to achieve a distribution of defect free fine crystals in the amorphous matrix. This has been shown by Professor Inoue's group for Al-RE-TM alloys and (2) finding an alloy slightly away from the glass forming composition such that the softer primary phase forms on solidification and rejects the solute in the remaining liquid. As the solute is rejected to the liquid its composition reaches to the glass forming composition. Thus, this gives rise to an in-situ amorphous/crystal composite. The high density of multiple shear bands results in a considerable increase in ductility both in compression and tension [35–37].

Keeping the above in view, it is pertinent to conclude that compositional variation based on thermo-dynamical aspects and phase stability as well as the topological calculation for better glass forming ability are a few challenges needed to be dealt with in future endeavors. In addition, the process modification so as to achieve the step by step consolidation of undercooled and viscous liquid for the synthesis of bulk glassy materials can play a determining role in the development of bulk amorphous metallic materials. The processing route should comply to the principle that the sole imperative is to not only produce bulk amorphous materials but a composite of amorphous and crystalline phases.

A few research works [24–32, 38–40] have demonstrated the development of bulk amorphous, nanocrystalline or a combination of amorphous-nanocrystalline-crystalline materials by spray forming. Although, the materials processed by this route are mainly based on Al-, Fe- and Mg-alloy systems, spray forming has already shown its efficacy in the synthesis of such materials. Therefore, in the following sections a detailed review on the recently reported works and the future directions of the process has been brought out, focusing, however, on the Bulk Metallic Glasses, whose compositions allow obtaining the amorphous state under the cooling conditions prevailing in the spray forming process.

14.3 Fundamentals of Bulk Metallic Glasses

14.3.1 Glassy Alloys

An important feature of the bulk metallic glass (BMG) alloys is their thermal behavior. In contrast to the amorphous alloys produced in 1960s, the BMGs show on heating a pronounced glass transition temperature (T_g) before any crystallization. The distinct thermal behavior of the normal amorphous alloys and of the bulk metallic glasses can be observed in Fig. 14.3, which presents DSC traces for both alloy compositions. On heating the amorphous structure, the amorphous metal (left in Fig. 14.3) presents no T_g before the crystallization; on the other hand the metallic glass (right) presents a T_g before crystallization temperature (T_x), which is the temperature of the onset of crystallization. The interval between T_g and T_x is called "supercooled liquid region" (ΔT_x) and is highly important for the further processing or shaping of the material. This is due to the fact that in this temperature range the viscosity drops strongly from around 10^{14} Pa.s to 10^7 Pa.s, allowing change in the material's geometry and ease in near net shaping; and at the same time keeping the amorphous structure unaffected. This supercooled liquid region can have a temperature range as high as 100 K [41]. BMGs can deform homogeneously with formability as good as polymers in their supercooled liquid region temperature

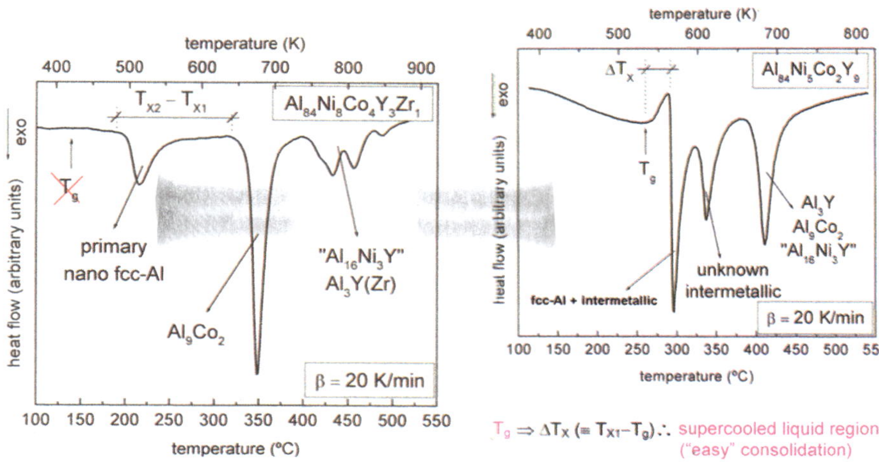

Fig. 14.3 DSC traces of amorphous metals and metallic glasses. *Left*: Alloy $Al_{84}Ni_8Co_4Y_3Zr_1$ presents an "amorphous metal" behavior, i.e., no T_g and primary nanocrystallization; *Right*: Alloy $Al_{84}Ni_5Co_2Y_9$ presents a "metallic glass" behavior, i.e., on heating the alloy shows T_g before T_x (the onset of the crystallization process), adapted from [24]

window. For instance, bulk metallic glasses can now be shaped by using the viscous flow of the supercooled liquid, with elongations that can exceed 15,000% [42] and the production of precision parts without dimensional changes due to almost no contraction during the transition of the supercooled liquid to vitreous state.

Turnbull and coworkers [43] illustrated the similarities between metallic and non-metallic glasses. They observed a glass transition, generally manifested in conventional glass-forming melts, in rapidly quenched metallic glasses also. Turnbull predicted that the reduced glass transition temperature ($T_{rg} = T_g/T_m$) could be used as an indicator for determining the glass-forming ability (GFA) i.e. ease of vitrifying a liquid on cooling. According to Turnbull's parameter, a liquid with $T_g = (2/3)T_m$ shows sluggish crystallization and can crystallize within a very narrow temperature range. Such stable liquids can be easily undercooled to a low temperature into the glassy state i.e. highly viscous solid with liquid structure. This indicator for the suppression of crystallization in undercooled melts has remained one of the best parameters for predicting the GFA of any liquid and has played a pivotal role in the development of bulk metallic glasses (BMGs). This concept was further extended by other authors, who proposed in addition to Turnbull's indicator the nowadays highly acceptable parameters to estimate the glass forming ability of Bulk Metallic Glasses, such as: the supercooled liquid region $\Delta T_x = T_x - T_g$ [44] and the γ parameter ($T_x/(T_l + T_g)$) [45]. There are several such parameters proposed by investigators which have been summarized in Ref. [21]. Despite the fact that none of these parameters can be considered a universal parameter, some important correlation between the critical cooling rate for overcoming crystal nucleation and growth, the T_g/T_m ratio and the maximum sample thickness for fully amorphous structure, have been observed in many

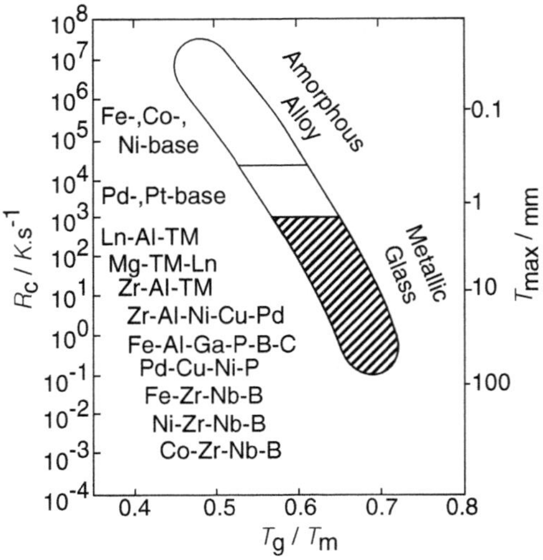

Fig. 14.4 Representative compositions of amorphous metals and metallic glasses. R_c means the minimal cooling rate necessary for the production of the amorphous state and t_{max} the thickness that can be obtained for the specific alloy composition. From [44] with permission

alloy systems. [44]. Figure 14.4 shows representative compositions, minimal cooling rates for obtaining the amorphous structures and the attainable dimensions for the specific alloy classes. The glass forming ability (GFA) of a given alloy composition can be evaluated by the minimal cooling rate necessary for obtaining the amorphous state (R_c) or the maximal thickness of the amorphous samples (t_{max}).

The remarkably sluggish crystallization kinetics in the undercooled liquid state has permitted formation of some bulk metallic glasses with high glass forming ability. The detailed studies of the liquid properties from the liquidus temperature down to the glass transition have been performed by investigators. The first complete TTT diagram between T_l and T_g was developed in the bulk metallic glass Vit1 [46]. This TTT diagram shows a typical 'C' shape due to the fact that the thermodynamic driving force for crystallization increases with increasing undercooling, while the atomic mobility in the liquid reduces. In a subsequent study, a complete TTT diagram was obtained for $Pd_{40}Cu_{30}Ni_{10}P_{20}$ [46], which is shown in Fig. 14.5. The TTT diagram has its 'nose' at 50 s and 680 K. The critical cooling rate of this alloy was measured to be 0.33 K/s.

The Vit1 (Zr-based) and the $Pd_{40}Cu_{30}Ni_{10}P_{20}$ alloys represent a class of materials with extremely high glass forming ability that allowed a deeper understanding of the glassy behavior and were discovered following a trial and error methodology. The most widely used parameters proposed to estimate the GFA of BMGs are (1) the reduced glass transition temperature ($T_g = T_m$), (2) the supercooled liquid region $\Delta T_x = T_x - T_g$ and (3) the γ parameter ($T_x/(T_l + T_g)$). GFA can be determined once the composition is known. The composition of the alloy needs to be first selected and tested for GFA. The lack of a reliable criterion for determining the glass forming ability led several authors to experiment and propose better parameter for determining a priori the glass forming ability, and the significant results of these investigations are outlined in the next section.

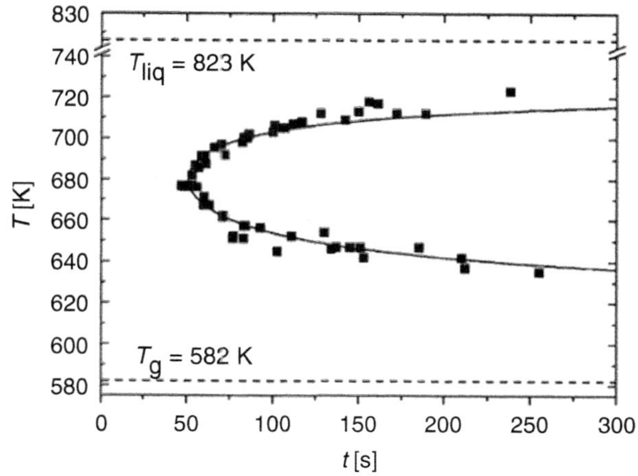

Fig. 14.5 Time-temperature transformations (TTT) diagram of Pd40Cu30Ni10P20. The solid line is a fitting using the inverse function of t(T), from [47] with permission

14.3.2 *Criteria for Glass Forming Ability*

There is no strict definition for glass forming ability; however, from an engineering point of view, it is possible to say that the lower the critical cooling rate for amorphous phase formation and the larger the critical thickness of the amorphous phase are, the higher the glass forming ability of a metallic glass will be. Thus, it is useful to indicate the glass forming ability of an existing composition with easily measurable parameters, as pointed out in the previous section.

Despite the huge advancement in the knowledge of bulk metallic glasses since their discovery in the 1980s the design of a BMG forming composition is still a complex task. Most of the existing criteria are based on three different factors: structural and electronic, thermodynamic, equilibrium diagram; and very often a combination of them.

Based on experimental observations of a series of glass forming compositions, a first attempt of a criterion was proposed by Inoue [48] stating the famous Inoue's empirical rules:

1. Multi-component systems with more than three components: equilibrium diagram, normally deep eutectics or low Liquidus temperatures,
2. Large difference in atomic size of the alloying elements: structural and electronic factors, and
3. Large negative heat of mixing in the liquid: thermodynamics.

Such a criterion is useful as a starting point to find elements that should be used to design a glass forming composition; however, it gives no information about the proportion among the elements; therefore, the development is dependent on a series of experiments where composition is changed step by step. These ideas were developed further in order to propose criteria for designing BMG compositions before doing any experimental trials, and, indeed some of them were already used in the 1970s to find amorphous compositions, such as using equilibrium diagram to find deep eutectics.

Before going into details of the existing criteria, it is useful to have a look on the fundamentals of crystallization theory, which is, basically, a competing process between the liquid phase and the derived crystalline phases, it says, the relative thermodynamic and kinetic stabilities of these phases. If a liquid is cooled below its liquidus temperature (T_l) the free energy difference between the liquid and the crystal (decreased energy, proportional to r^3, considering r as the radius of a spherical growing crystal) provides a driving force for crystal nucleation, while the creation of the liquid–crystal interface (increased energy, proportional to r^2) creates a positive interfacial energy that makes more difficult nucleation. This results in an energy barrier that a local composition fluctuation needs to overcome in order to form a nucleus, it says, a stable crystalline region with a radius higher than a critical value, where the decrease in the free energy difference among crystal and liquid equals the interface energy. To enable the growth of such a nucleus, atoms within the liquid need to be rearranged. The rate of such atomic transport is described by the atomic diffusivity, D, or viscosity. The resulting crystal nucleation rate, I, per unit volume is the product of a thermodynamic term, which depends on the probability of a fluctuation to overcome the nucleation barrier, and a kinetic term, which depends on atomic diffusion (or viscosity). This phenomenon was equated by Turnbull's classical nucleation and growth theory. In this framework, the homogeneous nucleation rate, I, of a crystalline phase formed from an undercooled liquid can be expressed by [49]:

$$I = \frac{10^{30}}{\eta} \exp\left[\frac{-16\pi}{3} \frac{\alpha^3 \Delta S_f T^2}{R(T_l - T)^2}\right] \tag{14.1}$$

where η is the viscosity, α is a factor which depends on the atomic arrangement at the interface and has a value close to unity, ΔS_f is the change in entropy per mole of alloy due to melting, T is the temperature of the melt and R is the universal gas constant. This equation was modified by Pariona et al. [50] to include the surface energy and the catalytic efficiency $g(\theta)$ for heterogeneous nucleation, including the latent heat of fusion $\Delta H_f = \Delta S_f \cdot T_l$:

$$I = \frac{10^{30}}{\eta} \exp\left(-16\pi\sigma^3 g(\theta) \middle| 3RT\Delta H_f^2\right) \tag{14.2}$$

According to theses above equations, the important parameters governing the GFA are $\eta(T)$, the kinetic factor that increases with decreased temperatures, and $T/(T_l - T)$, influencing the thermodynamic factor, which increases by decreasing the temperature. In addition, the higher the surface energy and the lower the latent heat of fusion are, the lower the nucleation rate is. Therefore, BMGs are compositions that present very high viscosity (three orders of magnitude higher than crystalline alloys) and high entropy of melting with a very high surface energy between liquid and crystalline phases, which increase both the thermodynamic barrier and the kinetic factor for nucleation of the crystal. On the other hand, heterogeneous nucleation has to be avoided as it lowers the critical energy barrier for nucleation, it says, special care must be taken with the purity of the melt. These parameters are influenced by the chemistry of an alloy and are defined by the above mentioned empirical rules of Inoue. In the following, the main proposed criteria up to date, which consider one or more of the influencing parameters, will be summarized.

It is possible to predict GFA by looking at deep eutectic points, as several studies carried out since the 1970s have showed that the composition of eutectics and the ones with good GFA are similar [51–53]. Basically, a deep eutectic composition shows a high stability of the liquid phase and difficulties to crystallize. Based on this, Cheney and Vecchino [54] proposed an α parameter to describe quantitatively the depth of an eutectic in order to evaluate GFA. The α parameter is a measure of the depth of an eutectic by comparing the relative, weighted liquidus temperatures of the isolated components to the eutectic temperature of the melt T_m. The calculation of α parameter as proposed by Cheney and Vecchino is shown in the following equation:

$$\alpha = \frac{\sum_{i=1}^{n} xiTi}{Tm} \qquad (14.3)$$

where x_i is the atomic fraction of element i, T_i is the melting temperature of element i, and n is the number of elements. In the case of a complex eutectic presenting multi-component phases, a better approach would be to use the melting temperature of the isolated phases instead of the elemental components. A eutectic will generate an α value higher than unity and a deep eutectic will produce a high α value. Considering that a deep eutectic composition shows a stable liquid phase to a temperature well below the melting temperatures of the isolated components or multi-component phases, it is reasonable to suppose that such composition will present a higher chance to be kept in the amorphous state upon cooling. In addition, the competition among the different phases, whose formation depends on the diffusion of the elements, makes the whole process very complex facilitating amorphous phase formation.

Experimentally, it is found that the glass-forming composition is located in the region with an α parameter larger than 1, and the best glass formers may have a α parameter larger than 1.5. However, the α parameter alone does not suffice to find a bulk metallic glass composition. An example is presented by the system Al-Au that shows a deep eutectic without forming amorphous phase, even when subjected to high cooling rates during melt spinning, as per the report of Egami et al. [55]. Further, Cheney and Vecchino [56] combined the Liquidus-based model with a model depicting the chemical short-range order (CSRO) to evaluate GFA of various metallic glass systems. They found that metallic glass compositions tend to locate at or near to deep eutectics, but also have to present an optimized structural topology.

Thermodynamic calculations provide a meaningful method to predict glass forming ability. It is based on the Miedema's model, which is an empirical theory for calculating enthalpy of mixing in various binary systems both for the liquid and solid state [57, 58]. Since the metallic glass formation process is controlled by thermodynamic factors, this theory was firstly used to predict the composition range of amorphous binary transition metal alloys [59, 60]. A simple method was proposed by Xia et al. [61] calculating the formation enthalpy of amorphous phase (ΔH_{amor}), solid solutions (ΔH_{SS}) and intermetallic compounds (ΔH_{inter}) and relating these values according to a γ^* parameter:

$$GFA \propto \gamma^* = \frac{(\varDelta H\text{amor})}{(\varDelta H\text{amor}) - (\varDelta H\text{inter})} \tag{14.4}$$

The glass forming ability increases for higher γ^* and it depends on a higher absolute value of ΔH_{amor}, and/or a small difference among ΔH_{amor} and ΔH_{inter}. According to the results of Xia et al., this criterion works well for binary Zr-Cu system. However, according to the results of Oliveira et al. [62] the criterion fails in calculating the glass forming ability of ternary alloys of the system Al–Ni–Y. This result was ascribed by the authors to uncertainty in calculating ΔH_{inter} using the Miedema's model and to the fact that the criterion does not consider the entropy.

One of the main rules to design alloys that form bulk metallic glasses is to select multi-component systems with more than three components having large difference in their atomic sizes. The BMGs were found to have new type of glassy structure with high degree of dense randomly packed atomic configurations. They also have new local atomic configurations, which are significantly different from those of the corresponding crystalline phases. Density measurements show that the density difference between BMG and fully crystallized state is in the range of 0.3–1.0% [63, 64], which is much smaller than the previously reported range of about 2% [65] for ordinary amorphous alloys. Such small differences in values indicate that the BMGs have highly dense randomly packed atomic configurations. The density of atomic arrangement is governed by two factors: atomic packing and atomic bonding. Atomic packing involves some geometrical parameters, such as atomic size ratio and size of clusters, while atomic bonding is more related to the electronic structures, such as the electronegativity and Fermi surface-Brillouin zone interaction. Based on the high density of glass forming compositions Miracle [66, 67] proposed a comprehensive efficient cluster packing model for metallic glasses consisting of $\leq$4 topologically distinct atomic clusters comprised of a central solute atom surrounded in the first coordination shell by solvent atoms. All atoms were idealized as hard spheres. This model attempts to describe the structure of bulk metallic glasses and predict the better composition for attaining the amorphous state. The model was successful in predicting high glass forming ability compositions, and despite considering solely structural factor it was essential for the understanding the essence of these structures.

Further, concerning structural factor, in 1984 Egami and Waseda [68] proposed the "λ-criterion" based on the concept of topological instability of a stable crystalline structure to justify the amorphization of binary solid solution alloys. In this criterion, the topological destabilization of a crystalline structure was associated with a critical solute concentration necessary to exceed the mean elastic volume strain in supersaturated solid solutions in binary systems. Based on this, Sá Lisboa et al. [69] extended this criterion to predict the thermal behavior of Al–TM–RE alloys. The λ parameter is calculated according to the equation:

$$\lambda = \sum_{i=B}^{Z} C^{i} \cdot \left| \frac{r_i^3}{r_A^3} - 1 \right| = C^{B} \cdot \left| \frac{r_B^3}{r_A^3} - 1 \right| + \ldots + C^{Z} \cdot \left| \frac{r_Z^3}{r_A^3} - 1 \right| \tag{14.5}$$

Where, A is aluminium, r_A = atomic radius of Al, r_i = atomic radii of the other elements from B to Z, C^i = atomic concentration of element i. If: $\lambda > 0.1 \rightarrow$ bulk metallic glassy behaviour; $\lambda < 0.1 \rightarrow$ amorphous behaviour. This criterion was extended later by Kiminami et al. [70] applying it to several glass forming compositions and showed that best glass-forming compositions will lie within fields of mutual and simultaneous topological instability of all crystalline phases competing with glass formation.

Finally, there is a consensus among the researchers that a unique criterion to predict glass forming ability does not exist so far. Most of the presented criteria can be applied for a particular system with relative success, however, failing to predict GFA in other systems. Based on the above discussion, in can be concluded that the bulk metallic glass-forming liquids are the alloys with typically 3–5 metallic components that have a large atomic size mismatch and a composition close to a deep eutectic. They are dense liquids with small free volumes and high viscosities which are several orders of magnitude higher than those in pure metals or previously known alloys. An electronic configuration leading to a certain value of conduction electron density (e/a) add another stabilization effect to the glassy state. In the microstructure, they have unique atomic configurations which are significantly different from those for conventional metallic glasses. Thermodynamically, these melts are energetically closer to the crystalline state than other metallic melts due to their high packing density in conjunction with a tendency to develop short-range order. These factors lead to slow crystallization kinetics and thus high glass-forming ability of BMGs.

14.3.2.1 Main Systems and Existing Alloys

Despite the non-availability of a comprehensive and universal criterion for selecting a specific composition with a high glass forming ability, the reported literatures show a huge knowledge about compositions capable of producing bulk metallic glasses. The first developed glass former compositions, early in the 1970s, were based on highly expensive Pd, Pt and Au elements. Nowadays, a series of new compositions are available based on more common and cheaper elements such as Cu, Mg, Al, Ni and Fe, enabling a widespread and commercial utilization of such materials. Table 14.1 shows the main systems and the year of their introduction.

Table 14.2 shows the glass forming ability of some known BMGs indicating their composition, glass transition temperature, T_g, onset temperature of

Table 14.1 Main BMG systems and year of introduction

System	Year	Reference
Introduction: Early and expensive systems		
Pd–Cu–Si	1974	[71]
Pt–Ni–P	1975	[72]
Au–Si–Ge	1975	[72]
Pd–Ni–P	1982	[73]
Breakthrough BMG: BMG with less expensive elements		
Ln–Al–TM	1989	[74]
Al–Y–Ni–Co	1990	[75]
Zr–Ti–Ni–Cu–Be	1993	[76]
Widespread: BMG with more common elements		
Cu–Zr–Ni–Ti	1995	[77]
Mg–Y–Cu, Mg–Y–Ni	1995	[48]
Ni–(Nb, Cr, Mo)–(P, B)	1999	[78]
Fe–Nb–B	2000	[18]
Fe–Co–B–Si–Y	2004	[79]
Fe–Cr–Co–Mo–Mn–C–B–Y	2004	[80]
Mg–Zn–Ca	2010	[81]
Binary systems		
Ca–Al	2004	[82]
Pd–Si	2003	[83]
Cu–Zr	2004	[84]
Cu–Hf	2004	[85]

Table 14.2 Glass forming ability of some known BMG [86]

Composition (at.%)	T_g (K)	T_x (K)	T_m (K)	T_{rg}
$Mg_{65}Ni_{20}Nd_{15}$	459	501	743	0.62
$Mg_{75}Ni_{15}Nd_{10}$	450	482	717	0.63
$Zr_{46.75}Ti_{8.25}Cu_{7.5}Ni_{10}Be_{27.5}$	622	727	909	0.68
$Zr_{44}Ti_{11}Cu_{10}Ni_{10}Be_{25}$	625	739	917	0.68
$Pd_{40}Ni_{40}P_{20}$	590	671	877	0.67
$Pd_{40}Cu_{30}Ni_{10}P_{20}$	586	678	744	0.79
$Cu_{60}Zr_{30}Ti_{10}$	713	763	1110	0.64
$Cu_{54}Zr_{27}Ti_{9}Be_{10}$	720	762	1090	0.66
$Cu_{60}Zr_{20}Hf_{10}Ti_{10}$	754	797	1189	0.63
$La_{55}Al_{25}Ni_{20}$	491	555	711	0.69
$La_{55}Al_{25}Ni_{10}Cu_{10}$	467	547	662	0.71
$La_{55}Al_{25}Ni_{5}Cu_{10}Co_{5}$	465	541	660	0.70
$[(Fe_{0.6}Co_{0.4})_{0.75}B_{0.20}Si_{0.05}]_{96}Nb_{4}$	825	875	1398	0.59
$[(Fe_{0.5}Co_{0.5})_{0.75}B_{0.20}Si_{0.05}]_{96}Nb_{4}$	820	870	1390	0.59
$(Fe_{44.3}Cr_{5}Co_{5}Mo_{12.8}Mn_{11.2}C_{15.8}B_{5.9})_{98.5}Y_{1.5}$	775	835	1270	0.61
$Fe_{60}Cr_{10}Mo_{9}C_{13}B_{6}Er_{2}$	808	848	1442	0.56
$Fe_{64}Cr_{10}Mo_{9}C_{15}Er_{2}$	803	850	1443	0.56
$Al_{84}Ni_{5}Co_{2}Y_{9}$	527	562	975	0.54

crystallization, T_x, melting point, T_m, and glass-forming ability represented by reduced glass transition temperature, T_{rg}.

14.4 Spray Forming Process

Spray forming is a well established process [87–95], therefore, it is briefly described and discussed in this section, particularly touching those aspects, which have direct relevance to the present subject i.e. producing amorphous or nano-crystalline materials. A schematic of the spray forming process is shown in Fig. 14.6. Spray forming comprises integral processes of liquid metal atomization into a spray of droplets and its deposition onto a substrate. The atomization produces a wide size range of liquid droplets that experience high cooling rate of the order of 10^3–10^6 Ks^{-1}. The cooling rate, however, varies depending upon the size of the droplets, which determines the specific surface area of droplets. The high cooling rate renders a rapid solidification effect on the droplets, which leads to high undercooling of the droplets prior to their deposition on the substrate. As the cooling rate and the undercooling of the melt, as mentioned before in the section on the 'fundamentals of bulk metallic glasses', are the major concern for the amorphization of a given glass forming composition, the atomization and deposition process becomes a potential route for the synthesis of bulk amorphous materials. This is highly advantageous, particularly when the droplets are deposited on the substrate in highly undercooled or viscous state. The degree of undercooling of

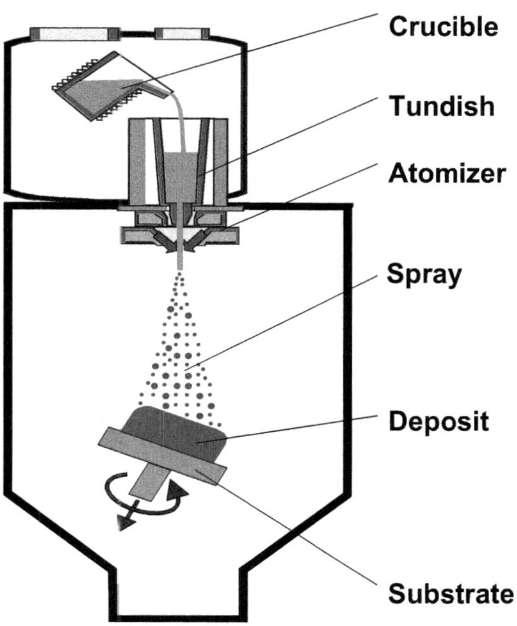

Fig. 14.6 A schematic of the spray forming process (from [97] with permission)

the droplets is also governed by the presence of potential catalytic nucleants that may arise due to the impurities in the liquid metal. One of the advantages of the atomization process is the fact that during disintegration of the melt stream the number of potential nucleation agents is divided into variously sized droplets leaving large fraction of droplets free of nuclei. This fraction of nuclei-free droplets increases as the mean droplet size decreases [28, 96], thereby increasing the possibility of highly undercooled spray, with large fraction of nuclei free droplets, prior to deposition. Therefore, a high undercooling in the spray can be achieved either by increasing the melt purity or reducing the mean droplet size, with small spread in size distribution, by manipulating the process parameters of atomisation. However, the control of the droplet sizes and the spread of the size distribution depend upon the nozzle design and process parameters used for the atomization. A little consideration will indicate that a narrow size distribution of droplets may ensures similar cooling conditions for a large fraction of undercooled droplets.

Figure 14.7 shows two similar sized particles of $Al_{85}Nd_8Ni_5Co_2$, where one particle is fully amorphous and the other depicts primary phase in a featureless nano-crystalline or amorphous phase. It is expected that a similar cooling condition would be experienced by the droplets due to their almost same size. However, it can be speculated that their structural differences might arise due to smaller undercooling experienced by the particle with crystalline phases, which may be

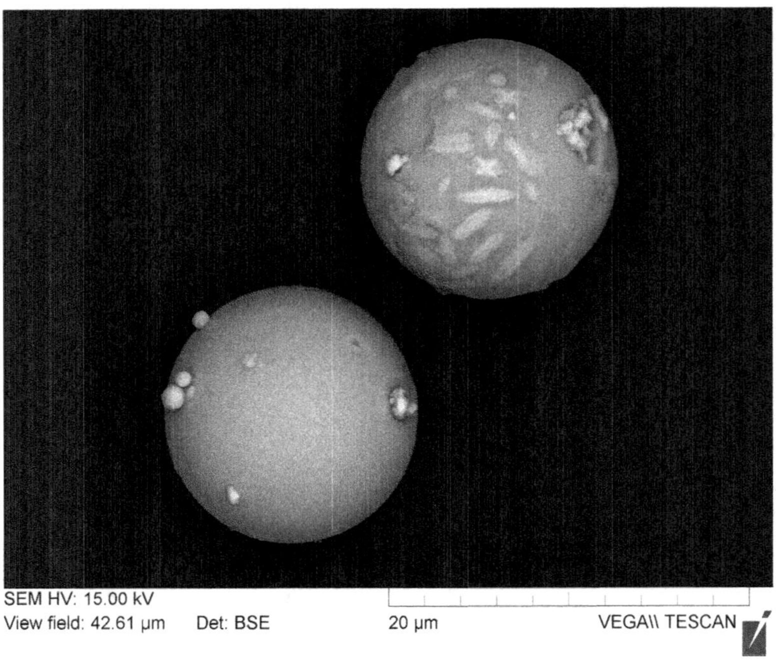

Fig. 14.7 SEM-BSE of a $Al_{85}Nd_8Ni_5Co_2$ glass former composition with similar sized particles showing different structures

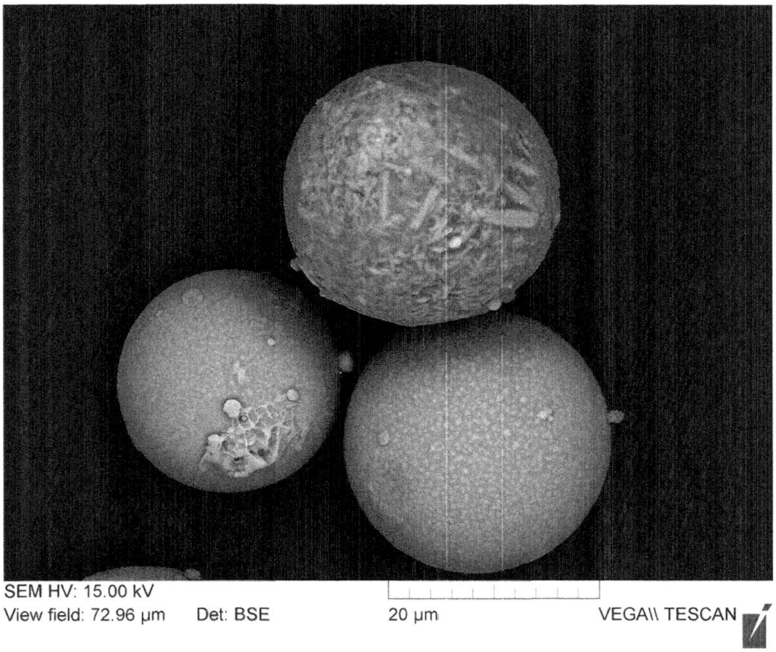

Fig. 14.8 AlYNiCo system showing particles with coarse primary crystals (*upper particle*) and small secondary crystals embedded in featureless surfaces (both *bottom particles*) (from [31] with permission)

attributed to the presence of potential catalytic nucleant. Similarly, the SEM picture in Fig. 14.8 shows the back scattered electron image of $Al_{85}Y_8Ni_5Co_2$ system showing particles with coarse primary crystals, small secondary crystals and featureless surfaces. In this case also, the cooling history and the achieved undercooling seem to determine the structural features of the particles.

Figure 14.9 shows the microstructural features of the $Al_{85}Y_8Ni_5Co_2$ system spray deposited in the form of a plate. This reveals large size prior solidified particle embedded in a relatively crystalline matrix. The particle seems to have been deformed under the influence of the impact force during deposition; however, keeping the prior particle boundaries intact. It is obvious from the figure that even when primary crystallization commences during the flight of the droplet, the remaining undercooled liquid may be kinetically arrested and further crystallization or growth would be restricted after deposition on the substrate due to a chilling effect on the substrate. The deposition of droplets with such a varying thermal history, as evident from the microstructure of the overspray powders, engenders unprecedented microstructure of the deposits, which is generally a combination of crystalline, nanocrystalline and amorphous phases coexisting together. The individual fractions of these phases depend to a large extent on the process parameters employed during spray forming. The spray condition, which is generally used for the deposition of non-glass forming alloys to get pore-free, dense and refined

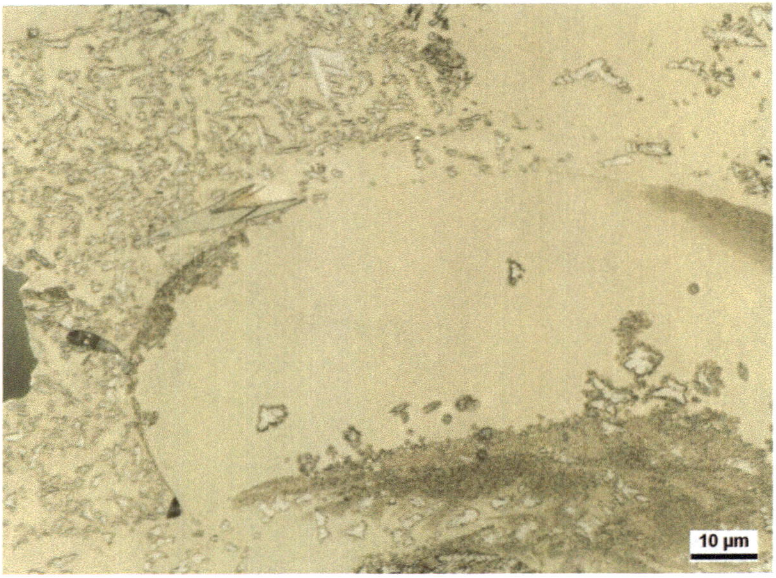

Fig. 14.9 Spray deposited AlYLaNiCo: the boundary of the central big particle remained intact after deposition. The primary crystallization that started during the flight of the droplet may be kinetically arrested and further crystallization or growth would be restricted after deposition on the substrate

microstructural feature may not be suitable for the synthesis of uniform amorphous/nano-crystalline phases.

In the last 15 years, several studies have been reported pertaining to the synthesis of amorphous and/or nanocrystalline materials by using spray forming. In these studies, the alloy systems reported are mainly devoted to Al-, Fe-, La- and Mg-based alloys. It has been invariably observed that the amorphization does take place irrespective of the glass forming alloy systems and varies from 20% to 98% of the amorphous content. The amount of amorphous fraction changes with the alloy content and the process parameters employed. In some of the alloys having comparatively high glass forming ability, fully amorphous phase could be achieved. In the next section, therefore, the details of these studies have been brought out and have been discussed in terms of composition, deposit size, phases evolved and parameters used.

14.5 Investigations on Spray Forming of Glass Forming Alloys

In this section, various reports available on the studies of the amorphous and/or nano-crystalline materials by employing spray forming process has been presented and discussed. As majority of the glass forming alloys that have been studied by using spray forming are Al-based and Fe-based, separate sections have been devoted for these two alloy systems. Other reported alloys have been grouped together in a separate section.

14.5.1 Aluminum Based Alloys

The first report on the development of amorphous materials by spray deposition was published, to the best of the authors' knowledge, by Oguchi et al. in the year 1990 [23, 98] on $Al_{84}Ni_{10}Mm_6$ system. In these experiments, a high pressure gas atomization process was employed to atomize the melt and the deposition was accomplished on a high speed rotating copper substrate. A fully amorphous sheet of 7 mm was produced that showed the characteristics similar to melt spun ribbons of the same composition. This experiment showed the efficacy of the 'two stage liquid quenching' technique i.e. rapid cooling of droplets during atomization and quenching of the undercooled droplets on a substrate. After a gap of 10 years, following the first report by Oguchi, Afonso et al. [24, 25] attempted spray forming of $Al_{85}Y_8Ni_5Co_2$ and $Al_{84}Y_3Ni_8Co_4Zr_1$ systems using a high gas to melt (G/M) mass flow ratio. The $Al_{85}Y_8Ni_5Co_2$ alloy was proposed by Inoue et al. [75] in 1990. Ribbons with 250 μm thickness, produced by melt-spinning were totally amorphous and presented good bend ductility with high tensile strength of 1250 MPa. By increasing the thickness to 300 μm the alloy presented an amorphous structure containing nanoscale fcc-Al phase with higher tensile strength of 1330 MPa. These results showed that the Al-based alloy does not present a good glass forming ability and can be classified as a marginal glass former. Thus, the formation of the amorphous phase depends on a rapid liquid quenching process which has to be obtained. The Bulk materials were produced by hot extrusion of amorphous powder (<25 μm diameter) in 1993 [99], which rendered the material a crystalline structure with a tensile strength of 900 MPa [100]. The later showed an amorphous matrix + nanoscale fcc-Al particles, by consolidation at 533 K, and a compressive strength of 1130 MPa. A bulk amorphous alloy was consolidated at 480 K (supercooled liquid region) and presented a compressive strength of 1030 MPa. Afonso et al. [24, 25], during spray forming of $Al_{85}Y_8Ni_5Co_2$ and $Al_{84}Y_3Ni_8Co_4Zr_1$ alloys, employed two ratios of volumetric gas flow rate to metal flow rate (G/M). For the former y G/M of 6.4 and 10.0m^3/kg and for the later a unique G/Mof 8.7 m^3/kg was used. The deposits weighted from 2 to 7 kg and the overspray powder weighted about 19–30% of the starting charge. The details of the processing and characterization of overspray powders and deposit have been presented elsewhere [24, 25].

Figure 14.10 shows the overspray powder of the $Al_{85}Y_8Ni_5Co_2$ alloy processed by the G/M = 10 [24]. This micrograph shows a large area of featureless particles indicating a high fraction of amorphous phase (84% as confirmed by DSC analysis). The size range of the particles (from few microns up to 100 μm, not in this picture) and the mass mean powder diameter (about 24 μm) for the three different G/M ratios were practically the same. These results are attributed to nozzle design, as it

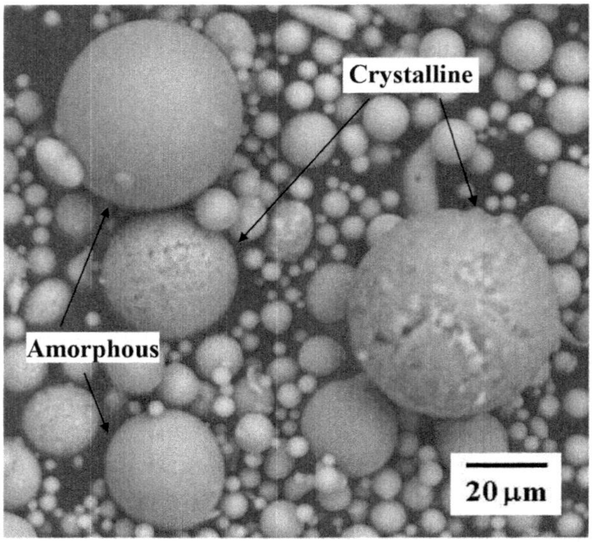

Fig. 14.10 Overspray powder of the $Al_{85}Y_8Ni_5Co_2$alloy processed by the G/M = 10, adapted from [24]

can minimize the effect of G/M ratio on the size of the droplets formed on the periphery of the spray cone, the source for most of the overspray powders. The spherical nature of the $Al_{85}Y_8Ni_5Co_2$ overspray powder is apparent in Fig. 14.10. Also, the difference between the featureless (amorphous or nanocrystalline, smooth surface) powders and those that are crystalline (rough surface) can be easily observed. EDS analysis of these "featureless" particles indicated that their compositions were approximately similar to the nominal composition of the alloy. The $Al_{84}Y_3Ni_8Co_4Zr_1$ powders appeared similar. The volume fractions of the featureless particles in the $Al_{85}Y_8Ni_5$Co2 powders were 82 and 41% (as determined by DSC analysis) for the high and low G/M, respectively, and 40% for the$Al_{84}Y_3Ni_8Co_4Zr_1$ powders, consistent with the increasing cooling rate with increasing G/M ratio and indicating a poor glass forming ability for the Zr-containing alloy. It is worth to point out the difference in the solidification behavior of the two biggest particles that is revealed in Fig. 14.10. Despite the fact that they have approximately the same size (~40 μm) and several satellite particles attached on their surfaces, one solidified amorphous and the other one crystalline. It appears that these distinct behaviors are closely related to two important aspects of solidification that involve kinetic competition: avoidance of crystallization upon cooling of the liquid and control of crystallization upon heating of the glass or the presence of catalytic potent nucleant in the crystalline particle.

Considering that both the particles are decorated with satellites, this could induce heterogeneous nucleation. If the solidification would be triggered by these satellites, both big particles would be crystalline. As this is not the case, it is to be assumed that the distinct behavior is associated with the different undercooling experienced by both the particles, dictated by different cooling/reheating rates they suffer or inherent heterogeneous nucleation.

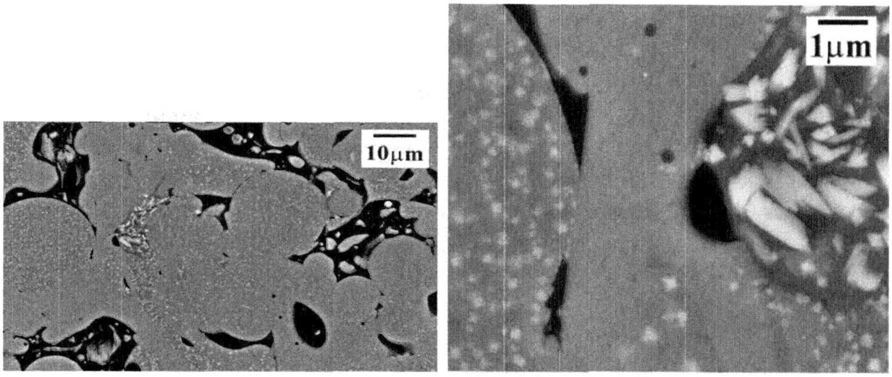

Fig. 14.11 Scanning electron micrograph (backscattering) of the $Al_{85}Y_8Ni_5Co_2$ alloy deposit, processed with G/M = 10.0 m^3/kg. *Black* regions are porosities, featureless regions are amorphous, *gray/white* nanoregions are nanoAl-fcc and the coarse white phases are intermetallics, from [25] with permission

It is possible to observe this distinct behavior for different particles size overall when observing the overspray powders of glass formers alloys. Moreover, there are also many particles presenting a mixed crystal/glass phase structures indicating that the transition is not sharp and occurs over a range of cooling rates reflecting the kinetic competition [31]. It is useful to note that the glass transition is not a phase transformation in a thermodynamic sense, but it is a kinetic manifestation of the slowing down of atomic transport in the liquid with cooling.

Although the overspray powders of the two compositions presented considerable amount of amorphous phase, only the $Al_{85}Y_8Ni_5Co_2$ deposit, processed using the higher G/M ratio (10.0 m^3/kg), contained a high fraction of about 76% of the amorphous phase (Fig. 14.11); the other deposits were fully crystalline [24, 25]. The presence of the amorphous phase only in this deposit indicates that a deposition with high G/M ratio is necessary for amorphous phase formation, and most important is to avoid excessive reheating and consequent crystallization during deposit build-up. In this case, the atomized droplets were subjected to a high cooling rate and the deposition on the substrate proceeds with high volume fraction of completely solid particles, both amorphous and crystalline, and this condition failed to reheat the deposit above its crystallization temperature. The irregular morphology of the porosity (ca. 9%) observed in this deposit is consistent with a high solid fraction of the droplets during deposition. However, other important features of deposit build-up arise and must be mentioned, since it is not possible to see the small amorphous particles that are observed in the overspray powder. Instead of this, we observe amorphous, featureless regions and spherical particles with nano fcc-Al crystallization. From this observation and from the material's flow, that is evident in the picture, it can be concluded that the featureless regions of the deposit were formed by the participation of a great numbers of amorphous particles that were kept in the supercooled liquid region for a sufficient time in order to allow bonding of one to other and viscous flow under the influence of the gas pressure. On the other hand, we can suppose that the particles that hold their geometry and that contains the nano fcc-Al embedded in the amorphous matrix attained a temperature beyond T_x, which promoted the partial crystallization of the material. The regions that presented the big intermetallic phases should be formed by the particles that hit the deposit in a crystalline condition and had time to allow the growth of the intermetallics. Unfortunately, due to intricate relationship between processing parameters and temperature control of the deposit by spray forming along with the restricted supercooled liquid region that these Al-based glass forming alloys present (less than 50 K), it is not possible to avoid/control the crystallization process. However, the microstructures reported in these papers [24, 25] confirm that spray forming, even not being a rapid solidification process in true sense, presents features of rapid solidification processes since the attained microstructures are similar to that one presented by melt spinning of thick ribbons (300 μm).

More recently, Srivastava et al. [31, 32, 38, 101, 102] presented similar results for $Al_{83}La_5Y_5Ni_5Co_2$ and $Al_{83}Y_8Ni_5Co_2$ alloys, with a high volume fraction of amorphous phase in the spray formed plate deposits of thickness 8–12 mm. Their experimental arrangement succeeded in maintaining the temperature of the plate below 290 °C during deposit build-up. It is possible to estimate that this maximal temperature is within the temperature range of the supercooled liquid region for these alloy type. The resulted deposit presented both scarce, pre-solidified dendrite arms and round particles, heterogeneously distributed and embedded in a very high fraction of a layered amorphous phase. Alloy $Al_{83}Y_8Ni_5Co_2$ presented a T_g and can be considered a BMG former; and it has a marginal glass forming ability and the maximal thickness to be obtained for amorphization should be approximately 1 mm.

Srivastava et al. have found that the TEM studies reveal a mixture of amorphous as well as nanocrystalline structures. This has been observed almost invariably in many of the Al based Al-RE-TM ally systems. Figure. 14.12 (a and b) shows the TEM bright field images, along with the SAD patterns, of AlLaYNiCo [32] and AlLaCeNiCo systems, respectively, clearly indicating the coexistence of nanocrystalline regions and the amorphous regions together. Considering that the amorphous matrix was obtained throughout the deposit one can suppose that it was produced layer by layer, where the incoming layer released the heat to the previous ones, thus, experiencing a chilling effect. This may also lead to softening and pressing of prior amorphous layers by attaining a temperature within the supercooled liquid region, which may lead to the bonding of the amorphous particles and result in a layered matrix due to the viscous flow of the material. In addition, the microstructure showed as well nanoparticles, possibly resulting from the partial crystallization of the matrix, an indication that the temperature reached in the deposit attained values slightly beyond T_x.

Guo et al. [26, 103] demonstrated an amorphous phase fraction of 36% in $Al_{89}La_6Ni_5$ flat product of thickness 2–3 mm. A number of nano-crystalline intermetallic second phases were observed along with unknown phases. In one of the

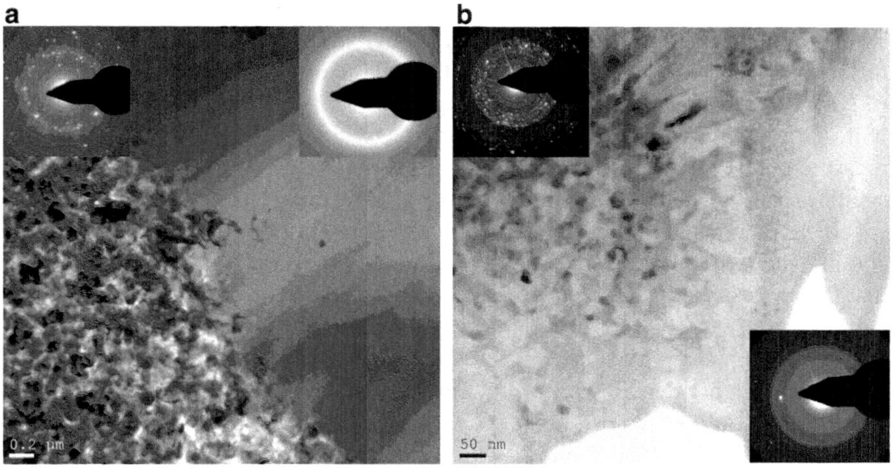

Fig. 14.12 TEM bright field images from the central regions of spray formed (**a**) AlLaYNiCo [32] and (**b**) AlLaCeNiCo systems

similar experiment, Guo et al. [104] produced 30 mm thick billet of $Al_{85}Nd_5Ni_{10}$. It was found in this study that the billet was fully crystallized. However, the oversprayed powder revealed an amorphous phase fraction of 63%. Although a liquid nitrogen cooled copper substrate was used in this experiment, so as to engender fast cooling of the deposit during and after deposition, amorphous phase was not observed in the deposit.

Similarly, with a little modification in the composition, Zhuo et al. [105] showed 91.7%, 78% and 54.3% of amorphous phase fraction in the bottom, middle and top of a 12 mm thick plate, respectively, of $Al_{86}Si_{0.5}Ni_{4.06}Co_{2.94}Y_6La_{0.5}$ system. In this case, melt was atomized with argon gas at 0.6 MPa. Similarly, Yan et al. [106] attempted to get amorphous powders of the $Al_{86}Ni_6Y_{4.5}Co_2La_{1.5}$ system and concluded that in such compositions the first phase to form was Al_2Y instead of fcc-Al, and this is the leading crystalline phase during solidification followed by precipitation of fcc-Al.

14.5.2 *Iron Based Alloys*

Saito et al. [107] reported in 1998, a fully amorphous phase, with a limited amount of $Fe_{14}Nd_2B$ intermetallic phase, in a 1–3 mm thick spray formed $Fe_{77}Nd_{15}B_8$ system. The deposition in this study was carried out using a subsonic atomizer at a high pressure of 6.0 MPa and a nozzle diameter of 2 mm that gives rise to a low melt mass flow rate and thereby increases the gas to melt mass flow rate ratio. In this study, the oversprayed powders showed fully crystalline state, which was observed by other investigators too for different alloy systems [28, 31, 32]. The authors concluded that the cooling rate of the deposit could be higher than that experience by the droplets during gas atomization. It is obviously evident that there must be a different cooling condition as the droplets impinge on the substrate. Afonso et al. [30] spray deposited $Fe_{83}Zr_{3.5}Nb_{3.5}B_9Cu_1$ alloy and found fully crystalline deposit with large fraction of intermetallic second phases due to the fact that a low gas pressure of 1.0 MPa with a G/M = 0.23 was employed. The low gas pressure and small gas to metal mas flow ratio lead to a large mean size of droplets; a high superheat may lead to high heat content in the spray prior to deposition, thus slow cooling and larger droplet size. However, in such cases it is important to point out that the process parameters must be such that efficient atomization takes place. As mentioned earlier, glass former alloys present high viscosity and the metal flow can be easily stopped by clogging the nozzle. However, all these parameters are not favorable for a high cooling rate, and also engender a high enthalpy transfer to the deposit.

Based on Fe–Nb–(Si)–B alloys with high glass forming ability and good soft magnetic properties, such as $Fe_{70}Nb_{10}B_{20}$ proposed by Ma et al. [108] and $(Fe_{0.75}Si_{0.10}B_{0.15})_{100-x}Nb_x$($x = 2$ and 4), proposed by Inoue et al. [109], Afonso et al. [110] proposed the new alloy composition $Fe_{63}Nb_{10}Al_4Si_3B_{20}$. Amorphous ribbons of this composition, presenting good soft magnetic properties for as-quenched amorphous and annealed nanocrystalline states, had been obtained by melt spinning of low cost Fe–Nb and Fe–(Si, Al)–B master alloys. In a further work, this alloy was processed by spray forming through nitrogen gas atomizer. Low cost Fe–Nb and Fe–(Al–Si)–B commercial master alloy were prepared to melt the alloy in an induction furnace under vacuum atmosphere. Additions of pure Fe, Al and Si were made to adjust the composition to the nominal one. The melt was superheated to 300 K over the melting temperature (1250 °C) to homogenize the composition of the alloy and to avoid clogging of the nozzle. A G/M ratio of 0.25 was used [29]. The resulting billet, of a mass of about 1.2 kg, as well as the overspray powder of about 1 kg, which was collected in the bottom of the atomization chamber, sieved and separated in the granulometric size ranges of 5–20, 20–30, 30–45, 45–75, 75–106, 106–150, 150–180, >180 μm, were characterized by different techniques.

The overspray powder presented a mass median powder diameter of about $d_{50} = 120$ μm and more than 80% amorphous phase was observed for particles below 75 μm. Even the coarsest particle size range analyzed, 150–180 μm, which should correspond to a cooling rate of approximately 10^3–10^4 K/s, presented more than 50% amorphous phase.

Despite of this it was not possible to maintain the high percentage of the amorphous phase throughout the deposit as it did by the aluminum alloy. Amorphous phase were observed only for the thinnest parts (below 1 mm thickness) of the deposit (Gaussian-like geometry). It seems that by the iron-based alloy it was not possible to drive the heat way as fast as it is possible by the aluminum alloy$Al_{85}Y_8Ni_5Co_2$, and as a consequence the amorphous structure of the previous layer crystallized under the heat effect of the incoming layers, which can be ascribed to the lower heat conductivity of the iron-based alloy. Thus, having a similar quantity of an amorphous structure throughout an iron-based deposit as it is possible in an aluminum-based deposit relies on a comparative higher glass forming ability for the iron-based alloy due to its lower heat conductivity.

In a recent report, Catto et al. [111] reported 95% amorphous fraction, up to 4 mm of a 16 mm thick deposit, in $[(Fe_{0.6}Co_{0.4})_{0.75}B_{0.2}Si_{0.05}]_{96}Nb_4$ system. This $[(Fe_{0.6}Co_{0.4})_{0.75}B_{0.2}Si_{0.05}]_{96}Nb_4$ alloy was proposed by Inoue et al. [79] in 2004 and the authors reported a high strength of over 4000 MPa. Rods of fully glassy alloy were produced with 4 mm diameter by ejection casting method in copper mold and showed $T_g = 552$ °C, $T_X = 602$ °C ($\Delta T_x = 50$ °C), denoting a very high glass forming ability with critical cooling rate in the range of 10^1–10^2 K/s and a BMG behavior for this Fe-based composition. This could account for the lower thermal conductivity of the Fe-based alloys allowing similar results to be obtained as it was for the $Al_{85}Y_8Ni_5Co_2$ alloy. In this experiment also, parameters were similar to that used by Afonso et al. [29]. A charge of the alloy with 1.6 kg was spray formed at 1693 K by using nitrogen as the atomizing gas and a G/M ratio of 0.94 m^3/kg, which led to a deposit with 1.3 and to 0.3 kg of overspray powder. The analyses presented for all particle size range amorphous band and almost absence of crystalline peaks. DSC traces indicated a high amount of amorphous phase formation for all range of particle size analyzed, over 95%, thereby, confirming the high glass forming ability of this composition. This high volume fraction, as mentioned before, was kept up to 4 mm Thickness in the deposit. This again emphasizes that the composition plays an important role during synthesis of bulk amorphous materials. Secondly, amorphous phase formation only up to 4 mm of the deposit indicates that the incoming droplets do not get enough quenching effect so as to transform to an amorphous phase during subsequent deposition of droplets. This is ascribed to the heat accumulation in the deposit as the thickness increases, a consequence of the low heat transfer coefficient of the Fe-based alloy.

Some features of the behavior of this BMG alloys should be mentioned as they render the final deposit important characteristics as, for example, low porosity. SEM micrographs of the powders show the glassy structure for all particle size ranges, see Fig. 14.13. It is worth to point out the behavior of what seems to be a consequence of the existence of the large supercooled liquid range for this alloy: the coarse range of the particles were deformed and not appeared round as the finest ones. We can suppose that these particles hit the wall of the atomization chamber within the supercooled liquid region being deformed by the impact. This behavior had a consequence during deposit build-up, leading to low levels of porosity, less than 2.3% whereas aluminum-based alloys presented porosity levels of 10%. Aside of this, the high glass forming ability of the composition led to a high volume percent of amorphous phase formation up to 4 mm thickness of the deposit, a similar value obtained for this alloy by copper mold casting.

Recently, Cava et al. [112] processed the $Fe_{43.2}Co_{28.8}B_{19.2}Si_{4.8}Nb_4$ (at.%) alloy by spray forming in an semi-industrial facility with the aim of investigating the formation of amorphous phases. The material was atomized using a "free fall"

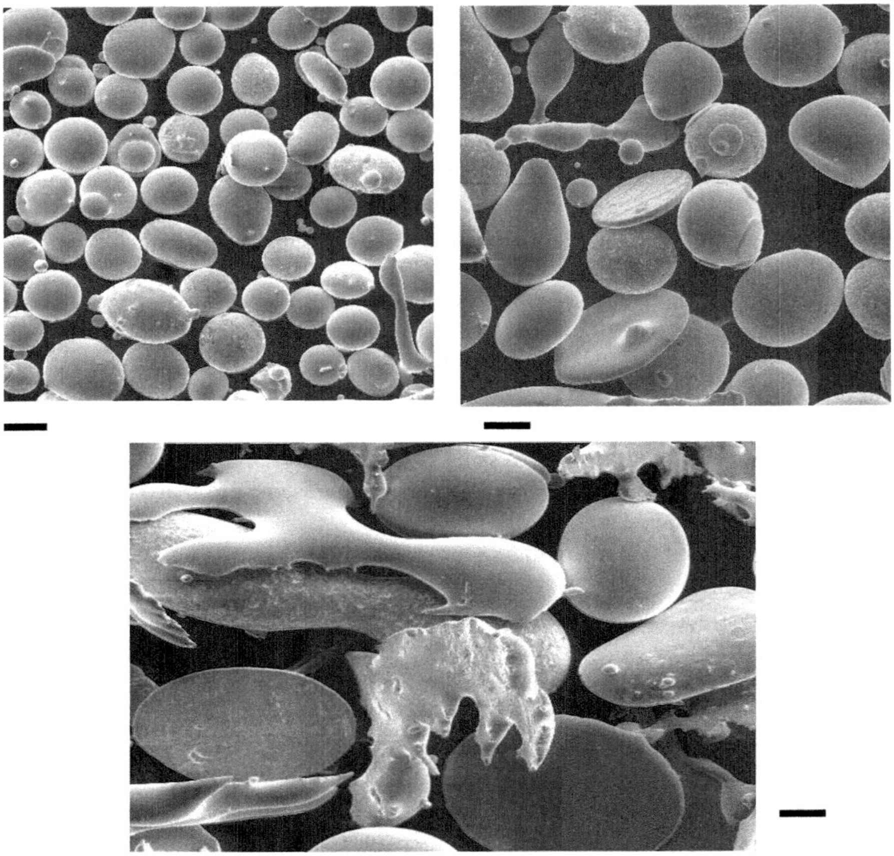

Fig. 14.13 $[(Fe_{0.6}Co_{0.4})_{0.75}B_{0.2}Si_{0.05}]_{96}Nb_4$ overspray powder. Particle size range: *Upper*: From *left* to *right*: particle size range of **+20–45 μm, −76 + 45 μm; *bottom*, respectively. *Bottom*: +106–150 μm.** Featureless aspect of all particle size range analyzed, confirming the high glass forming ability of this composition. Note the deformation of the coarse range of particle sizes. *Black bar* in all pictures is 50 μm long. From [111] with permission

nozzle with nitrogen gas, which was scanned on the substrates at a frequency of 15.2 Hz within a scan angle of ±4°. For deposition, it was used a preheated copper tube substrate (500 °C) with 770 mm length, 100 mm diameter and 10 mm thickness wall moving longitudinally in one direction against the spray cone at three speeds (10, 5 and 3 mm/s), allowing to obtain different deposit's thicknesses, respectively 5, 10 and 15 mm, see Fig. 14.14. The thicker layers (5–15 mm) presented partial/fully crystalline microstructure with onset of crystallization of the amorphous phase at Tx = 595 °C. On the other hand, thinner layers (below 2.5 mm) presented fully amorphous structure, and glassy behavior with glass transition (Tg) at 555 °C. The thermal stability and the crystallization kinetics of the alloy were studied using the isothermal DSC curves, measured at different

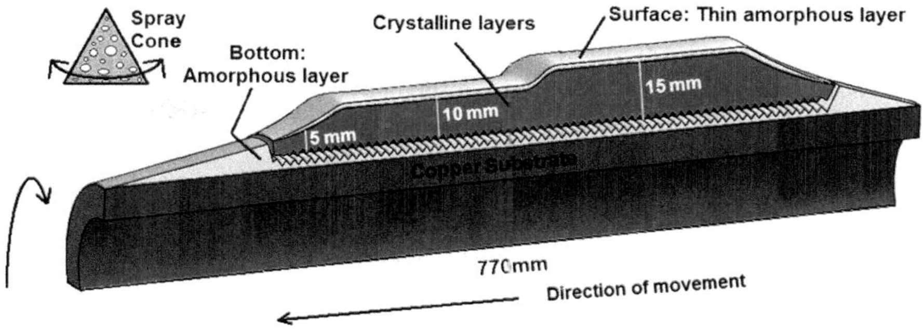

Fig. 14.14 Spray forming of a tube under quasi-industrial conditions [112]

heating rates and temperatures. Despite its good glass-forming ability (GFA) and high stability against crystallization, its incubation time for crystallization is almost zero. This behavior avoided the maintenance of the amorphous structure along the thickness of the tube.

14.5.3 Other Alloys

Although spray forming has been recognized as a process for amorphisation of various alloys, the major chunk of the work has been carried out on Al- and Fe-based systems. However, a few studies have been on Mg-based alloys due to the fact that these alloys have high glass forming ability and gives rise to high strength to weight ratio. The amorphous materials based on Mg show a small critical cooling rate and a cooling rate of 100 K/s has been observed for Mg–Cu–Y system. Replacing Y with Gd results in better GFA and up to 8 mm diameter amorphous material can be synthesized for $Mg_{65}Cu_{25}Gd_{10}$ by copper mold casting [113–115]. Chang et al. [116, 117] also attempted spray forming of the $Mg_{65}Cu_{25}Gd_{10}$ and produced 12 mm thick deposit, using a nozzle diameter of 6 mm and gas to melt flow rate ratio of 2.4 m^3/kg on a 5 mm thick copper substrate. This led to the formation of amorphous phase in the bottom (0–2 mm) but fully crystalline structure in the top region (7–12 mm). The difficulty arises as to why the spray forming route did not result in larger fraction of amorphous structure even when the glass forming ability of this system is high enough for large diameter copper mold casting. The reason might be that the high surface temperature with increasing deposit thickness may give rise to a hotter deposition surface for the subsequent incoming droplets, thus, reducing the second stage quenching effect.

14.6 Mechanisms of Microstructural Evolution

As discussed in section 14.4, the spray forming process can be divided into three distinct but integral processes (1) melt atomization and generation of spray of droplets direct towards the substrate under the influence of high velocity gas jets where the droplets experience high cooling rate (2) deposition of high velocity liquid/semi-solid droplets on the substrate/growing deposit and (3) the cooling of the deposition layer by convective heat transfer under high velocity gas or conduction through the deposit to the substrate or radiation. An understanding of all these process steps and the effect of different cooling conditions is extremely necessary to develop an insight into the mechanism of microstructural evolution of any alloy system. This insight may further enhance the understanding on the suitable process parameters required for the amorphisation/nanocrystallisation of materials by spray forming process. Spray forming is typically characterized to give rise to a droplet cooling rate of 10^3–10^4 K/s and the deposit cooling rates of and 10^0–10^1 K/s, respectively [88, 91, 118, 119]. However, Meyer et al. [40] in their attempt to understand the efficacy of spray deposition for the development of amorphous materials proposed three sub-process, namely, droplets in flight, droplets on deposition and post deposition cooling.

14.6.1 Process Control for Amorphous/Nanocrystalline Materials

14.6.1.1 Atomization and Droplets in Flight

Different atomization techniques can be used to atomize a melt into a spray of droplets. The most common atomization processes are free-fall [87, 88, 120] and close-coupled atomizers [87, 88, 121, 122], but also other atomizers have already been used in spray forming processes [123–126]. During the flight of droplets, cooling is achieved by releasing heat to the ambient by convection and radiation. The major influencing parameter with regard to the cooling rate of the droplets is their diameter and the gas jet velocity. To achieve a high cooling rate, formation of small droplets are favored and for the uniformity of the structure a narrow size distribution becomes an imperative, particularly when the prime concern is the synthesis of spatially uniform bulk amorphous and nano-crystalline materials.

14.6.1.2 Droplets on Deposition

When a droplet impacts and spreads, under the effect of high momentum, on the substrate or deposit surface, it releases heat mainly by conduction to the deposit [127] until it reaches in thermal equilibrium with the deposit [40]. A little consideration will show that a small size droplet in the spray will have lower thermal energy than the deposit surface and it would reheat to the equilibrium temperature [91]. The conductive heat transfer between the droplet and the deposit surface depends on their temperature difference as well as the thermal contact resistance between them. Both are dependent on the thermal state of the droplet and the surface. The thermal contact resistance depends on the phenomenon droplet spreading on the deposit on impact. A cold deposit surface leads to a non-uniform spreading of droplets due to the fact that the droplets' rapid solidification and release of a large amount of heat to the deposit [128–130]. This can then lead to an irregular contact surface and hence to an increased contact resistance. This phenomenon is well known in spray forming, when cold substrates are used a porous layer forms in the vicinity of the substrate. If a high heat transfer between the deposit and the substrate is necessary, too cold a deposit surface conditions of the substrate and the deposit should be avoided. It is reported [31] that the average amorphous fraction of overspray powder is lower than that found in the deposits. This indicates that the cooling rate during deposition must be higher than that during droplet cooling. As post-deposition cooling is known to be slow in spray forming, the droplet deposition stage must be a key cooling process for the production of bulk amorphous materials by spray forming.

14.6.1.3 Deposit Cooling

When a splat reaches in thermal equilibrium with the deposit surface, it enters the post deposition cooling stage. The growth of a deposit can be understood as layer by layer deposition of droplets [31]. The deposit cooling rate, therefore, depends upon a balance between incoming and outgoing heat content for a given deposited layer. Heat is transferred to the deposit by new layers of deposit while heat is released by convection and radiation as well as by conduction to the substrate [40]. An imbalance of heat transfer leads to an accumulation of heat in the deposit and may lead to crystallization of amorphous phase formed during deposition. This is in contrast to the fact that a continuous rapid cooling is needed. Therefore, as the cooling after deposition is also critical, conditions that affect this cooling stage should be chosen carefully.

The convective heat transfer from the deposition surface to the ambient depends upon the difference between the deposit and the ambient temperature, the heat transfer coefficient and the surface area. The surface and ambient temperature difference cannot be tailored so as to bring any drastic change in convective heat transfer. Only a moderate influence can be exercised for heat transfer coefficient by flow situation and atomization gas flow rate [119, 131]. Therefore, a strong post-deposition cooling effect can be achieved mostly by a change of the deposit geometry to increase the surface area. It should be noted that this effect is limited by the coverage of the gas flow field from the atomizer. The main parameter to change the convective heat transfer is, therefore, the change of the deposit surface area to volume ratio e.g. making large diameters tube or flat products. The same considerations are applicable for the heat transfer by radiation.

The conductive heat extraction from the deposit to the substrate is of utmost important when the thermal resistance between the deposit and the substrate is low and the substrate has enough heat capacity to digest heat from the deposit before both reach the equilibrium. Copper as substrate material is one of the best suited material due to its high density, specific heat and high thermal diffusivity. A thick substrate is generally preferred so that the equilibrium temperature between the substrate and the deposit remains below the glass transition temperature of the alloys composition. It is important to pre-heat the substrate to achieve a low thermal resistance between the deposit and the substrate; and hence a high heat transfer rate. This becomes more crucial as the deposit thickness increases and more heat requires to be conducted through the first layers into the substrate [28, 31, 101].

In addition, it is not only necessary to control the heat flow away from the deposition layers but also to control the incoming heat to the deposit. Therefore, it is important to use the atomization parameters such that most of the heat is released during flight of the droplets and at the same time maintaining the highest undercooled state of the droplets prior to deposition. This can either be achieved by reducing the droplet diameter or, if this is not possible, by reducing the deposit growth rate; hence, increasing the growth time for a given layer thickness so that the deposited layer gets enough time for heat release [40]. The melt mass flow rate plays a pivotal role in influencing the thermal history of a deposit. However, commonly used free-fall and close-coupled-atomizers usually cannot be run with the low mass flows required, therefore, there is a need for the implementation of alternative atomizers, such as pressure-gas-atomization or impulse atomization, for the spray forming process, particularly for the development of amorphous or nanocrystalline materials.

14.6.2 Microstructure Constitution of the Spray

The spray of droplets experiences a high cooling rate during flight and therefore undercools before crystallization. The spray of glass forming alloys, which becomes highly viscous as the temperature decreases, does speak a different story compared to other alloys systems. Generally, a droplet solidifies in a sequence as given in Fig. 14.15. A large sized droplet undercools and solidifies giving rise to a crystalline structure. In contrast, small size droplet does not crystallize and solidifies amorphous. The figure also shows the constitution of the spray in terms of the structure development and the thermal conditions associated with different stages of cooling during flight. The primary crystallization takes place, latent heat is released and solidification continues till a fully crystalline solid particle is formed. A small sized droplet, due to high surface area to volume ratio, may experience such a high cooling rate that the crystallization is fully avoided, leading to fully amorphous particles. In the alloys, the undercooling of droplets may show a change

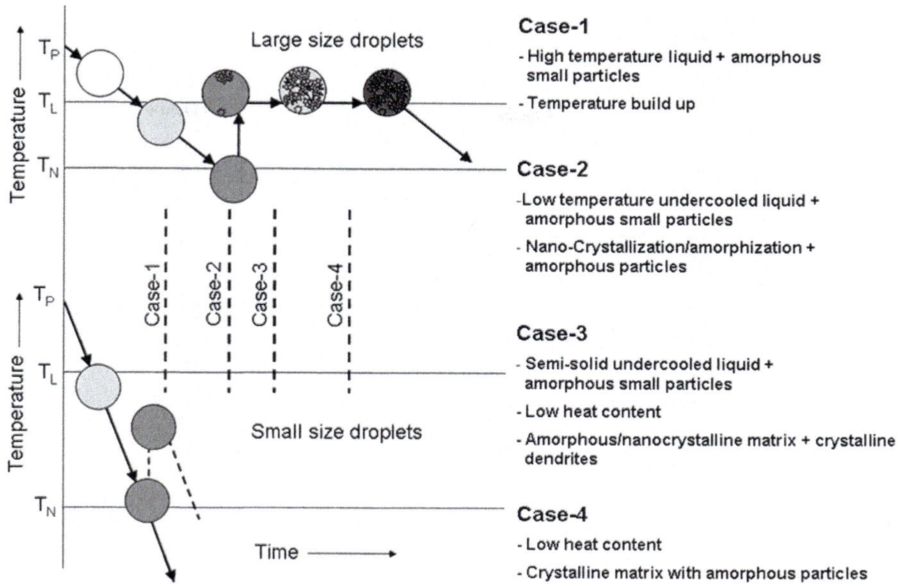

Fig. 14.15 Schematic illustration showing different solidification and thermal states of small and large size droplets. The figure shows the possible thermal states (cases: 1, 2, 3 and 4) of droplet prior to deposition and its effect on the final deposit structure (from [31] with permission)

Fig. 14.16 Micrographs of $Al_{85}Y_8Ni_5Co_2$ system (**a**) As-cast (**b**) oversprayed powders (**c**) deposit

in the solidification sequence depending upon the thermodynamic and kinetic aspects of the solidifying phases. The droplets which reach the substrate may have different solidification states. These include fully amorphous small particles, partially crystallized droplets having large fraction of undercooled liquid, fully crystalline small particles that might form due to the presence of heterogeneous nuclei and fully liquid high temperature droplets. Such a condition gives rise to a situation where the deposition of the spray brings in an amalgamation of several different phases. As a typical example, Fig. 14.16 shows the micrographs of as-cast alloy, atomized powders and the deposit of $Al_{85}Y_8Ni_5Co_2$ system together so as to reveal different structural features. As-cast alloy shows large intermetallic phases along with eutectic, the oversprayed powder particles reveal the presence of amorphous particles, crystalline as well as partially crystalline particles with

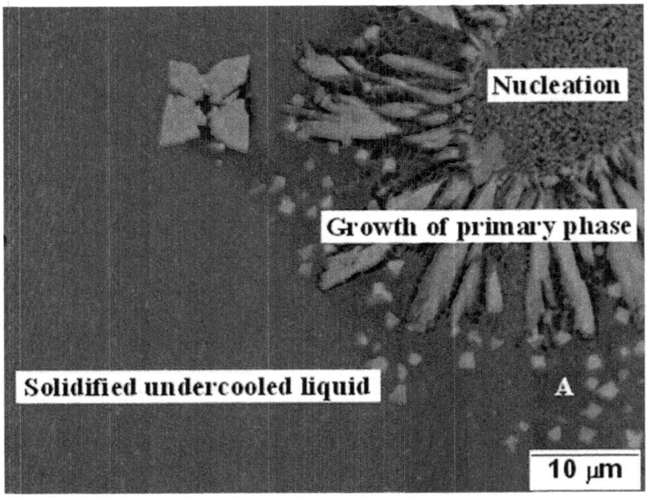

Fig. 14.17 Alloy $Al_{85}Y_8Ni_5Co_2$, partially crystallized and fragmented droplet on deposition (from [102] with permission)

nano-crystalline featureless regions and fully crystalline particles. In contrast, the deposit structure shows flow lines of amorphous/nanocrystalline regions obtained due to deposition highly undercooled liquid droplets, partially crystalline pre-solidified particles and fully crystalline pre-solidified particles.

Srivastava et al. [31, 32, 101, 102] observed a change in solidification sequence of powder particles for $Al_{85}Y_8Ni_5Co_2$ system. Instead of large size AlYNi based intermetallic phases as primary crystals, it was observed that the primary phase to form was Al_2Y in small particles in particular. A very high undercooling leads to full amorphisation without giving ways to crystallization. When the imperative becomes the development of amorphous or nanocrystalline materials, the aim should be to obtain maximal benefit of depositing larger droplets in highly undercooled state. This is due to the fact that amorphous particles would not be high in spray content and the crystallized particle will not help achieve the goal. Figure 14.17 shows the micrographs revealing the high magnification picture of a partially crystallized and fragmented droplet on deposition [32]. This clearly indicates that the liquid ahead of the solidification front is featureless and can be amorphous or nanocrystalline. It can be inferred from this observation that in these investigations the amorphous phases would have been generated mainly due to the deposition of partially crystalline state. Therefore, due to a wide structural inhomogeneity in the spray constitution a highly inhomogenous structure of deposit is produced. However, a narrow size distribution of droplets can be expected to ensure uniformity in the solidification state as well as increased fraction of amorphous content.

14.6.3 The Transient Layer on Deposition

The present chapter analyzes the efficacy of the spray forming process in the synthesis of amorphous or nano-crystalline bulk deposit. Therefore, droplets in the spray which have already witnessed crystallization are not of high importance as some crystalline fractions also come in the deposit. Therefore, the effort should be made to achieve the droplet deposition in undercooled state containing minimum heat. Figure 14.18 depicts a schematic of cooling path of a liquid droplet. A slowly cooled droplet in the spray undercools and finally crystallizes and becomes fully crystalline (cooling stage 1, as shown in Fig. 14.19, column 'A'), if there is no hindrance in the droplet's trajectory. However, a droplet with high cooling rate may cross the glass transition temperature without witnessing the crystallization event. If a moderately undercooled droplet, which generally experiences slow cooling, is to be brought to the glassy state a high cooling rate is required. This can be achieved by depositing the undercooled droplet on a relatively cold surface/substrate, say at point A (beginning of the 2nd cooling stage, as shown in Fig. 14.19 column 'B').

However, the solidification state (or flight distance) at which the droplet splats the surface governs the total heat content transferred to the substrate. A deferred impact of the droplet ensures low heat content, but the possibility of the droplet becomes crystallized increases. An early deposition of droplets with small

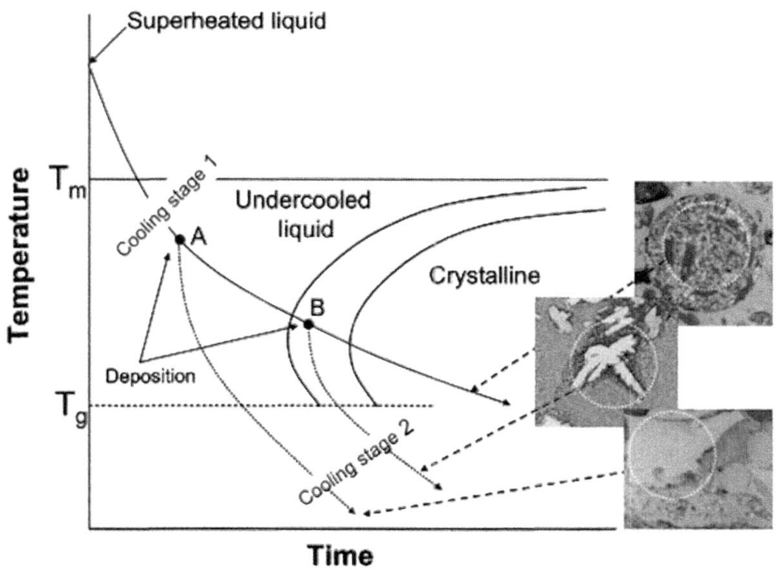

Fig. 14.18 Different stages of cooling for undercooled or partially crystallized liquid drops. Second stage cooling of undercooled drops can directly engender amorphous phase, whereas, partially crystallized drops may lead to a mixture of crystalline and amorphous phase. Amorphous phase formation is due to the quenching of remaining undercooled liquid (from [102] with permission)

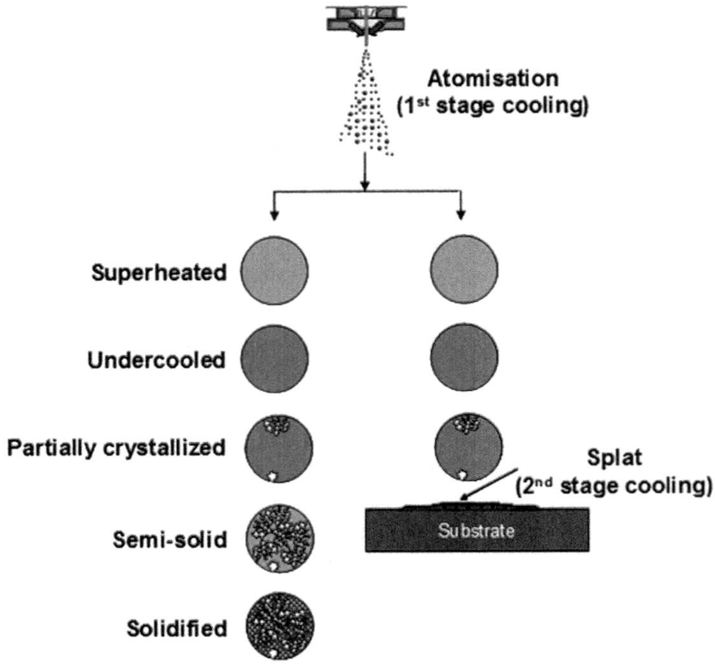

Fig. 14.19 A schematic showing that a liquid drop which crystallizes at a given cooling condition can give rise to higher fraction of amorphous or nano-crystalline structure when cooled in the 2nd stage (from [102] with permission). Colum A: *left*, Column B, *right*

undercooling may lead to heat accumulation and detrimental to the already amorphized deposit as crystallization may commence at this stage. On the other hand, if the crystallization commences and the droplet enters the two phase region (point B in Fig. 14.18), a large fraction of droplet will be in undercooled state mainly due to high viscosity of the liquid. If the droplet impinges the substrate in this thermal state (say point B), the undercooled liquid will immediately solidify amorphous along with already crystallized primary phase. The effect of this phenomenon can be seen in Fig. 14.17, already discussed. The featureless regions, which are optically irresolvable, is revealed in the spray formed deposit. This irresolvable region consisted of a large fraction of amorphous phase. The undercooled and highly viscous droplets impinge and flatten on the substrate and solidify depicting a featureless zone. The partially crystallized droplet, which would have still been left with remaining undercooled liquid prior to deposition, impinge on an already deposited droplet. The viscous liquid flows, with increased surface area, and solidifies in amorphous or nano-crystalline state, as indicated in Fig. 14.19. Therefore, it can be concluded that the droplets experience two stage of cooling during spray deposition; and bulk amorphization and/or nano-crystallization is the result of this phenomenon. The presence of higher fraction of crystalline phase in the oversprayed particles is also one of the indicators showing higher cooling rate during deposition or formation of bulk deposit.

14.7 Perspectives on Future Developments

Based on the theory and experiments presented throughout this chapter, it is possible to outline some considerations concerning the production of metastable phases by spray forming process. Basically, (a) the cooling rate in the deposit must be higher than that critical to avoid crystallization and (b) the temperature reached by the deposit during processing may not exceed the crystallization temperature in order to maintain the metastable structure. Considering the spray forming process and its processing parameters one should proceed by as cold as possible conditions, i.e., high G/M ratios, large flight distances, high heat extraction, keeping in mind that one must produce a workable deposit. On the other hand, the alloy should present the highest glass forming ability available; here the new knowledge about BMG forming alloys would be very helpful in selecting the most appropriate compositions with low minimal critical cooling rate (Rc $< 10^2$ K/s) and large supercooled liquid region(>50 K). Comparing the results for the aluminum-based and iron-based alloys it seems that the thermal conductivity of the alloy plays a key role, the higher the heat conductivity the easier to maintain the temperature of the deposit below the temperature of crystallization.

The results showed for glass former $Al_{85}Y_8Ni_5Co_2$ and $[(Fe_{0.6}Co_{0.4})_{0.75}B_{0.2}Si_{0.05}]_{96}Nb_4$ indicated special features of the solidification during the build-up of the spray formed deposit leading to the attainment of metastable phases as amorphous and nanocrystalline phases with a comparable performance as other fabrications processes. Despite the formation of amorphous phase, which was observed in the overspray powders of all alloys studied, amorphous phase was more difficult to be observed in the large size billet deposits. However, many of the alloy compositions studied by other investigators [28, 31, 32] showed higher fraction of amorphous phases in the deposit compared to overspray powders.

Since the cooling rate experienced by the atomized particles depend on their size, both amorphous and well-developed crystalline microstructures can be found in an atomization batch if an alloy with a glass forming composition is processed. The maintenance of the amorphous state will depend on the glass forming ability of the alloy considered and the temperatures reached during the deposition process. The different examples reported illustrated this situation. The billet of the spray forming $Al_{85}Y_8Ni_5Co_2$ alloy processed using high G/M ratio (10.0 m^3/kg) contained about 76% volume fraction of the amorphous phase. Processing the same alloy by a G/M ratio of 6.7 led to a fully crystalline deposit. It is important to point out here that there is no report so far in the literature showing a copper cast rod in this composition with more than 1 mm diameter of fully amorphous structure. Another processing using low G/M ratio (0.25 m^3/kg) of the $Fe_{83}Zr_{3.5}Nb_{3.5}B_9Cu_1$ alloy resulted in a fully crystalline deposit with irregular porosity due the high fraction of solid particles that hit the substrate when the deposition stage was obtained. The literature reports glassy structure for this alloy only for melt spun ribbons. The production of fully or partially amorphous deposits through spray forming requires very high gas-to-metal ratio that guarantee a deposition with high volume fraction of highly undercooled droplets prior to deposition, and it also requires a high glass forming ability (GFA) of the alloy composition. In addition, the cooling system of the process should be changed to match the needs of the alloys, even leading to some economic penalties. Here, the simulation of solidification carried out by Fritsching et al. [132, 133] for an Cu-based alloy (high conductivity) and a bearing steel (low conductivity) help understanding the results for the amorphous alloys reported before, namely the Al-based $Al_{85}Y_8Ni_5Co_2$ alloy (high conductivity) and the Fe-based $[(Fe_{0.6}Co_{0.4})_{0.75}B_{0.2}Si_{0.05}]_{96}Nb_4$ alloy(low conductivity). The calculations showed that the gradient temperature of a billet is strongly dependent of the materials properties. The calculated overall temperature distribution of a spray-formed 100Cr6 bearing steel billet at different times is compared with the ones of the CuSn6 billet, using similar standard process boundary conditions. Although such similarity in the process conditions, the results showed that because of the low thermal conductivity of 100Cr6 steel (30 W m^{-1} K^{-1}) the temperature difference from the top of the billet to the bottom as well as along the radius is relatively high. These calculations should also be made for flat as well as tube products which are plausible to give more surface area and small thicknesses. Even when the convective heat transfer coefficient decreases at the end of the spraying period, a non-uniform radial temperature distribution is observed. One can see that after spraying (>360 s), the billet cools down slowly and the residual liquid is enclosed by the totally solidified material. In this case, if shrinkage is suppressed, residual stress may rise and initiate hot cracks. In both cases, for the CuSn6 alloy and the 100Cr6 steel, higher temperature gradients are located at the base of the billets at the beginning of the process. At the top of the billets the temperature gradients are lower due to the high enthalpy contained in the mushy zone in the form of latent heat of solidification. In both the cases, it is shown that cooling rates of 10^2–10^3 K/s can be found in the very beginning of the process and decreases to 10^0–10^1 in the following cooling process. Besides changing the cooling conditions of the produced billets, efforts must be direct towards analyzing the nozzle design to achieve the desired droplet size distribution, the scale and size of deposits, cooling conditions of deposit and the overall process parameters. It is realized that the conventional spray forming parameters to achieve low porosity, structural refinement and homogeneity will not suffice to produce homogeneous partially amorphous or nano-crystalline materials.

A spray formed deposit of a viscous melt is expected to engender higher porosity. However, this can be consolidated at above the glass transition temperature to achieve nano-crystalline or partially amorphous materials. If one considers the spray deposition of glass forming systems in totality, several factors would come into picture and parameters such as alloy composition, purity of materials, nozzle design, substrate condition, in-flight cooling etc. needed to be considered for a successful synthesis of nano-crystalline/amorphous materials by spray forming. In summary, the development of bulk amorphous/nano-crystalline materials is of great importance and is considered to be of high promise. The spray atomization and deposition process shows some light towards a new beginning in the synthesis of such materials. However, a good understanding of the spray forming parameters and related modification in the process is necessary to achieve the desired properties.

14.8 List of Symbols

Latin

Symbol	Description
C^i	Atomic concentration of element i
$g(\theta)$	Catalytic efficiency of heterogeneous nucleation
I	Homogeneous nucleation rate of a crystalline phase formed from an undercooled liquid
n	Number of elements
R	Universal gas constant
r_i	Atomic radius of element i
T	Temperature of the melt
T_i	Melting temperature of element *i*
x_i	Atomic fraction of element *i*

Greek

Symbol	Description
α	Factor depending on the atomic arrangement at the interface
η	Viscosity
ΔS_f	Change in entropy per mole of alloy due to melting

References

1. Klement, W., Wilens, R. H., & Duwez, P. (1960). *Nature, 187*, 869.
2. Inoue, A., Ohtera, K., Kita, K., & Masumoto, T. (1988). *Japanese Journal of Applied Physics, 27*, L2248.
3. Inoue, A. (1998). *Progress in Materials Science, 43*, 365–520.
4. Johnson, W. L. (1999). *Materials Research Society Symposium Proceedings, 554*, 311–339.
5. Turnbull, D. (1961). *Transactions of AIME, 221*, 422.
6. Chen, H. S., & Turnbull, D. (1968). *The Journal of Chemical Physics, 48*, 2560.
7. Turnbull, D., & Fisher, J. C. (1949). *The Journal of Chemical Physics, 17*, 71.
8. Johnson, W. L. (2002). *JOM, 54*(3), 40–43.
9. Corner, R. D., et al. (2000). *International Journal of Impact Engineering, 24*, 435.
10. Ashby, M. F. (1992). *Materials Selection in materials design* (pp. 86–91). Oxford: Pergamon Press.
11. Johnson, W. L. (1996). *Materials Science Forum, 225–227*, 47.
12. Salimon, A. I., et al. (2004). *Materials Science and Engineering A, 375–377*, 385–388.
13. Schroers, J., Nguyen, T., O'Keeffe, S., & Desai, A. (2007). *Materials Science and Engineering A, 449–451*, 898–902.
14. Schroers, J. (2005). *JOM, 57*, 35.
15. Wesseiling, P., Nouri, A. S., & Lewandoski, J. J. (2005) In *TMS Annual Meeting*, San Francisco CA, February 16, 2005.
16. Johnson, W. L. (1999). *MRS Bulletin, 24*(10), 42.
17. Basu, J., & Ranganathan, S. (2003). *Sadhana, 28*(3–4), 783–798.

18. Inoue, A. (2000). *Acta Materialia, 48*, 279–306.
19. Johnson, W. L. (1999). *MRS Bulletin, 24*, 42–56.
20. Fan, G. J., Choo, H., & Liaw, P. K. (2007). *Journal of Non-Crystalline Solids, 353*, 102.
21. Jindal, V., Srivastava, V. C., & Uhlenwinkel, V. (2009). *Journal of Non-Crystalline Solids, 355*, 1552–1555.
22. Senkov, O. N., Senkov, S. V., Scott, J. M., & Miracle, D. M. (2005). *Materials Science and Engineering A, 393*, 12–21.
23. Oguchi, M., Inoue, A., Yamaguchi, H., & Masumoto, T. J. (1991). *Journal of Materials Science Letters, 10*, 289–291.
24. Afonso, C. R. M., Bolfarini, C., Kiminami, C. S., Bassim, N. D., Kaufman, M. J., Amateau, M. F., Eden, T. J., & Galbraith, J. M. (2001). *Journal of Non-Crystalline Solids, 284*, 134–138.
25. Afonso, C. R. M., Bolfarini, C., Kiminami, C. S., Bassim, N. D., Kaufman, M. J., Amateau, M. F., Eden, T. J., & Galbraith, J. M. (2001). *Scripta Materialia, 44*, 1625–1628.
26. Guo, M. L. T., Tsao, C. Y. A., Huang, J. C., & Jang, J. S. C. (2005). *Materials Science and Engineering A, 404*, 49–56.
27. Golumbfskie, W. J., Amateau, M. F., Eden, T. J., Wang, J. G., & Liu, Z. K. (2003). *Acta Materialia, 51*, 5199–5209.
28. Srivastava, V. C., Surreddi, K. B., Uhlenwinkel, V., Schulz, A., Eckert, J., & Zoch, H.-W. (2009). *Metallurgical and Materials Transactions A: Physical Metallurgy and Materials Science, 40*, 450–461.
29. Afonso, C. R. M., Bolfarini, C., Botta Filho, W. J., & Kiminami, C. S. (2007). *Materials Science and Engineering A, 449–451*, 884–889.
30. Afonso, C. R. M., Bolfarini, C., Botta Filho, W. J., & Kiminami, C. S. (2004). *Materials Science and Engineering A, 375–377*, 571–576.
31. Srivastava, V. C., Surreddi, K. B., Scudino, S., Schowalter, M., Uhlenwinkel, V., Schulz, A., Eckert, J., Rosenauer, A., & Zoch, H.-W. (2010). *Materials Science and Engineering A, 527*, 2747–2758.
32. Srivastava, V. C., Surreddi, K. B., Scudino, S., Schowalter, M., Uhlenwinkel, V., Schulz, A., Eckert, J., Rosenauer, A., & Zoch, H.-W. (2013). *Journal of Alloys and Compounds, 578*, 471–480.
33. Kawamura, Y., Mano, H., & Inoue, A. (2001). *Scripta Materialia, 44*, 1599–1604.
34. Zhong, Z. C., Jiang, X. Y., & Greer, A. L. (1997). *Materials Science and Engineering A, 226–228*, 531–535.
35. Hofmann, D. C., Suh, J. Y., Wiest, A., Duan, G., Lind, M. L., Demetriou, M. D., et al. (2008). *Nature, 451*, 1085–1089.
36. Hofmann, D. C., Duan, G., & Johnson, W. L. (2006). *Scripta Materialia, 54*, 1117–1122.
37. Das, J., Tang, M. B., Kim, K. B., Theissmann, R., Baier, F., Wang, W. H., et al. (2005). *Physical Review Letters, 94*, 205501.
38. Zhuo, L., Yang, B., Wang, H., & Zhang, T. (2011). *Journal of Alloys and Compounds, 509*, L169–L173.
39. Catto, F. L., Yonamine, T., Kiminami, C. S., Afonso, C. R. M., Botta, W. J., & Bolfarini, C. (2011). *Journal of Alloys and Compounds, 509*, S148–S154.
40. Meyer, C., Ellendt, N., Srivastava, V. C., & Uhlenwinkel, V. (2012). *International Journal of Materials Research, 103*, 1090–1095.
41. Inoue, A., & Zhang, T. (1996). *Materials Transactions, JIM, 37*, 185.
42. Inoue, A., Zhang, T., & Takeuchi, A. (1998). *Materials Science Forum, 269-272*, 855.
43. Turnbull, D. (1969). *Contemporary Physics, 10*, 473–488.
44. Inoue, A. (2000). *Acta Materialia, 48*, 279–306.
45. Lu, Z. P., & Liu, C. T. (2002). *Acta Materialia, 50*, 3501–3512.
46. Mukherjee, S., Zhou, Z., Schroers, J., Johnson, W. L., & Rhim, W. K. (2004). *Applied Physics Letters, 84*(24), 5010–5012.
47. Loefler, J. F., Schroers, J., & Johnson, W. L. (2000). *Aplied Physics Letters, 77*(5), 681–683.

48. Inoue, A. (1995). High strength bulk amorphous alloys with low critical cooling rates: Overview. *JIM Materials Transactions, JIM, 36*(7), 866–875.
49. Suryanarayana, C. (1999). *Non-equilibrium processing of materials [M]* (pp. 96–99). New York: Elsevier Science.
50. Pariona, M. M., Bolfarini, C., dos Santos, R. J., & Kiminami, C. S. (2000). *Journal of Materials Processing Technology, 102*, 221–229.
51. Donald, I.W., & Davies, H.A. (1978). Prediction of glass forming ability for metallic systems. *Journal of Non-Crystalline Solids, 30*(8), 77–85.
52. Turnbull, D. (1969). Under what conditions a glass can be formed? *Contemporary Physics, 10*, 473–488.
53. Ma, H., Hi, L. L., Xu, J., Li, Y., & Ma, E. (2005). Discovering inch-diameter metallic glasses in three-dimensional composition space. *Journal of Applied Physics Letters, 87*, 181915.
54. Cheney, J., & Vecchino, K. (2007). Prediction of glass-forming compositions using liquidus temperature calculations. *Materials Science and Engineering: A, 471*, 135–143.
55. Egami, T., et al. (2010). Why metallic glasses form and why they fail? In *WPI Advanced Institute for Materials Research:WPI-AIMR Annual Workshop*. Tohoku University.
56. Cheney, J., & Vecchino, K. (2008). GFA evaluated from liquidus-based model with a model for CSRO. *Journal of Alloys and Compounds*. doi:10.1016/j.jallcom.2008.03.071.
57. Miedema, A. R., Boom, R., & De Boer, M. R. (1975). On the heat offormation of solid alloys. *Journal of the Less-Common Metals, 41*, 283–298.
58. Boom, R., De Boer, M. R., & Miedema, A. R. (1976). On the heat of mixingof liquid alloys. *Journal of the Less-Common Metals, 45*, 237–245.
59. Van Der Kolk, G. J., Miedema, A. R., & Niessen, A. K. (1988). On thecomposition range of amorphous binary transition metal alloys. *Journal of the Less-Common Metals, 145*, 1–17.
60. Coehoorn, R., Van Der Kolk, G. J., Van Den Broek, J. J., Minemura, T., & Miedema, A. R. (1988). Thermodynamics of the stability of amorphous alloys of two transition metals. *Journal of the Less-Common Metals, 140*, 307–316.
61. Xia, L., Fang, S. S., Wang, Q., Dong, Y. D., & Liu, C. T. (2006). Thermodynamic modeling of glass formation in metallic glasses. *Applied Physics Letters, 88*, 171905.
62. De Oliveira, M. F., Aliaga, L. C. R., Bolfarini, C., Otta, W. J., & Kiminami, C. S. (2008). Thermodynamic and topological instability approaches for forecasting glass-forming ability in the ternary Al–Ni–Y system. *Journal of Alloys and Compounds, 464*, 118–121.
63. Inoue, A., Negishi, T., Kimura, H. M., Zhang, T., & Yavari, A. R. (1998). *Materials Transactions, JIM, 39*, 318.
64. Wang, W. H., Wang, R. J., Zhao, D. Q., Pan, M. X., & Yao, Y. S. (2000). *Physical Review B, 62*, 11292.
65. Wang, W. H., Wei, Q., Friedrich, S., Macht, M. P., Wanderka, N., & Wollenberger, H. (1997). *Applied Physics Letters, 71*, 1053.
66. Miracle, D. B. (2004). Efficient local packing in metallic glasses. *Journal of Non-Crystalline Solids, 342*, 89–96.
67. Miracle, D. B. (2004). A structural model for metallic glasses. *Nature Materials, 3*, 697–702.
68. Egami, T., & Waseda, Y. (1984). *Journal of Non-Crystalline Solids, 64*, 113.
69. SáLisboa, R. D., Bolfarini, C., Botta F, W. J., & Kiminami, C. S. (2005). *Applied Physics Letters, 86*, 211904.
70. Kiminami, C. S., SáLisboa, R. D., de Oliveira, M. F., Bolfarini, C. & Botta, W. J. (2007). *Materials Transactions, 48*(7), 1739–1742.
71. Chen, H. S. (1974). *Acta Metallurgica, 22*, 1505.
72. Chen, H. S., Krause, J. T., & Coleman, E. (1975). *Journal of Non-Crystalline Solids, 18*, 157.
73. Drehman, A. L., Greer, A. L., & Turnbull, D. (1982). *Applied Physics Letters, 41*, 716.
74. Inoue, A., Zhang, T., & Masumoto, T. (1989). *Materials Transactions, JIM, 30*, 965.
75. Inoue, A., et al. (1990). *Materials Transactions of JIM, 31*, 493.
76. Peker, A., & Johnson, W. L. (1993). *Applied Physics Letters, 63*, 2342.
77. Lin, X. H., & Johnson, W. L. (1995). *Journal of Applied Physics, 78*, 6514.

78. Leonhardt, M., Loser, W., & Lindenkreuz, H. G. (1999). *Acta Materialia, 47*, 2961.
79. Inoue, A., Shen, B. L., & Chang, C. T. (2004). *Acta Materialia, 52*, 4093.
80. Lu, Z. P., Liu, C. T., Thompson, J. R., & Porter W. D. (2004). *Physical Review Letters, 92*.
81. Gu, X. N., Zheng, Y. F., Zhong, S. P., Xi, T. F., Wang, J. Q., & Wang, W. H. (2010). *Biomaterials, 31*(6), 1093–1103.
82. Guo, F. Q., Poon, S. J., & Shiflet, G. J. (2004). *Applied Physics Letters, 84*, 37.
83. Pu, J., Wang, J. F., & Xiao, J. Z. (2003). *Nonferrous Metals Society, 13*, 1056.
84. Tang, M. B., Zhao, D. Q., Pan, M. X., & Wang, W. H. (2004). *Chinese Physics Letters, 21*, 901.
85. Inoue, A., & Zhang, W. (2004). *Materials Transactions, 45*, 584.
86. Wang, W. H., Dong, C., & Shek, C. H. (2004). Bulk metallic Glasses. *Materials Science and Engineering R, 44*, 45–89.
87. Lavernia, E. J., & Wu, Y. (1996). *Spray atomization and deposition* (pp. 155–260). West Sussex, England: John Wiley and Sons.
88. Grant, P. S. (1995). *Progress in Materials Science, 39*, 497.
89. Srivastava, V. C., Mandal, R. K., & Ojha, S. N. (2004). *Materials Science and Engineering A, 383*, 14.
90. Lavernia, E. J., Ayers, J. D., & Srivatsan, T. S. (1992). *International Materials Review, 37*, 1.
91. Grant, P. S. (2007). *Metallurgical and Materials Transactions A, 38*, 1520.
92. Srivastava, V. C., Mandal, R. K., Ojha, S. N., & Venkateswarlu, K. (2007). *Materials Science and Engineering A, 471*, 38.
93. Srivastava, V. C., Uhlenwinkel, V., Schulz, A., Zoch, H.-W., Mukhopadhyay, N. K., & Chowdhury, S. G. (2008). *Zeitschrift fuer Kristallographie, 223*, 711.
94. Srivastava, V. C., Mandal, R. K., & Ojha, S. N. (2001). *Materials Science and Engineering A, 304–306*, 555.
95. Srivastava, V. C., Mandal, R. K., & Ojha, S. N. (2001). *Journal of Materials Science Letters, 20*, 27.
96. Shukla, P., Mandal, R. K., & Ojha, S. N. (2001). *Bulletin of Materials Science, 24*, 547–554.
97. Srivastava, V. C., Huttunen-Saarivirta, E., Cui, C., Uhlenwinkel, V., Schulz, A., & Mukhopadhyay, N. K. (2014). *Journal of Alloys and Compounds, 597*, 258–268.
98. Oguchi, M., Inoue, A., Yamaguchi, H., & Masumoto, T. (1990). *Materials Transactions, JIM, 31*, 1005–1010.
99. Kawamura, Y., et al. (1993). *Materials Transactions of JIM, 34*, 969.
100. Inoue, A., et al. (2001). *Scripta Materialia, 44*, 1599.
101. Srivastava, V. C., Surreddi, K. B., Scudino, S., Schowalter, M., Uhlenwinkel, V., Schulz, A., Rosenauer, A., Zoch, H.-W., & Eckert, J. (2009). *Transactions of the Indian Institute of Metals, 62*(4–5), 331–335.
102. Srivastava, V. C., Ellendt, N., Meyer, C., & Uhlenwinkel, V. (2014). *Materialwissenschaft und Werkstofftechnik, 45*(8), 744–757.
103. Guo, M. L. T., Tsao, C. Y. A., Chang, K. F., Huang, J. C., & Jang, J. S. C. (2007). *Materials Transactions, 48*, 1717.
104. Guo, M. L. T., Tsao, C. Y. A., Huang, J. C., & Jang, J. S. C. (2006). *Intermetallics, 14*, 1069.
105. Zhuo, L., Yang, B., Wang, H., & Zhang, T. (2011). *Journal of Alloys and Compounds, 509*, L169.
106. Yan, M., Wang, J. Q., Schaffer, G. B., & Qian, M. (2011). *Journal of Materials Research, 26*, 944.
107. Saito, T., Takahashi, S., & Kuji, T. (1998). *Journal of Materials Science Letters, 17*, 1007.
108. Ma, L., Wang, L., Zhang, T., & Inoue, A. (1999). *Materials Research Bulletin, 34*, 915.
109. Inoue, A., & Shen, B. (2002). *Materials Transactions, JIM, 43*, 766.
110. Afonso, C. R. M., Bolfarini, C., BottaFilho, W. J., & Kiminami, C. S. (2004). *Journal of Metastable and Nanocrystalline Materials, 22*, 93.
111. Catto, F. L., Yonamine, T., Kiminami, C. S., Afonso, C. R. M., Botta, W. J., & Bolfarini, C. (2011). *Journal of Alloys and Compounds, 509*, S148.

112. Cava, R. D., Aliaga, L. C. R., Trivenõ Rios, C., Uhlenwinkel, V., Ellendt, N., Kiminami, C. S., et al. (2014). Microstructure characterization and kinetics of crystallization behavior of tubular spray formed Fe43.2Co28.8B19.2Si4.8Nb4 bulk metallic glass. *Journal of Heat Treament and Materials, 69*(5), 312–321. doi:10.3139/105.110236.
113. Xi, X. K., Wang, R. J., Zhao, D. Q., Pan, M. X., & Wang, W. H. (2004). *Journal of Non-Crystalline Solids, 344*, 105.
114. Men, H., & Kim, D. H. (2003). *Journal of Materials Research, 7*, 1502.
115. Yuan, G., & Inoue, A. (2005). *Journal of Alloys and Compounds, 387*, 134.
116. Chang, K. F., Chen, F. H., Fan, S. K., & Tsao, C. Y. A. (2008). *Advances in Materials Research, 51*, 57.
117. Chang, K. F., Guo, M. L. T., Kong, R. H., Tsao, C. Y. A., Huang, J. C., & Jang, J. S. C. (2008). *Materials Science and Engineering A, 477*, 58.
118. Fritsching, U. (2004). *Spray simulation. Modelling and numerical simulation of spray forming of metals*. Cambridge: Cambridge University Press.
119. Bergmann, D., & Fritsching, U. (2004). *International Journal of Thermal Sciences, 43*, 403.
120. Ellendt, N., Stelling, O., Uhlenwinkel, V., von Hehl, A., & Krug, P. (2010). *Materials Science and Engineering Technology, 41*, 532.
121. Kasama, A. H., Bolfarini, C., Kiminami, C. S., & Botta Filho, W. J. (2007). *Materials Science and Engineering A, 449–451*, 375.
122. Kasama, A. H., Mourisco, A. J., Kiminami, C. S., Botta Filho, W. J., & Bolfarini, C. (2004). *Materials Science and Engineering A, 375–377*, 589.
123. Ellendt, N., Schmidt, R., Knabe, J., Henein, H., & Uhlenwinkel, V. (2004). *Materials Science and Engineering A, 383*, 107.
124. McHugh, K., Uhlenwinkel, V., & Ellendt, N. (2008). Density of sprayformed materials. In *Proceedings of PM2008* Washington, 8.-12. June.
125. Chen, W. Z., Song, X. P., Qian, K. W., & Gu, H. C. (1998). *Materials Science and Engineering A, 247*, 126.
126. Achelis, L., Uhlenwinkel, V., Lagutkin, S., & Sheikhaliev, S. (2007). *Materials Science Forum, 534–536*, 13.
127. McDonald, A., Moreau, C., & Chandra, S. (2007). *International Journal of Heat and Mass Transfer, 50*, 1737.
128. Abedini, A., Pourmousa, A., Chandra, S., & Mostaghimi, J. (2006). *Surface and Coating Technology, 201*, 3350.
129. Dhiman, R., & Chandra, S. (2005). *International Journal of Heat and Mass Transfer, 48*, 5625.
130. Dhiman, R., McDonald, A., & Chandra, S. (2007). *Surface and Coating Technology, 201*, 7789.
131. Meyer, O., Schneider, A., Uhlenwinkel, V., & Fritsching, U. (2003). *International Journal of Thermal Sciences, 42*, 561.
132. Meyer, O., Fritsching, U., & Bauckhage, K. (2003). Numerical investigation of alternative process conditions for influencing the thermal history of spray deposited billets. *International Journal of Thermal Sciences, 42*, 153–168.
133. Cui, C. S., Fritsching, U., Schulz, A., Tinscher, R., Bauckhage, K., & Mayr, P. (2005). Spray forming of homogeneous 100Cr6 bearing steel billets. *Journal of Materials Processing Technology, 168*, 496–504.

Index

H. Henein et al. (eds.), *Metal Sprays and Spray Deposition*,
DOI 10.1007/978-3-319-52689-8

T

U

MIX
Papier aus verantwortungsvollen Quellen
Paper from responsible sources
FSC® C105338

Printed by Books on Demand, Germany